For Reference

Not to be taken from this room

McGRAW-HILL YEARBOOK OF SCIENCE & TECHNOLOGY

2011

REF.
Q
121
.M31
2011

Comprehensive coverage of recent events and research as compiled by the staff of the McGraw-Hill Encyclopedia of Science & Technology

New York Chicago San Francisco Lisbon London Madrid Mexico City
Milan New Delhi San Juan Seoul Singapore Sydney Toronto

The McGraw-Hill Companies

On the front cover

Conceptual design of a large neutrino detector, consisting of an array of light sensors embedded in transparent Antarctic ice. Also shown is the blue Cerenkov cone produced by a muon that was, in turn, produced when a neutrino that had traveled through the Earth interacted with an atomic nucleus in the ice. (*IceCube Research Center*)

ISBN 978-007-176371-4
MHID 0-07-176371-6
ISSN 0076-2016

McGRAW-HILL YEARBOOK OF SCIENCE & TECHNOLOGY
Copyright © 2011 by The McGraw-Hill Companies, Inc.
All rights reserved. Printed in the United States of America.
Except as permitted under the United States Copyright Act of 1976, no part of this publication may be reproduced or distributed in any form or by any means, or stored in a data-base or retrieval system, without prior written permission of the publisher.

The following articles are excluded from McGraw-Hill Copyright:
Animal prions; DNA barcoding in fungi; Positive train control; Wave slamming; Wood supply and demand.

1 2 3 4 5 6 7 8 9 0 DOW/DOW 1 5 4 3 2 1 0

This book was printed on acid-free paper.

It was set in Garamond Book and Neue Helvetica Black Condensed by Aptara, New Delhi, India. The art was prepared by Aptara. The book was printed and bound by RR Donnelley.

Contents

Editorial Staff

Mark Licker, Publisher

David Blumel, Senior Staff Editor

Stefan Malmoli, Senior Staff Editor

Jessa Netting, Senior Staff Editor

Renee Taylor, Editorial Coordinator

Charles Wagner, Manager, Digital Content

Jonathan Weil, Senior Staff Editor

Editing, Design, and Production Staff

Roger Kasunic, Vice President—Editing, Design, and Production

Frank Kotowski, Jr., Managing Editor

Consulting Editors

Dr. Milton B. Adesnik. *Department of Cell Biology, New York University School of Medicine, New York.* Cell biology.

Prof. William P. Banks. *Department of Psychology, Pomona College, Claremont, California.* General and experimental psychology.

Prof. Vernon D. Barger. *Department of Physics, University of Wisconsin-Madison.* Classical mechanics.

Prof. Rahim F. Benekohal. *Department of Civil and Environmental Engineering, University of Illinois, Urbana-Champaign.* Transportation engineering.

Dr. James A. Birchler. *Division of Biological Sciences, University of Missouri, Columbia.* Genetics.

Robert D. Briskman. *Technical Executive, Sirius XM Radio, New York.* Telecommunications.

Prof. Wai-Fah Chen. *Department of Civil and Environmental Engineering, University of Hawaii at Manoa, Honolulu.* Civil engineering.

Prof. J. John Cohen. *Department of Immunology, University of Colorado Medical School, Aurora.* Immunology.

Prof. Glyn A. O. Davies. *Senior Research Fellow in Aerostructures, Department of Aeronautics, Imperial College London, United Kingdom.* Aeronautical engineering and propulsion.

Prof. Peter J. Davies. *Department of Plant Biology, Cornell University, Ithaca, New York.* Plant physiology.

Prof. M. E. El-Hawary. *Associate Dean of Engineering, Dalhousie University, Halifax, Nova Scotia, Canada.* Electrical power engineering.

Barry A. J. Fisher. *Director, Scientific Services Bureau (Retired), Los Angeles County Sheriff's Department, Los Angeles, California.* Forensic science and technology.

Dr. Richard L. Greenspan. *The Charles Stark Draper Laboratory, Cambridge, Massachusetts.* Navigation.

Prof. Joseph H. Hamilton. *Landon C. Garland Distinguished Professor of Physics, Department of Physics and Astronomy, Vanderbilt University, Nashville, Tennessee.* Nuclear physics.

Dr. Lisa Hammersley. *Department of Geology, California State University, Sacramento.* Petrology.

Prof. John P. Harley. *Department of Biological Sciences, Eastern Kentucky University, Richmond.* Microbiology.

Prof. Terry Harrison. *Department of Anthropology, Paleoanthropology Laboratory, New York University, New York.* Anthropology and archeology.

Dr. Yassin Hassan. *Department of Nuclear Engineering, Texas A & M University, College Station.* Nuclear engineering.

Dr. Jason J. Head. *Department of Biology, University of Toronto, Mississauga, Ontario, Canada.* Vertebrate paleontology.

Dr. Ralph E. Hoffman. *Yale Psychiatric Institute, Yale University School of Medicine, New Haven, Connecticut.* Psychiatry.

Dr. Hong Hua. *College of Optical Sciences, University of Arizona, Tucson.* Electromagnetic radiation and optics.

Dr. S. C. Jong. *Senior Staff Scientist and Program Director, Mycology and Protistology Program, American Type Culture Collection, Manassas, Virginia.* Mycology.

Prof. Edwin C. Kan. *Department of Electrical & Computer Engineering, Cornell University, Ithaca, New York.* Physical electronics.

Dr. Bryan P. Kibble. *Independent Consultant, Hampton, Middlesex, United Kingdom.* Electricity & electromagnetism.

Prof. Robert E. Knowlton. *Department of Biological Sciences, George Washington University, Washington, D.C.* Invertebrate zoology.

Prof. Chao-Jun Li. *Canada Research Chair in Green Chemistry, Department of Chemistry, McGill University, Montreal, Quebec, Canada.* Organic chemistry.

Prof. Donald W. Linzey. *Wytheville Community College, Wytheville, Virginia.* Vertebrate zoology.

Dr. Dan Luss. *Cullen Professor of Engineering, Department of Chemical and Biomolecular Engineering, University of Houston, Texas.* Chemical engineering.

Prof. Albert Marden. *School of Mathematics, University of Minnesota, Minneapolis.* Mathematics.

Dr. Ramon A. Mata-Toledo. *Professor of Computer Science, James Madison University, Harrisonburg, Virginia.* Computers.

Prof. Krzysztof Matyjaszewski. *J. C. Warner Professor of Natural Sciences, Department of Chemistry, Carnegie Mellon University, Pittsburgh, Pennsylvania.* Polymer science and engineering.

Prof. Jay M. Pasachoff. *Director, Hopkins Observatory, and Field Memorial Professor of Astronomy, Williams College, Williamstown, Massachusetts.* Astronomy.

Prof. Stanley Pau. *College of Optical Sciences, University of Arizona, Tucson.* Electromagnetic radiation and optics.

Dr. William C. Peters. *Professor Emeritus, Mining and Geological Engineering, University of Arizona, Tucson.* Mining engineering.

Dr. Donald Platt. *Micro Aerospace Solutions, Inc., Melbourne, Florida.* SPACE TECHNOLOGY.

Dr. Kenneth P. H. Pritzker. *Professor, Laboratory Medicine and Pathobiology, and Surgery, University of Toronto, and Pathology and Laboratory Medicine, Mount Sinai Hospital, Toronto, Ontario, Canada.* MEDICINE AND PATHOLOGY.

Prof. Justin Revenaugh. *Department of Geology and Geophysics, University of Minnesota, Minneapolis.* GEOPHYSICS.

Dr. Roger M. Rowell. *Professor Emeritus, Department of Biological Systems Engineering, University of Wisconsin, Madison.* FORESTRY.

Dr. Thomas C. Royer. *Department of Ocean, Earth, and Atmospheric Sciences, Old Dominion University, Norfolk, Virginia.* OCEANOGRAPHY.

Prof. Ali M. Sadegh. *Director, Center for Advanced Engineering Design and Development, Department of Mechanical Engineering, The City College of the City University of New York.* MECHANICAL ENGINEERING.

Prof. Joseph A. Schetz. *Department of Aerospace & Ocean Engineering, Virginia Polytechnic Institute & State University, Blacksburg.* FLUID MECHANICS.

Dr. Alfred S. Schlachter. *Advanced Light Source, Lawrence Berkeley National Laboratory, Berkeley, California.* ATOMIC & MOLECULAR PHYSICS.

Prof. Ivan K. Schuller. *Department of Physics, University of California, San Diego.* CONDENSED-MATTER PHYSICS.

Jonathan Slutsky. *Naval Surface Warfare Center, Carderock Division, West Bethesda, Maryland.* NAVAL ARCHITECTURE AND MARINE ENGINEERING.

Dr. Arthur A. Spector. *Department of Biochemistry, University of Iowa, Iowa City.* BIOCHEMISTRY.

Dr. Anthony P. Stanton. *Tepper School of Business, Carnegie Mellon University, Pittsburgh, Pennsylvania.* GRAPHIC ARTS AND PHOTOGRAPHY.

Dr. Michael R. Stark. *Department of Physiology, Brigham Young University, Provo, Utah.* DEVELOPMENTAL BIOLOGY.

Prof. John F. Timoney. *Maxwell H. Gluck Equine Research Center, Department of Veterinary Science, University of Kentucky, Lexington.* VETERINARY MEDICINE.

Dr. Daniel A. Vallero. *Adjunct Professor of Engineering Ethics, Pratt School of Engineering, Duke University, Durham, North Carolina.* ENVIRONMENTAL ENGINEERING.

Dr. Sally E. Walker. *Associate Professor of Geology and Marine Science, University of Georgia, Athens.* INVERTEBRATE PALEONTOLOGY.

Prof. Pao K. Wang. *Department of Atmospheric and Oceanic Sciences, University of Wisconsin, Madison.* METEOROLOGY AND CLIMATOLOGY.

Dr. Nicole Y. Weekes. *Department of Psychology, Pomona College, Claremont, California.* NEUROPSYCHOLOGY.

Prof. Mary Anne White. *Department of Chemistry, Dalhousie University, Halifax, Nova Scotia, Canada.* MATERIALS SCIENCE AND METALLURGICAL ENGINEERING.

Dr. Thomas A. Wikle. *Department of Geography, Oklahoma State University, Stillwater.* PHYSICAL GEOGRAPHY.

Prof. Jonathan Wilker. *Department of Chemistry, Purdue University, West Lafayette, Indiana.* INORGANIC CHEMISTRY.

Article Titles and Authors

Preface

Events in our natural environment often captured the headlines this past year. Severe winter weather in some areas was followed by record summer heat and drought affecting food prices. Floods and earthquakes devastated populated regions. To what extent, if any, were the effects of these events exacerbated by human activities? What actions, if any, should be taken to mitigate future events? The answers to such questions require sound scientific information. As further examples, the debate over the reality of global warming and its causes continues, as do controversies in other areas such as the best ways to ensure our future energy supplies, and over stem cell research. An informed, nontechnical understanding of the trends and developments in science and technology is increasingly important not just for leaders in industry and government but also for an informed citizenry to have opinions based on current knowledge and to be able to make rational decisions about the direction of society. Thus, the 2011 edition of the *McGraw-Hill Yearbook of Science & Technology* continues its nearly half-century mission of keeping professionals and nonspecialists alike abreast of important research and development with a broad range of concise reviews invited by a distinguished panel of consulting editors and written by leaders in international science and technology.

In this edition, for example, we report on the rapid advances in cell biology and genetics with articles on topics such as breakthrough research on the reprogramming of human somatic cells into pluripotent stem cells, which promises to revolutionize regenerative medicine; new technologies that are drastically decreasing the time and cost of DNA sequencing; our understanding of epigenetic mechanisms that give rise to heritable characteristics through alterations to chromosomes that do not involve changes in their DNA sequence; and the intense scientific interest in so-called long noncoding RNAs, which are emerging as important regulatory components in the cell. In the neurosciences, we report on new findings on genetic influences on human language development; the effect of abuse on the brain; the genetics of memory; and the connection between anxiety disorders and the brain structure known as the amygdala. In biomedicine,we review the modeling of climate change on allergic disease; recent advances in the understanding of celiac disease; Type I diabetes; H1N1 influenza; and research on how animal prion diseases such as scrapie, "mad cow disease (BSE)," and chronic wasting disease (CWD) in deer and elk may infect humans. In engineering and technology, we chronicle developments in high-speed rail transportation; low-power radio links; microbial fuel cells; mobile WiMAX; active materials and smart structures; electric generation ancillary services; picoprojectors; and ramjet engines. In physics and astronomy, we review Einstein's mass-energy equivalence principle; the flow behavior of viscoelastic fluids; quantum coherence in photosynthesis; the search for a supersolid; ultrafast x-ray sources; wobbling motion in nuclei; the IceCube neutrino observatory; asteroid extinction; space flight; supermassive black holes; and water on the Moon. In chemistry, we report on click chemistry in polymer science; protein crystallization; harder-than-diamond carbon crystals; and molecule-based superconductors. And in the earth and environmental sciences, we review the 2010 Haiti earthquake; research on the understanding of crustal melting processes; advances in tornado observation; computing the fluid dynamics of air toxics; and tropical glacier monitoring. All in all, we chronicle advances in sciences from astronomy to zoology.

Each contribution to the *Yearbook* is a concise yet authoritative article authored by one or more authorities in the field. The topics are selected by our consulting editors, in conjunction with our editorial staff, based on present significance and potential applications. McGraw-Hill strives to make each article as readily understandable as possible for the nonspecialist reader through careful editing and the extensive use of specially prepared graphics.

Librarians, students, teachers, the scientific community, journalists and writers, and the general reader continue to find in the *McGraw-Hill Yearbook of Science & Technology* the information they need in order to follow the rapid pace of advances in science and technology and to understand the developments in these fields that will shape the world of the twenty-first century.

Mark D. Licker
PUBLISHER

A–Z

Active materials and smart structures

Active materials and smart structures offer a wealth of new opportunities for human ingenuity and engineering design. Whereas smart structures have the attributes of adaptability, flexibility, and even "intelligence," the active materials are the enabling factors that make smart structures possible.

Active materials. Active materials (also known as smart materials or intelligent materials) have the characteristic of responding in a controlled fashion to external stimuli (such as temperature, stress, voltage, magnetic field, moisture, and pH) by changing at least one of their properties (such as shape, color, or physical constants).

Piezoelectric materials. These materials produce a voltage when stress is applied (the direct piezoelectric effect) or display a change of shape when a voltage is applied (the reverse piezoelectric effect). Piezoceramics, such as $PbZrTiO_3$ (PZT), have high stiffness and good piezoelectric coefficients for generation of both strain and voltage; they are used in both sensors and actuators. However, a drawback of piezoceramics is their brittleness and high density. Piezopolymers, such as polyvinylidene fluoride (PVDF), are lightweight and conformable, but have low stiffness and hence a low strain-generating piezoelectric coefficient. However, their voltage-generating piezoelectric coefficient is excellent; they are extensively used in sensing applications, for example, on keyboards. On the other hand, electroactive polymers (EAPs), also known as artificial muscles, with a seeded fluid in pores, can strain up to 15% under a voltage gradient.

Magnetostrictive materials. These materials are similar to piezoelectric materials, but they respond to magnetic fields instead of electric fields. Magnetostrictive materials expand when a magnetic field is applied because their magnetic domains tend to align with the field lines. A commonly used magnetostrictive material is Terfenol (TbDyFe). Although the unbiased magnetostrictive response is quadratic, commercial applications of magnetostrictive materials use bias fields to obtain a linearized response that resembles piezomagnetic behavior. Also of interest are photoresponsive materials, which change shape upon exposure to light.

Shape memory alloys (SMAs). These are thermally activated materials that change shape through a phase transformation between low-temperature martensite and high-temperature austenite. Upon stretching at low temperatures, the martensite phase undergoes a quasiplastic deformation (around 8%), which is accommodated through detwinning of the crystalline lattice. Upon heating, martensite changes into austenite and the initial shape is recovered (that is, shape memory). The transition temperatures can be adjusted by changing the alloy composition. A common shape memory alloy is Nitinol (NiTi). Another phenomenon enabled by the austenitic-martensitic transformation is that of superelasticity (also known as pseudoelasticity), in which the shape memory alloy can undergo very large strains (around 8%) and fully recover upon unloading. Magnetic shape memory alloys are materials (for example, NiMnGa) that change their shape in response to a magnetic field. Shape memory polymers (SMPs) are polymeric materials that exhibit the shape memory effect through temperature-induced change from a temporary shape back into the permanent shape. Several shape memory polymer formulations exist, most notably polyurethane, polyethylene terephthalate (PET), and polyethylene oxide (PEO). Light-activated shape memory polymers exhibit the shape memory effect upon illumination with ultraviolet light of certain wavelengths that either activate or deactivate photo-crosslinked chemical bonds.

Chromogenic materials. These materials change color in response to electrical, optical, or thermal changes. Electrochromic materials change color or opacity on

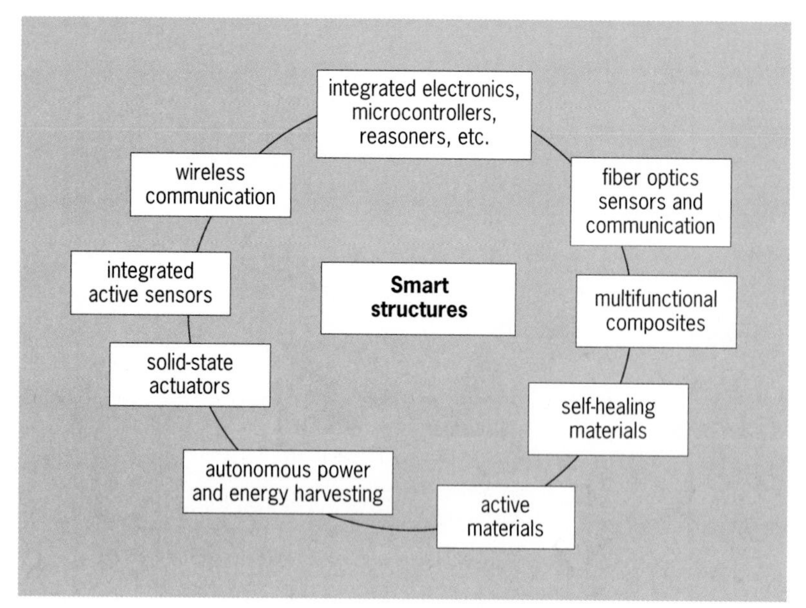

Fig. 1. The smart structures constellation: functional attributes and technology enablers.

the application of a voltage (for example, in liquid-crystal displays). Thermochromic materials change color depending on temperature. Photochromic materials change color in response to light (for example, light-sensitive sunglasses that darken when exposed to bright sunlight). Pressure-sensitive materials change color under stress and are used for taking "pressure footprints" of adjoining interfaces.

Electrorheological (ER) and magnetorheological (MR) fluids. These materials display dramatic changes in their viscosity when activated by electrical or magnetic fields and change from a fluid state to an almost solid state. This response is very fast (of the order of milliseconds) and reversible by the removal of the applied field. A similar behavior is displayed by ferrofluids, although the latter have a finer particle dispersion than magnetorheological fluids.

Smart structures. The concept of smart structures (also known as adaptive structures or intelligent structures) is usually derived through analogy with living organisms that can sense the environment, interpret the information sensed from the environment, and react to it appropriately. For example, sensors in the human skin can sense the pricking of a rose's thorns and send the information to the nervous system, which will interpret it and instruct the hand to let go and retreat. In order to achieve these functions, living organisms possess sensing, data processing, and actuation capabilities embedded into their complex bodies. Similarly, a bio-inspired smart structure would be equipped with sensing, data processing, and actuation capabilities (**Fig. 1**). A smart structure is a multifunctional system with capabilities well beyond the mundane load-bearing mission of conventional structures. Enabling technologies include active materials, integrated active sensors, fiber optics sensors, fiber optics communication, solid-state actuators, autonomous power and energy harvesting, multifunctional composites, self-healing materials, integrated electronics, reasoners, and microcontrollers.

Smart-structure concepts have been developed for many engineering fields. A smart building or bridge would feel an earthquake and "brace" itself to sustain it better; afterwards, it would quickly inspect itself to assess and control damage, if any. A smart aircraft would feel the effect of flight loads, the operational environment, or enemy fire on its load-bearing capability, and would take corrective actions to arrive safely at its destination. A smart space antenna would adapt its shape to maintain its focusing accuracy under uneven solar heating and micrometeorite impacts. A buried smart pipeline would monitor its state of corrosion, detect leaks, report its state of structural health, and even attempt self-repair. A smart automobile structure would adapt its suspension impedance according to the road type, making it stiffer for high-speed travelling on the highway, while making it more compliant for cross-country excursions. A smart machine tool would adapt the tool holder, pressure, feed, and impedance to optimize the machining process by increasing material removal rate while reducing chatter and vibration and minimizing energy usage.

Recent advances in smart structures research and implementation can be traced along three major directions: morphing structures, self-bracing structures, and self-monitoring structures.

Morphing structures. These are aircraft structures whose actuation is used to change an aircraft's shape, much as birds spread or retract their wings when perching, loitering, diving, or landing. Piezo-actuated trailing-edge flaps have been experimentally installed on helicopter blades in order to reduce helicopter noise and vibration through smart sensing and control. Morphing crewless air vehicles that can continuously change wing aspect ratio and planform (sweep, wing extension, wing folding, and so forth) have been built and tested. **Figure 2** shows how a morphing aircraft would cover the flight envelope much better than a fixed-geometry aircraft and even one with variable sweep or variable airfoil. The radial coordinate is a measure of performance efficiency, that is, how much of the flight envelope requirements are covered at a given flight regime: the closer to the perimeter, the better.

Self-bracing structures. These structures use structural actuation to change the stiffness, damping, or even mass of the structure in order to better sustain an earthquake event, an atmospheric gust, a violent storm, and so forth, much as one stiffens one's muscles. For example, shape memory alloy wires embedded in concrete, when activated, can close concrete cracks and increase overall stiffness. (The austenitic elastic modulus is up to three times larger than the martensitic modulus.) Magnetorheological dampers, when activated, can significantly change the structural damping and even act as structural stiffeners (**Fig. 3**).

Self-monitoring structures. These structures are equipped with structural health monitoring (SHM)

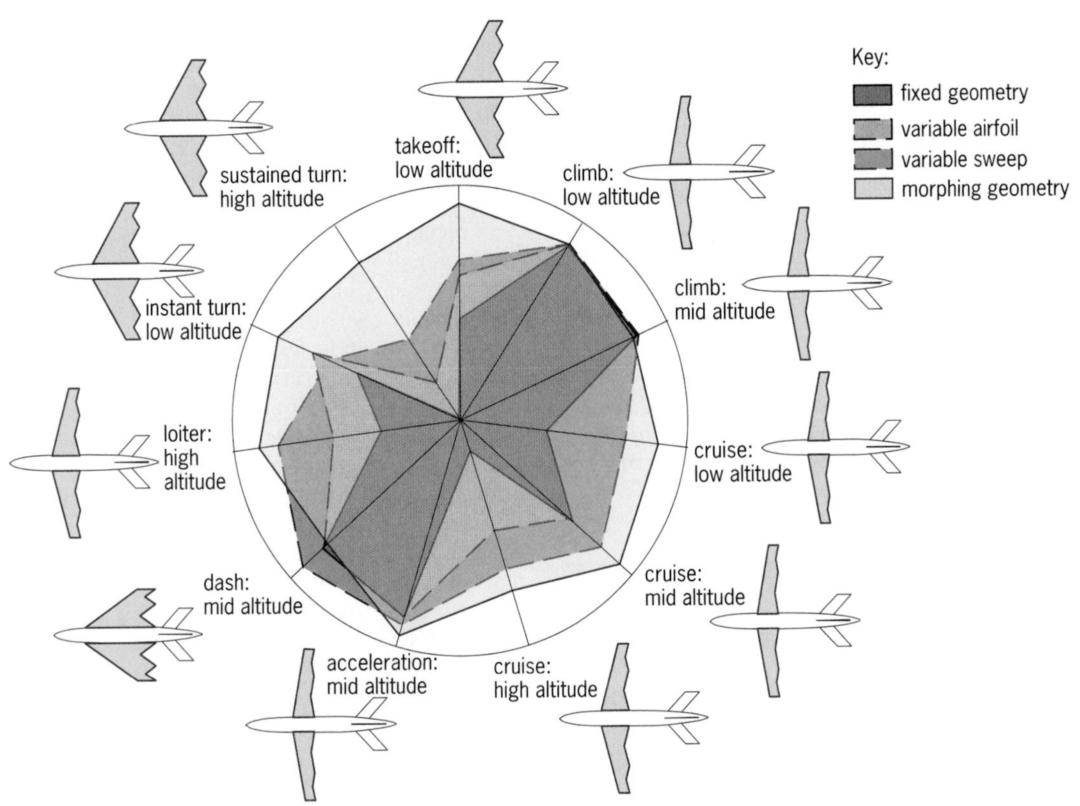

Fig. 2. **Advantage of morphing flight structures: Flight-envelope comparison between fixed geometry, variable airfoil, variable sweep, and fully morphing geometry. Radial coordinate is a measure of performance efficiency. (*Courtesy NextGen Aeronautics, Inc., reproduced with permission*)**

sensors and advanced data-processing algorithms capable of structural diagnosis and prognosis. The goal of structural health monitoring is to develop a monitoring methodology that is capable of detecting and identifying various damage types during the service life of the structure, monitor their evolution, and predict the remaining useful life with a continuously updating structural model. Structural health monitoring can be broadly classified into two categories, passive and active. Passive structural health monitoring techniques (such as acoustic emission, impact detection, and strain measurement) are relatively more mature; however, their utility is limited by the need for continuous monitoring and the indirect way in which damage existence is inferred. Active structural health monitoring techniques aim at directly interrogating the structure on demand using guided-wave ultrasonics and other methods. Active structural health monitoring resembles conventional nondestructive testing and evaluation, except that the sensors are permanently attached to the structure and interrogated automatically without human intervention. Historical structural health monitoring data would allow projection of damage progression trends and estimation of remaining useful life. In critical situations, on-board processing of structural health monitoring data would allow adaptive mission planning to ensure safe return to base. The opportunities offered by self-healing materials are also being considered.

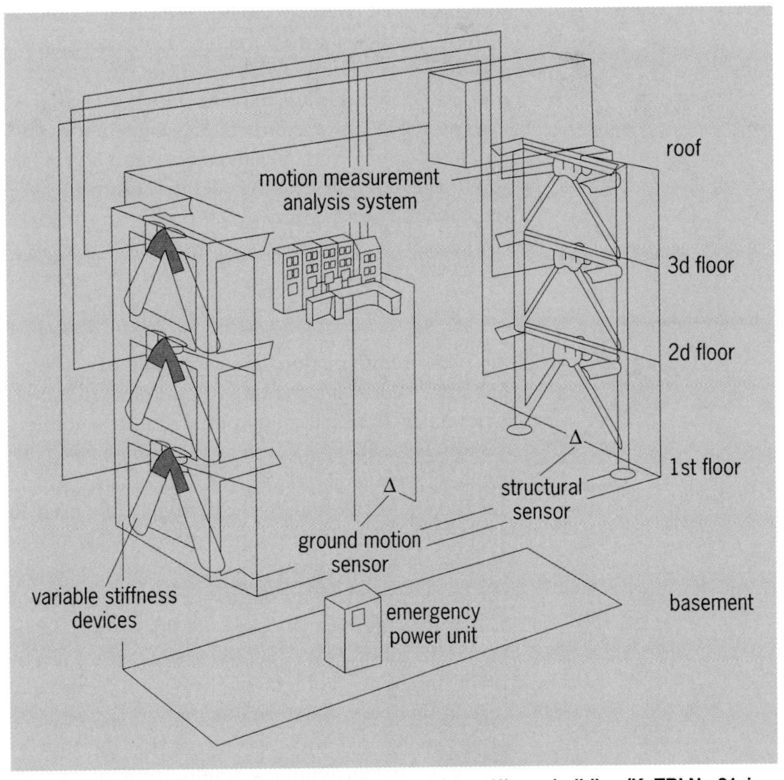

Fig. 3. **Schematic of an earthquake-resistant, variable-stiffness building (KaTRI No.21, in service in Tokyo, Japan), an example of a self-bracing structure. (*V. Giurgiutiu, Encyclopedia of Vibration, Elsevier, 2004*)**

For background information *see* ACOUSTIC NOISE; ADAPTIVE WINGS; FIBER-OPTIC SENSOR; MAGNETOSTRICTION; NONDESTRUCTIVE EVALUATION; PIEZOELECTRICITY; SHAPE MEMORY ALLOYS in the McGraw-Hill Encyclopedia of Science & Technology.

Victor Giurgiutiu

Bibliography. C. Boller, F.-K. Chang, and Y. Fujino (eds.), *Encyclopedia of Structural Health Monitoring*, Wiley, Hoboken, NJ, 2008; B. Culshaw, *Smart Structures and Materials*, Artech House, Boston/London, 1996; V. Giurgiutiu, *Structural Health Monitoring with Piezoelectric Wafer Active Sensors*, Elsevier Academic Press, New York, 2008; H. Janocha (ed.), *Adaptronics and Smart Structures*, 2d ed., Springer Verlag, Berlin, 2007; M. Schwartz (ed.), *Encyclopedia of Smart Materials*, Wiley-Interscience, New York, 2002.

Aeroacoustics

Many engineering devices, such as jet engines, helicopters, cooling fans, and wind turbines, generate sound as a result of the flow of air. Aeroacoustics is the study of this aerodynamic noise. The ultimate goal of aeroacoustics is the development of reliable predictive models that enable noise-reduction strategies to be identified and that allow noise to be accurately factored into engineering, environmental, and economic decisions.

Measures of sound. For the most part, flow-generated sound is important because of its direct impact on human beings. If the noise from a wind turbine or a jet engine were not audible, it would be of little engineering importance. This is because a sound field generally carries very little energy, many orders of magnitude less than the flow that produced it. The human factor is directly reflected in the most common engineering measures of sound level. The sound pressure level (SPL) is defined in decibels by the equation below. Here p_{ms} is the mean square of

$$\text{SPL (dB)} = 10 \log_{10} \left[\frac{p_{ms}}{(20\ \mu\text{Pa})^2} \right]$$

the pressure fluctuation associated with the sound. The dimensional constant of 20 micropascals (μPa) is nominally the smallest amplitude of pressure fluctuation that can be detected by the human ear. Thus a sound pressure level of 0 dB is a sound right at the threshold of audibility. The logarithm used in the definition of the sound pressure level allows for the extraordinary dynamic range of the human ear. The loudest sounds that our ears can stand (without pain) have a mean-square pressure fluctuation that is some 10^{13} times larger than the softest we can hear.

The frequency content of the sound is also important, not only in determining its impact on a listener, but also as a means of identifying the source. We therefore talk about the sound pressure level associated with specific frequency bands, which may be assembled to reveal the entire frequency spectrum. Common choices are to divide the spectrum into 1-Hz bands (which, for example, would give the sound pressure level of sound generated between 400 and 401 Hz) or into one-third-octave bands. An octave is a doubling of the frequency, so the ratio between the upper and lower frequencies of a one-third-octave band is $2^{1/3}$. The octave-band convention recognizes the fact that people also sense the frequency of sound logarithmically. A second integrated measure of sound level, which accounts for its frequency content, is the A-weighted sound pressure level, which is designed to reflect the varying sensitivity of the human ear to different frequencies.

Lighthill's theory. Compared to aerodynamics, aeroacoustics is a relatively new subject. Significant progress in the quantitative understanding of aeroacoustics began in middle of the twentieth century. A major pioneer in this effort was Sir James Lighthill, who in 1952 published a ground-breaking theory, known as the acoustic analogy, which quantitatively related sound sources to aerodynamics. He found that the Navier-Stokes equations, which describe all air flows under normal conditions, could be recast as an acoustic wave equation with sources defined by the aerodynamic variables. Lighthill's theory and the scientific developments it stimulated (in particular the work of N. Curle in 1955, including the effects of solid surfaces, and J. E. Ffowcs Williams and D. L. Hawkings in 1969, allowing arbitrary motion of those surfaces) led to a fundamental understanding of the types of sound sources that can exist in a flow, and how that sound will propagate to a listener.

Basic aerodynamic sound sources. The basic sources are the monopole, dipole, and quadrupole. Monopole sources are the result of volume changes, in the most obvious case produced by pulsation in solid boundaries (such as occurs with a loudspeaker) or unsteady flow in and out of a porous surface. In still air a monopole source radiates sound equally in all directions (**Fig. 1***a*). Dipole sources are produced when the air is subjected to a fluctuating force, such as when a helicopter rotor blade experiences an unsteady lift as it passes through the wake shed from a preceding blade. Sound radiates away from a dipole in two lobes aligned with the axis of the unsteady force (Fig. 1*b*). Roughly speaking, the sound level from a dipole source varies as the sixth power of the flow velocity. The most important quadrupole sources result purely from motion within the flow, specifically as a result of the rate of change of shear

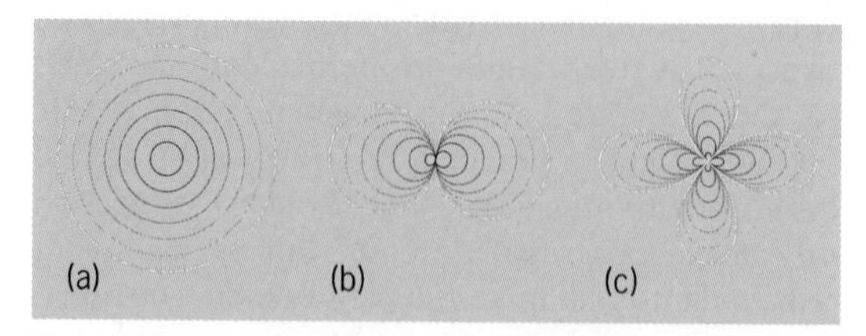

Fig. 1. Sound waves produced by (*a*) monopole, (*b*) dipole, and (c) quadrupole sources in an otherwise stationary uniform medium. Shading is proportional to the wave magnitude. Adjacent lobes of the dipole and quadrupole are noticeably out of phase.

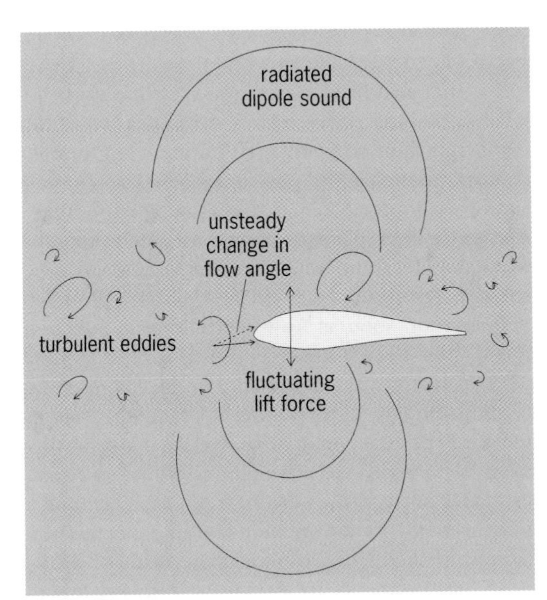

Fig. 2. Generation of leading-edge noise. Blade is traveling from right to left.

stress fluctuations associated with turbulence. These can be particularly intense in high-Mach-number turbulent jets, for example. Quadrupole sources ideally have a four-lobed directivity pattern (Fig. 1*c*) and produce sound that increases in intensity approximately as the eighth power of the velocity. Sound intensity therefore rises rapidly with velocity. A doubling of flow velocity can be expected to increase the sound intensity by a factor of 64 for a dipole source and a factor of 256 for a quadrupole source. The sound intensity from all these types of sources decays as the inverse square of the distance from the source.

Distributed sources. The distribution of sources throughout a flow can be complex and time-varying. This is particularly true when turbulence plays a role in the noise generation. An example of this is leading-edge noise, which can be an important source whenever a lifting surface cuts through a turbulent flow (**Fig. 2**). The eddies that comprise the turbulence are associated with vertical velocity fluctuations. When the leading edge of the blade encounters an eddy, these fluctuations locally change the angle of attack, creating a fluctuating lift force on the blade. This is a dipole sound source. Just like the eddies that produce them, these dipoles occur randomly in strength, in time, and in spanwise position along the blade. It is the integrated effect of all these sources that is experienced as leading- edge noise.

Effects of relative motion. The way the sound waves from different sources add up at the ear of a listener can be important in determining not only the intensity but also the nature of the sound that is heard. This is particularly true when relative motion is involved. Consider the disturbance produced by the rotating blade of a stationary helicopter, as perceived by a listener some distance away. **Figure 3** shows the situation as the blade advances toward the listener. What is heard at a given instant includes a combination of the pressure disturbances radiating from the leading and trailing edges of the blade. However, the disturbance from the trailing edge was produced earlier than that from the leading edge because it has a greater distance to travel, and was therefore produced when the blade was farther from the listener. From the listener's perspective it is as though the blade chord c were elongated by a factor $1 + v\Delta t/c$ where v is the blade speed and Δt is the difference in propagation time. From trigonometry we see that, if a is the sound speed and θ is the angle from the blade path to the listener, then $a\Delta t = (c + v\Delta t)\cos\theta$. From this it follows that the elongation factor is $1/(1 - M\cos\theta)$, where M is the Mach number, v/a. By the same argument, the blade appears foreshortened as it retreats from the listener, by a factor $1/(1 + M\cos\theta)$. With θ varying with time, the net sound field experienced by the listener mimics that which would be produced if the blade volume were fluctuating, and so a monopole source is heard, known as thickness noise. Effects of this type, in which there is a difference in the emission time (also known as retarded time) of the sound heard from different parts of the flow, can be important for all types of flows and sources. The size of the effect grows both with the Mach number and with the frequency of the sound being considered.

Reducing sound by reducing speed. As discussed above, the sound generated by aerodynamic sources in general increases rapidly with flow speed. Apart from moving away from a source (and thus taking advantage of the inverse-square-law decay), the simplest way of reducing aeroacoustic noise is often just to lower the flow speed. A famous example of this is the effect of the bypass duct in jet engines. The first commercial jet aircraft (the De Havilland *Comet*) used turbojet engines in which the entire air flow is mixed with fuel and burned to generate thrust. This results in a very high exhaust velocity that can be good, for example, for supersonic flight, but for commercial applications is inefficient and very noisy. Much of the noise is quadrupole noise generated by turbulent eddies that form as a result of the shearing between the high-speed jet and the

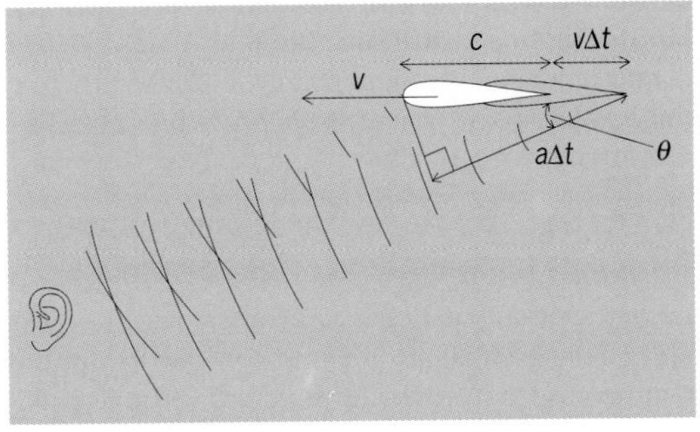

Fig. 3. Combination of acoustic disturbances from a moving helicopter blade.

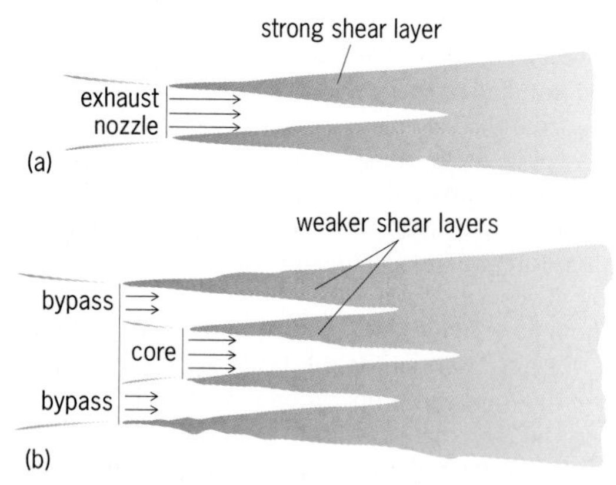

Fig. 4. Differences in the exhaust configuration for (*a*) turbojet engine and (*b*) turbofan engine.

surrounding air (**Fig. 4**). Modern commercial aircraft use turbofan engines in which some of the power of the engine is used to propel air through a duct surrounding the core of the engine, which bypasses the high-pressure and combustion stages. This is done by means of a large fan at the inlet that is attached to the rotating shaft of the engine. Because the power of the engine is now being directed into a much larger body of air, the exhaust velocities are reduced and so is the jet noise, roughly as the eighth power of the velocity. Furthermore, the velocity difference with the surrounding air is now shared between the shear layers of the core exhaust and the bypass, further reducing noise. The last 40 years have seen an impressive reduction of over 20 dB (that is, a factor greater than 100) in the noise of comparably sized commercial aircraft, in large part because of increases in the bypass ratio of their engines.

For background information *see* ACOUSTIC NOISE; AERODYNAMIC SOUND; AIRCRAFT PROPULSION; MULTIPOLE RADIATION; NAVIER-STOKES EQUATION; NOISE MEASUREMENT; SOUND; SOUND PRESSURE; TURBOFAN; TURBOJET; TURBULENT FLOW in the McGraw-Hill Encyclopedia of Science & Technology.

William J. Devenport

Bibliography. N. Curle, The influence of solid boundaries upon aerodynamic sound, *Proc. R. Soc. Lond. Ser. A. Math. Phys. Eng. Sci.*, 231:505–514, 1955; J. E. Ffowcs Williams and D. L. Hawkings, Sound generation by turbulence and surfaces in arbitrary motion, *Phil. Trans. R. Soc. Lond. Ser. A. Math. Phys. Eng. Sci.*, 264:321–342, 1969; M. S. Howe, *Acoustics of Fluid-Structure Interactions*, Cambridge University Press, 2008; M. J. Lighthill, On sound generated aerodynamically. I. General theory, *Proc. R. Soc. Lond. Ser. A. Math. Phys. Sci.*, 211:564–587, 1952.

Air toxics computational fluid dynamics

Air toxics computational fluid dynamics (CFD) is an emerging scientific method for calculating human exposure to toxic air pollutants. The U.S. Environmental Protection Agency (EPA) defines toxic air pollutants, also known as hazardous air pollutants, as those pollutants that are known or suspected to cause cancer or other serious health effects, such as reproductive effects, birth defects, or adverse environmental effects. Most air toxics originate from human-made sources, including mobile sources (for example, cars, trucks, and buses) and stationary sources (for example, factories, refineries, and power plants), as well as indoor sources (for example, some building materials and cleaning solvents). Some air toxics are also released from natural sources, such as from volcanic eruptions and forest fires. People are exposed to toxic air pollutants in many ways that can pose health risks, including directly by breathing contaminated air or indirectly by consuming or contacting materials that have been exposed to contaminated air. Accurate estimates of exposures can be supported through a measurement of the contaminated air concentration being breathed or surrounding the material. However, measurements are expensive and very few in number. As a result, measurements are not possible everywhere that they are needed.

It is important to determine the concentrations of air toxics along the air pathway from their many sources. Relationships with the sources make it possible to develop ways to decrease exposures to contaminated air below unhealthy levels. Having a model of these relationships also helps to estimate potential exposures to proposed sources of air toxics and the strength of their emissions. However, most models are not sufficient to do this for the wide range of air toxics. Air toxics computational fluid dynamics is a model simulation method that resolves scientific first principles of fluids and the laws of motion with chemistry in space and time. Thus, air toxics computational fluid dynamics provides improved accuracy, precision, and representativeness of contaminated air concentrations.

Air toxics computational fluid dynamics. Air toxics CFD is computational technology that enables reliable predictive modeling of air transport that is influenced by natural and built environments. CFD applications supporting air transport are now able to include replica geometry of the natural or built environment. Chemistry for toxic pollutants is resolved simultaneously with conservation equations for the pollutants within each CFD solution volume. Single- or multistep chemical reaction schemes are applied, depending on the required complexity of the modeled system.

Evolution. The predictive science of fluid flow began with the physical equations derived from the classical physics developed by Isaac Newton in the seventeenth century, first with its laws of motion and later using laws of conservation that may be found in any classical physics textbook. Others followed, describing the motion of fluids mathematically. The Navier-Stokes equation, which is the mathematical basis for modern CFD, was first derived by C. L. M. H. Navier in 1822. This equation is the general basis for all CFD applications, from weather prediction to vehicular aerodynamics design. The Navier-Stokes equation is nonlinear, and any solution will depend on the initial conditions at the problem boundaries. Numerical solutions require that the domain for the fluid flow being studied be divided into finite volumes. Practical solutions require some simplifying assumptions and numerical approximations. A model of a fluid-flow problem is developed by defining the boundary conditions and making some simplifying assumptions. Although reasonable models can be developed for most of the physical processes that influence air transport in the atmosphere, implementing these models into a numerical model is reliable only down to the model scale defined by the size of the finite volume. Numerical solutions of the Navier-Stokes equations began in the first half of the twentieth century with months of hand calculations, later supported by simple mechanical desk calculators. The modern age of electronic computers evolved rapidly during the second half of the twentieth century and continues into the twenty-first century. One of the first major applications was computational weather prediction, which began in 1950. Numerical air-contaminant transport models evolved within the framework of numerical meteorological (weather) models. Initial interest in modeling air-contaminant transport began in the 1950s for the long-range transport of materials from planned and unplanned radioactive sources. More important, in the 1970s, as the EPA was being established, there was a regulatory need to examine air quality in statewide regions. Current regional-scale air-transport models produce solutions on a gridded mesh of the order of 10 km. Fluid-flow and chemistry processes at smaller scales must be represented by subgrid-scale models, which are reliable only if the subgrid variations are small. As computing capacities advance, the practical scales of these regional-scale models may be reduced. CFD model applications supporting air transport are now able to include replica geometry of the natural or built environment. CFD modeling provides solutions at scales of 1-10 m for domains of 1-10 km^2. Over the next few years, implementation of the next-generation petascale [1000 trillion floating-point calculations (flops) per second] computing systems, followed by exascale (1 million trillion floating-point calculations per second) computing systems, will lead to the blending of fine-scaled regional models with larger-domain CFD applications.

CFD applications. Applications of CFD are built on classical science and the mathematical equations of fluid dynamics. Numerical solutions build on the fundamental computational methods of fluid dynamics applied to conservation laws and laws of motion with boundary and initial conditions in mathematical discretized form to estimate field variables quantitatively on a discretized grid (or for volumes) spanning the flow-field domain.

Traditionally, fine-scale CFD software uses the finite-volume numerical method to best accommodate complex geometries, unstructured grids, and chemistry. The CFD modeling process, like most numerical modeling processes, involves three main steps. The first step is to set up the problem being modeled. This includes setting up a virtual environment to replicate the real domain being studied, meshing the domain into finite volumes in which solutions will be calculated, and selecting the physical processes to be included in the model. For example, the modeling of urban environments requires geometry for all the buildings and meshing with finer mesh near building and terrain surfaces (**Fig. 1**). For air toxics CFD, the selected processes need to support the air-flow transport and chemistry that is being studied. In the second step, solutions are produced using computers to complete the numerical computations. Sufficient computing resources are needed to meet both the memory and speed requirements. In the third step, the produced solution data need to be examined and applied. This final step includes visualizations of the virtual environment to be able to view the solutions of interest.

Fig. 1. Example vertical profile of mesh.

Figures 2 and **3** show the complex wind patterns that influence the air transport and material transported from a point source near Madison Square Garden in New York City, respectively. Both commercially licensed and open-source general-purpose CFD software codes are available for public use. This software may be customized for specific applications. Some organizations develop their own customized CFD software, which is not available for public use. There is much information on the availability and applications of CFD software in the open literature. For starting points, see http://www.cfd-online.com.

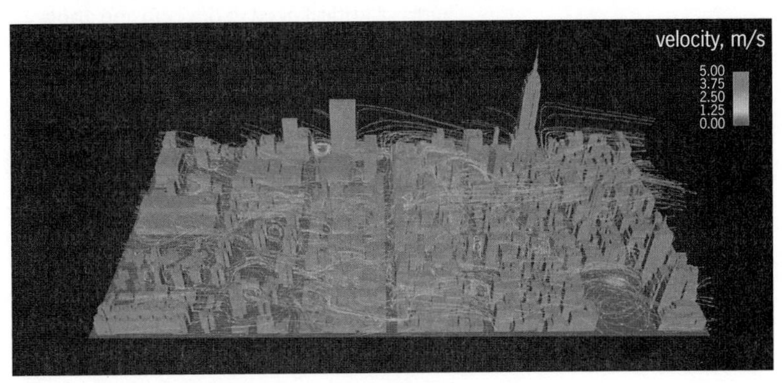

Fig. 2. Example velocity streamlines (wind patterns).

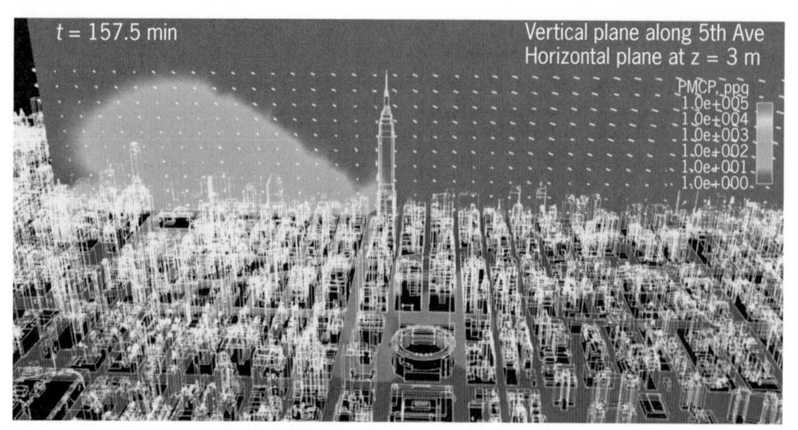

Fig. 3. Example air transport of material.

Emerging air toxics with computational fluid dynamics. Chemistry for air toxics are merged with CFD by coupling their material conservation equations within each CFD solution finite volumes. Air toxics CFD models will continue to evolve to support increased complexity for chemistry and increased spatial and temporal resolution for the fluid flow. Accurate solutions require resolutions to be resolved to scales smaller than scales that have significant gradients. For some secondary air toxics pollutants, which are produced by chemical reactions between other pollutants (for example, mixtures of volatile organic compounds and nitrogen oxides in sunlight react to form ground-level ozone), observations have shown that both temporal and spatial gradients are small relative to those used by present routine air quality models. For local air quality with local sources of air toxic pollutants, there are many situations where air toxics CFD would be necessary to provide reliable solutions.

Most chemical organizations recognize that detailed chemical kinetic modeling needs to be incorporated into the CFD model. Presently, applications for air toxics CFD are being improved because of developments in effective algorithmic design to reduce computational costs, efficient reduction of single- or multistep chemical reaction schemes of complex systems, acquisition of stable solutions within design-cycle time constraints, the handling of coverage-dependent surface reactions (catalysis), and multicomponent molecular-transport properties. There is a critical need to develop and apply air toxics CFD to meet the need for well-informed environmental decisions, especially involving human and ecological health.

For background information *see* AIR POLLUTION; COMPUTATIONAL FLUID DYNAMICS; ENVIRONMENTAL ENGINEERING; ENVIRONMENTAL FLUID MECHANICS; FLUID-FLOW PRINCIPLES; MODEL THEORY; NAVIER-STOKES EQUATION; NUMERICAL ANALYSIS; SIMULATION in the McGraw-Hill Encyclopedia of Science & Technology.

Alan H. Huber

Bibliography. R. A. Anthes and T. T. Warner, Development of hydrodynamic models suitable for air pollution and other mesometeorological studies, *Mon. Weather Rev.*, 106:1035–1078, 1978; G. K. Batchelor, *An Introduction to Fluid Dynamics*, Cambridge University Press, 1967; J. H. Ferziger and M. Peric, *Computational Methods for Fluid Dynamics*, 3d ed., Springer-Verlag, 2001; D. A. Vallero, *Fundamentals of Air Pollution*, Academic Press, 2008.

Anderson localization

The electrical transport properties of solids are strongly affected by disorder caused by the presence of impurities or imperfections in the crystal lattice. In 1958, in an article titled "Absence of diffusion in certain random lattices," P. W. Anderson demonstrated that disorder can spatially localize the electron charge carriers; that is, they become unable to carry an electric current because of their quantum-mechanical wave character. He simultaneously formulated the problem of electron localization in a disordered metal, made the link between localization and electrical transport, and gave the first quantitative estimate of the critical disorder for the disorder-driven metal-insulator transition.

Metals and insulators. The difference between metals and insulators is usually linked to the presence in a metal of negatively charged particles, that is, electrons, which can move easily through the metal. In insulators, these so-called conduction electrons are absent. In metals, there is a very high density of conduction electrons that generate an electric current when a voltage is applied across the metal. According to Ohm's law, the current grows linearly with the applied voltage. The ratio between voltage and current is the electrical resistance of a metal. Whether a material is metallic or insulating depends on its chemical composition, that is, on the constituent atoms. When copper atoms form a solid, a shiny metal with low electrical resistance is obtained. On the other hand, the arrangement of carbon atoms in a diamond crystal gives rise to an electrically insulating and optically transparent state.

For pure metals such as copper, the electrical resistance decreases when the temperature is decreased below room temperature and becomes independent of temperature at lower temperatures (**Fig. 1**). This temperature dependence can be linked to the fact that the copper atoms start to vibrate at higher temperatures and scatter the conduction electrons, giving rise to electrical resistance. The residual constant resistance at low temperatures can be linked to the presence of defects and impurity atoms that disturb the regular crystal structure. For a disordered metal, where the amount of impurities or defects is enhanced on purpose, it can be seen that the resistance first increases at lower temperatures before saturating at very low temperatures (Fig. 1) that can be reached only by cooling with liquid helium (around −269°C or −452°F). While for a perfect insulator the resistance should be infinitely large, real insulators have a very large but finite resistance as a result of thermal activation of electron

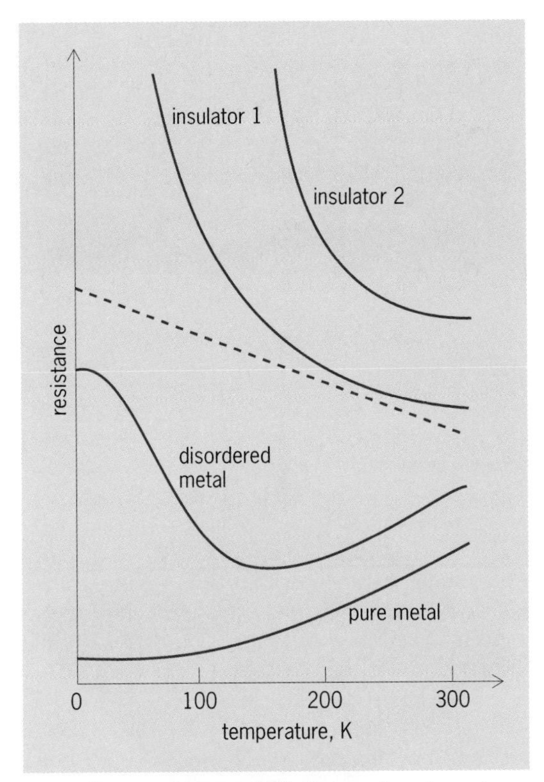

Fig. 1. Changes that occur in the temperature dependence of the electrical resistance of a metal as it becomes more and more disordered. In the metallic phase, the resistance remains finite at all temperatures, while in the insulating phase, the resistance becomes infinitely large at sufficiently low temperatures. The dashed line represents the border between the metallic phase and the insulating phase. The temperature scale is in kelvin (K). 0 K corresponds to −273.15°C (−459.67°F), while 300 K is close to room temperature.

charge carriers at higher temperatures. In contrast to that of a metal, the electrical resistance of an insulator strongly increases with decreasing temperature and becomes infinitely large at sufficiently low temperatures (Fig. 1). The stronger "insulator 2" has a faster increase in its resistance at low temperatures when compared to the weaker "insulator 1." We introduce the following operational definitions of a metal and an insulator: metals are electrical conductors at all temperatures, and their resistance curve levels off and reaches a disorder-dependent, finite value at very low temperatures; insulators are very poor electrical conductors with a resistance that decreases at higher temperatures and becomes immeasurably large at sufficiently low temperatures.

Disorder-driven metal-insulator transition. According to quantum mechanics, electrons will behave differently from particles that obey Newton's classical mechanics. In particular, the electrons should reveal a wavelike behavior. Anderson's work on the motion of electrons in disordered materials relied on solving the Schrödinger equation, that is, on calculating the shape of the electron waves for different energies of the electrons. Anderson's major finding was that, when the amount of disorder in a metal is increased, there exists a critical value for the disorder above which the metal becomes an insulator. This can be linked to the fact that for sufficiently large disorder, the electron waves that carry the electric current can no longer diffuse through the metal and become spatially localized, implying an insulating behavior. According to the intuitive Ioffe-Regel criterion, the spatial localization will start to dominate when the distance over which the electrons travel between subsequent scattering events by defects and impurities becomes smaller than the electron wavelength.

Experimental searches. Experimentally, investigators have looked in many disordered materials for the disorder-driven metal-insulator transition that Anderson predicted. In these experiments, a sufficient amount of disorder is introduced into a metal by substituting atoms of a nonmetallic element (such as silicon or oxygen) for the metal atoms. In this way, the critical value for the disorder can be reached. In Fig. 1, the metal-insulator transition is represented by the broken line. Below the transition, that is, for smaller amounts of disorder, the electrical resistance reaches a finite value at very low temperatures: the material is in the metallic phase. Above the transition, that is, for sufficiently strong disorder, the electrical resistance rapidly increases at lower temperatures: the material is in the insulating phase.

Quantum phase transition. The Anderson metal-insulator transition can be treated as a phase transition. In contrast to, for example, the phase transition between ice and water, which is driven by temperature, the Anderson transition is driven by disorder, and to observe its intrinsic properties, the transition needs to be investigated close to the absolute zero of temperature (−273.15°C or −459.67°F). Such a zero-temperature transition is referred to as a quantum phase transition, reflecting the fact that it is dominated by the quantum-mechanical wave character of the electrons.

Complicating effects. While many experimental results indeed point to the existence of a metal-insulator transition with increasing disorder, simply measuring the temperature dependence of the resistance for a series of samples with increasing disorder does not provide unambiguous evidence for the presence of Anderson localization. Indeed, several other mechanisms that occur when a metal is mixed with a nonmetallic element can modify the electrical conduction process. In particular, the electrical conduction can be strongly affected by changes in the Coulomb repulsion between the electrons (repulsion between electric charges having the same sign) and by percolationlike electrical transport, where spatial inhomogeneities in the material composition give rise to a metallic path that percolates through an insulating matrix.

Scaling theory and interference effects. A major breakthrough with respect to properly identifying Anderson localization was provided by the work of E. Abrahams, Anderson, D. C. Licciardello, and T. V. Ramakrishnan, who in 1979 published their "scaling theory" for the Anderson transition. This theory predicted that a continuous transition from "weak

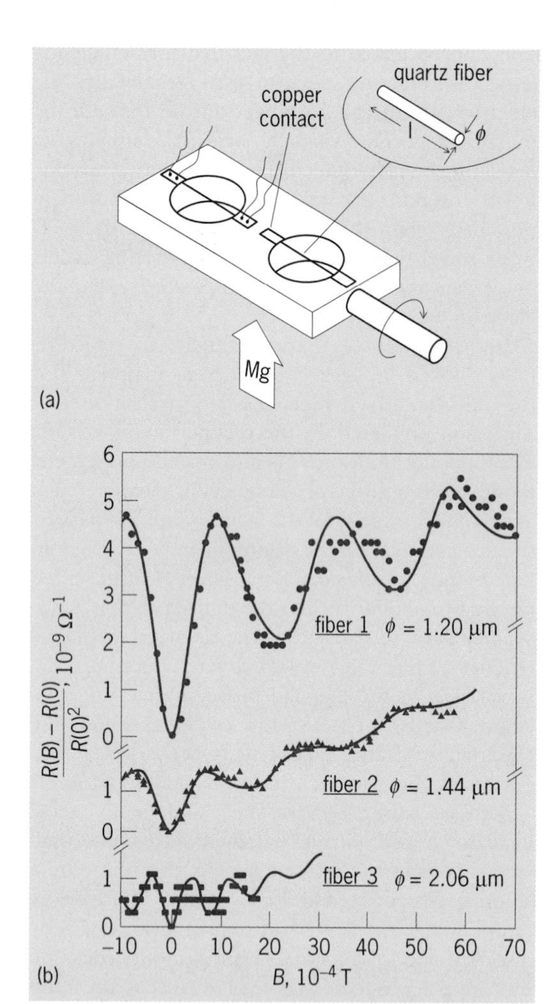

Fig. 2. Aharonov-Bohm resistance oscillations in thin-walled metal cylinders. (*a*) Fabrication of thin-walled magnesium cylinders with length *l* of 3–5 mm and micrometer-size diameter ϕ. A very thin film of magnesium is deposited by thermal evaporation onto a rotating quartz fiber that is connected to copper metal contacts. (*b*) Experimental observation of the oscillations at a temperature of 1.5 K: the resistance oscillations, which occur when a magnetic field *B* is applied parallel to the cylinder axis, are strongly damped for larger diameter ϕ because of the destruction of the quantum interference effects for electrons that travel around the cylinder. *R(B)* is the resistance of the cylinder at the magnetic field *B*, given on the horizontal axis, and *R*(0) is this resistance when *B* = 0. (*From M. Gijs, C. Van Haesendonck, and Y. Bruynseraede, Resistance oscillations and electron localization in cylindrical metal films, Phys. Rev. Lett., 52:2069–2072, 1984*)

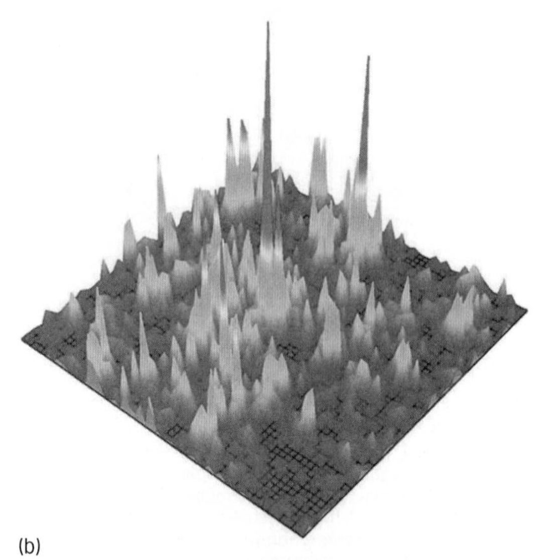

Fig. 3. Anderson localization of ultrasound in a disordered network of aluminum beads. (*a*) Zoom-in image of a network of 4-mm-diameter aluminum beads that are brazed together to form disk-shaped slabs with thicknesses ranging from 8 to 23 mm. (*b*) Lateral spatial distribution of the ultrasound wave intensity of 2.4-MHz waves that are emitted by a source on one side of the slab and, after transmission through the slab, are measured using a movable detector on the opposite side of the slab. The lateral size of the distribution, where the spatially separated transmission peaks are a direct result of the Anderson localization, is about 15 mm. (*From H. Hu et al., Localization of ultrasound in a three-dimensional elastic network, Nature Physics, 4:945–948, 2008*)

localization" toward "strong localization" of the electron wave functions occurs with increasing disorder. The weak localization gives rise to an increasing resistance for a disordered metal at lower temperatures (as seen in the curve labeled "disordered metal" in Fig. 1). On the other hand, the weak localization effects become much more pronounced in thin metal films and long metal wires.

Building further on Anderson's original work, the transition from metallic to insulating behavior could be linked to the presence of interference between the scattered electron waves, similar to the interference effects that occur between light waves that have traveled along different paths. The interference between the electron waves can be conveniently tuned by applying a magnetic field via the so-called Aharonov-Bohm effect. For cylindrically shaped metal films, the tuning gives rise to periodic oscillations of the resistance, where a minimum (maximum) resistance corresponds to the destructive (constructive) interference that in the case of an optical interference pattern gives rise to dark (bright) fringes. The experimental observation of the Aharonov-Bohm oscillations in hollow metal cylinders (**Fig. 2**) confirms that the Anderson metal-insulator transition can be directly linked to the quantum-mechanical wave character of the conduction electrons, and that interference effects are the driving mechanism behind the transition.

Localization of classical waves. In order to circumvent the unavoidable complications related to the influence of Coulomb repulsion on the Anderson localization of electron waves, considerable interest in classical-wave localization emerged in the early 1980s. Many experimental efforts focused on the localization of electromagnetic waves—in particular, optical waves—in strongly scattering media where wave absorption becomes an important obstacle.

Many questions remain concerning the strong localization of classical waves. Nevertheless, clear signatures of Anderson localization have been observed, for example, in ultrasound transmission through random networks of aluminum beads that are brazed together (**Fig. 3*a***). For frequencies in the range from 0.2 to 3 MHz, where the wavelength is comparable to the bead and pore sizes, the frequency dependence of the scattering makes it possible to probe both sides of the transition. For a frequency of 2.4 MHz, one has entered the regime of strong Anderson localization. This results in the presence of spatially separated transmission peaks (Fig. 3*b*) that are caused by the spatial localization of the ultrasound in the disordered network of aluminum beads.

Prospects. After a little more than 50 years, Anderson's original work continues to inspire many experimental and theoretical researchers, who try to further improve and extend our understanding of the Anderson localization phenomenon and the associated quantum phase transition physics in disordered electronic systems. To state it in Anderson's words, in his 1977 Nobel Lecture, "Localization, very few believed it at the time, and even fewer saw its importance. Among those who failed to fully understand it at first was certainly the author." A striking example of the continuing research efforts is the "topological insulators," which are insulating materials with electrically conducting surfaces. In order to properly understand the appearance of the surface conductance, it is essential to take into account that the electron charge carriers also carry a magnetic moment that results from their spin, that is, their fast intrinsic rotation (spinning). In this way, there emerges a close link with spintronics, another very lively field of modern solid-state physics that focuses on the use of spin rather than of charge to develop electronic components with strongly reduced dimensions.

For background information *see* AHARONOV-BOHM EFFECT; ELECTRIC INSULATOR; ELECTRICAL CONDUCTIVITY OF METALS; ELECTRICAL RESISTIVITY; ELECTRON SPIN; INTERFERENCE OF WAVES; NONRELATIVISTIC QUANTUM THEORY; PHASE TRANSITIONS; QUANTUM MECHANICS; SCHRÖDINGER'S WAVE EQUATION in the McGraw-Hill Encyclopedia of Science & Technology.

Chris Van Haesendonck;
Yvan Bruynseraede

Bibliography. H. Hu et al., Localization of ultrasound in a three-dimensional elastic network, *Nat. Phys.*, 4:945–948, 2008 A. Lagendijk, B. Van Tiggelen, and D. S. Wiersma, Fifty years of Anderson localization, *Phys. Today*, 62(8):24–29, August 2009; X. L. Qi and S. C. Zhang, The quantum spin Hall effect and topological insulators, *Phys. Today*, 63(1):33–38, January 2010.

Animal prions

Prion diseases have been recognized in several animal species and include scrapie in sheep, bovine spongiform encephalopathy (BSE or "mad cow disease") in cattle, chronic wasting disease (CWD) in deer and elk, and transmissible mink encephalopathy (TME) in mink. In humans, the most common form of prion disease is Creutzfeldt-Jakob disease (CJD). A new form of CJD in humans called variant CJD (vCJD) has been linked with exposure to BSE-contaminated materials. Thus, animal prion diseases have become a significant public health issue and have raised concerns about food safety. In light of the fact that BSE has caused a new form of prion disease in humans, as well as continuing uncertainty as to whether or not other animal prions may do the same, it is critical to understand how animal prions infect humans.

Prion diseases. Prions are a novel class of infectious agents that are responsible for a devastating group of rare but fatal diseases known as the transmissible spongiform encephalopathies (TSE or prion diseases). Prion diseases are defined by the damage that prions cause in the central nervous system, where infection can lead to the development of small holes or vacuoles in the brain, giving a very distinctive "spongiform" appearance to the brain tissue (**Fig. 1**). In most animal prion diseases as well as in vCJD and kuru (a prion disease that was once endemic in the cannibalistic Fore tribe of New Guinea), prion infection occurs via ingestion of contaminated food or other materials. Human infection also can occur via direct inoculation of infected tissues during certain medical procedures or, in the case of vCJD, via blood transfusion. Following infection, disease progresses slowly over a period of months or even years. There are no preclinical diagnostic tests or treatments available and, following the onset of symptoms (including imbalance, memory loss, and dementia), prion diseases are always fatal.

Prion spread between individuals within a species depends upon the type of prion (see **table**), but spread from one species to another can occur via the ingestion of prion-contaminated materials. The relative resistance of one species to infection with

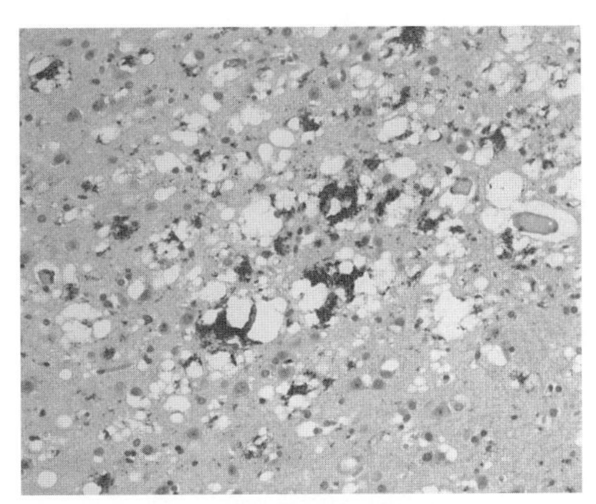

Fig. 1. Brain tissue from an individual with Creutzfeldt-Jakob disease (CJD) shows the spongiform appearance and deposition of PrP^{Sc} aggregates (dark areas) that are characteristic of prion diseases.

Major prion diseases of animals and humans

Prion disease	Infected tissues	Animal-to-animal spread	Host range
Bovine spongiform encephalopathy (BSE)	Brain, spinal cord, distal ileum	No	Bovine species, cats, humans, primates, goats
Scrapie	Brain, spinal cord, lymph nodes, spleen, placenta	Yes	Sheep, goats
Chronic wasting disease	Brain, spinal cord, lymph nodes, feces, spleen, urine, saliva	Yes	Deer, elk
Transmissible mink encephalopathy	Brain, spinal cord, lymph nodes, spleen	Unknown	Mink
Creutzfeldt-Jakob disease	Brain, spinal cord, cornea	No	Humans
Variant Creutzfeldt-Jakob disease (BSE origin)	Brain, spinal cord, lymph nodes, spleen, blood, appendix	Yes (blood transfusion)	Humans

prions from another species is known as the prion disease species barrier. Species barriers to BSE are relatively weak since goats, domestic cats, captive large cats (such as the cheetah), and exotic zoo animals (such as the greater kudu) developed BSE after being exposed to BSE-contaminated feed. Individuals who developed vCJD also were exposed presumably to BSE-contaminated food, although the type of material has never been determined. By contrast, there is no evidence that sheep scrapie has ever crossed species barriers to cause disease in people, even though humans consume products derived from sheep. In order to understand how prion species barriers are controlled, it is first necessary to understand how prions replicate.

Prion replication. Most infectious agents contain DNA or RNA nucleic acid genomes that encode the characteristics of that agent and are required for its replication and spread. However, unlike other infectious organisms such as viruses, bacteria, or fungi, prions do not appear to contain a nucleic acid genome. Rather, they appear to be composed primarily of an abnormal form of a protein known as prion protein (PrP). Abnormal PrP, known as PrP^{Sc} for PrP "scrapie," is found only in prion diseases and is believed to be the primary component of the infectious prion. PrP^{Sc} is not degraded easily by the cell, and is capable of replicating itself and accumulating over time into large, insoluble aggregates. Formation of these aggregates is what leads to the characteristic spongiform appearance of the brain (Fig. 1) and fatal prion infection.

The key to understanding how a prion can replicate itself in the absence of any nucleic acid genome is the fact that the primary component of the infectious particle, PrP^{Sc}, is directly derived from a form of prion protein that is naturally found in the infected host. The gene for PrP is highly conserved and is found in all mammalian species, including humans. The function of normal cellular PrP, known as PrP^{C} for PrP "cellular," is unknown, but it is expressed on the surface of most cells of the body. Unlike PrP^{Sc}, it is degraded easily and thus does not accumulate into large aggregates. When an individual is exposed to prion infectivity, PrP^{Sc} in the infectious material binds to the host's own PrP^{C} molecules and, through a poorly understood sequence of events, converts it into PrP^{Sc}. The newly made PrP^{Sc}, which is now derived from the infected host's PrP^{C}, binds and converts PrP^{C}, and the process repeats (**Fig. 2**, line 1). This replication process, which is dependent upon the host PrP^{C} molecule, is known as seeded polymerization.

All proteins are composed of long chains of amino acid building blocks that fold into a specific three-dimensional structure, known as the protein's conformation, which helps to determine its function. The process of seeded polymerization involves essentially a protein being forced to assume a different shape, which can change its properties and alter its function. This is somewhat analogous to what happens if one takes two identical pieces of paper and folds one piece into a paper airplane and crumples the other piece into a ball. Even though the starting material is exactly the same, the way the paper is folded differs, leading to final products with very different properties. At its most basic, the conversion of PrP^{C} to PrP^{Sc} can be explained in the same way.

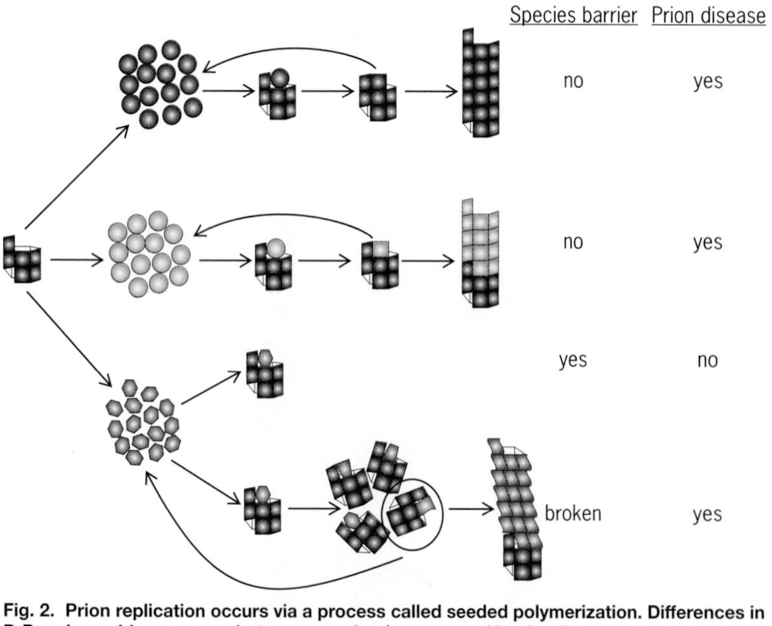

Fig. 2. Prion replication occurs via a process called seeded polymerization. Differences in PrP amino acid sequences between species (represented by the different shades) help to determine whether or not a prion can cross species barriers by influencing the final structure of PrP^{C} and PrP^{Sc} (represented by different shapes). Squares = PrP^{Sc}; circles and hexagons = PrP^{C}.

Because PrP^{Sc} is directly derived from the host's own PrP^{C} molecules, both forms of PrP are composed of the same amino acid starting material. However, when PrP^{C} is converted to PrP^{Sc}, it is forced to fold into a different three-dimensional structure. It is this new structure that enables the unique properties of PrP^{Sc} and distinguishes an infectious prion from normal prion protein.

PrP amino acid sequence and prion species barriers. Although PrP is found in all mammals, its amino acid sequence varies between different species. Since the conversion of PrP^{C} to PrP^{Sc} involves a change in the manner that PrP is folded, even small differences in its amino acid sequence can have significant effects on PrP^{Sc} formation and prion replication. This suggests a molecular basis for species barriers to prion infection based upon amino acid mismatches between PrP^{C} in the host and PrP^{Sc} in the infectious prion. Researchers have studied the importance of the PrP amino acid sequence in maintaining species barriers using mice that are susceptible to mouse prions, but highly resistant to hamster prions. Mice and hamster PrP^{C} differ at multiple amino acid residues. However, when mice are genetically engineered to express hamster PrP^{C} (known as transgenic mice), they become fully susceptible to hamster scrapie. Thus, the host prion protein amino acid sequence is a major determinant of whether or not a prion can cross a species barrier.

If the host PrP^{C} and incoming PrP^{Sc} do not have matching amino acid sequences, the prediction would be that PrP^{Sc} will be unable to replicate and infection will not occur. However, this is not always the case. For example, BSE prions can infect humans, even though the amino acid sequences of human and cattle PrP are different. This suggests that only certain amino acid differences will have a significant impact on the conversion of PrP^{C} to PrP^{Sc} and thus prion replication. Multiple studies in transgenic mice as well as in cell culture have used PrP^{C} molecules with amino acid sequences from more than one species to show that even a single amino acid difference between PrP^{C} and PrP^{Sc} can be sufficient to prevent species-specific PrP^{Sc} formation. Moreover, only certain amino acid residues are important in this process and the specific residues can vary from species to species. These specific but critical differences in amino acid sequence between PrP^{Sc} and PrP^{C} provide a mechanism for how prion disease species barriers are controlled. When the amino acid sequences of the host PrP^{C} and PrP^{Sc} match, PrP^{Sc} can replicate itself and disease occurs (Fig. 2, line 1). If the amino acid sequences do not match at certain critical positions, PrP^{Sc} does not form and disease does not occur (Fig. 2, line 3). A species barrier is crossed if the differences in the PrP^{C} and PrP^{Sc} amino acid sequences do not occur at amino acid residues critical for the formation of PrP^{Sc} within that species (Fig. 2, line 2).

PrP^{Sc} structure and species barriers. A protein's conformation is dependent on its amino acid sequence, but some amino acids will have a greater effect on the final structure than others. The fact that some species barriers to prion infection can still be crossed even though the PrP amino acid sequences do not match, as well as the fact that only certain amino acid residues appear to affect significantly PrP^{Sc} formation, suggests that it is the structure of PrP that is the final determinant of whether or not PrP^{Sc} can replicate efficiently in a new species. This in turn means that simply comparing the amino acid sequences of PrP^{C} from different species cannot be used to predict susceptibility to prion infection. Experimental techniques, such as those described above using transgenic mice expressing PrP^{C} molecules from different species, are currently in use to determine the susceptibility of one species to infection with prions from another species.

Unfortunately, prion species barriers are not absolute. When a prion from one species infects a new species, it is possible, even if there are significant amino acid differences between PrP^{Sc} and PrP^{C}, for a small amount of PrP^{Sc} to be formed over the lifetime of the host. If material from this disease-free animal is given to another animal of the same species, then that small amount of PrP^{Sc}, which now has the same amino acid sequence as the host PrP^{C}, will efficiently trigger new PrP^{Sc} formation, prion replication, and potentially prion disease. If this process is repeated several times, a prion from one species can eventually cause disease even in a resistant species (Fig. 2, line 4).

This process provides a possible explanation for the origin of BSE. Changes in the way that cattle and sheep were processed into meat and bone meal (MBM) may have allowed sheep scrapie to contaminate the MBM feed. When the MBM was fed to cattle, sheep PrP^{Sc} could have inefficiently converted a small amount of bovine PrP^{C} to PrP^{Sc}. Infected but clinically normal animals then were processed into MBM, which was fed back to cattle. The small amount of bovine PrP^{Sc} triggered another, more efficient round of prion replication. Repetition of this process over the course of years may have led eventually to the new form of prion, which caused BSE in cattle. Since sheep scrapie does not infect people but BSE does, one of the lessons of the BSE epidemic is that a prion crossing a species barrier not only can adapt to a new species but also can develop an expanded host range. Thus, it remains extremely important to continue to monitor domestic and wild animal populations for any indication that new types of prions have emerged that might cross species barriers to cause disease in humans.

For background information *see* BRAIN; INFECTIOUS DISEASE; NERVOUS SYSTEM DISORDERS; PRION DISEASE; PROTEIN; PUBLIC HEALTH; SCRAPIE; ZOONOSES in the McGraw-Hill Encyclopedia of Science & Technology. Suzette A. Priola

Bibliography. S. J. Collins, V. A. Lawson, and C. L. Masters, Transmissible spongiform encephalopathies, *Lancet*, 363:51–61, 2004; D. A. Harris (ed.), *Mad Cow Disease and Related Spongiform Encephalopathies*, Springer-Verlag,

Berlin/Heidelberg/New York, 2004; R. A. Moore, L. M. Taubner, and S. A. Priola, Prion protein misfolding and disease, *Curr. Opin. Struct. Biol.*, 19:14–22, 2009; C. J. Sigurdson, A prion disease of cervids: Chronic wasting disease, *Vet. Res.*, 39:41, 2008; C. J. Sigurdson and M. W. Miller, Other animal prion diseases, *Br. Med. Bull.*, 66:199–212, 2003.

Anxiety disorders and the amygdala

Everyone experiences, at least on occasion, feelings of anxiety—for example, pangs of dread before giving a public speech, worries about relationships or success at work, or a swelling sensation of fear when walking down a dimly lit street. These common experiences, however, are qualitatively different from clinical anxiety disorders, which are more than just exaggerated manifestations of these common experiences. Rather, individuals with anxiety disorders are burdened frequently by invasive thoughts and overwhelming emotions that can come on without warning, cause severe personal distress, and interfere with an individual's day-to-day functioning and personal relationships. As many as 40 million adults in the United States suffer from clinical anxiety disorders, which fall primarily into six diagnostic categories: panic disorder (PD), generalized anxiety disorder (GAD), social phobia or social anxiety disorder (SAD), specific phobia, posttraumatic stress disorder (PTSD), and obsessive-compulsive disorder (OCD).

There are striking differences among these disorders with regard to the circumstances that induce anxiety and the symptoms associated with each. In spite of these differences, though, all anxiety disorders appear to share at least one common element: the extreme, maladaptive, and seemingly unreasonable anticipation of negative events that might take place in the future. Such excessive anticipation can take the form of heightened fear of a specific object of immediate, certain danger to the individual. Alternatively, the anticipated threat can be more distal to the individual and its occurrence can be less certain or predictable. Appropriately, then, our understanding of the neural basis of anxiety disorders has benefited from extensive research on the amygdala, a brain region involved in responding both to certain, immediate threats as well as to conditions of uncertainty that require heightened vigilance and environmental monitoring.

Amygdala research. Tucked away deep within the medial temporal lobe of the brain, the amygdala is not actually a singular structure, but rather a collection of discrete cell bodies with distinct characteristics. The heterogeneous nature of this structure suggests that abnormalities within specific subregions may help explain the diverse characteristics of different anxiety disorders. Also of key relevance in its contributions to anxiety is the fact that the amygdala forms myriad connections with brain regions involved in movement, sensation, motivation, reward, and cognition.

Early research in primates demonstrated that destruction of the medial temporal lobe including (but not limited to) the amygdala resulted in dramatic changes in emotional function, particularly related to the expression of fear. These early observations set the stage for decades of research that characterized the amygdala as a "fear center" in the brain. Extensive research in rodents established the importance of distinct amygdala subregions in the acquisition of conditioned fear and the expression of fear responses, and research in humans has provided additional support for the role of the amygdala in learning associations involving feared or aversive stimuli.

In recent years, however, it has become increasingly clear that the amygdala is not a dedicated fear center, but a more dynamic region that responds to a wide range of experimental conditions. This region shows enhanced responses to salient stimuli in the environment, even when they are not aversive or threatening, and is involved in forming new associations involving rewarding stimuli. A number of human and animal studies have demonstrated increased amygdala activity under conditions of unpredictability or uncertainty, particularly for aversive events, but also in cases involving unpredictable neutral stimuli. An updated perspective on the function of the amygdala emphasizes its role in alerting the individual about novel, changing, or threatening aspects of the environment and engaging other brain regions that encourage responses or behaviors that promote safety, reward, and survival.

Neuroimaging studies. Research on animal models of anxiety, including fear-conditioning paradigms, lesion models, and pharmacological studies, has established a likely role for amygdala involvement in anxiety disorders. Subsequent research investigating amygdala dysfunction in humans with clinical anxiety disorders has been bolstered by the emergence of functional brain imaging techniques, which provide a noninvasive means of assessing brain activity with ever-improving temporal and spatial resolution. Chief among these methods are functional magnetic resonance imaging (fMRI) and positron emission tomography (PET), both of which provide indirect measures of neural activity in awake, responding humans.

The first wave of functional imaging studies in anxiety disorders utilized symptom provocation paradigms, in which disorder-specific stimuli are presented to clinically anxious patients while they undergo brain scanning. A general (although not always replicated) finding of these studies is that of greater amygdala activity in patients relative to healthy individuals. This hyperactivity has been observed in SAD patients taking part in a public speaking task or viewing threatening faces, PTSD patients presented with written or recorded scripts based on personal traumatic experiences, and individuals with specific phobias presented with images or videos relevant to their feared object.

This line of research showed that disorder-relevant stimuli robustly engage the amygdala in anxiety, but did not address whether elevated amygdala responses were specific to disorder-relevant stimuli or whether they reflected a more general pattern of aberrant amygdala activity. To address this issue, a number of studies have been conducted using more general "emotional stimuli," including pictures of threatening faces as well as aversive images and videos, which commonly elicit robust amygdala responses even in nonanxious individuals. These studies again revealed, across diagnoses, a general pattern of amygdala hyperactivity in response to emotionally evocative stimuli.

A number of recent functional imaging studies have provided additional insight into a broader role for the amygdala than just showing heightened activity to emotionally evocative stimuli. Consistent with the central role of excessive anticipatory activity across anxiety disorders, heightened amygdala activity has been observed during the anticipation of public speaking in SAD patients and anticipated peer judgment in adolescents with anxiety disorders. Additionally, the anticipation of aversive as well as neutral images in patients with GAD is associated with heightened amygdala activity. These studies provide evidence that amygdala hyperactivity in anxiety disorders may not reflect increased negative affect or heightened fear responses per se, but rather increased environmental monitoring or hypervigilance in preparing for potential threats or negative outcomes.

Connections with other brain regions. It is important to note that the amygdala does not function in isolation, and its abnormal activity in anxiety is likely related to altered activity in connected brain regions. In individuals with PTSD and GAD, the degree of amygdala activation in response to fearful faces is inversely related to activity in parts of the anterior cingulate cortex (ACC) and prefrontal cortex, which are brain regions involved in emotion regulation and conflict monitoring. These studies raise the possibility that heightened amygdala activity may be the result of abnormally low regulatory activity by prefrontal regions. Consistent with this hypothesis are the findings of recent studies using diffusion tensor imaging (DTI), a technique that provides a measure of the structural integrity of white matter pathways in the brain. In individuals with GAD and SAD, structural abnormalities have been observed in the uncinate fasciculus, a pathway connecting the amygdala with parts of the prefrontal cortex.

Neuroimaging genetics. A promising, yet nascent field of research investigates genetic influences on brain activity and structure that may contribute to anxious pathology. Of particular interest is a gene that codes for a protein involved in the transport and regulation of serotonin, a neurotransmitter with a substantial influence on amygdala function. Healthy individuals with the low-expressing form of this gene show increased amygdala responses while viewing emotional faces. These individuals also show a relative "decoupling" of activity between the amygdala and ACC. Similar to findings in GAD and SAD, this low-expressing genetic variant also is associated with poorer structural integrity of the uncinate fasciculus. It is hypothesized that this genetic variant results in enhanced amygdala reactivity under stressful conditions and thus may represent a risk factor for the development of anxiety disorders. Taken together, this research on the serotonin transported gene provides but one example of the ways in which neuroimaging genetics may bolster our understanding of biological factors that contribute to anxiety through mechanisms involving the amygdala.

Unanswered questions and future work. Although the development of new technologies and creative research methods have led to increased knowledge of the neural basis of clinical anxiety, questions regarding causal mechanisms remain largely unanswered. It is unclear, for example, whether abnormally elevated amygdala activity is a causal factor in the formation of anxiety disorders or whether it is a consequence of living with these disorders. To address such questions, researchers could use prospective research designs in healthy individuals with elevated probabilities of developing clinical anxiety—for example, subjects at high risk based on family history or genetic markers, or those likely to encounter traumatic circumstances that may lead to PTSD. Collecting neuroimaging data from these individuals before and after they develop disorders would add to our understanding of neurally based causal mechanisms or risk factors related to anxiety.

Finally, as acknowledged earlier, anxiety and the amygdala are both complex, heterogeneous entities. As evidenced by the cursory treatment of these complexities, it is not well understood how functional alterations in distinct amygdala subregions might map onto different manifestations of clinical anxiety. There is a rich literature in rodent models of anxiety suggesting that medial versus lateral nuclei play a role in responding, respectively, to more immediate threats versus threats that are more distal and unpredictable. How this maps onto anxiety in humans is unclear, and the relatively coarse spatial resolution of fMRI and PET has made such anatomical differentiation in humans difficult. Advancements in the understanding of these distinct roles in humans will contribute to a refined knowledge of the role of amygdala dysfunction in clinical anxiety and perhaps to a biologically informed system of classification, diagnosis, and treatment.

For background information *see* ANXIETY DISORDERS; BRAIN; MEDICAL IMAGING; NEUROBIOLOGY; OBSESSIVE-COMPULSIVE DISORDER; PHOBIA; POSTTRAUMATIC STRESS DISORDER; STRESS (PSYCHOLOGY) in the McGraw-Hill Encyclopedia of Science & Technology.

Daniel W. Grupe; Jack B. Nitschke

Bibliography. M. Davis and P. J. Whalen, The amygdala: Vigilance and emotion, *Mol. Psychiatr.*, 6:13–34, 2001; A. Etkin and T. D. Wager, Functional neuroimaging of anxiety: A meta-analysis of emotional processing in PTSD, social anxiety disorder, and specific phobia, *Am. J. Psychiatr.*, 164:1476–1488, 2007; J. B. Nitschke et al., Anticipatory activation in the amygdala and anterior cingulate in generalized anxiety disorder and prediction of treatment response, *Am. J. Psychiatr.*, 166:302–310, 2009; P. J. Whalen and E. A. Phelps, *The Human Amygdala*, Guilford Press, New York, 2009.

Asteroid extinction

It is generally recognized that the impact of a 6-to-10-km-diameter (4-to-6-mi) asteroid 65 million years ago contributed significantly to the extinction of the dinosaurs and many other species living at the time. Fortunately, these types of events are very rare, but there is growing recognition that impacts of smaller objects can cause significant regional disasters and that these events can happen relatively frequently. Humanity may now have the technology to divert these smaller objects, but we need to know where they are and to decide how and when to take action. This article overviews the nature of the threat, describes options for deflecting a threatening object, and discusses uncertainties that will affect the decision to act.

Nature of the threat. In 2003, an object estimated to be 2 m (6.5 ft) in diameter entered the atmosphere and exploded, sending fragments through the roof of a home near Chicago. In June 2006, a meteor exploded over a remote area of Norway with a force equivalent to an atomic bomb. A similar event in April 2010 caused a brilliant explosion over southwestern Wisconsin and showered debris on farms below. Nuclear detonation–detecting satellites recorded 136 atmospheric blasts in the megatons-of-TNT range over the period from 1975 to 1992. These types of events occur every few months.

In 1908, an object approximately 10 times larger—perhaps 30 m (100 ft) in diameter—entered the atmosphere and exploded over a remote area of Siberia. The blast leveled over 2000 km^2 (800 mi^2) of forest, an area larger than the Washington, D.C., metropolitan area. This type of event occurs less frequently, once every few centuries, but can cause devastation in a local area. Impacts of objects hundreds of meters in diameter probably occur at somewhat longer intervals (see **table**).

Finding hazardous objects. Potentially hazardous asteroids (PHAs) are defined to be objects larger than about 150 m (500 ft) in diameter that could come close to Earth. In 1998, the U.S. Congress authorized an effort to discover, by 2008, 90% of asteroids and comets larger than 1 km (0.62 mi) in diameter whose orbits cross that of Earth. Since that effort began, approximately 147 large, potentially hazardous asteroids have been found (**Fig. 1**), and there is a general belief that the 90% goal for objects larger than 1 km in diameter has been reached. The discovery effort has also found nearly 1000 potentially hazardous asteroids smaller than 1 km in diameter, and some believe that the total number of potentially hazardous asteroids could rise to as many as 20,000 if the discovery effort is broadened to include 90% of objects larger than 140 m (460 ft) in diameter.

Although no object larger than 1 km in diameter was discovered that poses a hazard in the next 100 years, the survey has found two smaller objects that do: a 300-m-diameter (1000-ft) object known as Apophis will pass very close to Earth (closer than geostationary weather satellites) in 2029 and has a probability of 1 in 250,000 of impacting Earth in 2036; and 2007 VK184, a 130-m-diameter (430-ft) object with a current impact probability in 2048 of 1 in 3000. The probabilities for both will likely drop as more observations become available and orbits are refined, but should either one hit the Earth, it would be a bad day for the planet and its inhabitants.

Options for defense. It is inevitable that action will be required to prevent an impact, and it is possible that action could be required in this century. Should 2007 VK184, Apophis, or another object warrant such action, what are the options?

Generally speaking, the desire is to move an approaching object intact to an orbit away from Earth. Working with an intact object assures that no fragments will be left in the original orbit to remain a hazard and also allows additional deflection efforts against a known target, if required. Minimizing the velocity change, ΔV, required to move an object dictates that the ΔV be applied as many years before the projected impact as possible (**Fig. 2**).

Deflection options fall into two basic categories: slow-push techniques, which apply a small force

Impact frequency and effects for objects of various sizes

Characteristic diameter of impacting object	Type of event	Approximate impact energy, MT*	Characteristic diameter impact interval, years
25 m (80 ft)	Airburst	1	200
50 m (160 ft)	Local scale	10	2000
140 m (460 ft)	Regional scale	300	30,000
300 m (1000 ft)	Continent scale	2,000	100,000
600 m (2000 ft)	Below global catastrophe threshold	20,000	200,000
1 km (0.6 mi)	Possible global catastrophe	100,000	700,000
5 km (3 mi)	Above global catastrophe threshold	10,000,000	30 million
10 km (6 mi)	Mass extinction	100,000,000	100 million

* MT stands for the chemical energy released by 1 million tons of TNT (5×10^{15} J).
SOURCE: National Research Council, *Defending Planet Earth: Near-Earth Object Surveys and Mitigation Strategies: Final Report*, The National Academies Press, 2010

over a long time to slowly move an approaching object to a nonthreatening orbit; and impulsive techniques, which move the object away by applying one or more instantaneous impulses. Some candidate slow-push techniques include: attaching a long-burning propulsion system to an asteroid or comet; attaching mass drivers to mine and accelerate asteroid material away at high speed, thereby providing thrust; shining a laser or focusing solar energy on a spot on the object, with departure of heated material providing a small thrust; and parking a spacecraft very close to the object and utilizing the gravitational attraction force between the objects to tow the object to a new orbit.

Impulsive techniques include kinetic impact—striking the object with one or more "bullets" (which could be spacecraft or other objects) travelling at very high relative velocities—and explosive techniques such as conventional and nuclear explosives that might be activated below, on, or above the surface of the object.

To date, the only technique that has actually been tested is kinetic impact: the Deep Impact mission used a 10.2-km/s (6.3-mi/s) impact of a 370-kg (820-lb) impactor to expel material from the comet Tempel 1 for analysis, and this impact provided a very tiny nudge (0.0001 mm/s) to the approximately 6-km-diameter (4-mi) body.

Need for research. The basic concept of applying a force to an object is easy to understand, but there are complications that will affect the success of a deflection mission. First, each comet and asteroid is unique. Some are likely to be solid, relatively homogenous bodies, others are accumulations of a few or large numbers of fragments held together by gravity; some are formed from a relatively low-density material, and others have high concentrations of nickel and iron; they may have very irregular shapes, be tumbling, or have an accompanying moon (**Fig. 3**) that could also be of sufficient size to be a threat. Some of these characteristics may be unknown or have a high degree of uncertainty without a data-gathering mission preceding a deflection attempt.

Uncertainties in these properties, along with inexact estimates of the object's size and density, create large uncertainties in the object's mass, a quantity that is critical in the design of a deflection mission and in predicting the response of the object to a deflection attempt. For example, a deflection technique that requires attachment to the object must be compatible with a range of possible material properties and must be designed to provide a force in the desired direction even if the object is tumbling. Similarly, an explosive device might fragment an object and make subsequent deflection attempts much more difficult.

Research is clearly needed to improve our understanding of the nature and variability of asteroids and comets. Research and possibly demonstration missions are required to assure confidence in deflection techniques and their effects.

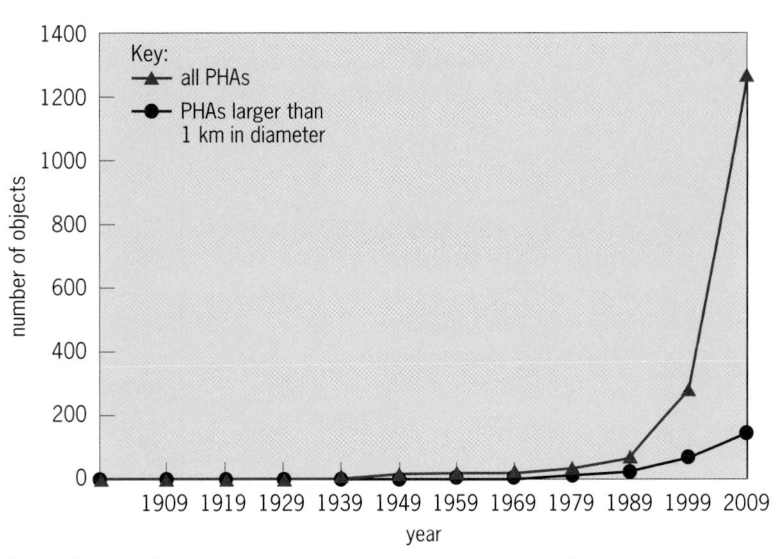

Fig. 1. Number of known potentially hazardous asteroids and number of potentially hazardous asteroids larger than 1 km (0.62 mi) in diameter from 1900 to May 2010.

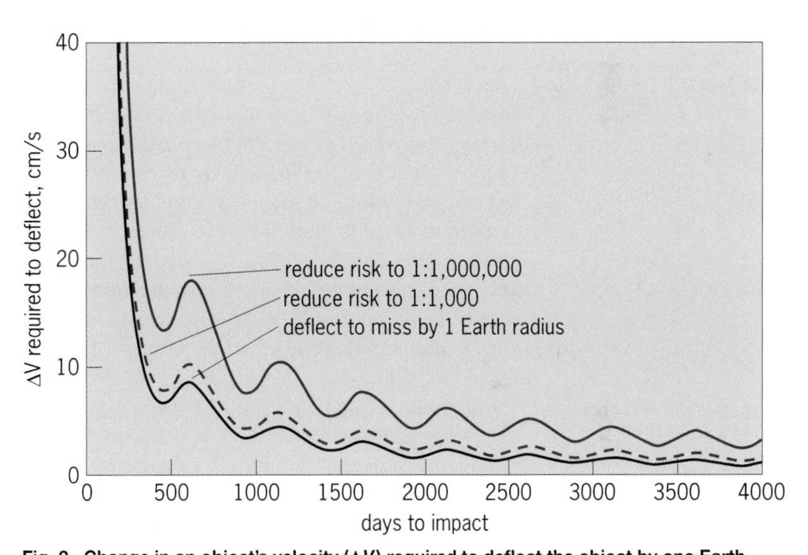

Fig. 2. Change in an object's velocity (ΔV) required to deflect the object by one Earth radius or to lower impact risk. The oscillation is due to the point in the object's orbit where the ΔV is applied.

Additional considerations. In addition to uncertainties in an object's physical properties, there are also uncertainties in an object's orbit that will affect predictions of impact. This phenomenon can be

Fig. 3. Asteroid Ida and its moon Dactyl. Ida measures 54 x 24 x 15 km (34 x 15 x 9 mi); Dactyl is 1.4 km (0.9 mi) in diameter. (*Photo courtesy NASA*)

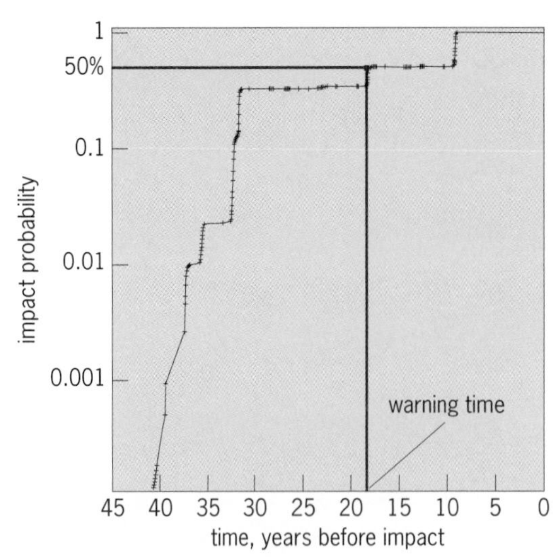

Fig. 4. Sample impact probability progression. (*From P. W. Chodas, NEO warning times for NEAs and comets, 2007 Planetary Defense Conference, George Washington University, Washington D.C., March 5–8, 2007, http://www.aero.org/conferences/planetarydefense/2007papers/S1-4-Chodas-Brief.pdf*)

illustrated by the example of a hypothetical object that was discovered 41 years before Earth encounter. Points on the curve in **Fig. 4** show where additional observations would be made to refine its orbit. Visibility from Earth is limited by the object's size and distance, and tiny forces due to solar heating and gravitational attraction from other bodies add uncertainties. In the example, over 20 years of observations would be required before the impact probability reached 50%, with the probability increasing to 90% only nine years before impact.

This characteristic will affect the decision process leading to a deflection attempt. Of course, the preference is to wait until the probability reaches near certainty before acting, but unfortunately, there are time and mission design constraints that must be considered. As illustrated earlier, the velocity change required to deflect an object increases as the object gets closer, meaning that the options available for deflection are reduced to the more energetic impulsive methods as time shortens. Second, sufficient time must be allocated to plan and coordinate a deflection campaign; to design, build, and launch the vehicles required; and for the vehicles to reach the oncoming object, to assess the effectiveness of the first wave, and to send a second wave if necessary. These factors point to a process in which decisions are made and funds are committed while recognizing that some deflection attempts may ultimately be found to be unnecessary.

In summary, discovery efforts are finding objects that might threaten in the future, and history shows that action will eventually be required. Techniques are being proposed that could be used to deflect an object, but their overall effectiveness for such a mission has not been tested, nor has our ability to develop a worldwide consensus that action is required, to define the nature of that action, or to commit the substantial resources that will be necessary. Increased efforts to discover smaller threatening objects, to speed the effort to refine the likelihood of the threat, to develop and test techniques that might be used to move an object, to focus on the design of deflection campaigns, and to develop and agree on the decision process are necessary.

For background information *see* ASTEROID; CELESTIAL MECHANICS; COMET; METEOR; METEORITE; NUCLEAR EXPLOSION; SPACE PROBE in the McGraw-Hill Encyclopedia of Science & Technology.

William H. Ailor

Bibliography. T. Gehrels, *Hazards due to Comets and Asteroids*, University of Arizona Press, 1994; J. S. Lewis, *Rain of Iron and Ice, the Very Real Threat of Comet and Asteroid Bombardment*, Helix Books, Perseus Publishing, 1996; National Research Council, *Defending Planet Earth: Near-Earth Object Surveys and Mitigation Strategies: Final Report*, The National Academies Press, 2010.

Asymmetric, pear-shaped nuclei

A basketball or an American football looks the same (ignoring its lettering) if one reflects its right and left sides. We say they are reflection symmetric. In the case of a pear whose right end is large and whose left end is small, reflecting the pear puts the large end on the left and small end on the right. Such a pear is reflection asymmetric.

For many years, nuclei were thought to have symmetric shapes: either spherical shapes, like those in the Mayer-Jensen spherical shell model (for which the Nobel Prize was awarded in 1963); or prolate (rugby football) or oblate (pancake) shapes (what are called quadrupole shapes), as predicted in the Bohr-Mottelson collective model (for which the Nobel Prize was awarded in 1974). However, in 1952 Aage Bohr proposed that intrinsic nuclear symmetry could be spontaneously broken and nuclei could have a stable, octupole deformation, that is, pear shapes. However, for several decades, searches for nuclei with such permanent pear shapes were made without success. In the mid-1980s, evidence for pear-shaped asymmetric nuclei was found around nuclei with atomic number $Z = 88$, the element radon. Two sets of bands of excited energy levels connected by strong, enhanced electric dipole electromagnetic radiation (gamma rays) were found in radium-221 ($^{221}_{88}$Ra), for example, as predicted by theory. However, only one such set was found in the neighboring even-Z, even-N (neutron number) nuclei, and not the two sets predicted by theory. Only recently have the theoretically predicted two sets of levels been found in even-even nuclei, this time in another region where pear-shaped nuclei were predicted, mainly in cerium-148 (^{148}Ce) and barium-144 (^{144}Ba). This discovery completes the experimental verification of stable octupole deformation that gives a nucleus a pear shape, as shown in **Fig. 1** for ^{144}Ba.

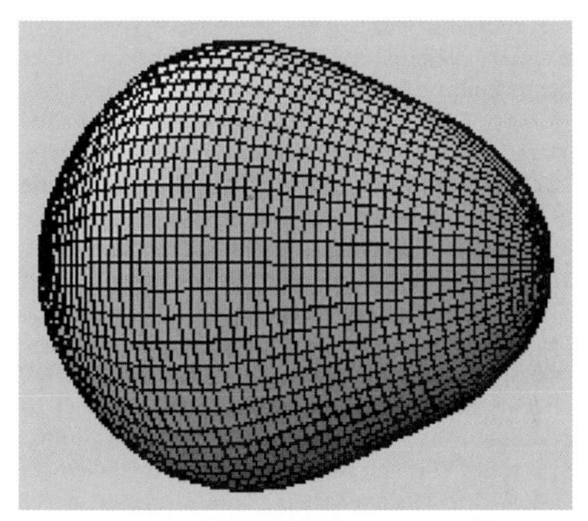

Fig. 1. Computer calculation of the barium 144 (^{144}Ba) pear-shaped nucleus with $\beta_2 = 0.16$ and $\beta_3 = -0.13$.

Theory and observation of reflection asymmetric nuclei. Nuclei that have a football shape have what is called a quadrupole deformation characterized by a long symmetry axis and a deformation roughly given by the equation below, where R_3 is the radius of the

$$\beta_2 \sim \frac{R_3 - R_1}{\frac{1}{2}(R_1 + R_3)}$$

long symmetry axis and $R_1 (= R_2)$ is the radius of the short axis. So for $\beta_2 \sim 0.25$, the long and short axes differ by about 25%. Stable octupole deformation is characterized by a deformation parameter β_3. A nucleus with both quadrupole and stable octupole deformation would have a pear shape and look like that illustrated for ^{144}Ba in Fig. 1.

In addition to stable octupole deformation, theory predicted that symmetric nuclei could undergo excited vibrations about a pear shape. Early on, such vibrational pear shapes were found in quadrupole-deformed nuclei, but not stable, ground-state pear shapes (that is, stable octupole ground-state deformations). Octupole vibrations to excited nuclear states with pear shapes can occur when the nuclear potential energy has the form shown in **Fig. 2*a***, with a minimum energy at $\beta_3 = 0$. Stable octupole shapes can occur when the nuclear potential energy has a minimum at $\beta_3 \neq 0$, shown in Fig. 2*b*.

Theoretically, rigid asymmetric deformed nuclei are predicted to have two sets of energy levels. Successive levels in each set differ by one unit of increasing spin (angular momentum) and have opposite parities. Moreover, the two sets have opposite parities for levels of the same spin. For nuclei with even nucleon number A, where I is the nuclear spin and π the parity of the level, the lowest state in one set has zero angular momentum (spin) and even (+) parity, shown as $I^\pi = 0^+$. [Parity specifies how the wave function that describes a nuclear state acts under a parity operation (simultaneous reflection of all spatial coordinates through the origin). If the wave function is unchanged, we say it has even parity (symbol + sign given as a superscript); if the wave function changes sign, it has odd parity (−, minus sign).] The next state would have one unit of spin and opposite odd parity (1^-), the next two units of spin and even parity (2^+), and so forth. In the other set, the first state has zero spin and odd (-) parity (0^-), the second state has spin 1 and even parity (1^+), and so forth. In summary, we have one set with

$$I^\pi = 0^+, 1^-, 2^+, 3^-, \ldots$$

and one set with

$$I^\pi = 0^-, 1^+, 2^-, 3^+, \ldots$$

In odd-A nuclei, where the spins are determined by the half-integer spin of the odd particle, we have one set with

$$I^\pi = 1/2^+, 3/2^-, 5/2^+, \ldots$$

and one set with

$$I^\pi = 1/2^-, 3/2^+, 5/2^-, \ldots$$

The first experimental evidence for stable octupole deformation was discovered in the 1980s, when two sets of bands of states, each with alternating even and odd spins and with alternating opposite parities, connected by strong (enhanced) electric dipole (E1) electromagnetic radiation (which can connect states that differ by one unit of angular momentum and have opposite parity), and with the two sets having opposite parities, were discovered in nuclei like radium-221 (^{221}Ra), with $Z = 88$ and $N = 133$. However, only one such set was discovered in even-even nuclei like radium-224 (^{224}Ra), with $Z = 88$ and $N = 136$, and thorium-224 (^{224}Th), with $Z = 90$ and $N = 134$.

Theoretically, octupole shapes can be understood by calculating the energies of the single-proton and single-neutron energy levels (orbitals) as a function of β_3, just as earlier these energies were calculated in

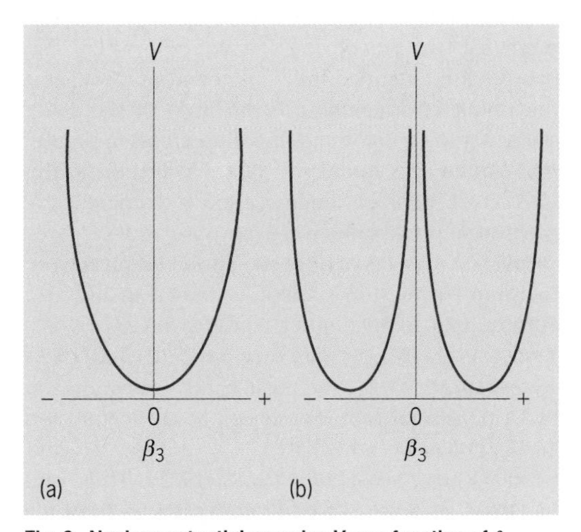

Fig. 2. Nuclear potential energies V as a function of β_3 (*a*) for octupole vibrational nuclei with a minimum at $\beta_3 = 0$, and (*b*) for nuclei with rigid, stable octupole deformation in their ground states with minimum at $\beta_3 \neq 0$.

the Nilsson model as a function of β_2. The spherical magic numbers for protons or neutrons: 2, 8, 20, 28, 50, and so forth in the Mayer-Jensen model occur when there are sizeable jumps in energy between lower-energy filled orbitals and the next available energy orbitals. In that case, the number of protons or neutrons that fill the lower orbitals up to 2, 8,. . . , give the nuclei special stability and are called magic; for example, lead-208 (^{208}Pb), with $Z = 82$ and $N = 126$, is spherical double magic. When such calculations were carried out as a function of β_3, there were gaps in the energy levels at $Z = 88$ and $N = 134$ when $\beta_3 \neq 0$. Thus, radium-222 (^{222}Ra) is a new type of double magic nucleus, but now for an asymmetric deformed shape. Earlier it had been pointed out that, for nuclei with quadrupole shapes, when Z and N have shell gaps at the same value of β_2, these gaps can reinforce each other and produce strongly deformed nuclei with $\beta_2 \sim 0.4$–0.6. As either N or Z move away by two to four particles from their deformed magic numbers, the superdeformation vanishes. These reinforcing shell gaps for Z and N are also critical for observing stable octupole deformation, which again vanishes as Z and N move away from the $\beta_3 \neq 0$ shell gaps.

Asymmetric nuclei around Z = 56 and N = 88. In 1984, G. A. Leander and coworkers predicted a new region of asymmetric, pear-shaped nuclei around $Z = 56$ and $N = 88$, and proposed the nuclide barium-145 ($^{145}_{56}$Ba$_{89}$) as the best candidate to observe the sets of doublet bands. Here, proton and neutron orbitals have shell gaps at $\beta_3 \sim 0.13$ for $Z = 56$ and $N = 88$, which reinforce each other to make $\beta_3 \neq 0$.

The first studies of ^{145}Ba found that the first few levels built on the ground state had a prolate shape with no octupole deformation. However, evidence for pear shapes was reported in barium-144 and -146 (^{144,146}Ba) and cerium-146 (^{146}Ce) based on strongly enhanced E1 transitions between the ground-state bands (which are a series of levels that result from the rotation of the ground state) and excited levels with opposite parities. However, the predicted two sets of doublets were not seen. The bands were extended in ^{144,146}Ba and in ^{146}Ce, and the predicted quenching (vanishing) of the rigid, stable octupole deformation with increasing angular momentum was found in ^{146}Ba. Nevertheless, the two sets of doublet bands needed to complete the theoretical predictions were not observed.

The first two sets of opposite-parity doublets were found in barium-143 (^{143}Ba), as shown in **Fig. 3*a*.** For example, in the first set, band (1) has $I^{\pi} = 9/2^-$, $13/2^-$, $17/2^-$, $21/2^-$,. . . and band (2) has $I^{\pi} = 15/2^+, 19/2^+, 23/2^+$, and these bands are connected by E1 transitions to form one set. In the second set, band (4) has $I^{\pi} = 13/2^+, 17/2^+, 21/2^+, \ldots$, and band (3) has $I^{\pi} = 11/2^-, 15/2^-, 19/2^-$. Thus, one has two sets, each set with alternating parities for the levels and each set with opposite parities; the first set has $13/2^-, 15/2^+, 17/2^-, 19/2^+$,. . . , and the second set has $13/2^+, 15/2^-, 17/2^+, 19/2^-$,. . . .

Earlier a nucleus had been found to have bands of excited states built on a spherical shape and other bands built on a well-deformed, prolate shape, providing evidence of a property known as nuclear shape coexistence. The levels of ^{145}Ba were reinvestigated with high-statistics data and a new type of shape coexistence was found in which we have bands of levels built on a symmetric prolate shape and other bands built on an asymmetric pear shape. The levels up to spins of about $11/2^-$ to $13/2^-$ are characterized by a prolate shape. Above those states, two sets of opposite-parity doublet bands emerge, characteristic of pear-shaped, asymmetric nuclei, to confirm the earlier theoretical predictions, as shown in Fig. 3*b*. The first few levels in bands labeled (2) and (3) in ^{145}Ba up to $11/2^-$ are characteristic of a collective prolate structure. At $13/2^-$ and above in band (2), the levels become the opposite-parity members of one set as expected for stable octupole deformation. So band (1) and the upper part of band (2) form one set, and the other opposite-parity set expected for stable octupole deformation is composed of bands (4) and (5). These two sets have opposite parities for levels of the same spin as theoretically predicted, for example:

bands (1), (2): $15/2^+, 17/2^-, 19/2^+, 21/2^-, \ldots$

bands (4), (5): $17/2^+, 19/2^-, 21/2^+, \ldots$

The upper levels of band (3), $15/2^-$ and $19/2^-$, are weakly populated and are connected to the prolate ground band.

To firmly establish the theoretical prediction of permanent ground-state pear shapes in nuclei with even Z and even N, it is critical to find both of the theoretically predicted sets of $\Delta I = 1$ bands with opposite parities for each set. Until recently, both sets had not been found. Now, using a higher-statistic data set with 100 times as many events as previously analyzed, the first example of both sets of opposite-parity bands in an even-even nucleus in this region was found in ^{148}Ce (**Fig. 4**). Subsequently, the second set of doublets in an even-even nucleus was found in ^{144}Ba. These two sets of $\Delta I = 1$ alternating parity bands complete the evidence for pear shapes in even-even nuclei. The pear shape for ^{144}Ba with the experimentally determined values of β_2 and β_3 is shown in Fig. 1. The levels in barium-140 through -142 ($^{140-142}$Ba), xenon-138 through -142 ($^{138-142}$Xe), cesium-139 through -144 ($^{139-144}$Cs), lanthanum-143 and -144 (^{143,144}La), and praseomdymium-149 through -152 ($^{149-152}$Pr) have been studied to map out this region of stable octupole pear shapes centered on the reinforcing shell gaps at $Z = 56$ and $N = 88$. The stable octupole deformation goes over to softer shapes as $\beta_3 \rightarrow 0$ to give rise to more vibrational octupole correlations as Z or N increase or decrease from 56 and 88, respectively.

For background information *see* MAGIC NUMBERS; NUCLEAR STRUCTURE; PARITY (QUANTUM

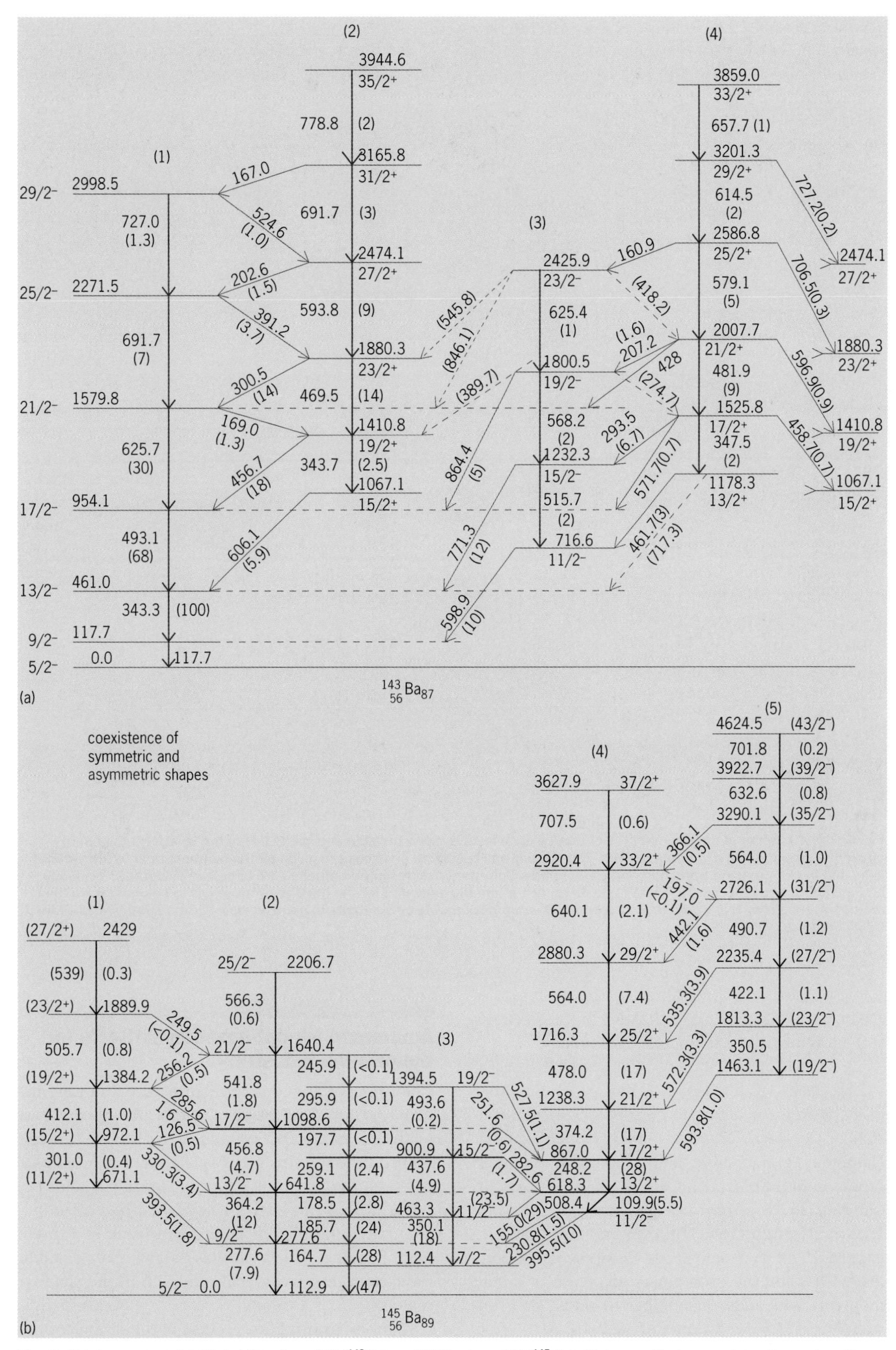

Fig. 3. Nuclear energy levels in (*a*) barium-143 (^{143}Ba) and (*b*) barium-145 (^{145}Ba). Above and below each level are given the spin and parity (I^{π}) of the level, and the energy above the ground state of the level in keV. The gamma-ray transition energies and intensities are given by the numbers next to the vertical lines and the E1 crossing transitions between bands. Spin-parity values in parentheses are based on systematic evidence and are not directly measured. Levels built on an asymmetric shape are in color in ^{145}Ba. (*From S. J. Zhu, et al., Octupole correlations in neutron-rich ^{143,145}Ba and a type of superdeformed band in ^{145}Ba, Phys. Rev. Rapid Comm, C60:051304, 1999*).

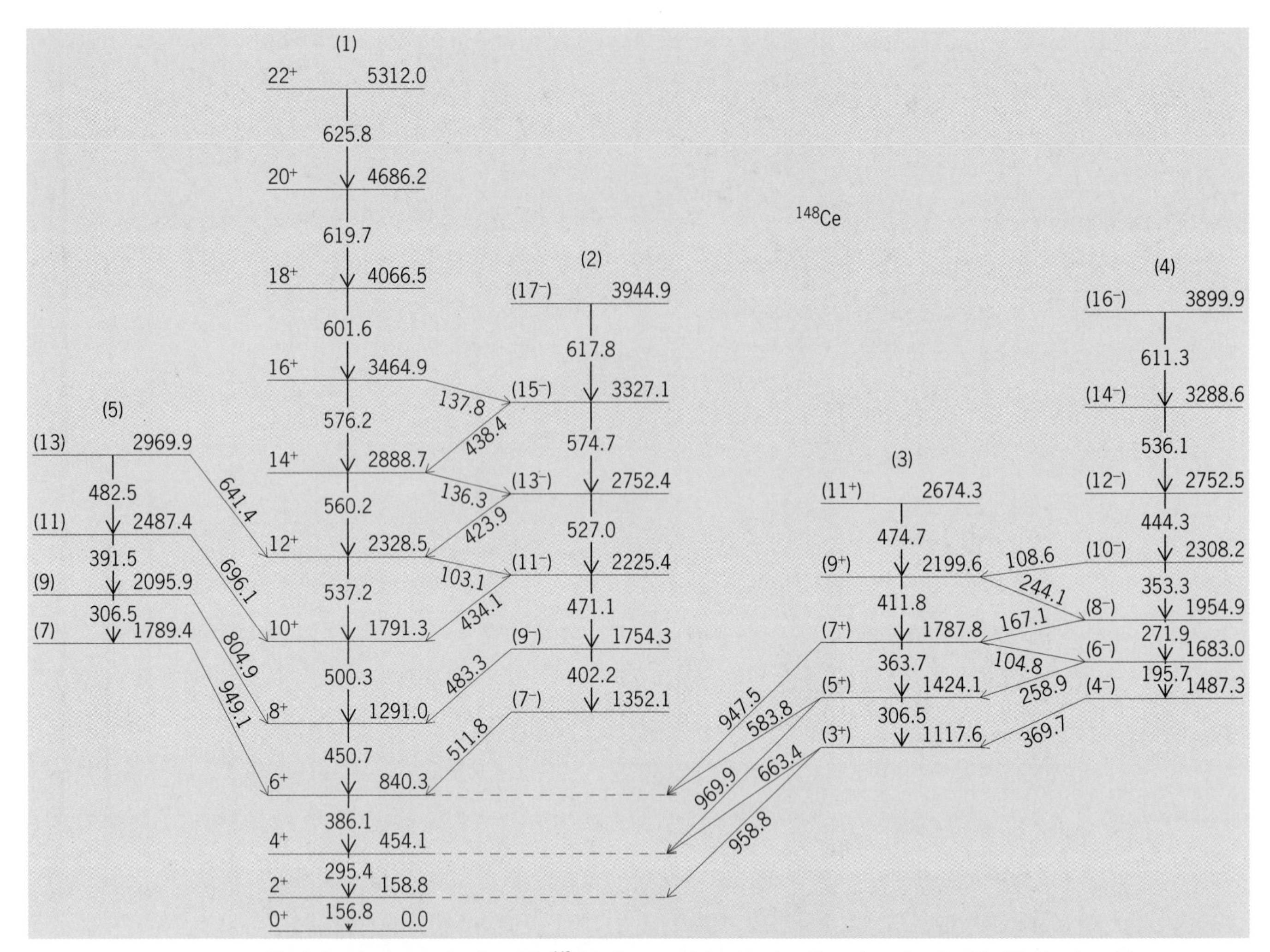

Fig. 4. Level scheme of cerium-148 (^{148}Ce). Above each level is given the spin and parity (I^{π}) of the level, and the energy above the ground state of the level in keV. The gamma-ray transition energies are given by the numbers next to the vertical lines and the E1 crossing transitions between bands. Spin-parity values in parentheses are based on systematic evidence and are not directly measured. There is no direct or systematic evidence for the parities of the levels in column (5), so they are not given. (*From Y. J. Chen, S. J. Zhu, et al., Search for octupole correlations in neutron-rich ^{148}Ce nucleus, Phys. Rev., C73: 054316, 2006*).

MECHANICS); RIGID-BODY DYNAMICS in the McGraw-Hill Encyclopedia of Science & Technology.

Sheng-Jiang Zhu; Yi-Xiao Luo; Joseph H. Hamilton

Bibliography. Y. J. Chen, S. J. Zhu, J. H. Hamilton, et al., Search for octupole correlations in neutron-rich ^{148}Ce, *Phys. Rev.*, C73:054316, 2006; J. H. Hamilton et al., Stable and vibrational octupole modes in Mo, Xe, Ba, La, Ce and Nd, in A. Corvello (ed.), *Sixth International Seminar on Nuclear Physics Highlights*, pp. 15–26, World Scientific, Singapore, 1999; Y. X. Luo et al., Octupole excitations in ^{141,144}Cs and the pronounced decrease of dipole moments with neutron number in odd-Z neutron-rich ^{141,143,144}Cs, *Nucl. Phys.*, A838:1–19, 2010; Y. X. Luo, J. H. Hamilton, et al., New level schemes and octupole correlations of light neutron-rich lanthanum isotopes ^{143,144}La, *Nucl. Phys.*, A818:121–138, 2009; S. J. Zhu, et al., Octupole correlations in neutron-rich ^{143,145}Ba and a type of superdeformed band in ^{145}Ba, *Phys. Rev. Rapid Comm.*, C60:051304, 1999.

Automated ink optimization software for the printing industry

Color images are commonly printed with cyan, magenta, yellow, and black (CMYK) inks rendered as halftone images and printed in register with each other. Although this process is known as four-color printing, the black ink is not technically a color because it does not have an associated hue (or dominant wavelength). The addition of black in a three-color printing system extends the range of printable tones and improves detail rendition in the shadow region.

Halftone printing is a process of rendering continuous tone images, such as photographs, with printing systems that only transfer a single thickness of ink. The image is divided into a grid of pixels that are small enough that observers cannot resolve the individual dots, typically 133 per linear inch or finer. Each pixel contains a halftone dot that is sized

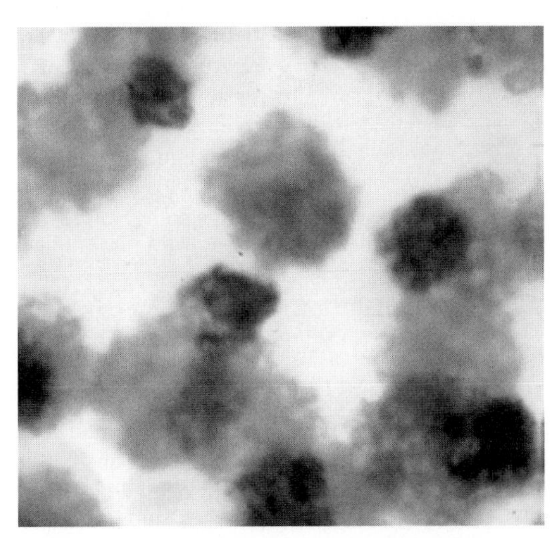

Fig. 1. Halftone dots in color image, shown here in grayscale.

according to the tonal value being conveyed by that pixel; small dots (about 2% of the pixel's surface area) represent highlight tones; mid-sized dots portray midtone values (image areas that are neither very dark nor very light); shadow dots (covering up to 98% of the pixel area) portray the darkest tones of the picture. Even though the dots are all equally dark, the human observer perceives the halftone image as consisting of a continuous gradation of tones.

Each pixel of an image has the potential for cyan, magenta, yellow, and black components. In complex color images, it is common for pixels to have some amount of all four inks, as seen in **Fig. 1**.

Theoretically, 100% coverage of all four process inks (400% coverage) would produce the darkest attainable shadow for a printing system. However, with modern printing methods, including offset lithography, it is impractical to apply 400% coverage to a substrate without incurring problems such as backtrapping, setoff, and ink show-through. Furthermore, shadow detail becomes muddy and less distinct. Backtrapping and setoff are printing problems that occur when heavy ink films are printed. Backtrapping occurs when a second ink is printed on top of a freshly printed ink film, and, rather than transferring all of the second ink to the paper, some of the first printed ink is lifted from the paper onto the printing blanket of the press. Setoff occurs when a printed area with heavy coverage is not sufficiently dried and it transfers ink to the next paper it touches (often the back of the sheet that follows it through the press).

Different industry segments and individual printers have set limits on the acceptable total ink coverage for their processes. The magazine industry, for example, through the Specifications for Web Offset Publications (SWOP) has established total dot area coverage limits of 300% for lower-quality coated paper and 310% for medium-quality coated paper. Similarly, the newspaper industry, through the Specifications for Newsprint Advertising Production (SNAP), limits total coverage to 260%. Commercial printers produce a more diverse range of products than do publication printers. They limit total ink coverage differently for different substrates and products. Those who adhere to the General Requirements and Applications for Commercial Offset Lithography (GRACoL) specifications are limited to 340% total area coverage for high-quality coated papers.

As photomechanical color separation was developed in the early twentieth century, the process of undercolor removal (UCR) was developed to limit the amount of ink printed in neutral shadows. During the exposure of the CMY color separation negatives, a photographic mask was placed in the image path that reduced the cyan, magenta, and yellow shadow densities. UCR was a forerunner to today's programmed ink reduction systems, but it was practiced only on neutral shadows and provided only modest savings in the use of CMY inks.

Gray component replacement (GCR). The gray component, a term first coined by J. Yule in 1940, of a CMYK pixel refers to the portion of the CMY inks that together would form a balanced gray and therefore do not add to the hue of the pixel; rather, they desaturate the color. P. Tobias in 1954 proposed a color separation system that would remove the gray component from each pixel and replace it with an appropriate value of black to match the color that would have been produced by an unaltered pixel. This system was never commercially developed.

With the advent of digital image processing in the early 1980s, it became feasible to evaluate the CMYK values of pixels and perform ink reduction by selectively reducing gray components and boosting the black ink as necessary. The process has been known by many names, including complementary color reduction, achromatic synthesis, polychromatic color removal, and integrated color removal, but the term gray component replacement is the most commonly used descriptor of the process. **Figure 2** shows a schematic representation of an unaltered pixel and one that has been subjected to GCR.

Figure 2 depicts the replacement of 100% of the gray component resulting in a pixel that contains only two process colors, plus black. However, the application of GCR can be partial, varying from 0–100%. In practice, GCR is rarely implemented at 100%.

The potential advantages to using gray component replacement include:

1. Savings based on reduced use of expensive process color inks in exchange for the slightly increased use of less-expensive black ink

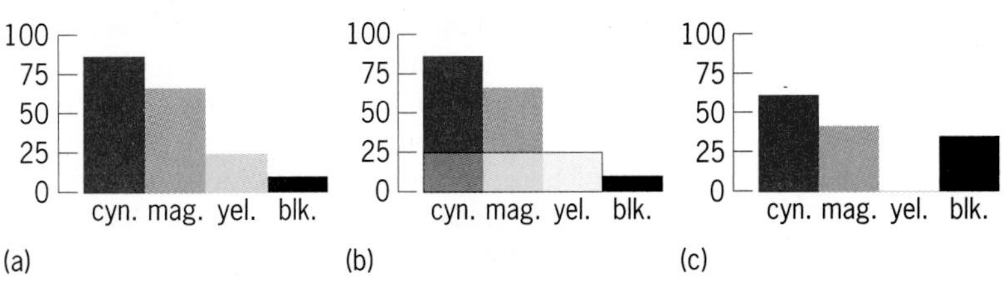

Fig. 2. CMYK values for (*a*) unaltered pixel, (*b*) gray component shaded, (*c*) gray component replaced with additional black.

Fig. 3. Composite pictorial image (shown here in grayscale) used for measuring ink consumption.

2. Greater color stability on press because less of the total color appearance is dependent on the balance of CMY inks

3. Fewer printing problems such as setoff, backtrapping, and ink show-through

4. Better drying characteristics because of the reduced total amount of ink on the sheet.

During the 1980s, numerous research papers on GCR were published in the Proceedings of the Technical Association of the Graphic Arts (TAGA). During the 1990s GCR was mentioned in several research studies, but typically not as the principal subject. The concept had become widely endorsed by the industry, and the work to refine the needed algorithms took place in proprietary research. During this period, GCR was more commonly found in the trade literature, where it was presented to a wider graphic arts audience.

Complicating factors. Several complicating factors make the implementation of GCR difficult enough that, after more than two decades of continuous developments, the process is only now gaining traction in the marketplace. One complication stems from the colorimetric impurities of printing inks. The cyan, magenta, and yellow inks are not perfect with respect to the portions of the visible spectrum that they absorb and transmit. Traditionally, densitometric calculations of hue error and grayness were used to quantify the relative impurities of the inks. Today, the colorimetric attributes of hue angle and chroma (color purity) provide more precise characterization. As a result of these impurities, equal amounts of CMY printed together do not result in neutral gray. The gray balance of a printing system refers to the relative values of CMY needed to produce a neutral scale. It is specific to the inks, substrate, and process being used, and it is not uniform throughout the tone scale.

Gray balance and, hence, GCR are influenced by factors other than the colorimetric impurities of the inks. The color, absorbency, and surface characteristics of the paper contribute to gray balance, as does the ink trapping of the printing system. Ink trapping is a measure of the relative efficiencies of ink transfer to previously printed ink films compared to the transfer to unprinted paper. With halftone printing, CMY inks rarely overlap in the highlights, partially overlap in the midtones, and completely overlap in the shadows, making the influence of ink trapping most pronounced on the gray balance of the darker areas.

The halftone structures of images also influence GCR. The mechanical and optical dot gains cause colors to appear darker than they should based on the dot area coverages in the digital file. Dot gain, the increase in size of a halftone dot when printed, is not uniform through the tone scale and varies with screen ruling and dot shape. It is also heavily dependent on the absorptivity and surface characteristics of the paper. There are pronounced differences in dot gain between conventional halftone patterns (amplitude modulated) and stochastic dot structures (frequency modulated).

Additionally, GCR must also allow for the additivity failure of printing inks, which, in turn, is influenced by the printing system. Additivity failure describes the phenomenon that the total density of an overprinted ink film is not equal to the sum of the individual ink densities.

All these concerns make it difficult to construct straightforward algorithms to perform gray component replacement, such that the colors of the corrected pixels perceptually match the colors of unaltered pixels. Early attempts at GCR in the 1980s were not widely adopted by printers largely because color matches were not deemed to be acceptable.

Undercolor addition. Although it is not practical to print too much ink in the darkest shadows of a reproduction, it is possible to print too little ink. Shadow areas require more than a single black ink film to achieve maximum density, so some amount of CMY undercolor is also needed. In areas of heavy implementation of GCR, such as shadow areas, the density of the gray component being removed can exceed the ability of the black ink to compensate for its removal. This leads to tone reversals in dark areas where saturated colors are imaged darker than neutral shadows, which should be the darkest parts of the image.

Some electronic scanners of the 1980s used separate undercolor addition (UCA) programs to reintroduce undercolor in the shadow areas. Scanner operators set three separate values to achieve ink optimization: undercolor removal, gray component replacement, and undercolor addition. As this technology has developed, these three functions have been incorporated into a single ink optimization setting.

Recent ink optimization study. Since 2000, widespread adoption of ink optimization software has begun, as evidenced by the number of vendors competing for the market. A. Stanton and colleagues participated in a collaborative study of ink optimization programs in 2009, including researchers from the Printing Industries of America.

Nine vendors participated in the study, which sought to answer two research questions:

1. Do the ink optimization programs reduce the amount of ink needed to print a series of test images, and to what extent does each program reduce ink usage?

2. Do the ink optimization programs cause unacceptable color shifts in the reproduction of test files compared to control files made through a default workflow?

The vendors were supplied with digital files and returned files with optimized ink reduction. Each vendor determined the level of ink reduction to apply to the files. A composite image (**Fig. 3**) was used as the basis for measuring reduction of ink consumption.

The photographs in the composite image presented a variety of challenging reproduction subjects. The composite image has high ink coverage and, therefore, offered ample opportunity for significant ink reductions.

This study examined two workflows (RGB to CMYK, and CMYK to CMYK), but only the CMYK-to-CMYK results are summarized here. The calculated ink reductions as percentages of total ink coverages ranged from 17 to 27%. With the use of ink optimization software, the mean reduction across vendors was 22%.

A separate test image (**Fig. 4**) was used to calculate the accuracy of color matches with and without the use of ink optimization software. Color matching accuracy was based on both objective measurements of color differences and subjective evaluations of acceptable color matches.

Kodak Spectrum™ proofs were made from the submitted files of each of the vendors, and a control proof was made using the default output of the images to the U.S. Web Coated (SWOP) v2 printing condition.

Judges visually compared the proofs made from vendor files against a control proof, rating the color matches on a scale of 1–5, with 5 being an excellent color match, 3 a commercially acceptable match, and 1 a poor match. The average scores for the vendors ranged from 2.6–3.6, with two of nine vendors falling below the commercially acceptable value of 3.0.

Color differences were measured using the Color Measurement Committee (CMC) color difference (ΔE) formula ΔE_{cmc}. The average color differences for the 24 patches of the X-Rite Color Checker Target ranged from 2.1 to 4.3. Linear regression between the measured color differences and subjective ratings showed an R-squared (predictive value of 0-1) value of 0.87. The graph indicated an average ΔE_{cmc} value of 3.25, corresponding to a subjective value of 3 (commercially acceptable match). This was considered to be a reasonable target based on research by A. Johnson and P. Green and H. Habekost and K. Rohlf.

Fig. 4. Test image for analysis of color matching (shown here in grayscale).

When the data for both workflows were considered, the ink optimization programs of three vendors did not provide acceptable color matches. The combined ink savings ranged from 26 to 16%. Two vendors were found to provide the best combination of ink savings and color matching. It was concluded that ink optimization software can provide both ink savings and reasonable color matches, but not all available software is successful at achieving these goals. Of the programs that are successful, there is substantial difference between the results that each will deliver.

Outlook. Today, the use of some level of GCR has become common; however, neither the precise extent of this use nor the levels being used are known. In the publication market, newspapers and magazines benefit substantially from ink savings and from the reduced drying problems with GCR. Lithographers serving the high-end market appreciate the enhanced shadow detail and improved stability on the press. It seems clear that ink optimization software has progressed from research labs to pressrooms, and it is rapidly becoming an integral part of electronic image processing. However, the software offerings on the market vary substantially in their effectiveness. Potential users should do careful comparisons before committing to implementation of ink optimization software.

For background information *see* COLOR; COLOR VISION; INK; PAPER; PRINTING in the McGraw-Hill Encyclopedia of Science & Technology.

Anthony P. Stanton

Bibliography. M. Habekost and K. Rohlf, The evaluation of colour difference equations and optimization of DE2000, *TAGA J.*, 4:149–164, 2008; A. Johnson and P. Green, The colour difference formula CIEDE2000 and its performance with a graphic arts data set, *TAGA J.l*, 2:59–71, 2006; A. Stanton, E. Neumann, and M. Bohan, The relative accuracy and effectiveness of automated ink optimization software, *TAGA Proceedings*, pp. 39–88, 2009; P. Tobias, A color correction process, *TAGA Proceedings*, pp. 85–90, 1954; J. A. C. Yule, Theory of subtractive color photography. III. Four-color processes and the black printer, *J. Opt. Soc. Am.*, 30(8):322–331, 1940, doi:10.1364/JOSA.30.000322.

Bacterial symbionts of farming ants

Leaf-cutting (attine) ants have an ancient mutualistic relationship with basidiomycete fungi, which they cultivate for food and defend from parasites with bacterially generated antibiotics (**Fig. 1**). The bacteria involved in these multipartite relationships are an ancient group called actinobacteria, notably *Pseudonocardia* species, which most attine ants shelter and feed in specialized structures on their cuticles called crypts.

Garden pests. Modern attine ant gardens (**Fig. 2**) are a complex microbial mix of fungi, yeasts, and bacteria—a living biomass dominated by the basidiomycete monoculture that attracts a diversity of transient and easily removed nonmutualistic "weed" fungi, including *Trichoderma* and *Fusarium*. However, the most dangerous and best-studied garden invader is the filamentous, ascomycetous fungi of the genus *Escovopsis*, which kill the food fungi, eat the dead remains, and leave the ants to starve. This is a catastrophe for ant communities, some of which house up to five million individuals in huge subterranean nests.

Escovopsis is a shape-changing fungus that appears to earn its entire living stealing food from leaf-cutter ants. Found only in ant gardens and associated refuse dumps, *Escovopsis* fulfill the postulates of pathogenicity elucidated by Robert Koch in the late 1800s, that is, when isolated from diseased gardens and reapplied to healthy gardens, they cause the same disease. Scientists think *Escovopsis* is vectored into ant colonies by neighboring invertebrates because founder queens carefully select disease-free garden patches following their mating flight, and this particular fungus does not produce airborne spores. Regardless of how it arrives into the nests, *Escovopsis* seems to be an integral member of the attine ant-microbe community and probably evolved into an obligate pathogen during its long association with the ants and their cultivar.

Bacterial weaponry. Actinobacteria are a ubiquitous, diverse, and very successful group of bacteria occupying a large number of different niches and taking many different forms, although all are Gram-positive (having tough outer coats that absorb the blue dye in Gram staining), which indicates that they were among the earliest inhabitants of Earth. *Pseudonocardia*, which resemble fungi in their filamentous form, are soil dwellers, and soil is a busy, competitive place—a perfect laboratory for the creation of small molecules, some of which have been adapted by humans, along with ants, for use as antibiotics. Indeed, four of the best-known human antibiotics—actinomycin, neomycin, streptomycin, and vancomycin—were derived from actinobacteria.

One of the *Pseudonocardia*-manufactured chemical weapons that attine ants use to protect their fungal gardens was described in 2009. Called "dentigerumycin" in honor of *Apterostigma dentigerum*, the ant that helped choreograph its evolution, this small molecule (molecular formula: $C_{40}H_{67}N_9O_{13}$) actively inhibits *Escovopsis*, but largely spares the ant's garden fungi. Dentigerumycin, a cyclic depsipeptide (a peptide with one or more ester bonds in addition to the amide bonds) containing unusual amino acids, also slows the growth of a multidrug-resistant strain of the human pathogen *Candida albicans* and may provide humans with an antifungal alternative to existing drugs.

Escovopsis, like *Candida*, can develop antibiotic resistance, but *Pseudonocardia* have millions of years of experience evolving new small molecules and usually outpace the fungus. In the event that this does not happen fast enough, the ants simply acquire new strains of actinobacteria, possibly from other symbiotic stock or the soil surrounding their nests. Researchers, however, are not yet able to fully explain how this is done. But, it appears that leaf-cutting ants tamed free-living strains of *Pseudonocardia* many times over during the course of their long evolutionary history together and that the acquisition is very selective; actinobacteria are the only clearly established symbionts of fungus-growing ants.

Understanding the chemistry underlying all that *Pseudonocardia* do for the ants is a major research issue and will ultimately yield important insights about how they and bacteria in general deal with the world. For example, biosynthetic pathways called "orphan pathways" are turned on in special and completely unknown circumstances and it is important to know how and why this happens. Moreover, it is unlikely that *Pseudonocardia* developed the small molecules that the ants use as antibiotics for the express purpose of warding off the ants' garden

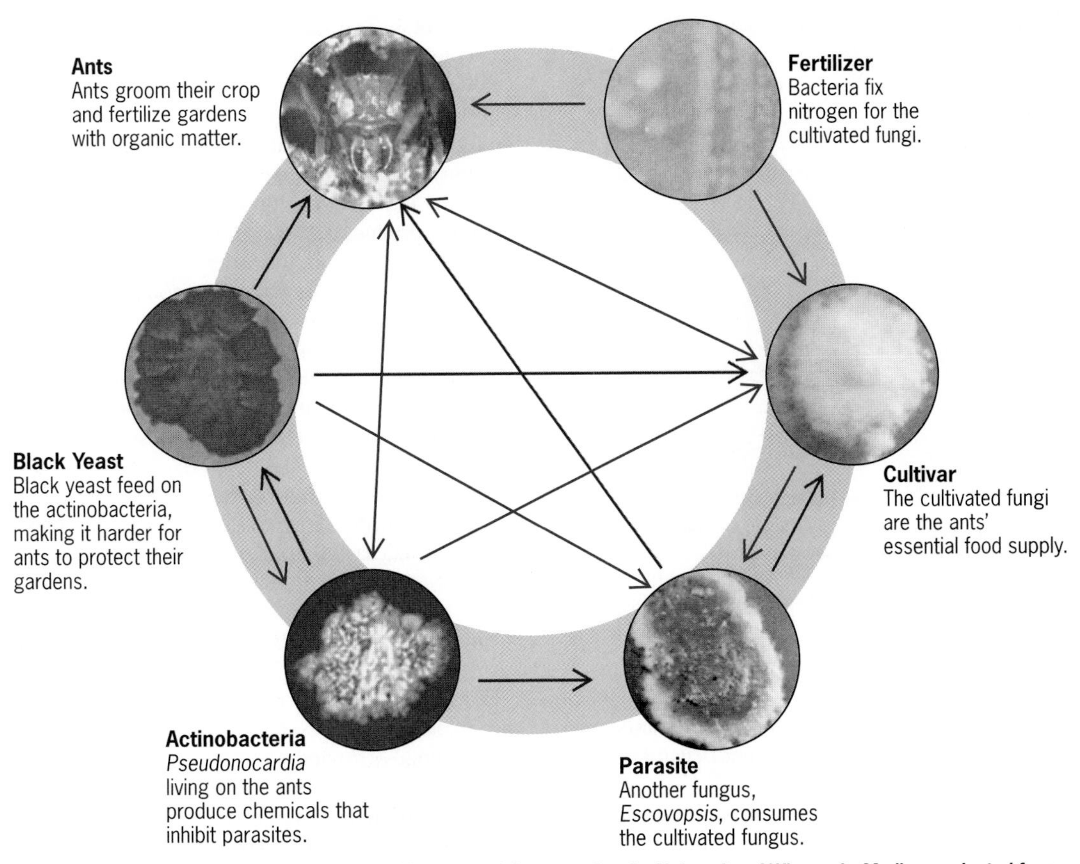

Fig. 1. The ant-microbe farming community. (*Courtesy of Cameron Currie, University of Wisconsin-Madison; adapted from an earlier version by Elsa Youngsteadt*)

Fig. 2. An *Acromyrmex octospinosus* ant tending its spongy fungus garden. (*Courtesy of Cameron Currie, University of Wisconsin-Madison; photo by Michael Poulsen*)

parasites; they probably have more peaceful uses in the wild. Bacteria tend to be gregarious, gathering in high-density populations, probably reflecting a safety in numbers strategy. Furthermore, they "chatter" continuously and their words are chemical. Thus, one important use for the actinobacterial small molecules is probably "quorum sensing," a bacterial monitoring language that tells them how many and what type of microbes are gathered in their neighborhood.

Genome sequencing is needed to define the small molecule-generating potential of *Pseudonocardia*; analytical chemistry is needed to isolate and characterize the molecules themselves; and functional assays are necessary to determine what these molecules actually do. The sequencing began in earnest in 2009, with three ant genomes and 14 ant-associated fungal and bacterial genomes under active investigation. Finalizing the genome sequences and annotation will take years, but results are being made available to the scientific community as they progress.

Black yeast prey on Pseudonocardia. Adding to the complexity of leaf-cutting ant communities, actinobacteria were discovered to have their own specialized fungal predators—a black yeast called *Phialophora*, which grows on and around the crypts housing the ants' antibiotic-producing bacteria. When the black yeast was put together with actinobacteria in a petri dish, *Phialophora* ate the bacteria, thereby robbing the ants of an important anti-*Escovopsis* defense. However, parasitism in this multipartite system is not considered a detrimental thing; it helps to keep the cooperators honest. For example, making antifungal agents is expensive, and selfish actinobacteria would be tempted to reproduce more and protect the ants less without the black yeast around. Thus, with the yeast, antibiotic production is also in the best interest of the threatened bacteria. Moreover, the black yeast undoubtedly adds value to the relationship because, like the cultivar, *Escovopsis*, and actinomycetes, the yeast

Fig. 3. Untimely death of an ant queen: *Acromyrmex octospinosus* queen from Panama with two *Ophiocordyceps stilbelliformis* stroma erupting from between the head and pronotum. Also note the fungus growing out the ends of her legs. (*Courtesy of David P. Hughes, Harvard University*)

has probably been part of the leaf-cutting ants' microbial balancing act since they first began farming about 50 million years ago.

Cordyceps infection of ants. Ants spend a great deal of time foraging about in the soil, and it should come as no surprise that they, as well as their fungal food gardens, are vulnerable to infection with pathogenic fungi. Principal among these are *Cordyceps*, which comprise at least 400 species and infect a wide range of insects and spiders in addition to ants. Ant queens are at particular risk of parasitic infections during the colony-founding stage because of the many energetic demands that must be met simultaneously. In addition, although it appears that vigorous general defenses against food crop fungi indirectly protect the ants from a range of parasitic fungi, an ant-associated genus of *Cordyceps*, called *Ophiocordyceps stilbelliformis*, has been found parasitizing leaf-cutter ants in Panama (**Fig. 3**). Fortunately for farming ants, this is a rare event because *Cordyceps* are a particularly nasty group of fungi that keep the infected ants alive just long enough to distance themselves from the nest and then become spore-making factories.

Attine ant-microbe evolution. The ancestors of fungus-growing ants probably transitioned from hunter-gatherers of arthropod prey, nectar, and plant juices to the farming life fortuitously, by taming the wild fungi growing on the walls of their nests in leaf litter or from a system of myrmecochory (spore or seed dispersal by ants) where specialized fungi used the ants for their own dispersal. Ants routinely ingest fungal spores and hyphal material and such infrabuccal contents are eventually expelled as pellets on nest middens (refuse dumps) and elsewhere, providing the fungi with a way of dispersing their spores and hyphae. Thus, the fungi were probably not passive symbionts that happened to come under ant control, but rather played a proactive role in the ant evolution from hunter-gatherer to fungus farmer.

Recent work using culture-independent genomic sequencing technologies is lending further weight to the long-standing scientific consensus that the ant-actinomycete association evolved for mutual benefit rather than as a coevolutionary arms race between antibiotic-producing *Pseudonocardia* and *Escovopsis* parasites. However, these new technologies must be used properly, statistically valid conclusions must be drawn, and wild as well as laboratory ant colonies must be tested in these endeavors. Importantly, the new tools and new data will attract further investigations by the growing community of researchers fascinated by the attine tribe of sophisticated farming ants.

For background information *see* ANTIBIOTIC; BACTERIA; CHEMICAL ECOLOGY; ECOLOGICAL COMMUNITIES; ECOLOGY; FOOD WEB; FUNGAL ECOLOGY; FUNGI; HYMENOPTERA; MUTUALISM; SOCIAL INSECTS; SOIL ECOLOGY; TROPHIC ECOLOGY in the McGraw-Hill Encyclopedia of Science & Technology.

Marcia Stone

Bibliography. C. R. Currie et al., Coevolved crypts and exocrine glands support mutualistic bacteria in fungus-growing ants, *Science*, 311:81-83, 2006; C. R. Currie et al., Fungus-growing ants use antibiotic-producing bacteria to control garden parasites, *Nature*, 398:701-704, 1999; O. Dong-Chan et al., Dentigerumycin: A bacterial mediator of an ant-fungus symbiosis, *Nat. Chem. Biol.*, 5:391-393, 2009; D. P. Hughes et al., Novel fungal disease in complex leaf-cutting ant societies, *Ecol. Entomol.*, 34:214-220, 2009; G. Yim, H. Huimi Wang, and J. Davies, Antibiotics as signalling molecules, *Philos. T. Roy. Soc. B.*, 362:1195-1200, 2007.

Betavoltaics

Betavoltaic devices are semiconductor energy conversion structures that are coupled with beta radiation sources. These devices are in many respects similar to solar cells and convert radiation directly into electrical energy.

Safety of radiation sources. Beta radiation, consisting of high-energy electrons, originates from nuclear reactions occurring in radioisotopes. Other types of radiation that can originate from radioisotope decay are alpha radiation (particles that are helium nuclei, consisting of two protons and two neutrons) and gamma radiation (high-frequency electromagnetic radiation similar in form to radio waves or visible light). Both alpha and beta radiation are easily stopped by a few millimeters (a fraction of an inch) of air or a layer of dead skin, while gamma rays can penetrate through the entire body. The amount of radiation exposure is determined by the total dose that the human body receives. The most dangerous type of radiation is alpha radiation, which has a significantly higher effective dose than either beta or gamma radiation. Beta radiation is the safest

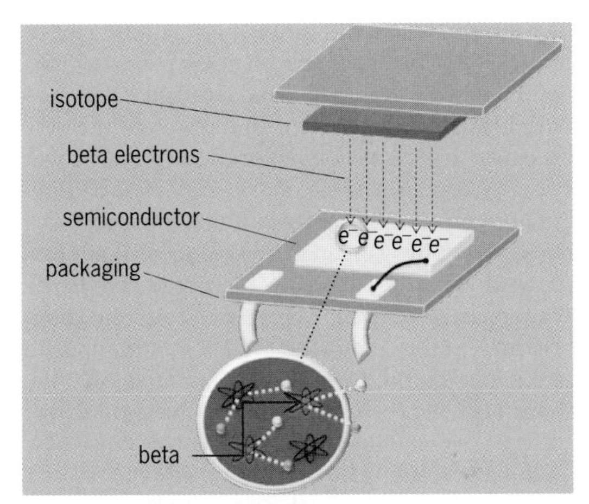

Fig. 1. Schematic of betavoltaic device.

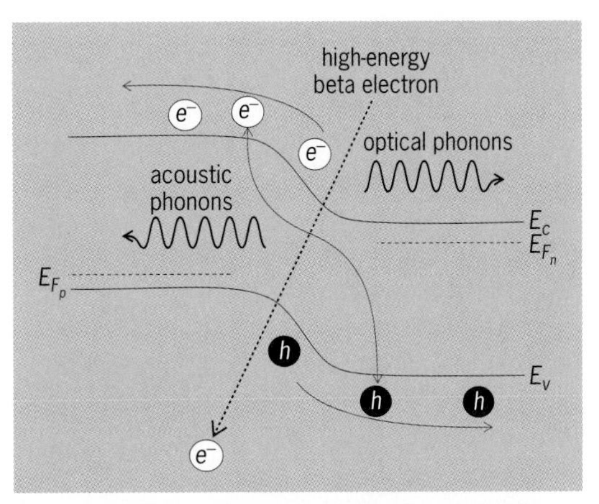

Fig. 2. Illustration of beta charge generation and collection in a *pn* semiconductor junction. e^- = electron, *h* = hole, E_v = energy of top of valence band, E_c = energy of bottom of conduction band, E_{Fp} and E_{Fn} are Fermi energies of positively and negatively charged particles, respectively.

radiation source from two perspectives: effective dose and penetration depth. The safety of beta sources allows them to be employed in high-volume applications such as smoke detectors and exit signs. Examples of beta radiation sources are nickel-63 (^{63}Ni), tritium (^{3}H), and promethium-147 (^{147}Pm).

Packaging. Betavoltaic devices find potential utilization in systems that require small amounts of power (0.01–100 μW) and operate in a small footprint (less than 10^{-2} cm^2 or 10^{-3} in.^{2}) for long time periods (2–25 years). **Figure 1** shows a schematic view of a betavoltaic device. In this realization, the isotope is integrated into the device package as a solid foil (gaseous sources can also be used). The package is an important part of the betavoltaic design, as it must position the isotope within about 100 μm (a few thousandths of an inch) of the semiconductor as well as provide the necessary shielding and physical durability.

History. Betavoltaic devices have a surprisingly long history. A commercial device, the Betacell, was produced in the early 1970s. The Betacell had an overall efficiency of 2.3%, and more than 100 people received heart pacemakers powered by it. At the time when the Betacell was manufactured, the majority of low-power applications required power levels that were too high for the Betacell to produce. With the continuing development of nanotechnology, the power requirements for electronics have significantly dropped, together with the size of individual devices and circuits. Betavoltaics now find a technology landscape in which the surface-to-volume ratio of the analog and digital devices and circuits has increased dramatically. In this environment, betavoltaics are increasingly becoming a competitive micropower option.

Theory of radiation cell operation. The betavoltaic energy conversion process begins with the emission of high-energy electrons from the radioisotope. The loss of the electron from the isotope would normally cause the radioisotope to acquire a positive charge. However, the radioisotope is connected to ground potential through the package, as shown in Fig. 1. The connection to ground provides a path for electrons to flow to the isotope, maintaining charge neutrality.

Betavoltaic devices use semiconductor *pn* junctions or metal Schottky junctions to collect charge. **Figure 2** illustrates some of the physical process that occurs during betavoltaic energy conversion in a *pn* semiconductor junction. The high-energy beta electron (10–200 keV in energy) is incident on the semiconductor and transfers its energy to the semiconductor via several sequential interactions. The first interactions produce plasmons (quanta of charge oscillations) and secondary electrons with lower energies. Subsequent interactions produce acoustic and optical phonons (quanta of crystal vibrations) as well as electron-hole pairs, shown in the diagram as open and closed circles.

The effects of this energy cascade are described by the heuristic parameter E_{e-h}, the mean electron-hole pair creation energy, which is the average energy required to produce one electron-hole pair. Using this parameter, it is possible to calculate the current flow out of the battery under a short-circuit condition (that is, when the battery terminals are connected together); this is referred to as the short-circuit current. The short-circuit current produced from a betavoltaic device is given by Eq. (1), where A

$$I_{sc}/A = J_{gen} = [J_\beta \cdot E_{\mathrm{mean}\,\beta} \cdot (1 - \eta)]/E_{e-h} \quad (1)$$

is the device area, J_{gen} is the net generated electron current density, J_β is the net flux of beta electrons from the radiation source, $E_{\mathrm{mean}\,\beta}$ is the mean beta-electron energy generated by the beta source, E_{e-h} is the mean electron-hole pair creation energy (which is 5.5 eV for silicon carbide, SiC), and η is the backscattering yield (the percentage of beta electrons backscattered at the semiconductor/air interface). For example, electrons with an energy of 5.68 keV (the mean electron energy for tritium) will produce around 1000 electron-hole pairs in SiC if the backscattering yield is 0%. For a flux of high-energy electrons equal to 5 nA/cm^2, a short-circuit

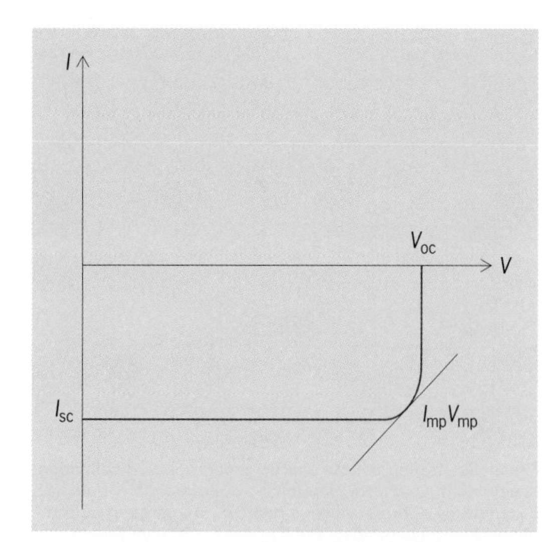

Fig. 3. I–V (current-voltage) plot of betavoltaic device.

current density of 5 μA/cm^2 will ensue. Equation (1) assumes 100% carrier collection efficiency of the beta electrons generated in the semiconductor.

A battery is defined by two parameters, the short-circuit current density (which has just been discussed) and the open-circuit voltage (the voltage when the battery's terminals are not connected). The open-circuit voltage can be calculated from the short-circuit current using well-known semiconductor device equations (also valid for solar cells). The open-circuit voltage is given in Eq. (2), where V_{oc}

$$V_{oc} = nV_T \ln(J_{gen}/J_{SS}) \qquad (2)$$

is the open-circuit voltage, n is the ideality factor, V_T is the thermal voltage (equal to 25.9 mV at a temperature of 300 K), ln is the natural logarithm function, J_{gen} is the current generated by the radioactive source, and J_{SS} is the reverse saturation current of the diode used in the cell. The ideality factor n is a measure of the quality of the semiconductor junction; the constant has a value between 1 (a high-quality junction) and 2 (a leaky junction). Reverse saturation current is the "leakage current" that flows internally inside the battery when the battery terminals are not connected. This leakage current can be affected by imperfections in the semiconductor material (such as electron traps or material dislocations), which will increase J_{SS} and therefore reduce V_{oc}.

The open-circuit voltage is determined by the electronic band gap of the semiconductor forming the pn junction (in addition to J_{SS}, as previously mentioned). **Figure 3** shows the plot of the betavoltaic $I - V$ (current-voltage) curve; the open-circuit voltage (V_{oc}), the short-circuit current (I_{sc}), and the maximum power point ($I_{mp}V_{mp}$) are indicated in the plot. The slope of the curve at the maximum power point is equal to the load resistance at maximum power.

Application of wide-band-gap semiconductors. Several semiconductor materials, such as silicon (Si), silicon carbide (SiC), gallium arsenide (GaAs), gallium phosphide (GaP), and gallium nitride (GaN), may be used to form the charge-separation junction. Wide-band-gap semiconductor materials produce betavoltaic batteries with the highest theoretical efficiencies. Semiconductor theory predicts that if a large-band-gap semiconductor is used to form the pn junction, the reverse saturation current will be very low (the "apparent" resistance that appears across the output of the device will be very high). This apparent resistance is referred to as the shunt resistance of the battery. Betavoltaic devices operate at very low output currents. It is important that the shunt resistance is very high, limiting the internal battery losses and increasing the battery efficiency. Silicon, the semiconductor industry workhorse, cannot realize sufficiently high open-circuit voltages or power-conversion efficiencies to be an optimal alternative for betavoltaic batteries. **Figure 4** compares the theoretical limit efficiencies of various semiconductor materials when excited by promethium [the measured efficiency (Fig. 4) for SiC betavoltaics powered by 63Ni]. The mean electron-hole pair creation energy E_{e-h} is similar for most of the material systems, which implies that the short-circuit current density for all the materials will be approximately the same. However, the larger-band-gap materials produce higher open-circuit voltages, which directly translate into devices with greater conversion efficiencies. Figure 4 does not account for backscattering losses or defects in the material, which could limit charge detection.

Betavoltaic sources. There are several candidate radioisotopes that can be inserted as a power source for betavoltaic batteries. Of these sources, nickel-63 (^{63}Ni), tritium (^{3}H), and promethium-147 (^{147}Pm) have been used in experimental devices, and promethium has been used in production devices as well. The energy spectra and absorption characteristics

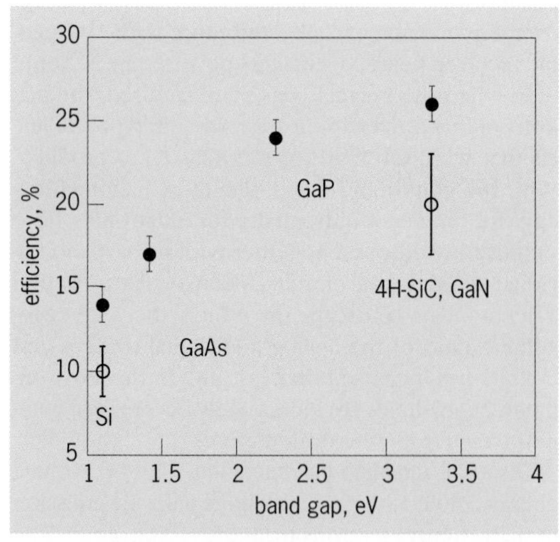

Fig. 4. Betavoltaic efficiency versus semiconductor band gap. The closed circles are the calculated efficiency of the device as a function of band gap. The open circles are measured values for betavoltaic devices. For the silicon (Si) device, ^{147}Pm is used as the radioisotope. For the silicon carbide (SiC) device, the data are taken with ^{63}Ni as the radioisotope.

Betavoltaic radioisotope source characteristics

Source	E_{max}, keV	E_{mean}, keV	Activity, GBq/mg	Half-life, years
^{63}Ni	67	17	0.37	92
^{3}H	18.6	5.685	37	12.32
^{147}Pm	230	73	15	2.62

of beta sources are fully defined by the mean energy E_{mean} and the maximum energy E_{max} of the beta source. The other important attributes of the radiation source are the isotope half-life and the isotope activity (which directly translates into beta-electron flux). The **table** compares the attributes of the popular beta sources.

If the energy of a beta electron is known, the penetration depth of the electron in a material of density ρ can be calculated using Eq. (3), where R is the range

$$R = 4E^{1.75}/100\rho \quad (3)$$

in micrometers, ρ is the density of the absorbing material in g/cm^3, and E is the energy of the electron in kiloelectronvolts. For a given source, the penetration depth determines how much radiation will be absorbed in the charge-separating semiconductor as well as how much radiation is "self absorbed" by the source. The amount of usable energy is limited to the number of high-energy electrons that are able to escape from the surface of the source. Because of the high absorbance of the beta source, only electrons from a very thin layer of radioisotope are extracted. In order to maximize source efficiency, it is necessary to form the isotope with a large surface-to-volume ratio. In the choice of betavoltaic sources, it is desirable to use sources with the highest mean energy; such isotopes will produce greater numbers of electron-hole pairs for the same activity. However, the down side is that the higher electron energies can produce radiation damage in the semiconductor, which ultimately limits the lifetime of the device. Also, at electron energies greater than 100 keV, the production of x-rays (bremsstrahlung radiation) becomes likely, increasing the shielding requirements of the package. Careful engineering is required to implement the optimal isotope for a specific application.

Applications of betavoltaic-powered systems. Betavoltaic batteries can deliver cost-effective continuous power in the range from 10 nW to 0.1 μW. The energy density of betavoltaic batteries is in the range of 3–200 W · h · cm^{-3} (depending on the isotope and the details of the design), which compares to 250–360 mW · h · cm^{-3} typical for lithium batteries. The half-lives of the important radioisotopes (*see* table) can be as long as 92 years or as short as 2.6 years. Therefore, betavoltaics find their greatest utility when there is need for a remote stand-alone application (where battery change-out is difficult), coupled with a need for long life. Further, electronic devices often need to operate in harsh environments where the temperature extremes preclude the use of conventional lithium batteries. In another type of systems implementation, betavoltaic batteries can be used to "trickle charge" thin-film lithium batteries (which are able to withstand harsh environments) or ultra-capacitors in order to be able to provide relatively large amounts of energy over a short period of time. This system topology can be used for applications that operate principally in the standby mode, such as antitamper circuits. Antitamper circuits need to have a small amount of continuous power to maintain security vigilance but a relatively large amount of available power to actuate alarms should there be a security breach.

Sensor networks are a ubiquitous class of applications that are remote in nature. The intelligent microprocessors that are at the heart of these networks are being developed using techniques that significantly reduce the operating energy requirements. Currently, digital circuits can idle at 0.36 μW of power and produce a 24-bit data packet every 3 s from a 1.8-μW power source. Low-power analog circuits are able to do 40 measurements per second from the same 1.8-μW power source. It is anticipated that both digital and analog circuits will continue to become more energy efficient. Medical applications represent systems that are poised to take advantage of the aforementioned low-energy circuits as well as new developments in MEMS (microelectronic machining). Low-power (1- to 100-μW) medical applications include implantable sensors (such as glucose monitors) and pacemakers.

For background information *see* BATTERY; BETA PARTICLES; BREMSSTRAHLUNG; HOLE STATES IN SOLIDS; MICRO-ELECTRO-MECHANICAL SYSTEMS (MEMS); PHONON; PHOTOVOLTAIC CELL; PHOTOVOLTAIC EFFECT; PLASMON; RADIATION DAMAGE TO MATERIALS; RADIOACTIVITY; SEMICONDUCTOR; SOLAR CELL; TRITIUM in the McGraw-Hill Encyclopedia of Science & Technology. Michael G. Spencer

Bibliography. C. Knight, J. Davidson, and S. Behrens, Energy options for wireless sensor nodes, *Sensors*, 8:8037–8066, 2008; J. Nelson, *The Physics of Solar Cells*, Imperial College Press, London, 2003; L. C. Olsen, Betavoltaic energy conversion, *Energ. Convers.*, 13(4):117–124, 1973.

Biorefinery (wood)

The term biorefinery relates to the concept of deriving multiple fuel, chemical, and material products from biomass in a way that is analogous to the operation of a petroleum refinery. The key feature that is common to both petroleum- and biobased refineries is the capability to generate a range of commercially useful products in a versatile manner that allows for rapid adjustment according to market requirements.

The origins of using wood or other forest biomass for biorefining date back to the eighteenth century, when forest chemicals, including pitch, pine tar, turpentine, and rosin, were first extracted for

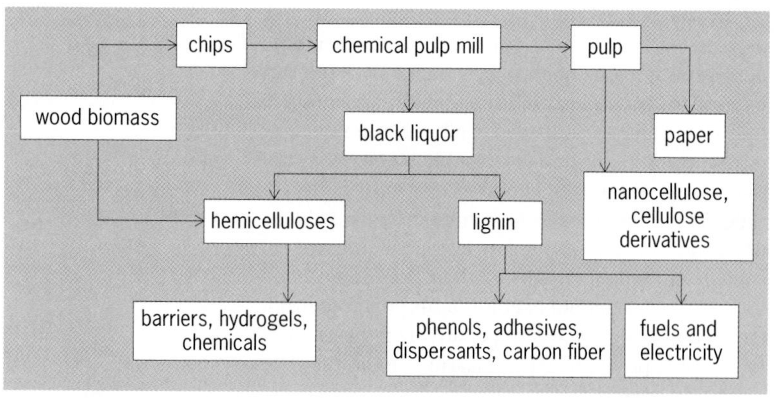

Fig. 1. Material flows and end products in a wood biorefinery concept based on a chemical pulp mill.

large-scale commercial use. Raw materials for biorefining can consist of (1) sugar-rich agricultural crops (such as wheat and corn) or agricultural residues and (2) forest biomass. These two routes are now broadly referred to as first- and second-generation biorefining, respectively. Just as the output of products, such as gasoline, diesel fuel, kerosene, and liquefied natural gas, from a petroleum refinery can vary in accordance with the type of incoming crude oil and its fractionation, so it is conceivable that biorefineries will have the capacity for production of different starch or cellulose product portfolios, depending on whether agricultural resources or woody biomass, respectively, are used as raw materials.

Regardless of the starting material, the end objective of biorefining is to displace fossil fuel–based products, namely fuels, chemicals, and materials, with renewable alternatives and with the added benefit of reduced environmental impact. Because there may be adverse effects on human food and animal feed supply chains from use of agricultural crops and the agricultural land base, there is increasing interest in second-generation biorefineries. However, it is worth noting that, in the case of agricultural crops, the range of biorefineries also might include those in which the output is a combination of food and feed (primary products) as well as secondary products (for example, fibers for use in composites). Depending on the technology platform, second-generation biorefineries offer the possibility of creating a wide range of fuels, chemicals, or fibers from the three main constituents of wood: cellulose, lignin, and hemicelluloses. Softwoods contain 40–45% cellulose, 26–34% lignin, and 7–14% hemicelluloses. Hardwoods generally contain slightly lower percentages of cellulose and lignin, but a higher percentage of hemicelluloses. The current global use of wood is approximately 3.6 billion m^3, of which more than half is employed in various forms of energy production. Estimates of the global forest biomass that is potentially available for biorefining, calculated by assuming that 25% of the total annual surplus forest growth could be used, are in the range of 0.7–1.2 billion m^3. One wood biorefining option is to build upon existing infrastructure within chemical pulp mills. For example, pulp and paper production at a kraft (sulfate) pulp mill can be complemented by the use of lignin and hemicelluloses extracted from black liquor (the thick, dark liquid by-product of the kraft pulping process) to derive various products as well as energy (**Fig. 1**). Alternatively, hemicelluloses may be pre-extracted by leaching from wood chips and converted to monosugars through hydrolysis, and these sugars then can be used to supplement cellulose in fermentation processes (for example, to generate ethanol if genetically modified yeasts or bacteria are used). Although there is evidence that this should be technically feasible, development of optimum fermentation processes for carbon-5 (C5) sugars continues to be an active area of research and development.

Biofuels. Increasing concerns about the future availability of oil, including its cost and security of supply, are driving biorefinery initiatives with a focus on production of ethanol as a fuel. In some countries, there is already a significant history of ethanol use as a fuel for road vehicles. One such country is Brazil, where ethanol is manufactured from sugarcane. In North America, fuel blends containing ethanol have been on the market since the late 1970s. For practical reasons related to air quality and engine performance, the addition of ethanol to gasoline in the United States is currently limited to a "blend wall" of 10%. In addition, the aviation industry has run successful experimental trials on ethanol as a fuel component.

To generate liquid fuels from wood, the first step involves a biochemical or thermochemical pretreatment. In the case of biochemical pretreatment, this step is designed to make the raw material more amenable to hydrolysis and is followed typically by fermentation of sugars to ethanol, separation of the sugar solution from residuals such as lignin, microbial fermentation of the sugar solution, distillation, and finally dehydration. Butanol also can be manufactured from fermentation of lignocellulosics and blended safely at relatively high levels with both gasoline and diesel fuel. Unlike ethanol, butanol can be transported in oil pipelines and has better fuel quality characteristics. Using the thermochemical route, liquid biofuels may be produced from forest biomass through fast pyrolysis processes that involve rapid heating to 500°C (932°F) in the absence of oxygen. The quality of oil obtained in this manner is dependent on the feedstock and process conditions, with the best yields of liquid products occurring at low feedstock moisture contents. At present, improvements in the quality of pyrolysis oils are necessary before adoption in transport fuel and combustion processes can be fully considered. Such improvements may be possible through variations in raw material or pyrolysis process conditions.

Bioderived platform chemicals. There is a considerable history of chemical coproducts from the chemical pulp industry. One example is tall oil, a mixture of rosins, sterols, and fatty acids, which has been converted to industrially useful derivatives. The U.S. Department of Energy has released a priority list of bioderived platform chemicals, some of which might be generated in a wood-biorefining operation (**Fig. 2**). Useful chemical intermediates such as glycerol, furfural, levulinic acid, and succinic acid are included on this list. Speciality chemical output from a wood biorefinery also could provide nutraceuticals, including xylitol and arabitol, and rare sugars, including xylose, arabinose, mannose, and ribose, which are derivable from wood hemicelluloses using industrial ion-exchange chromatography methods and crystallization from aqueous solution.

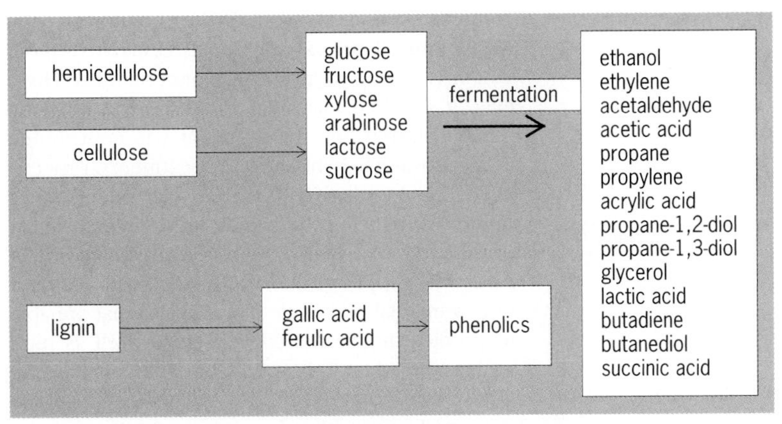

Fig. 2. Schematic outline showing examples of platform chemicals derivable from wood biomass in a wood biorefinery. (*Modeled on Fig. 3 from T. Werpy et al., Top Value Added Chemicals from Biomass, Volume 1, Results of Screening for Potential Candidates from Sugars and Synthesis Gas, U.S. Department of Energy, Washington, D.C., 2004*)

Biomaterials. Materials that may be obtained from a wood biorefinery can complement the liquid fuel and platform or speciality chemical portfolio. A future example could be the conversion of part of the cellulose stream to nanocellulose, either through mechanical homogenization of pulps to obtain microfibrillated cellulose (MFC) or through an acid hydrolysis procedure that results in crystalline cellulose nanowhiskers (nanoscale structures). Although originally considered for uses such as food and cosmetic additives, the potential of nanocellulose as a nontoxic and renewable polymer-reinforcing agent is now increasingly recognized, particularly in situations where fully bioderived composites can be obtained though combination with bioresins (**Fig. 3**). In addition to ethanol and butanol, the various products from hemicelluloses in a biorefinery include barrier materials, hydrogels, fiber additives, and composites. Of these possibilities, the favorable oxygen-barrier properties of hemicellulose films are of considerable interest, and commercial applications in packaging are therefore being explored. Hydrogels based on hemicelluloses are said to have potential uses in the health care and pharmaceutical industries. As well as its use for energy production, lignin from a wood biorefinery is one of the very few renewable sources for bulk aromatic chemicals (for example, phenols) and, as such, may be converted to commercial products, including adhesives, dispersants, and carbon fiber. The latter has good market-growth possibilities related to strong, lightweight composite materials in the automotive and aerospace industries. Lignin also has been used as an additive in thermoplastic composites, some of which are now in use commercially in applications as varied as loudspeakers, toys, and automotive parts.

Government policies. As a general rule, the policies of governments around the world in support of biorefining have focused particular attention on greater deployment of biofuels. Targets for the use of biofuels have now been introduced in at least 40 countries, although not all of these are mandatory. Because the targets are based largely on first-generation biofuels, some reviews of these targets are ongoing. Specific international policies that encourage second-generation biofuels include the U.S. Energy Independence and Security Act of 2007 and the European Union (EU) draft directive of 2008, which assume that second-generation biofuels will provide a substantial share of the total by 2020. As an incentive toward meeting national commitments in Europe, the contribution from biofuels obtained from wastes, residues, nonfood cellulosics, and lignocellulosics will count twice that of contributions from other biofuels. Estimates indicate that the policies in the United States and European Union could result in a 5–15% increase in the global uptake of second-generation biofuels by 2020, with a further increase in second-generation biorefining in subsequent years.

Future developments. While wood-based biorefineries centered on more diversified pulp mill operations appear quite achievable in the short term, the full development of second-generation biorefineries with ethanol as the core output will require continued investment in optimizing biochemical or thermochemical conversion technologies. Despite decades of research and development, this still is some years from full commercialization. The

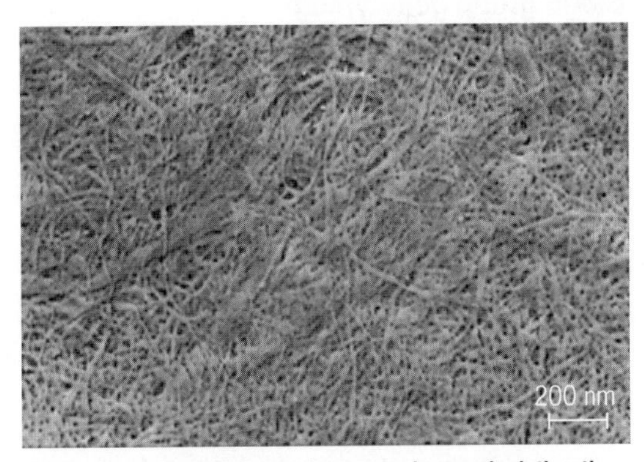

Fig. 3. Scanning electron microscopy image depicting the entangled nature of solvent-cast microfibrillated cellulose (MFC) nanofibers. Nanocelluloses such as MFC offer potential as a future high-value coproduct stream from wood biorefineries.

commercial risks are now further complicated by fluctuating oil prices and global financial turmoil, which is adding to investment uncertainties. However, it should be said that good progress in biorefining research and development has contributed new innovations in biomass pretreatments, processing technology efficiencies, and coproduct development, which will be commercially beneficial. At present, there also is an improved understanding of feedstock supply chains and a significant number of commercial pilot-scale biorefineries that are able to produce high-value coproducts as well as fuels such as bioethanol. It is widely expected that the full commercial uptake of second-generation biorefining will take another decade or so, but the policies designed to reward biofuel sustainability and environmental performance should ensure that second-generation biorefineries based on wood and other forest biomass will become considerably more significant in the commercial realm after 2020.

For background information *see* ALCOHOL FUEL; BIOMASS; CELLULOSE; CONSERVATION OF RESOURCES; HEMICELLULOSE; LIGNIN; PYROLYSIS; RENEWABLE RESOURCES; TALL OIL; WOOD ANATOMY; WOOD CHEMICALS; WOOD PROCESSING; WOOD PROPERTIES in the McGraw-Hill Encyclopedia of Science & Technology.

David V. Plackett

Bibliography. J. H. Clark, Green chemistry for the second-generation biorefinery—sustainable chemical manufacturing based on biomass, *J. Chem. Tech. Biotechnol.*, 82:603–609, 2007; A. Demirbas, *Biorefineries: For Biomass Upgrading Facilities*, Springer-Verlag, London, 2010; B. Kamm, P. R. Gruber, and M. Kamm, *Biorefineries—Industrial Processes and Products: Status Quo and Future Directions*, vol. 1, Wiley-VCH, Weinheim, Germany, 2006; P. Söderholm and R. Lundmark, The development of forest-based biorefineries: Implications for market behavior and policy, *Forest Prod. J.*, 59:6–16, 2009.

Blood group genotyping

Red blood cells express multiple molecules, including hundreds of different proteins and carbohydrates on their surfaces. These surface molecules vary among individuals. Some of these surface molecules are not present on some people's red blood cells. Other surface molecules exist in slightly different forms in different people. These variations are important during transfusion of red blood cells because cell surface molecules on the transfused red blood cells that are not present on a patient's red blood cells are "foreign" and can induce an immune response in the patient.

Two methods can be used to determine which protein and carbohydrate variations are expressed on an individual's red blood cells: serologic methods and genotyping (molecular) methods. While serologic methods have been used for decades, they are limited in accuracy, throughput, and cost. In many situations, genotyping methods can overcome these limitations and have the promise to change transfusion practices and improve safety.

Importance of red blood cell surface molecules (antigens). The most widely known red blood cell surface molecules are those of the ABO blood group system. These are carbohydrate structures, with O red blood cells lacking sugar residues that are present on A, B, or AB red blood cells. Each person's immune system will make antibodies to the A and B structures that their own red blood cells lack, and the antibodies will destroy transfused ABO-incompatible red blood cells in a process known as hemolysis. Transfusion of incompatible red blood cells and the resulting hemolysis is known as a hemolytic transfusion reaction and is often fatal.

There are many other molecules present on the surface of red blood cells that differ among people. Like the ABO sugar structures, many of these other molecular structures are known as antigens because they can be recognized by the immune system and induce an immune response directed at the specific molecular structure. However, unlike the ABO antigens, a person's immune system does not make antibodies to other red blood cell antigens unless the person has been exposed to those antigens, which usually occurs during pregnancy or from a transfusion.

The initial exposure to a foreign red blood cell antigen rarely causes clinically significant hemolysis, but it can induce an initial immune response. However, once a person's immune system has mounted such a response, subsequent exposures to the same red blood cell antigen will result often in a robust immune response. This secondary response can cause rapid hemolysis and a potentially severe or even fatal transfusion reaction. Additionally, if a woman has anti–red blood cell antibodies during pregnancy, those antibodies may be able to cause hemolysis in the fetus, resulting in severe and sometimes lethal fetal anemia.

Traditional approach to avoiding red blood cell destruction (hemolysis). Most efforts to avoid hemolytic transfusion reactions are focused on avoiding the transfusion of red blood cells that express antigens to which a patient has already made antibodies. Prior to a transfusion, each patient's blood is tested for the presence of anti–red blood cell antibodies. If antibodies are detected and identified, then donor red blood cells that lack the corresponding antigens are transfused.

With a few exceptions, blood banks usually do not take special precautions to prevent the development of new anti–red blood cell antibodies. In most cases, this is a safe practice because only about 5% of red blood cell transfusions will induce anti–red blood cell antibodies and most patients do not even have the ability to make such antibodies. The most common exception concerns the rhesus D [Rh(D)] protein, which has a high propensity to induce dangerous antibodies. Hence, most blood banks provide

Rh(D)-negative [Rh(D)-] red blood cells for Rh(D)- patients.

Blood banks test red blood cell antigens on donor red blood cell units that will be transfused and on patient's red blood cells. Donor red blood cell antigens are detected to determine which units to transfuse to patients who have antibodies or would otherwise benefit from blood that lacks specific antigens. A patient's red blood cell antigens are detected to facilitate identification of the anti-red blood cell antibodies that a patient has made and can make in the future.

Traditional methods to detect red blood cell surface antigens. Traditionally, antibodies are used to detect red blood cell antigens. While these methods work well in many situations, they can have significant limitations. Often the antibodies that are extracted from human blood and supplies can be limited and expensive. Antibody methods can be difficult or even impossible to interpret if preexisting antibodies are bound to red blood cells, as happens with some people. Antibody methods require pure red blood cells from the individual of interest. This can be difficult to obtain if the individual is a fetus or if a person has been transfused recently with red blood cells. Finally, antibodies recognize a small portion of a protein and might miss a variation in the protein that occurs at a different region of the protein. In this situation, antibody-based assays can give misleading or incomplete results.

Genes determine red blood cell surface molecules. Molecular variations on the surfaces of red blood cells are a result of genetic variations among individuals. Different forms of a gene, known as different alleles, are usually caused by variations at single DNA positions. These are termed single nucleotide polymorphisms (SNPs) and are responsible for most allelic variations in genes that encode red blood cell surface molecules.

Genotyping methods. Blood group genotyping detects the DNA alleles that encode for red blood cell surface proteins or for enzymes that modify those proteins. Theoretically, one could sequence the relevant genes, which is done sometimes for complex cases by a few reference laboratories. However, DNA sequencing is not currently feasible for widespread use at most blood banks. Instead, most genotyping methods detect the SNPs known to be responsible for most variations in red blood cell surface molecules. Several methods with similar strategies have been developed.

Genotyping methods take advantage of the fact that most allelic variations are caused by SNPs. The first step involves DNA extraction, usually from white blood cells because red blood cells do not have nuclei that contain DNA. Subsequent steps involve the polymerase chain reaction (PCR). During PCR, a DNA segment from a gene of interest is replicated multiple times. This is achieved by successive rounds of annealing (recombining strands of complementary DNA that were separated by heating or other means of denaturation), thereby creating small

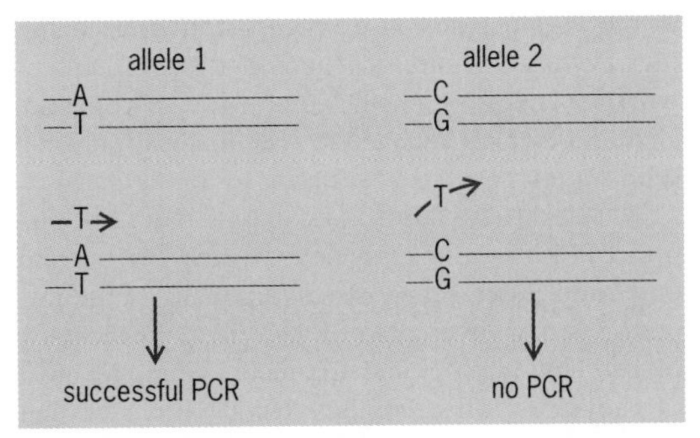

Fig. 1. The primer binds to allele 1, but not allele 2. Hence, PCR will be successful only with allele 1.

synthetic DNA fragments, known as oligonucleotide primers, which serve as the initial substrates for DNA extension catalyzed by a DNA polymerase.

PCR serves to amplify the target area of interest and to discriminate between blood group alleles. For each blood group gene of interest, at least one pair of oligonucleotide primers (short synthetic DNA fragments) is designed to amplify one of the alternate blood group alleles (**Fig. 1**). One of these primers binds specifically at an SNP that determines the allele. At specific salt and temperature conditions, this primer will bind only to DNA from the allele with which it is 100% complementary. Hence, the PCR reaction will successfully amplify only the specific allele targeted by the primer and will fail to amplify the alternate allele.

The most common approach to detect the PCR-amplified DNA segment involves fluorescence. Fluorescent conjugated deoxynucleotides are incorporated into the amplified allele by adding fluorescent deoxynucleotides into the PCR reaction. The fluorescent amplified allele is annealed to a DNA oligonucleotide that is attached to a solid substrate (**Fig. 2**). Fluorescence at a particular point in the solid substrate indicates that the PCR reaction occurred and the specific allele was present in the person's DNA. Alternate approaches, including those in which PCR products are detected in solution, can be used as well.

Multiple alleles can be analyzed simultaneously by adding multiple sets of oligonucleotide primers

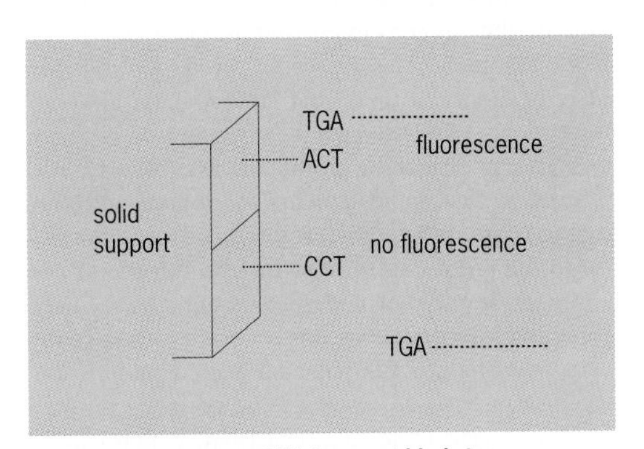

Fig. 2. The fluorescent DNA fragment binds to an oligonucleotide attached to a solid support only if it is perfectly complementary.

to the PCR reaction in a process known as multiplex PCR. Each fluorescent amplified DNA fragment will then anneal to specific complementary oligonucleotides that are attached to specific locations on a solid support.

Genotyping advantages. Genotyping that uses multiplex PCR can efficiently detect many alleles from one sample simultaneously. Additionally, some laboratories have been able to gain further efficiency by batching samples and automating some or most of the steps. This, combined with the fact that reagents can be produced relatively inexpensively in essentially limitless quantities, means that multiple blood group alleles often can be detected on many blood samples less expensively than with traditional antibody-based techniques.

Genotyping can be more accurate than traditional antibody methods because it is not subject to the same limitations. For example, genotyping can provide accurate results even if a person's red blood cells are coated with antibodies, and usually can provide accurate results even if a person has undergone a recent transfusion. Genotyping also can be used to identify rare alleles that may be difficult to detect with most antibody reagents.

Genotyping limitations. Genotyping is subject to potential erroneous results. As such, the genotyping systems need to be carefully designed and tested. In addition, because most genotyping techniques are designed to identify known allelic variations and do not directly detect the antigens on the red blood cell surface, they may fail to detect unusual genetic variations that may alter a surface protein or may even prevent a protein from coating the red blood cell surface.

Also, most blood genotyping platforms are designed to detect SNPs and are most useful when there are only a small number of allelic variations that exist for a particular gene. This is not the case for the ABO blood group system, which is encoded by multiple allelic variations. Also, many different alleles have been identified in the Rh genes, making accurate detection of the Rh alleles challenging in some cases.

Uses. Currently, blood group genotyping is being used to test donor blood, patients, and (in some cases) fetuses. Donor blood is tested to efficiently identify the antigens expressed on many donor red blood cell units. This allows for rapid identification of red blood cell units that lack specific antigens for patients who have made antibodies or who are at high risk of making antibodies. Genotyping also is used to test some patients for whom antibody methods are difficult or for whom it would be useful to minimize exposure to foreign red blood cell antigens. Genotyping and even sequencing are used sometimes for patients who are thought to have unusual red blood cell surface antigens. Finally, genotyping can be used to determine antigens on fetal red blood cells by testing fetal DNA, which can be extracted from the mother's blood or obtained from amniocentesis or chorionic villus sampling.

Future directions. Extensive blood genotyping could facilitate extended matching of multiple blood group antigens for many patients to minimize exposure to foreign antigens from transfusions. Just as most Rh(D)– patients receive Rh(D)– red blood cells to minimize the chances of developing anti-Rh(D) antibodies, patients that lack other antigens could receive red blood cells that lack those antigens. Because there are hundreds of antigenic molecules on the surface of red blood cells, it probably will be impossible to provide red blood cells that are antigen-matched perfectly for all patients. Future efforts are being made to prioritize antigens and patients for extended matching.

For background information *see* ALLELE; ANTIBODY; ANTIGEN; ANTIGEN-ANTIBODY REACTION; BLOOD; BLOOD GROUPS; DEOXYRIBONUCLEIC ACID (DNA); IMMUNOLOGY; POLYMORPHISM (GENETICS); RH INCOMPATIBILITY; SEROLOGY; TRANSFUSION in the McGraw-Hill Encyclopedia of Science & Technology.

Steven R. Sloan

Bibliography. B. Alberts, J. H. Wilson, and T. Hunt, *Molecular Biology of the Cell*, 5th ed., Garland Science, New York, 2008; G. Daniels, The molecular genetics of blood group polymorphism, *Hum. Genet.*, 126:729–742, 2009; A. M. Lesk, *Introduction to Genomics*, Oxford University Press, Oxford/New York, 2007; J. Roback et al. (eds.), *Technical Manual*, 16th ed., American Association of Blood Banks, Bethesda, MD, 2008; C. M. Westhoff and S. R. Sloan, Molecular genotyping in transfusion medicine, *Clin. Chem.*, 54:1948–1950, 2008.

Canine influenza

Influenza is a viral disease that strikes many different types of animals, including birds, pigs, horses, and humans. Canine influenza is considered an emerging disease, and resulted from a mutation in a previously existing equine influenza virus strain known as H3N8. Prior to this development, influenza A virus strains were unable to spread easily within canine populations. After the development of this new canine strain, outbreaks of canine influenza were reported for the first time in greyhound kennels and at greyhound racetracks in 2004. Serological studies indicate that the virus may have been present in greyhounds as early as 1999. Although canine influenza first appeared in greyhounds, it now causes respiratory illness in a variety of breeds, and all dogs are considered to be at risk of infection.

Background. Canine influenza is a respiratory illness that causes symptoms that resemble kennel cough. Kennel cough, also known as infectious tracheobronchitis, is caused by a number of different microbes, including both bacteria and viruses. Canine influenza, however, is caused by a specific strain of influenza virus. Early symptoms in milder cases include a low-grade fever, followed by a persistent cough and sometimes a purulent nasal discharge. Infected dogs appear lethargic and frequently have no appetite. In more severe cases, dogs have a high fever, with an increased respiratory rate and signs of pneumonia or bronchopneumonia. Fatalities have occurred in greyhounds, with autopsies revealing hemorrhaging in the respiratory tract. However, this does not appear to be frequent in household pets.

Influenza viruses are characterized on the basis of the type of protein antigens found on their surface: the hemagglutinin or H protein, and the neuraminidase or N protein. There are 16 different types of H protein and 9 different types of N protein. Typically, however, there are limited types that are found in each species of mammal. Currently, H3N8 is the main subtype found in horses. This virus subtype appears to have made the jump directly from equines to dogs (see **illustration**), and also appears to have diverged considerably from the original equine virus. As it can now be transmitted directly between dogs, it is considered to be a canine influenza virus. Analysis shows that there are four amino acid differences between the H protein in the equine virus and that of the canine virus. These differences are likely the reason that the virus can now be transmitted to dogs.

Between 2004 and 2006, canine influenza infections occurred in racing greyhounds in multiple locations in the United States, including the far south (Florida), as far west as Colorado, and as far north as Massachusetts. Since then, it has been reported in pet dogs in Florida, as well as establishing itself in Colorado and the New York City area, including New Jersey and Connecticut. Transmission appears to be similar in dogs as it is in humans and other mammals, occurring as a result of exposure to nasal discharges produced by the coughing or sneezing of an infected animal. Shedding of the virus can occur up to 7 days after clinical symptoms appear in a dog; 20–25% of the dogs can remain asymptomatic, but they can still shed the virus and spread the disease. The incubation period for canine influenza ranges from 2 to 5 days.

Because dogs as a whole have never had influenza, most of the population is expected to be susceptible. A new canine influenza vaccine was developed and released conditionally by the United States Department of Agriculture in June 2009. Trials with this vaccine for H3N8 showed that it reduces the incidence and severity of lung lesions, as well as the duration of coughing and viral shedding. In order to prevent transmission, disinfection of the kennels of infected animals should be performed with a bleach solution, detergents, or ethanol. Proper infection control practices should be implemented, with cages and bowls washed and disinfected between uses. Hand washing with soap and water by kennel personnel will minimize transmission, and gloves should be used when handling an animal that shows respiratory signs. The virus can survive on hands for up to 12 h, so proper hand washing is essential in the prevention of transmission.

Pathology and clinical signs. When influenza viruses enter the respiratory tract, they attach to epithelial cells by the hemagglutinin glycoprotein, which binds to sialic acid sugars on the surface of the host cells. The cell takes in the virus through endocytosis, a process in which the cell membrane wraps around the virus and engulfs it. The membrane forms an endosome, a membrane-bound vacuole, which becomes acidified. The hemagglutinin protein fuses the viral envelope to the endosome membrane, releasing the contents of the virus into the cytoplasm of the host cell, where it proceeds to take over the cell machinery to replicate its genome and proteins. New viruses are assembled and released from the cell. As they are released, the viruses attach to the cell surface sialic acid residues through the hemagglutinin protein. The neuraminidase protein that is found on the virus surface cleaves the residues, releasing the virus. After this, the host cell dies.

The replication of influenza virus in the cells of the respiratory tract causes rhinitis or nasal discharge, as well as coughing as a result of bronchitis and bronchiolitis. Because the cells lining the respiratory tract are killed by the infection, the underlying basement membrane becomes exposed and susceptible to secondary bacterial infections, which can increase the nasal discharge and coughing caused by the primary influenza infection. Exposure virtually guarantees infection with the virus. Approximately 20% of the dogs will remain asymptomatic, whereas 80% will develop clinical signs of disease. Most dogs exhibit the milder form of canine influenza. The most common clinical symptom is a cough with a duration of 10–21 days. The cough is soft and moist in most dogs, although some have a dry cough that resembles kennel cough (which is caused by *Bordetella bronchiseptica* or parainfluenza virus).

In severe cases, dogs may develop signs of pneumonia, including a high-grade fever of 104–106°F (40–41°C), and increased respiratory rate and difficulty in breathing. X-rays of the lungs may reveal lobe consolidation, a condition in which the lobe of the lung fills with fluid. Postmortem examination of fatal cases reveals hemorrhages in the lungs and pleural cavity. Signs of severe pneumonia may be present in the lungs, which may appear dark red to black. Histology may reveal tracheitis (inflammation of the trachea), bronchitis, bronchiolitis, and severe pneumonia.

Diagnosis and treatment. Clinical symptoms alone are insufficient to diagnose canine influenza, although it should be suspected in dogs with a persistent cough. Antibodies to canine influenza virus may be detected as early as 6–7 days after onset of infection. Using a reverse-transcriptase polymerase chain reaction (RT-PCR) assay, the virus may be detected in swabs taken from the nasopharynx of symptomatic dogs. However, RT-PCR becomes less accurate after the peak virus-shedding period ends, at approximately 4 days into the infection. In dogs that display suspicious symptoms of canine influenza, but have a negative result for virus when using RT-PCR,

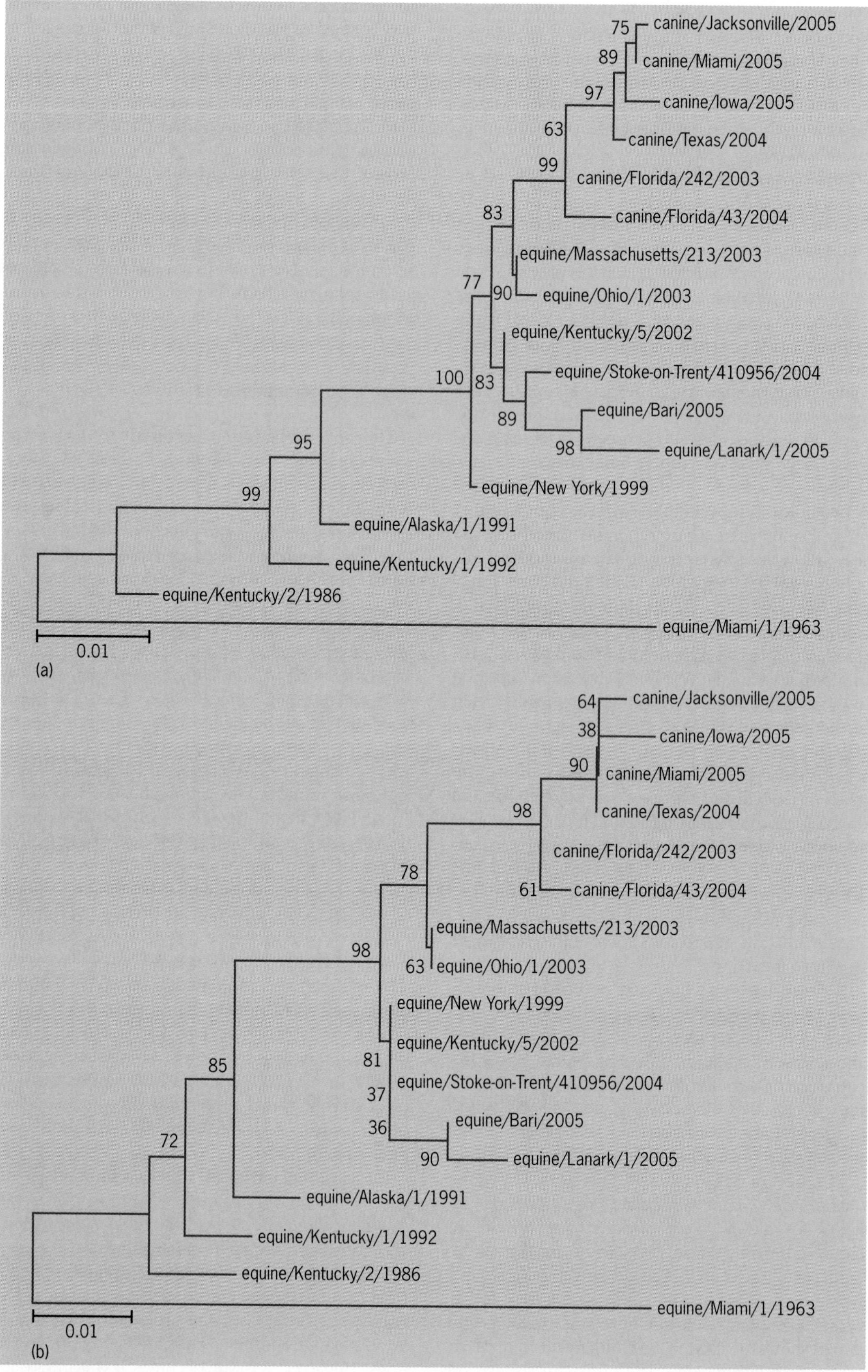

A study of the phylogenetic relationships among the hemagglutinin 3 (H3) genes. (*a*) Nucleotide tree of the canine influenza virus H3 genes with contemporary and older equine H3 genes. (*b*) Amino acid tree of the canine influenza virus H3 protein with contemporary and older equine H3 proteins. It is clear that the canine strain of influenza virus (H3N8) is most closely related to the equine strain from which it originated in 2004. Bootstrap analysis values greater than or equal to 80% are shown. Scale bars indicate nucleotide or amino acid substitutions per site. (*Source: Emerging Infectious Diseases; http://www.cdc.gov/eid/content/14/6/902.htm*)

serological techniques should be employed to measure antibody titers.

As canine influenza is a viral disease, treatment is limited largely to supportive therapy, such as administering intravenous fluids in severe cases and proper nutrition and care in milder cases. Pneumonia is usually a secondary bacterial infection and should be treated with appropriate antibiotics. A thick green nasal discharge usually indicates a bacterial infection of the nasopharynx and also should be treated with a broad-spectrum antibiotic. Although there are antiviral treatments for influenza in humans, these have not been tested in dogs and should not be administered. In most cases, dogs with the milder form of the disease recover fully within 2–3 weeks. More severe cases may require more time for recovery. Fatalities often show hemorrhaging in the respiratory tract; this most often occurs in greyhounds and is infrequent in pets.

Conclusions. Canine influenza virus appears to be an anomaly among influenza viruses because it crossed from horses directly to canines without undergoing antigenic recombination, also known as antigenic shift. As a result of its newness, dogs appear to be overwhelmingly susceptible as a species. Fortunately, a new killed-virus vaccine for H3N8 appears to induce excellent immunity in dogs, and a trial involving more than 700 dogs demonstrated that it does not cause side effects of any significance. As canine influenza virus damages the mucosa of infected dogs, it makes these animals highly susceptible to a host of canine bacterial infections, all of which can cause more severe disease and potential fatalities. Although canine influenza appears to be a mild disease in many dogs, immunization is important in preventing complications, such as bacterial pneumonia or rhinitis. As an emerging disease, canine influenza outbreaks should be carefully monitored to prevent the potential of transmission within pet populations. It also should be monitored to minimize the risk that the virus might undergo an antigenic recombination with a human virus, leading to the possibility of its spread within the human population.

For background information *see* ANIMAL VIRUS; ANTIBIOTIC; ANTIGEN; DOGS; EPIDEMIOLOGY; INFECTIOUS DISEASE; INFLUENZA; KENNEL COUGH; PNEUMONIA; VACCINATION; VIRUS in the McGraw-Hill Encyclopedia of Science & Technology.

Marcia M. Pierce

Bibliography. M. K. Cowan and K. P. Talaro, *Microbiology: A Systems Approach*, 2d ed., McGraw-Hill, New York, 2008; P. D. Kirkland et al., Influenza virus transmission from horses to dogs, Australia, *Emerg. Infect Dis.*, 16:699–702, 2010; P. R. Murray, M. A. Pfaller, and K. S. Rosenthal, *Medical Microbiology*, 5th ed., Mosby, St. Louis, 2005; E. Nester et al., *Microbiology: A Human Perspective*, 6th ed., McGraw-Hill, New York, 2008; S. Payungporn et al., Influenza A virus (H3N8) in dogs with respiratory disease, Florida, *Emerg. Infect. Dis.*, 14:902–908, 2008.

Cannon's conjecture

Cannon's conjecture attempts a group-theoretic generalization of W. P. Thurston's famous geometrization conjecture for 3-manifolds.

Geometric structures and the geometrization conjecture. The circle C locally looks like the real line $\mathbb{R}$. The real line can be wrapped around the circle infinitely often in such a way that two points r and s of $\mathbb{R}$ go to the same point of C if and only the difference $r - s$ is an integer. The resulting mapping p that takes $\mathbb{R}$ onto C is said to define a geometric structure on C, modeled on the geometry $\mathbb{R}$. This geometric structure on C allows the circle C to be studied by means of the algebraic and geometric properties of $\mathbb{R}$ and the mapping p. This particular geometric structure is heavily employed in number theory, in Fourier analysis, and in topology.

The advantages of a geometric structure become even more powerful in higher dimensions.

Thurston's geometrization conjecture claims that three-dimensional manifolds (those spaces that are locally like three-dimensional Euclidean space) also have geometric structures, in fact unique structures, modeled on one or more of eight well-understood three-dimensional model geometries, including Euclidean geometry, spherical geometry, negatively curved hyperbolic geometry, and five others. That is, after a canonical cutting into pieces, each piece is the (nicely folded) image of one of the eight model geometries.

Thurston's conjecture has, during the last 30 years, revolutionized the study of low-dimensional geometry and topology, in particular the study of two- and three-dimensional manifolds. His conjecture includes the famous Poincaré conjecture, which states that every compact, connected, simply connected 3-manifold M is homeomorphic with the three-dimensional sphere; indeed, spherical geometry is the only geometry among the eight that could serve as model for M. Thurston showed that the geometrization conjecture is true for the vast majority of 3-manifolds. Recently, Gregory Perelman amazed the mathematics community by proving the conjecture in its entirety, including the Poincaré conjecture, by combining methods from the topology of 3-manifolds, Riemannian geometry, and differential equations.

Extension to the theory of groups. Cannon's conjecture attempts to extend a major portion of the Thurston-Perelman geometrization conjecture from the study of manifolds into the mathematical theory of groups. In mathematics and physics, a group is an abstract algebraic object intended to model collections of rigid motions or symmetries of a geometric object. Motions or symmetries can be composed (repeated one after another) so that a collection of motions or symmetries forms an algebraic object with composition being the algebraic operation. In different contexts this composition might become, or be replaced by, some form of addition or multiplication. The Cannon conjecture claims that, if a group G satisfies certain natural abstract properties, then the group can be realized concretely as a group of rigid motions of three-dimensional negatively curved hyperbolic geometry; that is, G has a geometric structure modeled on hyperbolic space. Such rigid motions can all be expressed as 2×2 matrices with complex entries that act by rigid motion on hyperbolic space and by conformal mapping on the two-dimensional sphere. It would follow from the conjecture that such groups can be studied by means of matrix theory, noneuclidean geometry, and the theory of conformal mappings. A proof of the conjecture must either rely on the Perelman theorem or supply a new proof of the difficult hyperbolic portion of that theorem.

Precise statement and supporting definitions. Here is a precise statement of the conjecture followed by the supporting definitions.

Cannon's conjecture: Suppose that G is a group whose Cayley graph is Gromov-hyperbolic and whose space at infinity is the 2-sphere $\mathbb{S}^2$.Then G is a Kleinian group.

Cayley graph. Let G denote a group having a finite generating set $C = c_1, \ldots, c_k$, so that every element of G can be expressed as a finite product (or sum, or composition) of elements of C and their inverses. In fact, because, in a group, every element has an inverse, we lose no generality if we assume that the inverse of each element of c lies in C. Arthur Cayley associated with each such group G a connected graph $\Gamma = \Gamma(G, C)$ that gives a geometric picture of the group G and, in fact, supplies an abbreviated multiplication table for the group. The Cayley graph is defined as follows. The vertices of Γ are the elements of G. Each edge of Γ is an actual geometric interval joining two vertices g and $g \cdot c$ of the graph, where g is a member of G and c is a member of C. We refer to this interval, directed from initial vertex g to terminal vertex $g \cdot c$, by the symbol $e = (g, c, g \cdot c)$, and often picture that interval mentally as an arrow from g to $g \cdot c$ labeled by the generator c. The very same interval, directed in the opposite direction, is represented by the symbol $e^{-1} = (g \cdot c, c^{-1}, g)$. The graph is a physical object, a topological graph,and its members or "points" are its vertices and the points on its edges. This graph is not assumed to lie in some larger containing space. Each edge is assigned a length of one, and distance within the graph between points is path-length distance as measured within the graph, just as automobile distances are measured on the road and not as the crow flies. Hence, between vertices, path length simply counts the number of edges required to join one vertex to the other.

For example, the group $\mathbb{Z}$ of integers is generated by the integer 1 because every integer is a finite sum of 1s and -1s. (The number -1 is the inverse of the integer 1.) The Cayley graph can be viewed as a copy of the real line, with the integers representing the vertices of the graph and with the edges being directed from one integer n to the next integer $n + 1$, each labeled by the generator 1. The edges traversed in the opposite (negative) direction are considered to be labelled by -1. This graph is a topological space, with natural symmetries that reflect the structure of the group of integers. Addition of 1 corresponds to a translation of the line by one unit. Edge-path distance between points is the usual distance in the real line.

Gromov-hyperbolic group. Negatively curved noneuclidean geometry has thin triangles, a simple property not satisfied in Euclidean space. Here is the appropriate definition of thin triangles as applied to Cayley graphs. As noted above, the Cayley graph Γ has a natural notion of path length that views each edge interval as an interval of length 1. Between each pair, p, q, of elements of Γ, there will be one or more shortest paths that lie within the graph and are called geodesics. Distances are calculated only within the graph itself. A geodesic triangle Δ in Γ then consists of three points, a, b, and c, and geodesic paths, ab, bc, and ca, between them. It should be noted that these triangles are one-dimensional objects so that there is no interior to the triangle, contrary to what one would assume if the triangle were in the plane. If $\delta > 0$ is a positive number, then we say that the triangle Δ is δ-thin if the δ-neighborhood of each point of Δ intersects at least two of the paths ab, bc, and ca. The Cayley graph Γ and the group G are said to be Gromov-hyperbolic if there is a positive number δ such that every geodesic triangle in Γ is δ-thin. This property, unusual as it may seem, is actually generic in the sense that it is satisfied by the Cayley graphs of almost all groups that can be finitely described.

Space at infinity. Every Gromov-hyperbolic group G has a natural space at infinity defined as follows. Let 0 represent a point of the Cayley graph Γ. Each point at infinity is represented by an infinitely long geodesic ray R, which begins at $0 = r(0)$ and is parametrized by distance $t \in [0,\infty)$ from the point 0. Two rays R and S represent the same point at infinity if, for some positive bound B, $R(t)$ and $S(t)$ are within B of one another for all values of t. Points represented by rays R and S are said to be close to each other if $R(t)$ and $S(t)$ are close to one another for all t in a large initial interval $[0, T]$. The space at infinity is always finite dimensional, compact, and metrizable.

Kleinian group. According to a basic proposition of complex analysis, the 2×2 complex matrices define conformal mappings in the form of so-called linear fractional transformations, which apply not only to the complex plane but also on the 2-sphere $\mathbb{S}^2$, realized as the complex plane plus a point at infinity. A group G is Kleinian if it can be realized as a group of 2×2 complex matrices acting conformally on the 2-sphere $\mathbb{S}^2$ and by rigid motions on hyperbolic three-dimensional noneuclidean geometry, $\mathbb{H}^3$. There are some precise technical conditions that will not explained here, namely, the action on $\mathbb{H}^3$ should be isometric, properly discontinuous, and cocompact.

Difficulty in proving the conjecture. The difficulty in proving the conjecture is this: By hypothesis, the group G acts as a collection of symmetries on two objects, namely, on the Cayley graph and on the 2-sphere space at infinity. The conjecture requires one to put analytic coordinates on the sphere at infinity so that the symmetries preserve those analytic coordinates, in the sense of complex variables, by acting conformally. There are infinitely (uncountably) many nonequivalent ways of introducing analytic coordinates, and at most one of those ways can be compatible with the group. That is, a search is required for the (possibly existing) one and only one needle in an infinite haystack.

Attempts to prove the conjecture. Many approaches have been taken in attempts to prove the conjecture. Each approaches the creation of analytic coordinates on the sphere at infinity in a different way.

One approach uses a large "sphere" in the Cayley graph to approximate the sphere at infinity, optimizes the shape of that approximating sphere by means of a version of the classical Riemann mapping theorem from complex variables, and seeks to find a smooth limiting shape. Precise axioms have been worked out that are necessary and sufficient for the process to succeed, but the axioms require infinitely many calculations and are difficult to verify.

Another approach tries to modify the shape at infinity, compatible with the symmetries of the group, in such a way as to minimize that exotic measure of dimension called Hausdorff dimension. The Hausdorff dimension of a topological 2-sphere can be any positive real number equal to or greater than 2. The Hausdorff dimension depends on the shape of that 2-sphere. A Hausdorff dimension greater than 2 for the 2-sphere indicates that the 2-sphere is a fractal as defined by B. Mandelbrot. The Cannon conjecture is true if and only if the Hausdorff dimension can be reduced all the way to 2 in a way compatible with the symmetries of the group. Current work succeeds in reducing the dimension arbitrarily close to 2.

A third approach begins with a "visual" measure of distance at infinity and, modifying the distance measure, keeping this measure compatible with the symmetries of the group, attempts to make that measure as smooth and regular as possible.

It may be that the conjecture will finally be resolved by modifications of Perelman's approach via flow of Ricci curvature.

Related problems. The general version of the geometrization problem is this: Given an object defined without specific geometric shape, optimize the shape of that object geometrically. That is, discover what mathematical objects have a natural geometric realization or structure. For example, the flexible circle made out of string has natural geometric form as a round circle.

Thurston proved that the complement of a knotted circle in the 3-sphere has a unique geometric structure. For almost all knots, that structure seems to be based on hyperbolic geometry. According to the surveys of Jim Hoste, Morwen Thistlethwaite, and Jeffrey Weeks, there are 1,701,936 prime knots having 16 or fewer crossings, and only 32 of these are not hyperbolic.

The two-dimensional surfaces (the 2-sphere and the surfaces of the doughnut with one or more holes) also admit a geometric structure. The 2-sphere, of course, has a spherical structure. The one-holed doughnut (torus) has a Euclidean structure. All of the others have a hyperbolic structure.

Geometrization problems arise in the study of rational mappings of the complex plane. Every rational function of one complex variable defines a mapping from the 2-sphere $\mathbb{S}^2$, realized as the extended complex plane (that is, $\mathbb{C} \cup \{\infty\}$), to itself. This map is branched in the sense that, in local coordinates, it looks like a power map $z \mapsto z^k$, with k varying from point to point. A natural question asks, "If the local branching data is given for some self-map of the 2-sphere S^2, when is there a rational map having that branching data?" This question is a geometrization problem. The best answers to date have been given by Thurston.

Similar results have been proved in the three different studies of the geometrization problems for Kleinian groups (Cannon's conjecture), for knot complements, and for branched maps. Mathematicians seek precise connections.

For background information *see* COMPLEX NUMBERS AND COMPLEX VARIABLES; CONFORMAL MAPPING; EUCLIDEAN GEOMETRY; FRACTALS; GRAPH THEORY; GROUP THEORY; MANIFOLD (MATHEMATICS); MATRIX THEORY; NONEUCLIDEAN GEOMETRY; TOPOLOGY in the McGraw-Hill Encyclopedia of Science & Technology.

James W. Cannon

Bibliography. M. Bonk and B. Kleiner, Quasisymmetric parametrizations of two-dimensional metric spheres, *Invent. Math.*, 150:127–183, 2002; J. W. Cannon, Geometric group theory, in R. J. Daverman and R. B. Sher (eds.), *Handbook of Geometric Topology*, North-Holland, Amsterdam, pp. 261–305, 2002; J. W. Cannon and E. L. Swenson, Recognizing constant curvature discrete groups in dimension 3, *Trans. Am. Math. Soc.*, 350:809–849, 1998; M. Gessen, *Perfect Rigor*, Houghton Mifflin Harcourt, Boston/New York, 2009; W. Thurston, *Three-dimensional Geometry and Topology*, vol. 1, edited by S. Levy, Princeton Math. Ser., vol. 35, Princeton University Press, 1997.

Celiac disease

Celiac disease (CD) is an autoimmune condition of the small intestine that is believed to be T-cell-mediated and triggered by the consumption of gluten proteins found in foods containing wheat, rye, and barley. Symptoms can be overt and related to abdominal pain, diarrhea, weight loss, and other signs of intestinal malabsorption, or they can be subtle and present later in life as osteoporosis or iron-deficiency anemia. Conservative estimates place celiac disease at a prevalence of about 1 in 150 people in the United States, meaning that the majority of individuals remain undiagnosed. Screening is now simple with assays that detect autoantibodies against tissue transglutaminase (tTG) in patient serum. Diagnosis is confirmed by an intestinal biopsy showing villous atrophy, with implications of a life-long condition that can be treated only by eliminating dietary gluten. Once gluten is removed from the diet, the autoantibodies, intestinal injury, and symptoms resolve over time. If the patient is rechallenged with gluten, disease and symptoms will recur. Celiac disease is associated with other autoimmune diseases, notably type 1 diabetes and autoimmune thyroid disease, because of shared human leukocyte antigen (HLA) gene predisposition.

There is no cure for celiac disease, but it can be managed by maintaining a strict gluten-free diet. Even very small amounts of gluten (50 mg daily) can cause symptoms and lead to long-term complications. Adherence to a gluten-free diet is very complex and difficult, as it imposes a large number of lifestyle and financial restrictions on patients. The vast majority of processed foods contain gluten, because many preservatives and additives contain gluten or are contaminated by gluten. Unfortunately, there is no accurate and practical method to detect the presence of gluten in food. Failure to maintain a gluten-free diet can lead to increased morbidity and mortality in celiac disease patients of all ages. Gastrointestinal malignancies, particularly intestinal lymphoma, although rare, have been reported to occur with increased frequency in untreated adult patients. Patients who follow a strict gluten-free diet have a similar malignancy risk as the general population. Osteoporosis is another major complication in patients who have been previously undiagnosed, particularly in males.

Genetics. Major histocompatibility (MHC) molecules are heterodimers composed of an alpha chain and a beta chain designed to bind antigens composed of short peptide sequences. This is an important part of immune surveillance and necessary for T-cell recognition of antigen peptides. In the case of humans, they are known as HLA molecules. HLA molecules have tremendous variability and are often associated with risks of autoimmune disease. Virtually all celiac disease patients have either HLA-DQ2 (DQA1*0501, DQB1*0201) or HLA-DQ8 (DQA1*0301, DQB1*0302). These genes occur on chromosome 6 in the MHC region and encode for the MHC class II molecules. These molecules are expressed on professional antigen-presenting cells (APCs) that are critical for the adaptive immune system. (APCs are specialized cells that help fight off foreign substances that enter the body; they are divided into professional and nonprofessional categories, depending on the class of MHC molecules expressed.)

Although DQ2 or DQ8 is required for disease, this is not sufficient: 36% of the general population carry these high-risk HLAs, but only 5–10% of these will ever develop disease. Therefore, other factors, in addition to the required HLA molecules and dietary gluten, are needed for celiac disease to develop. Genome-wide association studies on celiac disease have identified numerous non-HLA loci that contribute to celiac disease risk, many of which have been shown to be shared by two or more immune-related diseases. These genes have functions that relate to immunological function and T-cell biology, suggesting that alterations in immune responses, rather than the intestinal epithelium or digestive enzymes, are associated with the pathogenesis of celiac disease.

Toxicity of gluten. Part of the pathology behind gluten is its direct toxic effect on enterocytes (cells that line the intestinal wall) through release of proinflammatory cytokines and activation of the innate immune system. Currently, it is unclear how gluten (which in wheat is found in the form of gliadin proteins) could have such a range of biological effects on innate immune cells and how it can bind to unrelated receptors. However, remarkable advances in the understanding of the role of the specific HLA molecules have taken place. The adaptive immune response in celiac disease is related as much to the gliadin peptides and the corresponding gliadin-specific T-cell receptors as it is to HLA-DQ2 or HLA-DQ8. When a person with celiac disease ingests gluten, gliadin peptides are recognized by the immune system and eventually bind DQ2 or DQ8 on the surface of APCs for presentation and activation of specific T cells. These gliadin-reactive T cells can be isolated from intestinal biopsy tissue from patients with celiac disease.

Biochemistry of pathogenesis. Gluten is a complex mixture of glutamine- and proline-rich proteins. In the case of wheat, it is in the form of gliadin, the wheat storage protein. The high proline content makes gliadin resistant to enzymatic degradation in the intestine. This can result in preservation of oligopeptides that persist in the gut lumen, leading to immunogenicity. These characteristics also make it an ideal substrate for the enzyme tTG, which converts the amide glutamine into the negatively charged glutamic acid, a common feature of the numerous gliadin epitopes identified in celiac disease. Deamidation (removal of the amide group) of specific amino acids of gliadin peptides results in negatively charged residues that can bind better into the pockets of HLA-DQ2 or HLA-DQ8 molecules, enhancing immunogenicity. The gliadin peptide/MHC complex on the surface of an APC is then presented to a specific T cell, leading to T-cell activation. Current understanding is that these gliadin-specific T cells play an important role in disease pathogenesis.

In an effort to identify physiologically stable epitopes of gliadin necessary for the T-cell receptor/MHC complex to initiate an inflammatory response, a recombinant representative α-gliadin was digested with gastric and pancreatic enzymes and analyzed. A relatively large 33-mer fragment (residues 57–89) was identified as one of the products. This 33-mer peptide was stable even after prolonged exposure to protease enzymes. It may remain intact throughout the digestive process and can act as a potential antigen for the T-cell proliferation. T-cell lines derived from celiac disease biopsies are able to recognize these epitopes. The 33-mer peptide also is found to be a high-affinity substrate for tTG. Because of its stability, ingesting even small amounts of wheat gluten can lead to the buildup of the 33-mer peptide in the intestine. This peptide, inert toward digestive breakdown, may be used in future vaccination, prevention, and treatment of celiac disease. But, there are other gliadin peptides outside this 33-mer region recognized by T cells. Thus, this 33-mer peptide may not be the sole source of antigenicity.

Recent biostructural data from the crystallization of DQ2 and DQ8 have led to a greater understanding of specific amino acid motifs required for gliadin to bind to DQ molecules during antigen presentation of gliadin peptides to T cells (see **illustration**). DQ8 has a basic pocket 9 in the peptide-binding groove secondary to a substitution of $\beta57$ aspartic acid for valine, alanine, or serine that disrupts a salt bridge between $\beta57$ aspartic acid and $\alpha76$ arginine, and this is important for pathogenesis in patients with DQ8. This pocket 9 has a strong preference for negatively charged residues, such as glutamic acid following specific deamidation. In fact, DQ8 prefers negatively charged residues at either or both anchor positions, pockets 1 and 9, which leads to higher binding affinity. The native sequence QQPQQPYPQ from one epitope, γ-gliadin, is selectively deamidated to become the highly immunogenic core sequence EQPQQPYPE (see illus. *a*). HLA-DQ2 has a preference for a proline at pocket 1 (which is abundant in gliadin) along with the negatively charged glutamic acid at position 4 or 6, such as in the prototypical α-1 gliadin epitope PFPQPQLPY (see illus. *b*). However, this peptide is practically nonstimulatory until it is deamidated to become PFPQPELPY. Therefore, enzymatic modification by tTG (deamidation) results in peptides that can bind to DQ molecules with even higher affinity, leading to enhanced immunogenicity. Gliadin contains a large number of peptides that are excellent substrates for tTG, bind to either HLA-DQ2 or HLA-DQ8, and can trigger T-cell responses.

In recent years, the identification of disease-relevant gliadin epitopes and the dissection of antigen presentation of these epitopes by DQ2 and DQ8 molecules have increased our knowledge of the HLA association in celiac disease. This understanding may lead eventually to treatment therapies by competing inhibitors of MHC class II–mediated peptide presentation to celiac disease (CD)–specific T cells.

For background information *see* ALLERGY; ANTIBODY; ANTIGEN; ANTIGEN-ANTIBODY REACTION; AUTOIMMUNITY; CELLULAR IMMUNOLOGY; FOOD ALLERGY; GASTROINTESTINAL TRACT DISORDERS; IMMUNITY; IMMUNOLOGY; WHEAT in the McGraw-Hill Encyclopedia of Science & Technology. Edwin Liu

Bibliography. Z. Hovhannisyan et al., The role of HLA-DQ8 beta57 polymorphism in the anti-gluten T-cell response in celiac disease, *Nature*, 456:534–538, 2008; K. A. Hunt and D. A. van Heel, Recent advances in celiac disease genetics, *Gut*, 58:473–476, 2009; J. See and J. Murray, Gluten-free diet: The medical and nutrition management of celiac disease, *Nutr. Clin. Pract.*, 21:1–15, 2006; L. Shan et al., Structural basis for gluten intolerance in celiac sprue, *Science*, 297:2275–2279, 2002; S. Tollefsen et al., HLA-DQ2 and -DQ8 signatures of gluten T cell epitopes in celiac disease, *J. Clin. Invest.*, 116:2226–2236, 2006; L. W. Vader et al., Specificity of tissue transglutaminase explains cereal toxicity in celiac disease, *J. Exp. Med.*, 195:643–649, 2002.

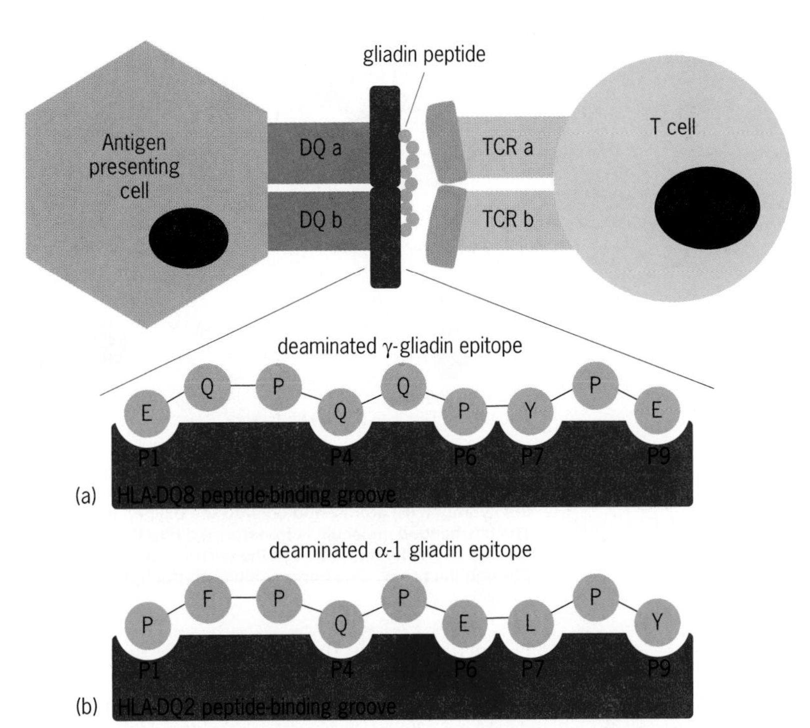

Antigen presentation of gliadin peptide to a CD4 T cell. Gliadin peptides bind to pockets P1, P4, P6, P7, and P9 of the (*a*) HLA-DQ8 and (*b*) HLA-DQ2 molecules.

Chirality in rotating nuclei

Rotating triaxial nuclei may acquire chirality. Similarities and differences between chirality of molecules and nuclei, as well as the experimental signatures for chirality, will be discussed.

Chiral molecules. Chirality (Greek "handedness") is a property of many complex molecules. Chiral molecules exist in two forms, one being the mirror image of the other. As with our hands, it is impossible to make the images identical by a suitable rotation. (For simple, achiral molecules, such as H_2O, this is possible.) The two forms are called left-handed and right-handed. They have the same binding energy because the electromagnetic interaction, which holds the molecule together, does not change under a reflection. Other properties that are insensitive to the geometry are also the same. The different geometry

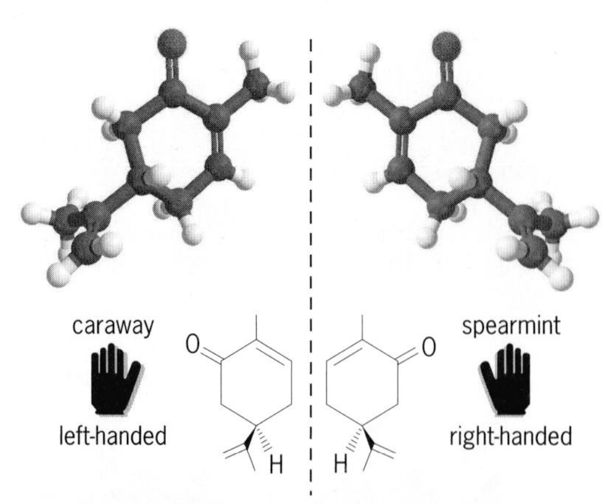

Fig. 1. The two chiral configurations of the carvon molecule. Black balls represent the carbon (C) atoms, white the hydrogen (H) atoms, and colored the oxygen (O) atom. The left-handed molecule is transformed into the right-handed by a reflection through the mirror plane that goes through the broken line perpendicular to the figure plane.

is the reason why the left-handed form turns the polarization plane of transmitted light in one direction by some angle while the right-handed form turns it in the opposite direction by the same angle. The differences between the two species may have dramatic consequences. **Figure 1** shows the carvon molecule. Its left-handed form smells of caraway and its right-handed form of spearmint. Organisms usually synthesize only the left-handed or the right-handed species of a molecule.

Shapes of atomic nuclei. Compared to molecules, nuclei are very compact. They resemble deformed droplets of liquid, with nearly constant interior density and a thin surface. This reflects the short range of the interaction that binds the nucleons together. Inside, the nucleons move freely on orbitals whose shapes are determined by the quantization of angular momentum. As an example, **Fig. 2** shows orbitals that carry an angular momentum of $j = (11/2)\hbar$ (where $\hbar$ is the quantized unit of angular momentum). The origin of nuclear deformation is the shape of the nucleonic orbitals, just as the shape of the electronic orbitals leads to the complex structure of molecules. However, there is a difference. The repulsive long-range Coulomb force between the electrons pushes the orbitals carrying $j = (1/2)\hbar$ or $(3/2)\hbar$ to the outside of the atoms, where they lobe out, making chemical bonds with neighboring atoms. In contrast, the short-range attractive force between the nucleons pulls the orbitals carrying $(1/2)\hbar \leq j \leq (11/2)\hbar$ as close as possible together, such that the different geometry of the orbitals nearly averages out. The resulting shapes are simple: spherical, axial symmetric (mostly football-like), sometimes pear-shaped. Some nuclei take the shape of a triaxial ellipsoid, that is, an ellipsoid whose three perpendicular axes each have a different length. Such simple shapes do not attain the property of chirality, which was considered to be an unusual concept by nuclear physicists for many years. *See* ASYMMETRIC, PEAR-SHAPED NUCLEI; WOBBLING MOTION IN NUCLEI.

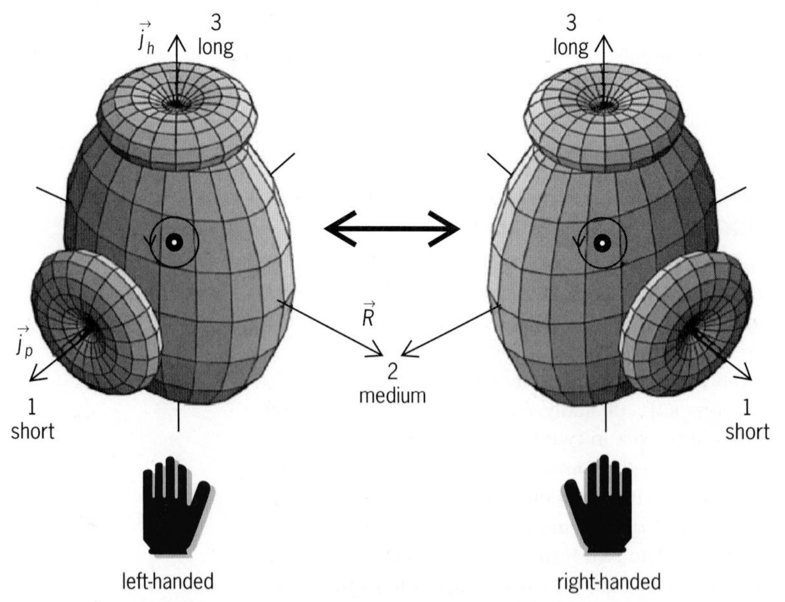

Fig. 2. Rotating triaxial nucleus with an unpaired particle carrying the angular momentum $\vec{j}_p$, an unpaired hole carrying $\vec{j}_h$, and collective angular momentum $\vec{R}$. All angular momenta are displayed as arrows. The total angular momentum $\vec{J}$, which is the axis of rotation, points into the direction of sight, and is depicted by the circle symbolizing the arrow tip. For purpose of illustration the orbitals are drawn outside the nucleus, whereas they are located inside.

Rotation of nuclei. As with molecules, deformed nuclei show rotational spectra. These are sequences of states of quantized energy and angular momentum that have the same structure, differing from each other only by the angular velocity of rotation. The dynamics of nuclear rotation is governed by the quantized motion of the individual nucleons, which makes nuclei react like a clockwork of gyroscopes, where the gyroscopes are the nucleonic orbitals carrying a fixed angular momentum $\vec{j}$. In even-even nuclei (nuclei with an even number of protons and an even number of neutrons), being in the lowest energy state, the nucleons occupy pairwise orbitals with opposite directions of $\vec{j}$. Because of this arrangement, it is energetically favorable that even-even triaxial nuclei rotate about the medium-length axis.

A new pattern evolves if not all the angular momenta $\vec{j}$ of the orbitals are paired off. Figure 2 illustrates such a situation for the case of an odd-odd nucleus (with an odd number of protons and an odd number of neutrons). The large shape represents the even-even "core" of the nucleus, which contains all the nucleons but the odd proton and has one extra neutron. The triaxial core generates the angular momentum component $\vec{R}$ along the medium-length axis. The additional odd proton occupies the vertical doughnut-like orbital. Its angular momentum $\vec{j}_p$ points along the short axis of the core. This orientation corresponds to the energy minimum of the short-range attractive interaction between the orbital and the core. The presence of the odd neutron can be seen as the horizontal doughnut-like orbital carved out of the even-even core (which has one neutron more). The angular momentum $\vec{j}_h$ of this hole orbital points along the long axis of the core. This orientation corresponds to the energy minimum of the short-range repulsive interaction between the hole orbital and the core. (The opposite sign of the hole-core interaction is analogous to the opposite sign of the gravitational pull on a bubble in water, its buoyancy.) The total angular momentum is the axis of rotation, which has components along all three axes, $\vec{J} = \vec{j}_p + \vec{j}_b + \vec{R}$. Consequently, it lies outside the three reflection planes of the triaxial shape, as shown in Fig. 2.

Emergence of chirality in nuclei. Stefan Frauendorf and Jie Meng first noticed that, in a case like that shown in Fig. 2, the rotating triaxial nucleus attains chirality. Looking in direction of the rotational axis from the tip of the vector $\vec{J}$ onto the nucleus, the upper half-axes are clockwise ordered in the right-handed configuration, and they are counterclockwise ordered in the left-handed configuration. In order to convert the right-handed into the left-handed configuration, one has to invert the directions of $\vec{j}_p$, $\vec{j}_b$, and $\vec{R}$, which select the three half-axes. Rotating the resulting configuration by 180^0 about the axis in the short-medium plane that is perpendicular to $\vec{J}$, one obtains the left-handed configuration shown in the figure. Inverting the direction of $\vec{J}$ is achieved by "time reversal," which means that the signs of all the velocities are changed (the nucleons run "backward").

The pictures of the two configurations in Fig. 2 look like mirror images. However, the left-handed configuration is not converted into the right-handed configuration by reflecting the positions of all constituent particles, as in the case of molecules. Such an operation both mirrors the nuclear shape and changes $\vec{J}$ ($= m\vec{r} \times \vec{v}$, where m is the mass, and $\vec{r}$ and $\vec{v}$ are the position and velocity vectors) into $-\vec{J}$. The combination of the two transformations does not change chirality.

The existence of two distinct configurations with opposite chirality gives rise to two bands of rotational states, where states with the same angular momentum are expected to have the same energy, and in turn to emit gamma radiation in the same way. Perturbations can change the energies somewhat (as will be further discussed) but not the electromagnetic transition probabilities, which are expected to remain the same for states of the same angular momentum. The states have the same parity quantum number, which indicates the symmetry under reflection. Candidates for rotational sequences of such chiral doublets were first found in odd-odd nuclei with approximately 134 nucleons and approximately 104 nucleons. The presence of chirality was best demonstrated in neodymium-135 (^{135}Nd) by the observation of equal emission of gamma radiation from the two rotational bands. This nucleus (**Fig. 3**) has an even number of 60 protons. In this case, two protons align their angular momentum $\vec{j}_p$ with the short axis. Yet another type of chiral configuration was found in the even-even nucleus ruthenium-112 (^{112}Ru), where two neutrons generate angular momentum along both the short and long axes.

Left-right coupling. The left- and right-handed configurations are coupled. For molecules, this coupling

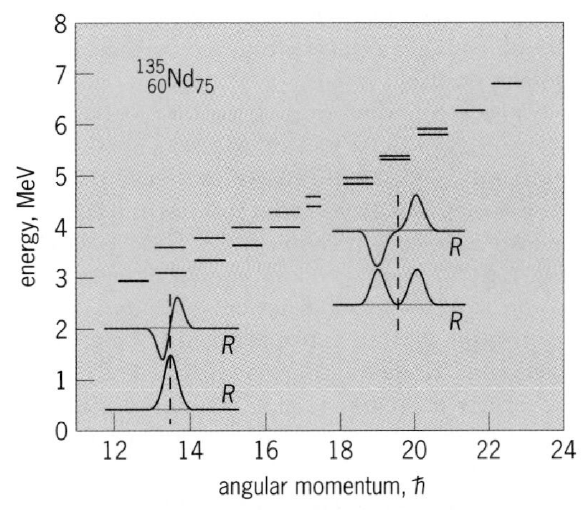

Fig. 3. The two energy sequences of rotational states in the nucleus neodymium-135 (^{135}Nd) that constitute chiral doublets. The insets illustrate how the left-handed and right-handed configurations are coupled. The broken lines depict the short-long plane perpendicular to the figure plane. The curves show the wave function of the component of the collective angular momentum $\vec{R}$ along the medium axis. The square of the wave function is the probability distribution of this component.

is usually so weak that, once formed, a right-handed configuration remains such for any time span of practical concern. In order to convert chirality, such a molecule has to pass through geometric arrangements of the nuclei that have very high energy, which generates a practically impenetrable barrier. There are exceptions: CH_3NHF oscillates with a frequency of 1000 GHz between right and left. In this case, the left- and right-handed configurations differ only by the places of one electron pair, which are separated by a low barrier.

For nuclei, the left-right coupling is substantial and depends on the total angular momentum. As seen in Fig. 2, the left-handed configuration differs from the right-handed configuration by the orientation of the component $\vec{R}$ along the medium-length axis. In order to convert chirality, the vector $\vec{R}$ must turn through the plane that contains the long and short axes. The energy is higher when $\vec{R}$ lies in this plane than when $\vec{R}$ is aligned with the medium-length axis. The energy difference, which increases with R, is the barrier that decouples the left-handed from the right-handed configuration. Since $R\left(= \sqrt{J^2 - j_p^2 - j_b^2}\right)$ increases with increasing J, the left-right coupling decreases with J.

The coupling generates two stationary states, which are the even and odd combinations of the left- and right-handed configurations. They have somewhat different energies, where the difference increases with the coupling strength. Figure 3 shows the sequence of such chiral doublets. Their energy distance reflects the decrease of the left-right coupling with increasing J. For $J = (27/2)\hbar$, the nucleus oscillates rapidly between left and right (chiral vibration). The wave functions are the ones of the ground and first excited states of a harmonic oscillator. For $J = (39/2)\hbar$, the chirality is strongest. The left- and right-handed components of the wave function are well separated, and the energies of the two bands approach each other. The nucleus turns three times around before it changes from left to right (static chirality).

For background information *see* ANGULAR MOMENTUM; MOLECULAR ISOMERISM; NUCLEAR STRUCTURE; RIGID-BODY DYNAMICS; STEREOCHEMISTRY; STRONG NUCLEAR INTERACTIONS in the McGraw-Hill Encyclopedia of Science & Technology. Stefan Frauendorf

Bibliography. S. Frauendorf, Spontaneous symmetry breaking in rotating nuclei, *Rev. Mod. Phys.*, 73:463–514, 2001; S. Frauendorf and J. Meng, Tilted rotation of triaxial nuclei, *Nucl. Phys. A*, 617:131–147, 1997; X. Y. Luo et al., Evolution of chirality from gamma soft Ru-108 to triaxial Ru-110, Ru-112, *Phys. Lett. B*, 670:307–312, 2009; S. Mukhopadhyay et al., From chiral vibration to static chirality in Nd-135, *Phys. Rev. Lett.*, 99:172501 (4 pp.), 2007; S. Zhu et al., A composite chiral pair of rotational bands in the odd-*A* nucleus Nd-135, *Phys. Rev. Lett.*, 91:132501 (4 pp.), 2003.

Click chemistry in polymer science

Polymers often have been built in highly defined molecular configurations and conformations. In many modern functional materials, the arrangement of their molecules dictates their properties and uses. As the structure of functional polymers becomes more complex, their directed synthesis is more elaborate, in particular aiming at the exact positioning of functional groups onto sites of medium and large molecules. Click chemistry, often described as being analogous to the fastening ("click") of two ends of a buckle, is a highly valuable tool for simple reactions between large molecules and molecule fragments, thus representing an ideal method for generating complex macromolecules and materials.

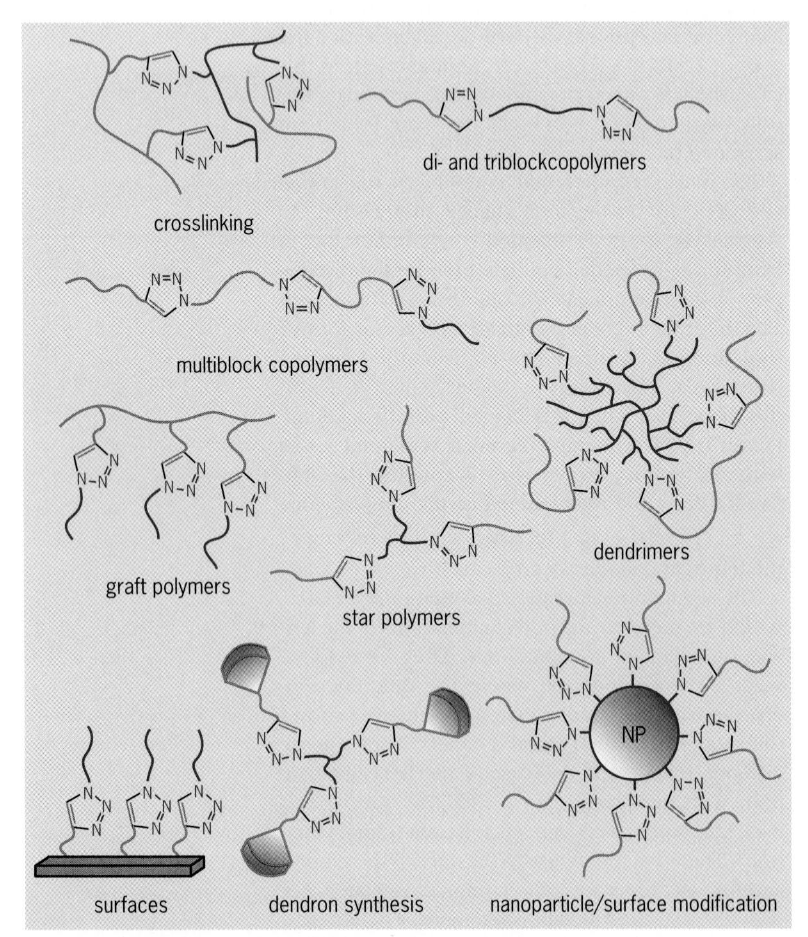

Fig. 1. Polymeric architectures available via click chemistry.

The synthesis of polymers often starts from controlled polymerization processes, followed by appropriate functionalization of the polymers' end or side-chain groups. With the enormous progress in the field of living polymerization reactions for which there is no termination step to stop chain growth, there is a need for efficient post functionalization methods of polymers and oligomers. In addition to conventional linear polymers, complex macromolecular architectures, such as star polymers, cyclic macromolecules, and hybrid polymers (incorporating peptides, proteins, and oligonucleotides) have gained importance, thus increasing the quest for highly efficient and universal functionalization chemistries (**Fig. 1**). As the solubility of many polymers often is restricted, post-functionalization methods need to reach a level of full conversion (100% yield) even under heterogeneous reaction conditions—an aim that had not been fulfilled before the advent of click chemistry.

The requirements for a highly efficient substrate- and solvent-independent linking method of molecular fragments onto polymers is fulfilled by click reactions, as defined by K. B. Sharpless in 2001. A click reaction is said to be wide in scope, featuring quantitative yields, being substrate- and solvent-insensitive, and singular in its reaction pathway (that is, no side products), mostly because of a high thermodynamic gain in energy (more than 20 kcal mol^{-1}). Examples of click reactions include the reaction of highly strained cyclic compounds (such as epoxides and aziridines) that can be quantitatively reacted with appropriate nucleophiles; cycloaddition reactions (1,3-dipolar cycloaddition reactions or [4+2] cycloaddition reactions) that can be highly straight forward, especially if reactive diene-ene pairs are used), or photochemical addition reactions (for example, thiol addition to double bonds) [**Fig. 2**].

The best known reaction among the known click reactions is the azide-alkyne click reaction (Fig. 2*a*), as discovered by the research groups of M. Meldal and K. B. Sharpless, which represents a metal-catalyzed variant of the Huisgen 1,3-dipolar cycloaddition reaction. The basic system for the azide-alkyne click reaction consists of the two main, to be coupled components (terminal azide and terminal alkyne), a metal salt [usually Cu(I)], and a cocatalyst (often an additional amino ligand) to accelerate the reaction. The reaction may be run either homogeneously (organic solvent or a water and organic solvent mixture) or heterogeneously

Fig. 2. Examples of click reactions. (*a*) Cu(I)-catalyzed azide-alkyne click reaction (also termed CuAAc). (*b*) Photochemical thiol-ene click reaction. (*c*) Copper-free 3,3-difluoro-cycloalkine click reaction. (*d*) Decarbonylative Diels-Alder reaction. (*e*) Thiocarbonyl click reaction.

(immiscible solvent mixture, liquid/solid interface, or insoluble reaction system), thus being an ideal reaction system for the often poorly soluble polymers. Most known solvents and biphasic reaction systems (mixtures of water and alcohol to water and toluene) can be used with excellent results. The copper (I) salts represent the major metal used to accelerate the reaction. Cu(I) sources, such as pure Cu(I) salts, (Cu(I)Br, or Cu(I)I, are used in amounts of approximately 0.25–2 mol% with respect to the azide or alkyne substrate. Aqueous regenerative systems (that is, Cu(II) salts/ascorbic acid) may also be used. Beside copper, other metals used include Ru-complexes (such as $CpRuCl(PPh_3)$, $[Cp^*RuCl_2]_2$, $Cp^*RuCl(NBD)$, and Au(I)-, Ni-, Pd- and Pt-salts, although with significantly less catalytic activity. Cocatalytic systems often used include simple amino bases such as triethylamine (TEA), 2,6-lutidine, N,N-diisopropylethylamine (DIPEA), or multivalent amines such as N,N,N′,N′,N″-pentamethylethylenetetramine (PMDETA), hexamethyltriethylenetetramine (HMTETA), or tris-(benzyltriazolylmethyl)amine (TBTA).

Click chemistry combined with living polymerization methods. Enormous interest has focused on combining click reactions with controlled polymerization processes. Because many known polymerization reactions require the absence of specific functional groups, there is considerable interest in the fixation of ligands onto polymers after a successful polymerization reaction has been completed. This is most important when living polymerization mechanisms are used, especially because the highly sophisticated chemical mechanism and equilibria of controlled/living polymerization reactions are often highly substrate specific and therefore strongly affected by even small amounts of functional groups or the respective coupling agents required for affixation.

As shown in **Fig. 3**, three different strategies can be employed to combine living/controlled polymerization methods with the azide-alkyne click reaction. Using these approaches, controlled radical polymerization reactions [such as nitroxide-mediated radical polymerization (NMP), atom transfer radical

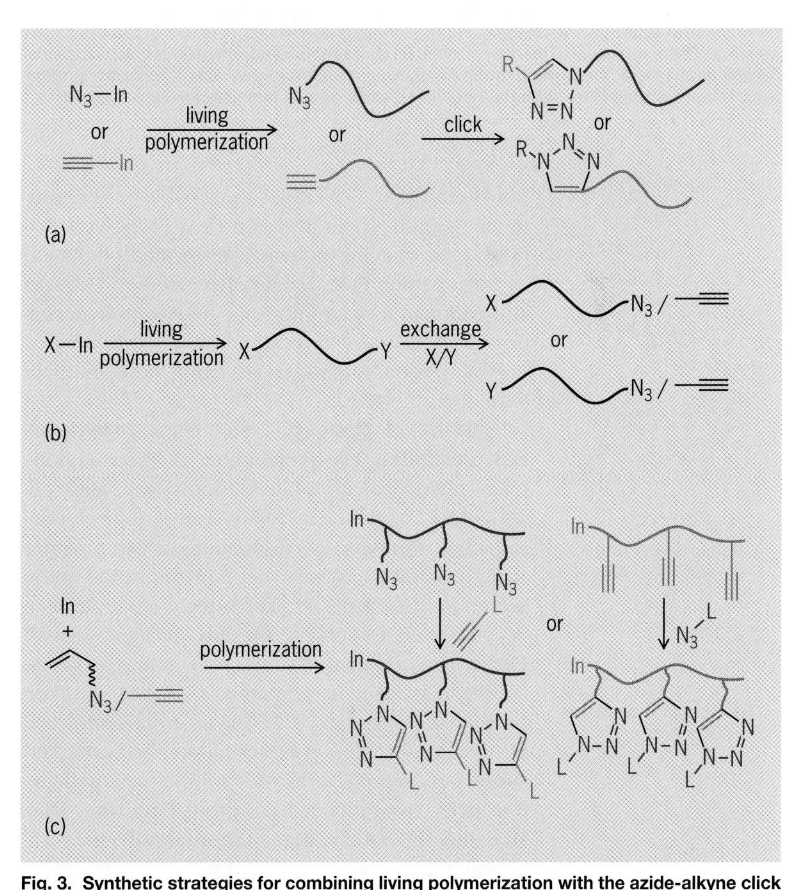

Fig. 3. Synthetic strategies for combining living polymerization with the azide-alkyne click reaction. (*a*) The incorporation of the azide and alkyne functional groups into the initiator structures used. This method implies that the subsequent polymerization method does not interfere with the presence of azide and alkyne groups. (*b*) Conventional initiation of a controlled polymerization, followed by subsequent exchange of the groups X and Y against a terminal azide or alkyne group. (*c*) The incorporation of the azide-alkyne group into the side chain of the monomeric structure. Again, this strategy implies noninterference between the azide-alkyne and the applied polymerization method. In this latter method, graft polymers can be generated, whereas in cases (*a*) and (*b*) linear, end-group modified polymers are obtained.

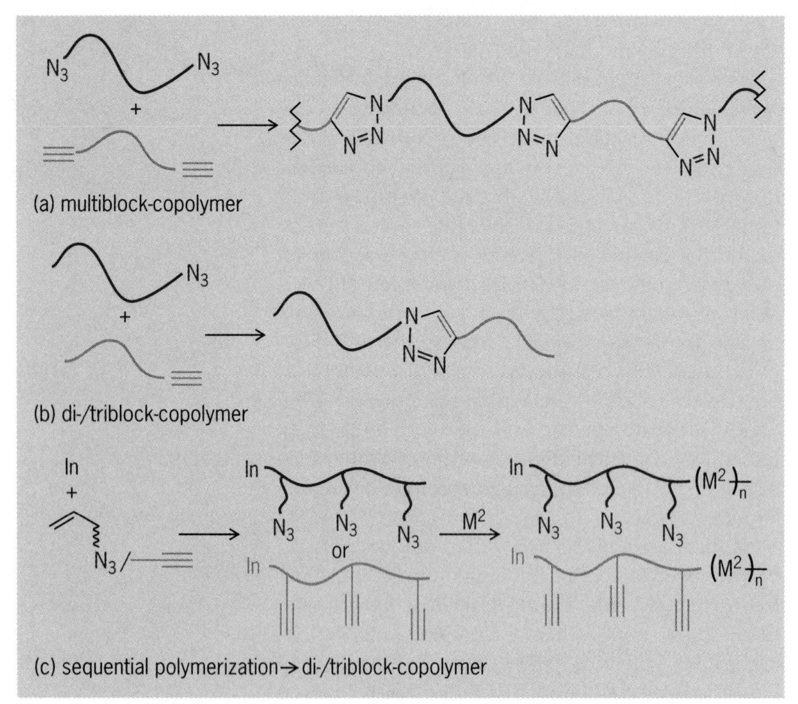

Fig. 4. Synthetic strategies for the preparation of block-copolymers via the azide-alkyne-click reaction. (*a*) The generation of multiblock copolymers by polyaddition of the respective azido-alkyne telechelic (bifunctional) polymers. (*b*) The use of the respective monofunctional telechelics, leading to defined di- or triblock copolymers. (c) Alternatively, the sequential block copolymerization of appropriate monomers yields the corresponding di- or triblock copolymers, with the respective azide-alkyne moieties for further grafting.

polymerization (ATRP), reversible addition-fragmentation chain transfer (RAFT) polymerization], ring opening polymerization methods [such as ring-opening metathesis polymerization (ROMP) and cationic and anionic ring-opening polymerization (ROP)], and living cationic and living anionic polymerization methods can easily be combined with click-chemistry.

Synthesis of block, star, star block copolymers, and dendrimers. The preparation of block copolymers, multisegmented block copolymers, and rod-coil block copolymers follows as a logical consequence from the corresponding linear polymer structures, prepared via living polymerization methods and click chemistry. Three main strategies can be employed to achieve the linkage as shown in **Fig. 4**. The preparation of star polymers is achieved in a similar fashion, preparing a central star-type polymer with pendant azide/alkyne groups prepared by living/controlled polymerization methods, and subsequent modification with another azide/alkyne telechelic (two functional end groups) polymer, thus attaching the outer sphere of the star polymer.

Highly branched polymers (such as dendrimers and hyperbranched polymers) are often prepared in tedious multistep syntheses, suffering from incomplete reactions and thus erroneous structures. Click methodology has been used extensively to prepare and functionalize dendrimers and hyperbranched polymers. The synthesis of dendrimers can be either achieved by convergent or divergent methods, using the azide-alkyne click reaction as an internal bond for the synthesis. This can lead to either hyperbranched polymers in one step or via sequential reaction. Additionally, whole dendron structures may be assembled via the azide-alkyne click reaction, using appropriately functionalized dendrons.

Synthesis of supramolecular polymers. Supramolecular polymer chemistry has been strongly influenced by the azide-alkyne click reaction, as it allows the affixation of supramolecular interactions at specific sites of a polymer chain. W. H. Binder's research group was the first to exploit the use of azide-alkyne click chemistry for the attachment of supramolecular entities onto the backbone of polymers prepared by living polymerization methods. One of the first examples concerned the combination of ROMP with click chemistry, thus achieving a controllable density of supramolecular entities in homopolymers, statistical copolymers, and block copolymers. The method represents a universal scaffold for the attachment of many supramolecular entities (such as the interaction of the Hamilton receptor on the polymer surface with barbituric acid-functionalized gold nanoparticles). Subsequently, the polymeric scaffolds can be used to take advantage of the microphase separation of block-copolymer and hydrogen-bonding interactions, which can be useful for attaching nanoparticles to the polymer surface via supramolecular recognition.

Click reactions on surfaces and nanoparticles. The chemistry on surfaces is a variation of the chemistry on polymers or other materials. An interesting aspect of the azide-alkyne click reaction lies in the fact that a reduced or enforced distance between the reaction partners leads to a strongly enhanced reaction rate. This effect has been demonstrated in the azide-alkyne click reaction within the pocket of enzymes (afinity-based protein profiling) by direct microcontact printing or via atomic force microscope (AFM) tips, thus opening the possibility for a sufficiently complete reaction at an interface.

In the case of self-assembled monolayers (SAMs), the use of appropriately azide- or alkyne-functionalized surfaces by direct ligand adsorption have been described. Alternatively, in-situ-generated terminal azides by bromide/azide exchange directly on the ω-bromoalkyl-functional monolayer can be done, eliminating the pressing instability of ω-azido-1-thioalkanes prior to the SAM formation process. Thus, a large variety of click reactions on SAMs, polymeric surfaces, or Langmuir-Blodgett layers (LbL-layers) have been reported.

Similar to SAMs, the surface of nanoparticles can be modified with the azide-alkyne click reaction. A large variety of nanoparticles (Au-, CdSe-, $Fe_2O_3^-$, SiO_2) as well as viruses and Au-nanorods have been surface-functionalized using this method. Mostly, the attached ligands serve as recognition sites to direct the location of nanosized objects onto materials via defined or nonspecific interactions. Selected examples of such recognition processes rely on hydrogen bonding, which directs the corresponding nanoparticles to a SAM surface, a polymeric surface, a liquid-liquid interface, or a block-copolymer phase or interface. In terms of the surface chemistry, an important point has been observed on comparing the Cu(I)-catalyzed reaction with the uncatalyzed, purely thermal click reaction on CdSe nanoparticles. Because copper ions interfere with the fluorescence properties of semiconductive nanoparticles, the use of Cu(I) species is not advantageous for their surface modification. Thus, without the use of the Cu(I) catalyst, the photoluminescence of the final, surface-modified CdSe nanoparticles remains nearly unchanged, whereas under Cu(I)-catalysis a significant drop in the quantum yields is observed. Therefore, the purely thermal azide-alkyne reaction may be sometimes advantageous over the metal-catalyzed click process. Compared to conventional surface-modification methods, the azide-alkyne methodology enables an elegant, fast, and efficient approach to functionalized nanoparticles.

Because of azide-alkyne click chemistry, polymer chemistry now approaches the level of small-molecule organic chemistry in terms of functional broadness, structural integrity, and molecular addressability. The synthesis of more complex molecules and materials can now be approached in cases, where in earlier times longer experiments and planning had been required.

For background information *see* ALKYNE; AZIDE; BRANCHED POLYMER; CLICK CHEMISTRY; CONTROLLED/LIVING RADICAL POLYMERIZATION; COPOLYMER; DIELS-ALDER REACTION; DENDRITIC MACROMOLECULE; HYDROGEN BOND; MACROMOLECULAR ENGINEERING; MONOMOLECULAR FILM; NANOPARTICLES; POLYMER; POLYMERIZATION; SUPRAMOLECULAR CHEMISTRY in the McGraw-Hill Encyclopedia of Science & Technology.

Wolfgang H. Binder; Florian Herbst

Bibliography. W. H. Binder and C. Kluger, Combining ring-opening metathesis polymerization (ROMP) with Sharpless-type "click" reactions: An easy method for the preparation of side chain functionalized poly(oxynorbornenes), *Macromolecules*, 37(25):9321–9330, 2004; W. H. Binder and R. Sachsenhofer, "Click"-chemistry in polymer and material science: An update, *Macromol. Rapid Comm.*, 29(12-13):952–981, 2008; W. H. Binder and R. Zirbs, "Click"-chemistry in macromolecular synthesis, in *Encyclopedia of Polymer Science and Technology*, Wiley, 2009; H. C. Kolb, M. G. Finn, and K. B. Sharpless, Click chemistry: Diverse chemical function from a few good reactions, *Angew. Chem. Int. Ed.*, 40(11):2004–2021, 2001; V. V. Rostovtsev et al., A stepwise Huisgen cycloaddition process: Copper(I)-catalyzed regioselective "ligation" of azides and terminal alkynes, *Angew. Chem. Int. Ed.*, 41(14):2596–2599, 2002; C. W. Tornoe, C. Christensen, and M. Meldal, Peptidotriazoles on solid phase: [1,2,3]-Triazoles by regiospecific copper(I)-catalyzed 1,3-dipolar cycloadditions of terminal alkynes to azides, *J. Org. Chem.*, 67(9):3057–3064, 2002.

Constraining slip and timing of past earthquake rupture

Why do earthquakes occur? What controls the recurrence of major earthquakes? Can earthquakes be predicted? And if so, how can they be predicted? These and similar questions have kept scientists as well as non-scientists occupied ever since the destructive potential of earthquakes was first witnessed. The recent earthquakes in Indonesia (moment magnitude M9.2, 2004), China (M7.9, 2008), Haiti (M7.0, 2010), and Chile (M8.8, 2010) dramatically exemplified the devastation and socioeconomic impact of major seismic events, underlining the scientific as well as public interest in knowing when a given fault is going to rupture again and how large the corresponding earthquake will become. A primary step toward addressing these questions is the identification of earthquake recurrence patterns in the existing seismic record. High-resolution light detection and ranging (Lidar) and interferometric synthetic aperture radar (InSAR) topographic data sets provide a powerful resource in this effort. They resolve current geomorphology as well as coseismic and postseismic deformation fields in unprecedented detail—valuable data for a deeper understanding of the mechanics of earthquakes and faulting. This article discusses both data sets and how they have been used in recent geomorphic, geodetic, and seismic studies.

Following the great M7.9 San Francisco earthquake of 1906, H. F. Reid (1910) formulated the elastic rebound theory of earthquakes that still forms a cornerstone for the understanding of seismic fault behavior. He proposed that fault behavior may be described by a seismic cycle consisting of essentially two phases: (1) a loading phase (interseismic period) in which plate-tectonic motion causes slow accumulation of elastic strain energy in the rocks on either side of a fault and (2) an unloading phase (coseismic period) in which the accumulated elastic strain energy is abruptly released during an earthquake via slip along these faults. While the cyclic nature of earthquake recurrence—with alternating periods of strain buildup and strain release—is well understood, it is currently not well known whether this behavior is also periodic. Periodicity, in the sense used here, refers to an essentially constant time interval between subsequent ruptures of the same magnitude along a given fault. How variable are rupture frequency and associated magnitude along a given fault through time? And what determines time and size of future earthquakes along a fault?

A primary step toward addressing these questions is the identification of earthquake recurrence patterns in the existing seismic record. Instrumental and historic records are of limited value in this regard, because earthquake recurrence times are large, with the completion of one seismic cycle usually taking more than a hundred years. Recording seismometers on the other hand have been in existence for only about 130 years and reliable historical records (which are sparse) generally cannot be obtained for

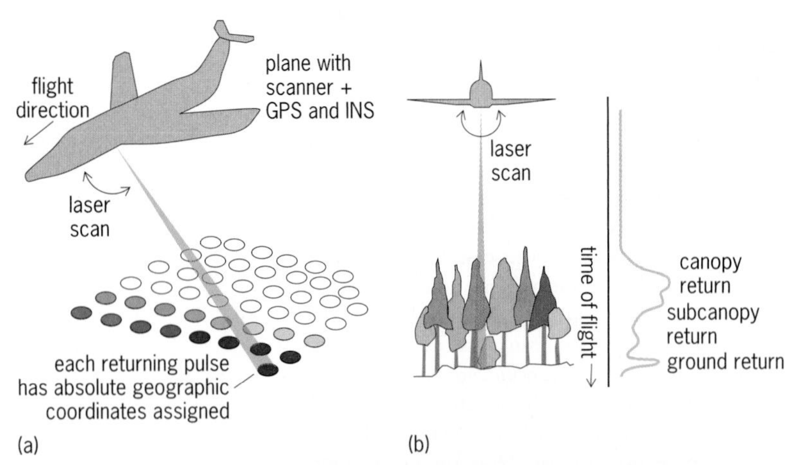

Fig. 1. Lidar data acquisition of (*a*) absolute geographic coordinates and (*b*) return pulses from the ground and vegetation. As the plane is flying along the area of interest, infrared laser pulses are shot (one at a time) from the scanner to the surface. A portion of the reflected pulse returns to the scanner where its time-of-flight is recorded. As the laser is sweeping the surface it generates a swath of surface elevation measurements. Each outgoing laser pulse may generate multiple returning pulses as the outgoing pulse may be partially reflected, for example, by vegetation cover. Filter algorithms automatically classify individual returns; for example, first return equals canopy height and last return equals surface elevation. (*Figures modified after GeoLas Consulting*)

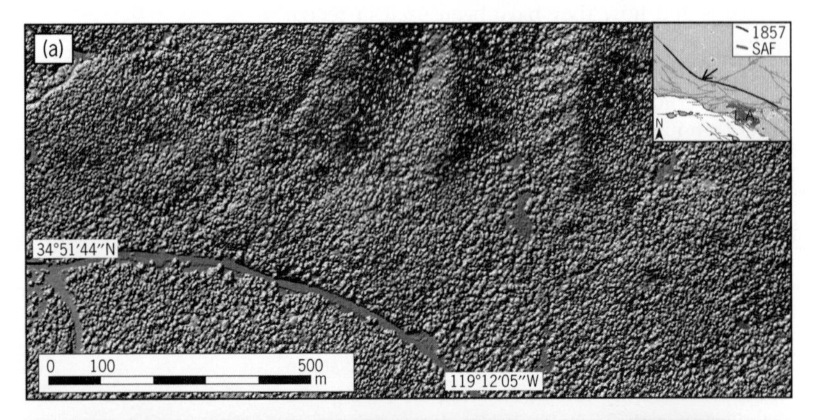

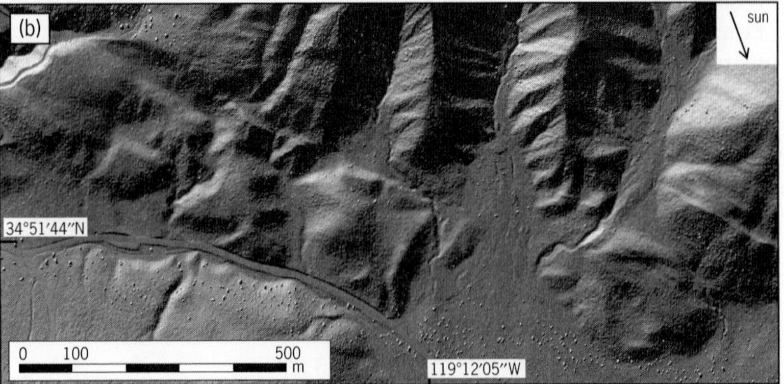

Fig. 2. Lidar-based DEMs of the south-central San Andreas Fault (SAF). (*a*) First pulse return, representing the top of the canopy and resembling the aerial photo imagery of this area. The fault-zone structure is well-hidden by the vegetation cover. (*b*) Last pulse return, representing the ground surface. The fault-zone structure is well expressed, with fault scarps and other lineaments running approximately from the upper left to the lower right corner (from WNW to ESE). As a result of the dense vegetation cover, resolution of these DEMs is 0.8 m, a rather coarse grid size. For sparsely vegetated areas, resolution may be as fine as 0.25 m. High spatial resolution in combination with the possibility of virtual deforestation makes Lidar data sets valuable resources in investigating fault-zone structure and slip amount of past earthquakes.

periods earlier than about 100 years before that. For many areas of the world, not a single complete seismic cycle has been documented instrumentally or historically, and for others only one is known, carrying no information on the temporal variability of earthquake frequency, size, and slip. Geologic and geomorphic evidence of seismic activity allow the existing seismic record to extend into past centuries and millennia since sufficiently large earthquakes disrupt and displace geologic and geomorphic units at or near the surface. These disruptions and displacements may be preserved in the geologic and geomorphic record, enabling size and age estimation for the causative earthquakes via geochronological methods and earthquake-scaling relationships. Two complementary approaches are commonly used: (1) paleoseismic excavation of the fault zone to study tectonically displaced stratigraphic units and soil horizons that constrain the slip and timing of past earthquakes at a point along the fault and (2) geomorphic offset measurement of tectonically displaced sublinear geomorphic features such as stream channels, fluvial terraces, or alluvial-fan edges, constraining the slip of past earthquakes and slip accumulation patterns along the investigated fault section. The latter approach commonly relies on field observation and the analysis of aerial photographs.

High-resolution topographic data sets, namely Lidar, also known as airborne laser swath mapping (ALSM) and interferometric synthetic aperture radar (InSAR), now present additional powerful means to further constrain the slip-per-event and slip accumulation pattern of past earthquake ruptures from offset geomorphic markers, as well as the ground deformation of recent earthquakes. Both data sets promise to contribute distinctly to a deeper understanding of the mechanics of earthquakes and faulting, thereby improving seismic hazard assessment and earthquake forecast models.

Lidar. Lidar data are acquired by a laser scanner, usually aboard an aircraft. During data acquisition, the scanner emits laser pulses that travel to the surface where they are reflected. Part of the reflected radiation returns to the scanner, is detected, and stops the time counter which was started when the laser pulse was sent out (**Fig. 1***a*). The intensity of the returning pulse is then recorded along with its time-of-flight. The latter is converted to a distance between scanner (source) and surface (reflector) by taking the speed of light into consideration. In combination with onboard GPS and an inertial navigation system (INS) this distance is then further converted to provide absolute geographic coordinates (latitude, longitude, and elevation) for the reflector (Fig. 1*a*). Note that each outgoing pulse may generate multiple returning pulses that differ in travel time and/or intensity, for example, because of partial pulse reflection by vegetation cover (Fig. 1*b*). Classification of those returning pulses by travel time and/or intensity permits virtual deforestation of the scanned area, resulting in a bare-earth depiction called a digital elevation model (DEM; **Fig. 2**). Generally, shot

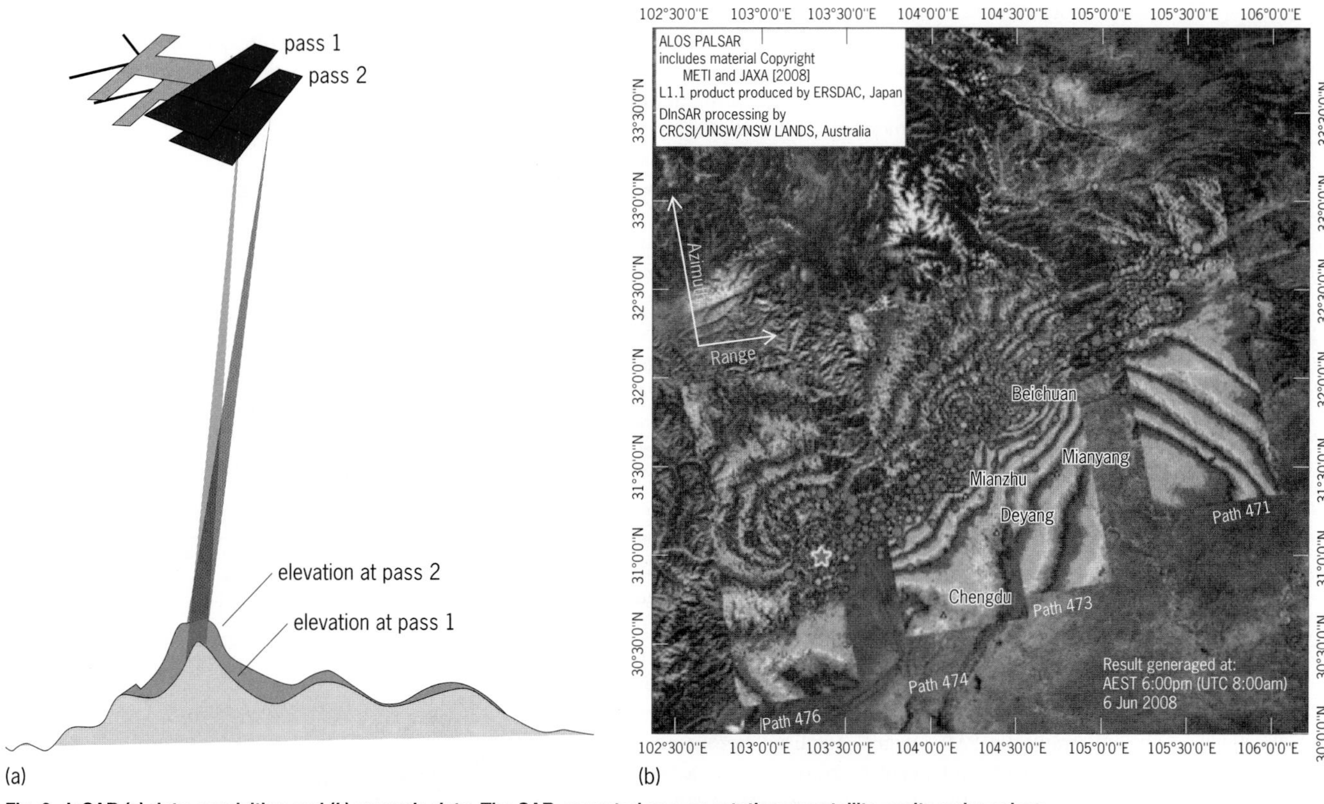

Fig. 3. InSAR (*a*) data acquisition and (*b*) example data. The SAR, mounted on a geostationary satellite, emits radar pulses that are reflected by the surface. If two SAR images of the same region but at different times are acquired (pass 1 and pass 2) and if other parameters remain essentially constant, then the difference in time-of-flight between pass 1 and pass 2 is due to a change in distance between the satellite and the surface. In (*b*) the SAR interferogram for the 2008 Sichuan earthquake in China, a single fringe equals an 11.8-cm displacement along light of sight. (*Image from http://cegrp.cga.harvard.edu/system/files/20080606_ALOS_PALSAR.jpg*)

densities (pulse returns per m^2), and thus DEM resolution, vary as a function of pulse frequency (2–400 kHz), aircraft elevation above ground, and vegetation cover. For example, average shot return densities of 3–4 per m^2 in the "B4" data set (available at http://www.opentopography.org) that covers the southern San Andreas Fault (SAF) allow generation of a DEM with <0.5 m grid size, permitting identification of sub-meter-scale tectono-geomorphic features. This data set is therefore useful for investigations of tectono-geomorphic studies concerning slip amount and slip distribution of past earthquake ruptures and slip accumulation patterns along the San Andreas Fault.

One of the first tectono-geomorphic studies that employed Lidar data was a study by K. W. Hudnut and colleagues on the M7.1 Hector Mine earthquake of 1999 along the Eastern California Shear Zone (ECSZ). The goal of this study was to use Lidar data to document the surface rupture of the 1999 earthquake. A recent study by O. Zielke and colleagues along the Carrizo section of the south-central San Andreas Fault used the "B4" data set to reevaluate the surface slip distribution associated with the most recent earthquake (the great M7.8 Fort Tejon earthquake of 1857) and preceding ground-rupturing events. This study showed that the average geomorphic slip along the Carrizo section during the 1857 event was 5–6 m, distinctly lower than the previously reported 8–10 m, eliminating a core assumption for a linkage between Carrizo segment rupture and the recurrence of major earthquakes along the south-central San Andreas Fault and lowering the average recurrence interval for Carrizo segment rupture from >250 years to approximately150 years. Another recent study that used a Lidar data set by G. E. Hilley and colleagues presented a method to automatically identify fault scarps and determine their relative ages, yielding information about the temporal development of the imaged fault zone. These and other studies that used Lidar data to investigate tectonic geomorphology, slip of past earthquakes, and fault-zone structure exemplify the great potential of high-resolution topographic data.

InSAR. Another approach in determining the slip and ground deformation of past earthquakes is to use InSAR. Similar to the approach for airborne Lidar, a satellite is carrying a SAR that emits radar pulses which are reflected by the surface. Based on the return signal, the distance from satellite to the ground is calculated, which in turn is converted to absolute geographic coordinates for each pulse return. If two SAR images of the same terrain made from essentially the same satellite positions are available, and if ground moisture is approximately the same for both images, then the difference in path lengths between the two images are attributable to changes in the position of the ground in the time between acquisitions of both images (**Fig. 3***a*).

The resulting depiction of ground deformation (usually displacements of 2.8 cm or more can be resolved) is termed a SAR interferogram and typically displayed as a contour plot in which each successive fringe represents an additional 2.8 cm of displacement (Fig. 3*b*). While the spatial resolution of InSAR is distinctly lower than Lidar (data are generally averaged to reduce noise, resulting in single displacement value for areas of approximately 100 by 100 m), its advantage lies in the large spatial coverage (a single SAR image covers 60 by 60 km), which allows for resolving the surface displacement field not only along a discrete fault but also in a wide area surrounding it. Additionally, InSAR does not present a static image of the current topography but instead the change in topography that occurred between acquisitions of the individual SAR images. It resolves this change at a very high resolution and across a very large area, therefore providing a detailed record of the spatial displacement field and the temporal variation thereof. As mentioned previously, generation of SAR interferograms requires multiple SAR images, acquired before and after a displacement event (earthquake). As a result, only recent earthquakes for which pre-earthquake SAR imagery exists can be investigated; that is, InSAR is not applicable for extending the existing record into the past. InSAR has been used for many recent earthquakes, including the M7.3 Landers earthquake of 1992; the M7.9 Denali earthquake of 2002; and the M7.9 Sichuan earthquake of 2008 (Fig. 3). Repeated acquisition of SAR images of these and other seismically active regions allows the investigation of the temporal variation of the corresponding displacement and deformation field during a seismic cycle. Studying the deformation field of a fault zone over a complete seismic cycle may reveal whether ground deformation is changing right before the occurrence of a major event, in which case InSAR may provide a means for earthquake forecasting.

For background information *see* EARTH INTERIOR; EARTHQUAKE; FAULT AND FAULT STRUCTURES; GEODESY; GEOMORPHOLOGY; PALEOSEISMOLOGY; PLATE TECTONICS; SEISMOGRAPHIC INSTRUMENTATION; SEISMOLOGY in the McGraw-Hill Encyclopedia of Science & Technology. Olaf Zielke

Bibliography. D. W. Burbank and R. S. Anderson, *Tectonic Geomorphology*, Blackwell Science Inc., Malden, MA, 2001; G. E. Hilley et al., Morphologic dating of fault scarps using airborne laser swath mapping (ALSM) data, *Geophys. Res. Lett.*, 37: L04301, 2010; K. W. Hudnut et al., High-resolution topography along surface rupture of the October 16, 1999 Hector Mine earthquake (Mw7.1) from airborne laser swath mapping, *Bull. Seism. Soc. Amer.*, 92(4):1570–1576, 2002; J. McCalpin, *Paleoseismology*, 2d ed., Academic Press, San Diego, CA, 2009; C. H. Scholz, *The Mechanics of Earthquakes and Faulting*, 2d ed., Cambridge University Press, Cambridge, U.K., 2002; O. Zielke et al., Slip in the 1857 and Earlier Large Earthquakes Along the Carrizo Plain, San Andreas Fault, *Science*, 327 (5969): 1119–1122, 2010.

Contagious cancers

Cancer arises independently in each individual as a result of genetic mutations that transform cells, leading to uncontrollable growth. However, there are two naturally occurring contagious cancers that have been detected. Devil facial tumor disease (DFTD) and canine transmissible venereal tumor (CTVT) each arose within a single individual, but then were able to spread from one individual to another. Moreover, both clonal cell lines lived on well beyond the normal life span of the host in which they arose. Although the two diseases are spread differently, with DFTD being spread between Tasmanian devils by biting and CTVT being transmitted during coitus in dogs, both appear to have arisen as a result of low levels of genetic diversity within the original host population.

Devil facial tumor disease. The Tasmanian devil (**Fig. 1**) is the world's largest remaining marsupial carnivore. Like other marsupials, devils give birth to altricial young at 3 weeks after conception. The young then spend 5 months in the pouch permanently attached to the teat and are fully weaned 10 months after birth. The Tasmanian devil is the size of a small dog [males: 10–12 kg (22–26.4 lb); females: 6–8 kg (13.2–17.6 lb)] and is a very effective scavenger. With a powerful jaw, devils have no problem eating almost every part of the carcass of prey. Previously, devils were found across mainland Australia, but they are restricted now to the island state of Tasmania. Devils have extremely low levels of genetic diversity as a result of repeated population bottlenecks, followed by population recovery.

DFTD was spotted initially by a wildlife photographer in 1996 in northeastern Tasmania. Since then, it has decimated Tasmanian devil populations. Tasmanian devils are now listed as endangered, with 70% of the species gone as a direct result of the disease. Some populations have lost 95% of their devils. Current estimates suggest that the disease will spread across the entire devil range, reaching northwestern

Fig. 1. A healthy Tasmanian devil. (*Photo by Sarah Peck*)

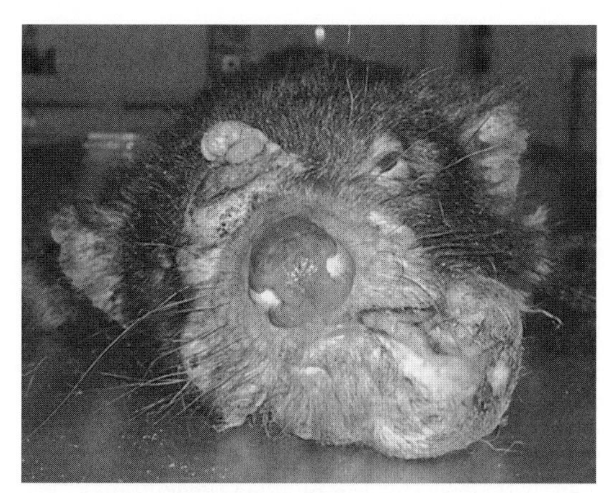

Fig. 2. Tasmanian devil affected by devil facial tumor disease (DFTD). (*Photo by Hannah Bender*)

populations within 5 years. Extinction in the wild is likely in approximately 25 years.

DFTD cells are believed to have originated from Schwann cells (which surround peripheral nerve axons, forming myelin sheaths of the neurilemma) of devil origin. The discovery that chromosomal rearrangements in DFTD cells taken from 11 biopsies from different devils were identical and grossly rearranged led to the conclusion that DFTD is spread as an allograft (a transfer of cells from a donor to a genetically dissimilar recipient of the same species).

When devils bite each other (which they often do during mating or fights for food), tumor cells can be transmitted from one animal to another. Tumors develop at the bite sites, primarily around the face and jaw, but then can metastasize (**Fig. 2**). All individuals who contract the disease die within 6 months as a result of starvation because of an inability to feed or as a result of organ failure from metastases.

Transmission of cells from one animal to another should lead to rejection of donor cells by the recipient's immune system. This occurs because the recipient's immune system recognizes foreign antigens, known as major histocompatibility complex antigens (MHC antigens), on the surface of the foreign cells. These MHC antigens are the cause for incompatibility and rejection during organ transplantation. The genes coding for the MHC antigens are the most polymorphic (variable) in the vertebrate genome. Each individual expresses a unique combination of MHC antigens that determines their "self" profile. Their immune system is trained to eliminate all cells that are "non-self" because they contain a different complement of MHC antigens. That is why donors and recipients are matched at MHC genes for organ transplantation.

Tasmanian devils have extremely low levels of genetic diversity across the genome, including the MHC genes. Tasmanian devils are essentially immunological clones, with their cells containing identical MHC antigens. Transmission of DFTD is possible because DFTD cells have the same MHC antigens as found in the affected devils. Therefore, affected devils do not "see" the tumor cells as foreign and do not mount an immune response, allowing them to grow unchallenged and under the radar of the immune system.

The recent discovery of MHC-disparate devils in the northwest region of Tasmania raises hopes that some animals may be resistant to the disease. The level of natural resilience in these populations remains to be determined. One key concern, however, is that DFTD has begun to evolve. More than a dozen DFTD strains, based on karyotype (chromosomal set) profiles, are known. It is feasible that DFTD will evolve immune evasion strategies in the face of strong selection pressure in MHC-disparate hosts.

Canine venereal transmissible tumor. CTVT is a sexually transmitted disease in dogs (**Fig. 3**), which also can be transmitted by licking or biting affected areas. It first was described in 1876 and is found now in dogs on six continents. Tumors grow on the genitalia. In immunocompromised individuals, the tumor can metastasize, but this is rare.

CTVT, like DFTD, is a contagious cancer that is believed to have evolved from a cell of the macrophage lineage. Genetic typing has shown that CTVT is clonal, and it is estimated to be between 250 and 2500 years old. CTVT is thought to have arisen in an inbred wolf population that lacked MHC diversity. Over time, CTVT evolved immune evasion strategies, which allowed it to affect MHC-disparate individuals and eventually cross the species barrier and affect dogs, jackals, and coyotes.

CTVT evades immune attack by downregulating cell-surface MHC antigens. This is a strategy that is used by a wide variety of human cancer cells to evade detection of MHC-bound cancer antigens. It also is

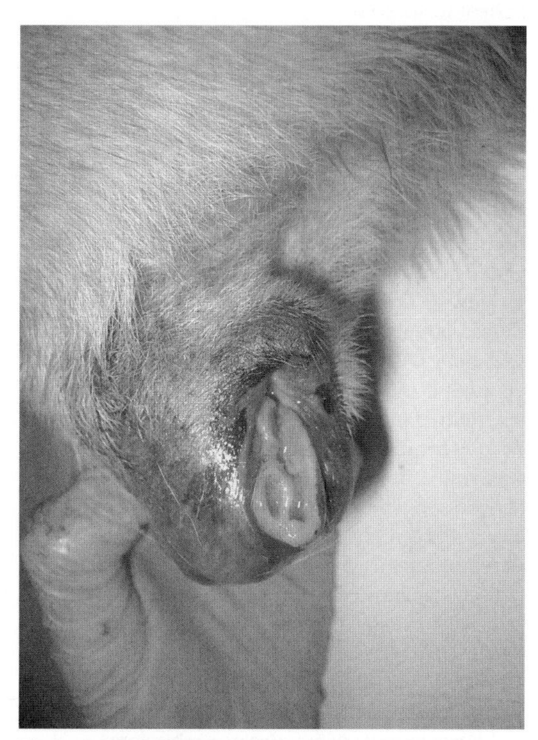

Fig. 3. A female dog affected by canine transmissible venereal tumor (CTVT). (*Photo by Elizabeth Murchison*)

seen at the fetal-maternal interface, in embryonic stem cells, and in neural progenitor cells.

Unlike DFTD, CTVT does not usually kill its host. During tumor growth, CTVT cells have very low levels of expression of MHC antigens and β2-microglobulin (β2-microglobulin is a protein that binds to and stabilizes MHC class I antigens). However, 3–9 months later, in healthy dogs, CTVT cells reexpress MHC antigens and β2-microglobulin at normal levels. As soon as cell-surface MHC antigens appear, the dog's immune system is able to destroy the tumor. The mechanisms behind this are not fully understood, but they appear to involve changes in expression of immunomodulatory cytokines (protein intercellular signals) that promote tolerance during tumor growth.

Transmission of other cancers. So far, DFTD and CTVT are the only naturally transmissible cancers found in nature. However, some cancers have been transmitted experimentally in the laboratory and rarely in medical cases.

Tumors can be transmitted readily between mice of the same strain, while a number of cancers can be transmitted through unrelated mouse strains. A sarcoma in Syrian hamsters is able to be transferred between individuals of an inbred laboratory colony by mosquitoes.

Cancers can be transmitted through organ transplantation. Hidden malignancies, such as melanoma, are particularly transmissible. Transmission of cancer cells from mother to young has been documented. In one instance, the cancer cells lost an entire MHC haplotype, not inherited by the child, to evade immune detection. There also is one documented report of a sarcoma being transmitted from patient to surgeon.

Will new contagious cancers emerge? The key similarity between DFTD and CTVT is that both diseases appear to have arisen in populations that lacked genetic diversity at MHC genes. These populations lacked the histocompatibility barriers needed to stop transmission of "non-self" cells. The possibility of other contagious cancers arising in the future cannot be ruled out, especially given the rapid loss of genetic diversity seen in many wildlife populations worldwide as a result of habitat fragmentation and repeated population bottlenecks. However, it is likely that both DFTD and CTVT have evolved other, as yet undiscovered, features, which allow them to be transmitted in such a stable fashion and for periods beyond the natural life span of their original host. Future studies need to focus on oncogenes and tumor suppressor genes, as well as mechanisms of immune evasion and regeneration of telomeres (the ends of chromosomes).

For background information *see* ANTIGEN; AUTOIMMUNITY; CANCER (MEDICINE); CELLULAR IMMUNOLOGY; DISEASE; DOGS; EPIDEMIOLOGY; HISTOCOMPATIBILITY; MARSUPIALIA; MUTATION; ONCOLOGY; TRANSPLANTATION BIOLOGY in the McGraw-Hill Encyclopedia of Science & Technology.

Katherine Belov

Bibliography. E. P. Murchison, Clonally transmissible cancers in dogs and Tasmanian devils, *Oncogene*, 27(suppl. 2):S19–S30, 2008; C. Murgia et al., Clonal origin and evolution of a transmissible cancer, *Cell*, 126:477–487, 2006; H. V. Siddle et al., Transmission of a fatal clonal tumor by biting occurs due to depleted MHC diversity in a threatened carnivorous marsupial, *Proc. Natl. Acad. Sci. USA*, 104:16221–16226, 2007.

Contaminants in recycled wood

Most people agree that wood should be recycled into new products. This can be done several times over and, if subsequent recycling is not economically viable, the wood can be burned to recover the energy that it contains. However, few people are aware of the technical difficulties associated with the simple idea of recycling old wood products into new ones. One aspect of this is minimization of the contamination levels in products made from recovered wood.

Recycling clean wood. Some wood products can be recycled with little or no concern for contaminants. Clean wood means wood (including lumber, tree and shrub trunks, branches, and limbs) that is free of paint, glue, filler, pentachlorophenol (PCP), creosote, tar asphalt, other wood preservatives, and treatments. While this definition specifically excludes treated wood, industry facilities that accept clean wood will inadvertently accept some treated woods, which need to be properly managed.

As examples, clean large wood beams from old barns and studs removed from old construction usually can be recycled directly. One concern from old wood is the possibility of mold growth on wood that has been damp or wet for some time during its use. Wood that has been dry during its lifetime has little or no strength loss and can be used in new construction without concern for loss of properties.

Recycling contaminated wood. Wood is rarely used in a pure form, that is, without interaction with other materials or products. It is normally joined with nails and screws or glued with one or more of a wide range of adhesives. Furthermore, it is possibly treated to enhance its resistance to fire and biological agents and is often finished with a paint, varnish, or stain. Consequently, when a wood product is recycled (see **illustration**), the materials and products associated with it are considered "contaminants" and must be separated from the wood. This is not an easy task as the contaminants are invariably intimately associated with the wood. Thus, some pass through the cleaning systems in place and contaminate the recovered wood stream.

Users of recovered wood need to know the level of contamination present to ensure that their own processes conform to regulations. For example, the European Panel Federation (EPF) imposes an industry standard on its members that defines maximum permissible concentrations of various heavy metals and preservatives. Likewise, an energy generator

***Left:* Waste wood mixed with metal. *Right:* Typical waste wood pile from an old house demolition.**

that burns recovered wood must have procedures in place to ensure that the metal content in the wood does not exceed the specifications set by the boiler supplier; otherwise, it might become clogged with metal deposits.

Contaminants in the form of metals and organic molecules are present in paints, varnishes, stains, and polishes. Practically all furniture has some form of coating applied to enhance and protect the aesthetic appearance of the wood. For example, door and window frames are painted, and garden furniture is often stained. The levels of contaminants in these finishing products are low and, as coatings applied to wood products, they certainly do not pose a health risk. However, the small quantities of contaminants in the finishes can add to those that are naturally present in the wood, with the natural levels being dependent on where the tree grew. In addition, there is contamination from the air, for example, exhaust fumes, and physical contact with soil and dirt while a product is used. Consequently, measurable quantities of heavy metals are present in most pieces of post-consumer wood.

Chromated copper arsenate (CCA) is a chemical wood preservative containing chromium, copper, and arsenic that protects wood from rot caused by insects and microbial agents. As a result, the use of CCA to pressure-treat wood can prolong the service life of the wood by 20 to 40 years beyond that without the preservative. CCA has been used to treat wood since the 1940s, and CCA-treated wood has been used extensively in residential applications since the 1970s. Wood treated with CCA produces no odors or vapors, and its surface can be painted or sealed easily. Wood products treated with CCA include lumber, timber, utility poles, posts, and plywood. Because of its ease in use and the effectiveness of its treatment, CCA-treated wood has been the most widely used type of treated wood in the United States, representing approximately 80% of the wood preservation market until about 2002. There are other preservative treatments for wood, including alkaline copper quaternary (ACQ), copper boron azole (CBA), ammoniacal copper borate (ACB), ammoniacal chromate arsenate (ACA), ammoniacal chromate zinc arsenate (ACZA), and copper azole.

To measure the heavy metal contents of wood, it is normally ground to a fine flour and then digested with strong acids to solubilize the metals. The concentrations of metals in the resultant solution are then measured by atomic absorption spectrometry (AAS) or inductive-coupled plasma (ICP) analysis. These analytical methods are accurate and can give precise measurements concerning the quantity of metal ions present. They cannot differentiate, however, between ions that were naturally present in the wood and those that were added by some chemical treatment. However, the level usually indicates whether or not some treatment had been applied. For example, the level of chromium (Cr) naturally present in wood is low, whereas the concentration of Cr present in wood treated with CCA is at least three orders of magnitude higher (see **Table 1**). The Cr content of chipped commercially treated wood has been found to be 2700 mg/kg, whereas that for various pieces of treated timber has been found to range from 400 to 11,000 mg/kg.

If a single piece of wood is analyzed, the metal concentration observed would indicate whether or not it had been subjected to some form of chemical treatment. Conversely, if the sample analyzed consists of particles that were chipped from, for example, demolition waste and the measured Cr content is, say, 20 mg/kg, then this indicates that some additional Cr is present. The additional Cr could be present because some of the chipped wood had

TABLE 1. Typical metal contamination levels found in virgin and CCA-treated wood (mg/kg)

	Copper		Chromium	
Description	Virgin	CCA-treated	Virgin	CCA-treated
Pine	0.5–17	1400–2700	0.03–0.5	2700–5200

TABLE 2. Maximum allowable concentrations of contaminants permitted in particle-board manufactured in accordance with European Panel Federation standards

Contaminant	Limit, mg/kg	Contaminant	Limit, mg/kg
Arsenic (As)	25	Lead (Pb)	90
Cadmium (Cd)	50	Mercury (Hg)	25
Chromium (Cr)	25	Fluorine (F)	100
Copper (Cu)	40	Chlorine (Cl)	1000
Pentachlorophenol (PCP)	5	Creosote	0.5

previously been treated with CCA. In fact, if only 0.75% of the wood by weight was treated with CCA, then this would be enough to explain the level of 20 mg/kg. However, the result could also be a consequence of the inclusion of steel, leather, soil, concrete, and pigments. For example, stainless steel contains 140,000 mg/kg of Cr. Thus, the 20 mg/kg measured in the sample could be the result of the inclusion of less than 0.02% by weight of stainless steel. Leather too has a high Cr content of approximately 21,000 mg/kg because of the tanning processes employed, so the result could be explained by the inclusion of a very small quantity (0.1% by weight) of leather.

Users of recovered wood normally specify that they will not accept preservative-treated wood. Certainly, this is true of American and European particleboard manufacturers, who are the main users of recovered wood. Limitations are also often placed on the proportions of the wood that is painted, stained, varnished, or colored in some manner. This is necessary because many pigments contain metals, leading to unacceptably high levels of metals being present.

Very high extraction values of 97%, 89%, and 93% have been claimed for copper, chromium, and arsenic, respectively, using a water-based extraction method. However, even though these extraction values are impressive, the process still leaves excessive quantities of metals in the particles: 50 ppm for copper; 400 ppm for chromium; and 200 ppm for arsenic. Particle-board manufacturers are not likely to use wood that still has this level of contamination. **Table 2** shows the maximum allowable content of contaminants in particle-board manufactured in Europe.

Recycled wood can also contain other wood preservatives, including creosote and pentachlorophenol (PCP). Creosote is a complex mixture of approximately 200 organic chemicals that is used primarily to preserve wood products such as railway ties and power poles. Under warm weather conditions, creosote tends to create odors and exude from the treated wood. Therefore, creosote-treated wood should never be used indoors and should not be used in outdoor areas frequented by people, especially children, or animals. Many other types of products, from wood packaging to heavy structural timbers, have been treated with PCP. Treatment processes include nonpressure treatment of millwork (ready-made products manufactured at a wood-planing mill or woodworking plant) and packaging; pressure treatment of lumber, timbers, and various types of poles; and hot- or cold-bath treatments of poles, notably those derived from Western tree species.

It is anticipated that any recycling program that uses preservative-treated wood would be burdened with the responsibility to ensure that the use of the ultimate, recycled product is consistent with all directives imposed by various environmental protection agencies for use of the original product. These include prohibitions against use in products in frequent or prolonged contact with bare skin; in residential, industrial, or commercial interiors (with some exceptions); in containers for storing animal food; and in cutting boards or countertops.

Landfill requirements. Wood contaminated with wood preservatives and other chemicals cannot be disposed in a normal landfill. According to new regulations on waste, all waste products consisting of wood that has been impregnated or coated with substances containing heavy metals, PCP, or creosote have been classified as "dangerous waste." Such types of wood waste are prohibited from being included in the "biomass" category. Thus, they must be separated from other types of waste and are subjected to special treatments before being disposed in designated toxic landfills.

Burning contaminated wood. Burning wood that has been impregnated or coated with oils or chemical products such as release agents, paints, varnishes, glues, or preservatives requires special precautions because of the toxicity of the fumes that are released during such combustion processes.

For background information *see* ENVIRONMENTAL TOXICOLOGY; LUMBER; RECYCLING TECHNOLOGY; WOOD ENGINEERING DESIGN; WOOD PROCESSING; WOOD PRODUCTS; WOOD PROPERTIES; WOODWORKING in the McGraw-Hill Encyclopedia of Science & Technology.

Roger M. Rowell

Bibliography. C. A. Clausen and W. R. Kenealy, Scaled-up remediation of CCA-treated wood, in *Proceedings of Environmental Impacts of Preservative-Treated Wood*, Florida Center for Environmental Solutions, Orlando, FL, 2004; European Panel Federation, *EPF standard for delivery conditions of recycled wood*, EPF position paper, 2002; European Panel Federation, *The use of recycled wood for wood-based panels*, EPF position paper, 2001; D. G. Humphrey, L. J. Cookson, and J. Gradev, *Determination and comment on CCA retentions in waste pallets made from treated pine*, Client

Report No. 907, Forestry and Forest Products, CSIRO, Australia, 2000; M. A. Irle, Biocomposites—Part 1: The cleanliness of recovered wood, Chapter 5, in R. M. Rowell, F. C. Jorge, and J. K. Rowell (eds.), *Sustainable Development in the Forest Products Industry*, Fernando Pessoa University Press, Oporto, Portugal, 2010; M. A. Irle, The impact of using recovered wood on heavy metal content in particleboards, in *Proceedings: Joint Wood Adhesive and Wood-Based Panel Research Group Symposium*, Tsukuba, Japan, 2005; M. A. Irle et al., The detection and measurement of metal contaminants in recycled wood intended for the manufacture of particleboards, in *Proceedings: Eighth European Panel Products Symposium*, Llandudno, U.K., 2004; K. E. Oskoui, Recovery and reuse of the wood and chromated copper arsenate (CCA) from CCA treated wood—a technical paper, in *Proceedings of Environmental Impacts of Preservative-Treated Wood*, Florida Center for Environmental Solutions, Orlando, FL, 2004; Waste and Resources Action Programme, *Assessment of types and levels of naturally occurring contaminants in virgin wood sources*, WRAP Final Report for Project WOO0035, 2005; Waste and Resources Action Programme, *Contamination levels in recovered wood products*, WRAP Annual Report, 2006.

Continuous plankton recorders in the North Pacific Ocean

All of the living marine resources that we enjoy, whether as food or for their intrinsic appeal, depend ultimately on the base of the food chain, the plankton. Scientists have long recognized that the open ocean is not an unchanging environment. Currents, topography, and climate combine to create variabilities in time and space in the physical, chemical, and biological characteristics of the oceans. Plankton have limited control over their movements and thus are subjected to the local conditions that prevail. These conditions, in turn, determine where various types of planktonic organisms occur and how fast and abundantly they grow. In this way, ocean climate changes are transferred up the food chain to the fish, marine mammals, and seabirds that are, to us, the most prominent product of the marine ecosystem. To understand and predict changes in living marine resources under future climate-change and ocean-use scenarios, we need information about the physical environment and the plankton. The question is how best to measure and monitor these properties over thousands of kilometers. Satellites have gone a long way toward enabling a wide view of surface ocean characteristics such as temperature, ocean color, sea-surface height, and the amount of phytoplankton (plant plankton), but zooplankton (animal plankton) cannot yet be viewed from space. Research ships cost a significant amount of money to operate, and any kind of comprehensive spatial coverage of the North Pacific Ocean plankton would be prohibitively expensive and time-consuming. When scientists at the North Pacific Marine Science Organization (PICES) were considering this problem, they turned to the Sir Alister Hardy Foundation for Ocean Science (SAHFOS) and the continuous plankton recorder (CPR) survey for help.

Continuous plankton recorder. The CPR was designed in the 1920s by Alister Hardy and first deployed in Drake's Passage in the South Atlantic on the *Discovery* expedition in 1927. Briefly, a CPR is a mechanical device that is towed behind a ship at about 7 m depth, where it filters the plankton from the seawater onto a continuously moving band of silk mesh with a pore size of about 270 μm. The movement of the CPR through the water drives a propeller that turns a take-up spindle in an internal cassette, drawing the filtering mesh across the entrance aperture (**Fig. 1**). A second piece of mesh covers the filtering mesh, sandwiching the plankton between them, and then the resulting sandwich winds onto the spindle in a tank containing preservative. Each cassette is preloaded with enough mesh to be towed for 500 nautical miles (926 km). Changing cassettes is a simple and quick procedure, so that greater distances can be covered using multiple deployments and cassette changes. Once the CPR has completed its tow, the band of mesh sandwich is unloaded and information from the ship's log (time and position of deployment and recovery) is used to reconstruct the voyage and divide the mesh into discrete samples, each of which covers 10 nmi (18 km) of the tow. Each sample is then viewed under a microscope, and the plankton are identified and counted. The CPR has changed very little since its first design and has been used in an almost unchanging fashion in the North Atlantic since the 1940s. Its principal advantages over other plankton samplers are that it is robust, needs no accompanying technician, and can be deployed from vessels at high speed. This makes it ideally suited to being towed behind commercial ships on their regular routes of passage, thus removing the need for costly research vessels.

Sampling of the North Pacific. The first routine CPR deployments in the North Pacific took place in 2000; since then, samples have been collected along two transects on an ongoing seasonal basis (**Fig. 2**). The longest transect is between British Columbia and Japan, following the Great Circle (shortest-distance)

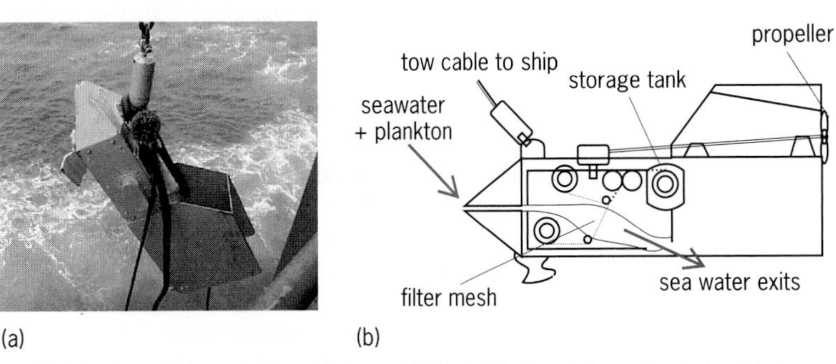

Fig. 1. Continuous plankton recorder (CPR). (*a*) About to be deployed. (*b*) Schematic illustration of how it works.

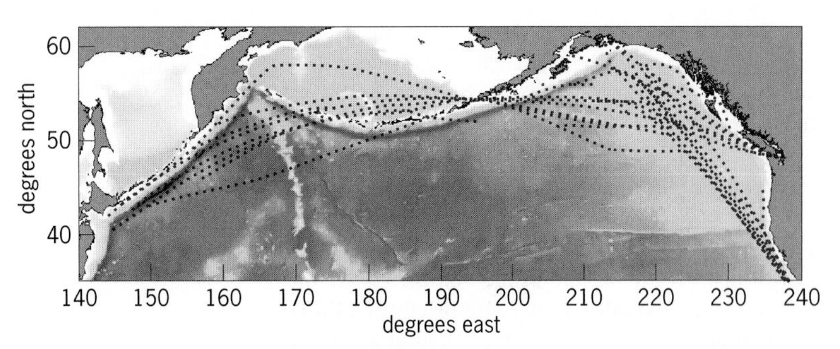

Fig. 2. Locations of all processed CPR samples collected from 2000 through 2009 (intermediate samples are archived). Ship tracks vary between months to avoid storms, for example, so the sample track coverage is quite wide.

route through the Aleutian Islands and the southern Bering Sea. The cargo vessel *Skaubryn* has towed the CPR on this route since 2000. It has a two-month turnaround time, so it tows the CPR three times a year between spring and fall, when plankton are most abundant. A second shorter transect runs between Puget Sound, Washington, and Cook Inlet, Alaska. This is towed by the container ship *Horizon Kodiak*. Prior to 2004, the transect was towed by oil tankers between northern California and Prince William Sound, Alaska. The recent shorter transect is towed roughly monthly between spring and fall. By the end of 2009, a decade of sampling had collected 146,950 nautical miles (272,151 km) of samples that have been archived. Funding limitations meant that only just over a quarter of the samples, spread over the whole survey, have been examined and counted. Nevertheless, 344 different taxonomic entities were identified and counted (many at the species level), of which slightly more than half were zooplankton. Some of these taxa are very rare, occurring only once or twice in the decade, while some are extremely common and can account for most of the plankton biomass at certain times of the year.

Effects of climate. During the decade of sampling, the ocean climate of the North Pacific has been quite variable. Alternating warm and cool periods have occurred particularly in the Northeast Pacific. These periods last 3–4 years, with abrupt shifts between them. For example, 2005 was one of the warmest years on record, whereas 2008 was the coldest in several decades. The zooplankton have responded to these changes in quite clear ways. Being cold-blooded, the rate at which zooplankton grow depends on the water temperature. One species of zooplankton, a copepod about 0.5 cm in length, is extremely abundant in subarctic oceanic waters in the spring. *Neocalanus plumchrus* spends much of the year dormant at depth (several hundred meters or more), but in the early spring juveniles migrate up to the surface, where they grow and molt through several life stages. At the subadult stage, they accumulate fat reserves and then descend to depth again, usually around June. They may be abundant in the surface for only three months of the year, but their large size, density, and fat content make them particularly attractive prey to fish, some seabirds, and even whales. Data from the CPR survey show that the timing of this surface maximum varies between years with a strong relationship to the temperature. This interannual variation has been as much as four weeks, a significant proportion of the short time spent in the surface waters. Furthermore, in cool years the peak tends to have a longer duration, whereas in warm years it is more abrupt. Predators that time their reproduction or migration in expectation of a rich food source are influenced by these changes.

Many species of plankton have a preference for, or tolerance of, certain temperature ranges. The CPR survey has found that in warm years, warm-water species extend farther north along the north–south transect (**Fig. 3**). Species also fare better or worse in warm or cool years, depending on their preference, so that a particular location may see a changing plankton community from year to year as a result of ocean-climate variability. These are just some of the ways that ocean climate can influence plankton communities by determining the distribution, growth, and timing of each species. In turn, this affects predators, whose diet is affected by the nutritional quality (different species may have different biochemical composition) and quantity of their prey.

Strengths and limitations. The CPR survey of the North Pacific is making a large contribution to our understanding of this extensive oceanic region, how it responds to ocean climate, and the resources that it supports, but it does have limitations. Many types of plankton are not well sampled by the small aperture and high speed of the CPR. For example, most gelatinous forms are too badly damaged during capture to be counted and even robust plankton are often too damaged to be identified to the species level. The CPR also samples at a fixed near-surface depth, giving little or no information about species that spend most of their time at greater depths. The samples each cover too great a distance to give detailed information at scales smaller than about 100 nautical miles (185 km). However, the CPR is a cost-effective and reliable tool that is complimentary to

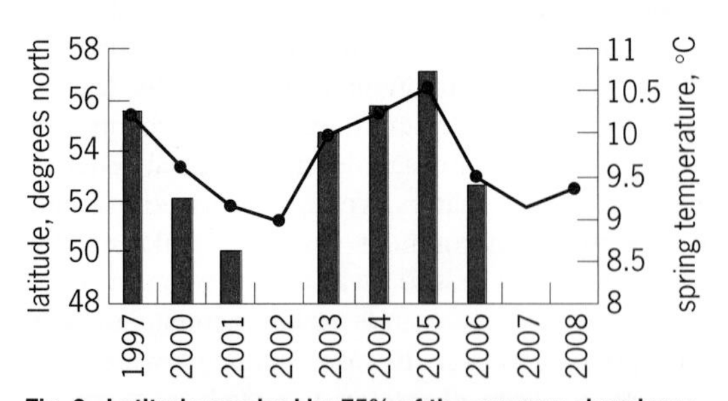

Fig. 3. Latitude reached by 75% of the summer abundance of subtropical copepods each year (bars) together with the spring temperature (line) as measured at Amphitrite Point lighthouse in British Columbia (http://www-sci.pac.dfo-mpo.gc.ca/osap/data/SearchTools/Searchlighthouse_e.htm). There is a strong positive correlation between temperature and latitude of the copepods. In cold years, such as 2002, 2007, and 2008, no subtropical copepods were found north of 48°N.

the local, more detailed surveys that are run by various research agencies around the rim of the North Pacific. Nearly a century after its invention, the CPR remains our best choice for sampling plankton in the open ocean.

For background information *see* GLOBAL CLIMATE CHANGE; MARINE BIOLOGICAL SAMPLING; MARINE ECOLOGY; MARINE FISHERIES; OCEANOGRAPHY; PACIFIC OCEAN; PHYTOPLANKTON; SEAWATER FERTILITY; ZOOPLANKTON in the McGraw-Hill Encyclopedia of Science & Technology. Sonia Batten

Bibliography. A. Hardy, *The Open Sea: Its Natural History: Part I: The World of Plankton*, Houghton Mifflin, Boston, 1971; T. Kiørboe, A Mechanistic Approach to Plankton Ecology, Princeton University Press, Princeton, NJ, 2008; C. B. Miller, *Biological Oceanography*, Wiley-Blackwell, Boston, 2003.

Coordinated Universal Time (UTC) scale

The precise measurement of time and time interval is a key enabler of modern technology in which a profusion of electronic devices and communications equipments depend on the synchronization of time and frequency in order to operate properly over distances of thousands of miles or more. Cell phones, mobile phones, the Internet, and satellite navigation, among other ubiquitous consumer services, all depend on clocks and oscillators that control their operation. Universal access by these and related services to a standardized reference time has benefited from the establishment in 1970 of a time scale called Coordinated Universal Time (UTC), and by the means to maintain that time scale both nationally and internationally.

Development of timekeeping. Efforts to measure the passage of time began with the recognition of diurnal and seasonal changes and progressed to the scheduling of human events. The rotation of the Earth was the basis for the definition of time scales until 1950. Apparent solar time, as read directly by a sundial, or more precisely determined by the elevation of the Sun over the horizon, defines the local time by the actual daily motion of the Sun. However, because of the tilt of the Earth's axis and the elliptical shape of the Earth's orbit, the time interval between successive passages of the Sun over a given meridian is not constant. The difference between mean and apparent solar time is called the equation of time. The maximum amount by which apparent noon differs from mean noon is about 16.5 min early around November 3, and as much as 14.5 min late around February 12. Until the early nineteenth century, apparent solar time determined the length of the day, but with the improvement in clocks and their growing use by ships at sea and by railroads, apparent solar time was gradually replaced by mean solar time.

The length of the day (LOD) as traditionally determined by astronomical observations determined Universal Time (UT), which also determined the length of the second, defined as 1/86,400 of a solar day. During the early part of the twentieth century it was well known that the LOD was irregular due the irregular rotational motion of the Earth. Another time scale was defined by the Conference on the Fundamental Constants of Astronomy held in Paris in 1950 to attempt to standardize timekeeping and to produce a more stable uniform time scale. The participants recommended that, instead of the period of rotation of the Earth on its axis, the standard of time ought to be based on the period of revolution of the Earth around the Sun, as defined by Simon Newcomb's *Tables of the Sun*, published in 1895. The measure of time defined in this way was given the name Ephemeris Time (ET), and Greenwich Mean Sidereal Time (GMST) was the name given to the mean Ephemeris Time measured with respect to the Greenwich meridian. GMST is used to derive the daily motion of the Sun and the LOD, known as Universal Time (UT1). The resulting effect on the second was that the General Conference on Weights and Measures (Conférence Générale des Poids et Mesures, CGPM) adopted in 1960 this definition: "the second is the fraction 1/31, 556,925.9747 of the length of the tropical year for 1900.0." The second so defined was subsequently known as the ephemeris second.

UT1 was initially determined using a worldwide network of transit telescopes, zenith tubes, and astrolabes. The techniques used today to estimate UT1 are very long baseline interferometry measurements of selected astronomical radio point sources, satellite laser ranging, and tracking of Global Positioning System (GPS) satellites. Even though ET was defined by the astronomical relationship of the Sun, it is actually derived from measurements of lunar positions. As improvements were made in how the time scale is defined and maintained, secondary time scales were defined, including UT0 and UT2. These secondary time scales may not have been actually realized; however, they denote different approaches to defining how the time is to be determined.

ET is more stable than UT in theory, but the development of atomic standards in the 1950s quickly provided the means to make it easier to generate a stable time scale. An atomic clock operating in a timing center offers a more stable and easily realizable method of generating a uniform and stable time scale; the problem then becomes getting the centers to agree on the time.

Atomic time. Atomic Time (AT) has become the basis of all modern time scales and has been maintained continuously in various laboratories since 1955, although not formally adopted until 1971 as an international time scale. The advent and operation of cesium atomic standards began in 1955 at the National Physical Laboratory in the United Kingdom. Worldwide broadcast systems such as LORAN, which were developed for navigation of ships, enabled accurate international comparison of these atomic standards at timing centers. In 1961 the Bureau International de l'Heure (BIH), which was responsible for maintaining UT1 internationally, began taking

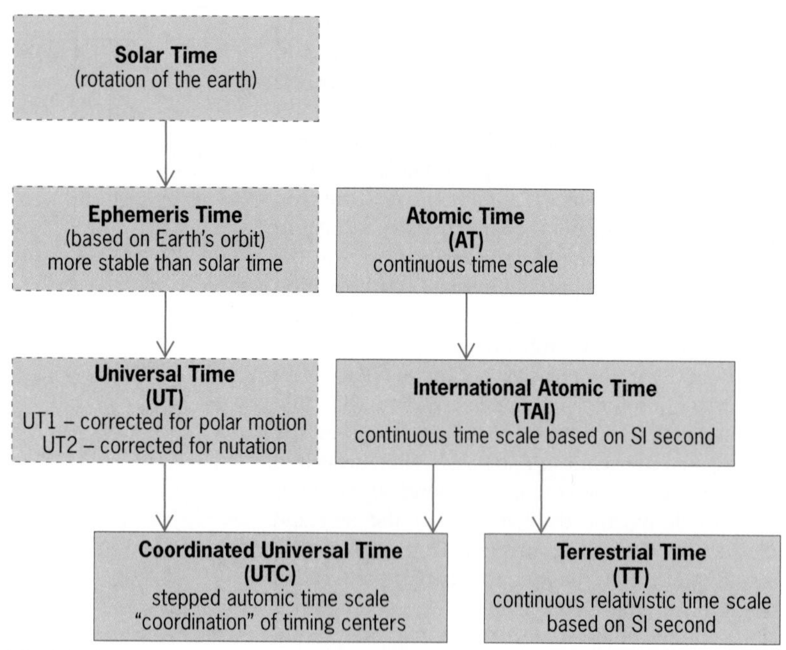

Fig. 1. Relationship of time scales and Coordinated Universal Time (UTC).

measurements from these long-range navigation systems, early satellite systems, and other broadcast media to coordinate time between the worldwide timing centers. This initial form of AT began as an informal arrangement between the United States and the United Kingdom, and was called Coordinated Universal Time. Initial international coordination was agreed through the International Radio Consultative Committee (CCIR) of the International Telecommunications Union (ITU) in 1962. With this agreement, timekeeping centers in other nations began to participate.

Only 7 years after the definition of the ephemeris second in 1960, the Thirteenth CGPM in October 1967 adopted the second of atomic time as the fundamental unit of time in the International System of Units (SI). The SI second was defined as the duration of 9,192,631,770 periods of the radiation produced by the transition between the two hyperfine levels of the ground state of the cesium-133 atom. This definition then decoupled the second from astronomically derived time.

The establishment of an atomic time scale on an international basis, to be known as International Atomic Time (Temps Atomique International, TAI), was recommended by a number of scientific unions and the CCIR in 1970. The Fourteenth CGPM meeting in 1971 adopted the establishment of TAI as a continuous time scale whose unit interval is the SI second as realized on the rotating geoid. It was to be determined by the readings of atomic clocks in various timing centers in accordance with the definition of the SI second. When TAI was established as a continuous time scale, decoupled from the rotation of the Earth, it needed to be related to the Coordinated Universal Time. The name Coordinated Universal Time (UTC) had been accepted by International Astronomical Union in 1967 and from 1961 to 1972 UTC was adjusted by both frequency offsets and subsecond steps to maintain agreement with UT2 (the version of UT1 corrected for Earth's seasonal variation) to within about 0.1 s. The coordination necessary to make the frequency and time adjustments required among a worldwide distribution of timing centers was very difficult. However, close coordination was considered necessary at that time because celestial navigation users required access to a time scale that directly related rotational time to the angle of the Earth rotation with an uncertainty of less than 1 s. Time received from radio broadcasts of UTC would then permit the celestial navigation users to determine their angular relationship to the Greenwich Meridian. The relationships of time scales in use today are primarily based on TAI; UT2 is no longer used (**Fig. 1**).

Definition of UTC. In 1972 the present UTC system was defined and adopted by the ITU with 1-s steps (leap seconds) and adjustments of the predicted differences of UT1 − UTC, or DUT1, to keep UTC within 0.9 s of UT1. It is in effect a "stepped" time scale. The rate of UTC is determined by TAI so that the basic time interval is the SI second. So, 1-s steps, either positive or negative, are then applied to maintain the difference with UT1 within 0.9 s (**Fig. 2**). Producing UTC this way is a compromise in providing both the SI second and an approximation to UT1 for celestial navigators to access by radio transmissions. Among the astronomical time scales only UT1 is actively maintained by the International Earth Rotation and Reference Service (IERS). The International Bureau of Weights and Measures (Bureau International des Poids et Mesures, BIPM) assumed responsibility for maintaining TAI, and consequently UTC, from the BIH in 1988.

TAI is also the primary time scale from which all the relativistic time scales are derived. To maintain accurate time on Earth and in the solar system, the celestial mechanics used to determine the position and motion of the planets and stars must take the theory of relativity into account. Clock time and time interval for spaceborne clocks must account for the variations induced by differences due to the relativistic effects between those in space and those on Earth. Consequently there are several relativistic time scales defined for this purpose. For practical everyday use, UTC is considered adequate and has been adopted for civil timekeeping. Terrestrial Time (TT) is a relativistic coordinate time scale derived from TAI as the fundamental time on the surface of the Earth. In relativity, coordinate time is the time kept by a clock in a specific coordinate frame over a region with little varying gravitational potential. TAI was defined by the Consultative Committee for the Definition of the Second (CCDS) in 1980 as a coordinate time scale on the rotating geoid with the SI second as the scale unit (the basic unit of measure). For use in nonterrestrial reference frames, relativistic corrections between time measured in those reference frames and TT would need to be applied since

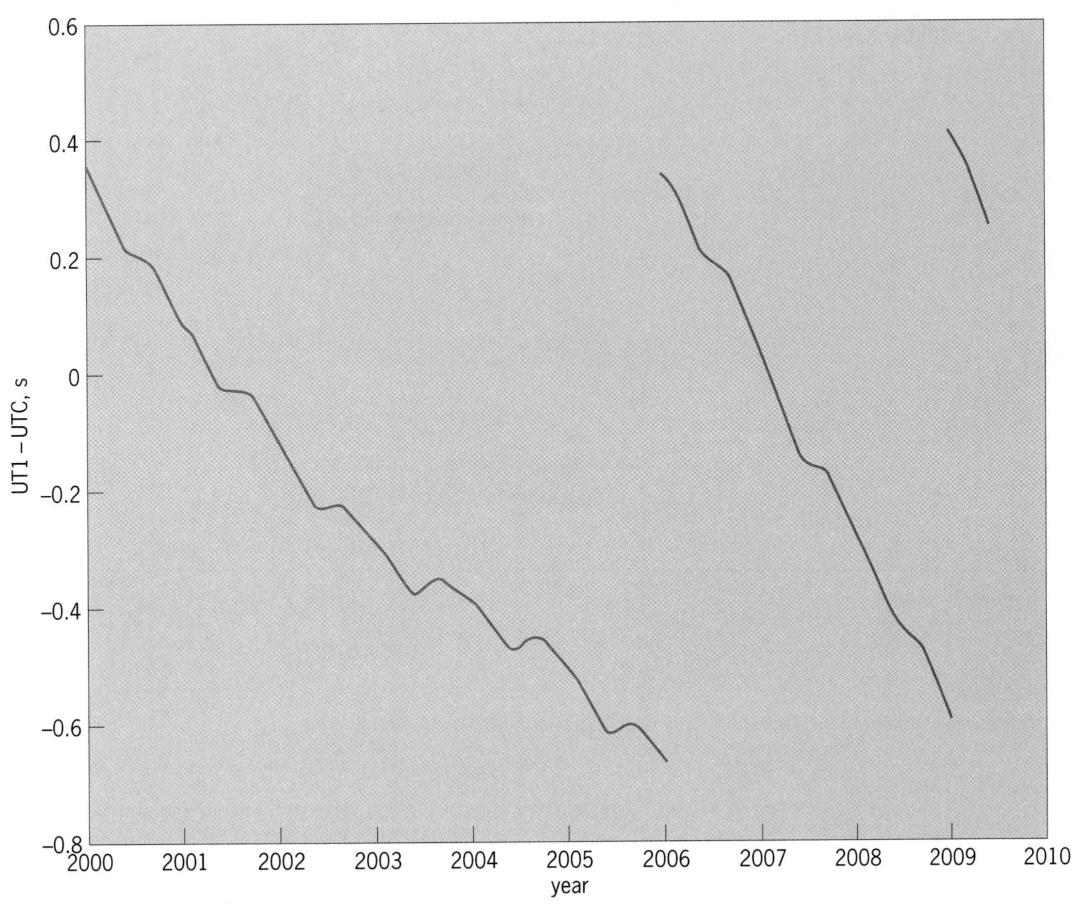

Fig. 2. Difference between UT1 and UTC in seconds since 2000.

there could be a significant difference between the two.

TAI is generated from clock comparison data supplied to the BIPM by participating timing centers and laboratories in a process using an algorithm known as ALGOS (**Fig. 3**). These clock comparison data are measurements made either using the GPS or by a technique known as two-way satellite time and frequency transfer (TWSTFT). This collection of continuous precise data provided to the BIPM compares the local realizations of UTC with local measurements of the individual atomic clocks themselves. This enables precise and accurate comparisons between the individual clocks on a worldwide basis.

Local realizations of UTC produced by a timing center are denoted as UTC(k), where k is the center producing the time estimate. UTC is determined from the collection of these clock and comparative link data. As is probably apparent from the need to collect and transmit comparison data, determining UTC is not a real-time process. UTC as determined internationally is a post-processed time. The international time scale is determined after the fact. The differences between the determined time and the local realizations at the timing center are published by the BIPM in a publication known as *Circular T*. To keep accurate time each of the timing centers then minimizes these differences between the final time and their approximation through a prediction process.

The first step in the BIPM process to determine UTC is to generate a free-running atomic time scale known as Echelle Atomique Libre (EAL) or Free Atomic Scale. EAL is then evaluated at the interval of the scale unit using data from contributing "primary" frequency standards—that is, cesium-based frequency standards that determine time intervals according to the SI definition of the second—and an optimum filter. TAI is determined from EAL by applying, if necessary, a correction to the scale interval of EAL to give a value as close as possible to the SI second. Correcting the scale unit is known as "steering" and is done infrequently. Having determined TAI, applying the appropriate number of leap seconds determined by the IERS will produce the final international value of UTC.

The accuracy of TAI is a primary consideration in maintaining the SI second and providing a reliable time scale in the long term. The optimization of the long-term stability is done at the expense of short-term accessibility. The calculation of TAI uses data over an extended period. Clock-comparison data are sent to the BIPM every 5 days (on days for which the Modified Julian Date ends in "4" and "9"). Blocks of data covering 60 days are used in the calculation of

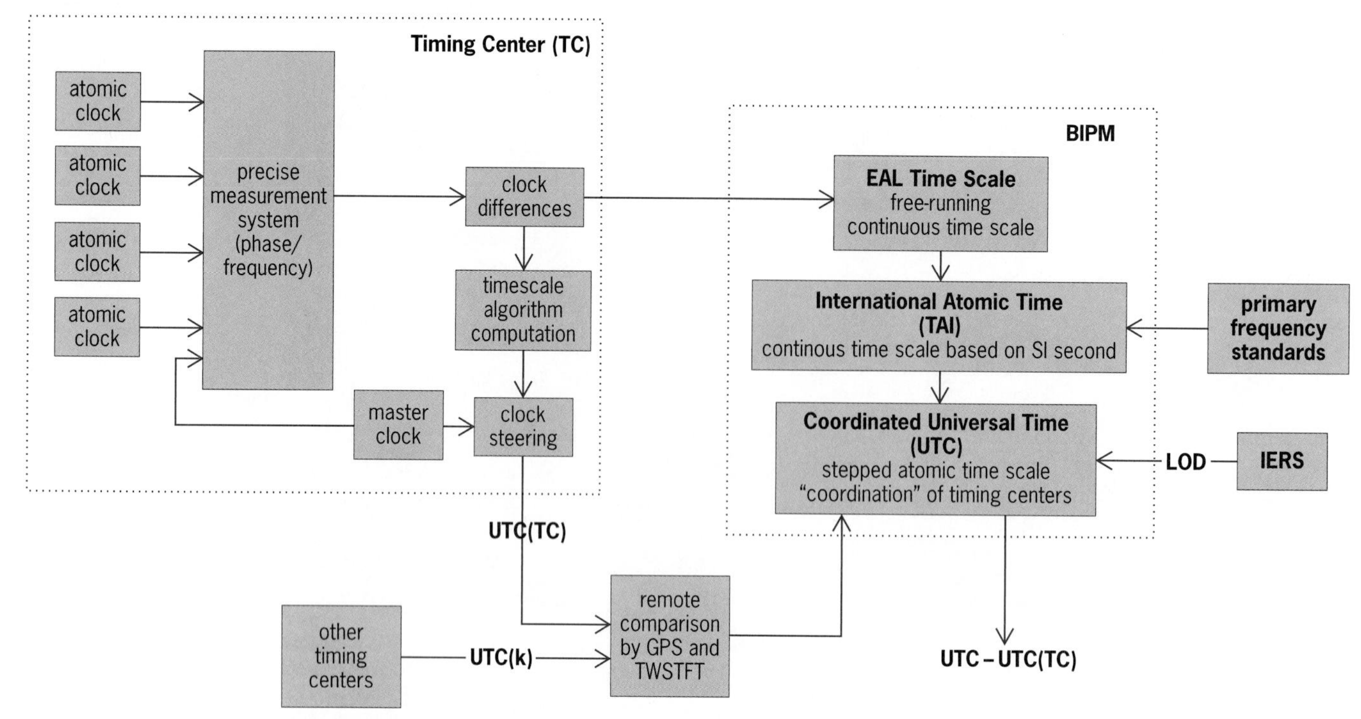

Fig. 3. Process of determining UTC, from timing centers to the BIPM.

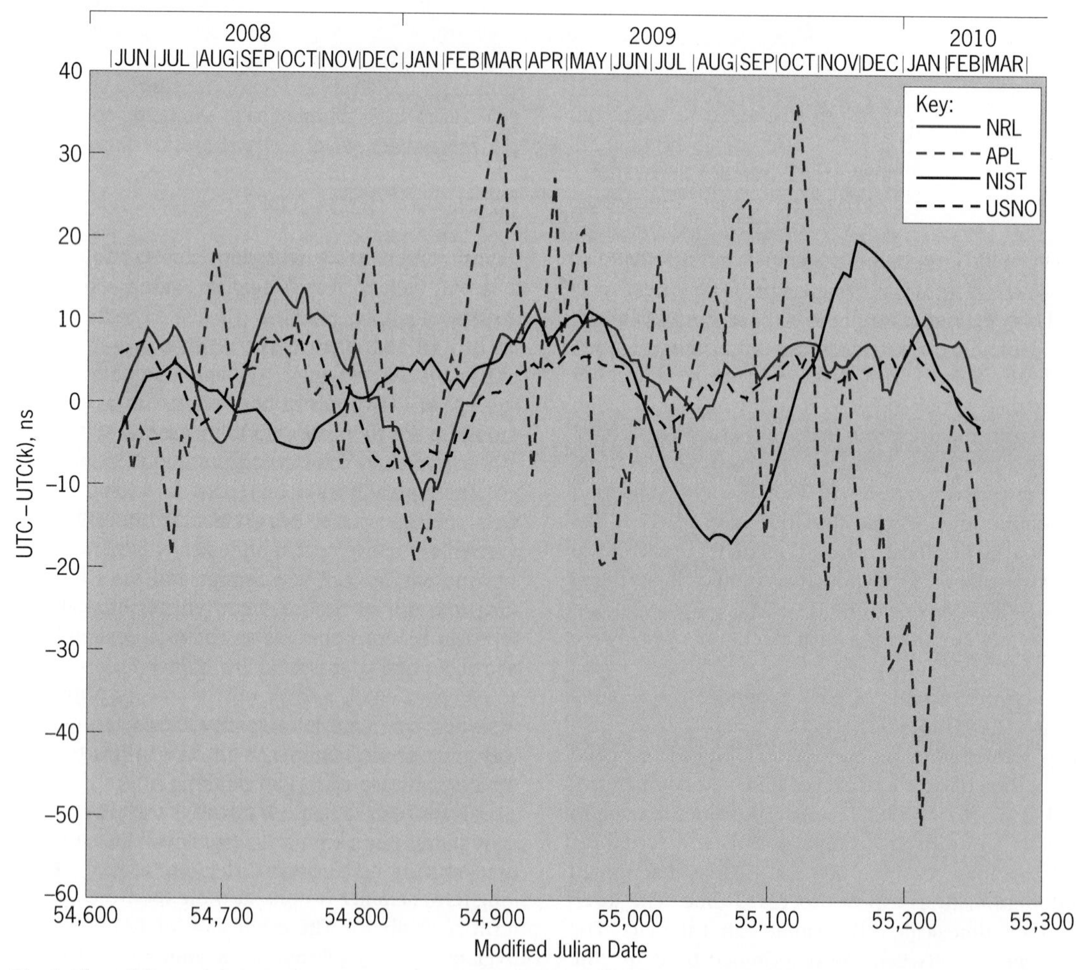

Fig. 4. Time differences between UTC and UTC(USNO), UTC(NIST), UTC(NRL), and UTC(APL), from data collected from BIPM *Circular T*. The respective timing centers are at the U. S. Naval Observatory (USNO), the National Institute for Standards and Technology (NIST), the U.S. Naval Research Laboratory (NRL), and the Johns Hopkins University Applied Physics Laboratory (APL).

the scale. A period of 60 days was chosen to place the effective integration time of the time scale at the effective stability limit of cesium clocks. Stability would therefore not be improved by a longer integration time. The period of 60 days is enough to smooth out the noise contributed by the time links, such as through the GPS, and the basic noise of the clocks. The monthly BIPM *Circular T* then alternates between provisional values, based on only 30 days of data, and the final definite values, based on the full 60 days of data. There are four centers in the United States that maintain realizations of UTC, and the difference between the values they generate in real time and the final UTC values determined after the fact are shown in **Fig. 4**.

Future of UTC. In 1999 issues were raised concerning the use of UTC as a stepped atomic time scale within the Radiocommunication Sector of the ITU (the ITU-R) and the Consultative Committee for Time and Frequency (the successor body of the CCDS). Since the ITU-R is responsible for the definition of the UTC, any major change could have a potentially significant impact on the synchronization of communications networks, navigation systems, and time distribution performance. Consequently, an ITU-R Special Rapporteur Group (SRG) was formed to investigate the possible future of UTC with particular emphasis on the leap second. The SRG was specifically tasked to study the requirements for globally accepted time scales for use in navigation and telecommunications systems and civil timekeeping. in addition, it was tasked to investigate the present and future requirements for the tolerance limit between UTC and UT1 and to explore whether leap second procedures satisfy user needs or if an alternative procedure should be developed.

The SRG held numerous meetings from 2000 to 2006 to attempt to make people aware of the issues and concerns, and held a general colloquium on the subject in Turin, Italy. Their activities prompted considerable discussion within the ITU-R and the timekeeping community but little outside that comparatively small international community. Analyses of Earth rotation by geophysical scientists were conducted which indicated that in the future multiple leap seconds per year will be required. The SRG completed its studies in 2006 and reported to the ITU-R, but technical discussions have continued within the ITU-R on eliminating the leap second so UTC would become a continuous time scale. A consensus on the proper course of action has yet to be reached.

For background information *see* ATOMIC CLOCK; ATOMIC TIME; DYNAMICAL TIME; EARTH ROTATION AND ORBITAL MOTION; EQUATION OF TIME; RADIO TELESCOPE; RELATIVITY; SATELLITE NAVIGATION SYSTEMS; TIME in the McGraw-Hill Encyclopedia of Science & Technology. Ronald L. Beard

Bibliography. E. F. Arias, The metrology of time, *Phil. Trans. R. Soc. London*, A363:2289–2305, 2005; E. F. Arias, B. Guinot, and T. J. Quinn, Rotation of the Earth and time scales, in Bureau International des Poids et Mesures, *Proceedings of ITU-R Special Rapporteur Group Colloquium on the UTC Time Scale*, Torino, Italy, May 28–29, 2003; C. Audoin and B. Guinot, *The Measurement of Time*, Cambridge University Press, 2001; Bureau International des Poids et Mesures, *BIPM Annual Report on Time Activities*, vol. 4, 2009; IEEE staff, *2008 IEEE International Frequency Control Symposium*, IEEE, 2008; J. Levine, Introduction to time and frequency metrology, *Rev. Sci. Instrum.*, 70:2567–2596, 1999; W. Lewandowski, Introduction, *Report of the 48th Civil GPS Service Interface Committee Meeting—Timing Subcommittee*, BIPM, 2008; R. A. Nelson et al., The leap second: its history and possible future, *Metrologia*, 38:509–529, 2001; *Proceedings of the 41st Annual Precise Time and Time Interval (PTTI) Applications and Planning Meeting*, U.S. Naval Observatory, Washington, D.C., 2009.

CRISPR-based immunity in prokaryotes

Recurrent viral attacks on prokaryotes have been a driving force for the evolution of immune mechanisms in these organisms that are crucial for their survival in the presence of these infectious elements. Viruses abuse bacteria to multiply, which often results in the death of the microbe. Therefore, bacteria have devised multiple defense strategies to resist phage predation. A relatively simple one is based on the elimination or modification of host proteins that are used by the virus as a surface receptor. Although this strategy prevents virus absorption and injection of the viral DNA, it may lead to an impaired fitness of the microbe. A second line of defense is formed by the well-studied restriction-modification system, which consists of an endonuclease enzyme and a methylase enzyme. While the host DNA is protected by methylation, invading DNA sequences without this methylation pattern are susceptible to cleavage by an endonuclease. However, the progeny of a single undetected virus particle will have a methylation pattern identical to the host and will no longer be recognized as an invader. Hence, an additional layer of defense against viral infection is based on clustered regularly interspaced short palindromic repeats (CRISPRs), which form the core of a sophisticated prokaryotic immune system that is both adaptive and inheritable.

CRISPR loci are present in 40% and 90% of the sequenced bacterial and archaeal genomes, respectively. A CRISPR locus consists of partially palindromic repeats (reading the same way forward as backward) of typically 30 nucleotides (nt) that are separated by similar-sized virus- and plasmid-derived sequences (spacers), which serve as a genetic memory of previously encountered invaders. The repeat sequence can differ between organisms as well as between different CRISPR arrays within one genome. The CRISPR array is transcribed into a long precursor CRISPR ribonucleic acid (pre-crRNA), which is subsequently cleaved in the repeat sequences to yield mature crRNAs. These small RNAs are used as a virus recognition tool to guide host defense; the presence of a spacer that matches the DNA sequence of an incoming phage renders the phage susceptible to detection and subsequent elimination. The size of one CRISPR locus ranges from a few to hundreds of spacer/repeat units. CRISPR loci evolve rapidly in natural ecosystems through frequent spacer acquisition and spacer loss. In fact, a CRISPR array can be seen as a record of the most recent infections that a lineage has overcome. The CRISPR-spacer hypervariability between closely related strains has been exploited for strain typing of the pathogenic bacteria *Mycobacterium tuberculosis* and *Corynebacterium diphtheriae* using a technique called spoligotyping.

CRISPR-associated (*cas*) gene clusters are located in close proximity of the CRISPR arrays and encode the protein machinery that carries out the various steps of the interference pathway. Multiple types of *cas* gene clusters have been recognized, with each containing from 4 up to almost 20 *cas* genes. Both the substantial differences in *cas* gene compositions between closely related species and the *cas* gene similarities between distantly related species suggest extensive horizontal gene transfer of *cas* genes. All CRISPR-harboring species contain the core genes *cas1* and *cas2* and usually a number of subtype-specific *cas* genes named after an organism in which they occur.

Mechanism of CRISPR defense. Despite the high degree of variation between the CRISPR/Cas systems, the biochemical processes that underlie the immune system are similar. The mechanism of the CRISPR/Cas immune response can be divided into three stages—(1) CRISPR adaptation: During this stage, new spacers are added to the existing CRISPR array. (2) CRISPR expression: The array is transcribed into a long precursor CRISPR RNA (pre-crRNA) transcript and processed to yield small crRNAs. These mature crRNAs, with the size of a single spacer and repeat, are retained by a complex of Cas proteins. (3) CRISPR interference: When infection takes place by a virus that is archived in the CRISPR blacklist, the crRNA-loaded protein complex uses the crRNA as a guide to recognize the invading DNA. The elimination of the viral DNA is carried out by the Cas protein complex alone or together with additional Cas proteins (**Fig. 1**).

CRISPR adaptation. The adaptation stage has been studied in most detail in lactic acid bacteria. After phage challenge, bacteriophage-insensitive mutants (BIMs) of *Streptococcus thermophilus* acquire resistance against subsequent infection with the same phage because of the presence of new spacer sequences in their CRISPR loci. Spacer acquisition

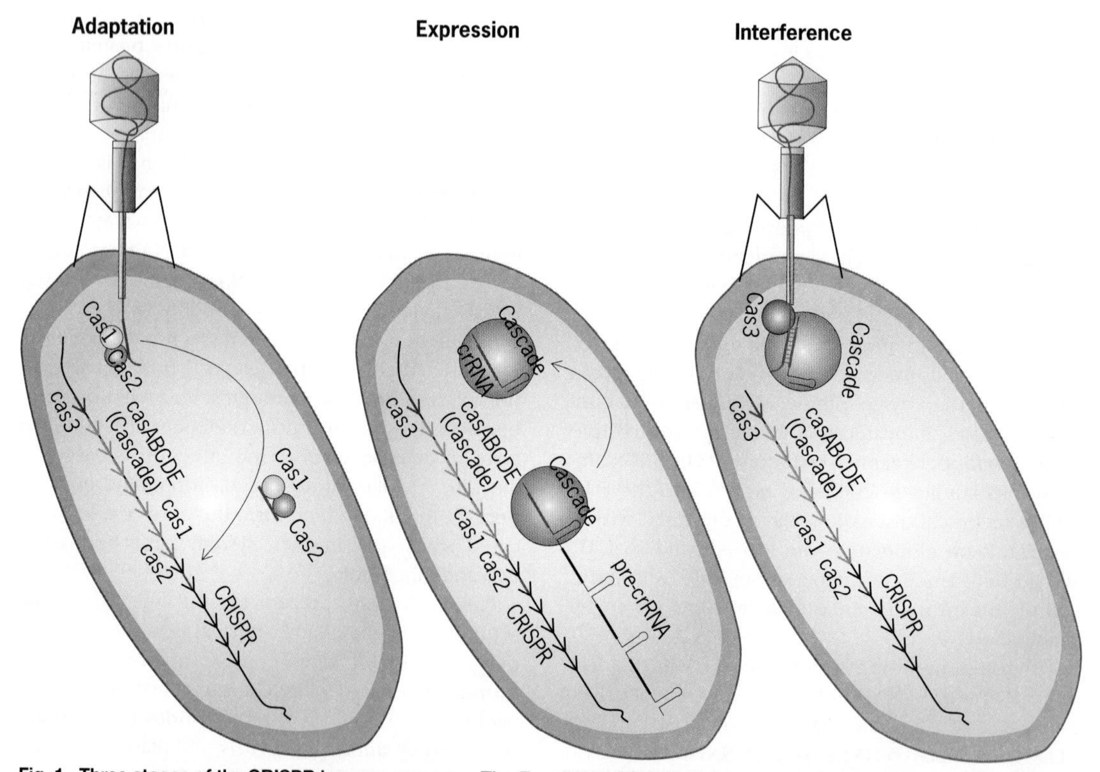

Fig. 1. Three stages of the CRISPR immune response. The *E. coli* K12 CRISPR/Cas locus consists of eight *cas* genes, which encode the machinery to carry out each of the three stages of the CRISPR immune response. On phage infection, Cas1 and Cas2 are thought to mediate the integration of new spacer sequences in the existing CRISPR array. During the subsequent expression stage, the Cascade protein complex is formed and the CRISPR locus is transcribed into a long pre-crRNA molecule. Cascade mediates the maturation of the pre-crRNA transcript by cleavage within the repeat sequence, at the base of each stem-loop. The mature crRNA remains bound to Cascade, and the crRNA-loaded Cascade specifically binds the double-stranded viral DNA during the CRISPR interference stage. In concert with Cas3, the elimination of the virus is accomplished, probably through cleavage of the viral DNA.

occurs in a polarized fashion; new spacers are added at one end of the CRISPR array, whereas spacers are lost at the opposite end of the array. Consequently, a CRISPR array forms a chronological record of bygone infections. Four *cas* genes reside adjacent to the most-active CRISPR locus of *S. thermophilus*, namely, *csn1*, *cas1*, *cas2*, and *csn2*. Inactivation of the species-specific *csn2* gene results in an inability to integrate new spacers, whereas gene disruption of *csn1* abolishes CRISPR interference regardless of the presence of a virus-derived spacer sequence.

Escherichia coli contains eight *cas* genes, of which only *cas3* and the subtype-specific *casABCDE* genes are required for the CRISPR interference stage (see below). The remaining two *cas* genes, *cas1* and *cas2*, are therefore anticipated to play a role in other stages, such as during spacer integration. Cas1 from *Pseudomonas aeruginosa* has been shown to be a DNA-specific endonuclease that generates approximately 80-bp (base pair) products, possibly the precursors of new spacers. The precise mechanistic details of spacer acquisition are still lacking.

CRISPR expression. The CRISPR array is transcribed into a long pre-crRNA transcript with a predicted stem-loop structure (shown in **Fig. 2**) due to the palindromic nature of the repeat sequences. The pre-crRNA is processed through single endonucleolytic cleavage within each repeat sequence at the base of the stem-loop structure. In *E. coli* K12, the pre-crRNA is cleaved by the endoribonuclease CasE, which is part of a Cas protein complex named Cascade (Cas-complex for antiviral defense). This protein complex consists of five different subunits, CasABCDE, and retains a single crRNA of 61 nt. In *Pyrococcus furiosus*, Cas6 is the pre-crRNA processing enzyme. Comparison of the crystal structures of CasE from *Thermus thermophilus* and Cas6 from *P. furiosus* shows that these proteins have a similar architecture, although CasE has a single catalytic histidine residue and Cas6 has a catalytic triad (His, Tyr, Lys).

In *P. furiosus*, additional 3′-trimming of the crRNA produces mature crRNA of approximately 40 nt. These mature crRNAs are transferred to a Cas protein complex that is composed of six proteins of the so-called RAMP (repeat-associated mysterious protein) family. The formation of ribonucleoprotein complexes composed of mature crRNA and several Cas proteins seems a common theme in CRISPR-based defense, even when variation exists in the Cas proteins and crRNA processing.

Little is known about the regulation of CRISPR defense. In *E. coli* K12, expression of the CRISPR and *cas* genes is repressed under normal growth conditions. Events associated with virus infection, including receptor binding and DNA injection, are prime candidates to instigate an induction of the CRISPR defense pathway. So far, only in *T. thermophilus* has the expression of *cas* genes been shown to be induced upon phage infection.

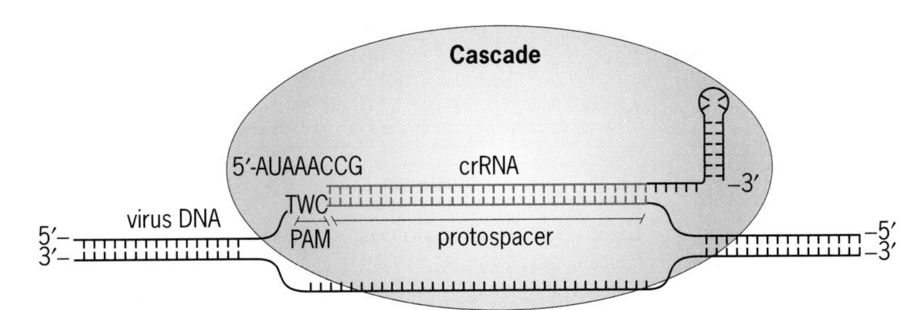

Fig. 2. ***E. coli*** **Cascade binding of viral double-stranded DNA. The mature crRNA consists of a spacer that is flanked by repeat sequences on either side. The stem-loop structure of the repeat is located at the 3′-end of the spacer sequence. Cascade binds a protospacer on the viral genome through base-pairing of the crRNA with the complementary viral DNA. The 3-nt protospacer adjacent motif (PAM) is located at the 3′-end of the protospacer and forms a single base pair with a residue of the 5′ crRNA-repeat sequence. The presence of the PAM is needed for CRISPR interference.**

CRISPR interference. Evidence is accumulating that some CRISPR/Cas systems target viral and plasmid DNA, while others target viral and plasmid messenger RNA (mRNA). In both *E. coli* and *S. thermophilus*, the presence of virus-derived spacer sequences confers immunity, and the complementary target sequence can be located either on the coding strand or on the template strand, which disqualifies mRNA as the potential target. In *Staphylococcus epidermidis*, the targeting of a noncoding sequence on a plasmid confers immunity. Although target DNA degradation has not been observed yet, these convincing lines of evidence point toward DNA as the prime target in these CRISPR/Cas subtypes.

The crRNA-loaded Cascade complex from *E. coli* is capable of binding double-stranded target DNA, but this is not sufficient for resistance (Fig. 2). In addition to Cascade, the predicted nuclease and helicase Cas3 is required for resistance, presumably to carry out target DNA cleavage after the target site has been located.

Although the above-mentioned findings have suggested that all CRISPR/Cas systems target DNA directly, a whole new variation on the CRISPR interference theme has emerged from studies done on *P. furiosus*. In vitro, the RAMP family Cas protein complex loaded with a mature crRNA is capable of cleaving complementary single-stranded RNA. The observed cleavage takes place at a well-defined position, 14 nt from the 3′-end of the crRNA, which is mechanistically similar to RNA interference (RNAi)–based defense against virus infections in plants.

During CRISPR interference, it is essential to distinguish self (the genomic CRISPR locus) from nonself (the target DNA). Recent analyses have revealed that this is determined by the sequence that flanks the complementary target DNA. The CRISPR/Cas system from *S. epidermidis* recognizes self DNA by base-pairing between the crRNA repeats and CRISPR repeats on the genome. In contrast, the target is identified by nonbase-pairing of the regions flanking the targeted sequence (protospacer) with the repeats of the crRNA (**Fig. 3**).

Other CRISPR/Cas subtypes require the presence of a 2- to 4-nt motif adjoining the protospacer, known as the protospacer adjacent motif (PAM) [Fig. 2].

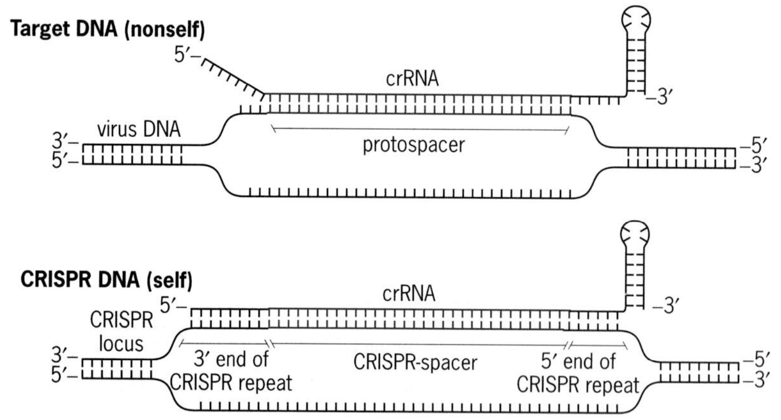

Fig. 3. *Staphylococcus epidermidis* self versus nonself discrimination during CRISPR interference. In *S. epidermidis*, the CRISPR locus is distinguished from target DNA through base-pairing of the repeat sequence of the crRNA with the residues that flank the complementary DNA sequence. CRISPR interference takes place only when no base-pairing occurs between crRNA and the flanks.

During CRISPR adaptation, new spacers are probably selected such that the targeted sequence contains a downstream PAM. Phages can evade CRISPR interference through single-nucleotide mutations in the protospacer or through mutation of the PAM sequence. Whether some phages contain more sophisticated systems to block CRISPR interference is currently unknown.

Applications. Apart from strain typing, CRISPR can be used for several applications. CRISPR-based BIMs of lactic acid bacteria are already being used in the food industry for the generation of dairy products. The fact that the incorporation of new spacer sequences occurs naturally after phage challenge makes these BIMs very convenient and legally allowed components of starter cultures in food fermentations. Pharmaceutical and chemical industries that rely on bacteria for the production of fine chemicals could also benefit from genetically engineered bacteriophage-resistant strains to prevent costly culture losses due to phage infections.

For background information *see* ARCHAEA; BACTERIA; BACTERIOPHAGE; CELLULAR IMMUNOLOGY; DEOXYRIBONUCLEIC ACID (DNA); FOOD FERMENTATION; GENOME; IMMUNITY; PLASMID; PROKARYOTAE; RIBONUCLEIC ACID (RNA); VIRUS in the McGraw-Hill Encyclopedia of Science & Technology.

Edze R. Westra; John van der Oost; Stan J. J. Brouns

Bibliography. S. J. J. Brouns et al., Small CRISPR RNAs guide antiviral defense in prokaryotes, *Science*, 321:960–964, 2008; P. Horvath and R. Barrangou, CRISPR/*Cas*, the immune system of Bacteria and Archaea, *Science*, 327:167–170, 2010; F. V. Karginov and G. J. Hannon, The CRISPR system: Small RNA-guided defense in bacteria and archaea, *Mol. Cell*, 37:7–19, 2010; L. A. Marraffini and E. J. Sontheimer, CRISPR interference: RNA-directed adaptive immunity in bacteria and archaea, *Nat. Rev. Genet.*, 11:181–190, 2010; J. van der Oost et al., CRISPR-based adaptive and heritable immunity in prokaryotes, *Trends Biochem. Sci.*, 34:401–407, 2009.

Cyber defense

Cyber defense is the practice of defending important computer-based information systems and networks from attack. Of primary concern is the protection of the information as it is stored and processed by computer systems and transmitted across networks. The three most important aspects of information that need to be protected, often referred to as the security triad, are its confidentiality, integrity, and availability.

Confidentiality means that only authorized individuals have access to the information. A credit card number is an example of information that most people would wish to keep confidential. When making an online purchase, a customer needs to transmit a credit card number to the merchant so that the merchant can charge the customer's card. The customer should be concerned about the confidentiality of the credit card information to be sure that no one else (perhaps eavesdropping on the transaction) can learn the credit card number and then use that information to make unauthorized purchases. In this situation, the confidentiality of the customer's information is typically protected by using a secure version of the Hypertext Transfer Protocol (HTTP), which uses a cryptographic algorithm, such as the Advanced Encryption Standard (AES), to encrypt the data sent between the customer and the merchant. Web pages that use the secure version of HTTP have URLs (uniform resource locators) that begin with HTTPS rather than HTTP. The encrypted channel between the customer and the merchant created by HTTPS ensures that it is not feasible for an eavesdropper to decrypt the messages and discover the customer's credit card number and other information.

In addition to confidentiality, the integrity of information is sometimes of great concern. Integrity means that only authorized individuals may modify the information. Blood type is an example of information that most people would not seek to keep confidential, but the integrity of which everyone should be concerned about. If someone were able to change a patient's blood type in a medical record, there is a good chance that the wrong type of blood would be used during a transfusion, with serious consequences. The integrity of information can also be protected with cryptography, as most cryptosystems not only render a message unintelligible to eavesdroppers, but also result in a message that will not decrypt properly if it has been modified in transit. Using a cryptosystem to protect both the confidentiality and the integrity of a message can be quite valuable. But as with our blood type example, there are situations in which only the integrity and not the confidentiality of information is required. In these situations, encryption is not necessary, and a different type of technique called a message digest (also called a cryptographic hash function) is used.

A message digest algorithm takes a message of arbitrary length and condenses it to a fixed-size digest (typically 128 or 256 bits). The digest has several important properties. First, given a particular message, it is easy to compute its digest. There are a number of widely accepted message digest algorithms [for example, MD5 (Message-Digest algorithm 5) and SHA-256 (Secure Hash Algorithm 256)], and using the same algorithm and the same message should always result in the same message digest. Second, given only the digest, it is not feasible to create a message that produces that digest. Third, it is not feasible to create a collision—two different messages that have the same digest. These properties of message digest functions protect the integrity of a message without necessarily protecting its confidentiality. For example, we can compute a digest using a patient's blood type and compare it to a stored digest. If the digests match, then the blood type information has not changed. If the digests do not match, then it has.

In addition to confidentiality and integrity, the availability of information can also be important. Availability means that information can be accessed by authorized users in a timely manner. In the case of someone needing an emergency blood transfusion, having no blood-type information (or information that takes several minutes to locate) could prove just as harmful as having information that was maliciously modified. The most common techniques used to protect the availability of information are redundancy and fault tolerance. However, even with these mechanisms in place, denial of service attacks designed to degrade the availability of information are notoriously easy to mount and difficult to combat.

As discussed earlier, when we talk about protecting information on computer systems and networks, we are usually concerned with maintaining appropriate levels of confidentiality, integrity, and availability. However, the goal of cyber defense is not to protect information for its own sake. The ultimate goal of cyber defense is to protect the people, institutions, and infrastructure that have come to depend upon digital information and the computer systems and networks that store, process, and exchange that information. There are numerous examples of how we now rely heavily on digital information, computer systems, and networks for such things as communications, financial transactions, and management and control of public utilities. A successful attack in one or more of these areas could cause a nuisance, a serious problem, or even a catastrophe. The goal of cyber defense is to prevent attacks from happening. Since it is not possible to prevent all possible attacks, cyber defense also seeks to minimize the damage when an attack does occur. Therefore, cyber defenders are those who undertake activities to prepare for an attack, protect the computer systems and networks they are defending, detect attacks, perform triage, and respond to attacks.

Preparation. Preparation is an important and ongoing component of cyber defense. Cyber defenders must regularly prepare themselves to deal with future attacks effectively. They must be knowledgeable about the computer systems and networks that they defend so that they understand how those systems and networks function, how they might be attacked, and how they can be repaired. In addition to accumulating knowledge about the technology in use, cyber defenders should acquire and practice using the tools and equipment they will use to combat an attack. If working as part of a team, a cyber defender should know what his or her role will be during an attack and be trained for that role. Team communication and cooperation will be critical during an attack, so team members should not only understand their own roles, but also know the roles of the other team members and when and how to contact them.

Protection. Another key function of cyber defense is the protection of the information, computer systems, and networks that are being defended. Proper protection can prevent many attacks, and it is often far easier to prevent attacks from occurring than it is to recover from these attacks after the fact. Like preparation, protection should be an ongoing process. Some of the steps that cyber defenders can take to protect their systems include using firewalls and other security appliances, educating users about common threats, and applying updates and patches to operating systems and software. It is particularly important that the last item be done automatically, as many updates and patches close serious security holes that have been discovered in products, and attackers usually target a vulnerability very quickly after a patch for it is released.

Detection. The actual handling of an attack usually cannot begin until the attack is detected. There are numerous mechanisms, some technical and some nontechnical, that can alert cyber defenders that an attack has occurred or is ongoing. Intrusion detection systems or other host and network monitoring devices are often used to detect attacks. Periodic reviews of the log files and other records of system use might uncover evidence of an attack. Occasionally, a media report, blog posting, or other information source will alert a cyber defender to the possibility of a new or emerging threat. One of the most common ways in which attacks are detected is by users of the system, who notice that something is not working properly or appears abnormal. Since even nontechnical users can be helpful in detecting attacks, they should be encouraged to report anything that seems suspicious, and their reports should be taken seriously. Some user reports will probably be false alarms, but others will be an early indication of serious problems.

Triage. In cyber defense, triage is the process of prioritizing attacks (when there are more than one) to allocate resources effectively. Triage may not always be necessary; for instance, there may be plenty of resources and therefore no need to neglect a lower-priority task in favor of a higher-priority one. However, cyber defenders often do not find themselves with an abundance of resources and will have to decide which attacks (and which symptoms of those attacks) to deal with first. Most of the time, cyber defenders will choose to focus first on mission-critical systems and services at the expense of other, less-important systems and services.

Response. Response encompasses all the actions that cyber defenders might take to contain an attack, repair the damage it has done, and restore operations to a properly functioning and secure state. Response actions may include technical activities such as improving security settings, recovering data that have been modified or deleted during the attack, and identifying and removing malicious processes and files from computer systems. Response actions might also include managerial actions such as changing a policy to prevent future attacks or terminating an employee who was found to be responsible for an attack. Some response activities might take the form of legal actions such as working with appropriate law enforcement agencies and prosecuting those responsible for the attack. While many cyber defenders will not have the legal training to lead such efforts, they may be relied upon to provide technical support for such things as tracking the attack back to an attacker, performing forensic analysis, or analyzing malware.

Cyber defense training exercises. In addition to deep technical knowledge, cyber defense requires many practical and organizational skills. Being able to work under pressure and as part of a team are vital. For this reason, several cyber defense exercises and competitions have been developed to allow cyber defenders to hone their skills in a realistic environment. One of the first was the Cyber Defense Exercise (CDX), which has been held annually since 2001. The CDX involves all five U.S. service academies (U.S. Air Force Academy, U.S. Coast Guard Academy, U.S. Merchant Marine Academy, U.S. Military Academy, and U.S. Naval Academy). Students are responsible for designing, deploying, and defending a set of computer systems and networks that are attacked by a team of security professionals.

In 2006, an annual Collegiate Cyber Defense Competition (CCDC) was begun so that all colleges and universities in the United States could compete, evaluate their cyber defense curricula, and give their students some cyber defense experience. In the CCDC, student teams do not design or build their network infrastructure or systems. Instead, the teams are given an existing, functional network and asked to improve its security (which is very weak to begin with) and defend it from attacks by a team of security professionals.

In February 2006, the U.S. Department of Homeland Security held the first of a biennial exercise series called "Cyber Storm," in which government employees were asked to defend against a simulated cyber attack on the nation's critical infrastructure. The simulation included a variety of attacks against communications, transportation, and public utilities. The U.S. military, as well as the international military community, has also created its own cyber defense exercises. Examples include the Black Demon exercise, which was created by the U.S. Air Force in 2000, and the semiannual International Cyber Defense Workshop, which is held each spring and fall. Another famous cyber defense exercise is the "Capture the Flag" competition, held each year since 1996 at the DEF CON® hacker convention.

Future developments. Attackers currently target corporations, organizations, and individuals, usually with the goal of profiting from the attack. These types of attacks are fairly common and can be quite damaging to the victims. Coordinated cyber attacks for political reasons have occurred (on Estonia in 2007, Georgia in 2008, and South Korea and the United States in 2009), but thus far they have caused only relatively minor damage. That may change in the near future, as cyber terrorism and cyber warfare are likely to occur. Cyber defense, and an appropriate number of properly trained cyber defenders, therefore, will become even more important.

For background information *see* COMPUTER SECURITY; CRYPTOGRAPHY; DATA COMMUNICATIONS; DIGITAL EVIDENCE; INTERNET; OPERATING SYSTEM; SOFTWARE; WORLD WIDE WEB in the McGraw-Hill Encyclopedia of Science & Technology.

Brett Tjaden; Robert Floodeen

Bibliography. Honeynet Project, *Know Your Enemy: Learning about Security Threats*, 2d ed., Addison-Wesley Professional, Boston, 2004; S. McClure, J. Scambray, and G. Kurtz, *Hacking Exposed: Network Security and Solutions*, 6th ed., McGraw-Hill/Osborne, Emeryville, CA, 2009; C. P. Pfleeger and S. L. Pfleeger, *Security in Computing*, 4th ed., Prentice Hall, Upper Saddle River, NJ, 2006; C. Prosise, K. Mandia, and M. Pepe, *Incident Response and Computer Forensics*, 2d ed., McGraw-Hill/Osborne, New York, 2003.

Decarboxylative couplings

A decarboxylative coupling is a chemical reaction in which a carboxylic acid extrudes a molecule of carbon dioxide (CO_2) and in its place forms a bond with an electrophilic carbon atom or heteroatom. The reaction can be employed, for example, in the regioselective synthesis of biaryls from aromatic or heteroaromatic carboxylates and aryl halides or pseudohalides, or for the analogous synthesis of aryl ketones from α-oxocarboxylates [reaction (1)].

$$\mathrm{R{-}C(=O){-}OH} + \mathrm{X{-}R^1} \xrightarrow[-\mathrm{HX}]{\text{catalyst (base, oxidant)}} \mathrm{R{-}R^1} + \mathrm{CO_2} \qquad (1)$$

In these couplings, the carboxylic acid serves as the source of a carbon nucleophile and reacts with carbon electrophiles. However, by the addition of a stoichiometric oxidant, the reactivity of the system can be turned around so that carboxylic acids may also undergo decarboxylative coupling with nucleophiles. Both the decarboxylation and the coupling steps are mediated by metal catalysts, such as palladium, copper, or silver complexes.

Decarboxylative cross-couplings can be used in place of cross-coupling reactions involving organometallic reagents. They are known as regiospecific, because the formation of the new bond occurs at the position predefined by the carboxylate group. However, the use of inexpensive and broadly available carboxylate salts instead of expensive and sensitive preformed organometallic reagents is ecologically and economically beneficial.

Carboxylic acids as sources of carbon nucleophiles. The principal aim of the development of this new synthetic method was to refine and extend the existing network of synthetic transformations. In recent times, it has become more important to integrate new raw materials, especially from renewable sources, and to improve the performance and sustainability of available synthetic steps to save raw material and energy resources. In this context, carboxylic acids represent particularly advantageous substrates, as they are widely available from natural sources or can be generated via sustainable methods. They are commercially available in a large structural variety, easy to store, and simple to handle. In recent years, various catalytic transformations have been developed that substantially extend the spectrum of applications that carboxylic acids have as building blocks in organic synthesis—for example, as substitutes of aryl-, alkyl-, or acyl halides and of organometallic reagents.

The decarboxylation step. In principle, any carboxylate salt can be turned into the corresponding organometallic reagent simply by an extrusion of CO_2. The difficulty connected to this process is that metal salts of carboxylates generally require extreme temperatures to lose CO_2, and that under such conditions, the resulting organometal species are likely to undergo fast protonation by the surrounding medium, giving the corresponding protonated products before a carbon-carbon or carbon-heteroatom bond is successfully formed.

Living organisms, which generally lack an air- and water-free environment, have long evolved to generate carbanion equivalents by straightforward enzymatic decarboxylation of ubiquitously available carboxylic acid derivatives, including the carboxylate-containing nucleotide orotidine 5′-monophosphate and certain pyruvates.

Protodecarboxylation. The simplest reaction that occurs under extrusion of CO_2 is the protodecarboxylation. In the absence of a catalyst, only a few carboxylic acids are known to decarboxylate at reasonable temperatures, for example, β-ketocarboxylic acids. Benzoic acids and benzoate salts of most metals are almost inert. Only when coordinated to Hg, Ag, or Cu cations do they lose CO_2 at elevated temperatures. In mechanistic studies of such protodecarboxylation reactions, M. Nilsson proved the intermediacy of an aryl-copper species (at 250°C) by capturing some of it with excess iodobenzene. This copper reaction-mediated precursor can be seen as the predecessor of a decarboxylative cross-coupling reaction (2).

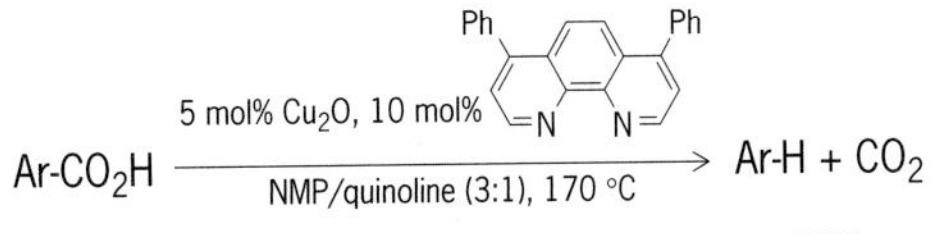

$$Ar\text{-}CO_2H \xrightarrow[\text{NMP/quinoline (3:1), 170 °C}]{\text{5 mol\% }Cu_2O\text{, 10 mol\% (Ph-phenanthroline)}} Ar\text{-}H + CO_2 \quad (2)$$

Ar = m-NO_2-C_6H_4 (89%); p-MeO-C_6H_4 (80%); p-NO_2-C_6H_4 (68%); p-CHO-C_6H_4 (65%); 2-NO_2-5-MeO-C_6H_3 (90%); p-HO-C_6H_4 (75%); p-MeC(O)-C_6H_4 (75%); p-MeC(O)N-C_6H_4 (76%); ...

Because of their model character for more complex decarboxylative couplings, protodecarboxylations have been studied intensively in recent years. The most generally applicable decarboxylation catalysts consist of phenanthroline-ligated copper complexes, and require temperatures of around 160°C. Silver catalysts are active below 120°C, but their substrate spectrum is limited to certain heteroarenecarboxylic and ortho-substituted benzoic acids. Palladium and rhodium catalysts operate under the mildest conditions, but have the strongest limitations with regard to their substrates.

Synthesis of biaryls via decarboxylative couplings. The scope of decarboxylative cross-couplings mediated by copper or silver alone is limited, because these metals are effective catalysts for the extrusion of CO_2, but much less so for selective cross-coupling processes involving oxidative addition and reductive elimination steps. In contrast, effective cross-coupling catalysts, such as palladium complexes, promote the decarboxylation of only a few particularly activated carboxylates. The first catalytic decarboxylative cross-coupling protocol involved a bimetallic catalyst system consisting of a decarboxylation and a cross-coupling catalyst. The **illustration** shows the proposed reaction mechanism.

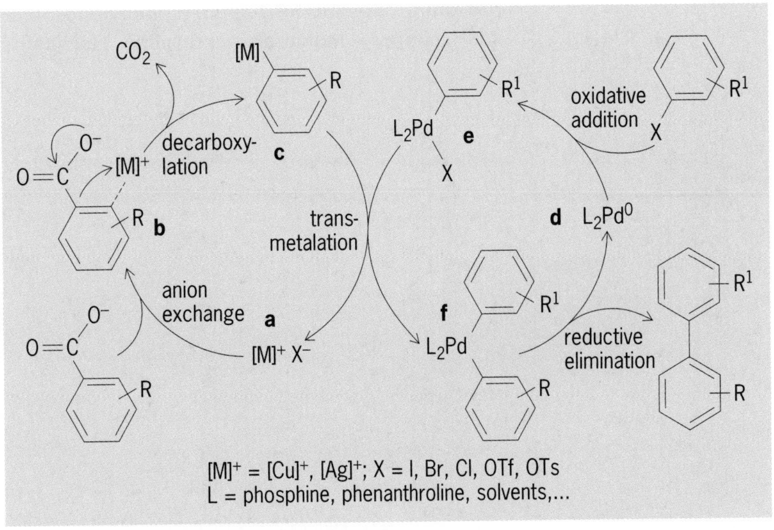

$[M]^+ = [Cu]^+, [Ag]^+$; X = I, Br, Cl, OTf, OTs
L = phosphine, phenanthroline, solvents,...

Decarboxylative cross-coupling. In the salt exchange reaction, the carboxylate group is transferred to the decarboxylation catalyst a, a copper or silver complex. By extrusion of CO_2, an organocopper (or silver) species b is generated. The carbon residue is then transferred onto an arylpalladium species e, generated in the reaction of the aryl electrophile with the second catalyst component, a low-valent palladium(0) species d. The carbon-carbon bond is formed in a reductive elimination from f, regenerating the original palladium species d. For further turnover of the decarboxylation catalyst to occur, the copper salt formed in the transmetallation step needs to undergo a salt metathesis (exchange) with the next carboxylate molecule.

The initial catalyst system consisted of 3 mol% copper(I) iodide, 5 mol% 1,10-phenanthroline, and 1 mol% $Pd(acac)_2$ [acac = acetylacetonate]. It allowed the coupling of ortho-nitrobenzoic acids with a number of bromo-, iodo-, and even some chloroarenes in the presence of potassium carbonate at 160°C in NMP (*N*-methyl-2-pyrrolidone) as the solvent. The reaction product water was removed by either azeotropic distillation or the addition of molecular sieves.

With the use of slightly higher loadings of the copper co-catalyst, a broader range of carboxylic acids, including ortho-substituted benzoic acids and heterocyclic derivatives, were effectively converted into the corresponding unsymmetrical biaryls [reaction (3)].

Subsequent mechanistic studies have indicated that these initial restrictions with regard to the carboxylic acid substrates are not intrinsic but a consequence of the strong affinity of the copper catalyst to the halide ions released in the cross-coupling step. This makes the salt metathesis (exchange reactions) unfavorable for nonactivated carboxylates. Two strategies have been considered to overcome this hurdle: one is to tune the decarboxylation catalyst to induce a preference for carboxylate over halide counter-ions, and the other is to use noncoordinating sulfonates as leaving groups at the electrophilic coupling partner.

Improved catalyst systems. The use of palladium complexes with customized phosphine ligands has made this reaction applicable to a broad spectrum of electrophilic coupling partners, including many aryl and heteroaryl halides and pseudohalides. Most common functionalities other than unprotected primary or secondary amines, alcohols, and thiols are tolerated. A catalyst system composed of palladium iodide/bis(*t*-butyl)-biphenylphosphine and copper iodide/phenanthroline facilitates, for example, the coupling even of notoriously nonreactive electron-rich chloroarenes in high yields. Another customized palladium catalyst consisting of $Pd(acac)_2$/tol-BINAP [tol-BINAP = 2,2′-bis(di-p-tolylphosphino)-1,1′-binaphthyl] allows the coupling of aryl triflate electrophiles. Because the triflate ($CF_3SO_3^-$) ions released in the process are unable to block the carboxylates from the coordination sphere of the decarboxylation catalyst regardless of their substitution patterns, a broad range of carboxylate substrates can be converted, including meta- and para-substituted benzoates. Two alternative protocols have been reported, one involving conventional heating (160°C for several hours) and the other microwave heating (190°C for 5–10 min). The latter is higher-yielding, particularly for deactivated carboxylates. Further improvement of the palladium catalysts allowed the use of the inexpensive but unreactive tosylates ($CH_3C_6H_4SO_3^-$) as the electrophilic substrates. The best results were obtained using 5 mol% $Pd(acac)_2$ and 7.5 mol% XPhos (2-dicyclohexylphosphino-2′,4′,6′-triisopropylbiphenyl) in combination with microwave heating (190°C/150 W/5 min) [reaction (4)].

The temperatures required by the abovementioned catalyst systems for decarboxylative cross-couplings remain high. Guided by density functional theory (DFT) studies, more active silver-based catalysts were developed that promote protodecarboxylations already at 80–120°C. Combining these systems with suitable palladium catalysts permitted a reduction of the reaction temperature to 120°C. However, because of the high affinity of silver to halides, only aryl triflates can so far be used as electrophilic coupling partners.

Applications in synthesis. Decarboxylative couplings have been employed successfully in large-scale syntheses of commercially important biaryls, for ex-

CO_2H R + Br R^1 —cat. $PdBr_2$, cat. Cu_2O/phenanthroline, K_2CO_3, 170°C, 24h, NMP/quinoline→ R^1 R

NO_2 R^1; Ac 69%; F 76%; CHO 61%; SO_2Me 42% (3)

R^1 = 4-Me (99%), 4-Cl (94%), 4-CHO (91%), 4-Ac (77%), 4-MeO (95%), 4-CO_2Et (96%), 4-MeS (98%), 4-CF_3 (93%), 4-CN (97%) (80% for Ar-Cl), 4-NO_2 (91%) (86% for Ar-Cl)

79%; S 62%

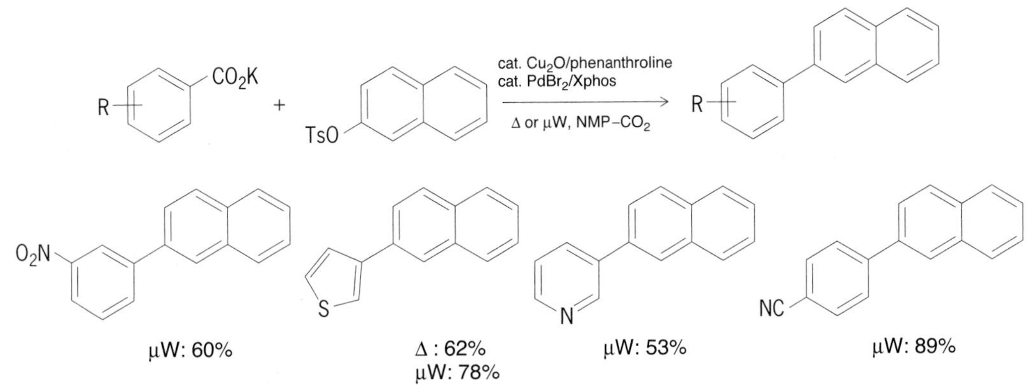

(4)

µW: 60% Δ : 62% µW: 78% µW: 53% µW: 89%

ample, the key intermediates in the syntheses of the agrochemicals boscalid [reaction (5)]

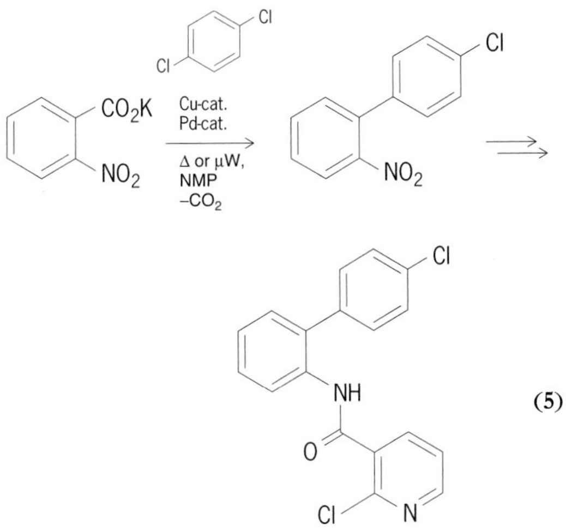

(5)

and bixafen. They have also been employed in concise syntheses of the angiotensin inhibitors valsartan and telmisartan.

Synthesis of aryl ketones. The reaction principle of decarboxylative cross-coupling is not restricted to the synthesis of biaryls. It can also be used for other substrate classes. A particularly interesting example is the decarboxylative coupling of α-oxocarboxylic acids with aryl halides or pseudohalides to give aryl ketones, as in reaction (6),

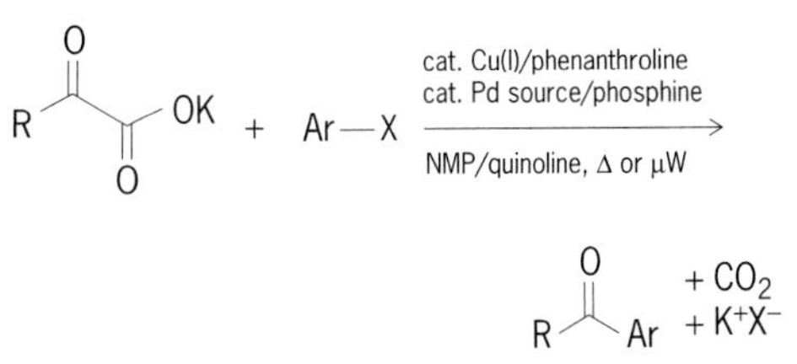

(6)

where X = halide or sulfonate. The novelty of this pathway is that it involves the generation and coupling of unprotected acyl anion equivalents, thereby reversing the polarity of the bond formation of traditional ketone synthesis.

Monometallic catalyst systems. As mentioned earlier, palladium complexes are capable of promoting the decarboxylation of some particularly activated carboxylates. It is thus possible to promote certain decarboxylative cross-couplings with palladium alone. For example, five-membered heteroarenes with carboxylate groups in the 2-position can be coupled with various aryl halides under formation of the corresponding arylated heteroarenes [reaction (7)].

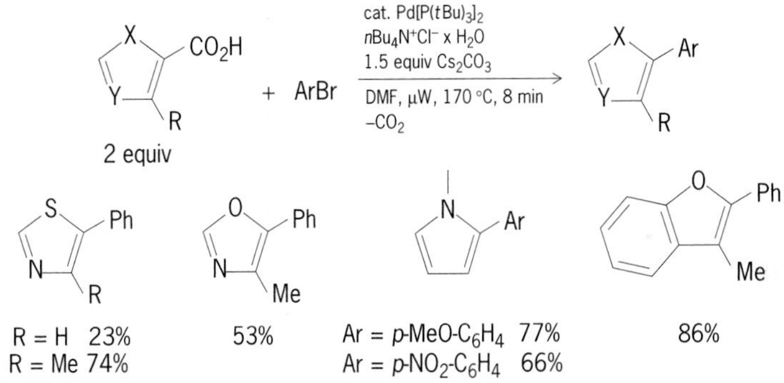

(7)

R = H 23% R = Me 74% 53% Ar = p-MeO-C_6H_4 77% Ar = p-NO_2-C_6H_4 66% 86%

An intramolecular reaction of this type was first used by W. Steglich in the total synthesis of lamellarin L (polycyclic pyrrole alkaloid). P. Forgione and F. Bilodeau developed a catalytic protocol, with general applicability also to intermolecular couplings of aryl halides with, for example, furan-, pyrrole-, thiophene-, oxazole-, and thiazole-2-carboxylates. The corresponding heteroarene-3-carboxylates cannot be converted.

Alkynylcarboxylate salts lose CO_2 under particularly mild conditions. They can thus be coupled with various aryl halides at 90°C to give the corresponding alkynyl arenes in the presence of $Pd_2(dba)_3$/dppf [tris(dibenzylideneacetone) dipalladium/diphenylphosphinoferrocene] and TBAF (tetrabutylammonium fluoride) in NMP.

Another class of carboxylates that require only palladium as the catalyst is oxalic acid monoesters. The decarboxylative ester synthesis is related to that of α-oxocarboxylates in that the extrusion of CO_2 gives rise to acyl anion equivalents. A catalyst consisting of 1-3 mol% $Pd(TFA)_2$

animals and penetrating the bone tissue. These animals, named *Osedax*, turned out to be siboglinids. They are gutless and associated with symbiotic bacteria that can break down the oils in whale bones. There are now five species described in the group, with many more that are currently being studied and awaiting description.

Discovery of *Swima*. A most recent surprise has been the discovery of a whole new group of deep-sea swimming worms that have been observed and collected with remotely operated vehicles. These annelids, which are 2–10 cm (0.8–4 in.) in length, swim with the help of long bristles that are used as paddles. Seven species have been recorded at depths of 1863–3793 m (6112–12,444 ft). Some live in close association with the sea floor, whereas others live in the free water column. Initially, a single new species, *Swima bombiviridis*, was described. The remaining undescribed six species have yet to be formally named. They are all currently referred to the genus *Swima* (**Fig. 1**).

The generic name *Swima* obviously has its origin in the swimming capabilities of these animals. All species are excellent swimmers and can move both forward and backward using metachronal waves (synchronized wavy movements) that originate posteriorly and are lengthy (involving at least eight segments). The single formally named species, *S. bombiviridis*, derives its name from the presence of four pairs of ellipsoid organs on the anterior segments. These organs (called bombs) can detach from the animals and then glow intensely for a number of seconds; thereafter, this green bioluminescence slowly diminishes. The species name *bombiviridis* comes from a combination of the Latin words *bombus*, meaning humming or buzzing, from which the English word bomb is derived, and *viridis*, meaning green.

Specimens of *S. bombiviridis*, when observed in situ, are usually positioned horizontally in the water column, with their long palps positioned forward and downward. Because of the light from the remotely operated vehicle, combined with the small size of the bombs [less than 1 mm (0.04 in.)], the actual release of the bombs could not be observed in situ. However, manual disturbance of the animals in the laboratory has resulted in the immediate release of bombs, which would immediately display bioluminescence. Also, manipulation in the laboratory, combined with the collection of specimens that had only some of their bombs released, shows that they do not discharge all bombs at a single event.

Fig. 1. A species of *Swima* from a depth of about 2000 m (6600 ft) at Juan de Fuca Ridge in the northeast Pacific. Four of the eight bombs have been discharged by this specimen. Length of the animal is approximately 4 cm (1.6 in.).

Bioluminescence. The ability to produce light is not uncommon in marine animals and is known also from a number of other annelids. In many cases, this production of light is associated with reproduction. In others, it is used as a defense mechanism. For example, the tube-living annelid *Chaetopterus* will produce a bioluminescent mucus when disturbed. However, the detachment of bioluminescent structures is a much rarer occurrence. Among annelids, it is known in a number of scale worms (Polynoidae). Mechanical disturbance stimulates the scales to glow and these then become detached from the body. Other known examples include a brittle star and a squid. The detachment of glowing parts may serve to distract a predator while the threatened animal escapes. With regard to *Swima*, the ability to produce bioluminescent bombs is not limited to the single described species, but is common for the group, including at least four of the undescribed species (**Fig. 2**). The actual biochemistry behind the luminescence in *Swima* remains to be elucidated.

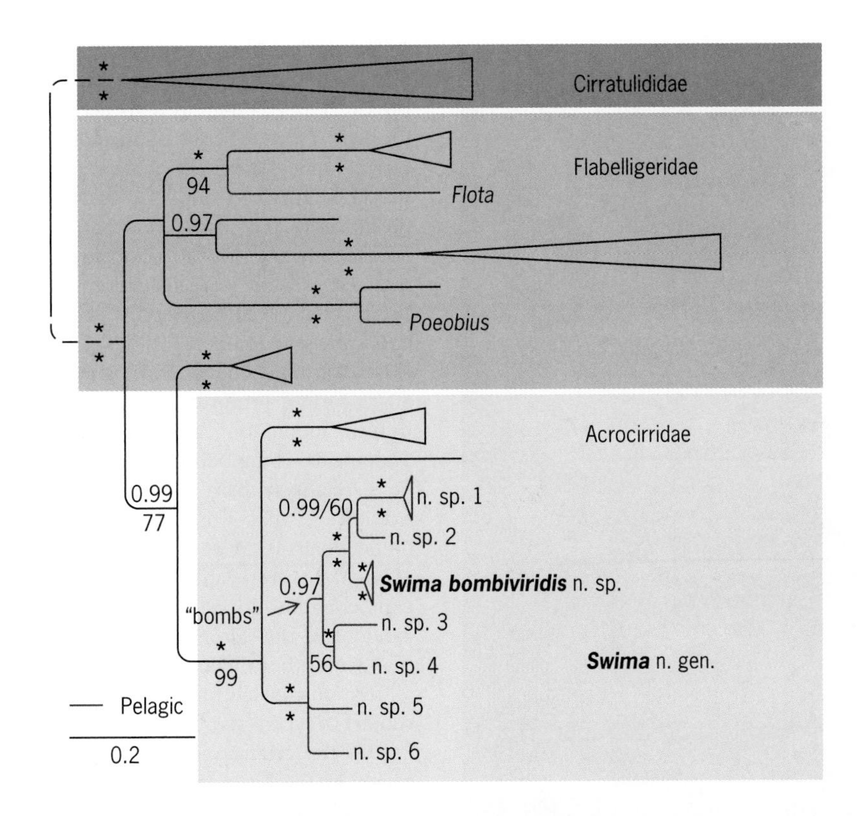

Fig. 2. Tree showing the phylogenetic relationships of *Swima* and related taxa based on genetic analyses of 18S rDNA, 16S rDNA, 28S rDNA, COI, and Cyt b. Three separate pelagic lineages are indicated: namely, the branches leading to *Flota*, *Poeobius*, and *Swima*. Support values (how well the data support a group) above the nodes indicate Bayesian posterior probabilities (0 to 1, with 1 as the strongest support); values below the nodes indicate bootstrap support from a parsimony analysis (0 to 100, with 100 as the strongest support). Asterisks indicate the highest possible support. (*From K. Osborn et al., Deep-sea, swimming worms with luminescent "bombs," Science, 325:964, 2009*)

Phylogeny. The phylogenetic relationships of *Swima* have been assessed using five genetic markers [18S rDNA, 16S rDNA, 28S rDNA, cytochrome oxidase I (COI), and cytochrome b (Cyt b)]. In addition to the seven new species, the phylogenetic analyses included a series of members of the families Cirratulidae, Flabelligeridae, and Acrocirridae. The results unequivocally show *Swima* to be a member of the Acrocirridae.

If the bioluminescent organs in *Swima* are compared with the anterior end of a traditional, benthic acrocirrid, for example, *Acrocirrus* (**Fig. 3**), then the latter animal is seen to be provided with branchiae (external gills), instead of bombs, at exactly the same position. Similar branchiae are present in other, related taxa. This positional similarity, combined with morphological similarities between the branchiae and the bombs, indicate that both organs are homologous and that the branchiae in an ancestor of *Swima* became modified into bombs.

The great majority of marine annelids are benthic (bottom-dwelling), although in many cases they have a larval stage that is pelagic (freely swimming or floating in the open ocean). However, there are a number of holopelagic groups (animals who are pelagic through their whole life cycle). Some, such as tomopterids and alciopids, have developed extreme adaptations to pelagic life and are completely transparent. There are also some pelagic polychaetes that are closely related to *Swima*. Within the sister family Flabelligeridae, there are two genera, *Poeobius* and *Flota*, that independently have made the transition from benthic to holopelagic life. *Poeobius* and *Flota* are usually referred to family Poeobiidae and Flotidae, respectively, although the analyses done with *Swima* (Fig. 2) have demonstrated both genera to be members of the Flabelligeridae (note, however, that Flabelligeridae on this phylogenetic tree is nonmonophyletic). This clearly indicates how a benthic-holopelagic transition could have happened several times independently within closely related groups. Within the acrocirrids, there are also two poorly known genera, *Chauvinelia* and *Helmetophorus*, that share some similarities with *Swima*. The swimming ability is unknown for these two genera, however, and the relationships between *Swima* and these taxa warrant further study. Finally, considering that the geographic areas that were covered at the discovery of *Swima* were very limited, many yet undescribed species in the group can be expected in the future.

For background information *see* ANNELIDA; BIOLUMINESCENCE; DEEP-SEA FAUNA; HYDROTHERMAL VENT; MARINE BIOLOGICAL SAMPLING; MARINE ECOLOGY; POGONOPHORA (SIBOGLINIDAE); POLY-

Fig. 3. *Acrocirrus validus* from a depth of a few meters off the east coast of Honshu, Japan. The spiral-shaped appendages that are situated on the head are the palps, with four pairs of branchiae following these on the anteriormost segments. The visible part is 5.5 mm (0.22 in.) in length, with the appendages excluded.

CHAETA; VESTIMENTIFERA in the McGraw-Hill Encyclopedia of Science & Technology. Fredrik Pleijel

Bibliography. K. Osborn et al., Deep-sea, swimming worms with luminescent "bombs," *Science*, 325:964, 2009; F. Pleijel, T. Dahlgren, and G. Rouse, Progress in science: From Siboglinidae to Pogonophora and Vestimentifera and back to Siboglinidae, *C. R. Biol.*, 332:140–148, 2009; G. Rouse and F. Pleijel, *Polychaetes*, Oxford University Press, Oxford, 2001.

Dicynodontia

Living mammals belong to a larger group of tetrapods called Synapsida. The synapsid lineage is over 300 million years old and includes numerous extinct animals that document the evolution of mammals from an ancient, lizardlike ancestor. Extinct nonmammalian synapsids are sometimes called "mammal-like reptiles," but all are more closely related to mammals than to any reptile. One of the most successful of these nonmammalian synapsid groups is Dicynodontia. Dicynodonts were herbivores known from the Middle Permian Period of Earth history [approximately 265 million years ago (Mya)] to at least the Late Triassic Period (approximately 215 Mya); a fragmentary specimen from Australia may imply that they survived until the Early Cretaceous (approximately 105 Mya). Although dicynodonts are not directly ancestral to mammals, they are important because they were major components of Permo-Triassic terrestrial ecosystems and survivors of the largest mass extinction.

Relationships. Dicynodonts and their closest relatives form a group called Anomodontia (**Fig. 1**). In turn, anomodonts are part of the synapsid group Therapsida. Anomodonts were long considered to be a relatively basal, or early diverging, therapsid lineage that was perhaps closely related to dinocephalians. More recent research suggests that anomodonts were later-diverging, sharing a more recent common ancestor with therocephalians and cynodonts, the latter of which include mammals, than with other therapsids.

Relationships within Anomodontia have been the subject of recent scrutiny, and a consensus is emerging on the arrangement of major anomodont lineages. At the base of the tree are a small number of nondicynodont anomodonts. *Eodicynodon* is the most basal dicynodont, and a series of predominantly Permian lineages constitute much of the rest of the dicynodont family tree. Most Triassic dicynodonts were members of Kannemeyeriiformes, but three emydopoid species occurred in the Early to Middle Triassic. Additional research is needed to clarify relationships between species within the major dicynodont lineages, and between the main groups of Triassic kannemeyeriiforms.

General appearance. Dicynodonts (**Fig. 2**) spanned a range of body sizes: The smallest species were marmot-sized, whereas the largest were rhinoceros-sized. Most dicynodonts had a large head, a stocky body, short limbs, and a short tail. The forelimbs and hindlimbs of early dicynodonts had a sprawling posture, but the hindlimbs of later species were more upright. Dicynodont skulls are characterized by a short snout and large temporal openings behind the eye sockets, which served as attachment areas for jaw musculature. The name "dicynodont" refers to enlarged tusks in the upper jaw that many species possessed, although some dicynodonts were tuskless. Most dicynodonts lacked teeth aside from the tusks (a few species retained small teeth in the upper and lower jaws); instead, a turtlelike beak covered much of the snout, the secondary palate, and the front of the lower jaw. Nondicynodont anomodonts generally were small, possessed teeth, lacked a beak, and had more gracile, longer-limbed bodies.

Life history and growth. The microscopic structure of dicynodont bone tissue is an important source of data on growth rates and growth patterns in the group. All dicynodonts that have been examined possess fibrolamellar bone, a rapidly growing tissue type, but most also show lines of arrested growth, indicating periodic pauses in growth perhaps associated with unfavorable environmental conditions. Although there is no evidence of a complete cessation of bone deposition in dicynodonts, implying that they grew throughout their lives, many species display a slowing of growth in adulthood, particularly large kannemeyeriiforms. Dicynodont bones are more highly vascularized (that is, they include a greater number of canals for blood vessels) than those of most contemporaneous therapsids. Because high vascularization is associated with higher rates of bone deposition, this may mean that dicynodonts grew more rapidly than other therapsids.

Studies of skull morphology and dimensions have concluded that adults of some dicynodont species were sexually dimorphic (exhibiting diagnostic morphological differences between the sexes). For example, it appears that *Diictodon* males were tusked, whereas females were tuskless, and presumed males of *Diictodon* and *Aulacephalodon* had larger nasal bosses (bony lumps located over the nostrils that were covered and exaggerated by the beak) than juveniles and presumed females.

Feeding system. The feeding system of dicynodonts was characterized by a propalinal (front-back) sliding movement of the lower jaw that was a critical component of their adaptation to a plant-eating lifestyle. Dicynodont skulls show many modifications for propaliny, including reduction and reorientation of the pterygoid bones, the presence of new muscle attachment areas on the skull and jaw, and changes to the jaw joint to allow sliding. Many differences in skull morphology among dicynodonts correspond to minor changes in the size and placement of jaw musculature. Recent mechanical studies indicate that variations in skull shape and construction also correlate with differences in how stress was applied to the skull during feeding. The height-to-width ratio of the back of the skull reflects the dominant movements of the head and has been used to hypothesize that dicynodont species differed in whether their preferred food sources were located above or below their heads.

Recent research has documented two evolutionary patterns related to feeding in anomodonts. First, propaliny likely evolved independently in the venyukovioid *Suminia* and in dicynodonts. Although this entailed similar changes to the jaw joint and musculature, *Suminia* had large teeth and no beak. Second, some Triassic dicynodonts reemphasized the orthal (up-and-down) component of their jaw movement. They accomplished this by changing the orientation of the jaw musculature so that it provided more vertical force, and by changing the shape of the jaw joint so that fore–aft sliding at the joint translated into vertical movements at the front of the jaw.

Ecology. Anomodonts were primarily terrestrial herbivores, but there was variation in that theme. Nondicynodont anomodont species differ in tooth shape and whether single or multiple tooth rows are present, presumably because of dietary differences. The skeleton of *Suminia* is the most thoroughly studied of all nondicynodont anomodonts and shows evidence of climbing abilities, making it the oldest known arboreal tetrapod.

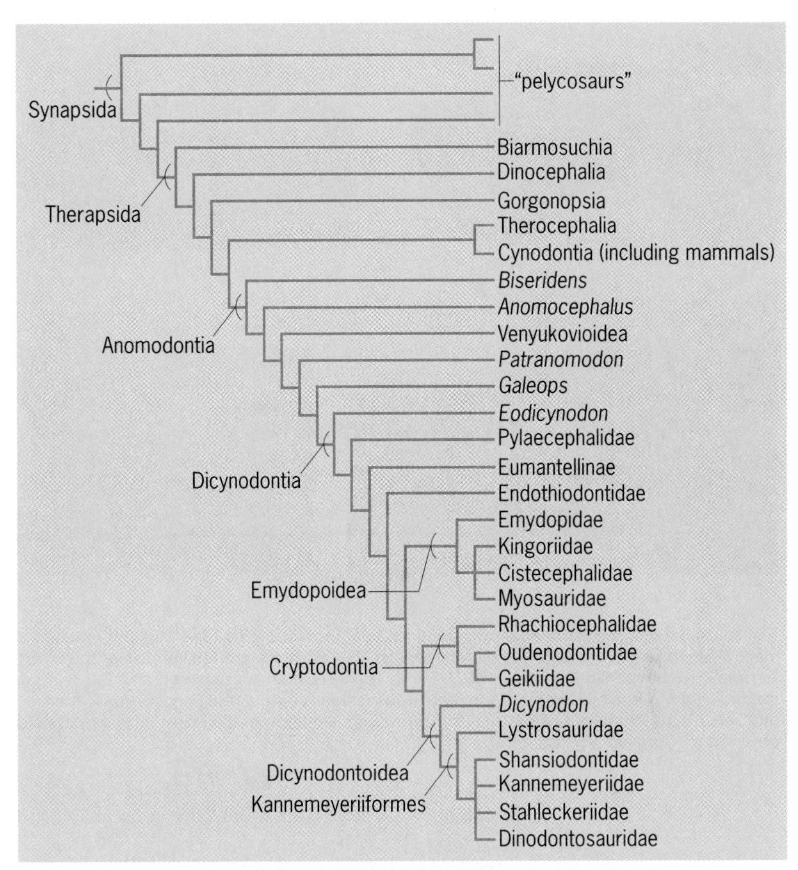

Fig. 1. Tree showing hypothesized relationships among major anomodont lineages, and between anomodonts and other synapsid lineages. Branches that are linked at their base represent lineages that are descended from a common ancestor. Groups are known only from the Permian Period of Earth history, with the following exceptions: Therocephalia, Cynodontia (including mammals), Kingoriidae, and Lystrosauridae are known from the Permian and Triassic; Myosauridae, Shansiodontidae, Kannemeyeriidae, Stahleckeriidae, and Dinodontosauridae are known only from the Triassic. Anomodonts mentioned in the text but not shown on the tree belong to the following groups: *Suminia* is a venyukovioid; *Diictodon* is a pylaecephalid; *Endothiodon* is an endothiodontid; *Kombuisia* is a kingoriid; *Aulacephalodon* is a geikiid; and *Lystrosaurus* is a lystrosaurid.

There has been much speculation about how variations in dicynodont beak and skull shape correspond to food preferences. The more conservative of these hypotheses are relatively well supported (for example, the jaws of *Diictodon* were likely specialized for slicing food, whereas *Lystrosaurus* emphasized crushing and grinding), but others (for example, *Endothiodon* as a specialist feeder on conifer cones) are unlikely to be confirmed without direct evidence (for example, preserved gut contents).

Digging was an important part of dicynodont ecology. All dicynodonts had strong forelimbs and flattened claws, which characterize digging animals, and *Diictodon* and *Lystrosaurus* remains have been found inside ancient burrow structures. Cistecephalids took this trend to an extreme and evolved a specialized burrowing lifestyle. Their skulls are heavily constructed, some members of the group had relatively small eyes, and their forelimbs were

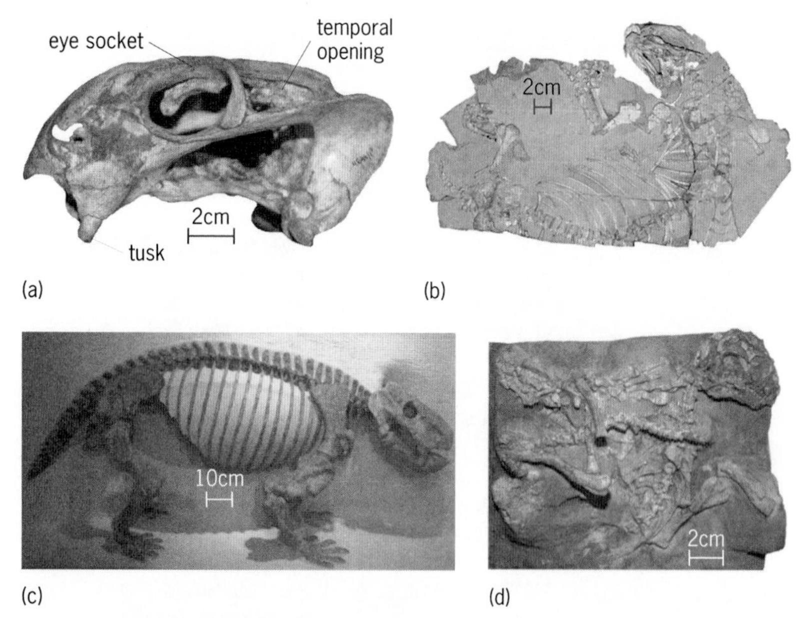

Fig. 2. (*a*) Skull of the Permian dicynodontoid *Delectosaurus* (PIN 4644/1) in left lateral view. Note the tusk and large temporal opening. (*b*) Nearly complete skeleton of the small Permian pylaecephalid *Eosimops* (BP/1/6674). The specimen is preserved as if it is lying on its back. (*c*) Mounted skeleton of the large Triassic dinodontosaurid *Dinodontosaurus* (MCZ 1670) in right lateral view. (*d*) Nearly complete skeleton of the Permian venyukovioid anomodont *Suminia* (PIN 2212/10).

molelike, with large attachment areas for powerful muscles.

The dicynodont *Lystrosaurus* is frequently hypothesized to have been semiaquatic because of its tall snout with nostrils positioned high on the face, the morphology of its limbs and feet, and its bone tissue structure. However, much of this evidence is equivocal. The tall snout of *Lystrosaurus* may be related to other aspects of its skull shape associated with increasing bite force, and similar bone tissue is widespread in dicynodonts that lack adaptations for a semiaquatic lifestyle.

Dicynodonts have broad ecological significance because they were the first highly abundant and diverse tetrapod herbivores. Dicynodont-dominated tetrapod assemblages in the Middle and Late Permian are among the first to show modern community structures, with abundant herbivores supporting smaller numbers of carnivores.

Biogeography. Dicynodont fossils are found on every continent. During the Permian and Early Triassic, dicynodonts were most common in middle to high paleolatitudes. Although some dicynodonts were endemic to specific areas, others were widespread, including *Diictodon* and *Lystrosaurus*. After the Early Triassic, dicynodont-bearing fauna spanned a wider paleolatitudinal range, but with more provinciality and fewer cosmopolitan species.

Recent debates have questioned whether anomodonts originated in Laurasia or Gondwana (ancient continents that fragmented and drifted apart to form eventually the present continents). Because the Russian venyukovioids were originally the most basal anomodonts known, the group was assumed to have a Laurasian origin. The discovery of *Patranomodon* and *Anomocephalus* in South Africa challenged that assumption, but the identification of the Chinese *Biseridens* as the most basal anomodont renewed support for the Laurasian-origin hypothesis. The geographic ranges of basal dicynodonts suggest that Dicynodontia originated in Gondwana.

Because dicynodonts were widespread, they have been used extensively in determining relative ages for the rocks in which they are found (biostratigraphy). The dicynodont *Lystrosaurus*, known from Antarctica, China, India, Mongolia, Russia, and South Africa, was an important early datum for plate tectonics because its distribution is difficult to explain using the modern continental configuration.

Taxonomic diversity and mass extinctions. Current compilations include 128 anomodont species in 68 genera, but these totals may change with the resolution of long-standing problems, such as the large number of potentially invalid species of *Dicynodon*. Dicynodonts underwent diversifications in the Middle Permian, Late Permian, and Middle Triassic. They experienced diversity declines in the late Middle Permian, at the Permo-Triassic boundary, and in the late Middle Triassic, although these patterns may be influenced by sampling of the fossil record.

Dicynodont diversity patterns have received attention in studies of the end-Permian extinction, the largest mass extinction in Earth history. The group's late Middle Permian diversity decline may correspond to the extinction pulse at the end of the Middle Permian in the marine realm, but age estimates of dicynodont-bearing rocks are currently too imprecise to be certain. Dicynodonts were affected by the extinction pulse at the Permo-Triassic boundary, with several Permian lineages becoming extinct. Only one dicynodont species (*Lystrosaurus curvatus*) crossed the Permo-Triassic boundary. Three other Triassic lineages (*Kombuisia*, Myosauridae, and Kannemeyeriiformes) have their closest relatives in the Permian, implying that members of these lineages also crossed the boundary, but they have not yet been found in the fossil record.

For background information *see* ANIMAL EVOLUTION; EXTINCTION (BIOLOGY); FOSSIL; HERBIVORY; PALEOGEOGRAPHY; PALEONTOLOGY; PERMIAN; REPTILIA; SYNAPSIDA; THERAPSIDA; TRIASSIC in the McGraw-Hill Encyclopedia of Science & Technology.

Kenneth D. Angielczyk

Bibliography. K. D. Angielczyk, *Dimetrodon* is not a dinosaur: Using tree thinking to understand the ancient relatives of mammals and their evolution, *Evol. Educ. Outreach*, 2:257–271, 2009; J. Fröbisch and R. R. Reisz, The Late Permian herbivore *Suminia* and the early evolution of arboreality in terrestrial vertebrate ecosystems, *Proc. R. Soc. B*, 276:3611–3618, 2009; T. S. Kemp, *The Origin and Evolution of Mammals*, Oxford University Press, Oxford, U.K., 2005; G. M. King, *The Dicynodonts: A Study in Palaeobiology*, Chapman & Hall, London, 1990; R. R. Reisz and H.-D. Sues, Herbivory in late

Paleozoic and Triassic terrestrial vertebrates, pp. 9–41, in H.-D. Sues (ed.), *Evolution of Herbivory in Terrestrial Vertebrates: Perspectives from the Fossil Record*, Cambridge University Press, Cambridge, U.K., 2000.

DNA barcoding in fungi

The kingdom Fungi, long thought to be primitive plants, are now known to be more closely related to animals and other metazoans, forming part of the larger group Opisthokonta (organisms with a single posterior flagellum). Fungi are a hugely diverse group of eukaryotes that includes chytrids, molds, mushrooms, lichens, rusts, smuts, and yeasts. They form close symbioses with plants (as mycorrhiza and endophytes) and algae (as lichens), break down woody and leafy matter on forest floors, play roles in weathering rocks, and parasitize animals, plants, and other fungi. Fungi straddle both the macroscopic and microscopic observational scales. Most people are familiar with mushrooms or cup fungi that grow on rotting logs and soil in spring or fall. However, the majority of fungal species remain hidden from view as microscopic structures. Although some species can be identified from macroscale structures, mycologists must rely on shapes, colors, and development of fungal cells observed with the microscope in order to study microfungi. Most fungi exist only as microscopic forms, growing as single cells (yeasts) or as a series of cylindrical cells (hyphae), periodically bearing structures for propagation (spores). Traditionally, great importance was placed on comparing fungal spore-producing structures, arising after either sexual or asexual division, for identification. Often, though, only a part of the fungal life cycle is observed. As an added complication, very similar structures sometimes occur in unrelated species. For such microorganisms, therefore, it is not surprising that scientists rely increasingly on DNA sequence comparisons to link different disconnected parts of the life cycle of a fungus and to distinguish between different species.

Mycologists have used DNA sequences for species identification in several ways since the early 1990s, but DNA barcoding, as it is known today, was proposed first by insect scientists at the University of Guelph, Canada, in 2003. The concept was to make species identification more efficient, employing a concept similar to Universal Product Codes used to identify retail products. DNA barcoding involves determining a string of DNA letters (500–700) from a single gene or region of DNA, standardized across large groups of life. These sequences then are compared with databases of reference sequences that have been firmly connected to accurately identified and preserved cultures or specimens, with associated data on geographical distributions and biology. The Barcode of Life Data Systems (BOLD) now includes close to a million DNA barcodes, representing more than 70,000 species; however, few of these are fungal.

How does DNA barcoding work? The first step in creating a DNA barcode is to select a defined DNA sequence, such as a gene or region, which will be unique for each species across a large group of organisms. For animals, the gene cytochrome *c* oxidase I (COI or *cox1*) was chosen as a marker. This gene is found in the energy production center of the cell, the mitochondrion, on a set of DNA molecules that is separate from the chromosomes of the nucleus. Each cell contains many mitochondria and it is easy to use a technique called the polymerase chain reaction (PCR) to select and amplify COI for analysis. In addition, only a small amount of tissue or cells is needed. In animals, COI varies enough to distinguish closely related species, but the sequences are also similar enough to allow statistical comparisons of sequences from very different species. One pitfall of using a single gene is that its rate of evolutionary change may differ from one group of organisms to another. As such, COI may be evolving too slowly in most plant species. However, in fungi, vast differences in length make it challenging to amplify using PCR. Hence, the gene area of choice in most fungi is a short structural gene (the 5.8S gene) that forms part of the nuclear ribosome, with two "spacer" regions [internal transcribed spacer 1 and 2 (ITS1 and ITS2)] on either side that do not encode for functional genes (see **illustration**). While the rate of change of the 5.8S gene is very slow, the two surrounding ITS1 and ITS2 spacers evolve much more quickly. The whole combined region is often referred to as the ITS. The ribosome is the cellular structure that functions to build proteins from DNA templates and therefore is integral to biological life. Ribosomal structural genes, located in the nucleus, exist in multiple copies and, similar to mitochondrial genes, are extremely easy to amplify by PCR. Other DNA areas also are used for species identification in fungi. For example, researchers working on yeasts (single cellular fungi that occur in several branches of the fungal tree of life) favor a variable stretch of another ribosomal structural gene, the 28S (LSU, or large subunit).

Organizations and databases. The ITS region was selected as the barcoding region to be recommended for fungi by a diverse group of mycologists at an international workshop organized by the Consortium for the Barcode of Life (CBOL) in 2007. CBOL is funded by the Alfred P. Sloan Foundation and has its offices

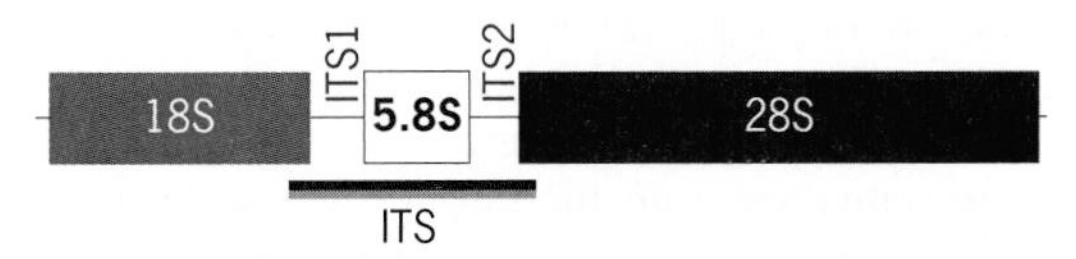

The ribosomal cassette of repetitive genes in fungi. The 5.8S gene is flanked by the 18S (small subunit) and 28S (large subunit) genes. The ITS region is indicated by the black bar.

at the Smithsonian Institution. The Consortium's mandate is to promote the use of barcoding technologies in all major biological disciplines, oversee the international standardization of barcoding markers for all groups of organisms, and provide technical protocols for scientists who wish to incorporate barcoding in their research. The Consortium includes 170 members from natural history collections and museums, government and nongovernmental organizations, biotechnology companies, and academia. The International Barcode of Life initiative, iBOL, is an international research network centered at the University of Guelph and is scheduled to begin its five-year mandate in October 2010. iBOL involves scientists from 25 countries, studying all groups of living organisms, except bacteria.

After the selection of standardized marker genes, barcoding requires the development of reliable databases based on sets of specimens of known identity and origins, which then can be used for identification purposes by other scientists. GenBank, located at the National Center for Biotechnology Information (NCBI) in the United States, and its international database partners, the European Molecular Biology Laboratory (EMBL) and the DNA Data Bank of Japan (DDBJ), house the majority of DNA sequences produced by researchers investigating fungal diversity. GenBank presently lists more than 150,000 sequences from the ITS in fungi, representing approximately 14,000 species. Although a careful verification process is performed for data quality, GenBank policies allow submitters to determine the identity of organisms associated with DNA sequences. Erroneously identified sequences complicate the identification of unknown samples. Therefore, NCBI is developing a well-validated set of reference sequences (RefSeq). An actively curated subset is already available for identification of bacterial sequences and this is now being expanded to fungi. Additionally, a number of sequence databases exist to aid fungal data gathering and identification with their own sets of reference sequences. The UNITE database is focused on well-verified root-associated fungi, but is expanding to other groups of fungi from soil. The Assembling the Fungal Tree of Life (AFTOL) database includes a diverse set that is representative of the entire fungal kingdom, including ITS sequences (also deposited at GenBank). Among economically important molds, there are online DNA sequence-based identification databases for the plant pathogenic genus *Fusarium* and the biocontrol genus *Trichoderma*. Several other specialized databases focusing on specific groups of fungi are housed at the Fungal Biodiversity Centre in the Netherlands as well as elsewhere.

Importance of DNA barcoding. The most common, conservative estimate for fungal species diversity is 1.5 million species. Given that approximately 100,000 species are now known to science, this means that more than 90% are undocumented. By enabling identification of fungi by anyone able to access the necessary technology, DNA barcoding will allow exploration of the full range of fungal diversity and the interactions that fungi have with other groups of organisms. New sequencing technologies have rapidly accelerated the amount of DNA sequence data that can be collected from the environment without collecting physical specimens. Mycologists continue to discover new diversity inside plants and rocks, and even at the bottom of the ocean. An understudied environment, the soil, is a particular focus for several research groups. One example is an enigmatic group of very common fungi known informally as "soil clone group I" (SCGI). Members of this group have been detected in soil on at least three continents and fall in the fungal phylum Ascomycota, but they are unrelated to any known class of Fungi. So far, they are only known by their unique DNA sequences and nothing is known of their biology or their appearance. In addition to addressing important gaps in our knowledge of the ecological roles of fungi in carbon cycling and other processes, there are several economical and practical implications to having an effective barcoding system in place for fungi. Governments are concerned with minimizing movement of pathogenic fungi as crops, lumber, and agricultural produce are traded around the globe. Hidden fungi, such as endophytes living inside plants, can be effectively detected and identified without labor-intensive culturing and microscopy. Last, barcoding will help the biotechnology sector increase the efficiency of discovery of novel fungi that can be used for biological control of harmful pests or pathogens and for production of industrial enzymes [as in bioconversion (including biofuels)], antibiotics, and probiotics, thereby benefiting human health.

For background information *see* BIODIVERSITY; BIOTECHNOLOGY; DEOXYRIBONUCLEIC ACID (DNA); FUNGAL ECOLOGY; FUNGAL GENOMICS; FUNGAL PHYLOGENETIC CLASSIFICATION; FUNGI; GENETIC CODE; GENETIC MAPPING; MYCOLOGY; RIBOSOMES in the McGraw-Hill Encyclopedia of Science & Technology.

Conrad L. Schoch; Keith A. Seifert

Bibliography. D. Begerow et al., Current state and perspectives of fungal DNA barcoding and rapid identification procedures, *Appl. Microbiol. Biotechnol.*, in press, 2010; D. L. Hawksworth, The magnitude of fungal diversity: The 1.5 million species estimate revisited, *Mycol. Res.*, 105:1422-1432, 2001; P. D. N. Hebert et al., Biological identifications through DNA barcodes, *Proc. R. Soc. Lond. B*, 270:313-321, 2003; T. Le Calvez et al., Fungal diversity in deep-sea hydrothermal ecosystems, *Appl. Environ. Microbiol.*, 75:6415-6421, 2009; T. M. Porter et al., Widespread occurrence and phylogenetic placement of a soil clone group adds a prominent new branch to the fungal tree of life, *Mol. Phylogenet. Evol.*, 46:635-644, 2008; K. A. Seifert, Progress towards DNA barcoding of fungi, *Mol. Ecol. Res.*, 9:83-89, 2009; K.A. Seifert et al., Prospects for fungus identification using *CO1* DNA barcodes, with *Penicillium* as a test case, *Proc. Natl. Acad. Sci. USA*, 104:3901-3906, 2007.

DNA sequencing

Deoxyribonucleic acid (DNA) is the common structure that forms the basic building blocks of life. Every nucleated cell in our body contains a copy of the 3 billion letters that comprise our unique DNA sequence. DNA sequencing is the process by which the precise order of the adenines (A), cytosines (C), guanines (G), and thymines (T), which are the nucleotide bases that comprise this DNA code, are determined. The first organism sequenced was the 5386 bases pairs (bp) of bacteriophage X174 by Frederick Sanger and colleagues in 1978. Since then, progress has been extremely rapid, with the generation of the sequences of viruses and numerous bacteria, culminating in the global effort to determine the sequence of the human genome.

DNA sequencing technologies. In the years since the completion of the human genome sequencing in 2000, there has been a silent revolution in the area of DNA sequencing technologies. Until recently, the vast majority of DNA sequencing was carried out using methods derived largely from those developed in the 1970s. One of the earliest sequencing techniques was developed by Allan M. Maxam and Walter Gilbert and was based on chemical modification and subsequent cleavage at specific positions along the DNA molecule. Initially, this method proved popular, but it quickly fell out of favor among molecular biologists with the invention of the more efficient and less toxic chain-terminator, or Sanger sequencing, method. The Sanger method relies on four separate sequencing reactions, each containing the DNA to be sequenced, a short oligonucleotide that binds to the DNA template for primer elongation during DNA synthesis, the four nucleoside triphosphates, and a small amount of a distinct DNA chain terminator. There is one terminator for each of the four bases (A, C, G, and T). These chain terminators truncate the synthesis of a complementary DNA (cDNA) strand as it is synthesized, generating a pool of fragments of different lengths when a chain terminator is incorporated. By labeling these fragments with a radioactive isotope or a fluorophore, the DNA sequence can be read following size fractionation using procedures that separate DNA chains that differ in length by only one nucleotide. Each DNA molecule to be sequenced by this procedure must first be cloned into an appropriate sequencing vector and introduced into bacteria for amplification in order to obtain enough DNA to give a robust signal after the sequencing reaction has been performed. These sequencing vectors often contain binding sites for universal sequencing primers. DNA for sequence analysis using this method may also be generated by a technique called polyermase chain reaction (PCR).

Unlike traditional sequencing techniques, in which each sequencing reaction takes place on a single DNA template, next-generation sequencing technologies attempt to parallelize millions of simultaneous sequencing reactions in order to vastly increase the throughput and yield of data. These new procedures have obviated the labor-intensive requirement of amplifying DNA fragments via transformation of bacteria by amplifying in parallel many individual DNA molecules, which in the process also attaches a sequence to each molecule that serves as a template for elongation from a universal primer during the sequencing reaction. In 2005, the 454 GS20 (now Roche 454™) was the first of a new generation of massively parallel sequencing technologies to come to market. The Roche 454 system is based on the idea of taking a large set of DNA fragments (a complete strand of DNA that has been fragmented into small, approximately 500-nucleotide lengths of DNA) and attaching individual DNA molecules to 20- or 28-μm beads, which are then added to a PCR reaction that amplifies them on the bead. After amplification, the beads are placed individually in one of several hundred thousand wells (more than a million in the newest machines) in a picotiter plate, which provides a fixed location where each sequencing reaction can take place. The picotiter plate serves as a flow cell over which individual nucleoside triphosphates are sequentially passed. After each cycle, light emitted by enzymatic detection of a pyrophosphate moiety is recorded on a charge-coupled camera device (CCD). Deconvolution of these flashes of light allows the sequence of the DNA molecule on the bead in each well to be read. In this "pyrosequencing" procedure, the intensity of the burst of light emitted in each well after the addition of each nucleoside triphosphate is proportional to the number of mononucleotides added to the growing cDNA chain, although accuracy is reduced when a string of 5 or more of the same base is added.

Subsequently, Applied Biosystems released their SOLiD™ sequencing platform in late 2007. The SOLiD system is based on similar technology to the Roche platform. Individual DNA fragments are distributed into separate wells for sequencing. The beads used are much smaller (1 μm rather than 28 μm), thereby allowing for a much higher bead density per sequencing run (typically of the order of 100 million beads per run). In addition, the SOLiD system does not involve the addition of individual nucleotides catalyzed by DNA polymerase, but rather the addition of short segments of DNA by ligation catalyzed by DNA ligase. In this procedure, after hybridization of the sequencing primer to bead-bound amplicons (small, replicating DNA fragments), 16 different probes (fluorescently labeled with one of four dyes, based on the nucleotides at the first two positions) are added successively. One of the 16 probes will ligate (join), depending on which one is complementary to the template strand around a bead, and the slide is imaged. Then, the probe is cleaved off, removing the fluorescent label and leaving 5 bases of the probe ligated to the DNA fragment. Several rounds of ligation and imaging are carried out to generate a color profile for every fifth dinucleotide. The unique feature of the SOLiD platform is that it is not possible to decipher the identity of any nucleotide without knowing the first base of the sequence, because it operates on a dinucleotide basis.

In late 2006, Solexa (now Illumina) launched the Genome Analyzer (GA). The technology behind the Illumina system is based on the idea of flowing template DNA fragments across a hollow glass slide (or flow cell), coating the interior of the flow cell with a random "lawn" of DNA molecules. The template DNA attached to the flow cell surface is isothermally amplified, resulting in clusters of DNA strands. The amplified clusters initially consist of double-stranded DNA; one strand is removed prior to sequencing. The flow cell, now carrying a lawn of clonally amplified single-stranded DNA clusters, is transferred to the sequencing machine, where the clusters undergo a sequencing-by-synthesis reaction. The machine operates in cycles, with one base being read from every cluster simultaneously at each cycle.

Computation and analysis. Perhaps one of the biggest impacts of next-generation sequencing has been the need for a complete reinvention of the computational methods used to analyze the raw sequence data. One of the biggest changes is that almost all of the next-generation sequencers are based on determining the presence of individual bases by detecting tiny flashes of light. Therefore, many of the methods commonly used in the area of image processing were adapted for DNA sequencing. Another major change has been the dramatic scale-up in the volume and types of sequence data produced by next-generation sequencing machines. Previously, sequence reads were of the order of 600–700 nucleotides in length, and up to 384 reads could be generated from each sequencing run. Next-generation machines can now generate sequence reads of several hundred nucleotides in length, but importantly each run can generate millions of reads, dramatically affecting data processing and storage.

An important application of new sequencing technologies is population-based sequencing, where there is already a high-quality reference genome (for example, the human reference genome). By aligning the short reads from multiple individuals to the reference genome, it is possible to discover genetic differences between individuals in a population (for example, to determine disease-causing genetic variants). In the past few years, new sequence alignment algorithms have been developed specifically for next-generation sequence data. The ability to map short sequence reads to the correct location depends on a number of factors, such as the complexity of the reference genome, the length of the sequence reads, error rates of the reads, and diversity of the individual or strain compared to the reference. By comparing reads to the reference in this way, it is possible to detect single nucleotide polymorphisms (SNPs) as well as short insertions and deletions. Many next-generation sequencing technologies can produce paired-end sequencing reads, in which each pair of reads is from opposite ends of a single DNA fragment with a known size range. This pairing distance information can be used to identify larger structural variants (SVs) by comparing the read mapping locations against the reference genome sequence. These SVs include large insertions, deletions, inversions, and translocations. For each of these structural rearrangement events, there is a specific pattern of how the reads align to the reference genome, which can be identified computationally.

As the evolution of DNA sequencing technologies has progressed, so too has the size and scale of experiments increased. In the early days of manual slab gel sequencing, experiments were focused on determining the DNA sequence of single genes and small viral genomes. The development of automated approaches in the early 1990s meant that researchers could begin to think of the possibility of determining the entire genome sequences of single-cell organisms, culminating in the publication of the first complete bacterial genome (*Haemophilus influenzae*) in 1995. In a similar vein, the advent of new sequencing technologies has meant that it is now possible to scale up to even larger sequencing experiments that were simply not possible just a few years ago. Sequencing is now truly moving into a new age, in which many hundreds or thousands of human individuals will have their complete genome sequence determined. One of the most notable projects in progress is the 1000 Genomes Project, whose goal is to "create the most detailed picture of human genetic variation to date." The pilot phase of the project was completed in late 2009 with the sequencing of approximately 180 individuals from four populations. The main phase of the project is planning to analyze over 1200 individuals from about 25 distinct human populations that have been drawn from all corners of the globe. As the focus shifts to medical-based sequencing, the results of this project will be essential to facilitate the identification of rare gene variants that contribute to human disease. Many other large population-based sequencing projects are also underway, including MalariaGEN (sequencing thousands of malaria isolates) and the Cancer Genome Project (sequencing cancer genomes).

For several years, one of the most essential tools for interrogating large numbers of gene expression levels in parallel has been the DNA microarray (an ordered array of DNA probes, providing a powerful reagent for analyzing the genome). One of the advantages of the high sequencing depth that can be generated using next-generation sequencing is that gene expression levels can be measured directly by sequencing RNA in the form of reverse-transcribed cDNA (also known as RNA-Seq). The subsequent sequencing reads are aligned back to a reference genome. The levels of gene expression are proportional to the sequencing depth across the transcript. Unlike array-based approaches, RNA-Seq is not subject to design biases or a fixed set of probes and thus allows for the detection of rare transcripts expressed at very low levels. A number of recent papers have demonstrated the reliability and reproducibility of RNA-Seq and found that it is possible to detect biological processes not observable with earlier methods for analyzing gene expression.

Outlook. There is intense research into faster, cheaper, and higher-throughput DNA sequencing technologies. The current target for next-generation sequencing vendors is the $1000 human genome. The recent dramatic reduction in sequencing costs is quickly making this goal seem realistic and achievable. With the continual unprecedented scale-up of sequencing throughput, it is possible to imagine a time in the not too distant future when a DNA sequencing machine will be found in every physician's clinic and the genome sequence of all living things will be determined.

For background information *see* BIOTECHNOLOGY; DEOXYRIBONUCLEIC ACID (DNA); DNA MICROARRAY; GENETIC CODE; GENETIC ENGINEERING; GENETIC MAPPING; HUMAN GENETICS; HUMAN GENOME; RIBONUCLEIC ACID (RNA) in the McGraw-Hill Encyclopedia of Science & Technology.

Thomas M. Keane; David J. Adams

Bibliography. M. Pop and S. L. Salzberg, Bioinformatics challenges of new sequencing technology, *Trends Genet.*, 24:142–149, 2008; D. J. Turner et al., Next-generation sequencing of vertebrate experimental organisms, *Mamm. Genome*, 20:327–338, 2009; K. V. Voelkerding, S. A. Dames, and J. D. Durtschi, Next-generation sequencing: From basic research to diagnostics, *Clin. Chem.*, 55:641–658, 2009.

Effect of abuse on the brain

A vast body of literature supports the idea that early adversities, such as childhood sexual abuse (CSA), physical abuse, and witnessing domestic violence, are major risk factors for psychopathology, accounting for 50–75% of the population-attributable risk for depression, suicide attempts, and drug abuse. The powerful relationship between childhood abuse and psychopathology may be best understood as a cascade. Exposure to early adversity alters trajectories of brain development, which in turn leads to social, emotional, and cognitive impairment, followed by the adoption of health-risk behaviors.

The extent to which the brain is affected by exposure depends on the timing of the exposure, the nature of the abuse, and an individual's gender, current age, and genetic risk factors. Brain regions that have been reported to be particularly susceptible to the effects of abuse include the hippocampus, amygdala, corpus callosum, prefrontal cortex, sensory cortex, cerebellum, and the fiber pathways to and from limbic regions. The hippocampus plays an important role in the encoding of spatial and emotional memories. It is one of the few brain regions to produce new neurons postnatally, and a disruption in this process has been associated with the development of depression. The amygdala is activated by fearful stimuli and operates in a reciprocal manner with the hippocampus in the storage of memories associated with emotional events. The brain regions are major components of the limbic system, involved in emotional expression, long-term memory, olfaction, and the initiation of behavioral reactions (such as fight, flight, or freeze). The corpus callosum is the largest nerve fiber tract in the brain and serves as the "information superhighway" between the left and right cerebral cortex. The prefrontal cortex is a slowly maturing part of the brain that is involved in working memory and executive functions. These higher-level processes include complex decision-making, prediction of outcomes from actions, inferring the mental state of others, and the inhibition of socially inappropriate behaviors. The cerebral cortex has specific regions where our primary sensory systems (vision, hearing, taste, smell, and touch) project, and these regions are largely responsible for our primary perception of sensory events. The cerebellum is largely known for its role in the smooth control and coordination of movement. Recent research also shows that it plays an important role in the regulation of attention and coordination of emotional responses.

Sensitive periods. David H. Hubel and Torsten N. Wiesel were awarded the Nobel Prize in Physiology or Medicine in 1981 for their discovery of the functional organization of the visual cortex and sensitive periods when the visual cortex was most influenced by early experience. These sensitive periods also apply to other regions of the brain. Moreover, because brain regions mature at different rates, they may differ in their susceptibility to the effects of early experience. The hippocampus, for example, attains approximately 85% of its adult size by 4 years of age. In contrast, the prefrontal cortex reaches its maximal size during adolescence.

Based on differential rates of maturation, researchers have hypothesized that specific brain regions should have differing periods of sensitivity to the effects of abuse. In one study, right-handed women, 18–22 years of age, who had three or more episodes of forced contact CSA, accompanied by fear or terror, were recruited. Controls were healthy women of the same age with no history of CSA or psychiatric disorders, raised in families with comparable socioeconomic status. Hippocampal and amygdala volume, frontal cortex gray matter volume (GMV), and midsagittal area of the corpus callosum were measured with magnetic resonance imaging (MRI) scans. Data were analyzed to assess the effect of CSA occurring during different years and developmental stages. Designated stages were preschool (3–5 years), latency (6–8 years), prepubertal (9–10 years), pubertal (11–13 years), and adolescent (14–16 years). Stages were selected to be relatively short, to make it possible to detect multiple distinct windows of vulnerability, as animal studies suggested that vulnerable periods may be brief.

Figure 1 illustrates the effects of CSA experienced during a given index age on hippocampal volume (Fig. 1*a*) and corpus callosum area (Fig. 1*b*) by comparing subjects who experienced CSA at that time to healthy controls. The hippocampus appeared to be

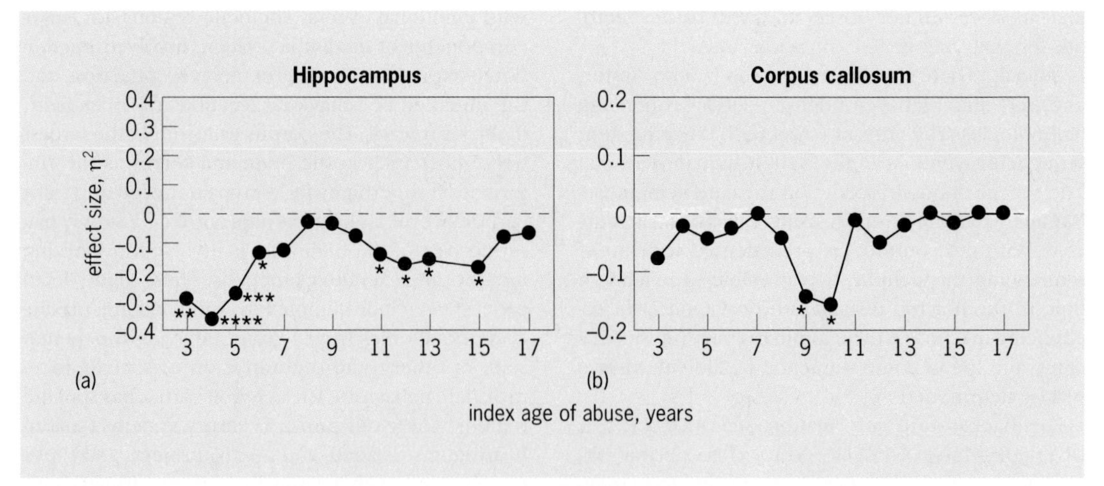

Fig. 1. Effects of exposure to CSA during specific ages on (*a*) hippocampal volume and (*b*) area of the rostral body of the corpus callosum. MRI scans from female subjects (18–22 years of age) retrospectively reporting sexual abuse during a given index age are compared to those of healthy controls of comparable age. Differences are indicated in terms of the effect size measure (eta squared), indicating the proportion of the variance in these measures that can be attributed to abuse at that age (p* < 0.05; ***p* < 0.01; ****p* < 0.001; *****p* < 0.0001).**

most vulnerable between 3 and 6 years of age. There also appeared to be a second period of vulnerability extending from 11 to 15 years. In contrast, the rostral body of the corpus callosum was vulnerable in the same group of subjects at 9–10 years of age.

Figure 2 shows the relationship between exposure during different developmental stages and MRI measures using a more comprehensive statistical model that accounts for the tendency of abusive experiences to carry over into subsequent stages. Hippocampal volume was reduced in association with CSA at 3–5 and 11–13 years. Corpus callosum was reduced with CSA at 9–10 years, and frontal cortex gray matter volume was attenuated in subjects with CSA at ages 14–16. The amygdala did not show any evidence for a sensitive period in these subjects.

The apparent vulnerability of the hippocampus to early stress fits with preclinical observations showing that exposure of the immature hippocampus to corticotropin-releasing hormone (CRH), a key limbic stress modulator, results in a delayed and progressive effect on cell survival and neuronal branching. Furthermore, there is a special population of cells in the immature (but not the adult) hippocampus that releases CRH in response to stress, potentially explaining the increased sensitivity of the hippocampus to

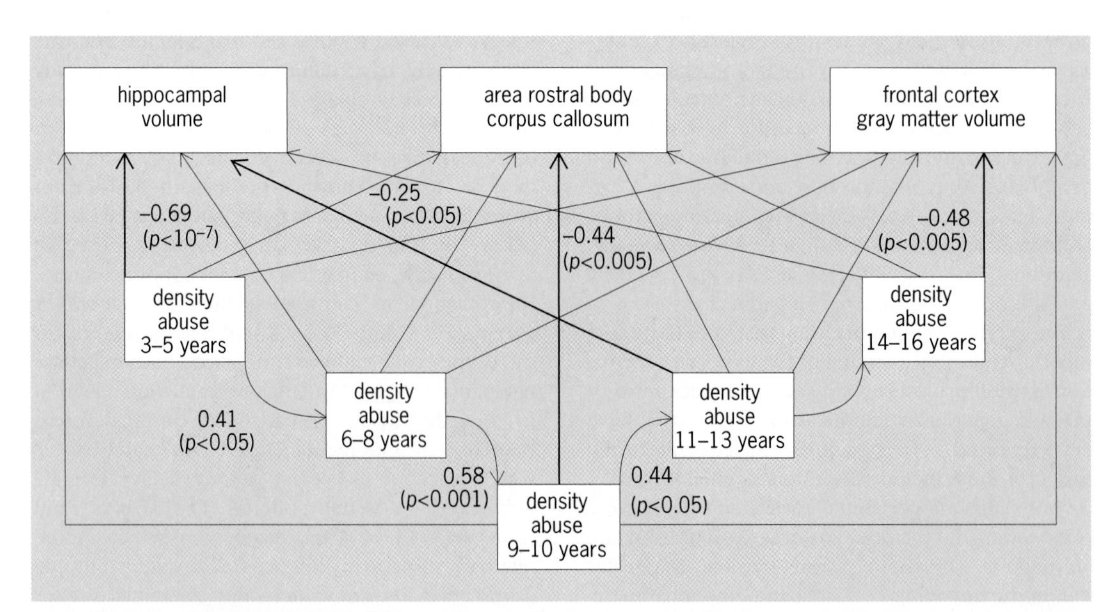

Fig. 2. Path analysis indicating relationships between density of abuse during different stages of development and covaried measures of regional brain size derived from structural equation modeling. Path analysis examined two main components. The first was that CSA during one period would predict CSA during the subsequent period. The second component examined the association between density of CSA during each stage and all morphometric measures. Numerical values represent standardized beta-weights (a measure of effect size) and their associated probability values. Light lines indicate paths that were evaluated in the model, but were not significantly predictive of any relationship between the variables. (*From S. L. Andersen et al., Preliminary evidence for sensitive periods in the effect of childhood sexual abuse on regional brain development, J. Neuropsychiatr. Clin. Neurosci., 20:292–301, 2008, with permission*)

abuse during early childhood. Early hippocampal vulnerability is consistent with evidence from humans and primates indicating that the hippocampus matures rapidly and is functional early in childhood. The amygdala may have an even earlier sensitive period than the hippocampus, as it has attained full adult size by 4 years of age. Amygdala abnormalities have been reported in imaging studies of children and adolescents who suffered very early emotional deprivation from traumatic experiences in orphanages. In contrast, prefrontal cortical functions mature during adolescence and early adulthood; this region grows at a slow rate until approximately 8 years of age, followed by a rapid growth spurt between ages 8 and 14. The lack of apparent effect of exposure to early CSA on frontal cortex volume is consistent with finding in rodents, which showed vulnerability of the hippocampus, but not prefrontal cortex, to early isolation stress. Corpus callosal vulnerability between 9 and 10 years of age is consistent with studies using diffusion tensor imaging (a technique of MRI), which showed that substantial changes occur in this structure between 8 and 12 years of age.

Silent period. Although the hippocampus may be particularly vulnerable at 3–5 years of age, there appears to be a "silent period" between exposure and manifestation of the consequences. Hence, studies that have examined hippocampal volume in children exposed to abuse have failed to detect a reduction in hippocampal volume. In contrast, reduced hippocampal volume has been consistently observed in studies examining adults who experienced early abuse. Likewise, preclinical studies have shown that exposure to early stress affects synaptic density in the hippocampus of developing rats. However, this effect does not become manifest until after they have passed through puberty into early adulthood. A silent period of approximately 9 years also has been observed between exposure to CSA and emergence of major depression and posttraumatic stress disorder, suggesting that these may be clinical correlates of delayed effects on the hippocampus.

Unique effects. Recent studies have shown that there may be unique effects associated with exposure to different types of abuse. In particular, brain regions that receive and process the adverse sensory input appear to be specifically affected. Young adults who experienced CSA or who visually witnessed domestic violence were found to have substantial reductions in gray matter volume in the visual cortex. Also, the nerve fiber pathway connecting the visual cortex to the limbic system was affected. In contrast, young adults exposed to high levels of parental verbal abuse as children were found to have an increase in gray matter volume in their left auditory cortex and a reduction in the integrity of the fiber pathway interconnecting Wernicke's area and Broca's area (regions of the brain for reception and expression of speech). Exposure to harsh physical punishment, conversely, was associated with a reduction in gray matter volume in the medial and dorsolateral prefrontal cortex, reduced integrity in cortical pain pathways, and reduced blood flow into major portions of the dopamine system (for example, caudate, putamen, accumbens, prefrontal cortex, and substantia nigra). Interestingly, young adults exposed to harsh physical punishment (but no other forms of maltreatment) had fewer symptoms of depression or anxiety than young adults exposed to sexual or emotional abuse; however, they used drugs and alcohol to a much greater degree.

Overall, these findings are consonant with the theme that the developing brain is shaped and molded by early experience. A key question is whether alterations relating to childhood abuse are the dysfunctional consequence of damage resulting from high levels of stress exposure during sensitive periods, or whether they serve a potentially adaptive purpose. The fact that sensory systems and pathways that receive, process, and relay the aversive inputs are affected suggests that the brain is modifying itself in specific ways to alter its perception or reaction to the exposure. The hypothesis that the brain is modified further in ways that may enable the individual to compete and reproduce more successfully in a malevolent environment requires further study.

For background information *see* AFFECTIVE DISORDERS; BRAIN; COGNITION; DEVELOPMENTAL PSYCHOLOGY; EMOTION; MEDICAL IMAGING; NEUROBIOLOGY; POSTTRAUMATIC STRESS DISORDER; PSYCHOLOGY; STRESS (PSYCHOLOGY) in the McGraw-Hill Encyclopedia of Science & Technology.

Martin H. Teicher

Bibliography. R. F. Anda et al., The enduring effects of abuse and related adverse experiences in childhood: A convergence of evidence from neurobiology and epidemiology, *Eur. Arch. Psychiatr. Clin. Neurosci.*, 256:174–186, 2006; S. L. Andersen et al., Preliminary evidence for sensitive periods in the effect of childhood sexual abuse on regional brain development, *J. Neuropsychiatr. Clin. Neurosci.*, 20:292–301, 2008; J. Choi et al., Preliminary evidence for white matter tract abnormalities in young adults exposed to parental verbal abuse, *Biol. Psychiatr.*, 65:227–234, 2009; M. H. Teicher, Scars that won't heal: The neurobiology of child abuse, *Sci. Amer.*, 286:68–75, 2002; M. H. Teicher et al., Neurobiology of childhood trauma and adversity, pp. 112–122, in E. Vermetten, R. A. Lanius, and C. Pain (eds.), *The Impact of Early Life Trauma on Health and Disease: The Hidden Epidemic*, Cambridge University Press, Cambridge, U.K., 2010; A. Tomoda et al., Childhood sexual abuse is associated with reduced gray matter volume in visual cortex of young women, *Biol. Psychiatr.*, 66:642–648, 2009.

Einstein's mass–energy equivalence principle

Einstein's mass–energy equivalence principle, $E = mc^2$, where E is energy, m is mass, and c is the speed of light, is probably the best-known formula in science. Despite its algebraic simplicity, however, it demonstrates an inherent fundamental difficulty for direct experimental verification: The comparison between energy and mass is realized by a factor of c^2, which magnifies any mass variation by 17 orders of magnitude. Accordingly, the energy equivalent of 1 kg (2.2 lb) of mass is comparable to the energy released by the largest nuclear explosion achieved so far. Therefore, the choice of physical system for experimental verification of Einstein's formula is limited to microscopic masses. Here the thermal neutron capture reaction appears to be the best candidate: A thermal neutron with a kinetic energy of a few millielectronvolts (meV) induces a nuclear reaction in which a new isotope with excitation energy of the order of 10 MeV is formed. The relative energy uncertainty of this reaction is therefore of the order of 10^{-9}, which would allow carrying out very precise experiments.

Basis of a direct test. The idea of an experiment based on a neutron-capture reaction consists of measuring the difference between the masses before and after the reaction and comparing the result to the energy released. The mass variation can be obtained from a careful measurement of all involved particle and atom masses. The energy released is given by the energy difference of the capture state (excited state) of the newly formed isotope and its ground state. The two states are connected via a sequence of gamma rays whose energies need to be measured. Einstein's energy–mass equivalence principle applied to the neutron-capture reaction of an isotope X with mass number L and forming the isotope with mass number $L+1$ can be written as in Eq. (1).

$$[A_r(n) + A_r(^{L}X) - A_r(^{L+1}X)]\, c^2 \frac{10^{-3}}{N_A h} = \sum_i \upsilon_i \quad (1)$$

The masses of atoms and particles are measured in relative mass numbers A_r as shown on the left-hand side of Eq. (1). [Thus, $A_r(n)$, $A_r(^{L}X)$, and $A_r(^{L+1}X)$ are the relative mass numbers, respectively, of the neutron, the isotope with mass number L, and the isotope with mass number $L+1$.] To convert relative mass numbers into real masses, they need to be multiplied by the so-called atomic mass unit u, which is defined within the SI system to be $u = (10^{-3}/N_A)$ kg. Here N_A denotes the Avogadro constant. To convert masses into energies, the left-hand side contains the factor c^2, which is exactly known. The right-hand side would contain a sum of gamma-ray energies. However, each gamma-ray energy E_i was converted into a corresponding frequency υ_i using the Planck relation, $E_i = h\upsilon_i$. Moving the Planck constant h to the left-hand side leaves a sum of frequencies on the right-hand side. The factor $N_A h$, named the molar Planck constant, can be extracted directly from its relation to a number of other fundamental constants, given in Eq. (2). All the constants on the right-hand

$$\frac{10^{-3}}{N_A h} = \frac{1}{c} \frac{2R_\infty}{\alpha^2 A_r(e)} \quad (2)$$

of this equation, namely, the Rydberg constant R_∞, the fine-structure constant α, and the relative electron mass number $A_r(e)$, have experimental uncertainties of 10^{-9} or better. Therefore the value of the molar Planck constant is assumed to be known with sufficiently high precision. The remaining task consists of the measurement of the relative atomic mass numbers and of the gamma-ray frequencies in Eq. (1).

Measurement of relative atomic masses. The relative atomic masses on the left-hand side of Eq. (1) can be measured using so-called Penning traps. The basic principle consists of confinement of a charged particle (such as an ion or molecule) with charge q in a strong homogeneous magnetic field B. This causes a circular cyclotron motion of the particle. The magnetic field results in confinement only in the horizontal plane, and electric fields are used for vertical trapping (**Fig. 1**). The cyclotron motion is typically superposed on other periodic motions; however, the detection and signal treatment electronics allow separation of the individual frequencies of these motions. For accurate measurements, all kinetic energy needs to be taken away from the particle before it is loaded into the trap. The cyclotron frequency ω_c is the signal that needs to be measured; it depends on

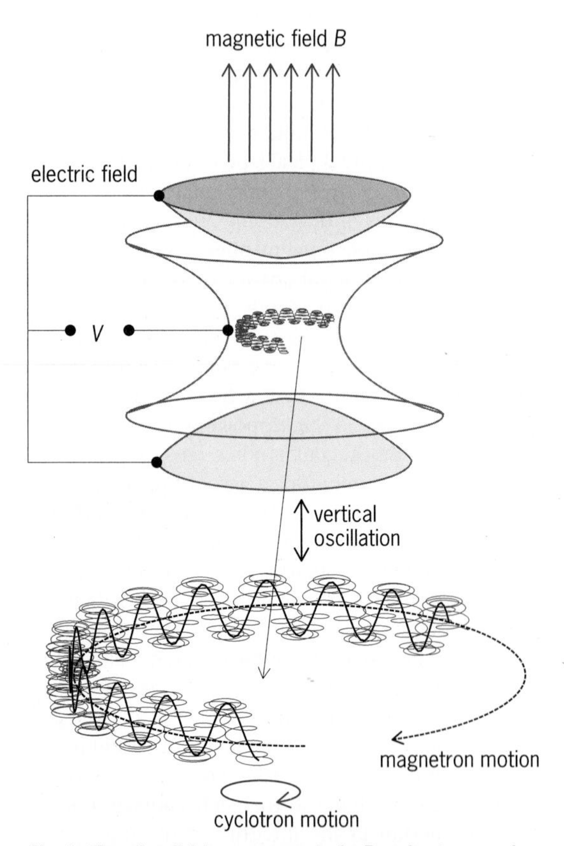

Fig. 1. Trapping field arrangement of a Penning trap, and trajectory of charged particle inside the trap. In the lower part of the figure, the superposition of different periodic motions (cyclotron motion, vertical oscillation, and magnetron motion) is shown. The signal from the cyclotron motion needs to be separated in order to measure its frequency.

the mass m of the particle as in Eq. (3). From this it

$$\omega_c = \frac{qB}{m} \tag{3}$$

follows that the mass ratio of two different charged particles, X and Y, precessing in the same magnetic field, is given by Eq. (4). As can be seen, this ratio

$$\frac{m(^{L}X)}{m(^{K}Y)} = \frac{A_r(^{L}X)}{A_r(^{K}Y)} = \frac{q(^{L}X)\omega_c(^{L}X)}{q(^{K}Y)\omega_c(^{K}Y)} \tag{4}$$

can be obtained from measurement of the ratio of the cyclotron frequencies. For precise values, a number of additional corrections need to be applied: Charged ions or molecules are missing electrons, and corrections need to be made for the corresponding mass deficits and binding energies. From the mass ratios as shown in Eq. (4), it is possible to deduce mass differences as required by Eq. (1).

It is worth noting that the expressions on the left-hand side of Eq. (1) have a particular numeric shape. In fact, the relative atomic mass of the neutron, $A_r(n)$, is of the order of one. The difference $A_r(^{L+1}X) - A_r(^{L}X)$ is also of the order of one. When these two expressions are subtracted, the remaining value is very small, but it still needs to be determined with very high accuracy. This puts rather strong demands on the Penning trap measurements. In fact, each relative atomic mass needs to be measured by a factor 1000 more accurately than the right-hand side of Eq. (1). Until 2004, the mass ratio was determined by loading single molecule species into the trap in a sequential way. In this measurement scheme, a mass comparison requires high stability of the magnetic field over time. This stability could not be realized with sufficient accuracy, which was the main limiting factor in the measurement. When the two atomic masses of interest are placed simultaneously in one trap, the magnetic field instabilities affect the cyclotron frequencies of both particles simultaneously. For the mass ratio the instability cancels out.

The left-hand side of Eq. (1) contains the relative atomic mass of the neutron. This is a neutral particle without electrical charge and cannot be held in a Penning trap. This problem can be solved by carrying out the neutron-capture experiment for at least two different elements. In this case, the measurement of the first element is used for the determination of $A_r(n)$. The value is then inserted into the equation for further experiments.

Measurement of gamma-ray frequencies. The measurement of gamma-ray frequencies in the right-hand side of Eq. (1) is realized via a wavelength measurement, as the speed of light (the propagation velocity of gamma rays) is known exactly. The wavelengths are determined via crystal diffraction: If a beam of electromagnetic radiation, such as gamma rays, passes a periodic arrangement of scattering centers such as atoms, a deviation of some fraction of the beam by a specific angle occurs. In practice, perfect silicon crystals are used for a periodic arrangement of atoms. The underlying physical law for this process is Bragg's law, given by Eq. (5). Here $n = \pm1$,

$$n\lambda_i = 2d \sin\theta_i^{(n)} \tag{5}$$

$\pm2, \ldots$, is an integer value called the diffraction order, λ_i is the wavelength of the ith gamma ray, d is the periodicity constant of the diffracting crystal, and $\theta_i^{(n)}$ is the diffraction angle for the ith gamma ray. The practical realization of the measurement is done via a double diffraction at two crystals consecutively (**Fig. 2**). The diffraction at the first crystal defines a direction in space very accurately. The second crystal is rotated successively into two different Bragg angles $\pm n$, and the angle between these two orders is measured.

The wavelength measurement is based on the knowledge of the lattice spacing d of the second crystal. The lattice spacing of the [220] orientation of the silicon crystals used in the measurement has been obtained with a relative uncertainty of 5×10^{-8} from comparative measurements with the currently best-known crystals (silicon crystals for the determination of the Avogadro constant). The precise measurement of the crystal diffraction angles is realized using optical interferometers. These

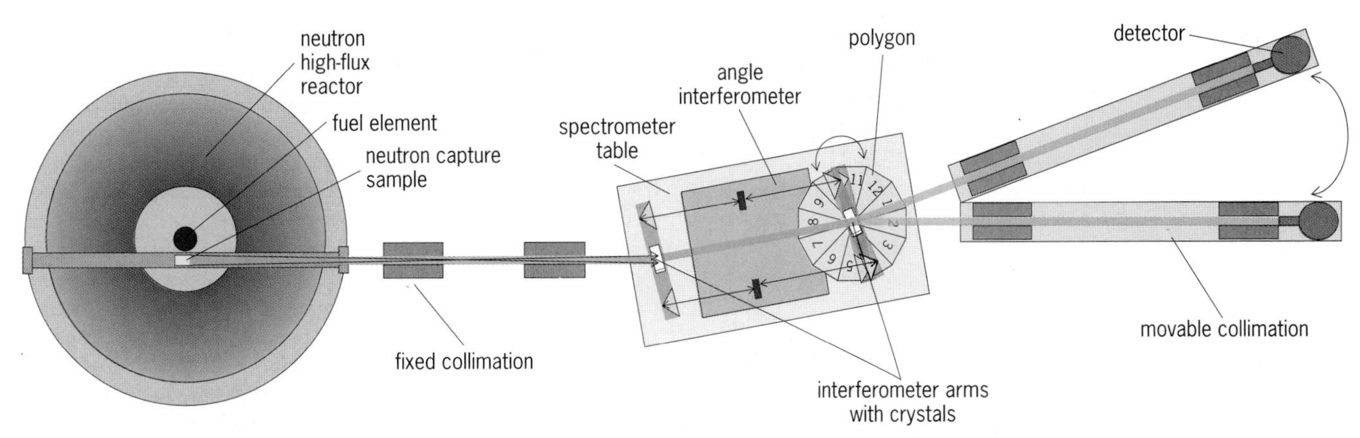

Fig. 2. Gamma-ray spectrometer GAMS4 at the high-flux reactor of the Institut Laue-Langevin. Samples are placed close to the reactor core in an intense neutron flux. The gamma-ray beam is then collimated toward the double-crystal spectrometer. Gamma rays are measured by double-crystal diffraction. The diffraction angle is measured as the difference of two diffraction orders of the second crystal.

interferometers provide good resolution of about 0.1 nanoradian over a range of about 20°. The angle interferometers need to be calibrated. This is done by measuring the 24 face angles of a 24-fold polygon. As the sum of these 24 angles has to be 360°, this procedure provides an absolute calibration.

Experiments and results. The experimental techniques discussed so far span a variety of criteria to yield their best possible performance. Following these criteria, three elements were chosen for the experiments: Hydrogen-1 and -2 (^{1,2}H, the proton and deuteron) were used for the determination of the neutron mass number $A_r(n)$, while silicon-28 and -29 (^{28,29}Si) and sulfur-32 and -33 (^{32,33}S) were used for the verification of the energy–mass equivalence principle. The Penning trap experiments were carried out at the Massachusetts Institute of Technology. The measurement of the gamma-ray energies was done in a collaboration of the National Institute of Standards and Technology and the Institut Laue-Langevin (ILL). Samples were placed in the neutron high-flux reactor of the ILL close to the core of the reactor. The gamma-ray wavelengths were measured at a distance of about 15 m (50 ft) using the crystal spectrometer GAMS4.

The combination of Penning trap and crystal diffraction experiments with hydrogen provided the most accurate measurement of the neutron mass so far, with $A_r(n) = 1.008\,664\,915\,97 \pm 0.000\,000\,000\,43$. The combination of experiments with sulfur and silicon yielded the most accurate direct verification of Einstein's mass–energy relation. With a relative uncertainty of 4×10^{-7} it was shown that in a thermal neutron-capture reaction mass and energy variation behave according to Einstein's prediction. This measurement improved the experimental validation by a factor of 50 compared to the most accurate previous measurements.

For background information *see* ATOMIC MASS; AVOGADRO NUMBER; FUNDAMENTAL CONSTANTS; GAMMA RAYS; NEUTRON; PARTICLE TRAP; PLANCK'S CONSTANT; RELATIVITY; REST MASS; RYDBERG CONSTANT; X-RAY DIFFRACTION in the McGraw-Hill Encyclopedia of Science & Technology.

Michael Jentschel

Bibliography. M. S. Dewey et al., Precision measurement of the ^{29}Si, ^{33}S, and ^{36}Cl binding energies, *Phys. Rev. C*, 73:044303 (11 pp.), 2006; S. Rainville et al., World Year of Physics: A direct test of $E = mc^2$, *Nature*, 438:1096–1097, 2005; S. Rainville, J. K. Thompson, and D. E. Pritchard, An ion balance for ultra-high-precision atomic mass measurements, *Science*, 303:334–338, 2004.

Electric generation ancillary services

The electricity market is evolving. For many years the electricity market was under rigid control. In the United States that rigid control was defined by the Public Utility Holding Company Act (PUHCA) of 1935, which prohibited nonutilities from participating in the market. The prohibition was accomplished by defining a utility as any entity that sold electricity for resale.

As a result of the PUHCA, any electricity wholesaler was subject to all state and federal laws regarding its operation, including the operation of nonutility businesses. Potentially, a refinery that needed steam for its operation and cogenerated surplus electricity could find its oil operations subject to regulation by the utility commission. Industrial plants that needed steam and could cogenerate electricity avoided being treated as a utility by forgoing low-cost cogeneration opportunities, operating without any connection to the rest of the grid, and installing switches that opened automatically when electricity flowed out of the plant to the grid.

The resulting inefficiencies were reversed by Public Utilities Regulatory Policies Act (PURPA) of 1978, which exempted qualifying facilities from PUHCA constraints, and thus new entrants appeared in the electricity market.

Electricity is a continuous service. Customers require energy and voltage regulation in variable amounts throughout the day. Historically utilities provided the energy and voltage regulation as a packaged service. The initial restructuring of the electric industry led to new entrants selling electricity at wholesale to the utilities. Subsequent restructuring allowed the new entrants to sell electricity at retail to residential, commercial, and industrial customers.

The new entrants often own power plants or control other discrete blocks of power. The discrete blocks of electricity under the control of nonutilities generally do not exactly match the requirements of continuous and variable electric service. Ancillary services were developed to provide the difference between the discrete blocks of electricity and the needs of the consumers. Ancillary services can also be sold to a utility, such as when an electricity generator sees a need of the utility that is not being served adequately. The prices for ancillary services may be determined by either a cost-based pricing model or a competitive pricing model. In the United States, ancillary services are regulated by the Federal Energy Regulatory Commission (FERC).

Exploring ancillary services. The electricity market is evolving. The ancillary services part of the electricity market is in its infancy, as is indicated by the structure of many ancillary service contracts. Most ancillary service contracts are for specific pieces of equipment that nominally address a single issue. Reactive power as an ancillary service is being sold increasingly in the United States, based on the capability of a generating unit to produce volt-amperes reactive (VARs), instead of having a competitive market for VARs integrated over time, variously abbreviated RVAH (reactive volt-ampere hours) or VARH (VAR hours). Similarly, some companies are installing devices to offset the effect of harmonic distortions. Again, the ancillary service contracts are for the individually installed devices rather than a competitive market. As the electricity market ma-

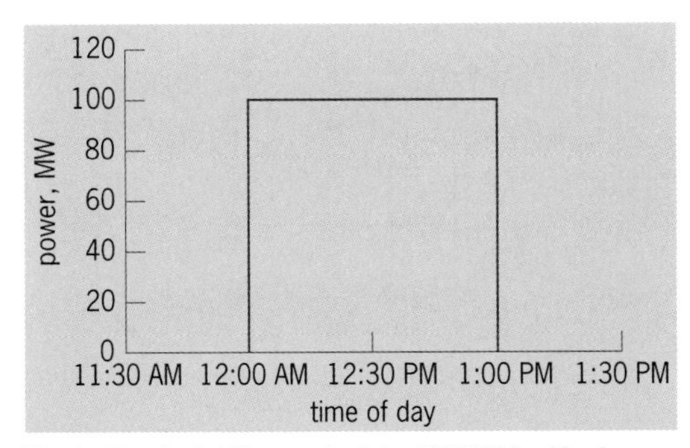

Fig. 1. Standard delivery schedule: 100 MW for 60 min.

tures, there should be competitive pricing for reactive power and harmonic distortions, but the ancillary services market has not yet matured enough in regard to these services.

Bulk electricity is typically sold through a forwards "energy-only" market. The market is energy-only in the sense that there is no provision with regard to reactive power or harmonic distortions. The continuous nature of the delivery of electricity results in the forwards markets being for blocks of active energy with a specified power level and a specified performance period, as is illustrated in **Fig. 1**. Though Fig. 1 is a line drawing, the output of generators is often measured discretely every few seconds, sometimes even within each second.

Figure 2 presents some ancillary service schedules that a generator might need to buy to be able to meet the varying load requirements of its customers:

Fig. 2*a*: 100 megawatts (MW) purchased for 1 min (shape A)

Fig. 2*b*: 100 MW purchased for the last 30 min of an hour (shape B)

Fig. 2*c*: 100 MW purchased for the first 30 min of an hour (shape C)

Fig. 2*d*: 100 MW purchased as a ramping up from 0 over a full hour (shape D)

Fig. 2*e*: 100 MW purchased as a ramping down to 0 over a full hour (shape E)

Choosing the on/off schedule of load shape B simplifies the arithmetic of the calculations for this article. The other load shapes can be analyzed in the same way. For instance, the single ramping schedule of load shape D has been analyzed financially in a filing with FERC. Reactive power is a confounding issue for ancillary services in that it must be utilized in varying amounts depending on the use of the active power, the power amounts shown in Figs. 1 and 2.

The value of electricity changes dramatically with slight imbalances on the system. The electric system in the United States is dominated by independent system operators (ISOs), which are essentially nonprofit superutilities that control the electric market in geographic regions of the country. The ISOs generally operate advanced markets in which buyers and sellers enter bids for the amount of electricity they want or are willing to supply to the market. The resulting transactions are forwards agreements

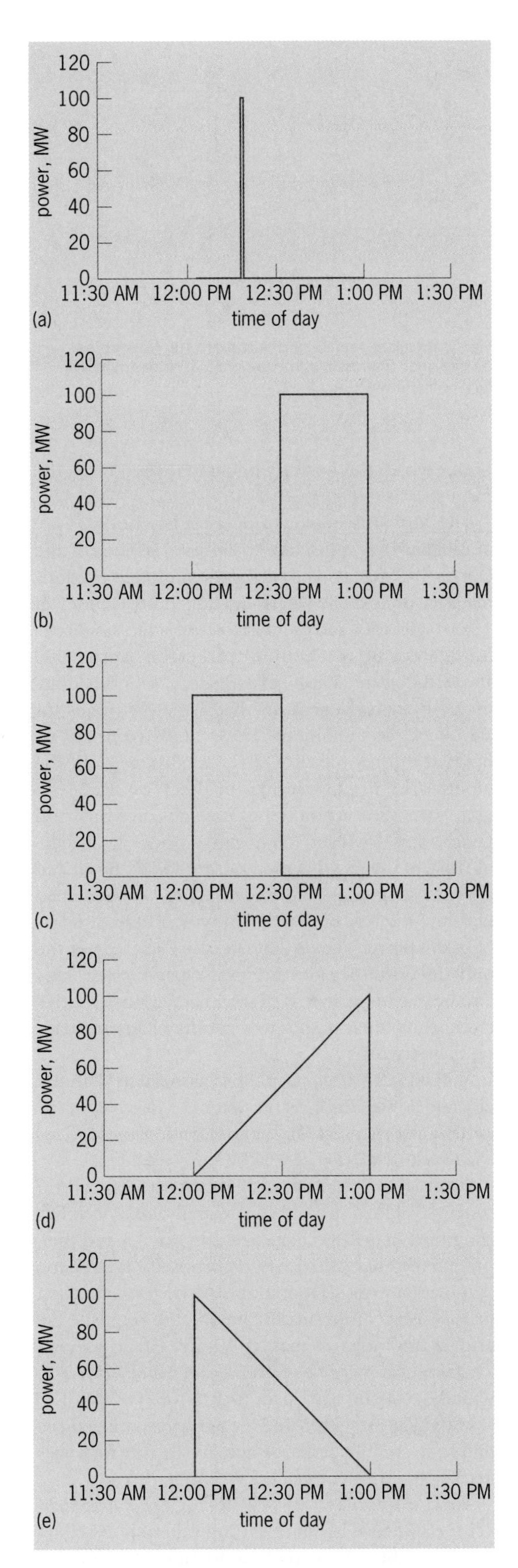

Fig. 2. Ancillary service schedules. (*a*) 100 MW for 1 min (shape A). (*b*) 100 MW for last 30 min of an hour (shape B). (*c*) 100 MW for first 30 min of an hour (shape C). (*d*) Ramp up for 60 min from 0 to 100 MW (shape D). (*e*) Ramp down for 60 min from 100 MW to 0 (shape E).

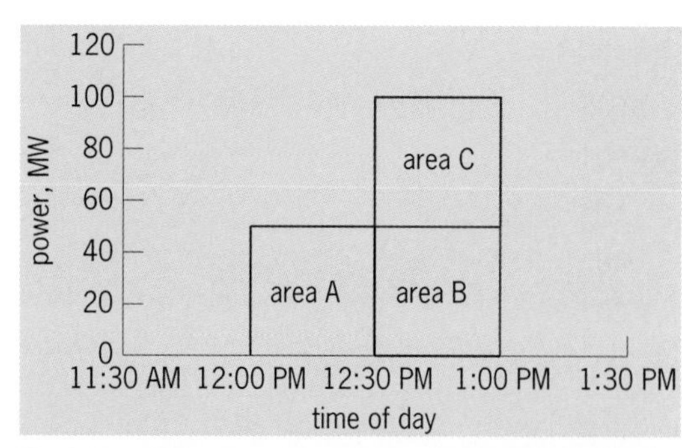

Fig. 3. Ancillary service normalization: 100 MW for last 30 min of an hour with purchase of 50 MW for 1 h in the forwards market.

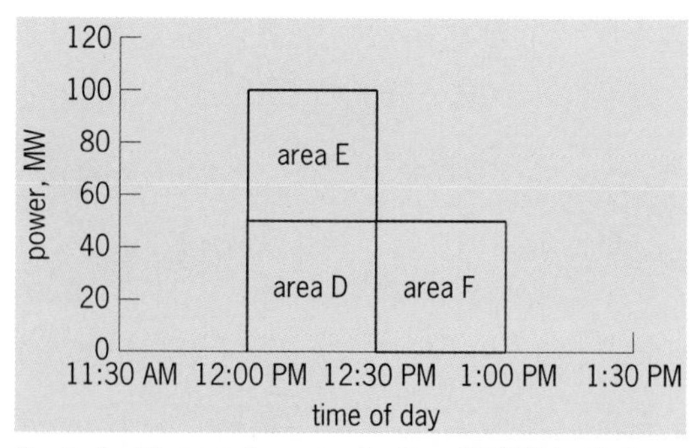

Fig. 4. Ancillary service normalization: 100 MW for first 30 min of an hour with purchase of 50 MW for 1 h in the forwards market.

under which the quantity being delivered takes the form shown in Fig. 1.

The forwards markets operated by the ISOs are nominally very efficient. An unusual aspect of the electricity markets is that they sometimes produce negative prices. The Electricity Reliability Council of Texas (ERCOT) serves most of the state of Texas. An analysis of the ERCOT market for April 2009 showed that the West Texas region, which is dominated by wind generators, had negative prices for about 23% of the month. These negative prices infiltrated the rest of ERCOT for less than 1% of the month. The negative prices in the West Texas region have been attributed to tax subsidies for wind generation. However, the negative prices are also indicative of the fact that electricity must be produced as it is consumed. That negative prices did indeed infiltrate the rest of ERCOT, in which there are lots of steam plants, shows that tax subsidies are not the only drivers of negative prices. Negative prices may also be due to the inertia of steam generators relative to changing their output to meet the changing loads on the network.

Matching the irregular shapes associated with the ebb and flow of the electric system requires both the rectangular shape of Fig. 1 and the ancillary services shapes shown in Fig. 2.

Financial analysis of an ancillary service. The on/off schedule of load shape B can be analyzed by normalizing load shape B to have zero net energy and then pricing the energy in the two portions of load shape B at market prices. Operationally, the normalization process is accomplished by buying the requisite energy in the forwards market. **Figure 3** shows such a transaction, with the purchase of 50 MW for 1 h, which is overlaid onto load shape B for 100 MW. The result is Fig. 3, with three areas representing three different combinations of activity in the forwards market and the spot market.

Area A: Purchase 50 MW in the forwards market for 30 min with an offsetting transaction selling 50 MW in the spot market for 30 min

Area B: Purchase 50 MW in the forwards market for 30 min

Area C: Purchase 50 MW in the spot market for 30 min

Areas B and C occur at the same time and complement each other. Thus, area C appears in Fig. 3 as being above area B.

Table 1 prices the energy presented in Fig. 3 using prices that can be considered to be typical for ERCOT, or at least not too extreme to be credible. The forwards market price is \$23/MWh, which is applicable to both area A and area B. The forwards purchase is a total of 50 MWh. Thus the generator would pay \$1150 for the net energy purchased in the forwards market. The spot price for dumping electricity is a negative \$25/MWh. Thus, for the right to dump the 25 MWh in area A, the generator would pay \$625. The spot price for taking electricity is \$170/MWh. Thus, the generator would pay \$4250 for taking the energy in area C. The total cost to the generator for 100 MW of load shape B is thus \$6025, or \$60.25/MW.

When there is a fair spot market for the unscheduled flows of electricity, some generators will have a contracyclical need for load shape B; that is, they will need 100 MW during the first 30 min and no power during the second 30 min. This is shown in Fig. 2*c*. **Figure 4** is the normalized version of Fig. 2*c*, much as Fig. 3 is a normalized version of Fig. 2*b*. **Table 2** shows that such a contracyclical need for load shape B would be worth \$3725 to the contracyclical generator, even after paying \$1150 in the forwards market for the energy. That is, taking ancillary service B would provide a windfall to the electricity generator. Figure 4 identifies the areas used in Table 2 for the analysis. Such a contracyclical generator might be a storage device, such as a flywheel, pumped storage hydro, or demand response.

The prices used in Tables 1 and 2 are extreme, but they did occur in the course of the normal operation of the ERCOT system for the forwards market in April 2009. Similar prices have occurred in the operations of some of the other ISOs in the United States. Some ISOs have imposed a floor of zero in the bids offered by electricity generators. That the other advanced markets have found the need for negative prices suggests that the floor prices are unduly restrictive.

The financial analysis of the energy ancillary services market can be interrelated to the ancillary services market for voltage regulation and reactive

TABLE 1. Total charge for an ancillary service: 50 MWh of energy in the forwards market, ancillary service of 100 MW for 30 min

	Area in Fig. 3	MWh	Price, $	Extension, $
Forwards purchase	A and B	50	23	1150.00
Spot dumping	(A)	−25	−25	625.00
Spot purchase	C	25	170	4250.00
Total/net		50		6025.00

TABLE 2. Total charge for an ancillary service: 50 MWh of energy in the forwards market, contracyclical ancillary service of 100 MW for 30 min

	Area in Fig. 4	MWh	Price, $	Extension, $
Forwards purchase	D and F	50	23	1150.00
Spot purchase	E	25	−25	(625.00)
Spot dumping	(F)	−25	170	(4250.00)
Total/net		50		(3725.00)

power. Voltage regulation can be effectuated by adding or absorbing reactive power. Because of limitations on the currents that a generator can handle, producing or absorbing reactive power may reduce the amount of active power that an electricity generator may be able to produce. Thus, some independent generators are able to sell reactive power as an ancillary service only by reducing their production of active power. In recognition, some ISOs pay for ancillary reactive power based on the generator foregoing revenue for active energy. Thus, the price of active power sometimes influences the price paid for reactive power on a real-time basis.

Conclusions. Electricity must be produced at the same time that it is consumed. This simultaneity forces some market participants to buy ancillary services to meet the moment-to-moment differences between the electricity they have under their control versus the electricity being used by their customers. In addition, market participants must deal with the reactive power needs associated with voltage control. The market concepts associated with reactive power are still being developed. The reactive ancillary services markets are generally priced on a nonmarket claimed investment cost of installed equipment. Ancillary services associated with the growing issue of harmonic distortions current follows the concepts being developed for reactive power.

The actual need for the various ancillary services is poorly known ahead of time. Hence, the market method evaluated in Fig. 3 is extremely simplified. Though Table 1 develops a very high price for a simple ancillary service, Table 2 shows that the same ancillary service can have a very different price (a credit using the exemplar data) for generators that have a contracyclical need for ancillary services.

An advantage of allowing prices to go negative is the incentive that such prices can provide to storage devices that help the system keep the lights on. In the case demonstrated in Table 2 using Fig. 4, the contracyclical generator earned $3725 for using the ancillary service. Allowing such prices will encourage more entrepreneurs to find solutions to the growth in unscheduled flows of electricity that require ever-growing supplies of ancillary services.

For background information *see* DISTORTION (ELECTRONIC CIRCUITS); ELECTRIC POWER SYSTEMS; ENERGY STORAGE; REACTIVE POWER; VOLT-AMPERE in the McGraw-Hill Encyclopedia of Science & Technology.

Mark B. Lively

Bibliography. M. B. Lively, The WOLF in pricing: How the concept of plug, play, and pay would work for microgrids, *IEEE Power Energy Mag.*, 7(1):61–69, January/February 2009; M. B. Lively, Thirty-one flavors or two flavors packaged thirty-one ways: Unbundling electricity service, *NRRI Quart. Bull.*, Summer 1996; North American Electric Reliability Council, *Interconnected Operations Services Reference Document*, Version 1.0, Princeton, NJ, November 1995; Order 888, Final Rule, in FERC Docket Nos. RM95-8-000 and RM94-7-000, *Promoting Wholesale Competition Through Open Access Non-discriminatory Transmission Services by Public Utilities; Recovery of Stranded Costs by Public Utilities and Transmitting Utilities*, issued April 26, 1996.

Epigenetic mechanisms in development

Transcriptional programming of multicellular development involves regulated interactions between DNA-binding transcription factors and specific DNA sequences in the regulatory regions of target genes. Through these sequence-specific DNA recognition events, transcription factors either activate or repress target gene transcription. In eukaryotic cells, genes are assembled into supercoiled complexes of DNA, histones, and other proteins, known as chromatin. Epigenetic mechanisms help to realize and reinforce the transcriptional decisions made when

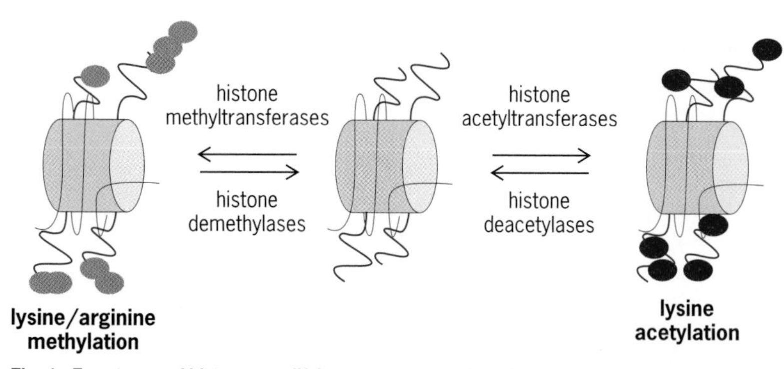

Fig. 1. Four types of histone-modifying enzymes regulate the distribution of methylation (ovals on the left) and acetylation (ovals on the right) on the N-terminal tails of core histones in chromatin.

transcription factors are recruited to specific genes, by introducing, removing, or perpetuating covalent modifications of chromatin structure. These covalent modifications are thought to create a molecular memory of decisions taken to activate or repress gene transcription and they may be inherited in specific cell lineages through successive cell divisions. Thus, epigenetic mechanisms can be defined as systems that facilitate the structural adaptation of chromosomal regions in order to record or propagate a change in the transcriptional activity of the genes contained within those regions. Accordingly, epigenetic mechanisms give rise to heritable phenotypes through alterations to chromosomes that do not involve changes in their DNA sequence.

Posttranslational modifications of chromatin. Posttranslational covalent modifications of the histones within chromatin, together with methylation of the DNA molecule itself, provide two of the best-characterized examples of epigenetic mechanisms that regulate gene transcription via their effects on chromatin structure. Four types of histone proteins, H2A, H2B, H3, and H4, form an octameric core particle around which DNA is coiled to make up the fundamental unit of chromatin called the nucleosome. Within nucleosomes, core histones can be modified covalently by various enzymatic processes.

The histone modifications about which most is known are acetylation of lysine residues and methylation of lysine or arginine residues, located within the histone N-terminal domains that protrude from the surface of the nucleosome. Acetylation of multiple lysines in core histones is catalyzed by histone acetyltransferases (HATs) and is frequently a hallmark of transcriptionally active genes, but these modifications can be removed by histone deacetylases (HDACs), which are associated with transcriptionally silent genes. By contrast, histone lysine methylation is catalyzed by residue-specific histone methyltransferases (HMTs) and this type of modification can signify either transcriptional activity or inactivity, depending on which lysine residues are modified. Thus, methylation of lysine 9 or lysine 27 in histone H3 is associated with transcriptional silencing, whereas methylation of lysine 4 in histone H3 can signify a state of permissiveness for the induction and/or maintenance of transcriptional activity of the associated target gene. As with acetylated histones, histone methylation marks can be removed by residue-specific demethylases. Therefore, the levels and distribution of histone acetylation and methylation marks that are associated with a specific gene depend both on the abilities of four different types of histone-modifying enzymes to be recruited to that gene and on the accessibility of the associated histones to these enzymes (**Fig. 1**). Once acetylation marks are in place, they act as recruitment signals for bromodomain-containing proteins that promote gene transcription. Once histone methylation marks are created, they can bind to a variety of proteins that may be involved in transcriptional repression or activation (these include chromodomain-containing and other core motif-containing transcription cofactors). In this way, histone modifications underpin the stable, albeit reversible, formation of multisubunit protein assemblies in chromatin that influence transcription of the associated genes. Precisely how individual histone marks contribute to determining whether an associated gene is transcriptionally active or inactive in a particular cell type depends on the variety and abundance of other histone modifications prevailing in the same region of chromatin. In other words, the molecular context is crucial to interpreting the transcriptional meaning of a particular individual histone modification.

As with histones, DNA can also be methylated by specific enzymes, known as DNA methyltransferases (DNMTs), which target cytosine residues within cytosine-guanine (CG) dinucleotides. Cytosine methylation is associated with transcriptional silencing and enriched at CG dinucleotides in chromosomal regions in which noncoding repeated DNA sequences and transposable elements are abundant. CG dinucleotides found in the transcriptional control elements of genes, such as promoters and enhancers, are relatively deficient in cytosine methylation. However, the cytosine methylation status of some CG dinucleotides within specific transcriptional control sequences of a small proportion of genes does change during multicellular development, as a consequence of the activity of DNMTs that are recruited to these genes in a cell-lineage-specific fashion. The resultant methylated CG-containing DNA sequences then act as recruitment signals for methyl-CG-binding proteins, which interact specifically with other components of the transcription silencing machinery, including histone-modifying enzymes, to stably repress the corresponding genes in particular cell types. The interplay between histone modifications and DNA methylation is thought to be crucial in allowing the transcriptional status of genes to be transmitted stably in proliferatively active cells, but it also provides multiple ways in which transcriptional status of a gene can be altered through the recruitment of enzymes with appropriate catalytic activities.

Epigenetic control of pluripotency in vertebrate embryos and embryonic stem cells. Development of a vertebrate embryo from a fertilized egg initially involves the production of a population of proliferating cells that acquire pluripotency, or the capacity to develop into several distinct cell types, while maintaining their capacity for self-renewal. In mammals, pluripotent cells, known as embryonic stem (ES) cells, arise within the inner cell mass of the early embryo. They can be cultured indefinitely in vitro, and they can differentiate into all cell lineages, including the germline. In lower vertebrates, for example, the zebrafish, similar pluripotent cells accumulate in the blastula stages around the time of zygotic gene activation at the maternal-zygotic transition. Analysis of the epigenetic status of genes both in mammalian ES cells and in zebrafish blastomeres at the maternal-zygotic gene transition has identified a population of developmental regulatory genes that are transcriptionally silent in pluripotent ES cells and blastomeres and associated with histone H3 molecules that are methylated on both lysine 4 and lysine 27 residues. The genes within this bivalently modified chromatin include a large proportion of key developmental regulatory genes that are required for the later specification and differentiation of specific cell lineages. When cell differentiation ensues and these genes become expressed, the associated histone H3 molecules lose their lysine 27 methylation marks, but retain those of lysine 4 methylation. Conversely, if the gene is stably silenced upon differentiation, the lysine 4 methylation marks can be lost, whereas those of lysine 27 methylation are retained. These findings indicate that bivalent modifications mark out genes within pluripotent cells that will be required for promoting the subsequent differentiation of multiple cell lineages, and they reveal an involvement of the epigenetic machinery in facilitating cell commitment to specific fates within the developing embryo.

Polycomb group and trithorax group functions in animals and plants. Formation and patterning of the primary axes of multicellular embryos require the correct spatiotemporal expression of *Hox* (homeobox) genes, which is governed in part by the functions of the *Polycomb group (PcG)* and *trithorax group (trxG)* of genes. *PcG* and *trxG* genes encode a variety of chromatin-associated proteins that play crucial roles in generating, recognizing, and interpreting covalent histone modifications within chromatin. PcG proteins maintain transcriptional repression of *Hox* and other developmental regulatory genes, by limiting the activities of trxG proteins that promote the stable transcription of *Hox* and other target genes. Thus, in *Drosophila*, the *PcG* gene *Enhancer of zeste* [*E(z)*] encodes a histone methyltransferase with specificity for histone H3 lysine 27, and the archetypal *PcG* gene *Polycomb* encodes a chromodomain protein that recognizes methylated lysine 27 of histone H3. Moreover, consistent with its role in promoting transcriptional activation, the defining member of the *trxG*, *trithorax*, encodes a histone methyltransferase with specificity for histone H3 lysine 4, which may facilitate poising of target genes for transcriptional activation. Similar functions have been defined for vertebrate homologues of these PcG and trxG proteins in embryonic patterning and the specification of cell fates, and evidence suggests roles for PcG and trxG functions in the creation of bivalent chromatin domains in pluripotent stem cells. In addition to these conserved functions, PcG proteins also are required for a wide variety of other biological processes. For example, in the roundworm *Caenorhabditis elegans*, the homologue of the PcG histone methyltransferase E(z), mes-2, is required for specification of the germline. Another example occurs in the plant *Arabidopsis thaliana*, where HMTs that are closely related to E(z) and methylate lysine 27 of histone H3 play essential roles in seed development and the regulation of flowering time by cold temperatures.

X-chromosome inactivation. One of the first examples of an epigenetic control mechanism to be documented was mammalian X-chromosome inactivation, a gene dosage compensation mechanism which ensures that the somatic cells of XY males and XX females express genes located on the X chromosome at similar levels. There are two distinct phases of transcriptional silencing to which the X chromosome is subjected during early development of female embryos. The first is called imprinted X-inactivation. It occurs shortly after fertilization and is targeted specifically to the paternally derived X chromosome of the zygote. In female embryonic preimplantation stages, this imprint is erased, leading to reactivation of the paternal X chromosome. Shortly after this, all somatic cells then randomly select either the maternally derived or paternally derived X chromosome for a second phase of stable, long-term X-inactivation, known as random X-inactivation. X-inactivation is remarkable because, in both its imprinted and random phases, it is initiated by a nonprotein-coding RNA (*Xist*) that is transcribed from a gene on the X chromosome called the X-inactivation center (XIC). The *Xist* RNA coats the X chromosome from which it was transcribed, which triggers gene silencing throughout the length of the chromosome. After random X-inactivation is initiated by *Xist* RNA in embryonic cells, PcG-dependent lysine 27 methylation of histone H3 molecules becomes widespread on the inactive X, and new DNA methylation is detected at CG dinucleotides within gene regulatory elements along the length of the chromosome.

Genomic imprinting. In mammals, another parent-of-origin-specific transcriptional silencing process known as genomic imprinting occurs at many locations throughout the genome. As a consequence of this, only one of the two alleles of a gene will be transcribed in somatic tissues. One such imprinted gene encodes the nonprotein-coding RNA *H19*, which is found in a cluster of imprinted genes on mouse chromosome 7. Expression of *H19* in somatic tissues is restricted to the maternally derived allele, while the

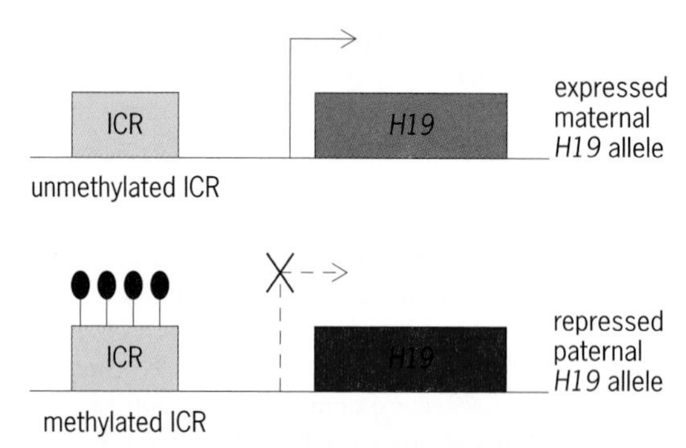

Fig. 2. Transcriptional repression of the *H19* gene is mediated by paternally imprinted DNA methylation at the adjacent imprinting control region (ICR). In the absence of methylation at the ICR, the maternal *H19* allele is expressed.

paternal allele is transcriptionally silent. This differential expression of *H19* paternal and maternal alleles is controlled by the adjacent imprinting control region (ICR). In the male germline, CG dinucleotides within the ICR are methylated, and this methylation status is inherited stably in the somatic tissues of developing embryos, causing transcriptional repression of the paternal allele of *H19* (**Fig. 2**). Conversely, in the female germline, the ICR is unmethylated and this state persists in embryonic somatic tissues, such that the maternal *H19* allele is transcribed. However, in the germline tissues of developing embryos, CG methylation marks within the ICR of the *H19* paternal allele are erased and reestablished during gametogenesis to ensure that the appropriate genomic imprints are created on gamete chromosomes and transmitted to the next generation. It remains unclear how the paternal *H19* ICR is specifically targeted for DNA methylation during male gametogenesis, although it seems likely that DNMTs are recruited to the ICR by a process triggered by ICR sequence-specific DNA-binding proteins.

For background information *see* CHROMOSOME; DEOXYRIBONUCLEIC ACID (DNA); DEVELOPMENTAL BIOLOGY; DEVELOPMENTAL GENETICS; DNA METHYLATION; GENE; GENETICS; GENOMICS; HISTONE; MATERNAL INFLUENCE; NUCLEOSOME; STEM CELLS; TRANSCRIPTION in the McGraw-Hill Encyclopedia of Science & Technology. Vincent T. Cunliffe

Bibliography. S. L. Berger, The complex language of chromatin regulation during transcription, *Nature*, 447:407–412, 2007; S. L. Berger et al., An operational definition of epigenetics, *Genes Dev.*, 23:781–783, 2009; A. Bird, Perceptions of epigenetics, *Nature*, 447:396–398, 2007; B. Schuettengruber et al., Genome regulation by polycomb and trithorax proteins, *Cell*, 128:735–745, 2007; N. L. Vastenhouw et al., Chromatin signature of embryonic pluripotency is established during genome activation, *Nature*, 464:922–926, 2010; J. R. Weaver, M. Susiarjo, and M. S. Bartolomei, Imprinting and epigenetic changes in the early embryo, *Mamm. Genome*, 20:532–543, 2009.

Establishing thermodynamic temperature with noise thermometry

Thermodynamic temperature T is a measure of the average energy content of the constituents of matter in equilibrium. The link between thermodynamic temperature and this energy content is the Boltzmann constant k. According to statistical mechanics, if the thermodynamic temperature of a given system is T, then the average energy of each of its accessible degrees of freedom is of the order of kT. (The precise definition of the Boltzmann constant takes into account other physical concepts, in particular, entropy, and will not be discussed here.)

The system of units accepted worldwide, the Système International d'Unités (SI), lists thermodynamic temperature as a base quantity; the corresponding base unit is the kelvin (K). The kelvin is defined as the fraction 1/273.16 of the thermodynamic temperature of the triple point of water. Instruments measuring the thermodynamic temperature are called primary thermometers, and are based on physical laws that relate, in a specified physical system, thermodynamic temperature to other physical quantities. Primary thermometers are complex and expensive physics experiments, typically carried out at national metrology institutions. The International Temperature Scale (ITS-90) provides a recipe that is accepted worldwide for realizing a temperature scale in a practical way, by explicitly stating the temperature of so-called fixed points, which are triple points (including the triple point of water) and phase transitions of given materials.

Modern metrology regards the present definition of SI units in terms of material properties or human artifacts as being increasingly unsatisfactory. In 2007, Resolution 12 of the 23d General Conference on Weights and Measures officially considered the possibility of redefining "the kilogram, the ampere, the kelvin, and the mole using fixed values of the fundamental constants at the time of the 24th General Conference (2011)." Although which of the constants would be employed in a future redefinition is a matter of present discussion, the Boltzmann constant is always included among the constants that would be so employed. This resolution, together with experimental evidence of deviations between thermodynamic temperature and ITS-90, has motivated increased efforts by national measurement institutions and international collaborations toward the realization of new primary thermometry experiments and new determinations of the Boltzmann constant.

Primary thermometry methods can be roughly divided into three classes if one limits the discussion to methods that can be applied at the triple point of water:

1. Methods based on physical laws of the ideal gas. Among these methods, the oldest is constant-volume gas thermometry, based on the ideal gas law $PV = nRT$ (where P is the pressure, V is the volume,

n is the number of moles of gas, and R is the molar gas constant); more recent methods are acoustic gas thermometry, dielectric-constant gas thermometry, refractive-index gas thermometry, microwave quasispherical cavity resonators, and density measurements.

2. Methods based on Planck's law of blackbody radiation: total radiation thermometry and spectral-band-limited radiation thermometry.

3. Methods based on the statistical properties of physical systems; these include Doppler-broadening thermometry and Johnson noise thermometry.

This article focuses on Johnson noise thermometry and its foundations, past and present developments, and future perspectives.

Johnson noise thermometry. Electrical noise was discovered during the development of vacuum-tube amplifiers. Despite technological improvements, it seemed that there was a limit on the ability of amplifiers to detect faint signals. In 1918, Walter Schottky linked the noise of electric currents to the discreteness of electric charge flow. In 1928, an even more fundamental source of electrical noise was identified in two papers that appeared together in *Physical Review*; John B. Johnson gave a thorough experimental study and Harry Nyquist a theoretical explanation of what is now called Johnson-Nyquist noise.

Nyquist theory explained the noise as a form of blackbody radiation in electrical conductors. Johnson-Nyquist noise is universal; that is, it is independent of the resistor's construction, shape, size, composition, or physical state. Given a resistor R in thermodynamic equilibrium at a temperature T, a voltage noise with a flat power spectrum appears at its terminals. If this noise is observed with a meter having measurement bandwidth B, its root-mean-square value V is given by $V^2 = 4kTRB$.

Johnson noise thermometry is the primary thermometry method based on the accurate measurement of Johnson-Nyquist noise. The main difficulty with its accurate implementation is the faintness of Johnson-Nyquist noise, which has to be amplified by a large gain factor (10^4–10^6) before meaningful voltage measurement methods can be applied.

We may call relative Johnson noise thermometers those where a periodic calibration is performed by measuring the Johnson noise of a resistor (the same or a different one) placed at a reference temperature, typically that of the triple point of water. Absolute Johnson noise thermometers are calibrated with electrical signals (of nonthermal origin) whose amplitude is known in purely electrical SI units. For a long time, relative Johnson noise thermometry has made important metrological contributions to the determination of the temperatures of fixed points, mainly at high temperatures, and to the establishment of ITS-90, but it cannot determine k.

Johnson's experiment of 1928 can be considered the very first example of absolute Johnson noise thermometry. It is presently employed in harsh environments where ITS-90 calibration of thermometry sensors is not feasible, such as in space installations or in nuclear power plants. In recent years, absolute Johnson noise thermometry has been considered to be a viable method for high-accuracy thermodynamic temperature measurements and (when performed at the triple point of water) as a new way of determining k.

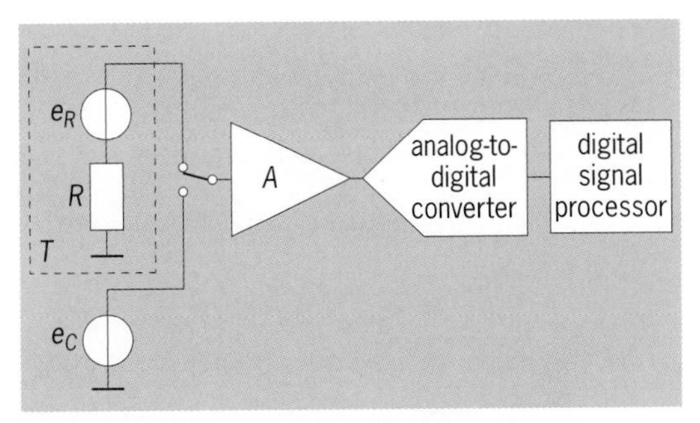

Fig. 1. Simplified schematic diagram of an absolute Johnson noise thermometer.

Absolute Johnson noise thermometry experiments. Absolute Johnson noise thermometry experiments are presently being conducted at the National Institute for Standards and Technology (NIST) in the United States and at the Istituto Nazionale di Ricerca Metrologica (INRIM) in Italy. The NIST experiment has been running since 2001, and the first prototype of the INRIM experiment has been operating since 2009.

A greatly simplified schematic of an absolute Johnson noise thermometry experiment is shown in **Fig. 1**. The Johnson-Nyquist noise signal e_R from a resistor R in a thermostat at temperature T is amplified by A and digitized by an analog-to-digital converter; the output codes are processed by a digital signal processor, where a fast Fourier transform is implemented. Apart from a proportionality factor dependent on the gain of A, the outcome of the measurement is the noise spectrum of e_R. The input of A is then switched to the measurement of a calibrating signal e_C, which is constructed to be of known amplitude and to have spectral properties matching those of e_R. The $e_R - e_C$ measurement cycle is repeated and the results averaged until an adequate precision is obtained. A more refined design would include a doubling of the signal acquisition channel in order to implement correlation algorithms to reject the noise added by A, analog and digital filters, transmission-line matching elements, and so forth.

In the NIST Johnson noise thermometer, e_C is generated with a waveform synthesizer given by a Josephson junction array driven by a programmable radio-frequency pulse generator (**Fig. 2**). The output of the Josephson synthesizer is a pseudorandom white noise, a deterministic signal having a comb spectrum of constant amplitude. This amplitude can be accurately calculated given the pulse generator rate and output codes (and is not dependent on the pulse waveform details). The synthesizer output is thus directly traceable to a frequency standard, and in principle has the same accuracy as the reproduction of dc voltage in national measurement institutes.

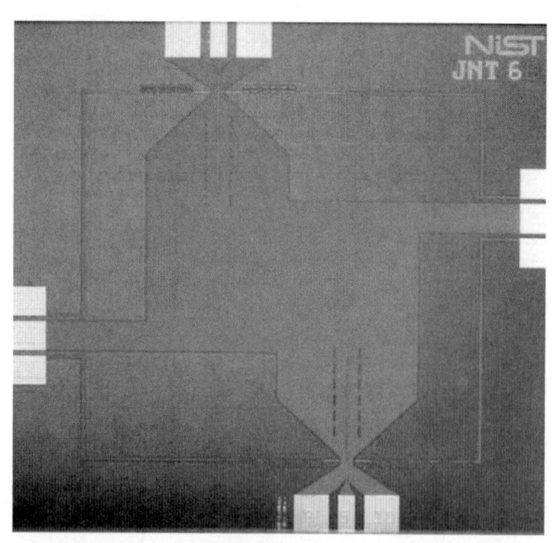

Fig. 2. Optical micrography of a Josephson junction array employed in the NIST Johnson noise thermometer. (*Courtesy of S. P. Benz, NIST*)

The NIST Johnson noise thermometer has been employed in the measurement of the thermodynamic temperature ratio between fixed points, with an uncertainty better than 5×10^{-5}. Although a redetermination of the Boltzmann constant has not yet been published, a target uncertainty of 6×10^{-6} has been estimated to be feasible.

In the INRIM Johnson noise thermometer, e_C is generated by the techniques of standard room-temperature electrical metrology. A flat (but uncalibrated) comb spectrum is generated at a high level (250 mV) with an electronic digital-to-analog converter. The root-mean-square voltage of the comb is measured with an ac-dc transfer technique based on a multijunction thermal converter. The signal is then scaled in amplitude with inductive voltage dividers. In contrast to Fig. 1, e_C is periodically injected in series with e_R to avoid mechanical switching (**Fig. 3**). Despite its short history, the INRIM Johnson noise thermometer has been tested at room temperature, where it gives results in agreement with

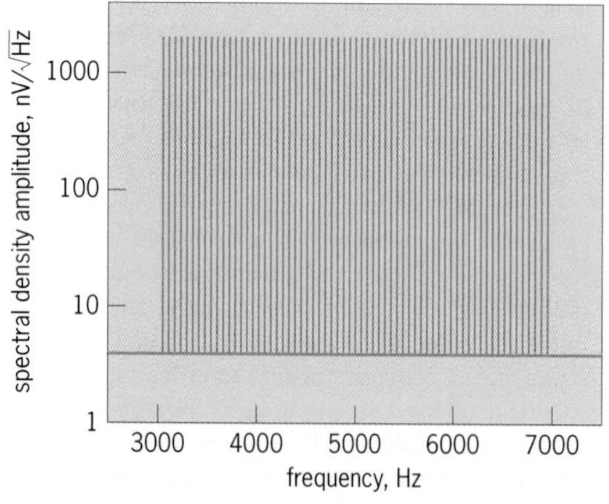

Fig. 3. Output of the INRIM Johnson noise thermometer during the calibration phase. In the power spectrum, the flat frequency comb e_C covers the measurement bandwidth. The spectrum baseline is the white Johnson-Nyquist noise e_R of the resistor *R*.

ITS-90 within a combined relative measurement uncertainty of 6×10^{-5}. Technical improvements that are under development should permit an uncertainty of better than 1×10^{-5} to be attained.

Perspectives. ITS-90 and the presently accepted value of the Boltzmann constant are based on experiments conducted in the 1980s. In the past few years, the metrology community has initiated new experiments based on both old and new primary thermometry methods. In the near future, these experiments will yield both high-accuracy, direct determinations of thermodynamic temperature (including temperatures far away from the triple point of water) and new determinations of the Boltzmann constant. Such results will be the subject of forthcoming discussions in the metrology community about the redefinition of the kelvin and the procedures for its realization. In any case, it is certain that they will soon have a deep impact on the foundations of thermometry.

For background information *See* BOLTZMANN CONSTANT; ELECTRICAL NOISE; GAS THERMOMETRY; HEAT RADIATION; JOSEPHSON EFFECT; LOW-TEMPERATURE THERMOMETRY; PHYSICAL MEASUREMENT; TEMPERATURE MEASUREMENT in the McGraw-Hill Encyclopedia of Science & Technology. Luca Callegaro

Bibliography. S. Benz et al., Electronic measurement of the Boltzmann constant with a quantum-voltage-calibrated Johnson noise thermometer, *Compt. Rendus Phys.*, 10:849–858, 2009; L. Callegaro et al., A Johnson noise thermometer with traceability to electrical standards, *Metrologia*, 46:409–415, 2009; B. Fellmuth, Ch. Gaiser, and J. Fischer, Determination of the Boltzmann constant—status and prospects, *Meas. Sci. Tech.*, 17:R145–R149, 2006; D. R. White et al., The status of Johnson noise thermometry, *Metrologia*, 33:325–335, 1996.

Ethnoarcheology

Since the 1970s, archeologists have carried out fieldwork in traditional societies to help answer questions regarding the interpretation of the archeological record and to develop analogies. This research strategy has been labeled ethnoarcheology and has transformed into one of the main sources of analogy. There are various definitions of ethnoarcheology, but it can be summarized simply as the acquisition of ethnographic data to assist archeological interpretation. It also can be defined as the study of the relationship between human behavior and its archeological consequences. Ethnoarcheology is differentiated from other actualistic studies in that it includes the systematic observation of living societies. It also is differentiated from other types of ethnography through its explicit focus on the intention to identify the archeological (material) context of human behavior. One of the most comprehensive definitions is provided by the archeologist Bill Sillar: "the study of how material culture is produced, used, and deposited by contemporary societies in relation to the wider social, ideological, economic, environmental, and/or technical aspects of the society concerned, and with specific reference to the problems of interpreting archeological material." Within the framework of ethnoarcheology, the key concept is that of analogy, which can be defined broadly as the transfer of information from one object or phenomenon to another based on certain relations of compatibility between them. Ethnoarcheology thus obtains information from a better-known source—living societies—in order to transfer this information to another, less well known subject—extinct societies.

Challenges. Ethnoarcheology has been viewed with mistrust because of the difficulties of extrapolating from contemporary to past societies, starting with the fact that the epistemological bases of how to conduct such extrapolations are not sufficiently developed. Many archeologists have cast doubt on the use of analogies to bridge the gap between the present and past. Nevertheless, the great majority of archeologists recognize the usefulness of analogical arguments in the process of interpretation or explanation of the archeological record and consider them indispensable tools.

Another point that has generated concern is that present-day indigenous societies—the source of analogies—have all had contact with Western society to a greater or lesser degree and are integrated in one form or another into the process of "globalization." It has been proposed that present-day societies cannot serve as analogical references for past societies. This criticism, however, is unjustified. The two elements of analogy (the source and the subject) need not be the same, but instead there should be certain conditions of comparability between terms. Analogy's strength does not lie in the degree of similarity between the source (in this case, the present-day society) and the subject (the past society as perceived through the archeological record), but rather in the logical structure of the argument and the similarity between the terms of the relation. Moreover, it is recognized that the power of a given analogy does not depend on the delimitation of which traditional or "pristine" group is the source of the analogy. Instead, it depends on its logical structure and the conditions of comparability.

Ethnoarcheology as ethnography. Ethnoarcheology attempts to formulate models that permit better understanding of the cultural patterns of human societies in both the present and the past. Essentially, it is a form of ethnography that takes into consideration aspects and relationships that are not approached in detail by traditional ethnographies. Ethnoarcheology is, in some ways, a view of ethnography through archeological eyes.

The use of ethnographic information to interpret the archeological record is neither new nor the exclusive domain of ethnoarcheology. What is new is that ethnoarcheological information has been obtained by archeologists, with the central objective of aiding comprehension of the archeological record. In the 1960s, upon the advent of processual archeology (a school of thought in archeology based on positivism and neo-evolutionism, with emphasis on the ecological relation of human beings and on the reconstruction of cultural process), the archeologist Lewis Binford became interested in ethnographic analogy in a systematic way. He developed his ethnoarcheological approach theoretically and conceptually in his pioneering work, *Nunamiut Ethnoarchaeology*. In addition to Binford's work, further contributions by John Yellen and Richard Gould in the late 1970s established the foundations of ethnoarcheology within the processual paradigm and transformed the subdiscipline into one of the most important producers of models used to interpret the archeological record of past societies.

Current ethnoarcheology. Contemporary ethnoarcheology emerged as a direct result of the testing of actualistic studies and the potential for such studies to explain the archeological record. Consequently, during the late 1970s and especially the 1980s, specific studies of living traditional societies were initiated by archeologists in several parts of the world, including Iran, Australia, Tanzania, India, the Philippines, the lowlands of South America, and the Andes. As such, a new approach was born—the search for general principles that connected human behavior to material culture, and the obtaining of conclusions that did not depend exclusively on sociocultural anthropological theory. Main research themes in this period included the production, use, and discard of pottery; animal bone transport, processing, and discard; lithic (stone tool) studies; herd management; and the relationship between settlement pattern and mobility. The initial belief by processual archeologists that human behavior was subject to laws (more or less similar to those of biology) pervaded ethnoarcheology and oriented its conceptual development in the 1970s. During these early years, there also was an underlying conviction that universal laws could be generated that related human behavior to material remains. In fact, archeologist Michael Schiffer presumed that, together with experimental archeology, ethnoarcheology would be the principal source for the production of these laws. It was within this tradition of research that several projects oriented toward the study of human ecology and foraging behavior started.

In the 1980s and 1990s, ethnoarcheology broadened its focus and began to be included within a postprocessual agenda. (Postprocessualism is a current movement of archeological research, initially led by British researcher Ian Hodder, which has a more idealist, historical, and postmodern orientation.) From within postprocessualism, the range of interests that ethnoarcheology incorporated was expanded. It widened its focus beyond techno-economic aspects, which dominated the previous years, to understanding greater levels of complexity, attempting to discern material correlates of the social and ideational. Principally, this new current of thought reconceptualized material culture, seeking

Fig. 1. Ethnoarcheologist recording an abandoned hut of the Fur in eastern Sudan. (*Photo by Gustavo G. Politis*)

Fig. 2. Awa Indians (Amazonian forest, Brazil), including one individual playing with the ethnoarcheologist's notebook during a hunting trip. (*Photo by Gustavo G. Politis*)

to determine the multiple dimensions in which it operates. In this sense, certain aspects, such as symbolism and the study of nonutilitarian dimensions of material culture within society, are emphasized. Ethnicity, gender, style, power, and resistance are among the new themes addressed. In addition to the pioneering work of Ian Hodder in Lake Baringo in Kenya, a good example of this trend is the research of Olivier Gosselain. Focusing on African pottery, Gosselain sought to develop general propositions concerned with the relationship between technological styles and aspects of social identity.

In parallel with these main trends, primarily Anglo-Saxon in origin, there is a Francophone ethnoarcheology that has its antecedents in the classic French ethnographic studies of material culture. This latter tradition is oriented toward the identification of technological processes (for example, pottery and metallurgy), paying attention to the broader social context. The influential research in technology of Pierre Lemonier called attention to the importance of the operational sequences that were involved, and not just the final product.

Currently, ethnoarcheological studies have multiplied and encompassed the analysis of all types of societies, and long-term ethnoarcheological projects have been initiated around the world (**Fig. 1**). In recent years, ethnoarcheological studies have been carried out among various societies in the Americas, Africa, and Oceania. In some cases, these studies have been undertaken by non-Western researchers (especially in Africa). Furthermore, these studies have not been limited to indigenous groups, but have also included Creole peoples, peasants, and Western urban societies. For some investigators (for example, A. González-Ruibal), the study of Western societies should be included as a detached form of ethnoarcheology called "archeology of the present."

Hunter-gatherers (**Fig. 2**) have been an important focus from the beginning of systematic ethnoarcheological research. Paradoxically, as interest grows in these studies and their contributions are valued as means of archeological inference, traditional societies are declining and the range of variation of analogous referents is diminishing. Consequently, the Westernization of indigenous societies notably reduces the availability of contemporary analogous referents that reflect some of the conditions of past societies or that are comparable in some way.

Outlook. Finally, ethnoarcheology has been active in the more general anthropological goal of understanding and exploring other forms of thought or cosmologies. Within this field, patterns of rationality and logical structures that differ from Western patterns are looked for. This kind of research, proposed by Spanish archeologist Almudena Hernando, is framed within the poststructuralism school of thought. In this use of ethnoarcheology, the correlation with material culture is secondary to the attempts to understand alternative cosmovisions and different logics independent of their material correlates. Obviously, one is not attempting to understand in depth extinct norms of thought, but rather to detect keys to its functioning and discern how and which ideological and social factors (as well as techno-economics) acted on the configuration of the material record.

For background information *see* ANTHROPOLOGY; ARCHEOLOGY; EARLY MODERN HUMANS; PALEOINDIAN; PHYSICAL ANTHROPOLOGY; SOCIOBIOLOGY in the McGraw-Hill Encyclopedia of Science & Technology.

Gustavo G. Politis

Bibliography. L. Binford, *Nunamiut Ethnoarchaeology*, Academic Press, New York, 1978; N. David and C. Kramer, *Ethnoarchaeology in Action*, Cambridge University Press, Cambridge, U.K., 2001; W. DeBoer and D. W. Lathrap, The making and breaking of Shipibo-Conibo ceramics, pp. 102–138, in C. Kramer (ed.), *Ethnoarchaeology: Implications of Ethnography for Archaeology*, Columbia University Press, New York, 1979; A. González-Ruibal, *La Experiencia del Otro: Una Introducción a la Etnoarqueología*, Editorial Akal, Madrid, 2003; O. Gosselain, Materializing identities: An African perspective, *J. Archaeol. Method Theory*, 7:187–218, 2000; W. Longacre, *Archaeology as Anthropology: A Case Study*, University of Arizona Press, Tucson, 1970; G. Miller, An introduction of ethnoarchaeology of the Andean camelids, Ph.D. thesis, University of California, Berkeley, 1979; J. F. O'Connell, Alyawara site structure and its archaeological implications, *Am. Antiquity*, 52:74–108, 1987; G. Politis, *Nukak: Ethnoarchaeology of an Amazonian People*, Left Coast Press, Walnut Creek, CA, 2007; B. Sillar, *Shaping Culture: Making Pots and Constructing Households—An Ethnoarchaeology Study of Pottery Production, Trade and Use in the Andes*, BAR International Series 883, Oxford, 2000; P. J. Watson, *Archaeological Ethnography in Western Iran*, University of Arizona Press, Tucson, 1979.

Floral volatiles

With over 250,000 described species of angiosperms, the flowering plants represent one of the most exuberant expressions of biological diversity on Earth. The explosive diversification of flowering plants is commonly attributed to coevolutionary relationships with animal pollinators, mediated through floral form, color, pattern, and fragrance. Many flowers are scented, and their odors vary markedly in their strength, complexity, and chemical composition. Furthermore, floral odors may vary spatially, at the scale of individual flower parts [for example, pollen and corolla (petal) odors typically differ], and temporally, with some fragrances showing pronounced diurnal or nocturnal rhythms as well as qualitative changes after pollination. Despite the predominance of insects as floral visitors and the acknowledged importance of chemistry to their foraging behavior, we are just beginning to appreciate the many ways in which floral volatiles can affect plant reproductive success. Improved analytical techniques, combined with a growing awareness of the general relevance of floral scent to pollinator function in agricultural as well as natural settings, promise to deepen our understanding of how fragrances contribute to floral function.

Volatile floral evolution. How did floral volatiles contribute to the evolution of pollination? The earliest flowers are thought to have attracted pollinators by presenting reproductive opportunities, such as rendezvous sites and warm, protected chambers for mating, and ovules and other floral tissues for feeding of their offspring. Water lilies, magnolias, soursops, and other ancient plant lineages that retain this pattern provide clues to the evolution of plant–pollinator interactions. The flowers and leaves of these plants emit strong odors, including compounds that repel insects and inhibit microbial growth. The first effective pollinators—mostly beetles, flies, and small moths—were probably those species that could tolerate such odors and use them to find flowers. In some cases, floral odors might have converged chemically with or provided a surrogate for the sex pheromones of their insect pollinators.

Nowadays, most examples of animal pollination generally involve a reciprocal transaction: flowers provide food substances (nectar and pollen) to floral visitors in exchange for pollen transfer between flowers of the same species. These transactions give rise to nectar or pollen markets, in which communities of flowering plants compete for pollinator services by advertising the presence of food with color and scent combinations that either appeal to the innate sensory preferences of the pollinators or can be learned by them to ensure floral constancy. Floral markets are universal features of modern plant communities from rainforests to alpine meadows, and several different animal groups have evolved to fill similar nectar- and pollen-feeding niches.

Multiple functions for fragrance. Floral volatiles (see **illustration**) serve many roles in floral markets, from providing pollinators with distance attraction and landing cues to offering species-specific blends that pollinators can easily learn in conjunction with nectar or pollen rewards. Conversely, fraudulent flowers lacking nectar or pollen rewards often are scentless or produce highly variable odors, making it harder for discriminating pollinators to learn to avoid them. However, floral scent plays additional roles beyond pollinator attraction. In some cases, distasteful or toxic compounds may reduce the number or duration of pollinator visits, or they may compel pollinators to move further between floral visits, which may reduce inbreeding through increased outcrossing distances. In other cases, floral odors may restrict or "filter out" ineffectual pollinators or floral enemies through repellence. Thus, the bouquet of floral volatiles from any given plant often includes "signal" components whose net impact is to improve the plant's reproductive success. Frequently, floral volatile blends also include "noise" components, which do not have a direct impact on the behavior of floral visitors, but are emitted as byproducts of the biosynthetic pathways that produce volatiles in plant cells. These also could be inherited as anachronistic traits common to that plant's ancestral lineage, just as modern whales retain the skeletal rudiments of a mammalian pelvis.

Information content in specialized systems. Among the best-understood odor-guided pollination systems are the bizarre cases of sexually deceptive pollination, in which male insects attempt to mate with orchid flowers as a result of sex pheromone mimicry, and brood site deception, in which female insects that lay eggs in dung or carrion pollinate flowers that mimic decaying organic matter. The floral volatiles that mediate these classes of deceptive

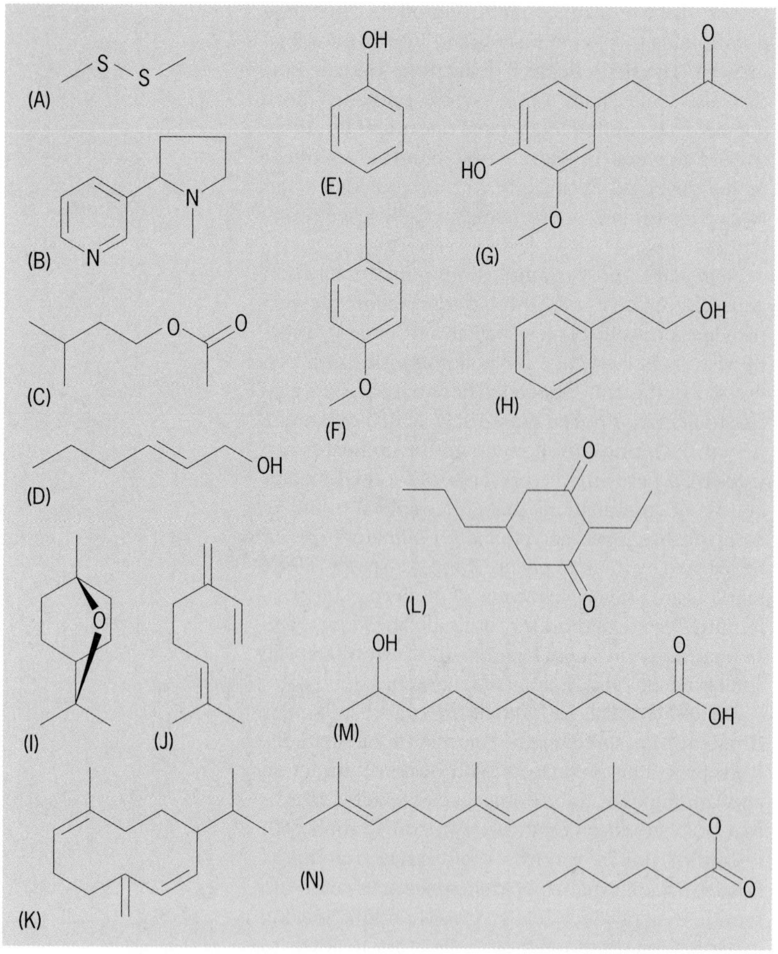

Representative floral volatiles with diverse biosynthetic origins and ecological functions. Pollinator specialization can be mediated by A (dimethyl disulfide) or E (phenol) in carrion and fecal mimicry, by F (4-methylanisole) or K (germacrene D) in fig pollination, by G (zingerone) or I (1,8-cineole) in aphrodisiac orchids, and by L (chiloglottone) or M (9-OH-decanoic acid) in sexually deceptive orchids. Compounds that enhance outcrossing by encouraging pollinators to leave flowers include B (nicotine), J (β-myrcene), H (2-phenylethanol), and N (farnesyl hexanoate). C (3-methylbutyl acetate) and D (*E*-2-hexenal) are common components of ripe fruit or wounded vegetation that occur in deceptive flowers. Compounds A, B, C, and E–H are derived at some level from amino acid metabolism, whereas D and L–N are fatty acid derivatives. I, J, K, and N are terpenoids.

flowers have highly specific information content, attracting precisely those insects for which such odors bear great reproductive significance. For example, phenol and cresol are universally present in plants of the Araceae and Apocynaceae that mimic herbivore feces, whereas dimethyl disulfide is present in all known cases of carrion mimicry investigated to date. Of course, sex-deceptive and brood site–deceptive flowers often require complex visual, tactile, and mechanical contrivances in order to manipulate pollinators into proper pollen removal and deposition. However, experiments with *Ophrys* and *Chiloglottis* orchids, as well as *Helicodiceros* trap inflorescences, have shown that highly specific odors are required for pollinator attraction in each system.

The other category of specialized flower-pollinator interactions driven by floral volatiles is "nursery" pollination, in which female insects pollinate flowers while inserting their own eggs into the ovary, allowing their larvae to eat the developing fertile seeds. Nursery pollination relationships have evolved in diverse flowering plant lineages worldwide and are best appreciated in the more than 700 known species of figs, each with its own obligate fig wasp pollinator. Female fig wasps, after mating and emerging from their natal figs, use odor to find new fig trees with receptive flowers, suggesting that fig odors should be species-specific, especially in forests with several sympatric fig species (sympatric species occupy the same range as another species, but maintain their identity by not interbreeding). Available data suggest that different fig species accomplish this either through unique blends of 15-carbon compounds (sesquiterpenes) or through the presence of unusual odors, such as 4-methylanisole. Unique odors also have been identified in nursery-pollinated *Yucca* and *Peltandra* plants, consistent with the prediction that obligate specialization should have its own fragrance.

Pleiotropic links to pigment and defense. Floral volatiles are products of multifunctional biosynthetic pathways that are familiar sources of phytohormones, drugs, polymers, and other structural components of plant cells. Volatile compounds can be synthesized anabolically from small molecules provided by cellular respiration and fatty acid metabolism, and they also can be derived catabolically from larger molecules, such as linolenic acid and beta-carotene. The structure and regulation of biosynthetic pathways suggest that floral volatiles have direct and indirect links to the production of floral pigments and defense compounds, and that abiotic stress, pathogenesis, and herbivore attack all have the potential to modify floral scent emissions. Indeed, herbivory on wild tomato species fundamentally changes floral volatile emission and nonvolatile chemistry, reducing floral attractiveness to bees. The physiological connections between plant defense and floral function are beginning to be explored, and they represent a frontier for further research. Finally, the impact of soil nutrients and fungal mycorrhizae on floral volatile biosynthesis merits closer examination, from the level of metabolic pathway dynamics all the way up to the ecosystem level.

For background information *see* AGRICULTURAL SCIENCE (PLANT); BOTANY; CHEMICAL ECOLOGY; ECOLOGY; FLOWER; PLANT; PLANT-ANIMAL INTERACTIONS; PLANT EVOLUTION; PLANT METABOLISM; PLANT PHYSIOLOGY; PLANT REPRODUCTION; POLLEN; POLLINATION in the McGraw-Hill Encyclopedia of Science & Technology. Robert A. Raguso

Bibliography. C. Chen et al., Private channel: A single unusual compound assures specific pollinator attraction in *Ficus semicordata*, *Funct. Ecol.*, 23:941–950, 2009; A. Jürgens, S. Dötterl, and U. Meve, The chemical nature of fetid floral odours in stapeliads (Apocynaceae-Asclepiadoideae-Ceropegieae), *New Phytologist*, 172:452–468, 2006; A. Kessler and R. Halitschke, Testing the potential for conflicting selection on floral chemical traits by pollinators and herbivores: Predictions and case study, *Funct. Ecol.*, 23:901–912, 2009; R. Raguso, Wake up and smell the roses: The ecology and evolution of floral scent, *Annu. Rev. Ecol. Evol. Systemat.*, 39:549–569, 2008; F. Schiestl, The evolution of floral scent and insect chemical communication, *Ecol. Lett.*, 13:643–656, 2010.

Flow behavior of viscoelastic fluids

Viscoelastic fluids are a common form of non-newtonian fluid. They can exhibit a response that resembles that of an elastic solid under some circumstances, or the response of a viscous liquid under other circumstances. Typically, fluids that exhibit this behavior are macromolecular in nature (that is, they have high molecular weight), such as polymeric fluids (melts and solutions) used to make plastic articles, food systems such as dough used to make bread and pasta, and biological fluids such as synovial fluids found in joints. The macromolecular nature of polymeric molecules along with physical interactions called entanglements lead to the elastic behavior (the fluids resemble a mass of live worms). Deformed molecules are driven by thermal motions to return to their undeformed states, giving the bulk fluid elastic recovery. This article attempts to provide a basic introduction to flow phenomena that are associated with the viscoelastic nature of fluids, such as rod-climbing, die or extrudate swell, entrance pressure losses, melt fracture, and draw resonance. In the case of polymeric fluids, these flow phenomena have a direct influence on the processing behavior and in some cases on the performance of the polymer in a given application.

In general, viscoelastic fluids are typically quite viscous, and the flow phenomena occur under creeping flow conditions in which the Reynolds number, Re, is less than 1.0. The primary dimensionless group that determines whether viscoelasticity is important is the Deborah number, De, which is the ratio of the relaxation time to the processing time, λ/t_{pr}. When De > 1, viscoelastic behavior is important. The basic rheological behavior of viscoelastic fluids determined in basic simple flows such as shear or extensional will first be reviewed, since this behavior accounts for the more complex flow behavior.

Fundamental rheological properties. The basic transport properties of polymeric materials that distinguish them most from other materials are their rheological behavior. There are many differences between the fundamental flow properties of a polymeric fluid and typical low-molecular-weight fluids such as water, benzene, and sulfuric acid, which are classified as Newtonian. Newtonian fluids can be characterized by a single flow property called viscosity (η) and their density (ρ). Polymeric fluids, on the other hand, exhibit a viscosity function that depends on shear rate, γ, or shear stress and are typically shear-thinning or pseudoplastic, meaning that the viscosity decreases with increasing shear rate (**Fig. 1**). At low shear rates the viscosity reaches a limiting value referred to as the zero-shear viscosity, η_0, followed by a region of shear thinning.

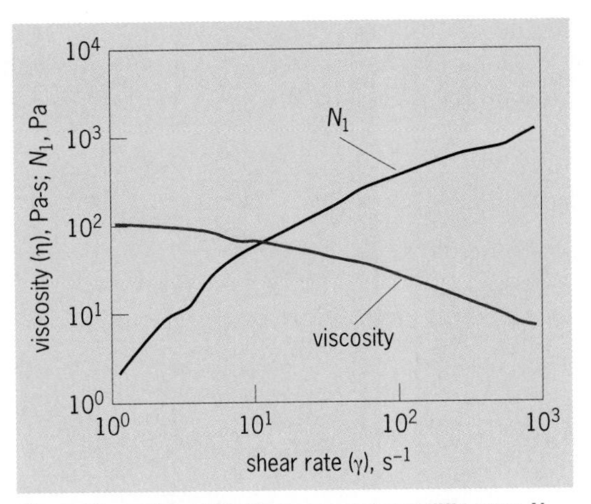

Fig. 1. Viscosity and the first normal stress difference, N_1, versus shear rate for a representative polymer melt, polystyrene, on a log-log plot. The viscosity shows a plateau followed by shear thinning, while N_1 is monotonically increasing with shear rate. (*From S.-P. Chang, Hole Pressure Measurements and Validation of the Higashitani-Pritchard-Baird Equation, M.S. thesis, Department of Chemical Engineering, Virginia Polytechnic Institute and State University, 1986*)

Other basic flow behaviors of polymeric fluids that distinguish them from Newtonian fluids include time-dependent rheological properties, viscoelastic behavior such as elastic recoil (memory), normal stresses in shear flow, and an extensional viscosity that is not simply related to the shear viscosity. The first normal stress difference, N_1, is related to tensions generated by stretching and orientation of the macromolecules along the flow direction, and is directly related to the elastic behavior of the fluid. (This quantity is shown in Fig. 1 and monotonically increases with shear rate.) The extensional viscosity of polymeric fluids, which is measured by the force required to stretch molten filaments, can rise to values higher than the so-called Trouton ratio, which is $3\eta_0$. These fundamental rheological properties are measured by means of rotational rheometers (cone-and-plate, parallel disks, and so forth); capillary rheometers, in which the pressure required to push a polymeric fluid through a capillary or tube is measured; and extensional rheometers (devices designed to stretch rather than shear a polymeric fluid). Because of these vastly different rheological properties, polymeric fluids are known to exhibit flow behavior that cannot be accounted for through merely a single rheological parameter such as the viscosity. Some of the differences in flow behavior include a nonlinear relation between pressure drop and volumetric flow rate for flow through a tube, swelling of the extrudate on emerging from a tube, the onset of a low-Reynolds-number flow instability called melt fracture, and gradual relaxation of stresses on cessation of flow. Some of the flow phenomena that are exhibited by viscoelastic fluids and are associated with these fundamental rheological properties will be described.

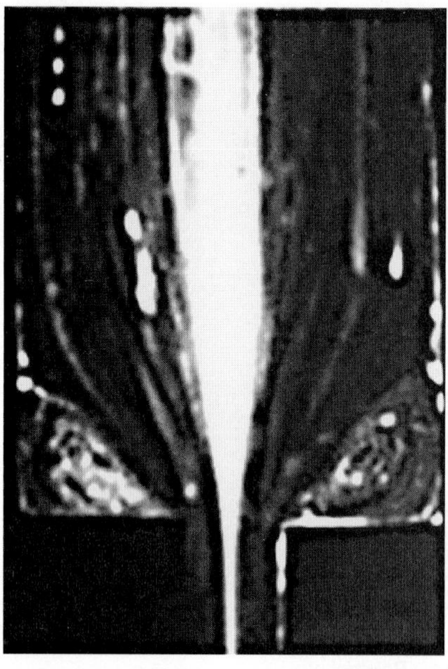

(a)

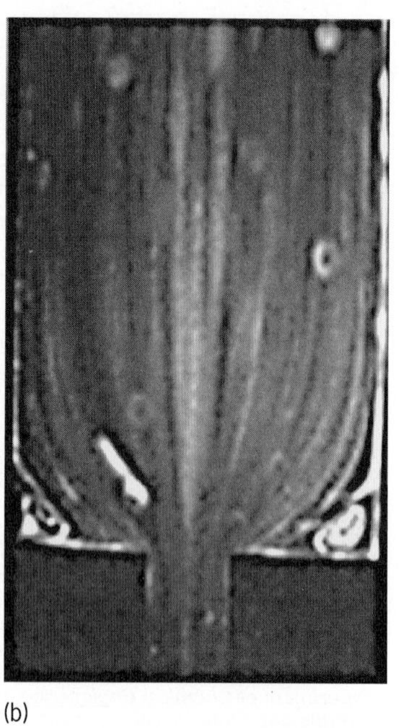

(b)

Fig. 2. Entry flow patterns in a planar geometry for polymer melts. (*a*) Pattern for polymer melt whose extensional viscosity increases with increasing extension rate (referred to as an extensional thickening fluid), low-density polyethylene (LDPE): shear rate (γ) = 60 s^{-1}, Deborah number (De) = 1.38. (*b*) Pattern for polymer melt whose extensional viscosity decreases with increasing extension rate, polystyrene: γ = 60 s^{-1}, De = 1.39. (*From S. White, Correlation of Entry Flow Behavior with Extensional Viscosity of Polymer Melts, Ph.D. thesis, Department of Chemical Engineering, Virginia Polytechnic Institute and State University, 1987*)

Rod-climbing. On stirring a fluid such as coffee in a cup, it is well known that if the stirring is too vigorous, the coffee will rise over the walls of the cup. This is associated with inertial forces that cause the pressure to rise at the wall. On the other hand, a polymeric fluid will climb the mixing rod no matter which direction the rod is rotated. In the case of a viscoelastic fluid the pressure is higher at the rod than at the walls due to the tensions along the streamlines caused by the first normal stress difference (Fig. 1).

Contraction flow. The pressure drop across a contraction, ΔP_{ent}, for a polymeric fluid can be quite large relative to the pressure drop across the capillary or tube. This pressure drop is thought to be due to the entry flow behavior of the polymer. In some cases, such as for low-density polyethylene (LDPE), the streamlines form natural entry angles (**Fig. 2***a*). The flow from the larger geometry into the smaller geometry is restricted well into the upstream region, which serves to effectively act as an extension to the capillary length and thereby increase the total pressure drop, ΔP. It is also observed that large vortices arise in the corners. For polymers such as polystyrene (PS; Fig. 2*b*), the streamlines become curved only a short distance from the contraction, and the vortices are quite small. In general, linear polymers such as polystyrene exhibit small regions of flow rearrangement and very small regions with vortices and, hence, lower values of ΔP_{ent}. On the other hand, Newtonian fluids at low Reynolds numbers exhibit no entrance vortices and very low values (about two decades smaller than polymeric fluids) of ΔP_{ent}.

Extrusion instabilities (melt fracture). The limiting factor in the extrusion rate of polymeric fluids is the onset of a low-Reynolds-number instability called melt fracture. The onset of melt fracture leads to varying degrees of imperfections that may affect only the surface appearance of a material on the one hand, while on the other hand may be so severe as to break up the polymer melt as it leaves a capillary. There are basically five types of melt fracture: sharkskin, ripple, bamboo, wavy, and severe. Three of these types of melt fracture are shown in **Fig. 3** for a linear low-density polyethylene (LLDPE). At the lowest apparent shear rate, γ_{a}, the extrudate is smooth (Fig. 3*a*), but at $\gamma_{\mathrm{a}} = 112\ \mathrm{s}^{-1}$ the extrudate exhibits a mild roughness called "sharkskin," which affects the appearance of the surface (Fig. 3*b*). This type of fracture is extremely detrimental to the manufacture of packaging films as they appear hazy rather than clear. As γ_{a} is increased, another form of fracture arises. At $\gamma_{\mathrm{a}} = 750\ \mathrm{s}^{-1}$, the fracture present is called "bamboo" (Fig. 3*c*). Finally, at $\gamma_{\mathrm{a}} = 2250\ \mathrm{s}^{-1}$, the fracture is severe (Fig. 3*d*). LLDPE does not seem to exhibit "wavy" fracture.

Devices that extrude polymeric fluids in order to shape the melt into useful products, such as fibers or flat films, are called dies, and the simplest shape of such a device is a capillary or tube. In manufacturing with dies, one may attempt to reduce the detrimental effect of melt fracture through die design or polymer modification. In order to do this, it is important to know the origin of melt fracture. The major sources for melt fracture are the capillary (or die in the general case) entry region, the capillary itself (called the die land in the general case), and the capillary (or die) exit. For a polymer such as LDPE, fracture originates at the capillary entry. As the extrusion rate is increased, the vortices no longer grow in size or intensity. Instead, the flow takes on a spiral motion in the capillary entry, sending sections of the nearly stagnant fluid into the capillary at regular intervals. This leads to regions of various flow histories passing through the capillary and leaving the capillary exit. When this type of fracture occurs, there is no indication of the flow problems in the pressure measured along the capillary walls. By streamlining the capillary entry or increasing the length of the capillary, it is possible to reduce the amplitude of the distortion, but the critical shear stress for fracture is unchanged. The critical wall shear stress for the onset of fracture for LDPE is of the order of 10^5 Pa (1 atmosphere). On the other hand, some polymers, such as high-density polyethylene (HDPE) and LLDPE, seem to slip in the capillary, leading to a "slip-stick" instability. There is a distinct flattening of the flow curve (shear stress versus shear rate), indicating a region where multiple flow rates are possible for the same wall shear stress. Eventually the flow curve appears to become normal again at high shear rates. When the slip-stick fracture occurs (which results in the ripple and then bamboo types of fracture), increasing the capillary length just makes the degree of distortion worse.

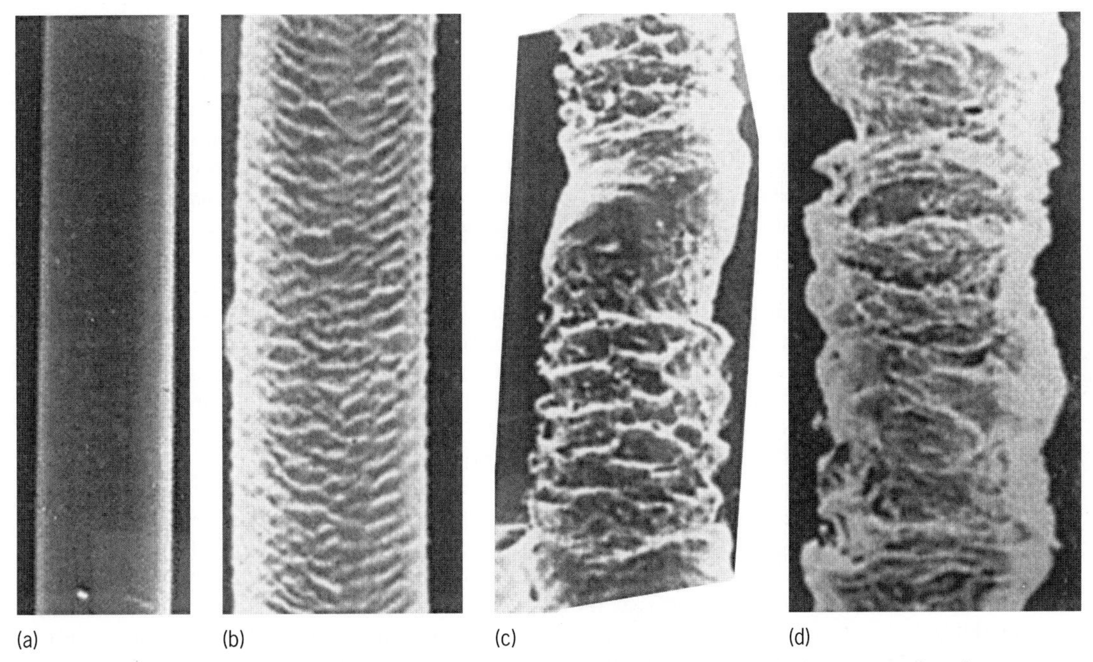

Fig. 3. Filaments extruded from a capillary and then quenched to solidify them to freeze-in the structure. Samples were obtained at different apparent shear rates γ_a, illustrating melt fracture types for a linear low-density polyethylene (LLDPE). (*a*) $\gamma_a = 37\ s^{-1}$; filament is smooth. (*b*) $\gamma_a = 112\ s^{-1}$; filament exhibits sharkskin fracture. (*c*) $\gamma_a = 750\ s^{-1}$; filament exhibits bamboo fracture. (*d*) $\gamma_a = 2250\ s^{-1}$; filament exhibits severe fracture. (*From R. Moynihan, The Flow Stability of Linear Low Density Polyethylene, Ph.D. thesis, Department of Chemical Engineering, Virginia Polytechnic Institute and State University, March 1990*)

Extrudate swell. The phenomenon associated with the increase of the diameter of an extrudate as a polymer leaves a capillary is known as die or extrudate swell. The present hypothesis is that extrudate swell is related to unconstrained elastic recovery (S_∞) following shear flow, which is the recovery in shape of a material deformed in shear once the stress required to deform it is removed. The quantity S_∞ is the ratio of N_1 to the shear stress (which is viscosity times shear rate), and gives N_1 a direct correlation to fluid elasticity. However, the relation of extrudate swell to S_∞ is only a rough estimate, since it varies from polymer to polymer even though they exhibit a similar range of values of N_1. Other factors that have been found to influence extrudate swell include the method of measurement, the length-to-diameter ratio of the capillary (L/D), the design of the entry region, and the extrusion rate as defined by the magnitude of the wall shear stress.

Draw resonance. In a number of industrial processes, stretching of the fluid is used to reduce the dimensions of the material. These processes include fiber spinning, a process for generating thin filaments used, for example, in textiles (shown in **Fig. 4** for the spinning of a single fiber); film casting, a process, for example, for coating paper with a thin film of polymer; and film blowing, a process for generating, for example, trash bags. In such processes, periodic fluctuations in the dimensions such as fiber diameter or film width can be observed when a critical draw-down ratio, D_R, is reached. Here, D_R is the ratio of the velocity of the fluid at the take-up device (Fig. 4) to the velocity of the fluid as it exits the melt-shaping device (capillaries in the case of fiber spinning and an annulus in the case of film blowing). By taking the fluid up faster than when it exits the capillary, for example, the diameter of the filament is reduced relative to that as it emerges from the capillary. This fluctuation in fiber diameter, which is referred to as draw resonance, appears as a sustained periodic fluctuation with a well-defined and steady period and amplitude of the cross section at the take-up device, and it occurs even when the flow rate and the take-up speed are constant. This instability should not be confused with the spinnability, since

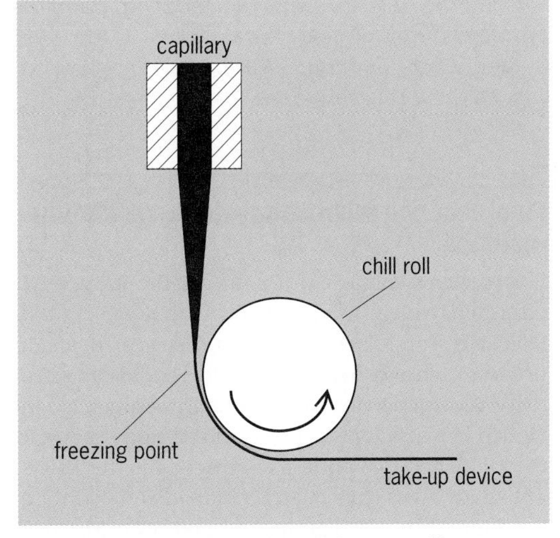

Fig. 4. Extrusion of a polymer melt from a capillary followed by passage over a chilled roll and then on to a take-up device. The filament is drawn by taking it up at a speed faster than it exits the capillary. The polymer melt is represented by the dark thread in the picture.

it has nothing to do with breakup of the filament. It appears in both purely viscous and elastic fluids, and there are two factors that reduce its effect: elasticity and nonisothermal conditions of spinning. For Newtonian fluids, the value of $D_R = 20.21$ is considered to be the critical draw-down ratio beyond which the flow becomes unstable. In general, it has been observed that for viscoelastic fluids the critical draw-down ratio for the onset of draw resonance is significantly lower than that of Newtonian fluids and can reach values as low as about 3.0. The nature of the extensional viscosity controls the onset of draw resonance for polymeric fluids, and those materials that are considered to be extensional viscosity thickening (that is, the extensional viscosity increases with extension rate) are more stable.

For background information *see* CREEPING FLOW; FLUID-FLOW PROPERTIES; FLUID MECHANICS; NEWTONIAN FLUID; NON-NEWTONIAN FLUID; PLASTICS PROCESSING; POLYMER; POLYOLEFIN RESINS; POLYSTYRENE RESIN; RHEOLOGY; VISCOSITY in the McGraw-Hill Encyclopedia of Science & Technology.

Donald G. Baird

Bibliography. D. Baird and D. Collias, *Polymer Processing: Principles and Design*, Wiley, New York, 1998; R. Bird, R. Armstrong, and O. Hassager, *Dynamics of Polymeric Liquids*, vol. 1: *Fluid Mechanics*, 2d ed., Wiley, New York, 1987; R. Larson, *The Structure and Rheology of Complex Fluids*, Oxford-University Press, Oxford, U.K., 1999; S. Middleman, *Fundamentals of Polymer Processing*, McGraw-Hill, New York, 1977.

Fluorescence microscopy: high-resolution methods

Fluorescence microscopy is one of the most powerful and widely used imaging techniques in biological research because it can visualize dynamic processes inside living cells with exquisite sensitivity and specificity. Major breakthroughs have included the development of confocal microscopy in the 1980s and two-photon microscopy in the 1990s. These advances have been accompanied by groundbreaking improvements in the ability to visualize specific proteins and organelles inside living cells using fluorescent proteins [for example, green fluorescent protein (GFP)] and transgenic technology. In addition, tremendous increases in computing power have enabled the acquisition and analysis of large imaging data sets and thus have further facilitated the ascendancy of fluorescence microscopy.

On the downside, all of these approaches fall short when it comes to resolving fine details of cellular substructures because they lack the required level of spatial resolution. Until a few years ago, the spatial resolution of light microscopy was thought to be fundamentally limited by the diffraction of light. According to a physical law put forward by Ernst Abbe in the 1870s, the smallest detail resolvable with a light microscope is on the order of half the wavelength of the used light, that is, 200–300 nanometers.

This is unfortunate because it has impeded the study of key cell biological processes and interactions that occur on the mesoscale of 10–200 nm. Whereas electron microscopy provides a spatial resolution down to a few nanometers, easily resolving cellular substructures, it provides no temporal information, which is crucial for many questions in cell biology. In addition, the labeling of particular proteins for electron microscopy is often problematic and sampling of cellular volumes [three-dimensional (3D) reconstruction] is very labor intensive. Conversely, fluorescence microscopy permits live-cell imaging over relatively large volumes with biomolecular specificity and relative ease.

However, as a result of a number of microscopy innovations in recent years, the classic limit of light microscopy has been resoundingly shattered. Although not abolishing diffraction per se, new nanoscopy techniques cleverly circumnavigate the limiting effects on spatial resolution. As a result, it now is possible to resolve details at the nanoscale (well below 100 nm) in biological specimens without forgoing the customary benefits of fluorescence microscopy (live-cell imaging, biomolecular specificity, and so forth).

Stimulated emission depletion (STED) microscopy was the first concrete concept that broke the classic diffraction limit of far-field optics. Since then, other powerful techniques, including photoactivatable localization microscopy (PALM), stochastic optical reconstruction microscopy (STORM), and structured illumination microscopy (SIM), have been developed for nanoimaging of fluorescent samples. These techniques rely on different principles and therefore come with specific strengths and weaknesses in terms of, for example, temporal resolution, depth penetration, multicolor imaging, and practical implementation. What they all have in common is that there is in theory no longer a hard resolution limit, and it is possible to achieve a resolution as high as a few nanometers as assessed under ideal conditions by using beads on a coverslip. However, in practice, they are limited by signal noise (from drift inherent in samples, particularly in living biological samples; detector noise; and other factors) to some tens of nanometers.

STED microscopy. The resolution of laser scanning microscopes (confocal and two-photon) is limited by the fact that a laser beam, even if it were perfectly collimated (with rays that are rendered parallel to a certain line or direction), is not focused by the microscope's objective to a infinitesimally small point but rather to a blurry intensity distribution, which is called the point-spread function (PSF) and which defines the spatial resolution of the optical system. The core idea of STED microscopy (see **illustration**) is to improve the spatial resolution by constricting the PSF beyond this classic limit (called the diffraction barrier). This is achieved by a second laser beam (called the STED beam), which can de-

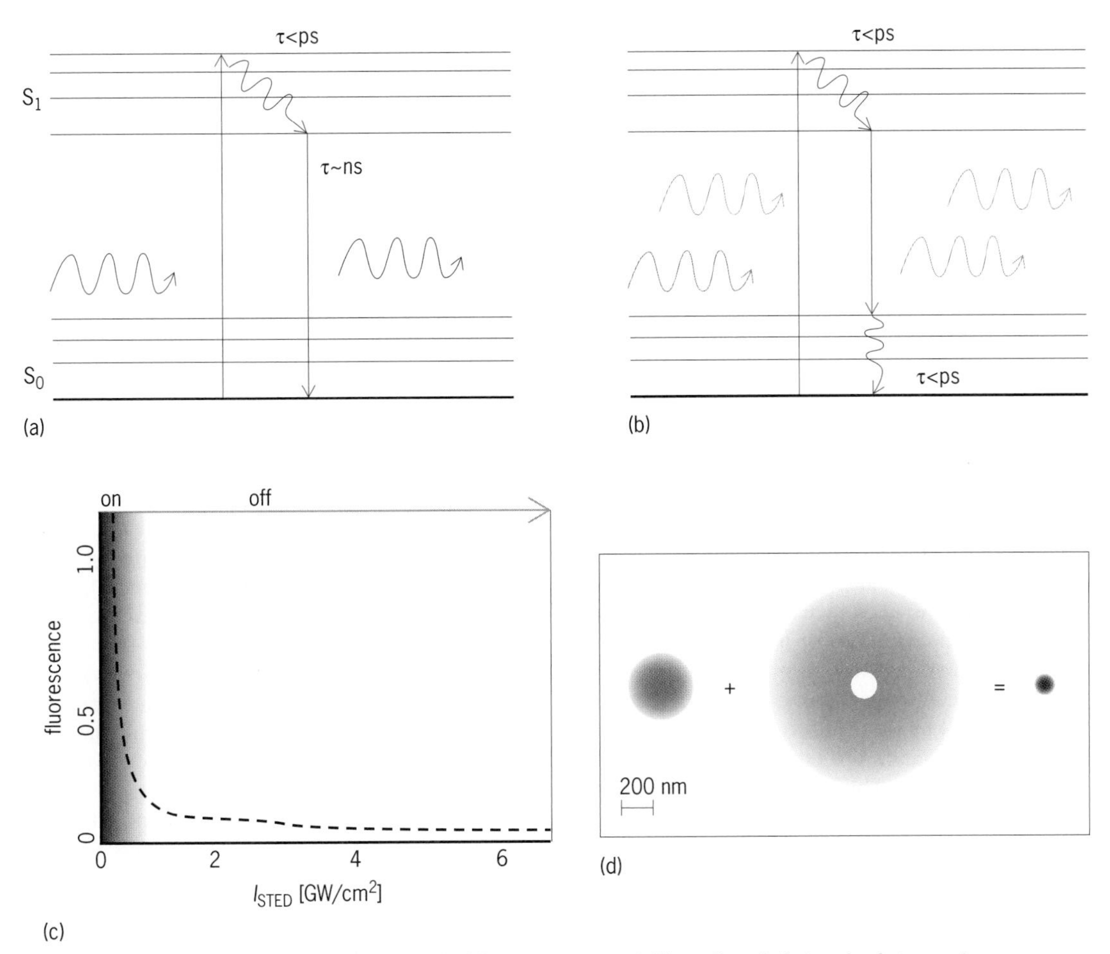

STED microscopy. (*a*) Energy diagram for a classical fluorescence event: Absorption of photons leads to spontaneous emission of fluorescence within typically a few nanoseconds, corresponding to the lifetime of the excited state of the fluorophore. (*b*) In STED microscopy, the excited state is quenched by stimulated emission: The passage of a longer-wavelength photon leads to the stimulated emission of a photon of the exact same wavelength and momentum as the incident photon, which can be spectrally separated from the fluorescence. (*c*) Turning off the fluorescence: The quenching of the fluorescence by the STED laser beam is a nonlinear process, which effectively turns the fluorescence off for higher STED intensities. (*d*) Confining the fluorescence spot (the point-spread function) in space: An annular intensity distribution ("doughnut") of the STED beam in the focal plane leads to a reduction in the spot size, with only the region in the center (the "null") being spared by the quenching process.

excite fluorescent molecules by stimulated emission at a wavelength that is longer than the fluorescence. By shaping the STED beam like a doughnut in the focal plane, it actively switches off the fluorescence on a circular rim around the PSF and thus only permits fluorescence to occur from the center of the doughnut (called the null). Currently, the fluorescence spot size can be reduced to about 10–20 nm, which is more than an order of magnitude smaller than what diffraction-limited light microscopy can achieve. The resolution (Δr) of a STED microscope is given by

$$\Delta r \approx \lambda/(2NA\sqrt{1+I/I_s})$$

where NA is the numerical aperture of the lens, λ is the wavelength, I is the intensity at the crest of the doughnut-shaped STED beam, and I_s is the STED intensity required to cut down the fluorescence probability of the molecules by half.

PALM and STORM. PALM and STORM are based on stochastic switching and computational localization of single fluorophores in wide-field illumination. As a general principle, PALM and STORM use the photoactivation properties of fluorescent proteins and organic dyes, respectively.

As a result of their ability to visualize single molecules, PALM and STORM techniques hold the current record in spatial resolution for fluorescence microscopy, routinely achieving a resolution better than 30 nm in the $x-y$ dimensions and 60 nm in the z dimension in biological samples. Whereas temporal resolution used to be notoriously low, acquisition times for two-dimensional (2D) images now can be on the order of minutes because of advances in fluorophores and image processing algorithms, improving the suitability of PALM and STORM for imaging dynamical events inside cells. In addition, it recently has become possible to map out rapidly the movements of many particles (hundreds of particles) by combining single-particle tracking with PALM (SPT-PALM). However, this is not imaging in the true sense of the word because only the tracks of a limited number of particles are localized, rather than producing an entire image. Although so far used mostly for imaging of monolayer tissue, reasonable depth penetration (for example, imaging a few cell layers below the surface of a brain slice) should in theory be possible, especially if combined with other imaging modalities such as two-photon laser scanning to improve optical sectioning. PALM and STORM can be used for two-color imaging, and its practical implementation is probably the least expensive and most straightforward among the nanoscopy techniques.

SIM. Structured illumination microscopy, another recent nanoscopy technique, relies on computationally restoring superresolved images from interference fringes (Moiré patterns) that have been caused by illuminating the sample with patterned light. In its nonlinear variant, which requires saturating illumination of the fluorophores, a lateral resolution of approximately 50 nm can be achieved by this technique. Relying on wide-field imaging, the temporal resolution is quite high (on the order of 10 Hz), being limited by the need to acquire multiple images of the Moiré patterns (10 or more). SIM poses no major constraints on imaging of multiple colors and thus it has been used for three-color imaging, which is a distinct advantage over the other techniques.

Conclusions. Nanoscopy techniques are bound to become central to cell biological research, providing new insights into dynamic processes inside living cells at the nanoscale. As many cellular functions are regulated on a scale that is just a little bit too small to be resolved reliably with conventional microscopy, the impact of these new techniques is expected to be broad and long-lasting. Because new superresolution methods have generated widespread enthusiasm for their potential to transform cell biology, it will be important to assess their scope and impact, by comparing them with ongoing refinement of traditional approaches such as electron microscopy and molecular biochemistry.

Certainly, the resolving capability of these new techniques is well established by now and they have come of age in many ways. However, future developments, in particular with regard to fluorescent probes and laser sources, are bound to broaden their scope and versatility and to improve the resolution even further.

For background information *see* ATOMIC FORCE MICROSCOPY; BIOELECTRONICS; BIOPHYSICS; CONFOCAL MICROSCOPY; DIFFRACTION; FLUORESCENCE; FLUORESCENCE MICROSCOPE; LASER PHOTOBIOLOGY; MICROSCOPE; MOIRÉ PATTERN; NANOTECHNOLOGY; OPTICAL MICROSCOPE in the McGraw-Hill Encyclopedia of Science & Technology. Urs Valentin Nägerl

Bibliography. E. Betzig et al., Imaging intracellular fluorescent proteins at nanometer resolution, *Science*, 313:1642–1645, 2006; M. G. Gustafsson, Nonlinear structured-illumination microscopy: Wide-field fluorescence imaging with theoretically unlimited resolution, *Proc. Natl. Acad. Sci. USA*, 102:13081–13086, 2005; S. W. Hell, Microscopy and its focal switch, *Nat. Methods*, 6:24–32, 2009; S. W. Hell and J. Wichmann, Breaking the diffraction resolution limit by stimulated emission depletion fluorescence microscopy, *Optics Lett.*, 19:780–782, 1994; S. T. Hess, T. P. Girirajan, and M. D. Mason, Ultra-high resolution imaging by fluorescence photoactivation localization microscopy, *Biophys. J.*, 91:4258–4272, 2006; U. V. Nägerl et al., Live-cell imaging of dendritic spines by STED microscopy, *Proc. Natl. Acad. Sci. USA*, 105:18982–18987, 2008.

Forensic firearms identification

On May 2, 1863, while returning from a scouting mission during the Battle of Chancellorsville, Virginia, Confederate General Thomas J. "Stonewall" Jackson was fired upon and wounded. Dr. Hunter McGuire removed a spherical ball 0.675 in. in diameter from Jackson's right hand and amputated his left arm. Although he survived the wounding, he developed pneumonia and died on September 10, 1863. Because the Union Army had abandoned the use of the smooth-bore musket that fired this type of ammunition the previous year, investigators were able to conclude that Jackson had been shot by his own troops, who were still using that firearm. Many believe this to be the first recorded instance of the use of firearms identification for forensic purposes.

Firearm identification. The Association of Firearm and Toolmark Examiners (AFTE) defines firearms identification as "A discipline of forensic science which has as its primary concern to determine if a bullet, cartridge case or other ammunition component was fired by a particular firearm." Forensic firearms identification is based on the theory that the methods used in the production of a firearm create random imperfections on the metal surfaces of that firearm which, in sufficient quantity and quality, create a unique pattern for that particular weapon. For example, virtually all firearm barrels start out as a solid steel rod. This rod is drilled to the approximate final diameter of the barrel, and then reamed using another tool to smooth out the interior of the barrel. The barrel is then rifled, a process that cuts or forms grooves, which spiral down the barrel. Next, the barrel is given a final smoothing. All of these processes require the use of tools that are, by their very nature, harder than the metal of the barrel. Any imperfections that are present on the tools as a result of their manufacture or subsequent wear will be randomly transferred to the surface of the barrel as marks, nicks, or scratches. Because these marks, known as striations or striae, are the result of the random wear of the tool and their random placement in the barrel, they create a unique pattern within that barrel (**Fig. 1**). Similar markings can be found on a firearm's firing pin, breech face (rear of the frame), and cartridge-case extractor and ejector; within the chamber of the firearm; and on any other metal surface that has been machined or formed.

Modern ammunition (the cartridge or shotgun shell) generally consists of one or more projectiles (bullet or shot) contained in a cartridge case that also

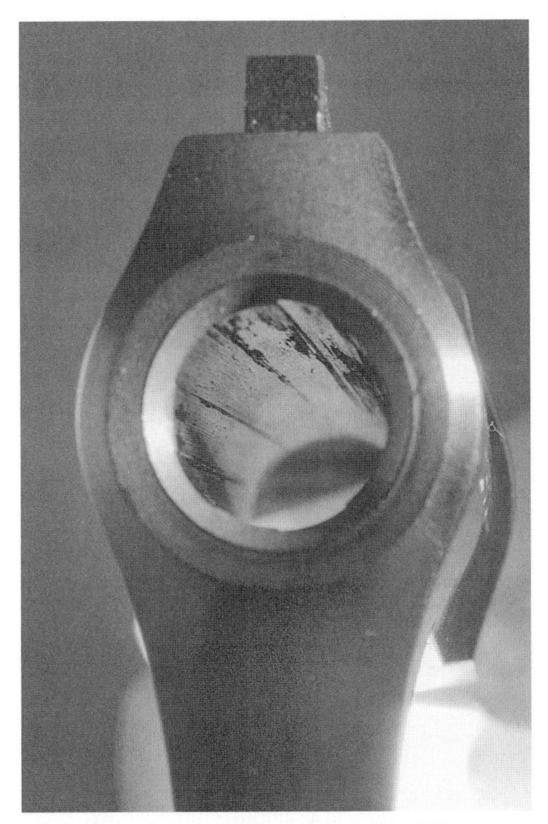

Fig. 1. Photo showing striations in a gun barrel. (*Larry Reynolds*)

Fig. 2. Striation markings on the surface of a fired bullet. (*Larry Reynolds*)

contains a predetermined amount of gunpowder. In the case of rifles and handguns, the bullet is inserted into one end of the cartridge case. In shotguns, small balls of lead, steel, or some other metal are contained within the case. The gunpowder is contained within the cartridge case in virtually all forms of ammunition. At the opposite end of the cartridge case, a contact explosive, known as a priming mixture, is either pressed into a cup (center fire ammunition) or spun into the rim of the cartridge case (rimfire ammunition). When discharging a firearm, pulling the trigger causes the weapon's firing pin to strike the rear of the cartridge case, detonating the priming mixture, which in turn ignites the gunpowder. As the gunpowder burns, the gases produced cause the pressure to build up within the cartridge. This pressure very quickly reaches the point where the bullet is forced out of the end of the cartridge case and into the barrel of the firearm. As the pressures continue to mount, the bullet, which is slightly larger than the diameter of the barrel, conforms to the rifling in the barrel; that is, as it is forced down the barrel, the bullet is in intimate contact with the surface of the barrel. At the same time, the cartridge case is slammed back into the breech face of the firearm.

This intimate contact between the bullet and cartridge case and the firearm results in the striations on the various parts of the firearm being impressed into the bullet and cartridge case. In essence, a "signature" of the barrel, breech face, and other machined parts of the firearm are imparted to the surface of the ammunition components (**Fig. 2**).

These signatures can be used by an expert to form an opinion as to whether a particular ammunition component has been fired from a particular firearm. This is accomplished by a direct comparison of the surface of the fired bullet or cartridge case to the surface of a fired bullet or cartridge case (known as a reference bullet or reference cartridge case) that has been fired from a known firearm. This comparison is dependent on the ability to produce reference material that ideally possesses only surface markings obtained as a result of having been cycled through the particular firearm in question. Reference bullets can be obtained by firing the weapon in question into some nonabrasive material dense enough to stop the bullet but not so dense as to damage it. Originally, this material consisted of rolls of absorbent cotton, or a soft grade of cotton waste. Although still in use in some places today, most laboratories now obtain reference bullets by firing into a tank filled with water. These tanks may be either vertical or horizontal and contain enough water to slow and stop the bullet's travel without damaging the surface.

Once the reference material has been recovered, the direct comparison is accomplished by using a comparison microscope (**Fig. 3**), a device that consists of two microscope bodies connected by an optical bridge, allowing the user to view portions of each microscopic field as if they were placed side by side. Although the comparison microscope has been known in one form or another since about

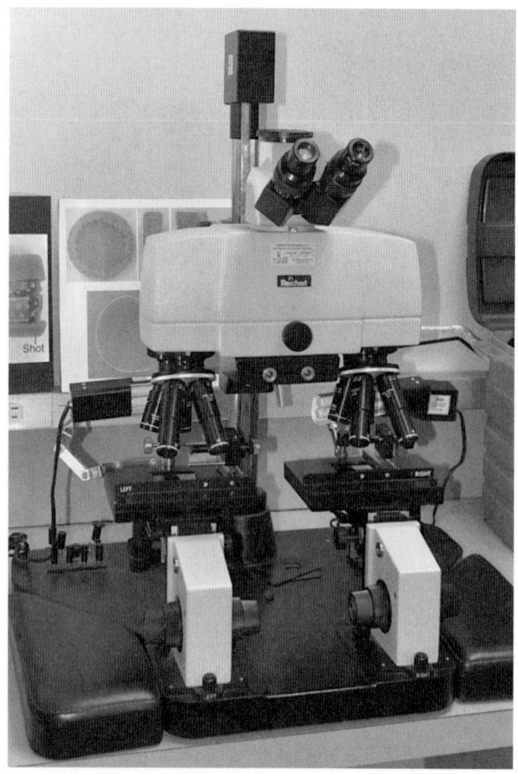

Fig. 3. Comparison microscope. (*Larry Reynolds*)

1886, it wasn't until 1925 that Philip O. Gravelle produced the forerunner of the modern firearms comparison microscope. This device's usefulness as an analytical tool became apparent in 1929, when it was used by Colonel Calvin Goddard in the evaluation of evidence from the St. Valentine's Day Massacre in Chicago, Illinois. Since then, although there have been advances in optics and design, the basic use of the firearms comparison microscope has not changed. For analysis, the surfaces of reference material and evidence are directly compared to one another in an effort to evaluate the correspondence or noncorrespondence of marks produced as a result of the firing process (**Fig. 4**).

Fig. 4. Two fired cartridge cases as viewed with a comparison microscope. (*Larry Reynolds*)

Today, forensic firearms examination encompasses much more than a determination of whether ammunition components were fired from a particular firearm.

Chemical analysis for the presence of gunshot residues. The process of discharging a firearm produces a number of residues, both visible and invisible, which can provide valuable information to the forensic firearms examiner. The detonation of the priming mixture results in the vaporization of the metallic components in the mixture. These vapors are blown out of any openings in the firearm, including the areas where the trigger and hammer are inserted into the frame, the cylinder gaps in revolvers, the ejection port in autoloaders, and the muzzle of the firearm. As these vapors cool, they form small spherical particles that contain combinations of the various metals present, particularly lead, antimony, and barium, which will deposit on any surface within a few feet of the firearm. Lead, antimony, and barium, while not particularly rare individually, are not generally found together in nature. As a result, their presence in a sample is indicative of gunshot residue. Although the presence of lead, antimony, and barium in a sample is not definitive proof of gunshot residue, their presence can be of investigative importance, particularly if these residues are found on samples taken from the hands of someone suspected of having discharged a firearm. Samples can be taken from any surface suspected of having gunshot residues on it either by swabbing the area with a cotton-tipped swab wet with dilute nitric acid, or by dabbing the area with an aluminum disc layered with adhesive. The samples are then analyzed by atomic absorption spectrophotometry (the cotton-tipped swabs) or scanning electron microscopy with energy dispersive x-ray analysis (the adhesive discs). Scanning electron microscopy has the added advantage of providing a visual confirmation of the particle's size and shape in addition to its elemental composition.

Distance determination. The incomplete combustion of gunpowder produces gaseous residues that are rich in nitrates and nitrites as well as particles of partially burned gunpowder. As these residues exit the muzzle of the firearm, they spread out in the atmosphere as they travel forward away from the muzzle. Depending on the type of firearm, residues can travel up to 8–10 ft (2.4-3 m) or more from the muzzle. The farther away from the muzzle the residues are, the more spread out they will be. Knowing this, a muzzle-to-target distance can be approximated by examining any residues present around a bullet entrance hole and comparing the density of the residues and the amount of spread to a series of targets fired from varying distances using the same firearm and ammunition. The visible residues, consisting mainly of unburned and partially burned gunpowder particles, are examined using either a lighted magnifier or a stereomicroscope. Most of the time this is sufficient for an estimation of the range of fire to be made; however, many laboratories will also chemically process the target in order to visualize the pattern formed by the combustion products, mainly nitrates and nitrites, which have also been deposited. Most of the common tests used for this purpose rely on the formation of colored diazo compounds (dyes).

Serial number restoration. Since 1968, virtually all firearms manufactured in, or imported into, the United States have serial numbers stamped on them, generally somewhere on the frame. Often, in an effort to impede the tracing of a firearm's ownership, the serial number will be obliterated in some way, such as by grinding, filing, or punching.

Most serial numbers on firearms are created by stamping the number into the metal with considerable force. This process deforms the structure of the metal underneath the stamped number. If, at some later time, the numbers are removed by filing or grinding, the deformed metal that was underneath will generally remain. Because the deformed metal is under stress, if the proper acid solution is applied the deformed metal will dissolve at a faster rate than the surrounding metal, revealing the removed numbers. Although there are many solutions that can be used for this purpose, the most commonly used consist of a mixture of dilute hydrochloric acid and copper chloride.

National Integrated Ballistic Information Network. In 1999, the Bureau of Alcohol, Tobacco, and Firearms established the National Integrated Ballistic Information Network (NIBIN), a database of ballistics imaging technology. Participants in this program acquire digital images of the markings made on fired bullets and cartridge cases using the Forensic Technology Integrated Ballistic Identification System (IBIS) and then upload them to the NIBIN database, where they are electronically compared to earlier entries. If a high-confidence match occurs, firearm examiners compare the original evidence using a comparison microscope to make the final decision regarding whether the items actually do match. Using this system, otherwise unrelated cases can be associated to one another.

For background information *see* ATOMIC SPECTROMETRY; CRIMINALISTICS; FORENSIC MICROSCOPY; OPTICAL MICROSCOPE; SCANNING ELECTRON MICROSCOPE in the McGraw-Hill Encyclopedia of Science & Technology. Ronald L. Singer

Bibliography. Association of Firearm and Tool Mark Examiners, *AFTE Glossary*, 1994; J. D. Gunther and C. O. Gunther, *Identification of Firearms*, John Wiley and Sons, New York, 1935; J. S. Hatcher, F. J. Jury, and J. Weller, *Firearms Investigation, Identification, and Evidence*, Stackpole Books, Harrisburg, PA, 1957; B. J. Heard, *Handbook of Firearms and Ballistics*, John Wiley and Sons, New York, 1997; R. Saferstein, *Criminalistics: An Introduction to Forensic Science*, 9th ed., Prentice Hall, Englewood Cliffs, NJ, 2007; T. Warlow, *Firearms, the Law, and Forensic Ballistics*, 2d ed., CRC Press, Boca Raton, FL, 2005.

FOXP2 gene and human language

Language is a key feature reflecting the evolution of the human brain. Arguably, it is the most unique feature that sets humans apart from all other organisms on Earth. Understanding the origin and mechanisms governing language production will provide insight into human evolution, as well as therapeutics for the many neurodevelopmental disorders involving language impairment (for example, specific language impairment, dyslexia, autism, and schizophrenia). A major advance into identifying the molecular mechanisms of language occurred when the first gene tied strongly to language function was uncovered. This gene encodes for the forkhead box P2 protein (FOXP2). FOXP2 belongs to the larger forkhead family of transcription factors. The role of transcription factors is to regulate the expression of other genes in a temporal or spatial manner. Because FOXP2 regulates gene expression, it is well poised to have a significant effect on the network of genes involved in the patterning and function of brain circuits involved in language and other elements of human higher cognition.

The expression of FOXP2 is also a critical indicator of its important function in cognition. FOXP2 is expressed in the developing human brain in areas known to be involved in human higher cognition, especially the prefrontal cortex and the basal ganglia, which together comprise the core of a frontal-striatal circuit. Disruption to circuitry connecting these regions occurs frequently in many neuropsychiatric diseases, supporting a role for FOXP2 in regulating gene expression signaling cascades that are affected in many disorders of cognition. The timing of the expression of FOXP2 is also an indicator of its relevance to the development of these critical circuits. The peak of FOXP2 expression occurs during mid-gestation in the human fetal brain. This is a time point in brain development when many neurons are born, migrate to their final location, and adopt their final specialized fate. Whether FOXP2 is involved in initiating neuronal differentiation, such as specific aspects of process outgrowth, or perhaps directs neurons toward becoming specific neuronal subtypes is an open question and an area of active research.

Function and mutations. The possibility of a gene involved in language was uncovered first through the study of a large multigenerational family, the KE family. This family demonstrated an autosomal dominantly inherited verbal dyspraxia (a motor speech disorder). In other words, this particular abnormality in speech vocalization was passed through every generation regardless of sex. The gene for FOXP2 was then identified when a single individual with a defined genetic abnormality displayed the same language symptoms as this unrelated large family. This individual had a translocation in chromosome 7 and also presented with verbal dyspraxia. By narrowing down the search for the mutated gene in the KE family to a particular region of chromosome 7, the researchers were able to discover a gene similar

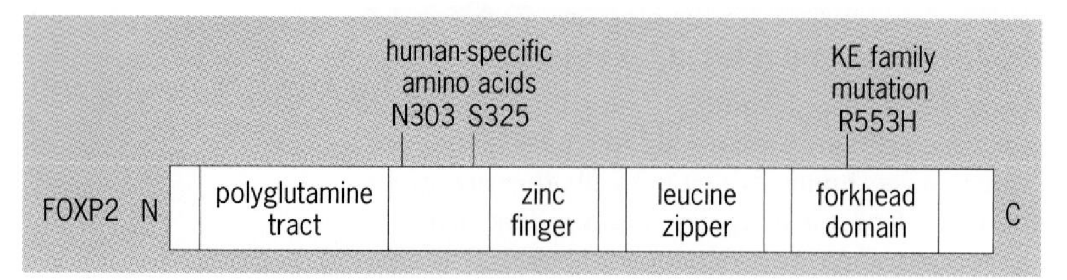

Fig. 1. Schematic representation of the FOXP2 protein.

to other forkhead family members and named it FOXP2.

FOXP2 has several key features, as depicted in **Fig. 1**. It has a DNA-binding domain that is highly homologous to the DNA-binding domains of other forkhead family members. It also has a leucine zipper (a common structural motif), which allows it to interact with other related transcription factors. Specifically, two molecules of FOXP2 can interact with one another (homodimerization), or one molecule of FOXP2 can interact with either one molecule of FOXP1 or one molecule of FOXP4 (heterodimerization) [**Fig. 2**]. Homo- or heterodimerization of FOXP2 is necessary for it to perform its function of regulating gene expression. FOXP2 also has a zinc finger domain and a polyglutamine tract, both of which may also contribute to protein-protein interactions and DNA binding.

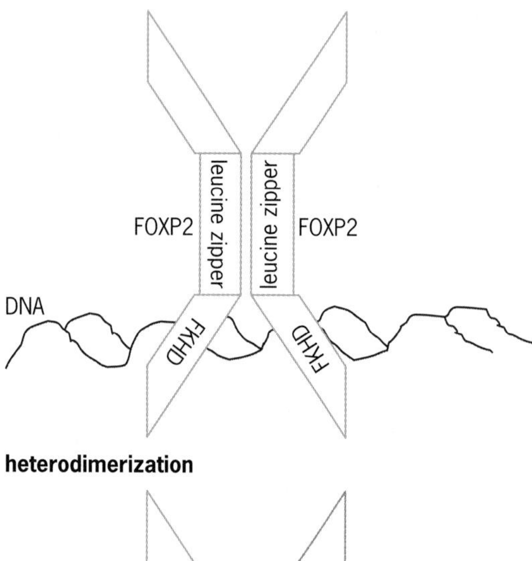

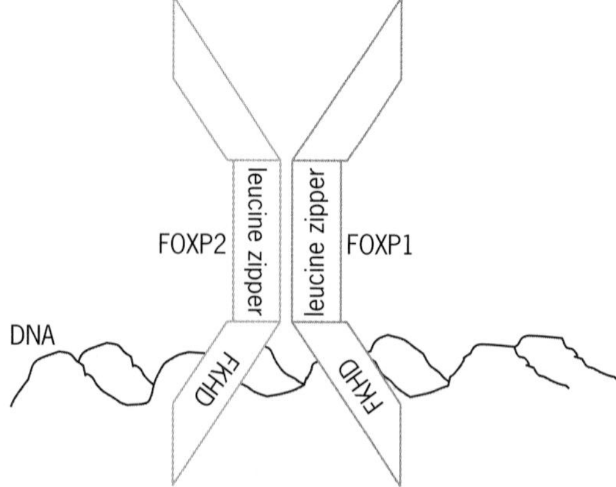

Fig. 2. Schematic representation of FOXP2 homodimers or heterodimers bound to DNA.

The inherited mutation in the KE family was determined to be an arginine-to-histidine mutation in the DNA-binding domain of FOXP2. A subsequent study using cell culture methods was able to demonstrate that this mutation in the DNA-binding domain completely prevents FOXP2 from binding to DNA and performing its function as a transcription factor. This mutation also might act as a dominant negative (a mutation resulting in a gene product that can interfere with the function of the normal gene product) by homodimerizing with wild-type FOXP2 and pulling it out of the nucleus. Imaging studies have demonstrated that the KE family has reduced gray matter in several areas of the brain, including the caudate nucleus and the inferior frontal cortex, where FOXP2 is expressed during development. The phenotype of these affected individuals is not limited to just the motor aspects of speech, that is, their distinct verbal dyspraxia, but rather the patients also demonstrate defects in writing, overall defects in motor learning, and below-average intelligence quotient (IQ). Several other individuals have been identified that have either mutations or truncations in FOXP2 and some form of language abnormality. Together, these reports solidify a role for FOXP2 in cognition and language. However, they clearly illustrate that FOXP2 is not a language gene per se, but rather is involved in development of circuits that, among other things, serve human language. This is supported by comparative analyses across different animal models (described below).

Evolution. FOXP2 is one of the most conserved mammalian proteins. When comparing the amino acid composition of FOXP2 throughout evolution, it is striking that the human form of FOXP2 only differs from the mouse form by three amino acids. Furthermore, two out of three of those different amino acids appear to have changed around the time that the common ancestor of humans and chimpanzees diverged. This recent evolutionary change in the FOXP2 protein is evidence of accelerated evolution. To examine whether the human-specific amino acid composition of FOXP2 confers a novel transcriptional function to FOXP2, either the human or chimpanzee form of FOXP2 was expressed in a human neuronal cell line. The two versions of the protein regulated different downstream targets consistent with a novel function for human FOXP2. In addition, these differentially regulated genes were also found to be differentially expressed in human and chimpanzee brain tissue. Thus, at least a subset of the differences in human and chimpanzee brains may be attributable to the differential function of FOXP2 during brain development.

Insight from animal models. The human brain was not built from scratch; rather, it expanded upon the structure and processes that evolution molded in its predecessors. The high conservation of the FOXP2 protein suggested that it might have some important conserved functions. To test this idea, as well as to learn more about the function of FOXP2 overall, several groups have generated animal models in which FOXP2 levels or features are manipulated. Mice with complete loss of Foxp2 die a few weeks after birth. These mice have severe motor defects and ultrastructural abnormalities, including reduced brain size and neurons with severely reduced neurite outgrowth. Mice lacking one copy of Foxp2 survive and are fertile, but also have some abnormalities, including motor defects. However, one of the more interesting observations was reduced ultrasonic vocalizations in distressed pups carrying only one Foxp2 gene. In addition, mice mutated to have either the KE family point mutation or the two human-specific amino acids also have abnormal ultrasonic vocalizations compared to controls. To investigate the divergent role of FOXP2 in evolution, the human version of FOXP2 was inserted into the region of the mouse Foxp2 gene. This resulted in a "humanized" mouse; in other words, the mouse expressed the human form of FOXP2 in areas where the mouse version would normally be expressed. These mice had various behavioral and cellular features that were different from normal mice, especially in the basal ganglia. However, perhaps the most intriguing finding was a change in ultrasonic vocalizations in mice expressing human FOXP2. This suggests a potential role for FOXP2 in circuits involved in vocalization in other mammals (**Fig. 3**).

Perhaps more instructive in this regard is work in songbirds, in which the pattern of FoxP2 expression has been shown to have remarkable parallels to human brain. Moreover, in zebra finches, a type of vocal-learning bird, FoxP2 levels change during song learning and song exposure in real time. Strikingly, when FoxP2 levels are reduced in vivo in brain nuclei important for song learning, finches display abnormal song learning. Thus, FOXP2 seems to have a conserved role in development and function of circuits that are involved in vocal learning mediated by auditory information. Along with the nonlinguistic deficits observed in humans with FOXP2 mutations, this suggests that these conserved circuits are critical for sensorimotor integration and motor learning in a broader sense than just language. However, the specialized functions of the human form of FOXP2 demonstrated by analysis of its transcriptional targets also suggest key human adaptations of this molecular circuitry that may be more specifically related to language in humans.

Human targets and disease. The elucidation of the function of FOXP2 in the human brain is more challenging from an experimental viewpoint. One way to begin to attempt this is by identifying the regions of DNA bound by FOXP2 and determining the gene that is regulated by the area of bound DNA. Initial studies using gene promoter microarrays identified a number of promoters specifically bound by FOXP2 in the human fetal brain compared to human fetal lung. What was remarkable about this list of promoters is that it was significantly enriched for genes having a known association with schizophrenia. More recently, FOXP2 has been shown to directly regulate expression of contactin-associated protein-like

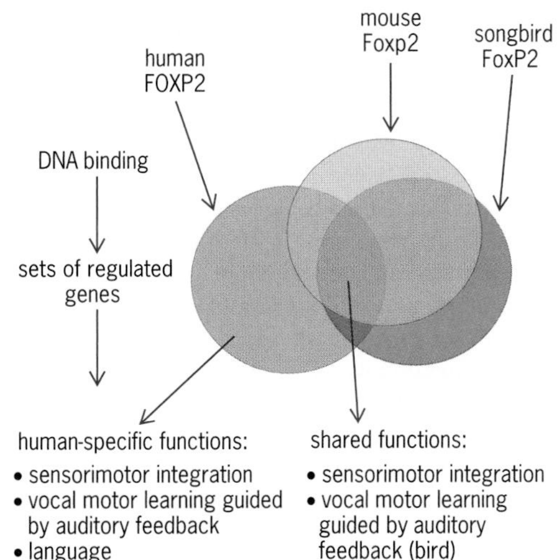

Fig. 3. FOXP2 regulation of gene expression in human, mouse, and bird brain.

2, CNTNAP2, an autism susceptibility gene. These studies suggest that the signaling pathways regulated by FOXP2 in the developing human brain are those that are uniquely vulnerable in many neuropsychiatric diseases and may link FOXP2-regulated genes to the disruptions in language and cognition in these disorders. Given the critical role of frontal-striatal circuits in the behavioral and cognitive problems associated with these disorders, this should not come as a surprise, Future work comparing the targets of human FOXP2 with those of other species should provide insight into the signaling cascades involved in these human-specific diseases.

For background information *see* BRAIN; COGNITION; DEOXYRIBONUCLEIC ACID (DNA); GENE; GENETICS; LINGUISTICS; MUTATION; NEUROBIOLOGY; SPEECH; SPEECH DISORDERS; TRANSCRIPTION in the McGraw-Hill Encyclopedia of Science & Technology.

Genevieve Konopka; Daniel H. Geschwind

Bibliography. S. E. Fisher and C. Scharff, FOXP2 as a molecular window into speech and language, *Trends Genet.*, 25:166–177, 2009; S. E. Fisher, C. S. Lai, and A. P. Monaco, Deciphering the genetic basis of speech and language disorders, *Annu. Rev. Neurosci.*, 26:57–80, 2003; S. Haesler et al., Incomplete and inaccurate vocal imitation after knockdown of *FoxP2* in songbird basal ganglia nucleus Area X, *PLoS Biol.*, 5:e321, 2007; G. Konopka et al., Human-specific transcriptional regulation of CNS development genes by FOXP2, *Nature*, 462:213–217, 2009; C. S. Lai et al., A forkhead-domain gene is mutated in a severe speech and language disorder, *Nature*, 413:519–523, 2001; E. Spiteri et al., Identification of the transcriptional targets of *FOXP2*, a gene linked to speech and language, in developing human brain, *Am. J. Hum. Genet.*, 81:1144–1157, 2007; I. Teramitsu et al., Parallel FoxP1 and FoxP2 expression in songbird and human brain predicts functional interaction, *J. Neurosci.*, 24:3152–3163, 2004; F. Vargha-Khadem et al., *FOXP2* and the neuroanatomy of speech and language, *Nat. Rev. Neurosci.*, 6:131–138, 2005.

Frustrated Lewis acid and base-pair reactions

In 1923, G. N. Lewis described electron donors as bases and electron acceptors as acids. Such Lewis acids and bases can combine to share electrons, forming a Lewis acid–base adduct. This fact leads to an understanding of the formation of main-group adducts and transition-metal coordination compounds, and it offers a rationale for many reactions in organic chemistry. In 2006, D. W. Stephan and colleagues uncovered a perturbation to this long-standing principle. By combining electron-rich and electron-poor species when the molecular structures preclude direct interaction of the electron acceptor and donor, they demonstrated that such sterically "frustrated" Lewis pairs (FLPs) provide unquenched Lewis acidity and basicity that is available for new reactivity.

In the initial study of FLPs, the species $[(C_6H_2Me_3)_2PH(C_6F_4)BH(C_6F_5)_2]$ (PH/BH) was prepared (Me = methyl). This is a rare example of a compound that contains both protic (H^+) and hydridic (H^-) centers. Upon heating to 150°C, this species liberates H_2 quantitatively, generating a red-orange solution of the phosphino-borane $(C_6H_2Me_3)_2P(C_6F_4)B(C_6F_5)_2$ (P/B) [reaction (1)].

$(C_6H_2Me_3)_2\overset{\oplus}{P}H-C_6F_4-\overset{\ominus}{B}H(C_6F_5)_2$ **(PH/BH)** colorless

150°C ↑ ↓ H_2 25°C (1)

$(C_6H_2Me_3)_2P-C_6F_4-B(C_6F_5)_2$ **(P/B)** red-orange

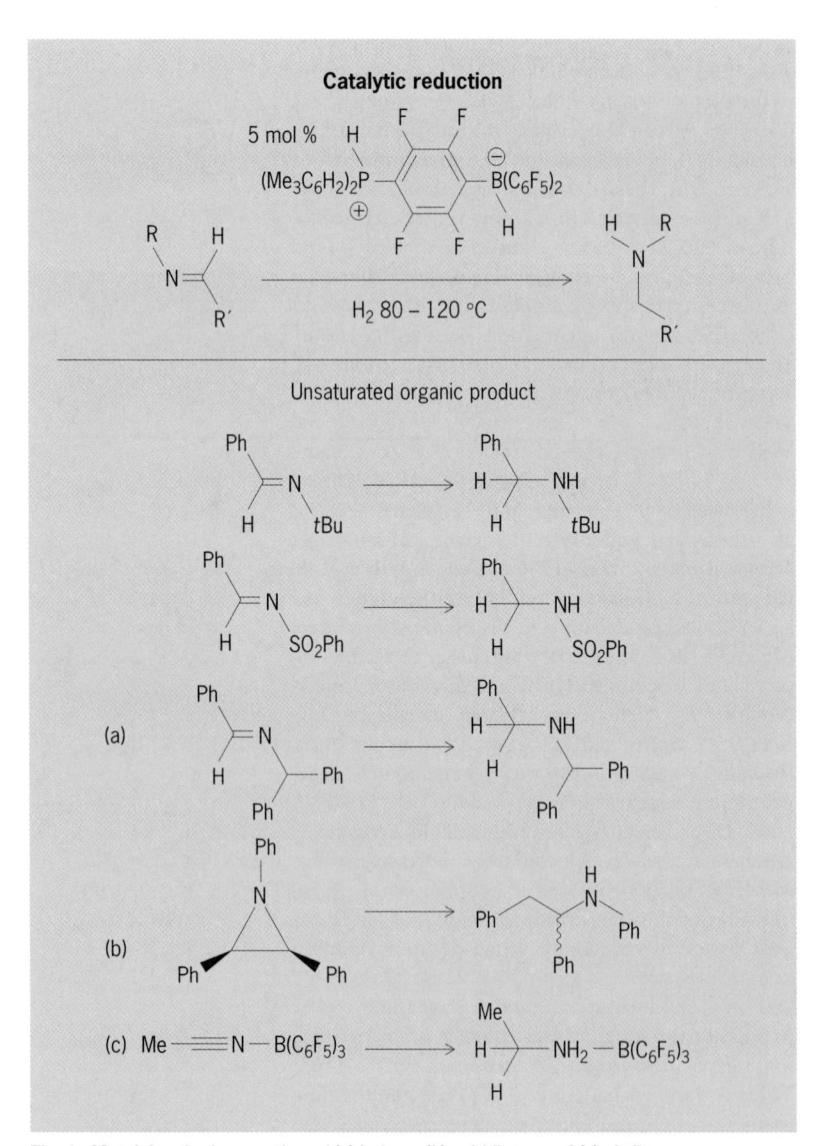

Fig. 1. Metal-free hydrogenation of (*a*) imines, (*b*) aziridines, and (*c*) nitriles.

Despite the fact that this P/B contains both a Lewis basic P center and a Lewis acidic B center, there is no evidence of intermolecular coordination in either solution or the solid state, consistent with a sterically "frustrated" Lewis pair. In an unprecedented finding, exposure of a solution of this neutral phosphino-borane to H_2 at 25°C led to the re-formation of the zwitterionic PH/BH salt. This easy interconversion of the PH/BH and P/B species was the first non-transition-metal system that was shown to release and take up hydrogen reversibly.

This unique reactivity was attributed to the presence of the frustrated Lewis pair. In further probing of this reactivity, simple combinations of $B(C_6F_5)_3$ with a variety of sterically frustrated phosphines were prepared. The crowded phosphines [where R = *t*-Bu (tert-butyl), $C_6H_2Me_3$] did not react with the Lewis acid $B(C_6F_5)_3$, confirming that the mixtures constitute FLPs. However, upon exposure to H_2 at 1 atm, there was an immediate reaction to form the corresponding salt $[R_3PH][HB(C_6F_5)_3]$, derived from the heterolytic activation of H_2 by the FLP. In contrast to the initial P/B system, these salts $[R_3PH][HB(C_6F_5)_3]$ were thermally stable with respect to loss of H_2.

Metal-free catalysis. Having observed the easy heterolytic cleavage of H_2 by FLPs, the potential of a metal-free hydrogenation catalyst was envisioned. To achieve this goal, a proton and a hydride must be transferred to an unsaturated organic molecule, thus regenerating the FLP, which will then be available for further activation with H_2. This cycle of transfer and H_2 activation will give rise to catalytic reduction of the substrate. This approach was initially demonstrated for the C=N double bonds of imine substrates. Treatment of such molecules with 5 mol% of the zwitterionic salt $[R_2PH(C_6F_4)BH(C_6F_5)_2]$

under a H_2 atmosphere effects the transfer of a proton and hydride to the C=N double bonds of imines to give the corresponding amines (**Fig. 1*a***). This catalysis proceeds cleanly and in high yield at temperatures between 80 and 120°C and H_2 pressures of 1–5 atm. This demonstration of the first metal-free hydrogenation catalyst proved effective for a number of imine substrates; however, these were restricted to imines with sterically demanding substituents on N. This limitation arises as less hindered imines bind tightly to the borane center of the P/B phosphino-borane, thus quenching further catalysis. This catalyst system also proved effective for the reductive ring opening of a three-membered CCN-aziridine ring (Fig. 1*b*). The mechanism has been shown to involve initial protonation of the N by the phosphonium center followed by BH attack at the adjacent carbon.

For imines with less bulky N substituents and nitriles (C≡N) [Fig. 1*c*], coordination to $B(C_6F_5)_3$ added steric congestion at N and permitted the catalytic reduction of the C=N or C≡N bonds, resulting in the formation of amine adducts of $B(C_6F_5)_3$. These latter reductions are thought to proceed via initial hydride transfer, followed by protonation to give the product amine-borane adduct.

A further extension of this metal-free catalysis recognized that sterically hindered imines can act as the Lewis base partner of $B(C_6F_5)_3$ forming an FLP. This permits the catalytic reduction of imine by simply employing a catalytic amount of $B(C_6F_5)_3$ in the presence of H_2.

The range of applications of such metal-free catalysis has been expanded. G. Erker's research group in Germany has developed related P/B catalysts that are effective for the hydrogenation of enamines and silyl enol ethers, species with C=C bonds adjacent to a N atom and $OSiR_3$ fragments (**Fig. 2**).

The mechanistic details of reactions of FLPs with H_2 continue to be the subject of intense theoretical study. Early computational studies support the view that the electrostatic approach of the Lewis acid and base is limited by steric demands. Nonetheless, the acid and base are associated, affording a species described as an "encounter complex." This species has a pocket between the base and acid in which H_2 is polarized for heterolytic cleavage. In this initial model, a linear arrangement of P···H–H···B atoms was suggested. More recent computations propose a perturbation of this model in which the H atoms are equidistant from B, giving a "side-on" interaction with B and "end-on" to P, as shown below in the theoretical model of activation of H_2 by phosphine and borane.

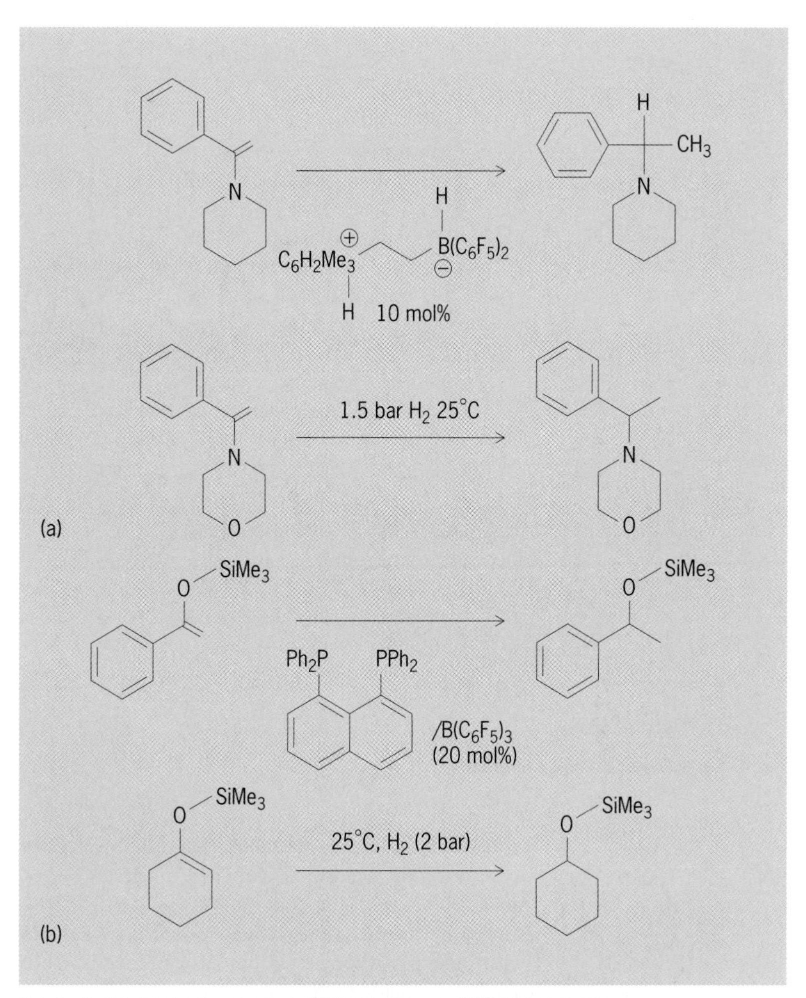

Fig. 2. Metal-free hydrogenation of (*a*) enamines and (*b*) silyl enol ethers.

This latter view is consistent with known interactions of H_2 with transition metals, and low-temperature matrix isolation experiments have indicated end-on interactions of phosphine with H_2.

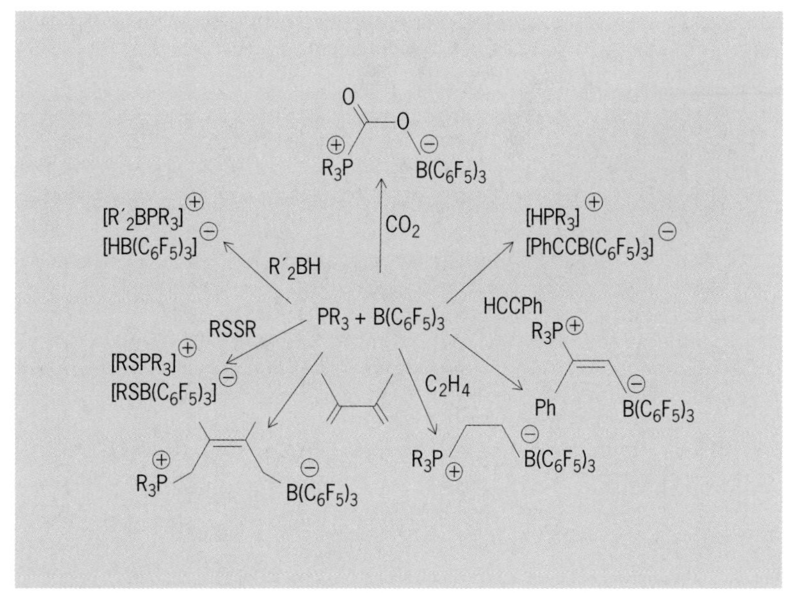

Fig. 3. Activation of small molecules by s.

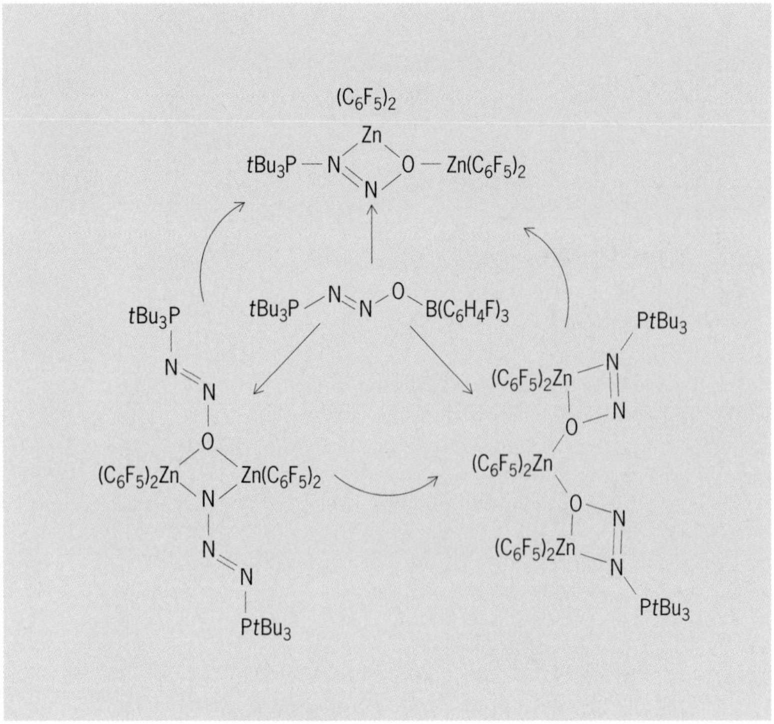

Fig. 4. Synthesis of Zn–N_2O complexes.

Activation of small molecules. The ability of FLPs to activate H_2 suggests the ability to activate other small molecules. This concept was initially demonstrated by the reactions of phosphine/borane FLPs with olefins. Although neither phosphines nor boranes react with olefins on their own, the combination results in the addition of phosphine and borane to the C=C double bond. In the case of ethylene, the product is $[R_3PCH_2CH_2B(C_6F_5)_3]$ (**Fig. 3**). For other terminal olefins, the additions proceed to give $[R_3PCH(R')CH_2B(C_6F_5)_3]$, in which the P base has added to the secondary carbon. Similarly, intramolecular reactions have been shown to give cyclic phosphonium zwitterions with a pendant $CH_2B(C_6F_5)_3$ fragment.

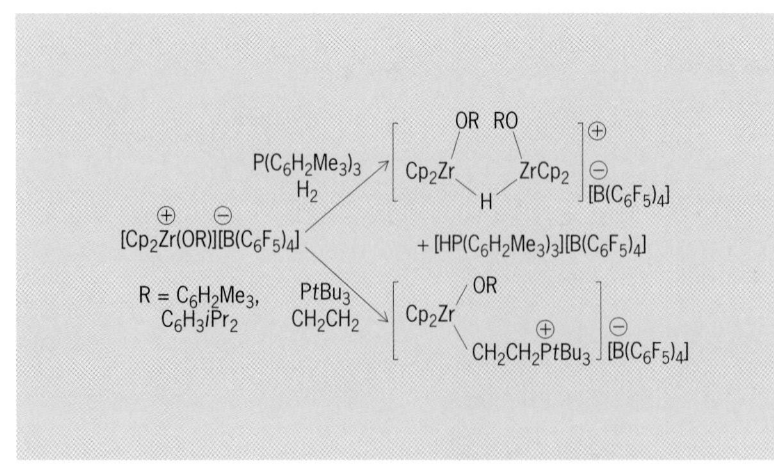

Fig. 5. FLP reactivity of Lewis acidic zirconocene Cp_2Zr or $C_5H_5(2Zr)$ cations.

FLPs also react with conjugated dienes (C=C–C=C). The *cis*-1,4-addition products, $[R_3PCH_2C(R')$=$C(R')HCH_2B(C_6F_5)_3]$ (Fig. 3) are the major products, although the reaction mixture contains other products thought to be either other isomers or 1,2-addition products.

Terminal alkynes also react with FLPs. If the sterically congested phosphine is basic, such as $t\text{Bu}_3P$, deprotonation of the alkyne results in formation of the salt $[R_3PH][RCCB(C_6F_5)_3]$. In contrast, for less basic phosphines, such as $(C_6H_2Me_3)_3P$, P/B addition to the C≡C occurs to give zwitterionic *trans*-$[R_3PC(R)$=$CHB(C_6F_5)_3]$ (Fig. 3).

FLPs have also been shown to effect other heterolytic bond cleavages. For example, reaction of $t\text{Bu}_3P$ (tri-tert-butyl phosphine) and $B(C_6F_5)_3$ with $(C_6H_4O_2)BH$ results in the cleavage of the B-H bond, affording $[(C_6H_4O_2)BPt\text{Bu}_3][HB(C_6F_5)_3]$ (Fig. 3). Similarly, treatment of the P/B FLPs with disulfides (RSSR) leads to heterolytic S-S bond cleavage to give the salts $[R'SPt\text{Bu}_3][R'SB(C_6F_5)_3]$ (Fig. 3).

Other Lewis acid and base combinations. Whereas much of the above chemistry has examined the range of small molecules that undergo FLP activation, other work has focused on the range of Lewis acid and base combinations that are capable of such reactivity. For example, combinations of carbenes and $B(C_6F_5)_3$ have been shown to be capable of H_2 activation. Sterically encumbered pyridines exhibited interesting behavior. For example, lutidine was shown to establish an equilibrium with $B(C_6F_5)_3$, providing both the classical Lewis acid–base adduct as well as FLP reactivity. Studies of the variations in the Lewis acids have included variations of B and related Al reagents.

Reactions with greenhouse gases. FLPs have also been shown to react with greenhouse gases. For example, $t\text{Bu}_3P$ and $B(C_6F_5)_3$ react with CO_2 to give the species $[t\text{Bu}_3PCO_2B(C_6F_5)]$ in reaction (1). This species is stable to 80°C, above which it loses CO_2. In a similar fashion, the linked FLP $[(C_6H_2Me_3)_2P(C_2H_4)B(C_6F_5)_2]$ also binds CO_2, forming similar P-C and O-B bonds, although this species readily loses CO_2 above −20°C in solution.

In an interesting study, researchers at Oxford combined CO_2 and H_2 in the presence of an FLP derived from a sterically hindered amine $(C_5H_6Me_4NH)$ and the borane $B(C_6F_5)_3$ at 160°C. After 4 days, methanol was formed. More recently, an FLP derived from $(C_6H_2Me_3)_3P$ and AlX_3 (X = Cl, Br) was also shown to react with CO_2. This combination of Lewis acid and base results in the isolation of $[(C_6H_2Me_3)_3PCO_2(AlX_3)_2]$, in which double activation of CO_2 was observed, as both oxygen atoms are bound to Al [reaction (2)]. This species proved to be much more thermally stable. Remarkably, this species reacts with NH_3BH_3 rapidly at room temperature to generate Al-bound CH_3O groups. Using ^{13}C-labeled CO_2 and quenching the reaction with water revealed the stoichiometric formation of methanol, which could be isolated in 50% yield [reaction (2)].

$$\text{Br}_3\text{Al}-\text{O}-\overset{\ominus}{\text{C}}(=\text{O}-\text{AlBr}_3)-\overset{\oplus}{\text{P}}\text{Mes}_3 \xrightarrow[\text{15min, 25°C}]{\text{NH}_3\text{BH}_3} [\text{Mes}_3\text{PH}][(\text{MeO})_n\text{AlBr}_{4-n}] + (\text{BHNH})_n \quad (2)$$

$[\text{Mes}_3\text{PH}][(\text{MeO})_n\text{AlBr}_{4-n}] \xrightarrow{H_2O} H_3COH$; $\xrightarrow{C_6F_5Br} PMes_3$

In related chemistry, $t\text{Bu}_3\text{P}$ and $\text{B}(\text{C}_6\text{F}_5)_3$ were reacted with the greenhouse gas N_2O to give the unprecedented complex $t\text{Bu}_3\text{PN}_2\text{OB}(\text{C}_6\text{F}_5)_3$. The P and OB fragments adopt a *trans*-disposition with respect to the formal N=N bond. This species is remarkably stable, evolving N_2 to generate $(t\text{Bu}_3\text{PO})\text{B}(\text{C}_6\text{F}_5)_3$ only above 140°C or under photolytic conditions. N_2O-FLP products were formed only using a strongly basic, sterically encumbered phosphine, but they proved to be tolerant of variations of the Lewis acid. Thus, several species of the form $t\text{Bu}_3\text{PN}_2\text{OBR}_3$ were obtained. In the case of $t\text{Bu}_3\text{PN}_2\text{OB}(\text{C}_6\text{H}_4\text{F})_3$, this species was shown to react with $\text{Zn}(\text{C}_6\text{F}_5)_2$ to give a series of Zn complexes of the form $[t\text{Bu}_3\text{PN}_2\text{OZn}(\text{C}_6\text{F}_5)_2]_2$, $[t\text{Bu}_3\text{PN}_2\text{OZn}(\text{C}_6\text{F}_5)_2]_2\text{Zn}(\text{C}_6\text{F}_5)_2$, and $[t\text{Bu}_3\text{PN}_2\text{O}(\text{Zn}(\text{C}_6\text{F}_5)_2)_2]$ (**Fig. 4**).

Outlook. Very recently, D. W. Stephan and colleagues described the use of combinations of early-transition-metal metallocene cations and sterically demanding phosphines in the activation of H_2, as well as additions to olefins (**Fig. 5**) and N_2O. These results suggest that this concept of FLP reactivity extends to organometallic chemistry.

In summary, the concept of combining Lewis acids and Lewis bases in which steric demands preclude classical Lewis acid-base adduct formation provides a new strategy for the catalysis and activation of a variety of small molecules. Research efforts are underway to exploit this fundamental finding for applications in chemical synthesis.

For background information *see* ACID AND BASE; CATALYSIS; COORDINATION CHEMISTRY; HYDROGENATION; ORGANIC SYNTHESIS; ORGANOMETALLIC COMPOUND; STERIC EFFECT (CHEMISTRY) in the McGraw-Hill Encyclopedia of Science & Technology.

Douglas W. Stephan

Bibliography. D. W. Stephan, Frustrated Lewis pairs: A concept to new reactivity and catalysis, *Org. Biomol. Chem.*, 6:1535–1539, 2008; D. W. Stephan and G. Erker, Frustrated Lewis pairs: Metal-free hydrogen activation and more, *Angew. Chem. Int. Ed.*, 49:46–76, 2010; G. C. Welch et al., Reversible metal-free activation of hydrogen, *Science*, 314:1124–1126, 2006.

Fungi in drinking water

Despite limitations in analysis methodology, knowledge about the occurrence and significance of fungi in drinking water has increased considerably in recent decades, and fungi are now generally accepted as drinking-water contaminants. However, the relevance of water-borne fungi for water quality and human health is poorly understood, and reports of effective treatment and disinfection against fungi in water are few. The occurrence of fungi in drinking water may be regarded as a potential health risk for particular vulnerable groups.

Filamentous fungi. Fungi are a diverse group of organisms that belong to the kingdom Eumycota, which comprises seven phyla according to the latest classification. For practical reasons, fungi have been divided into groups such as the filamentous fungi (molds), the yeasts, and the mushrooms. Some fungi are adapted primarily to aquatic environments and therefore are found naturally in water. Most filamentous fungi are adapted primarily to terrestrial environments. They are present in soil, organic material, air, and anything in contact with air, and these fungi can also enter drinking water from various locations.

Occurrence in drinking water. The first studies that reported fungi in drinking water, in the 1960s and 1970s, were undertaken primarily in response to a water-contamination problem, such as the presence of actinomycetes or cyanobacteria in water; thus fungi were never the main object of the analyses. However, significant amounts of microfungi were recovered from the water samples, leading to speculation about whether fungi could be causing the experienced problems. Since then, knowledge about filamentous fungi in drinking water has increased to the point at which fungi are now regarded as drinking-water contaminants. They have been reported from all types of water, from raw water to treated water, and from heavily polluted water to distilled or ultrapure water. Fungi have also been reported from bottled drinking water and even from water in chlorinated swimming pools. A wide diversity of fungal species has been reported from drinking water, including species of concern for human and animal health.

Analysis. There is no international standard method for the analysis of fungi in drinking water. The most frequently performed method is a three-step culturing procedure comprising isolation (for example, membrane filtration), quantification (colony counts), and identification (fungal identity). Such culturing methods are often time-consuming and require long experience in the field. Fungi have a tendency to be unevenly distributed in water, which may cause difficulty in ensuring representative samples. A frequent problem in both isolation and quantification of fungi is overgrowth of filters (see **illustration**), which can occur very quickly. Another problem is that not all fungi are able to

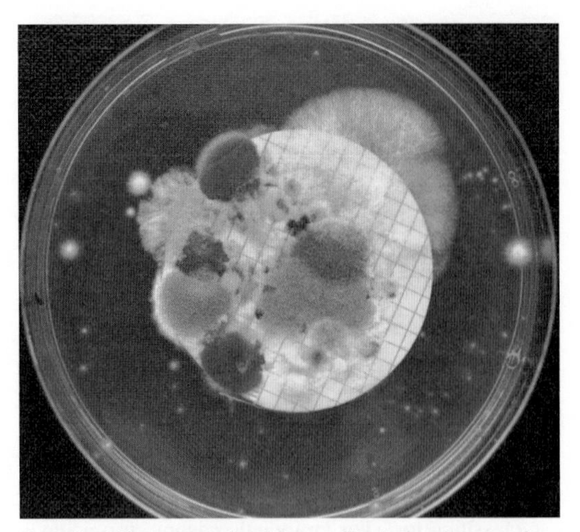

Growth of filamentous fungi from a filtrated water sample. (*Photo by G. Hageskal, National Veterinary Institute*)

grow under laboratory conditions. Therefore, the currently used methods may give a fair indication of the level of fungi present, but they will never give the precise number of fungi in the water. Another problem that is often experienced is that some fungi never sporulate and thus cannot be identified morphologically. However, DNA sequencing is not dependent on sporulating fungi, or even viable fungi, and is a good supplement to morphological identification. Moreover, the scant information about what levels of fungi are normal or acceptable in drinking water is an important limitation for the fungal water studies performed. Except for water regulations in Sweden, which include specifications on water-borne fungi, regulations for the occurrence of fungi in drinking water do not exist.

Significance for health and water quality. Filamentous fungi may have diverse effects on humans and animals, and several species may potentially be pathogenic, allergenic, and/or toxigenic. Along with the increasing frequencies of severely immunocompromised patients in hospitals, a dramatic increase in the number of cases of invasive fungal disease has occurred recently. Several studies have focused on analyzing hospital water systems with respect to the presence of fungi, and the results indicate that hospital water may contain a wide diversity of fungi, including potential pathogens. However, the link between water contaminated by fungi and human health is not fully understood, although it is hypothesized that fungi in water are aerosolized into air when water passes through installations such as taps and showers and thus are introduced to patients who are at high risk for fungal infections. Problems of resistance against antifungal drugs are also increasing, making the focus on prevention of exposure to pathogenic fungi more important than ever. With respect to allergy, a few case reports have been connected directly to water contaminated by fungi. Outbreaks of skin irritations associated with taking showers and baths and of hypersensitivity pneumonitis (also known as extrinsic allergic alveolitis) caused by elevated levels of filamentous fungi in water have been reported. Mycotoxin production by fungi in water has also been reported, although it is likely that concentrations will be low under normal circumstances. However, mycotoxin concentrations may increase occasionally if water is stored in cisterns or reservoirs, or even in bottles, for prolonged periods. Other chemical compounds produced by some fungal species may cause undesirable sensorial changes in drinking water, such as foul taste and odor. In addition, water contaminated by fungi may also act as inoculum of spoilage fungi into food and beverages.

Control strategies. Filamentous fungi survive water disinfection, and most current water treatment methods are not sufficient to eliminate fungi. The control strategies and efficiency of water treatment and disinfection against fungi have been studied to only a small extent, and the results are not consistent. Chemical coagulation has been reported to offer good removal of fungi in raw water. However, increasing levels of fungi have been reported in water distribution systems, and biofilms (microorganism communities) on pipes and water installations are probably an important causative factor for this increase. Fungi are known as excellent biofilm producers, and the biofilm life may both supply nutrients and protect the fungi against disinfectants and other threatening factors. Biofilms are difficult to remove. Most cleaning techniques have been found to be insufficient for the removal of fungi established in biofilms, although flow jetting has been reported to offer the most effective treatment.

Summary. Several studies have demonstrated that fungi are relatively common in drinking water. Species of pathogenic, allergenic, and toxigenic concern have been isolated from water, sometimes in high concentrations. Although the possible health efffects of fungi in water are still unclear, fungi in water may be aerosolized into air and introduced to severely immunocompromised patients. Sensorial changes have been associated with the occurrence of fungi in drinking-water systems, and waterborne fungi may also act as inoculum of spoilage fungi into food and beverages. Adequate water treatment could be a good solution, and monitoring water systems with regard to fungi may be required, especially in hospital water systems. Further studies are merited, particularly with respect to establishing accurate methodologies and investigating effects of water treatment against fungi in water. Epidemiological studies should also be conducted to determine the full health significance of drinking water that is contaminated by fungi. Because filamentous fungi may influence water quality in several ways and may constitute a potential health risk, the mycobiota of drinking water should be considered when the microbiological safety and quality of drinking water are assessed.

For background information *see* ENVIRONMENTAL TOXICOLOGY; FUNGAL ECOLOGY; FUNGAL INFECTIONS; FUNGI; MEDICAL MYCOLOGY; MYCOLOGY

MYCOTOXIN; PUBLIC HEALTH; WATER POLLUTION; WATER TREATMENT in the McGraw-Hill Encyclopedia of Science & Technology.

Gunhild Hageskal; Ida Skaar

Bibliography. G. Hageskal et al., Diversity and significance of mold species in Norwegian drinking water, *Appl. Environ. Microbiol.*, 72:7586–7593, 2006; G. Hageskal et al., Occurrence of moulds in drinking water, *J. Appl. Microbiol.*, 102:774–780, 2007; G. Hageskal, N. Lima, and I. Skaar, The study of fungi in drinking water, *Mycol. Res.*, 113:165–172, 2009; J. Kelley et al., *Identification and Control of Fungi in Distribution Systems*, AWWA Research Foundation and American Water Works Association, Denver, CO, 2003; R. R. M. Paterson and N. Lima, Fungal contamination of drinking water, in J. Lehr et al. (eds.), *Water Encyclopedia*, John Wiley & Sons, New York, 2005.

Galaxy evolution in the early universe

Galaxies are the fundamental building blocks of the universe. There are several hundred billion galaxies in the universe, each containing tens to hundreds of billions of stars. In the local universe, three general classes of galaxies are observed: spiral, irregular, and elliptical. Spiral galaxies, such as the Milky Way, are stellar systems characterized by a thin, rotating disk surrounding a central spherical bulge (**Fig. 1**). A small fraction of the matter in these galaxies (roughly 5%) is gas, located primarily in the thin disk and continuously forming new stars. Irregular galaxies are the less massive cousins of spiral galaxies; these systems have slightly larger gas fractions, little or no coherent spiral structure in their disks, and no central bulge. At the other extreme, elliptical galaxies are gas-poor, roughly spherical systems that resemble scaled-up bulges.

To understand how this galaxy menagerie was created, astronomers look backwards in time to the epoch of peak galaxy growth in the universe's history, during which the assembly of galaxies can be directly observed.

Fig. 1. Local spiral galaxy (NGC 1376). The galaxy center contains a spherical bulge and is surrounded by large spiral arms, in which gas is converted into new stars. [*NASA, ESA, Hubble Heritage Team (STScI/AURA)*]

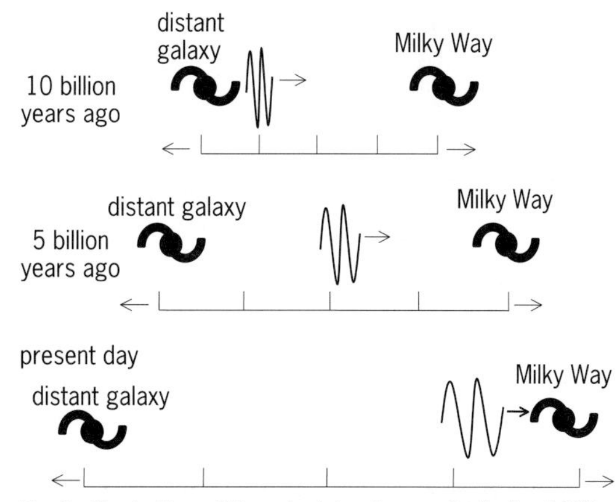

Fig. 2. Illustration of the principle of cosmological redshift. The length of time that light travels through the expanding universe (indicated here with the horizontal ruler) determines the amount that the light is "stretched." More distant objects, whose light spends more time traveling through space to reach the Milky Way, have larger redshifts.

Looking backward through the universe. Because the speed of light is finite, light from distant objects travels for substantial amounts of time before reaching Earth. When astronomers observe the nearest spiral galaxy to our Milky Way, the Andromeda Galaxy, at a distance of 2.5 million light-years (2.4×10^{19} km or 1.5×10^{19} mi), they are seeing light that has taken 2.5 million years to reach Earth and was therefore emitted by the Andromeda Galaxy 2.5 million years ago. Viewing increasingly distant galaxies is consequently equivalent to observing them as they were farther back in time.

The light from very distant galaxies takes billions of years to reach Earth, during which time it also gets "stretched" as it travels through the ever-expanding universe (**Fig. 2**). This stretching to longer, and therefore redder, wavelengths is referred to as cosmological redshifting, and the amount of redshifting a galaxy's light has experienced is a direct measure of the time that light has traveled through the universe. Galaxies with higher redshifts probe earlier epochs in the universe's history.

Through studies of galaxies at a variety of redshifts (denoted z), the broad context of galaxy evolution has been established. In particular, the star formation activity in the universe has been observed to have increased with time after the "big bang" for the first 2 billion years ($z > 4$), reached a plateau at a high value for the following 3 billion years ($z \approx 4$–1), and then dropped dramatically for the remaining 8 billion years of the universe's history ($z \approx 1$–0; **Fig. 3**). The epoch of most rapid activity in the universe's history therefore occured 2–5 billion years after the big bang; the stellar mass in galaxies correspondingly increased from around 15% to 50–75% of its current value during this era.

On cosmic scales, this behavior can be explained through the competing effects of the growth of structure in the universe and the formation of stars

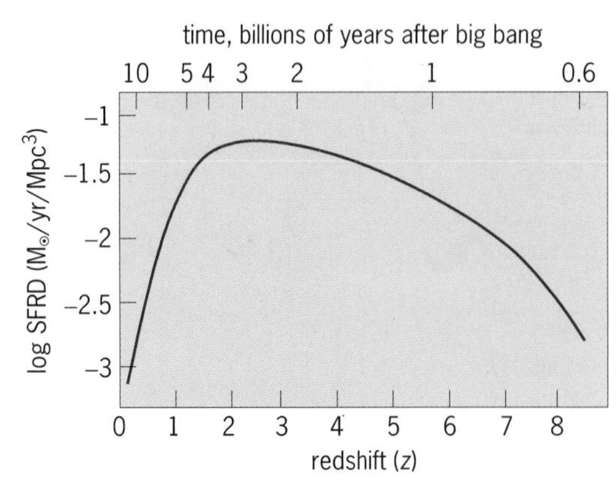

Fig. 3. Evolution of the volume-averaged rate of star formation with redshift (bottom axis) and time since the big bang (top axis). The cosmic star formation rate density (SFRD) is expressed in units of solar masses ($M_\odot$) formed per year per cubic megaparsec (1 Mpc = 3.09×10^{19} km = 1.92×10^{19} mi = the approximate radius of a group or cluster of galaxies).

from primordial gas. As the universe aged, matter concentrations within it gravitationally attracted nearby matter, leading to increasingly massive structures such as galaxies and clusters of galaxies. In the early universe, much of the mass accreted by galaxies was still in its primordial gaseous state and therefore contributed immediately to increased star formation rates inside these galaxies. As the masses of galaxies grew with time, the rate of star formation in the early universe rose (Fig. 3).

These high star formation rates eventually began to exhaust the available gas in the universe (after $z \approx 1$); material newly accreted by growing galaxies became increasingly gas-depleted and therefore unable to fuel further star formation. The rate of star formation in the universe correspondingly dropped dramatically (Fig. 3). Galaxies in the present-day universe ($z = 0$) are consequently much less active than they were 3 billion years after the big bang. To study the complex processes that drive galaxy evolution, the best laboratories are the most rapidly evolving systems, which are therefore not found in the local universe but instead were present 2–5 billion years after the big bang, at high redshift ($z \approx 1$–4).

The nature of galaxies in the early universe. Observations of the universe at $z \approx 1$–4 reveal that a much larger fraction of the galaxy population was undergoing rapid star formation than in the local universe. In this early epoch, the majority of galaxies were forming stars at a steady rate and were doing so much faster than their local kin.

The morphologies of high-redshift star-forming galaxies (**Fig. 4**) differ significantly from those of their star-forming analogs in the local universe, spiral galaxies (Fig. 1). At high redshift, individual morphological features within galaxies are larger, creating strongly asymmetric internal mass distributions. In particular, large star-forming gas complexes are observed, each with 1000 times the mass of local star-forming regions; collectively, these supermassive star-forming regions account for roughly one-third of the total mass in their host galaxies.

Kinematically, the star-forming gas in high-redshift galaxies is observed to reside in rotating disks, as in local spiral galaxies. The comparable rotation velocities and characteristic radii of high-redshift and local star-forming galaxies implies that they have similar masses. However, the dynamics of high-redshift galaxies differ from those of local spiral galaxies in two important ways. First, the supermassive star-forming clumps are significant dynamical perturbations whose effects are visible in the gas motions. Second, high-redshift galaxies have much larger random motions in their disks, indicating that their rotating, star-forming disks are substantially thicker in the vertical direction than their local counterparts. These unique morphological and dynamical features provide strong constraints on the processes that control the evolution of high-redshift galaxies and on the conversion of these galaxies into the galaxies observed today.

Drivers of galaxy evolution. The presence at high redshift of a significant population of galaxies with large, rotating disks and continuous powerful star formation indicates that the main driver of galaxy evolution in this epoch could not be galaxy mergers, whose gravitational torques produce short bursts of star formation and destroy large disks. Instead, the majority of high-redshift galaxies must have grown in a less destructive manner. The most probable source of fuel for their high star formation rates are smooth streams of inflowing gas, which bring gas into galaxies quickly and continuously without causing dynamical disturbances. Cosmologically, such streams are expected to exist between matter concentrations, creating a network of thin filaments that connect galaxies and galaxy clusters to one another. Recent

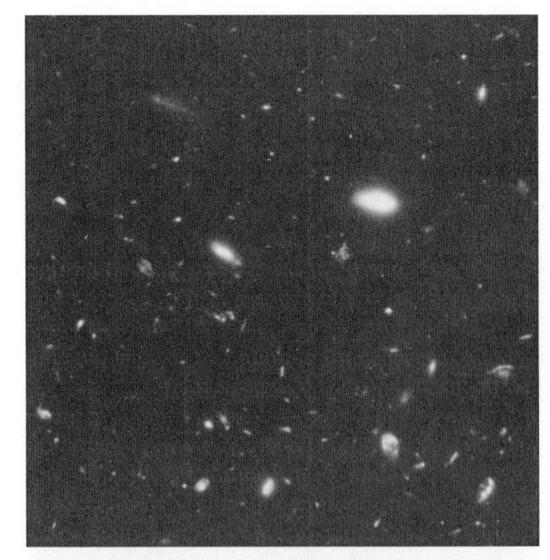

Fig. 4. Portion of the sky covered by the Hubble Ultra Deep Field, whose long exposure time has given astronomers an unprecedented view of distant galaxies. The clump-dominated morphologies of high-redshift galaxies are especially apparent in the extended systems at the lower right. (*NASA, ESA, S. Beckwith, HUDF Team*)

numerical models have shown that the gas in these filaments will be transported along the filaments and into the centers of galaxies, naturally linking the growth of galaxies to their cosmological context.

This scenario accounts for the high star formation rates and gas fractions (around 50%) observed in high-redshift galaxies, and such a significant amount of gas in turn explains the observed properties of high-redshift galaxies. In a rotating gas disk, the stability of gas against collapse under its self-gravity is determined by the rotational speed of the disk, the gas surface density, and the gas pressure. A disk that is unstable toward collapse will form structures on characteristic scales determined by these parameters. In the local universe, the gas in spiral galaxies collapses to form giant molecular clouds with masses a million times that of the Sun. At high redshift, the increased gas pressure (turbulence) and density leads to structures a thousand times more massive—namely, supermassive star-forming clumps.

These clumps are the direct result of galaxy evolution driven by massive and smooth inflows of gas. However, they are also important evolutionary drivers in their own right. Clumps are formed at large radii in their host galaxies, far from the galaxy centers, but this extended and asymmetric configuration is dynamically unstable. Observations and theoretical simulations show that dynamical friction on supermassive star-forming clumps in a galaxy will cause them to migrate to the galaxy center, carrying most of their mass inwards and forming a bulge similar to those of local galaxies. A small fraction of this mass will be stripped off at larger radii during this process and may account for the thick disk, a group of older stars that surrounds the star-forming thin disk in spiral galaxies. Moreover, the star formation occurring in supermassive star-forming clumps is an excellent candidate for producing the population of dense "globular" star clusters known to be associated with galaxy bulges and thick disks.

Linking galaxies across cosmic time. The gas-rich clump-driven phase in galaxy evolution may therefore be very important in creating some of the defining features of local spiral galaxies: bulges, thick disks, and globular star clusters. Observations of the dynamical structure of local elliptical galaxies have recently shown that this galaxy population also contains similar components, although the bulges are more massive and the thick disks smaller than in spiral galaxies. Comparison of the numbers of clump-dominated high-redshift galaxies and local galaxies suggest that the high-redshift population will eventually evolve into massive spiral galaxies and elliptical galaxies in the present day, and cosmological simulations support these conclusions. As a result, the current understanding of galaxy evolution at high redshift provides an elegant means to explain the presence of the primary substructures observed in local galaxies.

For background information *see* BIG BANG THEORY; COSMOLOGY; GALAXY, EXTERNAL; GALAXY FORMATION AND EVOLUTION; MILKY WAY GALAXY; MOLECULAR CLOUD; REDSHIFT; STAR CLUSTERS; STELLAR EVOLUTION; UNIVERSE in the McGraw-Hill Encyclopedia of Science & Technology. Kristen L. Shapiro

Bibliography. J. Binney and M. Merrifield, *Galactic Astronomy*, Princeton University Press, Princeton, NJ, 1998; R. Genzel et al., The rapid formation of a large rotating disk galaxy three billion years after the big bang, *Nature*, 442:786–789, 2006; L. Tacconi et al., High molecular gas fractions in normal massive star-forming galaxies in the young universe, *Nature*, 463:781–784, 2010.

Gene discovery in Drosophila melanogaster

At an elementary level, the development and homeostasis of an organism are the result of the differential expression of thousands of genes, the basic heredity units. Each gene consists of a portion of the DNA that contains sequences that code for a protein and noncoding sequences that determine its spatial and temporal expression pattern. The interregulations of a subset of an organism's genes form genetic hierarchies that dictate pathways that are responsible for the formation of the organism and its maintenance. The identification and analysis of genes in these networks allows scientists to understand the fundamental principles of development and identify how misexpression or mutations that eliminate gene function lead to genetic diseases.

Drosophila melanogaster as a research tool. In most cases, human experimentation to identify and study the role of genes involved in embryonic development is ethically impossible, leaving scientists to use model organisms as the primary means to discover and understand gene function. To the surprise of most nonscientists, many of the same genes and signals that function in the fruit fly, *Drosophila melanogaster*, also direct similar molecular pathways in humans. Thus, the simple fruit fly can be used to study complicated developmental pathways, including body patterning, organogenesis, and nervous system development. To determine the functions of conserved genes in these pathways, researchers studied the resulting physiological defects or phenotypes that occur when the genes are mutated or overexpressed in the fruit fly. By understanding the functions of genes in the fruit fly, insights into the activities of these same genes can be proposed for humans. In many cases, the defects that are observed when conserved genes are disrupted in fruit flies are similar to disease phenotypes that are observed when those same genes are mutated in humans. The success of using the fruit fly as a research tool to discover conserved genes in humans is seen by the number of vertebrate genes whose names are derived from the fruit fly mutants from which they were identified. These include notable regulatory genes such as *notch*, *sonic hedgehog*, *disheveled*, *son of sevenless*, and *eyes absent*, as well as *SMAD* and *Wnt* gene family members.

Through the implementation of the fruit fly as a model organism in the early 1900s, Thomas Hunt Morgan initiated the era of *Drosophila* genetics. Morgan was searching for an inexpensive organism that had a short generation time and could be reared in great numbers in a limited space. While breeding the fruit fly, Morgan identified a male with white eyes instead of the characteristic red. To test if this trait could be passed onto its offspring, this white-eyed male was crossed with normal red-eyed females. All of the resulting hybrid progeny had red eyes. Importantly, however, brother–sister mating between these hybrids resulted in one-quarter of the offspring with white eyes, all of which were males. Morgan concluded that the white-eyed trait was linked to sex chromosome (X) inheritance, thus confirming the chromosomal theory of inheritance. Morgan and his colleagues subsequently identified many spontaneous mutations that ultimately enabled them to establish behavior of the traits and localization of the genes on the chromosomes. For this groundbreaking work, Thomas Hunt Morgan was awarded the Nobel Prize for Physiology or Medicine in 1933.

Drosophila came of age in developmental biology with the work done by Christiane Nüsslein-Volhard and Eric F. Wieschaus (published in 1980) on mutations that affect early embryonic patterning. This was the first time that a phenotypic mutagenesis screen was performed to identify every gene involved in a developmental process. Up to this point, scientists had assumed that development was too complex to understand through genetics. Mutagenesis screens work in a backwards direction to identify and determine the function of genes in a specific process. Genes are identified, and the function of each gene can be determined by observing the resulting defect that occurs when it is absent or expressed at too high a level. To perform a simple mutagenesis screen, male fruit flies are fed a DNA-damaging agent that changes the gene sequence in all cells, including the germline cells that become sperm. These mutagenized males are crossed with females that contain a marked balancer chromosome (**illus.** *a*). The balancer chromosome is critical to the process because it prevents genetic crossing-over of the new mutation to the sister chromosome, it carries a dominant marker that allows researchers to follow the chromosome by observing a visible trait, and it contains a mutation that results in lethality when found in two copies. Crossing the F1 male offspring again with females that contain a balancer chromosome allows the establishment of fruit fly stocks that contain a single genetic mutation. By mating the F2 siblings to each other, fruit flies can be generated that carry two mutant copies of a gene, allowing researchers to study defects in a specific physiological process related to that gene.

In most cases, protein products encoded by multiple genes interact to regulate biological processes. Although simple mutagenesis screens identify many of these genes, in some cases the role of a gene in a pathway is only uncovered when its function is

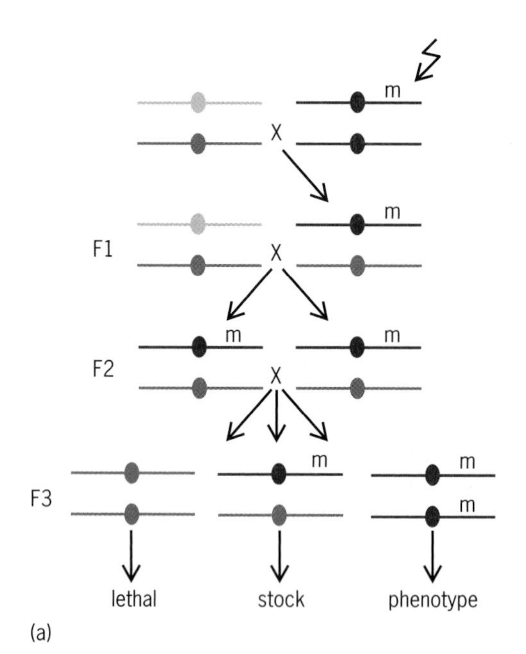

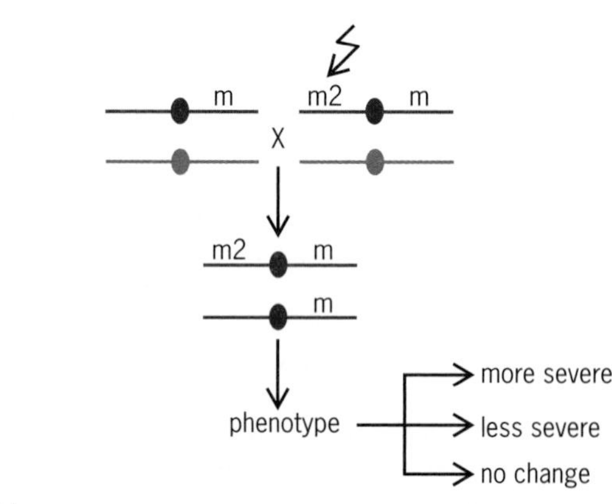

Traditional genetic screens performed in fruit flies. (*a*) Crossing scheme used to generate fruit flies containing mutant genes. Lines represent genomic DNA. Male fruit flies are fed a DNA-damaging agent or subjected to x-ray mutagenesis to induce random mutations (m) and then bred in mass with females containing a marked balancer chromosome (represented by dark black lines). Single F1 males are bred again with females containing a balancer chromosome to generate F2 male and female fruit flies carrying the same mutation. F2 siblings are backcrossed with each other to generate a balanced stock (50% of offspring) and produce F3 progeny that have a mutation in both copies of the gene (25%). Mutant progeny (m/m) are screened for defects. Fruit flies containing two copies of the balancer chromosome will die (25%). (*b*) Crossing scheme for modifier screens. Male fruit flies containing a previous mutation over a balancer chromosome are mutagenized as above and crossed with females carrying the same mutation. Offspring carrying a new mutation (m2) and mutant for both copies of the original mutation (m2; m/m) are compared to flies deficient for the original mutations (m/m) to determine if the defect is more severe (enhanced), less severe (suppressed), or has no effect. Enhanced and suppressed phenotypes suggest a genetic interaction between the two genes.

reduced in a sensitized genetic background. For example, if a weak mutation in a gene resulted in a small eye and a strong mutation in the same gene resulted in no eye, modifier screens can be performed in the weak sensitized background to identify mutations in other genes that result in an enhanced (no eye) or suppressed (normal eye) phenotype (see illus. *b*). Similarly, modifier screens can be performed to identify mutations that enhance or suppress phenotypes that occur when normal or mutant proteins are overexpressed. Using this approach, transgenic fruit flies expressing human disease proteins can be used to study the human disease phenotype in the fruit fly and modifier screens can be performed to identify other genes that participate in the pathogenesis of the disease. Thus, genes identified through modifier screens in the fruit fly can act as a blueprint to identify and test homologous vertebrate genes.

Genome sequencing. With the sequencing of the *Drosophila* genome in 2000, most genes in the fruit fly have now been identified by computer annotation. This is also the case for the human genome, the sequence being completed in 2004. The sequencing of these genomes reduced the need to use model organisms, such as the fruit fly, to identify new genes. Although one may predict that the sequencing of these genomes would bring the end to fruit fly research, just the opposite is happening. Sequence comparison has shown that 50% of fruit fly genes have human gene counterparts. Importantly, out of the 2753 genes that have been identified as causing diseases in humans, 1042 of these genes are found in the fruit fly. Many electronic database sites are helping scientists use the information gained through the study of fruit flies to decipher the molecular causes of human diseases and deduce the function of genes identified purely on the bases of sequence alone.

Researchers have used the information gained from sequencing the fruit fly genome to generate fly stocks that express RNA interference (RNAi) molecules, which can remove the function of specific genes. RNAi acts in a similar manner to classical mutations by specifically shutting down the expression of genes in the cell. However, unlike traditional mutations that affect every cell in the body, RNAi-mediated gene silencing allows genetic screens to be performed where the activity of a gene is only reduced in a specific target tissue, thus bypassing the lethality that is commonly associated with loss of gene function early in development. This technology allows the design of more elegant screens in the fruit fly to establish the function of a gene in a specific tissue or biological pathway at a defined stage of development.

Insights into human diseases. There are numerous examples of how research in the fruit fly has provided insights into possible molecular mechanisms associated with the pathology of human genetic diseases. Recently, the fruit fly is yielding insights into the most common cause of death and morbidity in humans, namely heart disease. Using an RNAi-based screen, a research group tested the requirement of 7061 evolutionarily conserved genes to regulate heart function in the adult fruit fly. Through this screen, 490 viable RNAi lines were identified that exhibited significant lethality when the heart was stressed by increased temperature. One of the identified molecular pathways was seen to involve the CCR4-Not complex. This complex had not been shown previously to influence heart function. Loss of the *not3* component of this complex caused dilated cardiomyopathy and myofibrillar disarray in the fruit fly. To determine if the role of *not3* was conserved in vertebrates, the mouse *not3* gene was disrupted using homologous recombination. Similar to the fruit fly, mice lacking one copy of the *not3* gene displayed impaired heart function and increased susceptibility to heart failure. Comparisons to published data regarding the function of the CCR4-Not complex suggested that NOT3 is required for chromatin modifications. Consistent with this idea, heart defects in *not3* mice were reversed by treatment with drugs that interfered with chromatin remodeling. Interestingly, some patients that display altered cardiac QT intervals (the QT interval is a measure of the duration of ventricular depolarization and repolarization) known to cause ventricular tachyarrhythmias (fast heart rhythms) carry a variant of the *not3* genes. Thus, using the fruit fly as a screening tool, it was possible to not only identify a new molecular component of heart development and disease, but also pinpoint a possible mechanism to treat the disease in humans. Ongoing discoveries using fruit flies will likely continue to provide valuable information into the understanding and treatment of many genetic diseases in humans.

For background information *see* CHROMOSOME; CROSSING-OVER (GENETICS); DEVELOPMENTAL BIOLOGY; DEVELOPMENTAL GENETICS; DIPTERA; DISEASE; GENE; GENETIC MAPPING; GENETICS; GENOMICS; HUMAN GENOME; MUTATION; RNA INTERFERENCE in the McGraw-Hill Encyclopedia of Science & Technology.

Robert B. Beckstead

Bibliography. S. Chien et al., Homophila: Human disease gene cognates in *Drosophila*, *Nucleic Acids Res.*, 30:149–151, 2002; G. G. Neely et al., A global in vivo *Drosophila* RNAi screen identifies *NOT3* as a conserved regulator of heart function, *Cell*, 141:142–153, 2010; Z. Orfanos, Transgenic tools for *Drosophila* muscle development, *J. Muscle Res. Cell Motil.*, 29:185–188, 2008; D. St. Johnston, The art and design of genetic screens: *Drosophila melanogaster*, *Nat. Rev. Genet.*, 3:176–188, 2002.

Genetics of memory

Most people have had the experience of almost but not quite remembering the name of a person that they have met, or knowing they have studied a topic but not remembering it well enough to answer that one last nagging question on the final exam. Since most of the time we move fluidly from idea to idea, it is most evident that there are memories when we realize that we should know an answer, but it is just not there. Therefore, how do we form memories? The main issue addressed here is how experience sometimes gives rise to a memory. Introspection also gives the inkling that not all memories are the same (that is, remembering a telephone number for a few

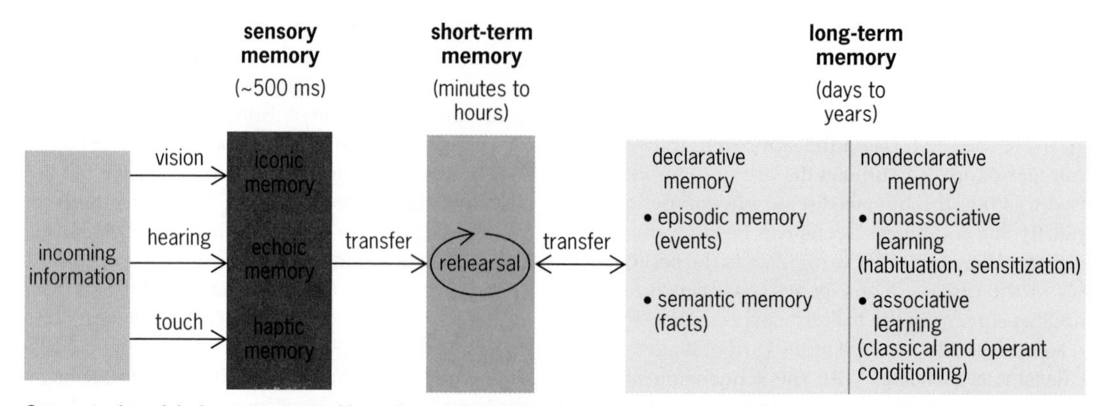

Conceptual model of memory types. Memories can be categorized by how long they last. Sensory memories last less than a second and are subdivided by the type of sensory information. Depending on how important the information or events are, memory items can be "transferred" to different phases, from short-term memory to long-term memory. Retrieval of long-term memories brings them back into a short-term-like memory state.

seconds is probably not the same as remembering the funeral of a favorite uncle). Thus, the next problem is how can we organize or conceptualize different sorts of memories. Finally, can we make some sense of the genes in the human genome that make it possible for us to learn and remember the many varied events that are experienced from day to day?

Conceptualization of different memory types. A prevailing view of memories is that they can be categorized by how long they last (see **illustration**). These include sensory, short-term, and long-term memories.

One type of memory is quite short-lived, lasting less than a second; it is called a sensory memory. Importantly, for a sensory memory to indeed be a memory, it must last longer than the duration of a sensory stimulus. Sensory memories last beyond the termination of a sensory cue, on the order of 500 milliseconds. These types of memories can come from different senses, too. For example, vision gives rise to iconic memories, hearing gives rise to echoic memories, and touch gives rise to haptic memories.

When we experience information or events that are important enough, it is possible to remember them for longer periods of time, giving us short-term memory (also sometimes referred to as working memory). This memory is stable over a range of minutes to hours. It is thought that humans can remember three or four items with working memory. Beyond this number, people begin to group information into chunks, and then remember again three or four chunks of information. Rehearsal of items can extend the duration that a working memory is maintained.

A memory that lasts days to years is called a long-term memory and can be further divided into declarative and nondeclarative memories. Declarative long-term memories can be about facts (semantic) or events (episodic) [for example, remembering the year that the U.S. Civil War ended or your last birthday party, respectively]. Nondeclarative memories are those that last for long periods of time but do not necessarily require active remembering. You can remember how to drive a car but do not necessarily need to think about how to turn the steering wheel to turn left. Furthermore, nondeclarative memories are sometimes categorized into those that arise from nonassociative and associative learning. Examples of nonassociative learning include habituation (a reduced behavioral response to repeated stimuli) and sensitization (an increased response after experiencing an intense or noxious stimulus). Associative memories arise from classical and operant learning, and probably also from mixed classical and operant learning. Classical learning is established when an animal receives a cue (for example, visual, aural, or tactile) that by itself does not evoke a given behavior, but which is paired in time with a stimulus that normally elicits a behavior (for example, food as a reward). The pairing of the cue with the reward can be used by an animal to make predictions, such as expecting food when that particular cue is perceived. Operant learning requires that an animal perform some behavior to receive a reward (or avoid some punishment). In this case, the behavior predicts the reward or punishment. Memories arising from associative and nonassociative learning also can be long-lasting (days to years).

Although the aforementioned memory categories appear linearly related, not all memories move from sensory to working to long-term phases. First, many memories are not behaviorally expressed continuously over time because of memory loss (forgetting), lack of access to the stored memories, or more complicated conditions. Second, from experiments in animals, it is clear that a memory need not be accessible at all times to influence behavior. That is, a memory can influence behavior in the minutes range, disappear for some time, and then reappear later to again influence behavior. Therefore, although memories can be conceptualized into various memory categories (see illustration), how the mechanisms that support memories at different times interact to influence behavior is still an area of ongoing investigation.

Genes important for memory. The approximately 23,000 protein-encoding genes in the human genome can influence memory formation in several ways. These genes determine how the brain develops; how neurons, glia, and brain structures are maintained over a lifetime; and the acute physiology of neurons and glia that are critical for memory. Importantly, it is possible to gain much insight into the mechanisms by which genes influence memory by examining functions in several animal models. An exhaustive list of genes critical for memory formation will not be detailed here. However, examples of genes and functions in development, maintenance, and physiology will be presented. The link between gene function in humans, mice, and the fruit fly *Drosophila melanogaster* are highlighted.

It is possible to identify genes that are critical for development of the nervous system and have strong effects on memory formation. Mutation of the gene encoding the Fragile X mental retardation protein (FMRP) [an RNA-binding protein] is associated with learning disabilities and is the most common cause of inherited mental retardation in humans (Fragile X syndrome). How does loss of FMRP lead to learning disabilities? Interestingly, examination of the brains of FMRP patients by noninvasive techniques reveals only subtle differences in different parts of the brain. Although there are severe changes in brain function, evident in learning disabilities, one only slowly learns about how FMRP is critical for memory formation by studying humans. Thus, one turns to animal models. Mutation of the FMRP gene in the fruit fly and the mouse leads to learning deficits, suggesting a common role for this gene across vastly different species. Also, looking in detail at the brains of FMRP-altered animals shows that they have deficits in neuronal structural specializations and altered physiology. Furthermore, the FMRP protein is critical for regulating expression of many genes. Therefore, it is possible to connect the role of a specific protein implicated in human mental retardation with molecular, cellular, and neural circuit functions.

Specific genes can also be critical for maintaining the function of the nervous system and the ability to form memories. Alzheimer's disease, with its associated age-dependent decline in the ability to form new memories, is a key example. Early insights into the genetic mechanisms of Alzheimer's disease came from a rare form called familial Alzheimer's disease. Even among the rather low number of early-onset Alzheimer's disease patients (those with signs of the disease who are younger than 65 years of age; approximately 6–7% of all patients with Alzheimer's disease), familial Alzheimer's disease accounts for only about 7% of these cases. Regardless of the frequency at which these cases are found, finding that mutation of the presenelins (PS) and amyloid precursor protein (APP) genes as the major cause for familial Alzheimer's disease pointed toward a common and more broadly applicable mechanism for disease onset and progression. Insights from animal models point to a role of mutant PS and APP in accumulation of a toxic extracellular APP peptide in large plaque deposits, which set off a series of events that leads to neuron loss. However, this simple model is complicated by the fact that PS-altered mice can induce age-dependent neurodegeneration in the absence of the APP peptide that is critical for plaque formation. Although complex, the molecular changes associated with PS and APP suggest that these proteins, and associated factors, are central to maintaining neuronal and nervous system integrity in a human degenerative disease that alters memory.

Furthermore, some genes are critical in altering the physiology of neurons during memory formation. One canonical signaling pathway that is important for memory formation regulates the levels of cyclic adenosine monophosphate (cAMP), a common second messenger in cells. The idea here is that a protein termed type-1 adenylyl cyclase (AC) makes cAMP only when it receives two signals at the same time—one coming from calcium (Ca^{2+}) [corresponding to some sensory stimulus] and the second from a guanine nucleotide-binding protein (G protein) [corresponding to a reward or punishment]. The cAMP signal then goes on to activate protein kinase A (PKA), which is followed by both short- and long-term changes in neuronal physiology that are important for memory formation. Mice and flies that are mutant for this type-1 AC do not make memories very well; indeed, their formation of memories is about half as good as that in normal animals. Importantly, at least in the fly, when the type-1 AC is mutated, the coincident dependent increase in cAMP and PKA activation in specific parts of the fly brain is lost. There are several other genes that are also important for regulating the physiology of neurons and memory formation (and these can be specific to the short-term and long-term memory phases).

Conclusions. We use our ability to form memories everyday. Sometimes, the memories are transient; other times, they are long-lasting. A set of genes in the human genome is responsible for establishing, maintaining, and regulating the physiology of the cells and circuits of the nervous system to allow us to form memories. Animal studies provide critical insights into what those genes are and how they function in the nervous system.

For background information *see* ALZHEIMER'S DISEASE; BRAIN; GENE; GENETICS; GENOMICS; HUMAN GENETICS; INSTRUMENTAL CONDITIONING; LEARNING MECHANISMS; MENTAL RETARDATION; NEUROBIOLOGY; SIGNAL TRANSDUCTION in the McGraw-Hill Encyclopedia of Science & Technology.

Daniela Ostrowski; Troy Zars

Bibliography. F. V. Bolduc and T. Tully, Fruit flies and intellectual disability, *Fly*, 3:91–104, 2009; N. Gervasi, P. Tchenio, and T. Preat, PKA dynamics in a *Drosophila* learning center: Coincidence detection by rutabaga adenylyl cyclase and spatial regulation by dunce phosphodiesterase, *Neuron*, 65:516–529, 2010; W. T. O'Donnell and S. T. Warren, A decade of molecular studies of Fragile X syndrome, *Annu. Rev. Neurosci.*, 25:315–318, 2002; J. Shen and R. J. Kelleher III, The presenilin hypothesis of Alzheimer's disease: Evidence for a loss-of-function pathogenic mechanism, *Proc. Natl. Acad. Sci. USA*, 104:403–409, 2007; L. R. Squire, Memory systems of the brain: A brief history and current perspective, *Neurobiol. Learn. Mem.*, 82:171–177, 2004.

Genome of the platypus

The platypus (*Ornithorhynchus anatinus*) [**Fig. 1**] is a true mammal, possessing fur and producing milk in females to nurture their young. Although the platypus lactates, it does not have nipples like other mammals. Instead, the young drink milk secretions from patches of abdominal skin. Moreover, platypuses possess many unique characteristics that have fascinated and perplexed biologists since their discovery. Most notable is the manner in which they reproduce. Unlike other mammals, members of the monotreme lineage (to which the platypus belongs) lay eggs. Platypus eggs hatch about 11 days after being laid, and the young are fed milk for the next 4 months of their development. Platypuses also have a unique system for determining the sex of their offspring. Whereas almost all mammals have an X-Y sex chromosome system, the platypus has five X and five Y chromosomes that share almost no similarity to the human X and Y. Furthermore, the platypus is the only known mammal with electroreception, which enables the platypus to sense electric fields generated by other animals. The platypus uses this ability to locate food sources, such as crayfish, while foraging underwater. In addition, the platypus is a venomous mammal. Although venom is not uncommon in other vertebrate lineages, it is rare in mammals. The male platypus has specialized spurs on its hind limbs that are used to deliver the venom. A dose of the venom, which is not lethal to humans, is so excruciatinly painful that the victim may be incapacitated. The exceptional features of its biology therefore make the platypus genome a particularly interesting and unique resource in its own right. However, the platypus is an even more important resource because of its unique evolutionary origin. The monotremes have been evolving independently from the rest of the mammals, including humans, for over 166 million years (**Fig. 2**), making them our most distant mammalian relatives. By analyzing the genome of the platypus, it is possible to gain important insights into the evolution and development of the human genome.

Fig. 1. Platypus, *Ornithorhynchus anatinus*. (Photo by Gerry Pearce, http://www.australian-wildlife.com)

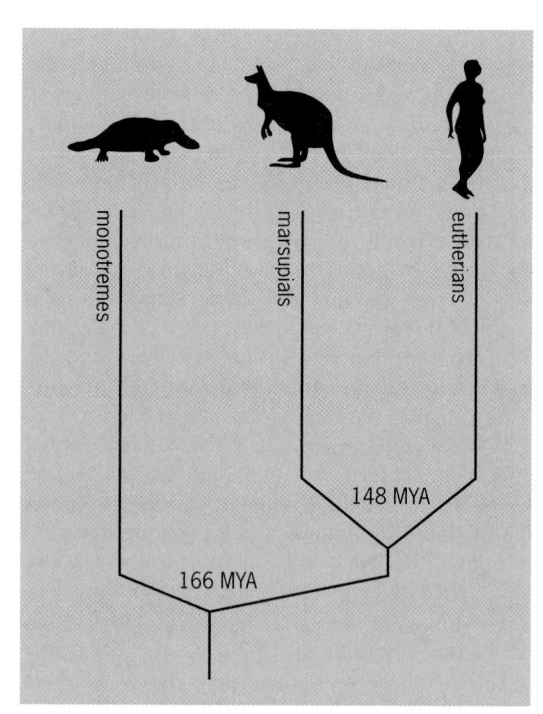

Fig. 2. Evolutionary tree of the extant mammalian lineages (MYA = millions of years ago).

Sequencing the platypus genome. The monotreme lineage contains only three living genera—the platypus and two echidna genera. Using the platypus genome, researchers have been able to determine that the platypus last shared a common ancestor with the echidna about 21 million years ago. The platypus genome is estimated to be approximately 2.3 billion base pairs, which is about three-quarters of the size of the human genome. It contains about 18,500 genes (roughly the same number as in other mammals), which are spread across its 52 chromosomes. About 82% of these genes are shared with the other mammals or the chicken. The remaining 18% represent platypus-specific and rapidly evolving genes. Some platypus genes are shared only with birds and reptiles, such as those required for egg formation, and do not exist in the other sequenced mammalian genomes. The platypus genome also contains lactation genes, which are specific to mammals. Thus, the platypus genome represents a mix of reptilian and mammalian characters.

Venom genes. Platypus venom contains at least 19 different protein components. These are clustered into three groups: the defensin-like peptides, C-type natriuretic peptides, and nerve growth factors. Genome analysis has shown that the venom proteins are derived from duplications of genes with very different functions. The defensins are a family of antibiotic-like peptides. The C-type natriuretic peptides typically control body fluid homeostasis and blood pressure, whereas nerve growth factors are important for the growth, maintenance, and survival of target neurons. Interestingly, comparisons with venomous snakes show that a similar set of genes are also used in reptiles to generate venom, although they have evolved independently. This suggests that the same selective pressures led to the development of venom from a conserved set of genes separately in the reptile and monotreme lineages by convergent evolution. These genes are useful because they help to create venom that can lower blood pressure, cause pain, and increase blood flow around the wound in the victim.

Tooth genes. Adult platypuses do not have teeth. Instead, they have a unique leather bill, which is better suited to foraging in an aquatic environment. However, developing platypus fetuses do develop an egg tooth, which is used to pierce the shell to help with hatching. This tooth is lost shortly after birth. Not surprisingly, the platypus genome does contain the genes required for tooth enamel. This is consistent with the development of a true tooth in the platypus fetus and also with the discovery of tooth-bearing platypus ancestors in the fossil record. This suggests that tooth development was lost relatively recently in monotreme evolution.

Immune genes. The array of immune genes in the platypus is different from that of other mammals. The biggest surprise is the large number of natural killer receptor genes in the platypus genome, which contains more than ten times as many of these genes in comparison to the human genome. The platypus genome also contains a diverse array of cathelicidin antimicrobial peptides similar to those found in the marsupial genome, but different from eutherian (placental) mammals (which typically have one). These differences in the immune system suggest that the platypus has an increased set of defense mechanisms against infection, which may be especially important for the protection of their naive young during early development.

Genomic imprinting. Another unique feature of the platypus genome is the apparent lack of genomic imprinting. All mammalians are diploid, meaning they carry two full sets of the genome in most cells. One set is inherited from the mother and the other from the father. However, these genomes are not equivalent. Some genes from each parent are inherently silenced by a mechanism known as genomic imprinting (**Fig. 3**). Genomic imprinting is present in the marsupial and eutherian mammal lineages (Fig. 2), but missing from monotremes. One theory is that imprinting evolved to regulate the amount of growth and nutrition of the fetus during development in the uterus of the mother. Since platypuses lay eggs, they have no need for such a mechanism. Analysis of the genome showed that the platypus has less repeat elements than marsupials and eutherians within the regions of the genome that became imprinted. The absence of imprinting in the platypus suggests that it may have evolved to help silence the expansion of

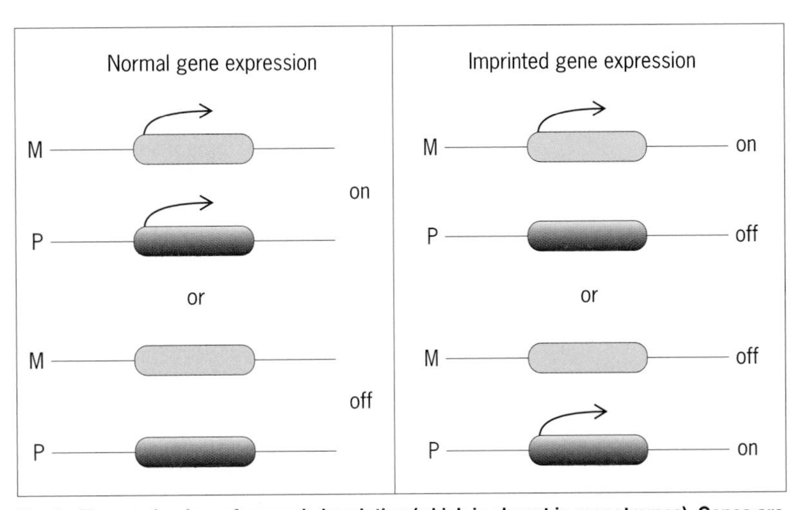

Fig. 3. The mechanism of genomic imprinting (which is absent in monotremes). Genes are represented by lightly tinted or darkly tinted ellipses. Arrows indicate an actively expressed gene. M indicates the maternal copy of the gene and P indicates the paternal copy of the gene. In the normal situation (left panel), both copies of the genes are either active or silent in the genome. However, genomic imprinting silences some genes that are inherited from the father (right panel, top) and some that are inherited from the mother (right panel, bottom), thereby resulting in expression of only a single copy.

foreign DNA elements that occurred in other mammalian lineages.

Sex determination. Mammals typically have an X–Y sex chromosome system, where XY individuals are male and XX individuals are female. Sex is determined by a gene on the Y chromosome called *SRY* (*Sex-determining Region Y*). *SRY* evolved from a gene on the X chromosome called *SOX3*. The platypus is unique in that it has five X and five Y chromosomes. In addition, there is no sign of the *SRY* gene in the platypus genome, and *SOX3* is autosomal (pertaining to a chromosome that is not a sex chromosome) and is not found on one of the ten sex chromosomes. Instead, a gene called *DMRT1* resides on one of the platypus X chromosomes. *DMRT1* is important for testis development in mammals, but is autosomal and thus does not determine sex. However, in birds, *DMRT1* resides on one of their sex chromosomes and has been shown to determine sex. Thus, the platypus appears to have a unique sex-determining system, perhaps more reminiscent of birds than mammals.

Conclusions. The genome of the platypus is smaller than that of other mammals, but contains a similar number of genes. These genes share identity with other mammals, birds, and reptiles, reflecting many of the unique characteristics of this mammalian species. Detailed analysis of the platypus genome has revealed many insights into the evolution of various components of the mammalian genome, including the development of the immune system, sex-determining mechanisms, and the origins of genomic imprinting. Future work will continue to unlock the complexities of platypus biology and uncover the common features of our evolutionary past.

For background information *see* DEOXYRIBONUCLEIC ACID (DNA); EUTHERIA; GENE; GENETIC MAPPING; GENOMICS; MAMMALIA; MARSUPIALIA;

MONOTREMATA; PLATYPUS; PROTOTHERIA; SEX DETERMINATION in the McGraw-Hill Encyclopedia of Science & Technology. Andrew J. Pask

Bibliography. A. J. Pask et al., Analysis of the platypus genome suggests a transposon origin for mammalian imprinting, *Genome Biol.*, 10:R1, 2009; W. C. Warren et al., Genome analysis of the platypus reveals unique signatures of evolution, *Nature*, 453:175–183, 2008; P. D. Waters and J. A. Marshall Graves, Monotreme sex chromosomes—implications for the evolution of amniote sex chromosomes, *Reprod. Fertil. Dev.*, 21:943–951, 2009; C. M. Whittington et al., Understanding and utilising mammalian venom via a platypus venom transcriptome, *J. Proteomics*, 72:155–164, 2009.

Genomics and relationships of Australian mammals

The continent of Australia is home to many exotic mammals, including hopping kangaroos, egg-laying platypus, spiky echidnas, carnivorous Tasmanian devils, and cuddly koalas. The ancestors of these present-day Australian mammals lived on the Gondwana supercontinent (the ancient landmass that later fragmented and drifted apart to eventually form the present continents). Splitting of the Gondwana landmass, starting about 145–200 million years ago, separated mammals in Australia from mammals in Africa, South America, India, and Antarctica. Vast oceans between the modern continents created geographical barriers, allowing the independent evolution of the unique mammalian fauna of Australia. It is possible to trace these historical events and understand the relationship of animals with the use of the molecule common to all living things: DNA.

Molecular clock. There are four nucleotide bases, namely, adenine (A), guanine (G), cytosine (C), and thymine (T), whose sequence provides the information content of DNA. Approximately 3 billion of these bases are arranged in a specific order on chromosomes to form the genome of mammals. The genome of an individual animal is replicated with amazing fidelity, and is passed on from generation to generation through the gametes (eggs or sperm). However, DNA replication during gamete formation is not absolutely foolproof, and slippage during gametogenesis can cause the DNA of the offspring to differ in several bases compared to the DNA of the parents. For example, let ACG[**T**]ACGTACGTACGTACGT represent a sequence in the genome of a mammalian male. If the fourth nucleotide base thymine (T) of this sequence is replaced by a guanine (G) base during spermatogenesis, then a mutated sequence ACG[**G**]ACGTACGTACGTACGT is passed on to his offspring. If the rate of mutation were consistent in subsequent generations, then one would expect 10 mutations in 10 generations, 100 in 100 generations, and 10 million in 10 million generations. If all the animals reproduced at the age of 20 years, then one also could project that 1 mutation occurred in 20 years, 10 in 200 years, 100 in 2000 years, and 10 million in 200 million years.

It is this simple "molecular clock" principle of DNA mutation that is utilized to date the divergence time of two species. However, in practice, complexity arises because the rate of mutation in the genome is not linear. Moreover, not all mutations are substitution mutations; some are insertions or deletions. Most significantly, the DNA sequence of the ancestor is not usually known beyond a few generations. Nevertheless, sophisticated algorithms and statistical procedures have been invented to make use of the DNA sequence of extant species to infer relationships and to deduce when and how mammal species evolved. Using these types of analyses, the genomics and relationships of Australian mammals can be investigated.

Evolution of mammals. Recent studies using the DNA sequence of 66 genes in over 4500 extant mammal species showed that all species of the class Mammalia diverged from birds and reptiles about 166 million years ago. The egg-laying monotreme mammals, forming the subclass Prototheria (meaning "first beasts"), were the first to diverge from therian (placental and marsupial) mammals. This was followed by divergence of the marsupials (infraclass Metatheria, meaning "sort-of beasts") from placental mammals (infraclass Eutheria, meaning "true beasts") approximately 148 million years ago. The presence of marsupials in South America and Australasia, as well as the fossil record, suggests that marsupials were present on Gondwana before Australia and South America were separated about 70 million years ago. Although monotremes are currently endemic to the Australian continent and neighboring islands, a single monotreme tooth has been found in Argentina, suggesting that they were present on Gondwana. However, the paucity of fossil records for monotremes and Australian marsupials does not enable a clear understanding of their evolutionary history and relationships. Hence, DNA-sequencing technology has proved to be critical in delineating the relationships of Australian mammals.

Relationships of monotremes. Monotremes are represented by five extant species: the platypus and four species of echidnas. They all share features that define mammals, including hair on the body, high rates of metabolism, and feeding of newborns via mammary gland openings in the skin. However, they also retain some of the anatomical features of reptiles, such as the presence of a cloaca, which is the opening duct for the urinary, defecatory, and reproductive tract systems. Monotremes lay eggs, as do reptiles and birds; however, in monotremes, the egg stays inside the mother for some time, during which she supplies nutrients to it. One of the most interesting aspects of the genome of monotremes is that their multiple sex chromosomes have homology to the sex chromosomes of birds rather than to those of marsupial or placental mammals. Platypuses are found on the east coast of Australia and in Tasma-

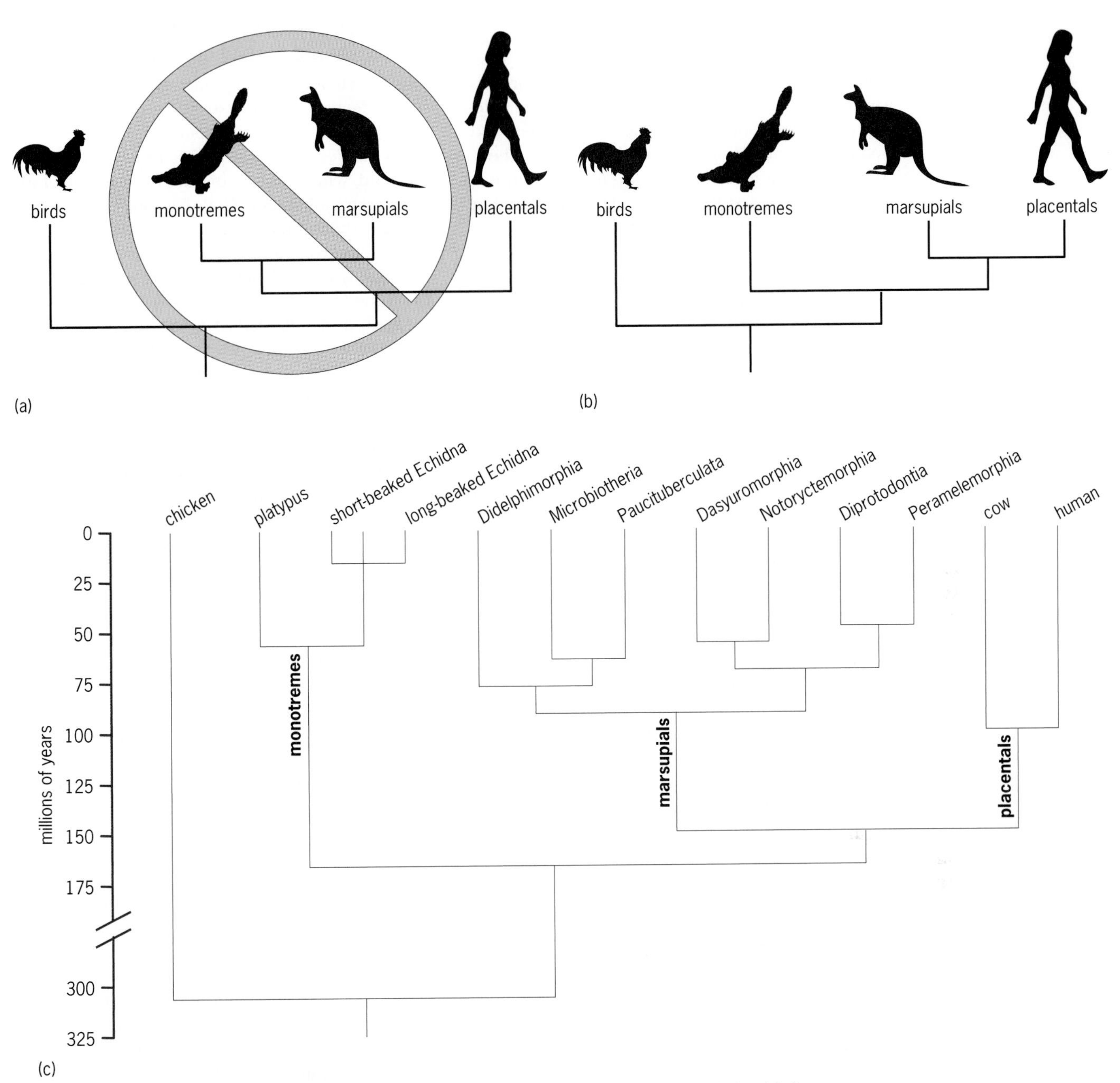

Schematic representation of mammal relationships. (*a*) Marsupionta hypothesis, showing a close relationship between marsupials and monotremes. (*b*) Theria hypothesis, showing the accurate relationship between marsupials and monotremes. (*c*) Relationship between different monotreme and marsupial orders. (Time scale is an approximate measure.)

nia, which were separated only recently by water. Short-beaked echidnas are endemic to mainland Australia and the surrounding islands, and three species of long-beaked echidnas are endemic to neighboring New Guinea, which also was separated only recently from the mainland by a body of water.

Comparison of the mitochondrial DNA sequence of monotremes, marsupials, and placentals suggested at first that monotremes and marsupials are more closely related to each other than either is to placentals (Marsupionta hypothesis; **illus**. *a*). However, comparisons of many nuclear genes give a clear indication that marsupials and placentals both form a distinct phylogenetic group and are more closely related than either is to monotremes (Theria hypothesis; illus. *b*). Marsupials and placentals diverged from monotremes about 166 million years ago (illus. *c*). These conclusions have been robustly confirmed by comparisons of the whole genome sequence of the platypus. Comparison of the DNA sequences within particular genes (the protamine P1 genes) of the platypus and echidna suggests that they diverged from each other approximately 60 million years ago. Within the echidna lineage, short-beaked echidnas diverged from the long-beaked New Guinea echidnas about 10 million years ago.

Relationships of marsupials. There are over 330 species of marsupials. Of these, more than 200 species from four orders (Dasyuromorphia, Diprotodontia, Notoryctemorphia, and Peramelemorphia) are found in Australia, New Guinea, and some neighboring islands to the northwest. About 100 species from three orders (Didelphimorphia, Microbiotheria, and Paucituberculata) are found in South America, and a single didelphid species [the Virginia opossum (*Didelphis virginiana*)] is found in North America. Of the Australian marsupials, Diprotodontia is the most diverse, encompassing nearly 125 species and including kangaroos, wallabies, potoroos, wombats, koalas, opossums, and gliders. Similarly, Dasyuromorphia encompasses quolls, dunnarts, and the extinct thylacine, whereas Peramelemorphia consists of bandicoots. Notoryctemorphia consists of a single species, the marsupial mole.

Fossil records indicate that marsupials originated from China, spreading first to North America in the early Late Cretaceous Period, spreading later into South America, and then dispersing to Australia via Antarctica before the continents drifted apart. Mitochondrial DNA sequence data and nuclear gene data have shown that Didelphimorphia were the first to diverge approximately 82.5 million years ago in South America and that all extant orders of marsupials were separate lineages about 60 million years ago. Molecular data support the monophyly (the development from a single common ancestral form) of each marsupial order and date their divergence to at least 66.8 million years ago in Australia.

Until recently, molecular studies used a limited set of loci from the nuclear or mitochondrial genome for identifying the relationships between Australian marsupials. However, with the availability of the genome sequence of the Australian model marsupial, the tammar wallaby (*Macropus eugenii*), and the South American model marsupial, the gray short-tailed opossum (*Monodelphis domestica*), more light will be shed on the evolution of this distinct class of mammals and their interrelationships.

For background information *see* ANIMAL EVOLUTION; AUSTRALIA; CONTINENTAL DRIFT; DEOXYRIBONUCLEIC ACID (DNA); EUTHERIA; GENETIC MAPPING; GENOMICS; MAMMALIA; MARSUPIALIA; METATHERIA; MONOTREMATA; MUTATION; PHYLOGENY; PROTOTHERIA; THERIA in the McGraw-Hill Encyclopedia of Science & Technology.

Hardip R. Patel; Jennifer A. Marshall Graves

Bibliography. O. R. P. Bininda-Emonds et al., The delayed rise of present-day mammals, *Nature*, 446:507–512, 2007; L. Bromham and D. Penny, The modern molecular clock, *Nat. Rev. Genet.*, 4:216–224, 2003; T. S. Mikkelsen et al., Genome of the marsupial *Monodelphis domestica* reveals innovation in non-coding sequences, *Nature*, 447:167–177, 2007; W. C. Warren et al., Genome analysis of the platypus reveals unique signatures of evolution, *Nature*, 453:175–183, 2008.

Harder-than-diamond carbon crystals

The search for ultrahard materials has interested both scientists and industry for decades, as many applications for these compounds are foreseen. One of the hardest natural materials known on Earth is diamond, which is made of pure carbon. Therefore, the search for harder crystals has long been focused on the carbon system. In addition, some research has been focused on boron nitride, which has a hardness just slightly lower than that of diamond and is easier to synthesize.

Basics of natural carbon system and polymorphs. The number of natural carbon polymorphs (that is, crystals with the same chemistry but with different structures) in nature is quite small. So far, only five or six natural carbon crystal types are known on Earth. The most well-known are graphite and its high-pressure polymorph, diamond. Some more exotic natural carbon minerals have also been described, such as chaoite, which was found in 1963 in the rocks of the Ries meteoritic impact crater in Bavaria, Germany. More recently, two unnamed minerals were found in the Popigai impact crater in Siberia, Russia, and within a highly shocked meteorite. The latter will be discussed in detail later in this article. Physical and chemical synthesis has produced two other carbon polymorphs: the fullerene types, which look like soccer balls, and carbon nanotubes. Fullerene was subsequently found in sediments from the Sudbury impact crater in Ontario, Canada. Based on these different natural samples, the search for harder materials than diamond can involve three approaches: theoretical prediction, experimental research, and natural sample evaluation.

Theoretical prediction. Theoretical prediction of ultrahard materials derived from carbon is based mainly on molecular dynamics, whereby the interactions of the different atoms in a crystal structure are modeled. As long as the potential function, which is the description of the terms by which the particles will interact in the simulation, is correct, this approach allows scientists to investigate a large range of physical parameters, such as pressure and temperature, and predict the different structures that will result and the different fields of existence of these structures.

Experimental research. Experimental research into the carbon system in the search for ultrahard compounds began with the study of the graphite-to-diamond transformation at high pressure and a range of temperatures. Following the discovery of new pure carbon polymorphs, which were not supposed to exist, the range of carbon material to study under compression broadened. These studies led to important discoveries of ultrahard materials. For example, aggregated diamond nanorods were discovered after compressing fullerene at 37 GPa and temperatures ranging from 300–2500 K (80–4040°F). So far, this compound is the hardest synthesized material known on Earth, as its hardness has a value of 310 GPa while that of diamond is only around

160 GPa, as determined by the Vickers hardness test. The Vickers hardness test consists of indenting the test material with a diamond tip for a certain duration and load and then observing the surface of the tested material. After this, a hardness number is attributed based on the mean length of the scratch on the surface. Since the test involves a force applied on a surface, the hardness can be expressed in GPa.

Natural sample observation. The study of naturally formed carbon samples is also very important when looking for ultrahard material. Natural geologic material can (1) contain pure carbon, usually as graphite, and (2) be submitted to ultrahigh pressure and temperature during, for example, a meteoritic impact event. Therefore, examining these geologic materials within either the meteorite itself (which has been compressed during its ejection from its parent body) or the impact crater is important when searching for ultrahard materials derived from natural events. This year, this approach led to the discovery of two new ultrahard phases that were predicted by neither theoretical calculation nor experimental research.

Discovery of two new ultrahard carbon phases. Ureilites, or meteorites with a high carbon content occurring as graphite, are considered to be the choice when looking for naturally occurring ultrahard materials. Like many meteorites, ureilites were submitted to a huge shock event during their ejection from their parent body, and part of the graphite was converted to diamond during the peak shock pressure. These diamonds are known to occur as small flakes inside the meteorite.

Investigation of one of these meteorites, and particularly the carbonaceous areas within it, provided evidence of new ultrahard phases. In order to examine the meteorite, it was cut and then polished with a diamond paste, a standard sample preparation procedure. However, researchers discovered that the section was not perfectly polished in the carbonaceous area, as some protruding materials were standing high above the silicate matrix (**Fig. 1**). This observation suggests the presence of a material that is harder than the polishing material and therefore harder than diamond. Unfortunately, no other hardness comparison was possible, as it could have damaged the structure of the material and consequently would have prevented any further research on it. Ultimately, two different ultrahard phase were found in two different carbonaceous areas within the sample, but they share similarities in terms of spatial arrangement and therefore of formation mechanism.

Fig. 1. Side view of one of the carbonaceous areas containing an ultrahard phase.

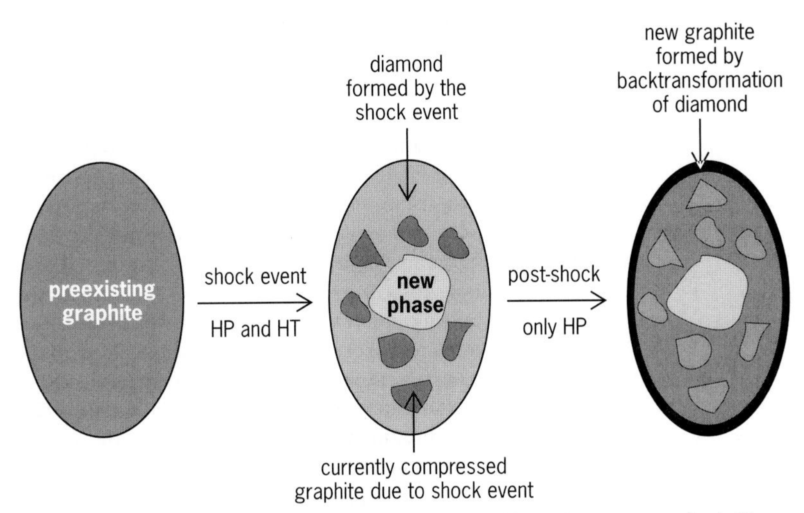

Fig. 2. Formation scenario for the new carbon phase and its carbon compound satellites. HP = high pressure. HT = high temperature.

A typical carbonaceous area shows zonation from the exterior to the interior of graphite, graphite mixed with diamond, and an ultrahard phase. This concentric arrangement suggests that the initial phase present was graphite, which was converted to either diamond or the new phase at the center during the shock event. This is supported by two observations: (1) the graphite mixed with diamond shows evidence of compression, presumably during the shock event, and (2) the graphite that is not mixed with diamond does not show any sign of compression. One has to interpret this feature as a back transformation from diamond to graphite following the shock event. During the pressure release, the temperature stays high enough to allow the back transformation of diamond to new and unshocked graphite (**Fig. 2**). This scenario is supported by the different measurements made on this phase using Raman spectroscopy.

Investigation of the structure of both new phases was done using x-ray diffraction analysis. In the two carbonaceous areas, the ultrahard phases were found to be different. In the first area, the unpolished region is composed of a mixture of diamond, graphite, and a polytype of diamond. Polytypes are variations of the same chemical compound that are identical in two dimensions but differ in the third. Two polytypes of diamond were previously known: diamond, which is the 3C (cubic) polytype, and lonsdaleite, a hexagonal polytype (2H). The new polytype found in the meteorite is a rhombohedral polytype called the 21R polytype. Its structure is very close to that proposed theoretically based on the similarities between the carbon system and the

SiC system. The other area gave a totally new carbon mineral that had never been found in nature, synthesized experimentally, or predicted by theoretical calculations. This new phase has a structure that is distinct from that of the 21R polytype, thus confirming the occurrence of a new carbon polymorph. The best fit for the new polymorph structure was obtained for a rhombohedral lattice (R3m space group) with the parameters $a = 3.5610(9)°$ and $\alpha = 90.2(2)°$. The unit cell contains four structural positions for carbon atoms: C1 (0.276, 0.276, 0.276), C2 (0.273, −0.256, −0.256), C3 (0, 0, 0), and C4 (0.5, 0.5, 0). Although the quality of the diffraction data was not sufficient for a full-profile structural refinement, processing of the diffraction pattern with fixed structural positions and optimization occupancy factors indicates that the C3 and C4 positions are only partially filled. The density of the new phase is therefore intermediate between the densities of graphite and diamond. The proposed crystal structure of the new carbon polymorph is closely related to the structure of diamond: all carbon atoms are tetrahedrally coordinated, but slightly shifted from their symmetric positions in the diamond structure, and two of the four structural positions are partially occupied. Particular peaks on the diffraction pattern may also indicate a significant degree of disorder of the crystal structure.

Outlook. There are a large number of potential applications for ultrahard materials, but the following questions must first be answered.

Can we synthesize it experimentally? In order to answer this, it is important to know the conditions of formation of these new materials precisely. At this point in time, it is very difficult to constrain the pressure, temperature, and strain orientation that gave birth to these new compounds.

If synthesis is possible, how much will it cost? There are a number of different applications that could result from the discovery of new ultrahard materials. They could be used, for example, to coat objects to protect them from scratching. Infusing car paint with such a material could put an end to annoying scratches. It might also be possible to incorporate ultrahard materials into LCD or cell-phone screens. If it were possible to synthesize large, single crystals, they could be used to cut other hard materials or used in scientific apparatus that requires ultrahard materials.

For background information *see* CARBON; CRYSTAL STRUCTURE; DIAMOND; GRAPHITE; HARDNESS SCALES; HIGH-PRESSURE CHEMISTRY; HIGH-PRESSURE MINERAL SYNTHESIS; HIGH-PRESSURE PHYSICS; METEORITE; POLYMORPHISM (CRYSTALLOGRAPHY) in the McGraw-Hill Encyclopedia of Science & Technology.

Tristan Ferroir

Bibliography. N. Dubrovinskaia et al., Nanocrystalline diamond synthesized from C_{60}, *Diamond and Related Materials*, 14:16–22, 2005; A. El Goresy et al., A new natural, super-hard, transparent polymorph of carbon from the Popigai impact crater, Russia, *Compt. Rendus Geosci.*, 335(12):889–898, 2003; F. J. Ribeiro et al., Hypothetical hard structures of carbon with cubic symmetry, *Phys. Rev. B.*, 74:172101, 2006; P. Scharff, New carbon materials for research and technology, *Carbon*, 36(5-6):481–486, 1998.

Health and environmental interoperability

In recent years, many natural and human-induced disasters have demonstrated to the public the interconnectedness and interdependence of critical infrastructures. The ease with which these connections and dependencies occur within systems is known as interoperability. From the Haiti and Chile earthquakes, to the volcanic ashes in Scandinavia, to the BP (British Petroleum) rig explosion and the oil spill in the Gulf of Mexico, to the financial crisis in Greece and southern Europe, the consequences and ripple effects of these interdependencies create many unintended consequences. Although disasters are never completely predictable, new tools are available to help to prepare and respond more effectively.

A key aspect of interoperability is to examine events holistically and systematically by incorporating the perspectives of multiple scientific, engineering, and other disciplines. As such, it offers ways to model systems and predict their behavior. Although the value of interoperability is dramatically evident in a disaster, it can also improve everyday life, especially by sharpening the focus on wellness and disease prevention. It provides a means to improve healthcare by focusing on the individual, rather than the institutions. That is, rather than looking for ways to improve hospitals, public health agencies, and research facilities, the focus is the wellness of the person. The current health-care system emphasizes diagnosis, treatment, and cure. However, such a focus is on the disease, not the person.

Transforming the current system into one that focuses on wellness and disease prevention requires that many new factors be considered, including many environmental factors that affect humans, including potable water, safe food, clean air, and many other factors.

Interoperability and some challenges. A good place to begin to apply interoperability is the way health information is recorded. The electronic health record (EHR) is distinguished from the personal health record (PHR). An EHR is found in a hospital or clinic, predominantly to document information performed in that place. It often lacks other important information, such as vaccine registries, mental health, and dental records. A PHR, on the other hand, incorporates all these types of information and numerous other data particular to an individual patient who is the owner of the PHR. It is more than a health record, since it includes procedures and many data that were recorded throughout the life of the patient by many hospitals and caregivers, which may be located in many different cities, states, and or countries.

The need for improved information about the individual must be balanced against other aspects of well-being, including privacy. For example, as molecular and systems biology has improved, so has the ability to identify individuals based on genetic and metabolic information (that is, genomics and metabonomics, respectively).

Another challenge to society in general is the field of scientific information management. Issues such as storage and retrieval of information that can be valuable to members of different communities may go undetectable just because members of one profession do not share the same information with individuals of other disciplines. Imagine, for example, that an astronomer developed an image-processing algorithm to analyze satellite images. A radiologist working in the area of nuclear magnetic resonance could use this algorithm for the purpose of breast cancer detection. Being completely unaware of the developments on the field of astronomy, the radiologist will never find out about the existence of this algorithm with our current uses of information and knowledge management.

Public health critical infrastructure. The scientific community has come to recognize that preparing for the provision of healthcare and erecting a viable public health system is not just a matter of converging heterogeneous technologies, it also requires the application of the social sciences as well. In this respect, enhancing the records of all parties involved in the areas needed to improve healthcare deliveries and public health infrastructures must undertake efforts to build in interoperability and consider the interdependence of the processes and systems at the forefront of all future considerations.

The current challenge is that (1) information that seems unrelated is frequently very much related; (2) healthcare, environmental, and public health information is connected, and yet functionally, it is disconnected; (3) a "systems" approach is crucial, as is the need to take a "holistic" view of the problem to be able to see the whole and not just discrete pieces and to help determine, for example, unintended consequences when needed, but absent; (4) it is necessary to integrate multidisciplinary and interdisciplinary orientations and activities when trying to understand the problem, and moving toward generating potential solutions, yet present approaches are grossly insufficient in this respect; and (5) the problems that are routinely encountered with respect to medical errors, such as adverse drug effects, occur and are perpetuated because of our lack of information exchange.

Public health infrastructure, that is, the state of information technology in state and local health departments and the Centers for Disease Control and Prevention (CDC), is currently incapable of meeting the collective needs for public health protections and services, and may likely worsen in the immediate future.

A major proposal related to safe food calls for an enterprise architecture that emphasizes four main conclusions:

1. Healthcare and public health as well as the food and agriculture critical infrastructures are highly interdependent, and that globalization has created an environment where every country depends on many others for a continuous year-round supply chain of food and related products.
2. Related adverse consequences affect the U.S. financial and national security, and potentially the security of countries with whom we import and export food products.
3. The United States lacks a national approach to protect this critical national infrastructure and needs a food product safety management enterprise architecture to enable food product safety.
4. Without a national food product safety management enterprise architecture, the United States will continue to experience contaminated food-product incidents that threaten food-product consumers and also threaten the U.S. financial and national security.

Dust and water: the case for interoperability. A large-scale exposure to the toxin ciguatera occurred in several U.S. cities in the winter of 2008, with a particularly severe outbreak in Washington D.C. One potential source of the toxin was the frequent occurrences of dust storms throughout the world. By linking to multiple sets of information related to individual health and public health, particularly global food sustainability and global security, researchers were able to arrive at plausible causes and responses. Such a shift in interoperability is needed to achieve not only information sharing and a common operational picture, but also to build cost-effective and efficient systems. In this example, the world's exponential population increases without sustainable environmental resources strains food supplying systems and other natural resources (such as supplies of drinking water and reliable sources of energy). Worsening contamination strains already vulnerable

Net primary productivity of ecosystems

Ecosystem type	Net primary productivity grams dry organic matter per square meter per year
Open ocean water	100
Coastal seawater	200
Desert	200
Tundra	400
Upwelling area	600
Rice paddy	340–1200
Freshwater pond	950–1500
Temperate deciduous forest	1200–1600
Cropland (cornfield)	1000–6000
Temperate grassland	≤1500
Cattail swamp	2500
Tropical rain forest	≤2800
Coral reef	4900
Sugarcane field	≤9400

SOURCE: R.M. Maier, I.L. Pepper, and C.P. Gerba, *Environmental Microbiology.* 2d ed., Elsevier Academic Press, Burlington, MA, 2009.

nations. Thus, the environmental conditions have a direct effect on public health and security.

Interoperability is also needed to investigate threats to global climate and ecosystems. Coral reef protection is an example. Reefs are among the most productive ecosystems on Earth (see **table**). Threats to the function and structure of coral-reef ecosystems have been associated with wastewater discharges to major rivers that ultimately drain into oceans, carrying nutrients and microbial populations. More recently, these discharges have been found to contain drug, personal care products, and antibiotics that are not degraded or incompletely degraded, leading to stresses on aquatic ecosystems. For example, the corals off the coast of Florida (the world's third largest barrier reef) have been highly stressed, with half of the live coral being lost in the past few years. An additional indication is that fish feeding on these corals are developing deformities and experiencing premature mortality.

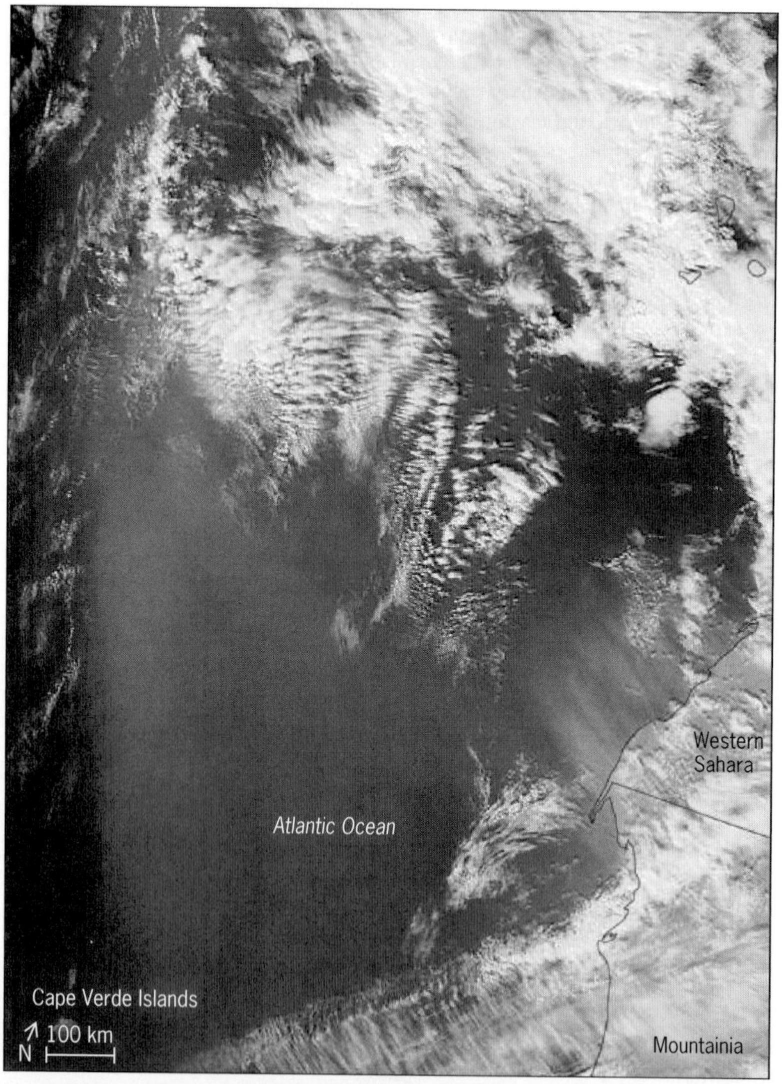

Plumes blowing off Africa's west coast on January 22, 2008. The plumes carry dust from the Sahara Desert across the Atlantic Ocean. The dust in the plumes contains bacteria, fungi, and their spores that have been associated with coral reef destruction. In this photo, the dust stayed airborne over the Atlantic Ocean for several days in mid-January 2008, before the National Air and Space Agency's (NASA) *Aqua* satellite captured this image using Moderate Resolution Imaging Spectroradiometer (MODIS). (*NASA's Earth Observatory*)

The threats to coral reefs come in many chemical and biological forms. The U.S. Coral Reef Task Force recently identified what it considers to be the most prominent threats that federal agencies and states must address to protect coral reefs in the United States. They are:

1. Pollution, including eutrophication and sedimentation from poor or overly intensive land use, chemical loading, oil and chemical spills, marine debris, and invasive alien species.
2. Overfishing and exploitation of coral-reef species for recreational and commercial purposes, and the collateral damage and degradation to habitats and ecosystems from fishing activities.
3. Habitat destruction and harmful fishing practices, including those fishing techniques that have negative impacts on coral reefs and associated habitats. This can include legal techniques, such as traps and trawls, which are used inappropriately, as well as illegal activities such as cyanide and dynamite fishing.
4. Dredging and shoreline modification in connection with coastal navigation or development.
5. Vessel groundings and anchoring that directly destroy corals and reef framework.
6. Disease outbreaks that are increasing in frequency and geographic range are affecting a greater diversity of coral-reef species.
7. Global climate change and associated impacts, including reduced rates of coral calcification, increased coral bleaching and mortality (associated with variety of stresses including increased sea surface temperatures), increased storm frequency, and sea-level rise.

Microbes appear to be transported long distances in winds aloft. For example, some of the invasive bacteria that threaten coral-reef habitats may be coming from Africa in the form of Saharan dust. Deserts commonly contain gravel and bedrock, along with some sand. The Sahara is the exception, with sand covering 20% of the spatial extent of the desert. This means that the Sahara often loses large amounts of dust by winds that advectively transport particles in plumes that can travel across the Atlantic Ocean, depositing dust along the way (see **illustration**). Saharan dust carries disease-causing bacteria and fungi that have been associated with the destruction of coral reefs in the Caribbean Sea.

Scientists from numerous disciplines are studying these phenomena and trying to evaluate the linkages, the threats, and possible interventions. Recently, this multidisciplinary venture has identified algae and bacteria as a much larger concern than had previously been thought. The bacteria are likely arriving in the coral-reef ecosystems by long-range atmospheric transport.

Coral reefs are threatened by an array of physical, chemical, and biological stressors. Ultraviolet (UV) radiation is an example of a physical stressor. UV radiation is a ubiquitous stressor that is very likely affecting human and ecological systems on a global scale. It has been implicated in observed shifts in polar plankton community composition, local and global declines in amphibian population abundance and diversity, coral bleaching syndrome, not to mention an increasing incidence of human skin cancer and other diseases. Disruption and loss of coral-reef ecosystem communities because of coral bleaching and disease leads to coral mortality, changes in reef persistence and formation dynamics, as well as cascading reef community interactions.

Biological agents can present an even greater and more ominous threat to coral reefs. The microbial ecology within a coral reef is complex. For example, some algae are symbiotic and others are parasitic to corals. Fungi are also widely distributed in calcium carbonate; that is, they live within corals (endolithic association). Bacterial diseases are also appearing. Thus, various biological taxa of microorganisms present different, but possibly synergistic threats to coral reef systems. This indicates that ecological and medical information systems need interoperability in the event that the dust is transporting human pathogens as it has coral pathogens.

Outlook. Linking disparate data and information is needed in "person-centered" health system. For example, it is not common for medical practitioners to access environmental databases, even though these are often rich in information needed for proper diagnosis, treatment, and prevention of diseases. Contaminants have been measured and modeled in many communities, so healthcare providers could glean information about exposures and risk. If a child shows symptoms of neurotoxicity, for instance, the provider would be wise in determining the proximity of a child's home to possible sources of lead and mercury. Such information could be readily input to geographic information systems and other tools to improve the medical care and to prevent future exposures.

Spatially relevant health information is scarcely available to a clinician at the time of a medical diagnostic encounter and certainly it is not a typical part of a comprehensive electronic medical record. Leveraged by a geographic information system, information on patients' potential environmental exposures can be delivered to the clinician while the patient is in the examination room, and influence future outcomes by modifying behaviors, for example. Using modern information technology in this way can go a long way toward helping the physicians and patients they serve.

As food and water sustainability become more of a commodity worldwide, the need to preserve these assets will progressively increase. At the same time, the quality of the air, water, food, and other environmental factors will permit, when combined with personal health records, the anticipation of problems for more effective and efficient prevention. Clearly, the future of combining genetics and molecular biology with health and environmental systems opens the doors for the future of personalized medicine.

[Disclaimer: The views expressed in this paper are those of the author and do not reflect the official policy or position of the National Defense University, the Department of Defense, or the U.S. Government.]

For background information *see* FUNGAL ECOLOGY; GEOGRAPHIC INFORMATION SYSTEMS; INFORMATION MANAGEMENT; INFORMATION TECHNOLOGY; MEDICAL INFORMATION SYSTEMS; MODEL THEORY; PATHOGEN; POLLUTION; PUBLIC HEALTH; REEF in the McGraw-Hill Encyclopedia of Science & Technology.

Luis Kun; Daniel Vallero

Bibliography. W. Boddie and L. Kun, Enhancing the healthcare/public health and the food and agriculture critical infrastructures through an enabling food product safety enterprise architecture, L. Kun (guest ed.), Special issue on protection of the healthcare and public health critical infrastructure and key assets, *IEEE Eng. Med. Biol. Mag.*, 27(6):54–58; 2008; B. Davenhall, The Missing Component (Geomedicine), *ESRI ArcUser*, Winter 2010, http://www.esri.com/news/arcuser/0110/geomedicine.html; S. Golubic, G. Radtke and T. Le Campion-Alsumard, Endolithic fungi in marine ecosystems, *Trends Microbiol.*, 13:229–235, 2005; L. G. Kun, Editorial: Interoperability: the cure for what ails us, *IEEE-EMBS Mag.*, 26(1):87–91, 2007; L. G. Kun, Government affairs editorial: Global health and policy implications of dust storms: A disease transportation system, L. Kun (guest ed.), Special issue on protection of the healthcare and public health critical infrastructure and key assets, *IEEE Eng. Med. Biol. Mag.*, 27(6):74–79, 2008; L. Kun (guest ed.), Special issue on protection of the healthcare and public health critical infrastructure and key assets, *IEEE Eng. Med. Biol. Mag.*, vol. 27(6), 2008; R. Mathews and C. Spencer, National security strategy for US water, L. Kun (guest ed.), Special issue on protection of the healthcare and public health critical infrastructure and key assets, *IEEE Eng. Med. Biol. Mag.*, 27(6):42–53, 2008; W. R. Munns et al, Approaches for integrated risk assessment, *Human and Ecological Risk Assessment*, 9(1):267–273, 2003; National Oceanic and Atmospheric Administration, *National Coral Reef Action Strategy: Report to Congress on Implementation of the Coral Reef Conservation Act of 2000 and the National Action Plan to Conserve Coral Reefs*, 2002; J. L. Nitzkin and C. Buttery, Public health critical infrastructure, L. Kun (guest ed.), Special issue on protection of the healthcare and public health critical infrastructure and key assets, *IEEE Eng. Med. Biol. Mag.*, 27(6):16–20, 008; M. Schrope, Future of corals is going down the pan, *New Scientist*, 175(2355): 11, 2002.

High-speed rail

High-speed rail (HSR) is currently regarded as one of the most significant technological breakthroughs in passenger transportation developed in

(a) (b) (c)

Fig. 1. High-speed rail lines. (*a*) Tōkaidō. (*b*) TGV. (*c*) ICE.

the second half of the twentieth century. According to the definition of the International Union of Railways (UIC), high-speed rail is a type of passenger rail transportation capable of operating at speeds of 200 km/h (124 mi/h) for upgraded track and 250 km/h (155 mi/h) or faster for new track. During the past 50 years, the high-speed rail has developed rapidly to meet the increasing demand for passenger rail travel. The development process can be divided into three stages.

Stage 1 (1964 to 1990). In this stage, Japan, France, Italy, and Germany constructed high-speed railways and put them into operation. Japan was the first country in the world to build dedicated railway lines for high-speed travel, called Shinkansen. In 1964, the world's first contemporary high-speed rail, the Tōkaidō Line, went into operation in Japan (**Fig. 1***a*). In 1975, Japan constructed the Sanyō Shinkansen, and the two new lines, the Tōhoku Shinkansen and Jōetsu Shinkansen, were built in the 1980s. In France, the high-speed rail is called Train de Grande Vitesse (TGV). It was developed during the 1970s by GEC Alsthom and Société Nationale des Chemins de Fer Français (SNCF). The TGV began operation between Paris and Lyon in September 1981 (Fig. 1*b*). The success of this line led to the expansion of the network, with new lines built in the south, west, north, and east of the country. Italy opened what is often regarded as Europe's first high-speed rail route, the Direttissima, which from 1978 connected Rome with Florence. However, the major works were not finally completed until the early 1990s, long after France's faster TGV network was established. In Germany, the high-speed rail is called InterCityExpress (ICE). The Hanover to Würzburg high-speed railway was the first high-speed railway line for InterCity-Express traffic, starting as early as 1973. The line opened fully in 1991 (Fig. 1*c*).

Stage 2 (mid-1990s). In this stage, the rapid success of high-speed rail attracted more study and attention. Japan, France, Italy, and Germany started comprehensive planning of their railway networks. Spain introduced the technology of high-speed rail from France and Germany and constructed its first high-speed railway between Madrid and Seville in 1992. In 1994, the Channel Tunnel connected the United Kingdom to the European continent. In 1997, Eurostar connected France, Belgium, the Netherlands, and Germany. In France and Germany, the governments started to reconstruct existing lines, besides building new lines.

Stage 3 (mid-1990s to present). In this stage, China, Russia, South Korea, Australia, and the United Kingdom started to construct new high-speed railway lines. The Czech Republic, Hungary, Poland, and Austria reconstructed their existing lines and connected them to the European high-speed network (**Fig. 2**). By 2008, the total length of the high-speed railway lines had reached 9372 km (5823 mi).

By March, 2010, China had the longest high-speed railway line in the world. The total length of HSR in China is 3500 km (2175 mi). The main operator of regular high-speed train services in the People's Republic of China is the China Railway High-Speed (CRH) [**Fig. 3**].

In 2004, the Ministry of Railway (MOR), China, announced the Mid-to-Long Term Railway Network Plan, composed of a grid of eight high-speed rail corridors: four running north-south and four traversing east-west, totaling 12,000 km (7500 mi) [it was revised to 16,000 km (9900 mi) in 2008] (**Fig. 4**; not including Taiwan). Most of the new lines follow the routes of existing trunk lines and are designated for passenger travel only. They are known as passenger-designated lines (PDL). Several sections of the national grid, especially along the southeast coastal corridor, were built to link cities that had no previous rail connections. Those sections will carry a mix of passenger and freight, but are sometimes mislabeled. High-speed trains on PDLs can generally reach 300–350 km/h (186–217 mi/h). On mixed-use HSR lines, passenger train service can attain peak speeds of 200–250 km/h (124–155 mi/h). This ambitious national grid project was planned to be built by 2020, but the government's stimulus plan has expedited this schedule considerably for many of the lines.

Technology. High-speed rail is an integration of the most advanced techniques, and it has its own technical characteristics, mainly the highly regular track, a train control system based on on-board signals, and streamlined and sealed ac-dc-ac EMUs (electric

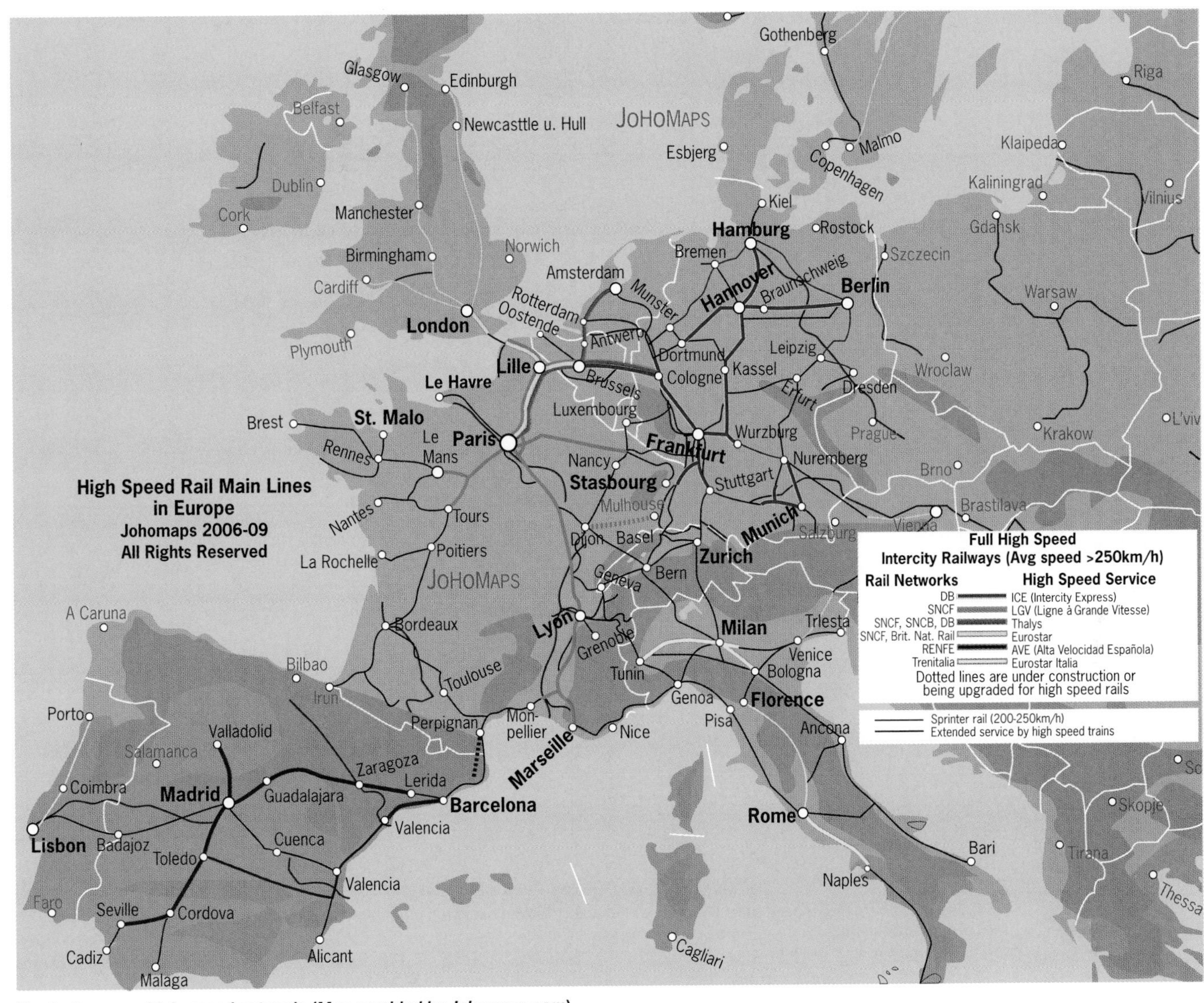

Fig. 2. European high-speed network. ***(Map provided by Johomaps.com)***

multiple units; a set of two or more electrically powered passenger rail cars that operate together).

High-speed rail requires higher density and stability for the track. Because its speed is faster than the ordinary rail, it demands higher regularity. To ensure high regularity, it is required that much stricter codes are adopted in the design and construction of the track, roadbed, and tunnel. In addition, it is necessary to enlarge the curve radius of the track because the train will produce a centrifugal force while passing through curves in direct proportion to the square of the velocity and in inverse proportion to the curve radius.

When a train is going forward at a speed more than 200 km/h (1234 mi/h), the driver is unable to control the train by trackside signal and must control the train by an on-vehicle signal system. The on-vehicle signal system can be divided into the grade speed control system and the continuous speed control system, the later is the development direction of train control system.

Fig. 3. China Railway High-Speed (CRH).

Because the speed of high-speed rail is faster, it encounters more air resistance, so the train is streamlined and made lighter to reduce the air resistance. Meanwhile, the high-speed rail adopts ac-dc-ac EMUs with high power. In order to shorten the braking distance, the high-speed rail should also improve the braking capacity.

At present, Japan, France, and Germany have the most advanced techniques and the richest

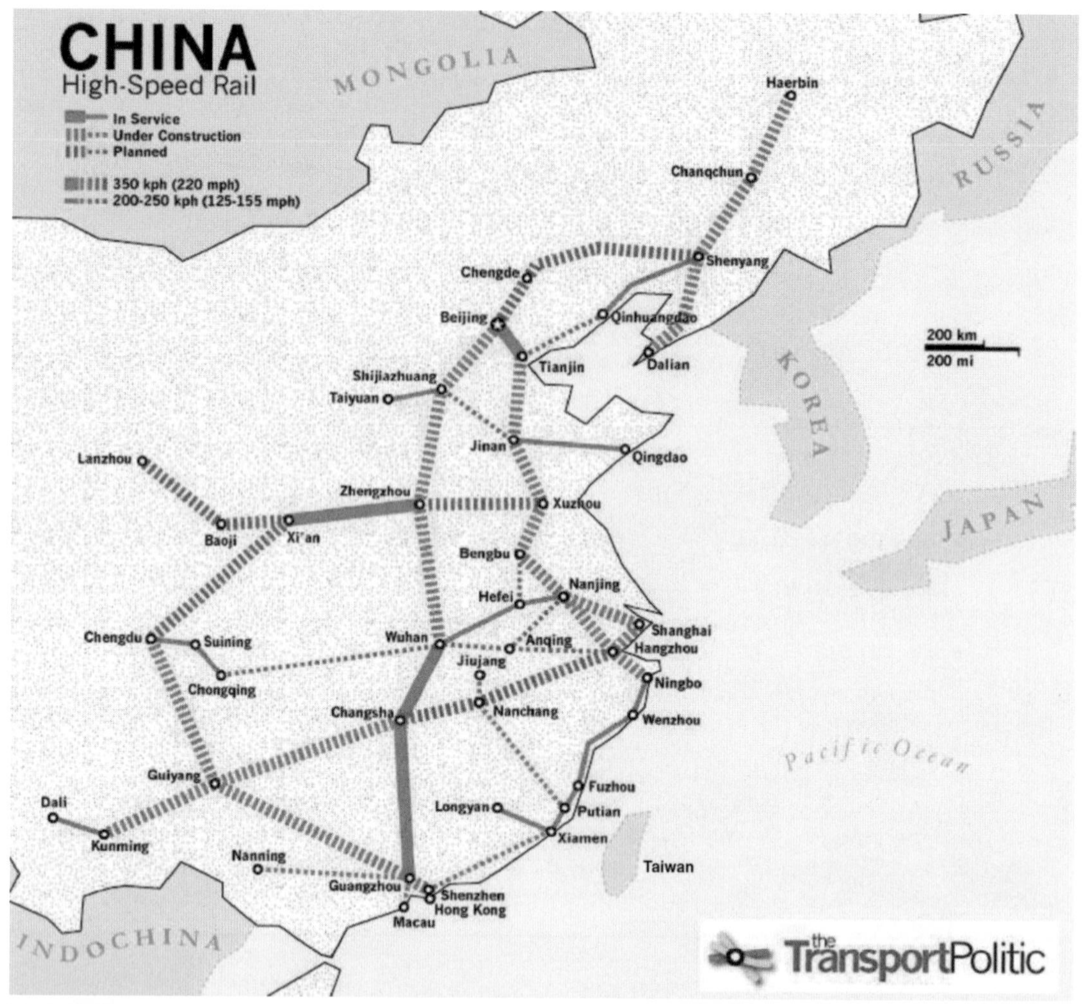

Fig. 4. China's high-speed rail program, not including Taiwan, showing eight high-speed rail corridors. (*Yonah Freemark/The Transport Politic*)

operating experience of HSR. Japan's Shinkansen sets mass rapid transit as its objective. Its power type is power distributed among the EMUs, which has a large passenger seating capacity. The targeted speed is only 240 km/h (149 mi/h). To improve the speed, the Shinkansen has to strengthen the function of its EMUs.

France has the world's fastest test speed (574.8 km/h or 357.2 mi/h) and overall trip speed (320 km/h or 199 mi/h). TGV adopts power-centralized EMUs. It has a short car body, thus a low train seating capacity and a high density. Its high-speed railway lines cooperate with the existing lines. The high-speed railway lines employ the ballasted track (for example, a crushed rock trackbed), and the maximum grade reaches 3.5%, which helps to keep the speed but reduce the cost.

The most significant characteristic of ICE is that the passenger train and the freight train are mixed and the high-speed railway lines and the existing lines are mixed. ICE adopts power-centralized EMUs. It employs a ballastless track and advanced techniques, and the maximum grade reaches 4.0%. Thus the construction cost is higher but the life-cycle cost is lower.

There is another type of high-speed rail, the maglev train. It is a system of transportation that suspends, guides, and propels the train using magnetic levitation from a very large number of magnets for lift and propulsion. Because of the lack of physical contact between the track and the vehicle, maglev trains experience no rolling resistance. This method has the potential to be faster, quieter, and smoother than wheeled mass- transit systems. There are three types of maglev technology: electromagnetic suspension (EMS), electrodynamic suspension (EDS), and magnetodynamic suspension (MDS). For EMS, electromagnets in the train attract it to a magnetically conductive track. EDS uses electromagnets on both the track and train to push the train away from the rail. And MDS uses the attractive magnetic force of a permanent magnet array near a steel track to lift the train and hold it in place. The highest test speed of a maglev train, 581 km/h (361 mi/h), was achieved in Japan in 2003. The Shanghai Maglev Train is the first operational high-speed conventional

Fig. 5. A maglev train coming out of the Pudong International Airport.

maglev railway in the world. It connects Pudong International Airport to downtown Shanghai. The highest speed achieved on the Shanghai track is 501 km/h (311 mi/h) [**Fig. 5**].

Construction. Nowadays, many countries are constructing or are trying to construct high-speed railway lines. In sum, the constructing modes can be listed as follows:

1. Newly constructed double-track high-speed railways, which deal exclusively with passenger transport, such as the Shinkansen and the TGV.

2. Newly constructed double-track high-speed railways, which deal with passenger transport and freight transport, such as the line connecting Rome with Florence; the speed of the passenger train is 250 km/h (155 mi/h), and the speed of the freight train is 120 km/h (75 mi/h).

3. Part of a newly built high-speed rail cooperates with part of an existing line, such as the line between Hanover and Würzburg.

4. A tilting train operates on an existing line, dealing with passenger transport and freight transport. It is a popular phenomenon is Europe, such as the X2000 type in Sweden.

Operation. Since the 1970s, railways have entered a period of decline in industrialized countries, mainly because of the challenge of road and air transportation. To remedy this situation, new technologies such as HSR, heavy-haul transport, and information techniques have been developed. Reforms of operational systems have been put forward and are beginning to play an important role in rail transportation. Now there are three modes for HSR: (1) using the existing railway administrative bureau (traditional mode); (2) separating operations from infrastructure (split mode); and (3) combining rail operations with infrastructure (aggregative mode).

The traditional mode would be the preferred option of the railway administrative bureau. Because both the conventional railways and the HSR would be operated by the same owner, it would be easier to coordinate the freight and passenger train operations. However, this mode is based on the planned economy system; thus it has many disadvantages in establishing a modernized enterprise system and moving HSR into a market economy environment.

The split mode separates operations from infrastructure. In this mode, infrastructure and operations would be managed independently. The infrastructure company (IC) usually provides support for the construction of HSR, while a passenger transportation corporation (PTC) would manage the operation of HSR (**Fig. 6**). The main aim of the separation of track from operations is to ensure competition in service provision and hence improve customer service at lower costs. This mode provides a clearly defined relationship between the companies and the government, releases railway transportation enterprises from the heavy burden of owning fixed infrastructure, and allows companies to compete fairly in the market place. But this mode requires the infrastructure company to invest sufficient funds to bear the cost of the project.

The aggregative mode combines rail operations with infrastructure. This mode combines the infrastructure and railway operations as a single entity (**Fig. 7**). All functions, such as construction, maintenance, and operation of HSR, would be undertaken by a high-speed railway group (HSRG). This would facilitate the establishment of an incorporated company and allow the HSRG to become a self-operating and self-profiting entity, and would also help to reduce internal exchange costs. But it does not allow the analysis of cost control, as compared with the split mode.

Performance. Compared to conventional railway transport, the high-speed railway transport has some advantages.

High speed. The greatest advantage of high-speed rail is its high speed (>200 km/h or 124 mi/h). The average speed of conventional railway is 140–160 km/h (87–100 mi/h) in China; for comparison, the speed of a Boeing plane is 700-1000 km/h (435–621 mi/h).

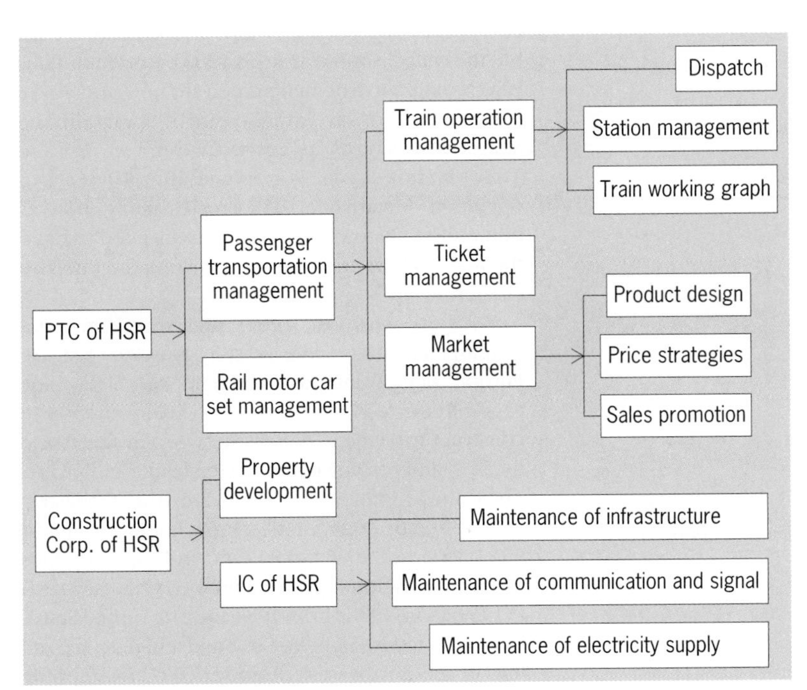

Fig. 6. Structure of split mode alternatives.

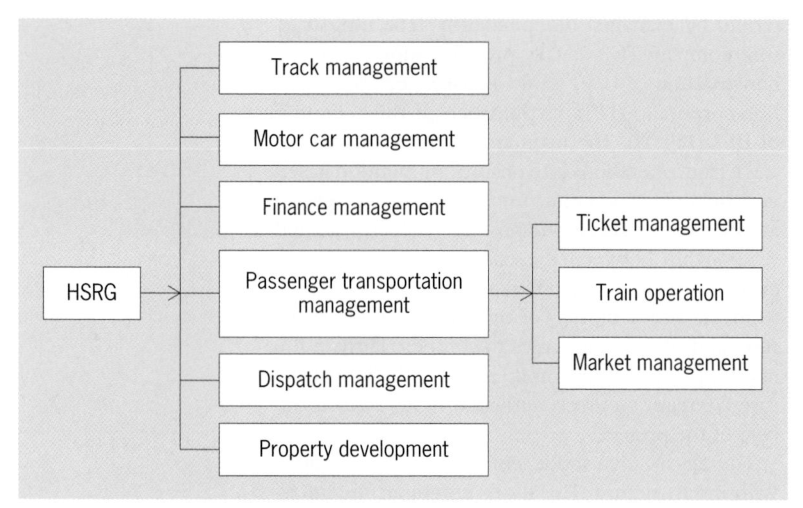

Fig. 7. Structure of aggregative mode.

Large-capacity. Compared to other transportation modes, rail has the largest transportation capacity. The Shinkansen can transport 518,400 tourists a day, far exceeding that of road transport and aircraft transport.

High safety. High-speed rail adopts advanced techniques. Its operation control system is an integration of electronic techniques and smart software. The Shinkansen has been in operation for over 40 years and has a zero accident record, which seems like a miracle in railway transport.

All-weather operation. Unlike aircraft transport and road transport, the HSR can run all the time, including in rain, snow, fog, or other bad weather.

Low energy consumption. The energy consumption of HSR is 136 kcal/km, while the automobile consumes 765 kcal/km, the airplane consumes 714 kcal/km. At the same time, HSR can use cleaner secondary energy sources (such as wind- or solar-generated), but the energy source of airplanes is petroleum. Thus HSR is more environmental friendly.

High economic returns. After 7 years of operation, the Shinkansen returns its entire investment. For the TGV, the investment is returned after 10 years of operation. Meanwhile, HSR creates many jobs, as numerous construction workers are needed to build the line, and the trade generated along the line also creates new job opportunities.

Since its birth in 1964, high-speed rail has developed rapidly and is still growing fast. In Europe, France intends to extend its railway line to 4000 km (2500 mi) by 2020. Spain intends to construct the largest railway network in the world by 2010 and replace France as the leader in Europe. The United Kingdom intends to construct the fastest railway line by 2025. In Asia, India intends to invest $54 billion in the next 5-8 years, and China intends to invest 1 trillion RMB (renminbi or Chinese yuan) [$146 billion] in the next 10 years. The United States intends to invest $8 billion to build a high-speed rail line in the northeast of America. We are stepping into the high-speed rail age.

For background information *see* MAGNETIC LEVITATION; RAILROAD CONTROL SYSTEMS; RAILROAD ENGINEERING; TRANSPORTATION ENGINEERING in the McGraw-Hill Encyclopedia of Science & Technology.

Zuyan Shen; Yuanqi Li

Bibliography. Q. H. Hai, Main technical characteristics of high speed railway and high speed EMUs, *Electric Drive for Locomotives* (in Chinese), 5:5–9, 2003; L. X. Qian, The level of development of high-speed railway in the world and China's technological development in high-speed railway, *Railway Purchase and Logistics* (in Chinese), 10:19–21, 2009; R. Vickerman, High-speed rail in Europe: Experience and issues for future development, *Ann. Reg.l Sci.*, 31:21–38, 1997; W. G. Wong and B. M. Han, Evaluation of management strategies for the operation of high-speed railways in China, *Transport. Res. Part A: Policy and Practice*, 36:277–289, 2002.

Holographic data storage

Digital data are ubiquitous in modern life, and the capabilities of current storage technologies are continually being challenged by applications such as the distribution of content, digital video, interactive multimedia, small personal data-storage devices, archiving of valuable digital assets, and downloading over high-speed networks. Current optical data-storage technologies, such as the compact disk (CD), digital versatile disk (DVD), and Blu-ray disk (BD) have been widely adopted because they provide random access to data, are inexpensive removable media, and rapidly replicate video and music content.

These traditional optical storage technologies stream data one bit at a time, and record the data on the surface of the disk-shaped media. In these technologies, the data are read by detecting changes in the reflectivity of the small marks made on the surface of the media during recording. The traditional path for increasing optical recording density is to record smaller marks that are closer together. These improvements in characteristic mark sizes and track spacing have yielded storage densities for CD, DVD, and BD of approximately 0.66, 3.2, and 17 Gb/in.^{2}, respectively. BD has decreased the size of the marks to the practical limits of far-field recording.

To further increase storage capacities, multilayer disk recording is possible but signal-to-noise losses and reduced media manufacturing yields make using more than two to four layers impractical.

Holographic data storage. Holographic data storage (HDS) breaks through the density limitations of conventional storage technologies by going beyond recording on the surface, to storing data in three dimensions (3D). Unlike serial technologies (optical and magnetic) which that record one data bit at a time, page-wise holography records and reads over a 1 million bits of data with a single flash of light, enabling transfer rates significantly higher than traditional optical storage devices. Page-wise holographic data storage has demonstrated the highest storage densities (712 Gb/in.^{2}) of any removable technology, and has a theoretically achievable density of around 40 Tb/in.^{2}. High storage densities, fast transfer rates, and random access, combined with durable, reliable, low- cost media, make page-wise holography a compelling choice for next-generation storage and content distribution applications.

Holographic technologies offer improvements in the performance and cost curves of storage that make increasingly large amounts of data accessible to users, while reducing the total cost of storing the data.

The value proposition for holographic data storage products includes:

1. Highest performance for removable storage, which combines a demonstrated data density of over 712 Gb/in.^{2}, random access (around 250 ms), and transfer rates capable of exceeding 120 MB/s.

2. 50+ years media archive life, requiring no special handling, refreshing, or environmental controls; and no wear from media contact with a read/write head.

3. Near-line random access to content, which is much faster than tape but not as fast as disk, making petabytes (10^{15} bytes) of data almost instantly accessible.

4. Smaller media formats with higher density.

5. Lowest cost per gigabyte for professional-grade media, making archiving affordable for terabytes (10^{12} bytes) to exabytes (10^{18} bytes) of data.

6. Improved data protection with a true (intrinsic) write-once-read-many (WORM) media format that ensures that the data retains its original state.

7. Lowest total cost of ownership, resulting from low media costs; reduced frequency of media migration; smaller media size (which reduces data center floor space requirements); and power savings, achieved by decreasing the use of hard disk drives to store infrequently accessed data.

8. Low-cost, high-speed media replication similar to CD or DVD.

9. Low-cost media using plastic substrates and easy manufacturing process similar to DVD.

In HDS, data are stored as shown in **Fig. 1**. Light from a single laser beam is split into two beams: the signal beam (which carries the data) and the reference beam. The hologram is formed where these two beams intersect in the recording medium. The process for encoding data onto the signal beam is accomplished by a device called a spatial light modulator (SLM). The SLM translates the electronic data of 0s and 1s into an optical "checkerboard" pattern of light and dark pixels. This is called a data page and typically is around 1000 × 1000 pixels. At the point where the reference beam and the data-carrying signal beam intersect, the hologram is recorded in the light-sensitive storage medium. A chemical reaction occurs causing the hologram to be stored. By varying the reference-beam angle, or media position, hundreds of unique holograms are recorded (multiplexed) in the same volume of material.

Many multiplexing techniques have been proposed. The following examples were developed by InPhase Technologies. The primary multiplexing method is angle multiplexing, which allows data pages to be superimposed by varying the reference-beam angle. For each reference-beam angle a different data page is stored. Each hologram or page is stored through the 3D volume of the media that is exposed. By changing the reference beam, the Bragg selectivity, which diffracts light along a partic-

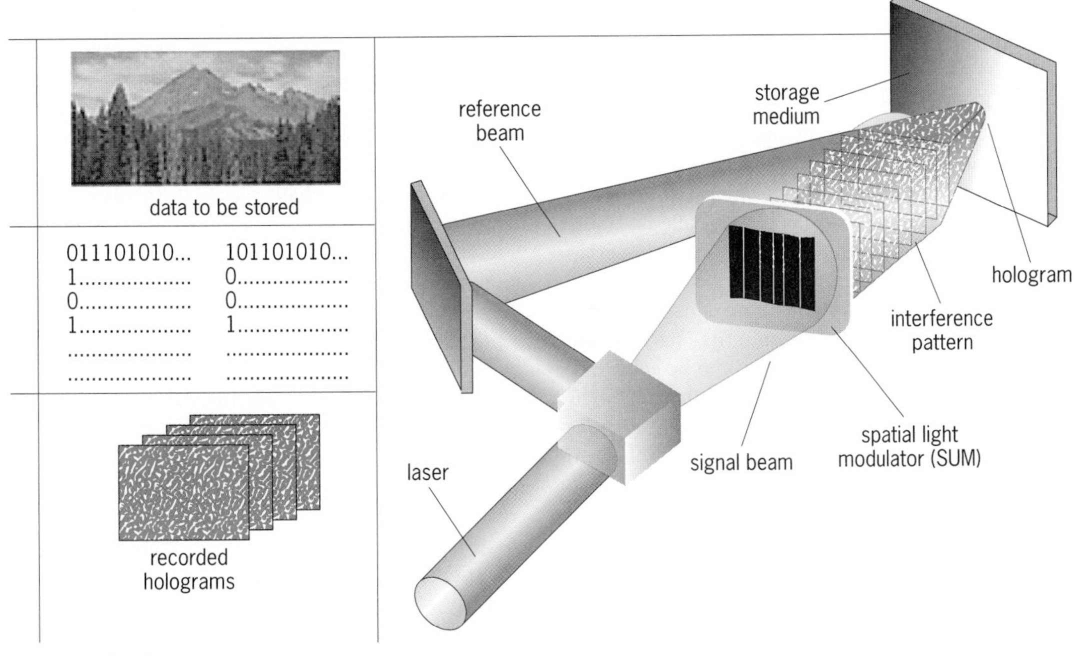

Fig. 1. Writing data.

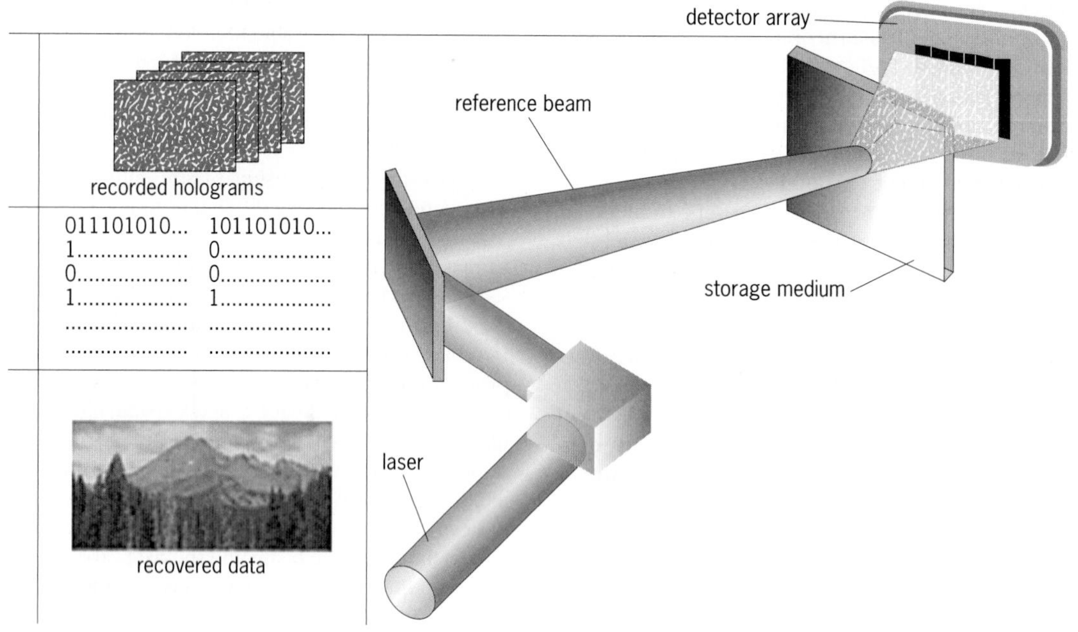

Fig. 2. Reading data.

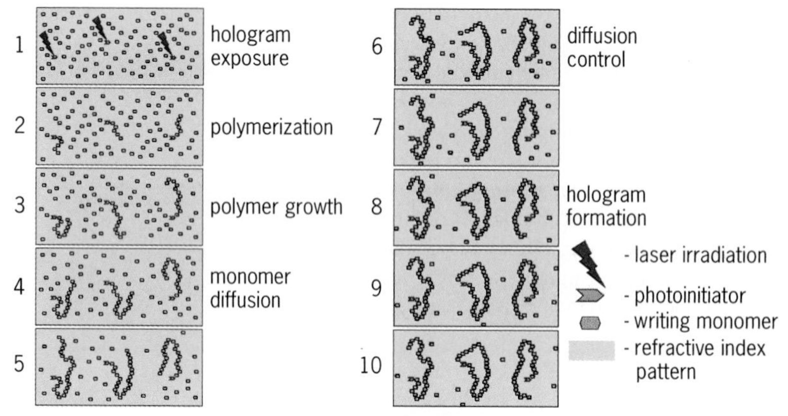

Fig. 3. Hologram recording mechanism for photopolymers.

ular angle of incidence of the thick media, allows for multiple holograms to be stored in the same volume with different reference beam angles. A group of multiplexed holograms (or pages) is called a book. Polytopic multiplexing allows for books to overlap in the media and therefore reach close to the two-dimensional (2D) diffraction-limited density on a per bit basis. Then by multiplexing multiple pages in the same volume much higher density can be achieved.

For data recovery, a reference beam (which is nominally a replica of the recording reference beam) is used to illuminate the medium. The hologram diffracts the probe beam, thus reconstructing a replica of the encoded signal beam. This diffracted beam is then projected onto a pixelated photodetector that captures the entire data page of over 1 million bits at once. This parallel read-out provides holography with its fast transfer rates. **Figure 2** illustrates a basic geometry for reading holograms. The detected data page images are then processed and decoded in order to recover the stored information.

In addition to angle multiplexing, InPhase invented polytopic multiplexing. In polytopic multiplexing, the book spacing is determined by the beam waist or signal bandwidth. Since the waists do not overlap, undesired reconstructions from neighboring books can be filtered out by introducing an aperture that passes only the desired beam waist. Writing is performed as in the nonpolytopic case, excepting that the aperture is present and the book spacing is determined by the polytopic criterion. Upon readout, the desired page and several of its neighbors may be simultaneously reconstructed since the probe beam must illuminate data-bearing fringes from several books at once. However, the physical aperture blocks the components diffracted from the neighboring books. This is particularly important when, as is required for high-density storage, a high numerical aperture and a thick medium are used.

One of the major challenges in the history of holographic data storage has been the development of a recording material that could enable the promise of this technology. Holographic storage requires a material that can rapidly image (capture) the complex optical interference patterns generated during the writing step, such that the imaged patterns are (1) generated with adequately high contrast, (2) preserved (both dimensionally and in their contrast) during subsequent, spatially co-located writing, (3) are unchanged by readout of the holograms, (4) are robust to environmental effects, (5) survive over long periods of time (many years), and (6) can be inexpensively manufactured in high volume.

Early efforts in holographic data storage sought to demonstrate the basic capabilities of the approach; that is, the ability to record and recover data pages with high fidelity and to multiplex to achieve storage density. Because of this initial focus on the feasibility of the technology, these efforts mostly used photorefractive crystals that were well known. Photorefractive materials translate optical patterns into refractive index gratings within the material through the electro-optic effect. While valuable for basic demonstrations of the technology, photorefractive crystals suffered from limited refractive-index contrast, slow writing sensitivity, and unwanted volatility where the readout of holograms led to erasure. Attention turned to alternatives such as photochromic, photoaddressable, and photopolymer materials in which, similarly, optical interference patterns could be imaged as refractive-index modulations. Photopolymers quickly became the leading candidates as, in addition to their imaging properties, they exhibited high contrast and recording speeds and were nonvolatile (retain information in the absence of power).

Photopolymers can meet the requirement for holographic storage and have become the primary focus of development worldwide. Photopolymers consist of a photoreactive, polymerizable system dispersed within a host. During holographic recording, the optical interference pattern initiates a pattern of polymerization in the photoreactive system—polymerization is induced in the light intensity maxima of the interference pattern while no polymerization occurs in the nulls. This patterned polymerization sets up a concentration gradient in the unreacted species. Unpolymerized species diffuse from the nulls to the maxima of the interference pattern to equalize its concentration in the recording area, creating a refractive-index modulation set up by the difference between the refractive indices of the photosensitive component and the host material. **Figure 3** shows the steps in the formation of the refractive index modulation.

The polymerization that occurs during the recording process results in changes in the dimensions and the bulk refractive index of the material. Changes such as these, if not controlled, can distort the imaging process, degrade the fidelity with which holographic data can be recovered, and ultimately limit the density of data the material can support. The design of the photopolymer media for holographic storage applications must therefore balance the advantages of photopolymers (photosensitivity and large refractive index modulations) against the impact of the changes that accompany the polymerization.

State of the art. Figure 4 shows a picture of a first-generation drive and removable media. It stores 300 GB per disk and reads and writes data at 20 MBytes per second (160 Mbits/s) through standard interfaces. The holographic recording media consists of 1.5 mm of InPhase Tapestry™ photopolymer sandwiched between two 130-mm-diameter, 1-mm-thick substrates. One of the substrates contains a tracking pattern for tangential positioning. Both substrates have an antireflecting coating that is optimized for 405-nanometer (violet) light (same as used in Blu-ray systems). The disk package has a magnetic hub for securing it to the drive spindle. The entire disk is encased in a light-tight cartridge.

(a)

(b)

Fig. 4. First HDS drive with *(a)* top of enclosure removed and *(b)* disk inside cartridge.

For background information *see* COMPACT DISK; OPTICAL INFORMATION SYSTEMS; OPTICAL RECORDING; VIDEO DISK; HOLOGRAPHY; POLYMER; ELECTROOPTICS; REFRACTION OF WAVES; INTERFERENCE OF WAVES in the McGraw-Hill Encyclopedia of Science & Technology. Kevin Curtis

Bibliography. K. Anderson and K. Curtis, Polytopic multiplexing, *Opt. Lett.*, 29:1402–1404, 2004; L. Dhar et al., Temperature-induced changes in photopolymer volume holograms, *Appl. Phys. Lett.*, 73:1337–1339, 1996; K. Curtis, et al., *Holographic Data Storage: From Theory to Practical Systems*, Wiley, 2010; K. Shimada, et al., High density recording using Monocular architecture for 500GB consumer system, *Opt. Data Storage Conf.*, Buena Vista Florida, 2009.

Homo heidelbergensis

Homo heidelbergensis is a species of extinct humans that lived in Europe and possibly also in Africa and Asia approximately 600–300 ka (that is, thousand years before the present) in the time period known as the Middle Pleistocene. The species was named by Otto Schoetensack in 1908, following the discovery of the Mauer mandible (lower jaw) by workmen in a sandpit in the village of Mauer, near Heidelberg, Germany. This specimen serves as the holotype, or type specimen, of this species; that is, it is the specimen on which *H. heidelbergensis* was described in the original publication. Therefore, the original definition of the species relies on mandibular features of a European hominin (fossil human). *Homo heidelbergensis* was among the earliest fossil human species to be recognized, following *H. neanderthalensis*, first discovered in Germany in 1856, and *Pithecanthropus erectus*, later renamed *H. erectus*, discovered in Java in the early 1890s.

Mauer jaw. The unusual features of the Mauer jaw were noticed and described early on. Particular attention was paid to the lack of a chin and to the large overall size of the mandible, both considered primitive, apelike features. On the other hand, the average size of the teeth and the humanlike character of their morphology were also pointed out. The Mauer mandible was considered more ancient than the newly discovered Neanderthals, and almost from the start it was thought to be either directly ancestral to or an early representative of the Neanderthals. Today, this mandible is considered to exhibit mainly primitive features and therefore its potential ancestral position relative to Neanderthals is not clear. Up until recently, the Mauer specimen was one of the earliest hominins known from Europe. Although its exact chronology is not entirely resolved, it is believed to be as old as 500 ka.

Further fossil evidence. Since the early twentieth century, fossil discoveries gradually produced a number of fossils of similar antiquity as the Mauer individual. These included several fossil mandibles that could be compared with the Mauer jaw. The discovery of the Arago mandibles in Tautavel, France, in the late 1960s was crucial because their similarity to the Mauer specimen suggested that they belonged to the same species. This was important in linking the lower jaw of *H. heidelbergensis* with the cranial fossil human remains recovered in Arago. For the first time, paleoanthropologists could literally give a face to *H. heidelbergensis*. From that point forward, fossils similar to the Arago crania could be considered as belonging to the same species as the Mauer mandible. This group would grow to include Middle Pleistocene specimens from Europe, and possibly Africa and Asia as well, although the inclusion of non-European specimens in *H. heidelbergensis* is currently debated (see below). Important European representatives of the taxon include Arago 21 (France), Petralona (Greece) [**Fig. 1**], Steinheim (Germany), and Cranium 5 (from the Sima de los Huesos site of the Atapuerca complex in northern Spain).

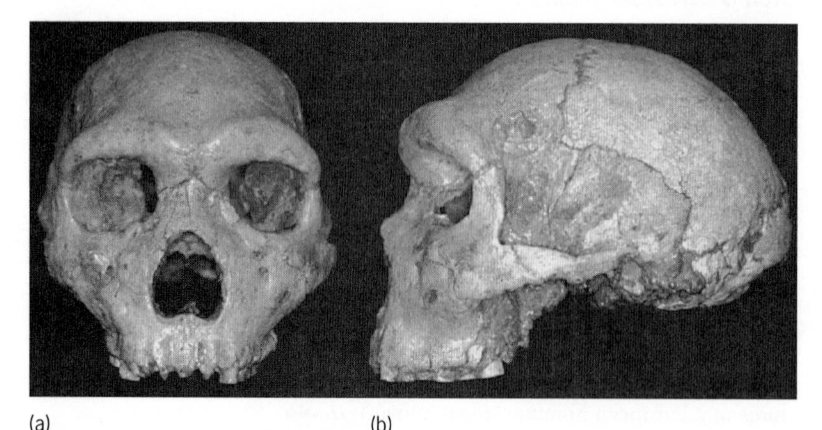

(a) (b)

Fig. 1. Two views of the Petralona cranium from Greece: (*a*) front view, (*b*) side view. The Petralona cranium is likely similar to what the Mauer cranium would have looked like. (*Photographs courtesy of Eric Delson*)

The anatomical definition of *H. heidelbergensis* has been somewhat elusive, mainly because this species does not appear to have many autapomorphic features (that is, features unique to itself). Fossils attributed to *H. heidelbergensis* show several morphological similarities to *H. erectus*, including broad and relatively forward-projecting faces, pronounced brow ridges, relatively low and long braincases, and heavy musculature markings at the back of the skull. However, they also exhibit derived (that is, specialized) cranial features, found in the later species *H. neanderthalensis* and *H. sapiens*, mainly related to an increase in brain size: increased convexity of the cranial bones (especially the frontal and parietals), a less angular posterior part of the skull, decreased forward projection of the face, and decreased overall muscular markings. These characteristics suggest affinities with later *Homo* species, that is, Neanderthals and modern humans.

Status of Homo heidelbergensis. The current status of *H. heidelbergensis*, its geographic range, and its relationship to later human species, including our own, are at the heart of current discussion in paleoanthropology. Strong anatomical similarities have been observed between European and African Middle Pleistocene fossil human specimens—for example, the Petralona cranium from Greece and the Kabwe (also known as Broken Hill) cranium from Zambia. These similarities are particularly striking in analyses of metric data, that is, statistical analyses of cranial measurements. The results of such analyses suggest that African and European specimens, and even some Asian specimens, from this time period should be included in the same species, *H. heidelbergensis*. In this view, *H. heidelbergensis* is a transcontinental species and is likely the last common ancestor of the two distinct evolutionary lineages leading to the two later species, *H. neanderthalensis* in Europe and *H. sapiens* in Africa.

An alternative approach to the study of the human fossil record involves the analysis of small-scale anatomical features that usually cannot be measured. These features are instead "scored" as present or absent, and intermediate conditions, when present, are noted. Such nonmetric analyses reveal the presence of many Neanderthal-like anatomical details in the European subset of these Middle Pleistocene specimens, suggesting that they were the direct ancestors of Neanderthals. If this interpretation is accepted, it follows that the name *H. heidelbergensis* should be applied only to the European Middle Pleistocene humans. As a direct ancestor of Neanderthals, the species *H. heidelbergensis* could be sunk into *H. neanderthalensis*, since the two represent the same lineage. On the other hand, it could be retained but recognized only as an arbitrary segment of the Neanderthal lineage, that is, a chronospecies. In either case, it can no longer be considered as the last common ancestor of both Neanderthals and modern humans. This last common ancestor then could

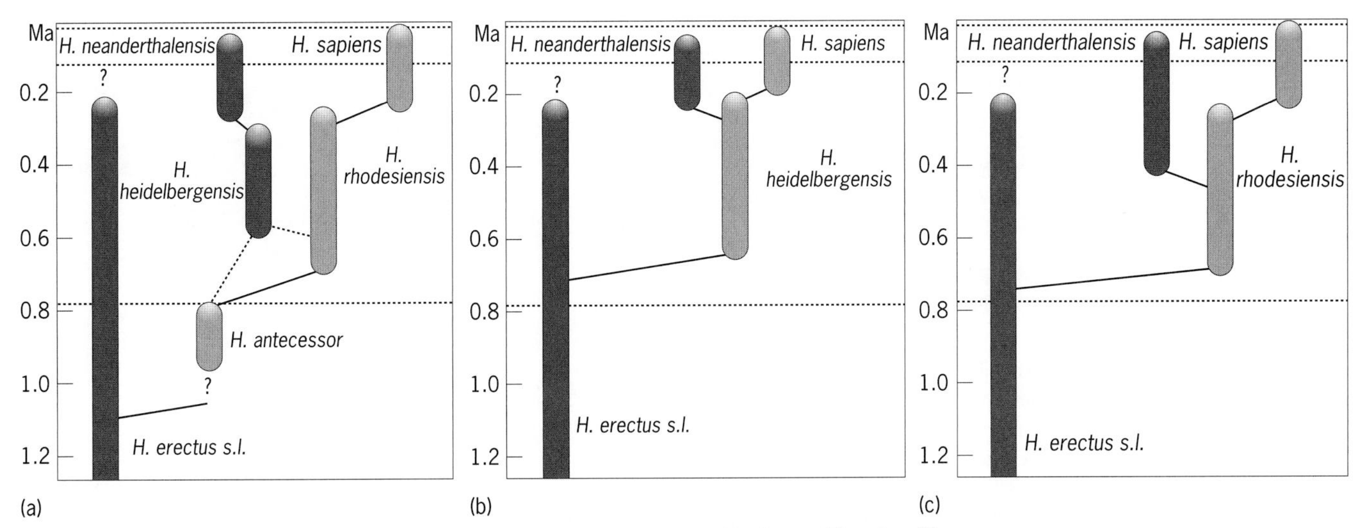

Fig. 2. Alternative evolutionary scenarios (*a–c*) for *H. heidelbergensis* as discussed in the text. Time, in million years ago (Ma), is indicated on the left side of each panel. The horizontal dotted lines signify the beginning and the end of the Middle Pleistocene. (*Adapted from J.-J. Hublin, The origin of the Neandertals, Proc. Natl. Acad. Sci. USA, 106:16022–16027, 2009*)

be represented by the African Middle Pleistocene specimens, which in this scheme would have to be recognized as another species, *H. rhodesiensis*. Alternatively, it could be represented by yet another, earlier hominin discovered in the 1990s in Europe, dating from as early as 1.2 million years before the present and named *H. antecessor*.

These alternative scenarios are summarized in **Fig. 2**. Figure 2*a* shows *H. antecessor* as the last Neanderthal-modern human ancestor, with *H. heidelbergensis* leading to *H. neanderthalensis* in Europe and *H. rhodesiensis* leading to *H. sapiens* in Africa. This view postulates an early separation of the lineages, which is difficult to reconcile with the pronounced morphological similarities of the African and European Middle Pleistocene humans. It also fits uncomfortably with the date of divergence estimated from genetic data for the Neanderthal and modern human lineages, the most recent of which suggest a population separation date of approximately 370 ka.

Figure 2*b* shows *H. heidelbergensis* as a direct descendent of *H. erectus* rather than of *H. antecessor* and as a transcontinental species, spanning a geographic range that includes both Africa and Europe. The European representatives of this species then give rise to *H. neanderthalensis*, whereas the African ones give rise to *H. sapiens*. This view is consistent with the estimated time of population separation, as well as the pronounced similarities between European and African Middle Pleistocene specimens. However, it does not explain the occurrence of Neanderthal-like characteristics among the earliest European Middle Pleistocene specimens, such as those from Sima de los Huesos, which are currently dated at approximately 600 ka. The last panel, Fig. 2*c*, proposes an intermediate time of population divergence and recognizes the European Middle Pleistocene specimens postdating 400 ka as belonging to the Neanderthal lineage. It thus assigns specimens previously recognized as *H. heidelbergensis* into *H. neanderthalensis*. Both *H. neanderthalensis* and *H. sapiens* are descendent from *H. rhodesiensis*, representing the African and the earliest of the European Middle Pleistocene samples. This view accommodates the estimated time of population divergence and the morphological similarities between the European and African human fossil record in this time period. However, as with the scenario in Fig. 2*b*, it also cannot properly account for the occurrence of Neanderthal features that could date as early as 600 ka in the Atapuerca Sima de los Huesos sample.

Whether they represent one species or more, the Middle Pleistocene humans commonly placed in *H. heidelbergensis* appear to share some important trends. Perhaps most interesting is the trend for increased brain size. Mean cranial capacity (as can be calculated based on a limited number of well-preserved specimens) is 1206 cm^3 (74 $in.^3$), compared with 973 cm^3 (59 $in.^3$) for the *H. erectus* mean, and this is well within the normal modern human range. This increase appears to be higher than what would be predicted from gradual evolution within the *H. erectus* lineage. Furthermore, estimates of brain size relative to body mass indicate that this increase cannot be attributed to greater body size alone. These larger brains, in turn, appear to be linked with technological advances, such as more sophisticated toolkits, as well as with hunting skills and hunting equipment.

For background information *see* ANTHROPOLOGY; ARCHEOLOGY; BRAIN; EARLY MODERN HUMANS; FOSSIL HUMANS; MOLECULAR ANTHROPOLOGY; NEANDERTALS; PHYSICAL ANTHROPOLOGY; PLEISTOCENE in the McGraw-Hill Encyclopedia of Science & Technology. Katerina Harvati

Bibliography. E. Delson et al., *Encyclopedia of Human Evolution and Prehistory*, 2d ed., Garland, New York, 2000; J.-J. Hublin, The origin of the

Neandertals, *Proc. Natl. Acad. Sci. USA*, 106:16022–16027, 2009; R. G. Klein, *The Human Career: Human Biological and Cultural Origins*, 3d ed., University of Chicago Press, 2009; A. Mounier, F. Marchal, and S. Condemi, Is *Homo heidelbergensis* a distinct species?: New insight on the Mauer mandible, *J. Human Evol.*, 56:219–246, 2009.

H1N1 influenza

The H1N1 strain of influenza A (also known as swine flu) has spread worldwide since March 2009, when it is believed to have originated in Mexico, and is now of concern in many countries. Interestingly, the average person is likely able to identify H1N1 specifically as an "influenza virus," an awareness that was never very prevalent before. Previously, most individuals were not aware that there are different strains or even different types of influenza viruses. The development of this pandemic has made the H1N1 strain a household term, familiar to adults and children alike.

Background. Influenza is an infectious disease caused by viruses in the family Orthomyxoviridae. These viruses infect mammals and birds and can be transmitted from one type of animal to another. Though influenza is often confused with the common cold, it is a more severe disease and can, in some cases, lead to mortality of those infected. Symptoms in humans include chills, fever, sore throat, muscle pain or myalgia, headache, coughing, weakness, and overall general discomfort. Influenza viruses are transmitted through the air by coughing or sneezing, or through contact with contaminated surfaces (for example, transmission can occur during handshaking).

The viruses that cause influenza are similar in structure, consisting of a roughly spherical protein capsid that is 80–120 nanometers (nm) in diameter, surrounded by a viral envelope (**Fig. 1**). Two main types of glycoproteins are associated with the envelope's surface: hemagglutinin, or H protein, and neuraminidase, or N protein. There are 16 different types of H protein and 9 different types of N protein; each strain is named for the type of H and N proteins present on the envelope of the virus. These proteins are of vital importance in the infection process of these viruses. Hemagglutinin is used by the virus to bind to sialic acid sugar residues that are found on the surfaces of epithelial cells in the human respiratory tract. This allows the virus to be taken into the cell by endocytosis. Neuraminidase is used by the virus after replication, allowing it to escape from the host cell by cleaving the same host sialic acid residues.

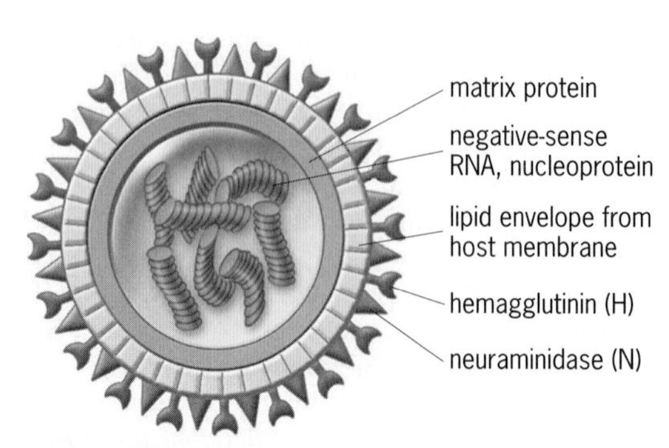

Fig. 1. Schematic drawing of influenza virus. ***(Courtesy of M. K. Cowan and K. P. Talaro, Microbiology: A Systems Approach, 2d ed., McGraw-Hill, New York, 2008)***

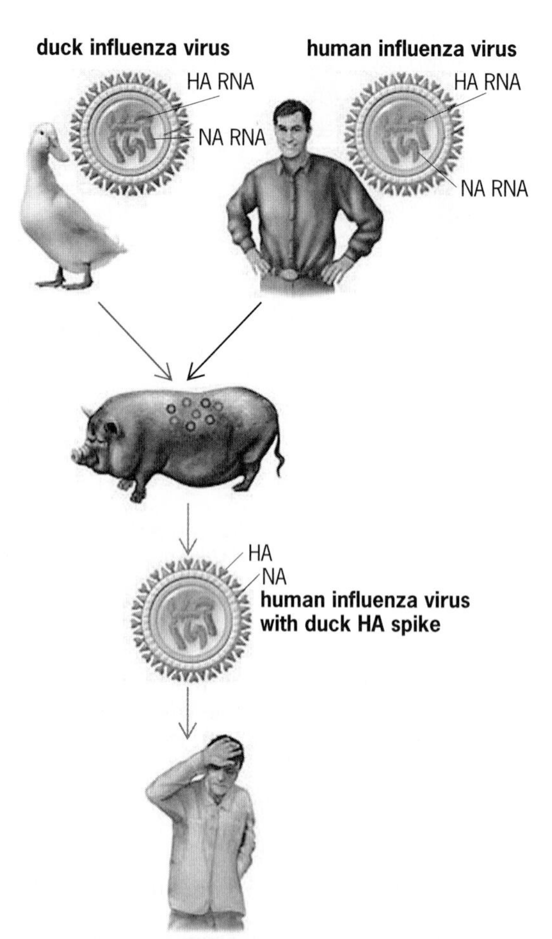

Fig. 2. Antigenic shift event in influenza virus. HA = hemagglutinin; NA = neuraminidase. ***(Courtesy of M. K. Cowan and K. P. Talaro, Microbiology: A Systems Approach, 2d ed., McGraw-Hill, New York, 2008)***

The internal core of the virus contains a genome that consists of seven or eight pieces of RNA that are negative-sense; that is, the RNA sequences must be converted by an RNA-dependent RNA polymerase enzyme to positive-sense RNA strands before they can be used by the host cell machinery. This segmentation allows for the mixing of genetic information when two different viruses enter the same host cell, leading to a sudden change in the type of antigen present in a virus. This allows the new virus to infect naïve hosts, who will lack immunity to the new type of virus. This process is known as antigenic shift and can result in new influenza pandemics (**Fig. 2**).

There are three genera of influenza viruses: influenza A viruses, which are the most virulent or deadly influenza viruses for humans; influenza B viruses, which infect primarily humans but are less dangerous than influenza A infections; and influenza C viruses, which infect humans, dogs, and pigs. Of the different genera, only influenza A viruses have the ability to undergo antigenic shift, in which different influenza strains recombine during infection of a host cell to produce a "new" strain of virus. Antigenic shift led to the development of a new influenza virus in the spring of 2009, which then spread quickly, resulting in the 2009 H1N1 pandemic.

The "Spanish flu" pandemic of 1918 also was caused by an H1N1 strain of influenza; this pandemic led to the deaths of an estimated 20–50 million individuals worldwide. This strain of influenza virus was also unusual in that most of the people who died from it were less than 65 years of age and more than half of those were adults between 20 and 40 years of age. Usually, influenza is more deadly in the very young, such as babies and toddlers, and in the elderly. The fact that the 2009 virus is also an H1N1 strain may account for some of the panic about this particular pandemic.

Development of the H1N1 pandemic. In April 2009, an outbreak of influenza-like illness occurred in Mexico. Shortly thereafter, the Centers for Disease Control (CDC) reported cases of type A/H1N1 influenza in the southwest region of the United States. It soon became clear that the two outbreaks were related, and the World Health Organization (WHO) issued a health advisory about influenza-like illness in the United States and Mexico. The disease spread rapidly in Mexico, causing over 2000 confirmed cases by early May 2009, despite the measures put in place by the Mexican government. These measures included thermal scanners at airports to detect passengers with fever and the distribution of thousands of face masks to the public. A little more than a month later, the WHO declared H1N1 to be a pandemic, the first to be declared since the 1968 Hong Kong flu.

H1N1 traveled from Mexico to the United States, from which it has spread worldwide. More than 214 countries and territories of countries have reported H1N1 cases (**Fig. 3**). More than 18,000 deaths have been attributed to the H1N1 pandemic flu. The 2009 H1N1 strain of influenza differs from the 1918 Spanish flu strain because it causes primarily more severe disease in the very young (children under 5, but especially those under 2 years of age), adults over 65, pregnant women, and those with underlying medical conditions, such as patients with asthma, diabetes, chronic lung disease, or heart disease.

Influenza outbreaks peak in the wintertime; because there are seasonal variations between the Northern and Southern hemispheres, there are two different flu seasons. Therefore, two different vaccines are developed in preparation for each hemisphere. A recent study showed that influenza virus survives longer on surfaces at colder temperatures and that aerosol transmission of the virus is higher at cold temperatures with low relative humidity. This may account for seasonal flu variation in temperate climates, where there is less of a temperature difference from season to season. As a result of the two flu seasons, the spread of H1N1 worldwide has affected the two hemispheres at different times.

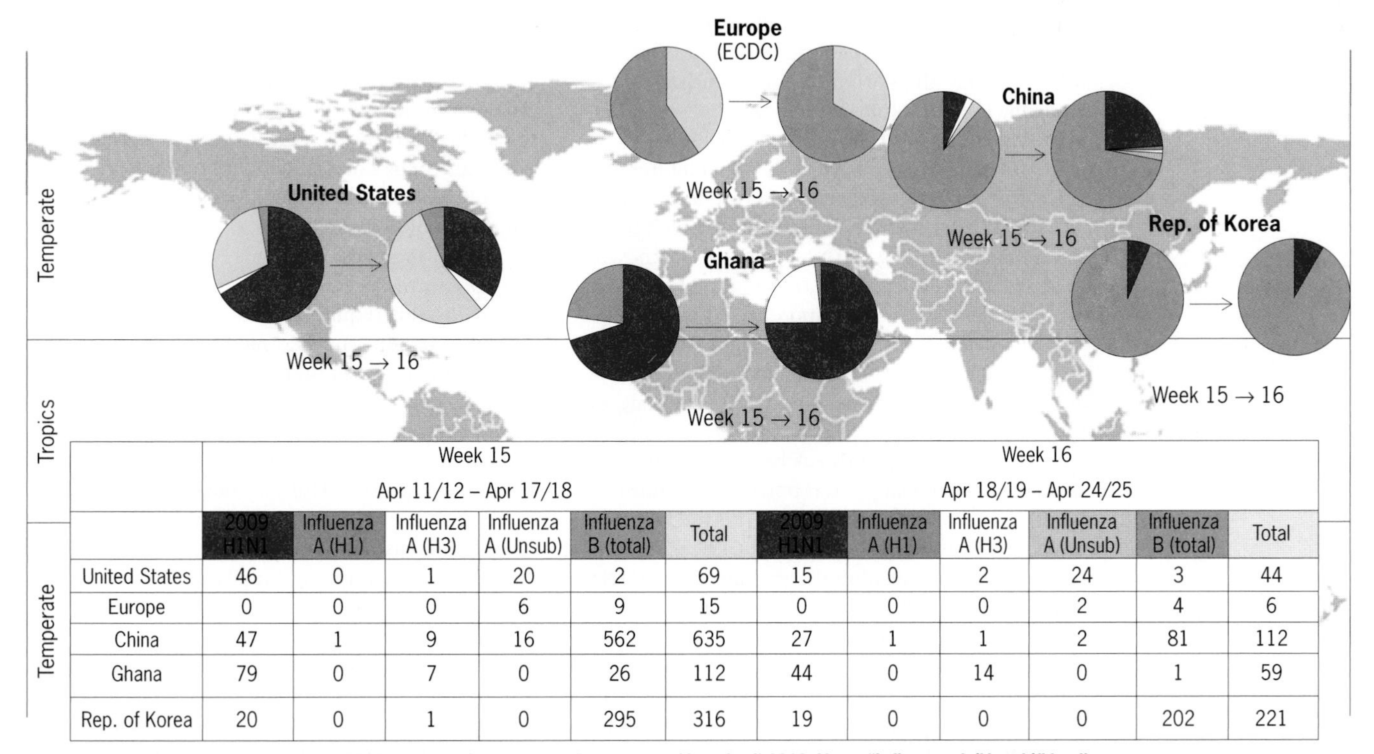

	Week 15 Apr 11/12 – Apr 17/18						Week 16 Apr 18/19 – Apr 24/25					
	2009 H1N1	Influenza A (H1)	Influenza A (H3)	Influenza A (Unsub)	Influenza B (total)	Total	2009 H1N1	Influenza A (H1)	Influenza A (H3)	Influenza A (Unsub)	Influenza B (total)	Total
Temperate												
United States	46	0	1	20	2	69	15	0	2	24	3	44
Europe	0	0	0	6	9	15	0	0	0	2	4	6
China	47	1	9	16	562	635	27	1	1	2	81	112
Ghana	79	0	7	0	26	112	44	0	14	0	1	59
Rep. of Korea	20	0	1	0	295	316	19	0	0	0	202	221

Fig. 3. International co-circulation of 2009 H1N1 and seasonal influenza as of late April 2010. Note: "Influenza A (Unsub)" is all unsubtyped influenza A viruses. It does not include unsubtypable influenza A viruses. (*Source: Centers for Disease Control*)

Prevention and treatment of H1N1 influenza. Influenza is prevented through vaccines that are produced as directed by the WHO, which predicts which virus strains are most likely to be circulating through the population in the upcoming season. As stated previously, two different vaccines are developed, based on the variations between the Southern and Northern hemispheres. After the 2009 H1N1 pandemic was declared, a vaccine based on its characteristics was rushed into production and distribution in an attempt to stem the number of cases that would occur in the United States. Two types of vaccines were developed: a live, attenuated virus vaccine that is administered using a nasal spray, and a killed-virus vaccine that is injected into the muscle of the arm. The use of the live vaccine is limited to immune-competent individuals; it is contraindicated in individuals with underlying medical conditions, because it can cause disease in these patients.

Severe cases of H1N1 influenza are treated using the neuraminidase inhibitors oseltamivir (Tamiflu®) and zanamivir (Relenz®) in individuals who have been symptomatic for no more than two days. A third neuraminidase inhibitor, peramivir, also was approved for use in hospitalized patients with 2009 H1N1 influenza. This treatment was approved for patients who did not respond to the other neuraminidase inhibitors and when the infection in those patients is life-threatening or critical. Amantadine and rimantadine are drugs used to inhibit the M2 ion channel that is required to allow viral uncoating after entry into the host cell; however, these drugs are both contraindicated in H1N1 infection, as the virus appears to be almost universally resistant to these antivirals.

Current status of the H1N1 pandemic. Presently, the most active areas of pandemic influenza are in West Africa, the Caribbean, and Southeast Asia. Pandemic influenza transmission is active in several countries of the tropical Americas, including Cuba and Guatemala. Cases of pneumonia in children under 5 years of age increased in Peru during March and April 2010; however, the extent to which these cases were caused by H1N1 viruses is unknown. In Southeast Asia, cases of H1N1 continue to increase in Malaysia and Singapore, whereas cases in Thailand peaked in March 2010 and have since steadily decreased. H1N1 activity in the United States and Europe has decreased since the beginning of April 2010.

When the H1N1 flu outbreak began in April 2009, the CDC began tracking and reporting lab-confirmed cases of H1N1 in the United States. This tracking was suspended in July 2009, when the CDC started using a system of estimating the prevalence of H1N1 cases by correcting for underascertainment of disease. Using this system, the CDC estimates that between 43 and 88 million cases of 2009 H1N1 infection occurred during the period from April 2009 to March 2010. Hospitalizations attributed to H1N1 infection within the same time period are estimated to have been between 192,000 and 398,000, and the number of deaths due to H1N1 is estimated to have been between 8720 and 18,050. Undoubtedly, case numbers, hospitalizations, and deaths from pandemic H1N1 will increase, especially as epidemiologists predict that the virus will return in the next flu season in both Europe and the United States.

For background information *see* ANIMAL VIRUS; ANTIGEN; DISEASE; EPIDEMIC; EPIDEMIOLOGY; INFECTIOUS DISEASE; INFLUENZA; PUBLIC HEALTH; VACCINATION; VIRULENCE; VIRUS; VIRUS CLASSIFICATION; ZOONOSES in the McGraw-Hill Encyclopedia of Science & Technology. Marcia M. Pierce

Bibliography. M. K. Cowan and K. P. Talaro, *Microbiology: A Systems Approach*, 2d ed., McGraw-Hill, New York, 2008; P. R. Murray, M. A. Pfaller, and K. S. Rosenthal, *Medical Microbiology*, 5th ed., Mosby, St. Louis, 2005; E. Nester et al., *Microbiology: A Human Perspective*, 6th ed., McGraw-Hill, New York, 2008.

Human altruism

Sociobiologist E. O. Wilson called altruism the "central theoretical problem" of the sciences. Humans exhibit a level and diversity of altruism that is unknown among other species. Thus, the study of human altruism has attracted attention from researchers in all the biological and social sciences. Growing empirical evidence increasingly points to the varieties of human altruism, leading to decreasing emphasis on once-standard models that assume narrow self-interest. Such models are now being replaced with new explanations for the emergence and existence of human altruism.

Altruism defined. Definitions differ across disciplines, with some researchers defining altruism according to behavioral or fitness outcomes and others defining it according to motivation. Generally, researchers in the biological sciences define altruism as any behavior that benefits another organism, possibly at a cost to oneself. Similarly, evolutionary scientists consider an altruistic behavior as any behavior that increases another's reproductive fitness at a cost to the organism's own fitness. These definitions do not require that the altruistic act be intended to benefit another. As discussed below, the motivation may be entirely egoistic.

Social psychologists are more likely to focus on the motivation underlying the altruistic act—specifically, whether the helper's goal was to benefit the other person. Thus, in much social scientific research, altruism refers to an action that is motivated to increase another's welfare, even at a cost to oneself. Importantly—and unlike strictly biological definitions—this definition excludes behaviors that may help another but that are motivated primarily by egoism (for example, expectations of praise or material rewards for helping). Note that a person may act out of altruistic motivation and then receive (unanticipated) material benefits as a result of that act. Given that these material benefits did not motivate the action, it would still be classified as altruistic according to the social psychological definition.

Explanations of altruism. There are three primary evolutionary explanations for the existence of altruistic behavior in humans. One is based on *kin altruism*, or actions that benefit one's genetic relatives at a cost to oneself. There is strong evidence that humans tend to favor kin over nonkin, in terms of both actual behaviors and the proximate mechanisms that drive prosocial behavior (for example, love). Cross-cultural laboratory experiments have demonstrated support for theories of kin altruism, showing that participants shoulder higher costs to benefit others as the genetic relatedness of the benefactor and beneficiary increases.

Kin altruism explains altruism toward genetically related others; *reciprocal altruism* can explain helping behavior between nonrelatives. Reciprocal altruists help those who have helped them (or who will help them in the future). Computer-simulation tournaments, field studies, and laboratory experiments show the robustness of reciprocal strategies for generating cooperation between individuals and groups. Furthermore, sociologists and anthropologists have shown that norms governing reciprocity are universal.

A limit of reciprocal altruism approaches is that they cannot explain altruism and cooperation beyond the level of dyads (pairs of individuals). Theories of indirect reciprocity and generalized exchange explain altruism on larger scales. Indirect reciprocity occurs when a person helps someone who has helped another in the past. Note that helping those who have helped others (and withholding help from those who have not) requires some level of knowledge of how someone in need has acted in the past. This may occur through direct observation (in small groups) or reputations that get transmitted via gossip.

The most controversial evolutionary explanation of altruism is based on *group selection*. Group selection approaches explain the emergence of altruism via the tendency for groups composed of altruists to be more apt to survive (and perpetuate altruism) than groups dominated by egoists.

Beyond biological evolution approaches, cultural evolution models (and gene-culture coevolution approaches) describe the processes through which humans learn and transmit altruism via observation and imitation. From this perspective, cultures that contain altruistic norms are more likely to persist than those that contain egoistic norms. Altruistic and prosocial behavior appears to vary substantially cross-culturally, suggesting that altruistic tendencies are in part culturally transmitted.

Evidence for the existence of altruism. There is much evidence for the existence of biological altruism in humans, as well as in other species. Thus, much of the debate about the existence of altruism has centered on the existence of psychological altruism. Given the obvious difficulties of "seeing" motivation, this debate has gone on for some time. However, an increasing variety of findings point to the existence of psychological altruism—that is, some people, under some conditions, are motivated solely to benefit others.

One of the longest-running research programs on psychological altruism is the empathy-altruism hypothesis, most closely associated with C. Daniel Batson and associates. According to this hypothesis, people will act altruistically when they experience empathy for a person in need. A large body of research supports the association between empathy and altruism, and Batson and colleagues have conducted a series of experiments to rule out several egoistic explanations for the empathy-altruism link. For example, one egoistic alternative suggests that helping is associated with empathy because empathizing with someone in need leads to feelings of distress, which can be alleviated by helping the target. However, in several experiments that manipulated both empathy and ease of escaping the situation, those who felt high empathy for someone in need were more likely to help, regardless of whether exiting the situation was easy or difficult. If people were acting solely to reduce their own distress at seeing someone in need, they should be more likely to use their ability to escape as a way of reducing distress when escape is easier than helping. However, rates of helping were equal, regardless of whether avoiding the situation was difficult or easy, as long as people felt empathy for the person in need.

Other research focuses more explicitly on "person × situation" approaches to altruism. In contrast to pure self-interest models (which predict that people will not help unrelated others when there are no reputational benefits for doing so), person × situation approaches predict that prosocials (or altruists) will exhibit greater behavioral consistency (in terms of their level of prosociality) across situations, whereas proselfs (or egoists) will tend to act prosocially primarily when there are situational or reputational incentives for doing so. Although there is some evidence that even subtle reminders of reputation can lead to increased prosociality, recent findings suggest that the altruism of prosocials cannot be explained by a greater sensitivity to reputation. In short, these studies show that at least some people are motivated to benefit others even when there is no incentive to themselves for doing so.

Recent research on neural responses to helping lends a different kind of evidence for the existence of altruistic motivations. Using functional magnetic resonance imaging (fMRI) to monitor brain activity, researchers have shown that helping others increases activity in reward centers of the brain, even when helping is mandatory. Voluntarily helping someone in need may cause one to experience a "warm glow" feeling for having helped. However, mandatory helpers who do not experience the "warm glow" appear to be responding solely to the fact that someone in need of help received it, rather than because they feel good about having played a (voluntary) role in helping. This is consistent with the existence for psychological altruism—being concerned solely for the welfare of someone in need, even when no rewards are gained by helping. It is not clear whether those who are more likely to experience these psychological rewards following mandatory helping are the same types who help even in the absence of reputational benefits.

When, why, and how human altruism occurs is an important area of study in a variety of scientific disciplines. New findings on human altruism go against formerly widespread assumptions that humans act purely out of self-interest. Findings increasingly show that some people, some of the time, will act to benefit others even at a cost to themselves.

For background information *see* BEHAVIOR GENETICS; BRAIN; COGNITION; MOTIVATION; NEUROBIOLOGY; PERSONALITY THEORY; PSYCHOLOGY; SOCIAL MAMMALS; SOCIOBIOLOGY in the McGraw-Hill Encyclopedia of Science & Technology.

Ashley Harrell; Brent T. Simpson

Bibliography. R. Axelrod, *The Evolution of Cooperation*, Basic Books, New York, 1984; C. D. Batson, *The Altruism Question: Toward a Social-Psychological Answer*, Erlbaum, Hillsdale, NJ, 1991; W. T. Harbaugh, U. Mayr, and D. R. Burghart, Neural responses to taxation and voluntary giving reveal motives for charitable donations, *Science*, 316:1622–1625, 2007; B. Simpson and R. Willer, Altruism and indirect reciprocity: The interaction of person and situation in prosocial behavior, *Soc. Psychol. Q.*, 71:37–52, 2008; E. Sober and D. S. Wilson, *Unto Others: The Evolution and Psychology of Unselfish Behavior*, Harvard University Press, Cambridge, MA, 2001.

IceCube neutrino observatory

It was only in the tenth century that, by building a large box camera, the astronomer Alhazen demonstrated once and for all that we "see" by capturing light from objects that emit or simply reflect light. Light is the messenger that brings the heavens to human eyes; eventually telescopes aided our eyes. For most of human history, that is how astronomy was done. Then astronomers discovered the astounding power of sensing the color (wavelength) of messenger photons. High-energy photons with tiny wavelengths, collected by satellites, revealed the most violent processes in the universe; gamma-ray bursts were discovered. Long-wavelength infrared light divulged spectacular views of star-forming regions. After the opening of the electromagnetic spectrum from radio waves to gamma rays, the idea of changing the messenger itself emerged half a century ago with the entry of the neutrino into astronomy. Neutrinos have the potential to be ideal cosmic messengers. Unfortunately, building neutrino telescopes has turned out to be a daunting technical challenge.

Invention and discovery of the neutrino. Matter is made of electrons (e), protons (p), neutrons (n), and neutrinos. Without neutrinos, nuclear interactions that change protons into neutrons and the reverse would be impossible. Stars are powered by nuclear reactors, and when the fuel runs out, they collapse in supernova explosions, releasing the chemical elements that we are made of. The neutrino was invented in the 1930s as the invisible agent that makes the transformation $n \Leftrightarrow p + e^-$ possible; the neutrino's only role was to balance energy between the two sides of the arrow. In order to do that, the neutrino could have no electric charge and no mass, so that it hardly interacted with conventional particle detectors.

A quarter of a century later, Clyde Cowen and Frederick Reines discovered the neutrino by placing 200 L (50 gal) of water a distance of 11 m (36 ft) from the Savannah River nuclear reactor. Despite the huge flux of neutrinos streaming through the detector, only a handful of neutrinos per hour interacted with nuclei in the water. Reines said that he instantly realized that his discovery was a new window on the universe that would open onto the Sun as well as onto the most violent and still enigmatic sources that create cosmic rays, by far the most energetic particles reaching us from space.

Detecting solar and supernova neutrinos. Raymond Davis proved Reines right by placing a railroad car filled with cleaning fluid 1.5 km (0.9 mi) underground in the Homestake Gold Mine in Lead, South Dakota. Neutrinos produced by the Sun's nuclear reactor would convert an atom of chlorine into an atom of argon at a rate of one per day. He found them, but he found only one of the three that he had expected, based on theoretical calculations by John Bahcall. Thirty years later, data from a dozen underground experiments on three continents revealed the origin of the discrepancy: on their way to the detector, the electron neutrinos produced by the Sun transform themselves into muon and tau neutrinos, which escape detection by the cleaning fluid. Oscillating neutrinos imply that neutrinos have mass, the first evidence that the standard model of particle physics is incomplete. Discovering new physics not in the laboratory but in the sky represented a remarkable and totally unexpected achievement for the new field of particle astrophysics.

In 1987, three of these experiments (Kamiokande, Irvine-Michigan-Brookhaven, and Baksan) detected some 20 neutrinos from a supernova explosion in the Large Magellanic Cloud. These 20 events were sufficient to provide the first close-up view of the formation of a neutron star and to constrain the properties of neutrinos themselves (such as their mass), improving on the constraints obtained over decades with accelerator experiments.

Detecting high-energy neutrinos. It should be clear from the foregoing discussion that detecting neutrinos is not easy, especially in the case of the extremely small fluxes of neutrinos from sources beyond the Sun, such as the cosmic accelerators that produce cosmic rays in our own Milky Way Galaxy and in the universe. The neutrinos that these sources emit will reveal nuclear processes far more energetic than those that can be created in earthbound laboratories. Throughout the cosmos, nature accelerates elementary particles to energies in excess of 10^{20} eV, equivalent to a macroscopic energy of 50 J carried by a single elementary particle. We have no idea where these particles originate or how they can be accelerated to such high energies. Because cosmic rays are electrically charged, their paths become scrambled by pervasive galactic magnetic fields, so the directions of their arrival at Earth do not reveal their exact origin. This is why the cosmic-ray puzzle persists almost a century after the discovery of radiation from space by Victor Hess. However, cosmic rays produce electrically neutral neutrinos in collisions with background microwave photons in the vicinity of their sources. If we can detect these neutrinos, the directions from which they came will point back to these sources. A simple calculation shows that it takes a kilometer-scale neutrino detector to observe them.

The problem of how to build such detectors was solved conceptually by the physicist Moisey Markov, who suggested the transformation of large volumes of natural sea or lake water into a particle detector. IceCube is about to realize this concept with a twist: the water is frozen.

IceCube design. The IceCube project transforms natural ultra-transparent Antarctic ice into a particle detector. The ice itself is the detector, a volume of 1 km (0.6 mi) on each side located between depths of 1500 and 2500 m (5000 and 8000 ft). The experiment looks for high-energy neutrinos that travel through the cosmos and mostly stream through the detector without leaving a trace. The roughly one in 10^6 neutrinos that makes a direct hit on a nucleus in the ice will create secondary particles that, unlike the neutrino itself, are recorded by optical sensors viewing the cubic kilometer of natural ice. The sensors detect the faint light pulse resulting from the nuclear interaction initiated by a single neutrino; its origin is the same as that of the characteristic blue glow of water shielding a nuclear reactor. The detailed light pattern observed by the detector reveals the direction of the incident neutrino, making neutrino astronomy possible.

Among the secondary particles produced by the neutrino interactions are muons. These can stream through the ultra-transparent ice for kilometers, leaving a trail of blue Cerenkov photons. (The effective volume of the detector thus exceeds the instrumented volume for muon neutrinos.) So-called Cerenkov radiation is emitted when the atoms, perturbed by the passing electrically charged muon, restore themselves to equilibrium. Because the high-energy muon travels through the detector at close to the speed of light c, it outstrips the Cerenkov photons, which propagate at the speed of light in ice. The latter is reduced to a value of c/n, where n is the index of refraction of ice with a value of 1.4. As a result, the light waves pile up just as water waves form a wake behind a speedboat. This "Cerenkov cone" reveals the direction of the muon that is aligned with

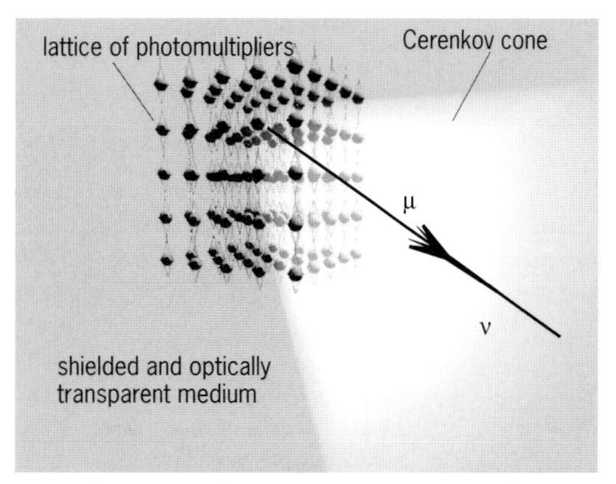

Fig. 1. Conceptual design of a large neutrino detector. A neutrino (ν), selected by the fact that it traveled through the Earth, enters from the lower right and interacts with a nucleus in the ice, producing a muon (μ) that is detected by the wake of Cerenkov photons (Cerenkov cone) that it leaves inside the detector (*IceCube Research Center*).

the neutrino within the resolution of the measurement. By observing muons that have penetrated the Earth, we make sure that IceCube records the arrival directions of neutrinos; no other particle can penetrate Earth and reach a detector at the South Pole from the northern skies (**Fig. 1**).

Suspended within the ice are arrays of light sensors called phototubes or photomultipliers that record photons one at a time. The signals detected by each sensor are transmitted to the surface over 86 cables to which the sensors are attached (**Fig. 2**). The photomultipliers are 25-cm (10-in.) light bulbs, although their functionality is reversed, detecting light and transforming it into electrical signals. The pulses are digitized and sent to computers on top of the ice that process the information and determine

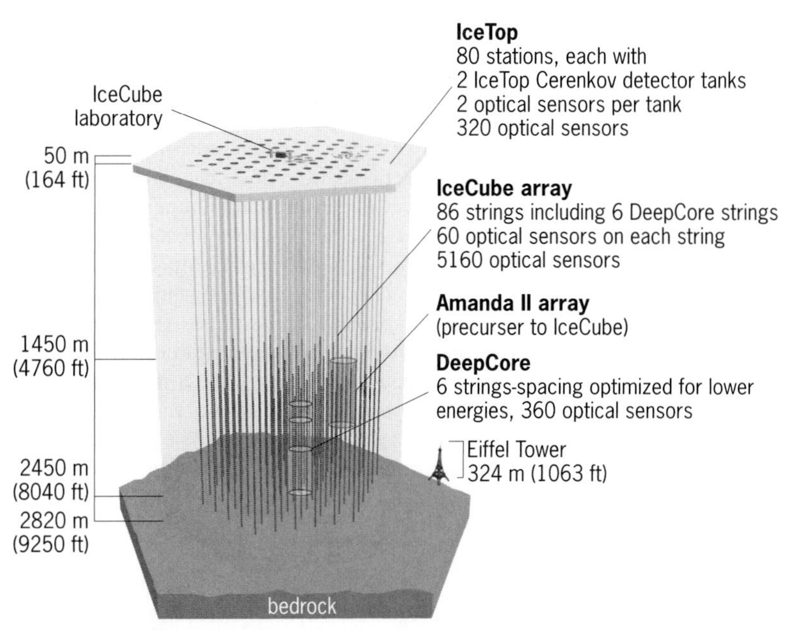

Fig. 2. Design of the completed IceCube neutrino detector, with 5160 optical sensors viewing a cubic kilometer of natural ice. Also shown are the AMANDA and the DeepCore detectors, which are surrounded by IceCube and optimized for low-energy neutrinos (*IceCube Research Center*).

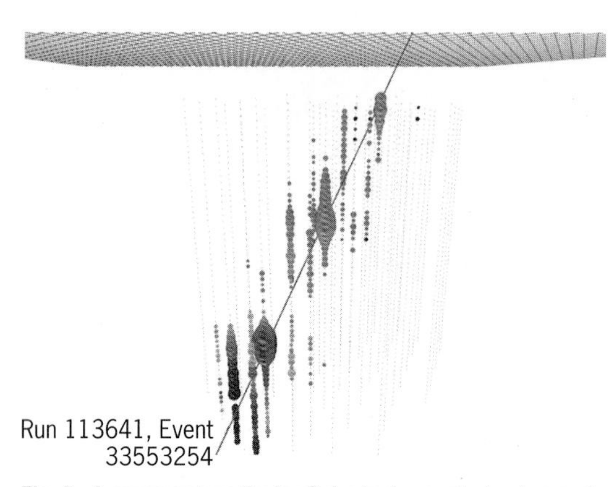

Fig. 3. A muon enters the IceCube instrumented volume at the bottom left and exits at the upper right. Shaded dots indicate light sensors that have detected photons emitted by the muon. The shading of the dots reflects the time of detection, with darker dots early and brighter dots late; their size indicates the number of photons detected. (*IceCube Research Center*)

the direction and energy of the neutrino. This information is then sent by satellite to the operations center located near the campus of the University of Wisconsin in Madison (**Fig. 3**).

IceCube capabilities. When IceCube is fully up and running by spring 2011, it will have a sensitivity to astronomical sources that is almost 100 times superior to that of the previous best telescope, IceCube's predecessor, AMANDA. IceCube should have the sensitivity to finally establish that supernovae are the sources of galactic cosmic rays, confirming the 1932 prediction of Walter Baade and Fritz Zwicky.

IceCube is expected to collect 10^6 events over the next decade, providing the data to study the neutrino itself, including its behavior at energies not within reach of accelerators. DeepCore, an infill of the IceCube array completed in January 2010, lowers the neutrino energy threshold to 10 GeV over a significant instrumented volume of very clear ice in the lower half of the IceCube detector. DeepCore will detect close to 100,000 atmospheric neutrinos per year, two orders of magnitude more than existing experiments like the one near Kamioka, Japan. While the IceCube investigators hope to reproduce the Kamioka detector's historic observation of the disappearance of muon neutrinos produced in the atmosphere, they also have the statistics to show that these reappear as tau neutrinos, thus closing the loop on the oscillations of atmospheric neutrinos.

Finally, IceCube is ready to be a major player in the ultimate astronomical event, the first twenty-first-century galactic supernova. The arrival of neutrinos and gravitational waves, the latter observed by the Advanced LIGO detector, will alert astronomical telescopes, readying them to catch the early light of the supernova a few hours later.

For background information, *see* CERENKOV RADIATION; COSMIC RAYS; GAMMA-RAY BURSTS; INFRARED ASTRONOMY; LIGO (LASER INTERFEROMETER GRAVITATIONAL-WAVE OBSERVATORY); NEUTRINO; NEUTRINO TELESCOPE; PHOTOMULTIPLIER; SOLAR NEUTRINOS; SUPERNOVA; TELESCOPE in the McGraw-Hill Encyclopedia of Science & Technology.

Francis Halzen

Bibliography. J. N. Bahcall and F. Halzen, Neutrino astronomy: The Sun and beyond, *Phys. World*, 9(9):41–45, September 1996; G. B. Gelmini, A. Kusenko, and T. J. Weiler, Through neutrino eyes, *Sci. Amer.*, 302(5):38–45, May 2010; F. Halzen, Neutrino astrophysics experiments beneath the sea and ice, *Science*, 315:66–68, 2007; F. Halzen and S. R. Klein, Astronomy and astrophysics with neutrinos, *Phys. Today*, 61(5):29–35, May 2008.

Induced pluripotent stem cells

Since the first derivation of embryonic stem cells (ESCs) from mouse and human embryos, much enthusiasm has been generated because the pluripotency of the cells (that is, their capacity to generate any cell type in the body) promised to revolutionize the field of regenerative medicine (in which diseased organs or tissues in a patient could be replaced or repaired by transplantation of cultured cells or an in vitro–generated tissue or organ). However, that promise has been often limited by the fact that, if used for transplantation, the origin of the cells from an allogeneic (genetically different but of the same species) donor blastocyst will most likely be recognized as "nonself" by the recipient immune system. In that regard, using ESC-derived cells or tissues for transplantation would be considered the same as using organs for transplantation from an allogeneic donor, with the need for lifetime immunosuppression and the associated complications. In addition, the fact that ESCs typically require the use and destruction of blastocyst-staged human embryos has elicited significant ethical considerations that have, in some cases, halted the progress of ESC research. All these obstacles have been solved by a single major discovery made by the team of Shinya Yamanaka in Japan. About a decade ago, Yamanaka started working on the possibility of converting adult somatic cells into pluripotent stem cells. To do so, he focused on a group of genes that were uniquely or highly expressed in ESCs. His team originally used retroviruses to overexpress 24 genes in mouse fibroblasts and amazingly found a few weeks later that colonies highly similar to normal mouse ESCs emerged in the culture plates. Unlike the starting fibroblast population, the resulting cells could be maintained indefinitely in culture and could function as ESCs. Moreover, they went further to demonstrate that only a cocktail of 4 out of the 24 genes, named *Oct4*, *Klf4*, *Sox2*, and *cMyc* (all of which encode proteins that function as transcription factors to regulate the expression of different sets of genes), were sufficient to induce nuclear reprogramming of somatic cells to bring them "back in time" to a primordial, pluripotent state. To distinguish these new cells from their embryonic counterparts, they were named induced pluripotent stem cells or iPS cells. It took a few months for the same team as well as other teams to show that the same principles could be applied to reprogram human somatic cells, a major step toward the long-sought use of pluripotent stem cells in regenerative medicine (**Fig. 1**). The functional demonstration that iPS cells were truly pluripotent was their ability to contribute to all types of tissues when injected into immunocompromised mice to form teratomas (tumors that contain tissues from all three primary germ layers). Most significantly, mouse iPS cells were capable of generating an entire mouse when injected into a mouse blastocyst.

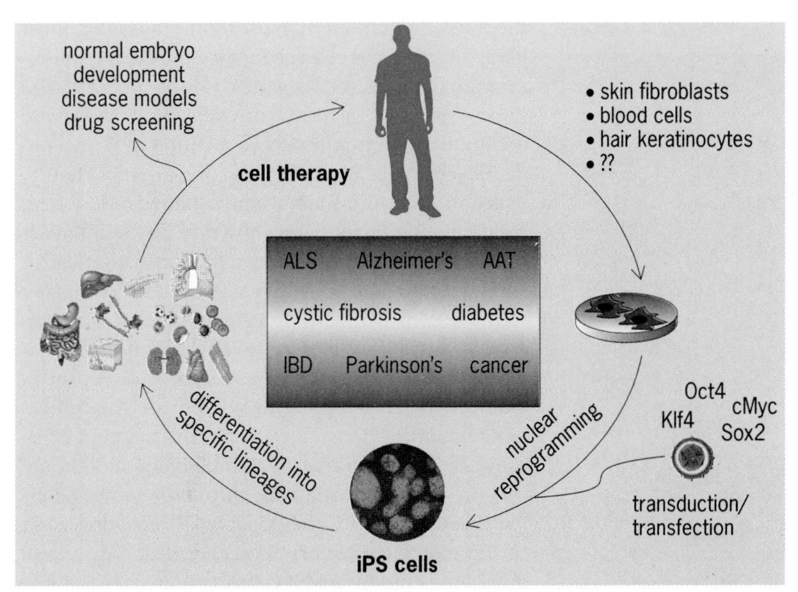

Fig. 1. The potential of iPS cell research. Adult somatic cells (such as skin fibroblasts and peripheral blood cells) can be obtained easily from any individual and can serve as the source for the generation of personalized pluripotent stem cells. Then, iPS cells can be employed for the generation of specific tissues or organs that can serve for the study of any disease, screen for new therapeutic drugs, or have use in cell replacement therapy. Representative target diseases that could benefit from this methodology are shown. AAT = alpha-1 antitrypsin deficiency; ALS = amyotrophic lateral sclerosis; IBD = inflammatory bowel disease.

Genes and cells of origin. Although the original study was performed using fibroblasts, researchers have shown that different types of cells, including cells from stomach, pancreas, liver, and even blood lymphocytes, are capable of being reprogrammed to become iPS cells. Using genetic tracing (for instance, a β-cell trace for reprogrammed pancreatic cells) or a defined (VDJ) rearrangement (of an antibody-encoding gene) that takes place during the development of B lymphocytes for reprogrammed lymphocytes, the source of cell origin as a fully differentiated somatic cell was categorically confirmed. These studies demonstrated that virtually any cell type could be potentially a target for reprogramming, ruling out the idea that a rare progenitor or stem cell was at the basis of iPS cell generation. In an attempt to understand the biology behind the reprogramming process, several groups have attempted to decipher the role of each of the individual reprogramming factors. Although it is still not fully understood, it appears that cMyc has a different function from the other three, Oct4, Klf4, and Sox2. During the earlier stages of nuclear reprogramming, cMyc is responsible for inducing downregulation of genes commonly expressed by differentiated somatic cells and at the same time triggering cellular metabolic changes. Oct4, Klf4, and Sox2 appear to be important in establishing and maintaining the stem-cell gene expression program. In this regard, however, several laboratories have shown that cMyc is dispensable for the reprogramming process, but at the expense of significantly lowering reprogramming efficiency. Moreover, when using specific cells such as neural stem cells as sources for reprogramming, iPS cells can be derived using two of the reprogramming factors or even a single factor, Oct4. The reason for this may well be the fact that these cells normally express high levels of endogenous Sox2, hence obviating the need for its exogenous expression. Regarding the dynamics of the reprogramming process, it is a relatively slow process, requiring at least 8–10 days of sustained expression of the reprogramming factors. During that time, the endogenous loci of genes that are important in establishing a stem-cell program are remodeled to become actively expressed, whereas the expression of the reprogramming genes introduced through the viral vectors becomes silent. This feature might be particularly important for appropriate differentiation of the iPS cells.

Methodology as the key for the future use of iPS cells in regenerative medicine. It is now widely accepted that the permanent presence of the reprogramming factor transgenes within the genome of the established iPS cells, in which the reprogramming genes are not controlled by their natural regulatory elements but rather by viral or other regulatory elements, represents a severe obstacle for the application of this technology to the clinical arena. The original studies used individual integrating retroviruses, and the generated iPS cells contained multiple viral integrations. Importantly, a high percentage of mice generated with those iPS cells developed tumors later in life, in most cases associated with the reactivation of the transduced oncogene, cMyc. Furthermore, the permanent overexpression of the reprogramming factors within iPS cells may interfere with their appropriate induced differentiation.

Several methodologies have been employed to improve the overall efficiency of reprogramming as well as to obtain iPS cells that are free of exogenous transgenes introduced by the retroviral vector. The use of nonintegrating adenoviruses to express all reprogramming factors was the first demonstration that nuclear reprogramming could be achieved without any integration, albeit at very low efficiency. Similarly, the use of excisable transposons (genetic elements capable of moving to a new chromosomal location) and the use of episomal vectors (extrachromosomal replicating units) have been reported as being capable of inducing reprogramming, but again at the expense of efficiency. The use of a single lentiviral vector (derived from a class of retroviruses such as HIV, which is capable of integrating into the genome of nonreplicating cells) that expresses all reprogramming genes from a single contiguous piece of DNA (stem-cell cassette) has proven to be the most efficient method for reprogramming, most likely because every transduced cell receives all reprogramming factors together. More recently, the same single vector has been excised using genetic techniques (such as loxP/Cre technology), generating iPS cells that are free of transgenes. Importantly, excision of the introduced genes that encode the reprogramming factors was shown to be essential for optimizing appropriate lineage differentiation, an attribute that will have significant implications for the future use of iPS cells and their progeny in regenerative medicine. Ideally, though, the induction of reprogramming by chemical means or direct introduction of purified proteins will ensure that no genetic modification occurs at all in the targeted cells. In this regard, different small molecules and chemicals (in most cases affecting chromatin remodeling) have been shown to be capable of replacing each of the individual factors. However, full reprogramming has yet to be achieved using only chemicals or small molecules. Two research groups have reported the generation of mouse and human iPS cells using transfection of purified proteins, but the efficiency of reprogramming was exceedingly low. Hence, its general application will depend on improving current technologies.

In summary, the choice of methodology to achieve appropriate nuclear reprogramming and generation of iPS cells will need to be made on the basis of several considerations (**Fig. 2**). Most likely, a balance between practical efficiency and the risk of inducing genetic modifications will need to be considered, depending on the individual goal of each laboratory. However, the speed at which this field has been evolving and the plethora of groups around the world that are now focusing on iPS cells will most likely shape their future applications and the ways in which results are achieved.

Clinical application of iPS cells. Several months after the initial discovery by Yamanaka and coworkers, further investigations demonstrated the potential application of iPS cells for medical therapy, using mouse models of sickle-cell anemia and Parkinson's disease. Similarly, phenotypic correction of murine hemophilia A was shown using iPS cell–based therapy. These studies, while serving as proofs of principle, exemplify the future potential applications of pluripotent stem cells in clinical therapy (Fig. 1). This holds especially true in the case of monogenic diseases, in which isolated fibroblasts or iPS cells derived from patients could be the target for gene correction before proceeding with differentiation and regeneration of the affected tissue. Several laboratories have already generated iPS cells from patients with a variety of human diseases, including amyotrophic lateral sclerosis (ALS), Parkinson's disease, Fanconi's anemia, severe combined immunodeficiency, muscular dystrophy, spinal muscular atrophy (SMA), diabetes, and Down syndrome. In most cases, these diseased iPS cells have been characterized as being pluripotent, but their specific differentiation toward a cell type or tissue of interest is still unknown, with a few exceptions worth noting. ALS-specific iPS cells have been shown to be capable of differentiating into spinal motor neurons, but no specific disease-like phenotype has been reported. Similarly, iPS cells have been generated from several patients suffering from Parkinson's disease. When these iPS cells were differentiated into dopaminergic neurons, they were undistinguishable from normal controls. Why there was no disease phenotype was not clear, but it was suggested that long-term

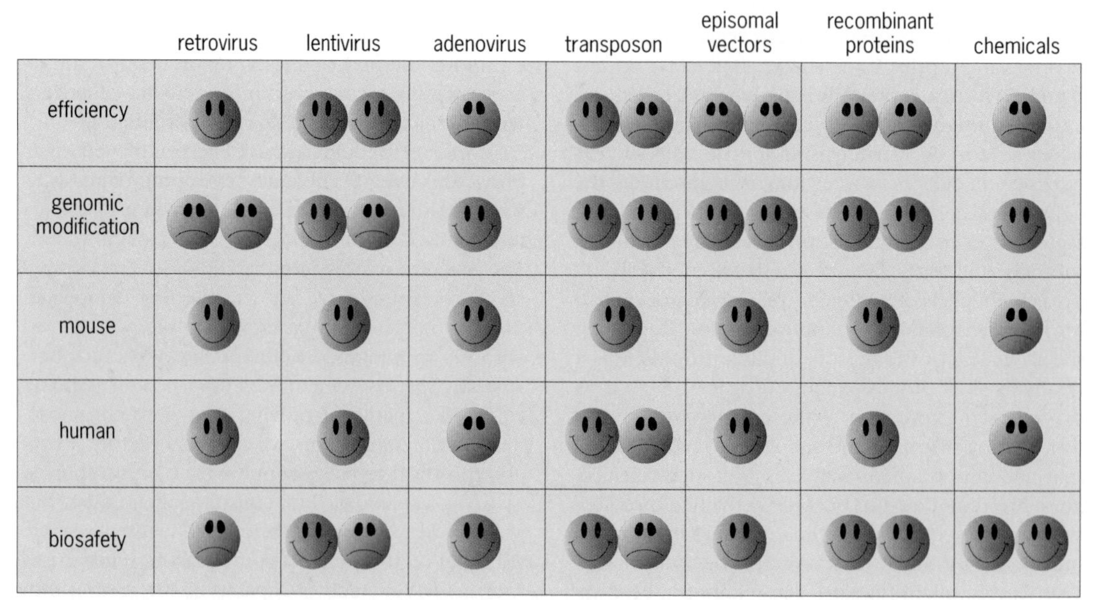

Fig. 2. Methodologies for the generation of iPS cells. A comparison of most methods published to date for the generation of iPS cells is shown. Their overall efficiency, potential to generate genomic modifications, ability to reprogram mouse and human cells, and their biosafety are summarized. Note that full reprogramming using only chemicals has not been reported yet.

cultures may be necessary to recapitulate the clinical effects that occur in chronic diseases such as Parkinson's disease. Alternatively, it may be related to the intrinsic biology of the reprogramming process. In contrast to these studies, others have shown that iPS cells generated from a patient with SMA were capable of recapitulating some of the cellular phenotypes characteristic of the disease upon differentiation into motor neurons, proposing that this would serve to model the pathology seen in a genetically inherited disease. It was further shown that iPS cells derived from a patient with familial dysautonomia (FD), an example of a monogenic disease, can be used to generate unlimited amounts of neural crest precursors that not only evidenced specific defects associated with the disease but also were capable of responding favorably, although in a partial manner, to a drug treatment. Finally, in a study of patients with Fanconi's anemia, fibroblasts were transduced with a lentivirus expressing a wild-type version of the *FANCA* gene that is usually affected in these patients. It was shown that correction of the mutation was a prerequisite for normal reprogramming to occur, and corrected iPS cells were capable of differentiating into hematopoietic lineages similar to normal controls.

Thus, the discovery that easily accessible somatic cells can be induced to turn their biological clocks back and become pluripotent has opened a new window for the future of regenerative medicine. For the first time, it feels like it will happen in our lifetime.

For background information *see* CELL DIFFERENTIATION; CELL LINEAGE; EMBRYOLOGY; EMBRYONIC DIFFERENTIATION; GENE; GENETIC ENGINEERING; REGENERATIVE BIOLOGY; SOMATIC CELL GENETICS; STEM CELLS; TRANSPLANTATION BIOLOGY in the McGraw-Hill Encyclopedia of Science & Technology.

Gustavo Mostoslavsky

Bibliography. E. Kiskinis and K. Eggan, Progress toward the clinical application of patient-specific pluripotent stem cells, *J. Clin. Invest.*, 120:51–59, 2010; N. Maherali and K. Hochedlinger, Guidelines and techniques for the generation of induced pluripotent stem cells, *Cell Stem Cell*, 3:595–605, 2008; K. Takahashi and S. Yamanaka, Induction of pluripotent stem cells from mouse embryonic and adult fibroblast cultures by defined factors, *Cell*, 126:663–676, 2006; S. Yamanaka, A fresh look at iPS cells, *Cell*, 137:13–17, 2009; S. Yamanaka, Ekiden to iPS cells, *Nature Med.*, 15:1145–1148, 2009.

Infectious disease control in animal shelters

Since the first animal shelter opened in 1866 in New York City, over 5000 organizations have been established across the United States to protect and promote the welfare of homeless animals. Each year, millions of animals find refuge in shelters prior to entering permanent homes, yet many animals still end their lives in these facilities as a result of causes that could be prevented. Veterinarians in the emerging specialty of shelter medicine are working to understand the precise needs of companion animals in the shelter environment. Because shelters are often limited in their resources, housing units may contain multiple animals, stress levels may be high, and there is often little positive human interaction; the result is an environment that harbors disease. Although research and consultations conducted over the past decade have led to marked improvements in the control of infectious disease in shelters, ongoing progress must be made in disseminating this knowledge to local shelters.

An effective population management plan decreases disease incidence. Sound population management is the key factor that underlies control of infectious disease in animal shelters. The goal of most shelters is to move animals through the facility as expeditiously as possible to either good permanent homes or appropriate rescue organizations. To accomplish this goal, shelters utilize a population management or flow-through plan. A well-run shelter with an effective population management plan has systems in place to ensure that animals are quickly processed, assessed, and directed to an appropriate end point. Ultimately, this decreases each animal's length of stay or time spent in the shelter, leading to fewer opportunities of exposure to disease and more lives saved.

In order to assess the success of population plans and evaluate trends, shelter personnel must collect detailed data on the animals entering and leaving the facility over an extended period of time. Shelter personnel calculate staffing levels and holding capacity to ascertain availability of resources in the form of both personnel and housing units to properly care for the animals entering their facility. The demands

Fig. 1. Crowded conditions at one animal shelter. Individual dog runs contain four or five dogs.

Canine core vaccines	Feline core vaccines
• distemper (CDV) • parvovirus (CPV) • adenovirus-2 (CAV-2/ hepatitis) • parainfluenza (CPiV) • *Bordetella bronchiseptica*	• feline herpesvirus-1 (feline viral rhinotracheitis/FHV-1) • feline calicivirus (FCV) • feline panleukopenia (FPV)

Fig. 2. Core vaccines administered at intake in animal shelters.

that the population places on staff and the facility must be evaluated; otherwise, the shelter risks creating an environment of severe crowding (**Fig. 1**), increased length of stay, and a diminished capacity to provide animal care. Calculation of the required stray holding capacity (RSHC) determines the number of appropriate housing units to be allocated to animals in their required stray holding period. This number is based on the average expected number of animals per day for any given month (monthly daily average; MDA) and the minimum expected holding period for that shelter (RSHC = MDA stray intake × number of required days for stray hold). Likewise, with accurate data, the adoption-driven capacity of a shelter can be calculated to determine the optimum number of housing units for adoption areas, using the average length of time that an animal will wait to be adopted based on previous data and the staff's ability to care for the animals. On the basis of these calculations and others, a shelter can optimally maintain its animal population and provide adequate care, thereby preventing infectious disease outbreaks.

Intake procedures act as the first line of defense. The point of entry to a shelter is also the first site of potential disease transmission. Intake procedures, or those procedures conducted immediately upon entry, are the first line of defense against the spread of infectious agents. Intake procedures should include a brief physical examination, so that animals with illness are flagged and appropriate steps taken to ensure that they are not introduced to the general population. After a general assessment, staff should check for ownership by scanning for microchips. Animals are then vaccinated and dewormed prior to housing.

When processing newly admitted animals, it is important for staff to be cognizant of transmission of disease between animals. Intake areas should be designated places that can be easily disinfected between animal visits. Ideally, there should be separate intake areas for cats and dogs to minimize stress. Intake areas should be properly equipped with refrigerators to keep vaccines at appropriate temperatures, microchip scanning wands, computer stations to input information, and disinfection supplies. Once the animal is processed, it should be housed immediately in a designated area. There should be no lag time between time of arrival and initiation of intake procedures. All animals should be vaccinated and dewormed prior to housing in a designated area.

Research has demonstrated that administration of an effective vaccine upon entry into a shelter decreases significantly the incidence of disease. In fact, some vaccines, such as that of canine distemper virus, may be effective within minutes of administration. If animals are coming to a shelter from known high-risk areas (that is, transfer from locations where disease is endemic), it may be best to vaccinate at the shelter of origin so that they are less likely to enter the facility harboring an infectious agent. Core vaccines administered at intake in shelters are presented in **Fig. 2**.

Vaccines are available in various forms: modified live virus (MLV), killed, or recombinant. Killed vaccines are not recommended in shelters because their efficacy requires a booster at 3–4 weeks after the initial vaccination. Because of the higher likelihood of exposure to infectious agents in a shelter than in the owner's home, the recommended dosing schedules for young animals is more frequent and extended than in private small-animal practice. A typical recommended dosing schedule is shown in **Fig. 3**.

Core vaccine and deworming schedule for shelter puppies under 5 months	Core vaccine and deworming schedule for shelter kittens under 5 months
MLV or Recombinant DHPP • Administer subcutaneously every 2 weeks, starting 4 weeks of age, up to 18–20 weeks	MLV FVRCP • Administer subcutaneously every 2 weeks, starting 4 weeks of age, up to 18–20 weeks
Intranasal MLV *Bordetella* • At intake or at 2–4 weeks	
Roundworm treatment • Give orally every 2 weeks, starting 2 weeks of age, up to 16 weeks • If coccidiosis is an endemic problem, treat all puppies at the time of intake	Roundworm treatment • Give orally every 2 weeks, starting 2 weeks of age, up to 16 weeks • If coccidiosis is an endemic problem, treat all kittens at the time of intake

Fig. 3. A recommended dosing schedule for core vaccines and deworming. MLV = modified live virus; DHPP = distemper, hepatitis, parainfluenza, and parvovirus (vaccine); FVRCP = feline viral rhinotracheitis, calicivirus, and panleukopenia (vaccine).

Appropriate medication at intake also decreases the incidence of parasite transmission. Standard intake deworming procedures include treatments for roundworms (pyrantel pamoate). If coccidiosis is an endemic puppy or kitten problem, effective preventive treatment is recommended (ponazuril). In addition, animals with heavy flea infestations receive flea and tapeworm medications. Decreasing parasite burden increases the likelihood that animals remain healthy during their shelter stay.

Appropriate facility design reduces stress and improves health. Inadequate housing can lead to infectious disease spread, increased length of stay, and increased need for euthanasia for medical reasons. Current studies in cats demonstrate that stress of

Fig. 4. A hole made between two stainless steel housing units [2 ft × 2 ft (0.6 m × 0.6 m)].

cramped housing units leads to upper respiratory tract infections (feline URI) by triggering feline herpesvirus. Cats housed in large housing units with ample room to display feline behaviors, such as stretching, scratching, and pouncing, and in which cat litter is separate from bedding and food, are less likely to develop feline URI. This simple realization is leading shelter personnel to reassess feline housing units and to create new and innovative ways of increasing space for cats. One novel interim solution for increasing existing space is the cutting of apertures between existing cages to create a two-compartment "suite," where cat litter can be kept separate from bedding (**Fig. 4**). Simple enrichment options, such as provision of hiding boxes for cats, further aid in decreasing stress and improving health.

Housing units should be easy to disinfect, but need not be completely disinfected daily. Ideally, spot cleaning to remove soiled materials should occur on a daily basis. Although daily deep cleaning might appear to be desirable to prevent disease, the stress that it places on cats actually increases risk for contracting illness. Feline housing units should be thoroughly disinfected only when occupants are changed or if completely soiled.

Canine housing units should be designed for increased ease of cleaning, while providing a comfortable environment. Single-occupancy units with double-sided runs connected by guillotine doors are ideal for most shelters that cannot accommodate larger "real-life" housing rooms. Housing units should be large enough to contain bedding that can be kept separate from the area for elimination. Double-sided guillotine runs can minimize disease spread by decreasing animal handling and potential contact between dogs. Dogs can be placed on one side of the run while the other is disinfected. It is important to ensure that dogs are housed in dry areas because moisture increases likelihood of pathogen multiplication. Appropriate disinfectants that target the primary agents affecting animal shelters should be used. This is important because many labeled disinfectants (including quaternary ammonium compounds) are not truly effective against some of the more resistant agents, including canine parvovirus.

If an animal does contract an infectious agent, it is important to ensure there is minimal exposure to the rest of the population. Appropriate isolation and segregation policies will minimize disease transmission. Animals diagnosed with infectious disease should be housed in a separate area with separate staff and cleaning supplies. In general, puppies and kittens should be housed apart from adult animals because they are more susceptible to infection. Separate staff should be assigned to these animals as well, or the most susceptible population should be handled prior to entering the general population. Facilities not equipped to handle serious infectious disease should have other procedures in place to ensure that ill animals are either transferred to emergency centers or humanely euthanized, thus minimizing exposure to other animals.

Although some infectious disease is inevitable in a shelter population, simple steps will minimize spread and reduce transmission. Recent understanding of the importance of population management has led to policies that are less reactive and more proactive in disease prevention. Carefully orchestrated intake procedures, improved facility design, and appropriate isolation facilities will improve the health of animals in shelters, thereby increasing their chances of adoption.

For background information *see* CANINE DISTEMPER; CANINE PARVOVIRUS INFECTION; DISEASE; DISEASE ECOLOGY; EPIDEMIOLOGY; FELINE INFECTIOUS PERITONITIS; FELINE PANLEUKOPENIA; INFECTIOUS DISEASE; KENNEL COUGH; VACCINATION in the McGraw-Hill Encyclopedia of Science & Technology.

Jyothi V. Robertson

Bibliography. C. Greene, *Infectious Diseases of the Dog and Cat*, 3d ed., Saunders, Philadelphia, 2006; K. Hurley and L. Miller, *Infectious Disease Management in Animal Shelters*, Wiley-Blackwell, Ames, IA, 2009; M. A. Kennedy et al., Virucidal efficacy of the newer quaternary ammonium compounds, *J. Am. Anim. Hosp. Assoc.*, 31(3):254–258, 1995; L. J. Larson and R. D. Schultz, Effect of vaccination with recombinant canine distemper virus vaccine immediately before exposure under shelter-like conditions, *Vet. Ther.*, 7(2):113–118, 2006; L. Miller and S. Zawistowski, *Shelter Medicine for Veterinarians and Staff*, Wiley-Blackwell, Ames, IA, 2004; I. R. Tizzard, Vaccination and vaccines, pp. 235–252, in I. R. Tizzard (ed.), *Veterinary Immunology*, 6th ed., Saunders, Philadelphia, 2000.

Inkjet-printed paper-based antennas and RFIDs

Radio-frequency identification (RFID) is gaining in popularity, especially as we head toward a world of ubiquitous computing and of everyday objects interconnected by wireless sensor networks (the Internet of Things). Automatic identification systems have become not only an important aspect of technology but also a part of daily life. We use RFID in cars, transportation systems, and access points, and for simple transactions; We also need RFID in our global supply chain and logistics systems, as well as in tracking and locating applications. Over the past few years, there has been an increasing focus on wearable electronics, including personal wireless body-area networks (WBANs) that provides medical, lifestyle, assisted-living, mobile-computing, and tracking functions for the user. Sensing and identification are the key functions of such conformal applications. RFID technology will play an important role in this technology, since most business and manufacturing identification needs have supported the development and adoption of smaller, cheaper, and longer-range RFID tags. Solutions compatible with RFID standards find quicker acceptance and see faster improvement. This article will identify the key technologies for realizing "green" paper-based antennas and RFIDs using inkjet-printing technologies.

Principles of operation of RFIDs. RFIDs use electromagnetic waves to transmit and receive information stored in a tag or transponder to and from a reader (**Fig. 1**). This technology has several benefits over conventional methods of identification (for example, bar codes), such as a higher read range, faster data transfer, the ability of RFID tags to be embedded within objects, the absence of any requirement for line-of-sight communication, and the ability to read an enormous number of tags simultaneously. A typical RFID system consists of RFID tags, an RFID reader, and middleware. Each individual RFID tag is identified by the reader over high-frequency (HF), ultrahigh-frequency (UHF), or radio-frequency (RF) bands in the radio spectrum. While RFID tags come in different shapes, have different dimensions, and may feature different capabilities, all RFID tags have the following essential components: tag antenna, integrated circuit (IC), and substrate (**Fig. 2**). Lately, there have been a variety of RFID-enabled wireless sensors in numerous applications ranging from "smart skin" and biomonitoring, to "smart energy" and geolocation applications.

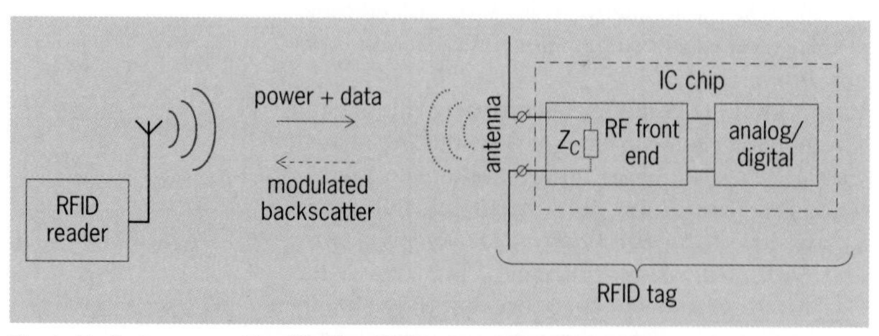

Fig. 1. Radio-frequency identification (RFID) system diagram, showing a reader with antenna, and sensor with antenna and integrated-circuit (IC) chip.

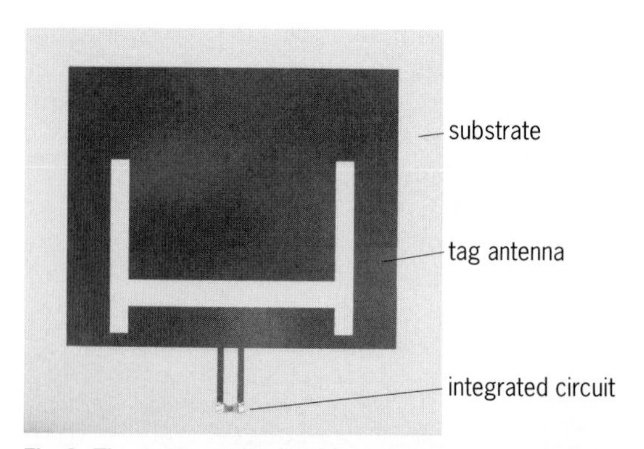

Fig. 2. The components of a passive metal-patch RFID tag. The tag antenna is inkjet printed on a flexible organic substrate (for example, paper), with an integrated-circuit chip assembled in the bottom center of the metal trace. These three essential components together make the RFID tag functional.

In RFID applications, the wireless signal emitted by the RFID reader travels outward and encounters (illuminates) the antenna element in the tag. This is very similar to radar systems that operate on the backscattering principle by detecting the reflection of the illumination wave by the target objects, especially those with dimensions greater than half the wavelength of the incident wave. An electromagnetic wave is transmitted from the reader antenna, reaches the tag antenna, and is transferred to its feeding lines in the form of an alternating-current (ac) sinusoidal continuous wave, thus requiring a conversion from ac power to direct current (dc). After rectification by diode topologies, this power can be used to "turn on" the tag's integrated-circuit chip. A proportion of the incoming radio-frequency energy is reflected by the antenna and reradiated outward. The amount of reflected energy can be influenced by altering the load connected to the antenna. In order to transmit data from the tag to the reader, a load impedance inside the integrated-circuit chip, connected in parallel with the antenna, is switched on and off in time with the data stream to be transmitted. The strength of the signal reflected from the tag can thus be modulated. Once the data in the integrated circuit are modulated and encoded, they are transmitted back to the reader. The reader then decodes and demodulates the modulated data and retrieves the required information. This mechanism of modulated backscatter coupling is widely used, especially for long-range UHF RFID systems.

Paper-based RFID. For wearable or mountable RFID applications, special requirements arise for the substrate material. The substrate needs to be flexible to fit the human body or the mounting topology and should preferably be organically compatible or inkjet-printed to allow for good integrability and ruggedness. Several aspects make paper an excellent candidate for low-cost and environmentally friendly RFID applications. As an organic-based substrate, paper is widely available. The high demand and the mass production of paper make it one of the cheapest materials in use, if not the cheapest. From a manufacturing point of view, paper is well suited for reel-to-reel processing. Thus, mass-fabricating RFID inlays on paper is feasible for an inkjet printing process for RFID antennas, followed by assembly of integrated circuits or other microelectronic devices.

Other qualities of paper contribute to its being the best inexpensive substrate. It can feature a low surface profile. When it is coated with certain plastics (such as those on photographic paper), it is commonly referred to as inkjet paper. In addition, paper is suitable for fast printing processes, such as direct-write methodologies or inkjet printing of electronics, instead of the traditional metal-etching techniques that are commonly used in radio-frequency and wireless electronics. As a fast process, inkjet printing can be used efficiently in conjunction with the appropriate bonding methods to produce multilayer electronics and modules on or in paper. This also enables critical wireless components such as antennas, integrated circuits, memory, batteries, and sensors to be easily embedded in paper modules. In addition, paper can be made hydrophobic and fire-retardant by adding certain textiles or thin films to it, a procedure that easily resolves any moisture-absorbing issues that fiber-based materials suffer from. Paper is one of the most environmentally friendly materials, and the proposed approach could potentially set the foundation for the first generation of truly green radio-frequency electronics and modules operating at frequencies up to at least 10 GHz.

There are a wide range of available paper types that vary in density, coating, thickness, and texture, thus differing in dielectric properties, such as dielectric constant (in the range of 2.8–3.2) and dielectric loss tangent. These dielectric properties are the foundation of radio-frequency parameters needed by the design engineer, so the characterization of paper substrates is an essential step in RFID-on-paper designs. Unlike etching, which is a subtractive method of removing unwanted metal from the substrate's surface, inkjet printing deposits a single ink droplet from the nozzle at the desired position. Therefore, no waste is created, resulting in an economical fabrication process. For example, after a silver nanoparticle droplet is driven through the nozzle, it is necessary for a sintering process to follow, whereby these particles are bound together by heating but not melting them, and excess solvent and material impurities are removed. Another benefit provided by the sintering process is an increase in the bond of the deposit with the paper substrate. The conductivity of the conductive ink has been observed to vary from 0.4 to 2.5×10^7 siemens per meter for frequencies up to 10 GHz, depending on the curing temperature and time. Inkjet printing accuracy down to 1 picoliter (pL or 10^{-12} L) has been achieved. **Figure 3** illustrates a UHF RFID tag fabricated by

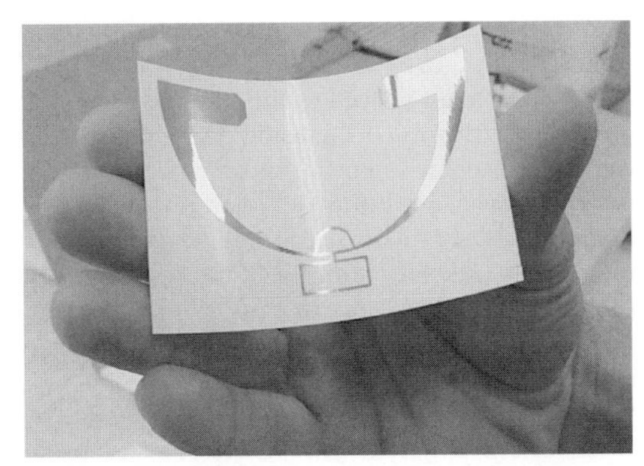

Fig. 3. Inkjet-printed RFID tag on flexible photographic paper-based substrate.

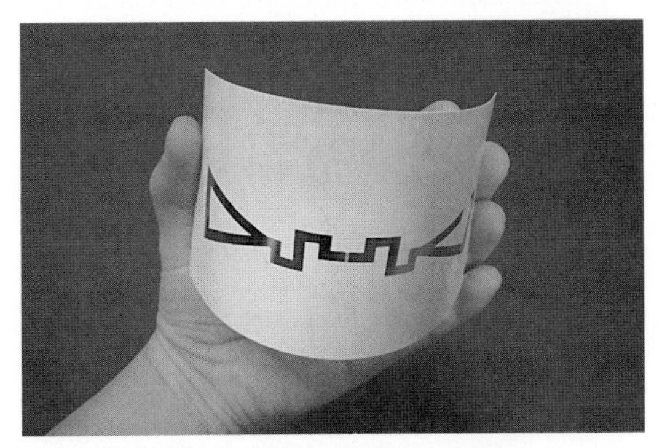

Fig. 4. Conformal paper-based RFID tag with an inkjet-printed carbon-nanotube film in the center.

an inkjet printing method on a photographic-paper substrate.

RFID-enabled sensors. One of the several unique features of inkjet-printed paper-based RFID technology is its ability to easily integrate or print nanostructures and additional functionalities, such as power scavengers and sensors. In this way, inexpensive and reliable RFID-enabled sensors can be easily printed and realized, enabling an ever increasing "cognitive intelligence" and potentially setting the foundation for the Internet of Things. **Figure 4** shows an example of an inkjet-printed RFID-enabled gas sensor. The tag is fabricated on a photographic paper substrate. A carbon-nanotube (CNT) thin film is inkjet-printed in the center of the RFID tag as a sensing section while wireless communication is enabled by an RFID antenna. Carbon nanotubes have shown sensitivity toward extremely small quantities of gases, such as ammonia (NH_3) and nitrogen dioxide (NO_2). Because of the distortion of the electron cloud of a carbon nanotube (from a uniform distribution in graphite to an asymmetric distribution around cylindrical carbon nanotubes), a rich π-electron conjugation forms outside of the carbon nanotube, making it electrochemically active. The electrical properties of carbon nanotubes are extremely sensitive to charge transfer and chemical doping effects by various molecules. When electron-withdrawing molecules (for example, NO_2) or electron-donating molecules (for example, NH_3) are absorbed by the semiconducting carbon nanotubes, they change the density of the main charge carriers of the nanotubes, resulting in changes of the conductance of the nanotubes, which are relayed to the RFID module and broadcast in the form of a modified reflection coefficient or resonant frequency. This behavior forms the basis for applications of inkjet-printed carbon nanotubes as gas sensors that can be integrated easily with paper-based inkjet-printed RFIDs. Such sensors can be deployed easily as an early warning system for poisonous gas leakage or wearable biomonitoring applications. *See* QUANTUM ELECTRONIC PROPERTIES OF MATTER.

Wearable RFID with energy scavenger. Most state-of-the-art long-range RFID tags are realized by an active RFID topology that requires a power supply. Until now, chemical-cell batteries have been commonly used, but their frequent replacement is costly, highly inconvenient, and environmentally detrimental, especially in the light of the increased use of RFIDs. Extraction of electrical energy from human body movements is a promising alternative. The human body is an underestimated energy source that is continuously generating steady thermoelectric, blood pressure, and motion energy outputs. For example, it has been estimated that an average of 67 W of power are generated in the heel movement of a 68-kg (150-lb) person walking at a brisk pace. Even a small percentage of this power would provide enough energy to operate many of the body-worn RFID systems produced today. For practical wearable RFID applications, the critical requirement for the energy sources is their capability to provide sufficient energy for a very short period of time (cycled-operation) to transmit a few packets of information bits.

A proof-of-concept example of a paper-based autonomous RFID tag involves a piezoelectric push button that was embedded in a shoe whose sole and the heel platform made an ideal test bed for exploring body-energy harvesting. **Figure 5** shows the self-powered RFID tag mounted on a shoe. The tag is fabricated on a paper-based substrate for better integration with the shoe's fabric. There is ample space for adding more RFID-enabled sensors, if desired. In this design, a logo printed on the shoes serves as the RFID antenna, eliminating the need for additional space for the typically large RFID antennas. Another obvious direct benefit brought by this integration is that the RFID bearer would not feel any obtrusive effects. To utilize the human walking stride power, a piezoelectric push button was chosen because of its compactness, simplicity, relative low cost in terms of power density, and energy requirement. Whenever the piezoelectric push button is pressed, an internal spring is compressed. When the pressure exceeds a threshold, a spring-loaded hammer is released, delivering a dynamic mechanical force to the piezoelectric element. Once the hammer strikes the piezoelectric element, a pressure wave is generated and reflected several times between the hammer and the element, creating a mechanical resonance. During this process, some fraction of the mechanical energy is converted by the piezoelectric element to electrical energy and effectively is scavenged. The generated output voltage features a waveform similar to an ac signal because of the dynamic polarization of the piezoelectric element.

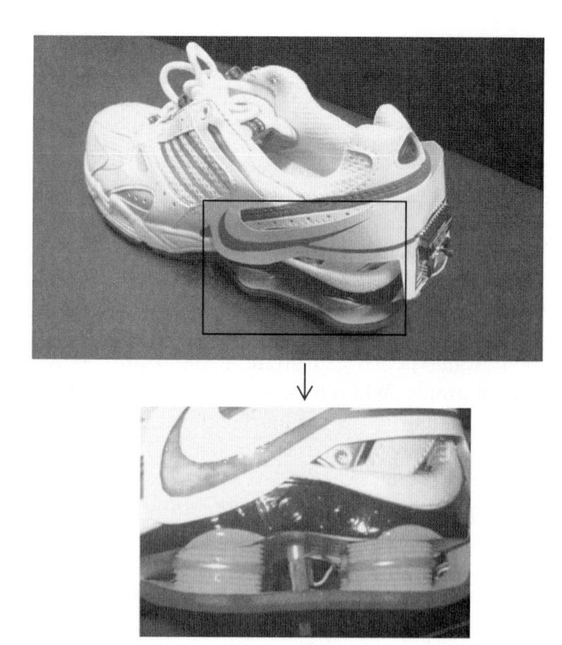

Fig. 5. Self-powered RFID shoes with mounted electronics.

The push button is 35 mm (1.4 in.) long and 5 mm (0.2 in.) in diameter, and has a deflection of 4.5 mm (0.18 in.) at a maximum force of 15 N (3.4 lb). Since piezoelectric elements typically produce high voltages at low currents, and radio-frequency transmitter circuitry requires low voltages at high currents, a step-down transformer is used for the better matching of the impedance to the latter circuitry. An amorphous-core device with a 25:1 turns ratio was chosen that transforms a peak of voltage close to 1000 V at the piezoelectric element down to tens of volts. The amount of electrical energy harvested from each walking step is 848.4 μJ, ample for powering the RFID tag to transmit 18 bits of digital word information. This technique can find direct applications in human positioning, biomonitoring, and tracking projects, and serve as a promising solution for other self-powered wearable paper-based RFID-enabled applications.

Outlook. As discussed, these inkjet-printed paper-based antennas and RFIDs could set the foundation for the ubiquitous implementation of "green" autonomous sensing/identification and widespread implementation of Internet of Things. The use of paper drastically reduces the cost and enables the mass implementation of multilayer UHF and wireless RFID-enabled modules for rugged applications that require conformal substrates. The proposed approach could be the first truly low-cost and green technology bridging RF electronics, nanotechnology, and sensing technologies.

For background information *see* ANTENNA (ELECTROMAGNETISM); CARBON NANOTUBES; CON-

DUCTION (ELECTRICITY); DIELECTRIC MATERIALS; IMPEDANCE MATCHING; INKJET PRINTING; INTEGRATED CIRCUITS; PAPER; PIEZOELECTRICITY; RADAR; SINTERING; TRANSFORMER in the McGraw-Hill Encyclopedia of Science & Technology.

Li Yang; Manos M. Tentzeris

Bibliography. V. Lakafosis et al., Progress towards the first wireless sensor networks consisting of inkjet-printed, paper-based RFID-enabled sensor tags, *Proc. IEEE*, 98:1601–1609, 2010; A. Rida et al., Conductive inkjet-printed antennas on flexible low-cost paper-based substrates for RFID and WSN applications, *IEEE Antenn. Propag. Mag.*, 51(3):13–23, 2009; R. Vyas et al., Paper-based RFID-enabled wireless platforms for sensing applications, *IEEE Trans. Microw. Theor. Tech.*, 57(5):1370–1382, 2009; L. Yang et al., A novel conformal RFID-enabled module utilizing inkjet-printed antennas and carbon nanotubes for gas-detection applications, *IEEE Antenn. Wireless Propag. Lett.*, 8:653–656, 2009; L. Yang et al., RFID tag and RF structures on paper substrates using inkjet-printing technology, *IEEE Trans. Microw. Theor. Tech.*, 55(12):2894–2901, 2007.

Innovative nuclear fuel cycles

Construction of new nuclear power plants is expected to restart shortly in the United States, and the number of nuclear plants worldwide is expected to jump beyond the 435 current plants producing 15% of the world's electricity. Most nuclear plants use nuclear fission of fissionable uranium-235 to produce large amounts of energy due to mass conversion into energy. That heat generation is removed by coolants which go on to produce electricity. The fissions generate fission products (FP) and neutrons absorbed into fertile uranium-238 to produce plutonium (Pu) and minor actinides (MA), consisting of americium (Am), curium (Cm), and neptunium (Np). There are two principal types of reactors: thermal and fast. Thermal reactors use a moderator to reduce their energy level, while fast reactors with no moderator operate at a much higher energy level and can better fission plutonium and minor actinides. Also, they produce more neutrons and can breed fissile material.

By far the dominant world fuel cycle employs thermal reactors cooled by water and utilizes the once-through fuel cycle (OTFC) [**Fig. 1**]. Nuclear fuel cycles identify all necessary steps. Starting from the raw material, uranium (U), its enrichment of uranium-235 from 0.7% to 3–5% of uranium-238 is followed by its fabrication into fuel rods and assemblies that are inserted into reactor cores to undergo fission and to be discharged as spent nuclear fuel (SNF) for site storage and permanent disposal into a deep geological repository or for eventual reprocessing and recycling. Reprocessing allows recovery of reusable materials, while recycling permits more generation of nuclear power. During operation, uranium is burned until the uranium-235 content drops to 0.8%. One-third of the fuel is discharged, typically after 18 months, and is stored in water pools to reduce its decay heat. Dry storage of spent nuclear fuel is possible after 4 or 5 years.

This article reviews innovative fuel cycles and their radioactive waste management. Innovation is measured in terms of improvements made to the once-through fuel cycle with regard to its ability to produce energy, its economics, its proliferation resistance, and reductions in waste volume and radioactivity.

Characteristics of the once-through fuel cycle. Spent nuclear fuel burnup is measured in megawatt days per metric ton of heavy metal (MWd/MTHM). Its current value is 50,000 MWd/MTHM. At equilibrium conditions it contains 92.5% uranium, 6% fission products, 1.3% plutonium, and 0.2% minor actinides. Irradiated uranium is not hazardous. Fissionable plutonium is important because it is useful in nuclear weapons and hazardous if inhaled. Fission products are responsible for spent nuclear fuel decay heat, which drops by a factor of 100 after 1 century due to the short half-lives of fission products. Spent nuclear fuel radioactivity remains high after 100,000 years due to actinides (plutonium and minor actinides). Other characteristics of the once-through fuel cycle are: its volume of high-level waste (HLW) to repository is highest; proliferation resistance is good; only 1% of available energy is used; its decay heat was the controlling factor at the Yucca Mountain repository; high radioactivity after 100,000 years could cause it to face the most generational opposition; health risks depend upon mobile fission products and repository design and provided engineered safeguards; and it is the most economical. In the United States, once-through fuel cycle costs were estimated at $8.67 per megawatt hour (MWh) in 2007 dollars, including $1/MWh for disposal of once-through spent nuclear fuel. In the United States, two changes are possible: a 20% increase in fuel burnup and an interim large centralized dry storage facility for 153,000 metric tons of spent nuclear fuel at a minimal cost of $0.1–0.2/MWh. Worldwide, 200,000 metric tons of spent

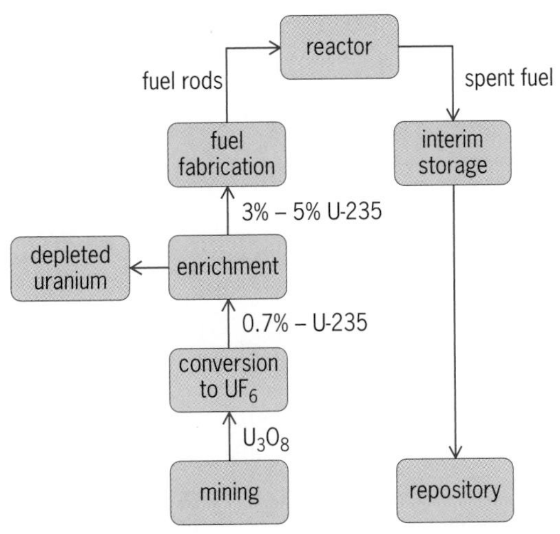

Fig. 1. Once-through fuel cycle.

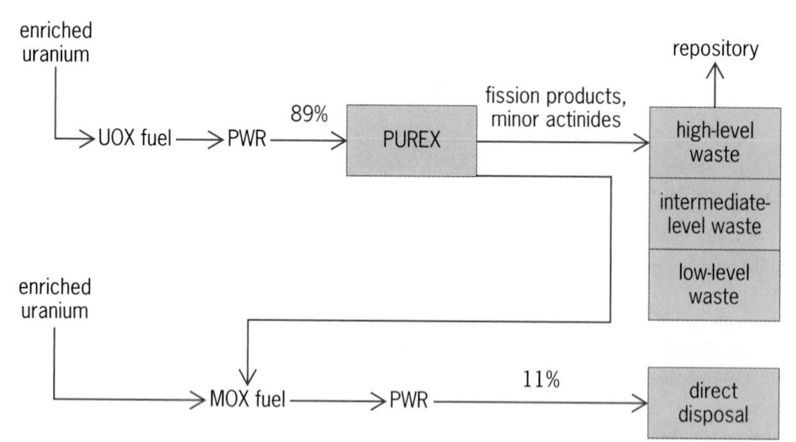

Fig. 2. **Conventional reprocessing fuel cycle with a light-water reactor, here a pressurized-water reactor (PWR). (*Organisation for Economic Co-operation and Development, Advanced Nuclear Fuel Cycles and Radioactive Waste Management, NEA no. 5990, 2006*).**

nuclear fuel had been discharged by 2008, and that inventory is growing by 10,000 metric tons/year, emphasizing the urgency of a solution.

Management of actinides in light-water reactors. In 1976, the presidential candidates in the United States agreed that separation of plutonium using plutonium-uranium extraction (PUREX) should be discontinued due to proliferation concerns, and this policy became United States law in 1977. Other countries chose to reprocess plutonium using a mixed oxide (MOX) of uranium and plutonium in light-water reactors (LWRs; **Fig. 2**). MOX fuel resembles uranium oxide (UOX) except for replacing one-third of the peripheral rods by MOX rods so that the plutonium consumed in MOX assemblies matches that produced in UOX assemblies. The presence of MOX fuel in light-water reactor cores impacts the ability to shut down the reactor, thereby limiting the number of positions MOX can occupy. Three plutonium concentrations by weight may be needed to avoid excessive local power peaking in peripheral plutonium fuel rods. Also, there are limitations in processing and plutonium fabrication facilities because some of the recovered fissile plutonium recovered from spent nuclear fuel has a half-life of 14.4 years and decays to form americium-241, which emits radiation harmful to workers. At present, only 5 years of plutonium cool down can be tolerated, explaining French shipment of dry spent nuclear fuel to reprocessing 4.5 years after discharge. MOX recycle is possible but it is limited to two or three recycles due to plutonium degradation and minor actinide buildup. The capacity of available facilities and worker constraints also play an important role.

Three innovative changes have been made to Fig. 2: The new Co Extraction (COEX) French process of handling uranium and plutonium avoids plutonium separation; depleted uranium is used in plutonium rods; and uranium from reprocessing, called RepU, is utilized in reloads to save uranium. The radioactivity of RepU is much higher than that of natural uranium due to the presence of uranium-232, -233, and -234. Blending with medium-enriched uranium is preferred by the Russians to reenrichment because they can separate uranium-232 and avoid radiation exposure to workers. The Electric Power Research Institute (EPRI) has found that RepU costs are lower when the cost of U_3O_8 exceeds $48/lb, which was its value in July 2009.

France has carried out intensive development programs to avoid plutonium disposal underground. The European Pressurized Reactor (EPR) can handle a full core of MOX fuel, and an advanced fuel assembly known as Core Reload After Initial Loading (CORAIL) can undergo several recycles. Also, separation and transmutation (S&T) is employed to separate curium and store it for 50–60 years, and to fabricate americium into targets for burning in fast reactors or accelerators. **Figure 3** shows the innovative MOX/enriched uranium (MOX/EU) cycle, also known as MIX, to reduce the actinides by a factor of 3.4 compared to the once-through fuel cycle. MIX plutonium is a secondary contributor to fission (1.3% plutonium in 3.4% enriched uranium), but it requires higher throughput of plutonium fuel than Fig. 2, increased costs, and enlarged facilities.

The net result of French activities is illustrated in **Fig. 4**, which shows the projected plutonium inventory in metric tons for current constant French nuclear capacity, which was at about 63,000 MWe electrical at end of 2008. Curve 3 corresponds to the once-through fuel cycle, while curve 2 is for a single MOX recycle. Curve 4 employs advanced CORAIL design and curve 1 uses MIX to level the plutonium inventory. Cores with 30% MOX are used until the EPR is available in 2020. All current standardized 900-MWe plants are to be replaced by EPRs by 2040, and significant upgrades of the reprocessing and plutonium fuel fabrication facilities are necessary in 2020. Issues of funding, SNF/MOX storage, waste-disposal volume, reduced repository radiotoxicity, and times to reach equilibrium are left out.

Another innovative light-water reactor fuel cycle developed by the Paul Scherer Institute (PSI) deserves mention. It employs an inert matrix fuel (IMF), zirconium oxide (ZrO_2), for burning plutonium and minor actinides, and duplicates the MOX configuration. The inert matrix fuel has very low solubility, is difficult to reprocess, and more suitable for

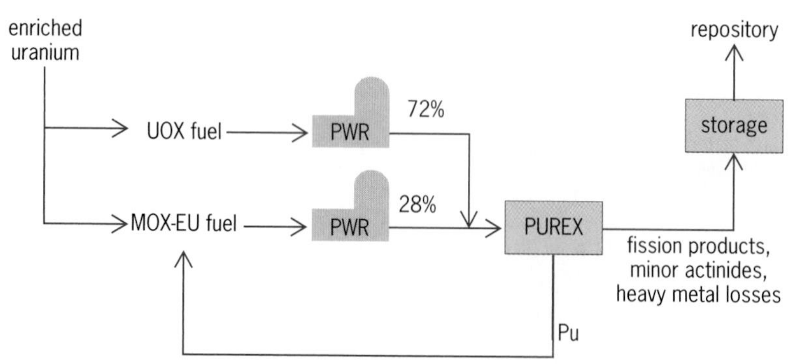

Fig. 3. **Plutonium burning in light-water reactors with enriched uranium (the MOX/EU or MIX cycle). (*Organisation for Economic Co-operation and Development, Advanced Nuclear Fuel Cycles and Radioactive Waste Management, NEA no. 5990, 2006*).**

geological disposal. It is undergoing irradiation tests by PSI.

In summary, the European Community has chosen to burn MOX fuel in light-water reactors because it is judged to be a proven, safe, and licensable process that yields 20–30% savings in uranium and enrichment costs. Those gains require intensive planning of required facilities and limiting concentrations of americium-241 and uranium-232 to avoid their radiation impact upon workers. Also, costs exceed those of the once-through fuel cycle by 6–10% of total nuclear power costs. Several other options have been developed for multiple recycling in light-water reactors over their duration of over 100 years, but their success, increased costs, and waste management require confirmation.

Liquid-metal fast reactor (LMFR) fuel cycles. Fast reactors received early attention due to their ability to breed. Because of a lack of a moderator, their core power generation per unit volume is high and requires high-heat-transfer coolants. Liquid sodium, lead, lead-bismuth, and sodium/potassium have been used, but liquid sodium remains the preferred choice. By comparison to light-water reactors, liquid-metal reactors can sustain much higher reactor exit temperatures. This circumstance considerably improves steam conditions at the steam turbines producing electricity and results in 30% improved thermal cycle efficiency. The consensus still remains that their capital cost is 10–20% higher than light-water reactors. Recently, the interest in liquid-metal fast reactors has shifted to burning the actinides (plutonium and minor actinides) produced in light-water reactors and to improving their waste management. Without actinides, light-water reactor spent nuclear fuel radiotoxicity drops after 100 years and falls below uranium toxicity after less than 1000 years. The coupling of the liquid-metal fast reactor and the light-water reactor was performed successfully at the Experimental Breeder Reactor II (EBRII) at the Idaho National Laboratory, simulating the Integral Fast Reactor (IFR) and its pyroprocessing technology. The Integral Fast Reactor is a liquid-metal fast reactor called Integral because of its ability to recover, recycle, and consume all actinides, and to resolve several nuclear waste-disposal concerns. This innovative coupling of a light-water reactor and the Integral Fast Reactor is illustrated in **Fig. 5**. The developmental uranium extraction + (UREX+) process in Fig. 5 does not separate plutonium and captures technetium (Tc) into zirconium cladding to reduce repository health risks. Pyroprocessing employs electromechanical separation in a molten salt to transport uranium, plutonium, and minor actinides to a cathode, leaving fission products at the anode. It is carried out in a hot cell and should cope better with multiple recycling of IFR fuel.

About 25 liquid-metal fast reactors have been built and operated around the world. The remaining issue with liquid-metal fast reactors is not feasibility but fuel recycling and capital costs. Two fast reactors are being constructed in South Korea and Japan to reduce LMFR capital costs. The Japanese prototype is scheduled for 2025 startup. According to Fig. 5, 36.8% of the power would be generated in the liquid-metal fast reactor, resulting in a severe cost penalty without a significant drop in LMFR costs. Another solution is to increase the burnup of IFR fuel beyond its demonstrated value of 20%.

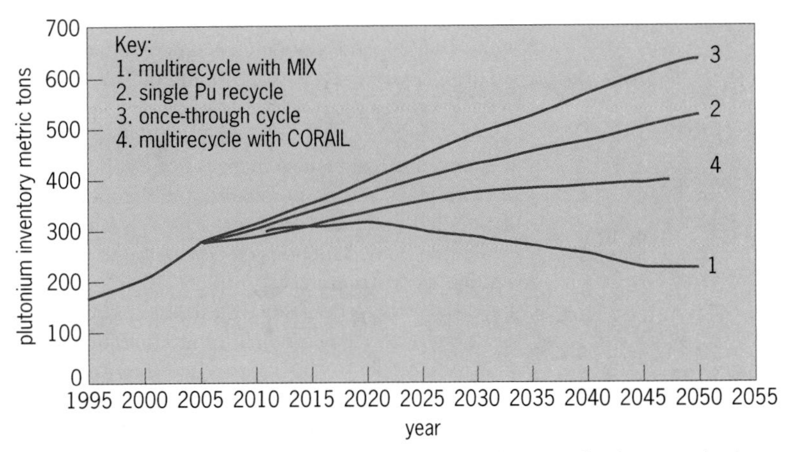

Fig. 4. Projected plutonium inventory from French plutonium recycling in pressurized water reactors.

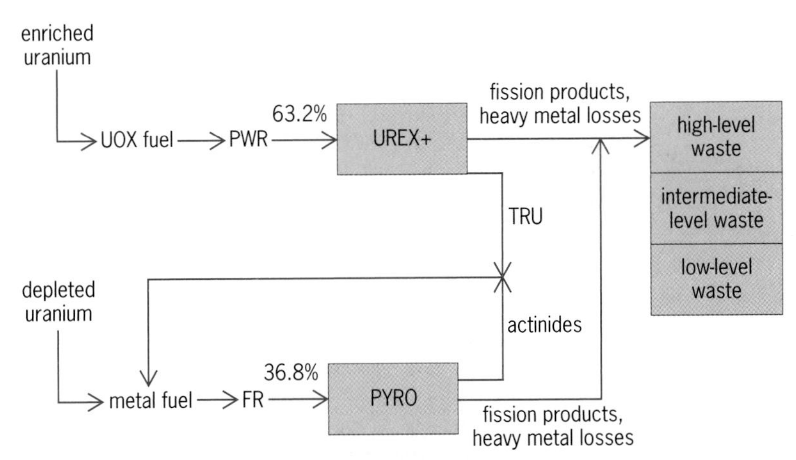

Fig. 5. Burning of actinides and transuranics (TRU) in a cycle coupling a pressurized-water reactor (PWR) and a fast reactor (FR). (*Organisation for Economic Co-operation and Development, Advanced Nuclear Fuel Cycles and Radioactive Waste Management, NEA no. 5990, 2006*).

In summary, the coupling of liquid metal fast reactor to light-water reactor will significantly reduce light-water reactor spent nuclear fuel radiotoxicity and lead to a factor of 50 to 100 savings in energy resources if multiple recycling of IFR fuel is successful. When the liquid-metal fast reactor and the light-water reactor can compete, a direct LMFR fuel cycle, which recycles only its own spent nuclear fuel, could replace light-water reactor fuel cycles.

Other fuel cycles. Numerous other fuel cycles have been proposed. The largest number of gas-cooled plants was in Great Britain where carbon dioxide (CO_2) coolant and a single plutonium recycle were utilized. That program is being replaced by light-water reactors, while waste management of CO_2 fuel plants relies upon geological repository disposal. There has been a trend to switch to helium high-temperature gas-cooled reactors (HTGR), which were first developed using pebble bed fuel to attain increased reactor exit temperature. A direct recycle using a gas turbine is under consideration to reduce capital costs, but it is developmental and the associated waste management is not established. In the United States, development of high-temperature gas-cooled reactors for process heat is underway, using pebble bed fuel or coated small fuel particles known as tristructural isotropic (TRISO) encapsulated by two layers of silicon carbide and graphite. An innovative TRISO deep-burn modular helium reactor (DB-MHR) has been proposed to reduce light-water reactor actinides by 60%, but field experience and disposal plans are lacking.

The use of thorium fuel is receiving attention because it can absorb neutrons to form fissile uranium-233. It has the advantage of not forming actinides. In the United States, several past applications of thorium were carried out but their pursuit was deferred because uranium was available. India, with large reserves of thorium, has an intensive thorium program but its waste disposal plans are undefined.

There are many other fuel cycles, such as gas-cooled fast reactors, or the DUPIC [Direct Use of spent PWR (pressurized-water reactor) fuel In CANDU] reactor, which has been shown by South Korea to save 40% of uranium but is applicable only in countries with light-water reactors and CANDU plants. An International Fuel Cycle Program to exchange information and avoid duplication would make great sense if more emphasis is put on waste management.

Deployment strategies. Deployment takes a very long time and is very dependent upon the peculiar conditions prevalent in each country. For illustrative purposes, in the Oak Ridge National Laboratory (ORNL) program for sustained nuclear energy, the United States nuclear program is assumed to increase the number of its reactors from 104 in 2010 to 108 in 2020, 116 by 2030, and 136 by 2060. The first reprocessing facility of 1000 metric tons per year comes on-line by 2030, and grows to 2000 metric tons per year by 2040, and 3000 metric tons per year by 2070. The plan proposes to reprocess the oldest fuel first so that, after 30 or more years, transmutation of long-lived actinides can be carried out in light-water reactors. Also, due to insufficient reactor shutdown ability, the ORNL plans may need a MIX program. Furthermore, the ORNL program fails to recognize that power producers have a firm Department of Energy contract to remove spent nuclear fuel, and will not agree to recycling unless compensated for increased costs.

Conclusions. Several conclusions may be drawn from this survey.

1. The search for an alternate geological repository should be initiated immediately because it is needed irrespective of selected fuel cycles. Minimal seismic and volcanic repository sites could accelerate licensing and acceptance of once-through spent fuel.

2. The low cost of a large centralized spent nuclear storage justifies its pursuit to provide time to consider advanced fuel cycles.

3. A pilot line to confirm UREX+ is needed to produce actinide fuel for irradiation in operating light-water reactors and international fast reactors.

4. Development of fast reactors is overdue. Advantage should be taken of experience with the Integral Fast Reactor and its pyroprocessing methodology.

5. Guarded nuclear park concepts containing several Integral Fast Reactors and their complete fuel-cycle facilities (reprocessing, fabrication, waste treatment, and storage) will lead to enforceable proliferation resistance because no fissile material will leave the site.

6. Continued research in advanced nuclear fuel cycles should be encouraged, preferably on an international integrated level.

For background information *see* NUCLEAR FUEL CYCLE; NUCEAR FUELS; NUCLEAR FUELS REPROCESSING; NUCLEAR POWER; NUCLEAR REACTOR in the McGraw-Hill Encyclopedia of Science & Technology

Salomon Levy

Bibliography. Oak Ridge National Laboratory (ORNL), *A Practical Solution to Nuclear Fuel Treatment to Enable Sustained Nuclear Energy and Recovery of Vital Materials*, ORNL/TM-2010/81, April 2010; Organisation for Economic Co-operation and Development, *Advanced Nuclear Fuel Cycles and Radioactive Waste Management*, NEA rep. no. 5990, 2006; Organisation for Economic Co-operation and Development, *Management of Recyclable Fissile and Fertile Materials*, NEA rep. no. 6107, 2007; Organisation for Economic Co-operation and Development, *Plutonium Management in the Medium Term*, 2003; The technology of the Integral Fast Reactor and its associated fuel cycle, *Prog. Nucl. Energ.*, 31:1–217, 1997; United States Department of Energy, *Global Nuclear Energy Partnership Programmatic Environmental Impact Statement*, DOE/EIS-0396, October 2008.

Interference from power-line telecommunications

Power-line telecommunications (PLT) means the use of existing electric power lines (mains wiring) for broadband communications purposes. To achieve data rates of the order of tens of megabits per second, a frequency range up to tens of megahertz (typically, 30 MHz) is used. Electric power cables (mains cables) were not originally designed to carry broadband communication signals and actually have a configuration and layout that cause them to act as antennas at high-frequency (HF; 2–30 MHz) radio frequencies. Therefore, there is potential for interference to HF radio systems near a PLT installation, as well as at large distances due to cumulative contributions from a large number of PLT installations.

HF radio spectrum. Radio frequencies in the HF range have the extraordinary property in that they can be used for communication or broadcasting over very long ranges without any kind of infrastructure or satellites. By "ground wave" the signals can travel along the Earth's surface for up to hundreds of kilometers, and by "sky wave" the signals can travel across continents thanks to reflections from the ionosphere. Users of the HF radio spectrum include short-wave broadcasters, radio amateurs, radio astronomers, aeronautical systems, maritime users, military users, and intelligence agencies.

Analog AM broadcasting has been used to reach distant areas for a long time, and in the last decade the digital HF broadcasting system Digital Radio Mondiale (DRM) has also been introduced. The HF radio band is also important in emergency situations where other communications infrastructure is nonexisting or knocked out. Some view the HF radio band as a limited natural resource that should be carefully protected.

PLT systems. For some time, power-line communications systems with low data rates have been used, for example, to transmit metering information. These systems operate at frequencies below 150 kHz and are uncontroversial. But since the late 1990s, high-data-rate PLT systems have been introduced. PLT is also known as power-line communications (PLC) or broadband over power line (BPL).

PLT systems are categorized as access PLT, used to get a broadband connection to a building, and in-house PLT, used for local networking within a building. Access PLT has not seen widespread application because the economics have in most cases turned out to be poor compared to alternatives such as digital subscriber line (DSL), optical fiber, and coaxial cable. But in-house PLT has gained popularity because it can be bought by anyone at a price comparable to the alternative, a wireless local-area network (WLAN), and is very easy to install. As an example, approximately 6 million in-house PLT modems had been purchased in Germany by February 2010. One application is to use in-house PLT modems for Internet Protocol television (IPTV), as the television receiver is often in a different room than the broadband 3Wmodem.

Efforts are underway to use significantly higher PLT injection frequencies than 30 MHz. Depending on the injected power level and spectrum rolloff, such PLT systems could pose interference to in-home FM radio reception.

Interference mechanisms. The interference mechanisms related to PLT are illustrated in **Fig. 1** for the in-house case. The signal from a PLT modem is injected differentially between the two live wires of a plug receptacle (mains socket). If the building wiring (mains) cable was a good twisted pair or coaxial cable designed for carrying communication signals, the radiated signal from the two wires would cancel so that resulting emission would be low. But building electric power (mains) wires are far from such ideal communication cables: They are not twisted, so that cancellation of radiation from the differential signal is suboptimal. They are poorly balanced, so that part of the signal appears as common mode (in phase) on the two wires relative to ground. For common-mode signals, radiation from the two wires adds together instead of cancelling. Branches and equipment plugged into other sockets create impedance discontinuities and a poorly terminated transmission line, which give rise to reflections and standing waves, and these phenomena also can cause radiation. To make things worse, the physical dimensions of building electric power (mains) wiring are comparable to the wavelength of HF radio waves, increasing the potential for standing waves and radiation ("antenna behavior").

Different interference mechanisms should be considered depending on the distance between the PLT modem and the radio receiver. A radio receiver in the same or nearby buildings may suffer interference conducted through the plug receptacle (mains socket) or from the radiated near field around the electric power (mains) wires carrying PLT signals. (The near field is the region so near a current-carrying cable that the equations for a propagating electromagnetic wave are not valid.) At ranges up to some kilometers, ground-wave propagation may cause interference. And at really long ranges, interference may be carried by sky wave. The transmission loss increases with range, but so does the number of potential interferers, and the total interference is the sum of the received power from all PLT modems within HF radio range (called the cumulative effect).

Radio emissions from a PLT modem can be quantified by several different measures: injected power spectral density from the modem onto the line,

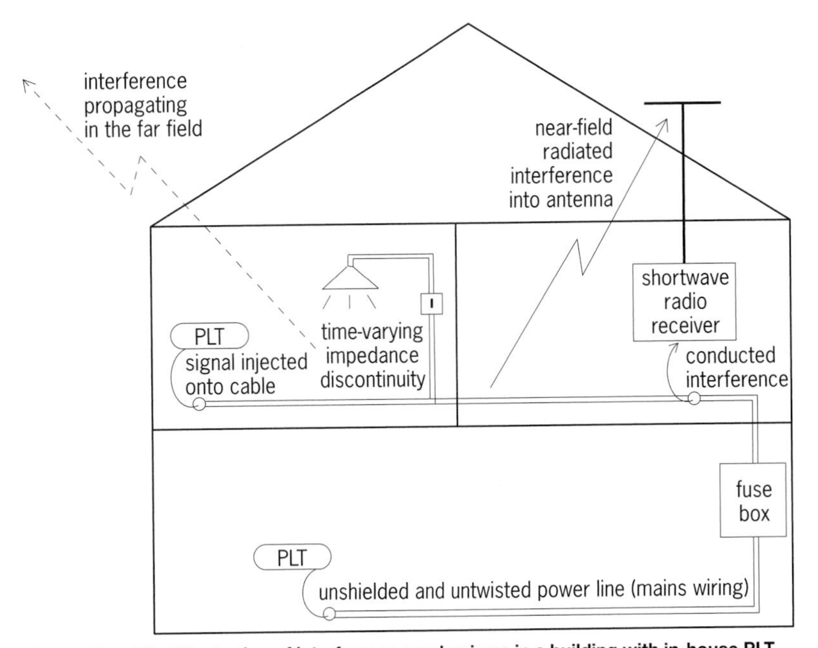

Fig. 1. Simplified illustration of interference mechanisms in a building with in-house PLT. The branch to the lamp is an example of an impedance discontinuity, which is time-varying as the light switch is turned on and off. The broadband connection to the building is not included in the figure.

electric or magnetic field strength at a particular distance from the line or building, or effective isotropic radiated power (EIRP) of the interference propagating in the far field. These measures will be a function of frequency, and different from one PLT installation to the next.

The interference potential of PLT may be assessed by estimating the received field strength (either from a few nearby PLT modems or from a large number of far-away PLT modems), and comparing this with the noise floor (the signal created by the sum of all noise sources) at a time prior to the introduction of PLT. If the PLT interference has field strength comparable to the noise floor (or higher), it will decrease the quality of radio reception.

Regulatory status. There is an inherent conflict when it comes to regulation of radio interference from PLT systems. Telecommunication authorities in charge of regulation are mandated to support proliferation of PLT to improve competition in the broadband market, but also to protect radio users from harmful interference. Although PLT vendors want emission limits to be as high as possible to be able to make useful systems, radio users want limits to be as low as possible to avoid interference.

One issue currently under discussion in Europe is whether to interpret the power-line (mains) connection of a PLT modem as a power-line (mains) port or as a communications port, because it serves both purposes. Different measurement setups (impedance termination networks) are defined for these types of ports as traditional communications cables have better and more well-defined characteristics than electric power (mains) wiring. It is therefore possible to inject a signal more than 30 dB stronger (with 1000 times more power) onto a communications port than onto a power-line (mains) port and still have measured values below the defined limits. In the United States, only radiated limits (and not conducted limits) apply to PLT systems, and for access PLT there are additional rules specifying exclusion zones, excluded frequency bands, and so forth.

The loudest protesters against PLT have been radio amateurs, with the result that PLT systems often include notches in the radio amateur frequency bands (**Fig. 2**). Only in the notched radio amateur bands is the level below the limit applicable to equipment connected to a power-line (mains) port. In principle, all HF radio users would want PLT systems to include notches in "their" band, but it is not practicable to notch the entire HF spectrum.

It is often difficult to compare different limit values on emissions from PLT systems because the limits are applied to measurements that are not directly comparable. Some consider the signal injected onto the line (often in combination with a requirement on line balance), and some consider the electric or magnetic field strength measured 1, 3, 10, or 30 m (3, 10, 30, or 100 ft) from the power line. Because these distances are mostly in the near field, no constant factor can be used to correctly convert between elec-

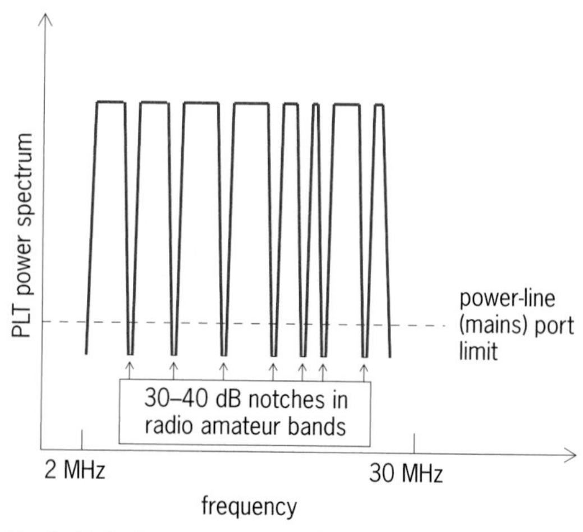

Fig. 2. Typical power spectrum injected onto the electric power (mains) line by a PLT modem.

tric and magnetic field strength or between different measurement distances. It is similarly difficult to extrapolate the emitted power in the far field, which is of interest for prediction of cumulative effects. Another issue is which measurement bandwidth to use and which kind of detector to use ("peak," "quasi-peak," or "average"), and it is also difficult to establish conversion factors between these detectors as they depend on the properties of the signal.

Measurements. A number of reports documenting measurements of emissions from PLT systems have been published. The measured power spectra typically resemble the shape in Fig. 2, where the power spectrum corresponds either to the signal conducted on the line, or to the electric or magnetic field near the line. A general observation is that the situation will be different from house to house, and in different locations in the same house. Also, measurements reveal that some PLT modems emit power continuously, regardless of whether data is being transmitted or not.

The typical antenna gain (ratio between EIRP and differentially injected power) of an in-house PLT system is estimated to be approximately −30 dB, based on reported measurements. The differentially injected power is typically about −50 dBm/Hz (10 nW in each hertz of bandwidth), so that the EIRP of each PLT modem is approximately −80 dBm/Hz. Cumulative PLT interference levels can be predicted by combining this number with estimated PLT modem density and propagation path loss models.

Danger of sneaking harm. A PLT system interfering with a nearby radio receiver may be identified in the event of an interference complaint (and interference complaint rates on PLT systems are on the rise). But cumulative effects will be harder to notice, as the incoherent addition of many sources will give noise-like characteristics, and the contribution of one additional source will not be measurable. The cumulative effects will therefore be possible to notice only in the long term. If the cumulative effect at some point in the future is proven to cause real harm to HF radio systems, it will be too late to remove the interference, as there will then already be a large installed base of PLT modems.

For background information *see* ANTENNA (ELECTROMAGNETISM); ELECTRICAL INTERFERENCE; ELECTRICAL NOISE; ELECTROMAGNETIC COMPATIBILITY; INDUCTIVE COORDINATION; POWER LINE COMMUNICATION; RADIO-WAVE PROPAGATION; TRANSMISSION LINE in the McGraw-Hill Encyclopedia of Science & Technology. Roald Otnes

Bibliography. A. Chubukjian et al., Potential effects of broadband wireline telecommunications on the HF spectrum, *IEEE Comm. Mag.*, 46(11):49–54, November 2008; P. S. Henry, Interference characteristics of broadband power line communication systems using aerial medium voltage wires, *IEEE Comm. Mag.*, 43(4):92–98, April 2005; ITU-R, Impact of powerline telecommunication systems on radiocommunication systems operating in the LF, MF, HF and VHF bands below 80MHz, draft report ITU-R SM.[PLT] Doc. 1/67-E, September 23, 2009; R. Marshall, Environmental effects of the widespread deployment of high speed power line communication—cumulative effects on signal/noise ratio for radio systems, *EMC J.*, 87:33–41, March 2010.

Internet-based simulation for earthquake engineering

Structural experiments play an important role in earthquake engineering research for understanding the behavior of buildings and other structures subjected to earthquake forces. As the scale and complexity of modern structural experiments increase, the existing earthquake engineering laboratories, which are often faced with limited resources (such as space and equipment), inevitably become incapable of satisfying the demand of various types of large-scale and complicated experiments.

Rather than endlessly increasing the capacity of each laboratory, a more practical and cost-effective approach has been developed. The core part of the approach is called the pseudo-dynamic test (PDT) method, which divides the test structure into separate physical and numerical parts. Only the physical parts need to be constructed and tested in a laboratory, while the numerical parts are modeled and analyzed using a computer. The key idea behind the PDT method is to minimize the construction of physical structures, requiring only the most critical or relevant parts of a structure to be tested experimentally in the laboratory, while the rest of the structure can be simulated analytically on a computer. This method greatly reduces the demand on the limited resources of the laboratory and allows for the testing of structures larger than the laboratory itself.

Because the physical and numerical parts can be connected using data communication through a computer network or over the Internet, the PDT method can be distributed. This allows for the integration of a number of geographically distributed earthquake engineering laboratories through the Internet in order to jointly conduct a single structural experiment. In the field of earthquake engineering, the software environment needed for supporting the distributed PDT experiments among geographically distributed laboratories has been called Internet-based simulation or, more recently, the distributed hybrid simulation (DHS) environment.

Distributed hybrid simulation environment. The DHS environment allows different laboratories to conduct a test simultaneously and collaboratively. The sharing and integration of resources among laboratories significantly reduces the investment and cost for tackling large-scale and complex earthquake engineering experiments. In addition, integration allows for opportunities to exchange experiences and technologies, and it assists with the further collaboration on research between these laboratories. Several DHS software environments have been developed in recent years. Examples include NCREE-ISEEdb and NCREE-ISEEap from Taiwan's National Center for Research on Earthquake Engineering (NCREE); UI-SimCor from University of Illinois at Urbana-Champaign; and OpenFresco from University of California, Berkeley. Both UI-SimCor and OpenFresco are now included in a United States multi-university national initiative project called the NEES (Network for Earthquake Engineering Simulation) project. The project is sponsored by the National Science Foundation and aims to explore the benefits of sharing and integrating laboratory resources, including equipment, experiment data, and simulation software through the network.

The DHS environment integrates experimental testing facilities at multiple sites, including software components for facility control, a numerical analysis engine, data acquisition, and data exchange through the Internet. This article uses NCREE-ISEEdb as an example for introducing the overall framework and the components of a DHS environment for facilitating collaborative pseudo-dynamic experiments among earthquake engineering laboratories.

A collaborative pseudo-dynamic experiment requires network interconnection among several participating programs. These programs may be placed at geographically distributed laboratories, with a need to exchange analytical and measured data continuously during the experiment. These participating programs could include an analysis engine (also known as a command generation module, CGM) which operates the numerical parts of the pseudo-dynamic experiment, and one or more facility controllers (or facility controlling modules, FCMs) which control the hydraulic actuators that deform the specimen (that is, the structures being tested). Usually each pseudo-dynamic experiment involves a different setup for the analysis engine, specimens, and facility controllers. Therefore, each experiment requires unique interconnections between the participating programs, leading to a complex software setup for every collaborative experiment. As such, it is tedious to modify, rewrite and

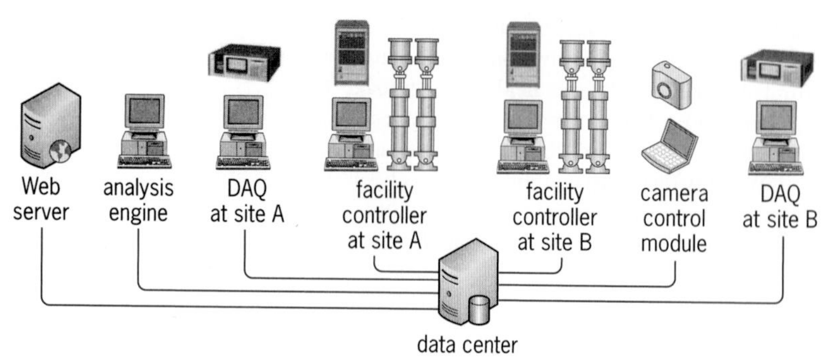

Fig. 1. Use of databases for exchanging information in NCREE-ISEEdb. DAQ = data acquisition.

test the source codes of each program whenever a new experiment is run.

NCREE-ISEEdb employs a database server to act as a data center to facilitate information exchange. This allows authorized geographically distributed programs to input and extract analytical data, measured data, and other user-defined information in the form of floating-point numbers. All participating programs (or modules) communicate only with the data center (**Fig. 1**), thus removing the need for coordinating communication between multiple programs.

The data center not only works as a data repository for experimental results, it also manages the interconnection relationship among the participating programs in each experiment. This is done prior to running an experiment and helps to minimize the work required to set up or modify participating programs for each individual experiment. In addition, the data center provides a simple Internet interface for researchers to easily create and set up a web page for each experiment, and allows users to browse the data of ongoing or completed experiments through tables and graphs.

The analysis engine generates the controlling commands of a collaborative experiment. It is also the first participating program to input data to the data center in each step. In a typical pseudo-dynamic experiment, the analysis engine is a numerical analysis engine that analyzes the numerical part of the structure, considering the interactions between the analytical responses and the experimental responses measured from the specimens. Such an analysis engine can be built using a general-purpose structural dynamic program or as a tailor-made program for a specific experiment.

A facility controller is a software component used to control an experimental facility in a laboratory. During a pseudo-dynamic experiment, the analysis engine simulates and calculates for each step the predicted structural responses (including nodal displacements, velocities, and accelerations). Once the predicted responses are calculated, the facility controller obtains data (for example, the calculated responses of the specimen from the analysis engine through the data center) and carries out the command of manipulating the facilities to deform the specimen in a displacement-controlled manner. The facility controller then returns the responses (for example, the measured resisting forces) to the analysis engine through the data center after carrying out the command. After receiving the dynamic resisting forces, the analysis engine calculates the responses of the system and the predicted responses of the next step.

Application example of networked pseudo-dynamic experiment. A transnational collaborative pseudo-dynamic experiment was carried out among NCREE, National Taiwan University (NTU), and Carleton University (CU) of Canada in 2006 to simulate the response of a multi-span bridge system (**Fig. 2**) subjected to a series of bilateral ground motions. The specimens of three double-skinned, concrete-filled tubular (DSCFT) piers, denoted as P1, P2, and P3, were located at CU, NCREE, and NTU, respectively. Each pier was controlled along two lateral directions with a pretensioned axial rod to simulate the bilateral responses and the gravity load. The fourth pier and the bridge decks were simulated by a three-dimensional (3D) numerical model. The NCREE-ISEEdb platform was used not only for the data repository and communication among laboratories, but also for data sharing on the Internet, video broadcasting, two-dimensional (2D) plots, and 3D visualization of bridge deformation.

Figure 3 shows the network configuration connecting the hardware and software components for the pseudo-dynamic experiment. Three piers, P1, P2, and P3, were controlled by the facility controlling modules (FCMs) at the laboratories where they were set up and tested. The FCMs complied with the commands (that is, bilateral target displacements at pier top) received from the data center, and sent the responses (that is, bilateral resisting forces at pier top) back to the data center. The analysis engine, after receiving and interpreting the specimens' responses, generated the commands for the next step by performing numerical dynamic analysis. A visualization module accessed the experimental or numerical analysis data in the data center, generated 2D plots (tables and graphs) and 3D figures of structure deformation, and sent the plots and figures to the Web server. The Web server then shared the data, plots, and figures on a Web page. The video modules captured video records of the experiment in progress and transmitted them as video streams through the Internet. During the experiment, the data, plots, figures, and videos were updated continuously. Participating researchers at remote sites

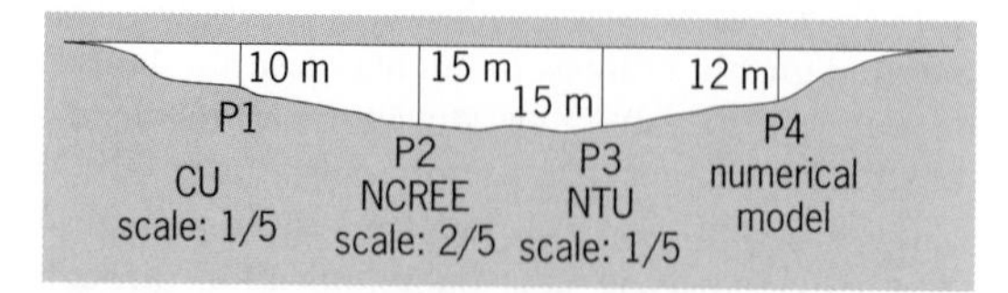

Fig. 2. Elevation of the multispan bridge system in the experiment. CU = Carleton University, Canada. NCREE = National Center for Research on Earthquake Engineering, Taiwan. NTU = National Taiwan University. P1–P4 = piers.

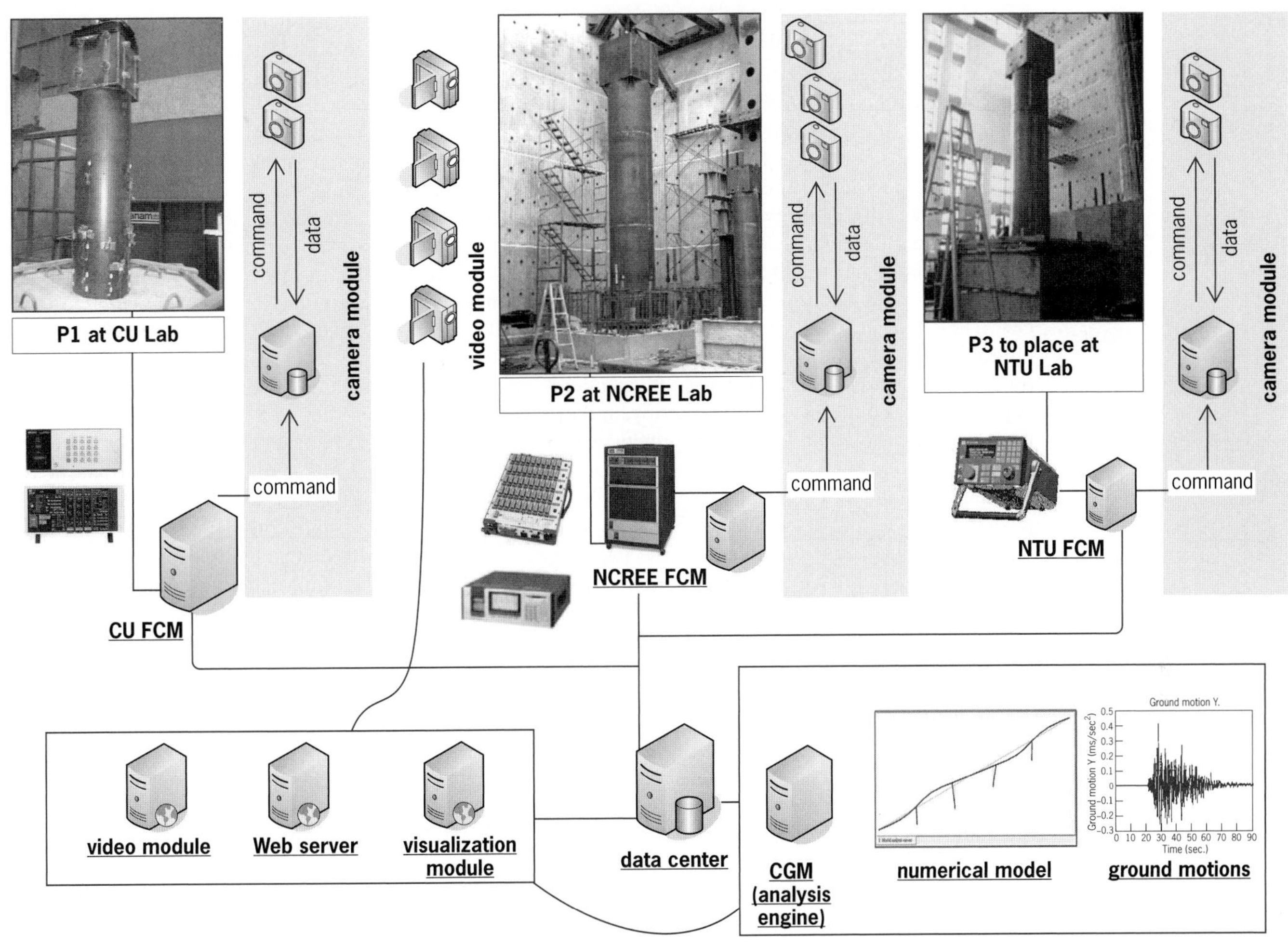

Fig. 3. Network configuration of the pseudo-dynamic experiment. CGM = command generation module. FCM = facility controlling module. P1, P2, P3 = piers.

or with Internet connections could observe the progress of the experiment and access the information. The camera modules took photos of the specimens after the FCMs completed their commands at each step. In the experiment, the camera images were stored off-line and were not shared in real time.

Outlook. As the DHS approach integrates the PDT method, the Internet, and other information technologies, engineers can now optimize the use of resources among geographically distributed laboratories around the world, with the joint goal of tackling increasingly large-scale and complex earthquake engineering experiments and for advancing knowledge in earthquake engineering. In early 2010, several earthquake engineering laboratories acquired the ability and experience to apply DHS, although further research is still ongoing to improve the robustness of the approach and the interoperability among laboratories with different DHS environments. The ultimate goal is the advancement of technology in order to tackle real-time distributed dynamic experiments.

For background information *see* EARTHQUAKE ENGINEERING; SIMULATION; DATABASE MANAGEMENT SYSTEM; NUMERICAL ANALYSIS; STRUCTURAL ANALYSIS; MODEL THEORY in the McGraw-Hill Encyclopedia of Science & Technology. Shang-Hsien Hsieh

Bibliography. V. Saouma and M. V. Sivaselvan (eds.), *Hybrid Simulation: Theory, Implementation and Application*, Routledge, Taylor & Francis Group, 2008; W. F. Chen and C. Scawthorn (eds.), *Earthquake Engineering Handbook*, CRC Press, Boca Raton, FL, 2002; Y. S. Yang et al., ISEE: Internet-based simulation for earthquake engineering, Part I: Database approach, *Earthquake Engineering & Structural Dynamics*, 36:2291–2306, 2007; K. J. Wang et al., ISEE: Internet-based simulation for earthquake engineering, Part II: The application protocol approach, *Earthquake Engineering & Structural Dynamics*, 36:2307–2323, 2007; O. S. Kwon et al., *UI-SIMCOR Users Manual and Examples for UI-SIMCOR v2.6 and NEES-SAM v2.0*, Department of Civil and Environmental Engineering, University of Illinois at Urbana-Champaign, 2007; Y. Takahashi and G. L. Fenves, Software framework for distributed experimental-computational simulation of structural systems, *Earthquake Engineering & Structural Dynamics*, 35:267–291, 2006.

In vivo 3D optical microendoscopy

Cancer remains one of the principal threats to human health. More than 7 million people die of cancer worldwide each year. The high cancer mortality is due mainly to the lack of early detection modalities, especially for internal organs. Computerized tomography (CT), magnetic resonance imaging (MRI), and ultrasound imaging are commonly used diagnostic tools, but they have issues of relatively low resolution, low contrast, radiation risk, or high cost. These imaging modalities typically provide resolutions of 0.1–1 mm. On the other hand, several optical imaging techniques, such as confocal microcopy, nonlinear optical (NLO) microscopy, and optical coherence tomography (OCT), provide much higher resolutions. Confocal and NLO microscopy can achieve submicrometer resolutions while OCT's resolution ranges from 1 to 15 micrometers depending on the light source employed. These high resolutions are achieved by spatial gating of a pinhole for confocal microscopy, spectral gating of harmonic generation or multiphoton absorption for NLO imaging, and coherence gating for OCT. With such cellular or even subcellular resolutions, early cancers can be detected.

However, the high resolution of these advanced optical imaging techniques is obtained at the price of imaging depth. For most biological tissues, confocal and NLO microscopy can image only up to 0.4 mm, while OCT can image up to 3 mm. Thus biopsy is needed for internal organs. For diagnosing diseases at an early stage, lack of visible evidence often leads to negative biopsies. Biopsy also introduces high risks, such as trauma. To fully utilize the capabilities of these advanced optical imaging techniques, miniature endoscopic probes with active optical scanning engines must be developed so that real-time, in vivo, noninvasive imaging for early internal cancer detection can be realized.

High-resolution endoscopic imaging is very challenging, since fast image scanning must be realized in very small imaging probes. The most critical subsystem for miniaturizing endoscopic imaging probes is that responsible for light-beam scanning. Some optical fiber–based probes have been reported that operate by rotating a long fiber, swinging the fiber tip, or rotating a pair of scanning graded-index (GRIN) lenses, but they suffer from slow speed, hysteresis, optical coupling uniformity, and other drawbacks.

Micro-electro-mechanical systems (MEMS) technology provides a viable solution and extends high-resolution three-dimensional (3D) optical imaging into internal organs. In vivo 3D endoscopic imaging has been demonstrated by several research groups. This new technology can be used to perform real-time, in vivo imaging without biopsy of various organs such as the lung, gastrointestinal tract, prostate gland, bladder, and pancreas for early cancer detection. It can also be used for real-time image-guided surgery. The instantaneous information that the endoscopic probes provide to the physician or surgeon has not been possible with the technology available up to now. The imaging probes are portable and inexpensive, and thus can be used in rural areas to allow more people to access medical care and lead healthier lives. The probes are disposable, and therefore minimize the risk of disease transmission. They are also applicable for bedside or point-of-care medical care. This article reviews MEMS technology and discusses the use of MEMS devices to miniaturize imaging probes and obtain 3D images of internal organs in vivo.

MEMS scanning mirrors. MEMS technology leverages integrated-circuit batch fabrication techniques and produces various functional devices with small size and low cost. MEMS have been applied in a wide variety of fields in both research and industry, for instance, DNA microarrays for gene sequencing and expression profiling, MEMS motion sensors in video games, and MEMS mirrors for portable projectors. For optical imaging, MEMS mirrors can be used to miniaturize imaging probes and extend the optical biopsy capability into the human body.

MEMS devices can be made by using surface micromachining or bulk micromachining processes. Most MEMS micromirrors are surface micromachined. However, for biomedical imaging

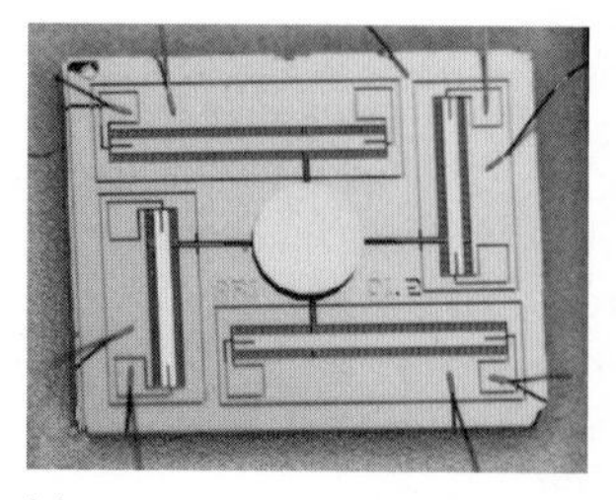
(a)

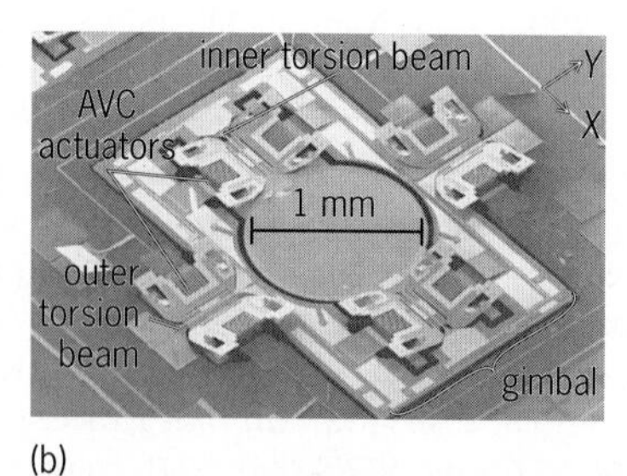

(b)

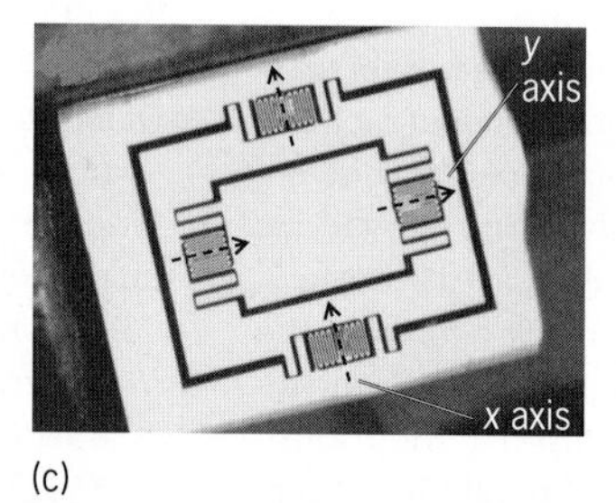

(c)

(d)

Fig. 1. Scanning electron microscopic (SEM) images of some 2D MEMS mirrors. (*a*) Electrostatic mirror (*V. Milanovic, Multilevel beam SOI-MEMS fabrication and application, J. Microelectromech. Syst., 13:19–30, © 2004, IEEE*). (*b*) Another electrostatic mirror. AVC = angular vertical comb (*W. Piyawattanametha et al., Surface- and bulk-micromachined two-dimensional scanner driven by angular vertical comb actuators, J. Microelectromech. Syst., 14:1329–1338, © 2005, IEEE*). (c) Electromagnetic mirror (*K. Kim et al., Two-axis magnetically-driven MEMS scanning catheter for endoscopic high-speed optical coherence tomography, Optics Express, 15:18130–18140, 2007*). (*d*) Electrothermal mirror (*L. Wu and H. Xie, A large vertical displacement electrothermal bimorph microactuator with very small lateral shift, Sensor. Actuator. Phys., 145:371–379, 2008*).

applications, relatively large mirrors (greater than 0.5 mm) are required. Therefore, bulk micromachining or combined surface-bulk micromachining processes are used to make large, flat micromirrors.

Micromirrors can be actuated electrothermally, electrostatically, piezoelectrically, or electromagnetically. Among these actuation methods, electrostatic actuation is most commonly used because it consumes low power and has a fast response. Electrostatic micromirrors with 0.5–1.0-mm aperture sizes, 5–10° scan angles, and 1–15-k Hz resonant frequencies have been demonstrated (**Figs. 1*a,b***). However, high drive voltages, of the order of hundreds of volts, are often required, which are not suitable for internal organ applications. The area fill factors (the mirror area divided by the whole device area) of electrostatic micromirrors are typically small (approximately 5%) because of the large area required for the interdigitated comb-finger drives, that is, pairs of beam arrays engaging each other like two overlapping combs. Piezoelectric actuators have low power consumption and high bandwidth, but piezoelectric actuation typically has small displacements, charge leakage problems, and hysteresis effects that often require a feedback control loop.

For endoscopic optical imaging applications, the MEMS mirrors must scan large angles at low drive voltages and have high area fill factors. In order to increase the scan range and decrease the drive voltage, electromagnetic micromirrors have been explored. Electromagnetic micromirrors with 0.5–1.0-mm aperture size can typically scan about 30° with a voltage level of less than 10 V. The electromagnetic micromirror shown in Fig. 1*c* has a fill factor of approximately 10%. However, the requirement for magnets for actuation not only complicates packaging but also limits application ranges because of the magnetic field. To further increase the scan range and fill factor, as well as to avoid the magnetic interference, electrothermal micromirrors with 0.8–2.0-mm aperture size, fill factors greater than 25%, and scan angles greater than 60°, but still at less than 10 V, have been developed (Fig. 1*d*). The low drive voltage and large scan range of electrothermal and electromagnetic micromirrors are achieved at the price of high power consumption. The electrical currents required for both electrothermal and electromagnetic micromirrors are of the order of 50 mA.

Miniature endoscopic probes. MEMS technology has been applied to optical coherence tomography (OCT), confocal microscopy, and nonlinear optical microscopy to develop miniature endoscopic imaging probes, enabling in vivo early disease diagnosis and surgery of internal organs.

Endoscopic OCT. The first MEMS endoscopic OCT imaging was demonstrated in 2001 using a one-dimensional (1D) scanning electrothermal MEMS micromirror, in which a 5-mm-diameter probe was directly inserted into the biopsy channel of a cystoscope. In vivo imaging of the human bladder has been performed with an imaging depth of approximately 2.5 mm and axial resolution of about 10 μm at a frame rate of about 5 frames per second. Many research groups subsequently explored this concept

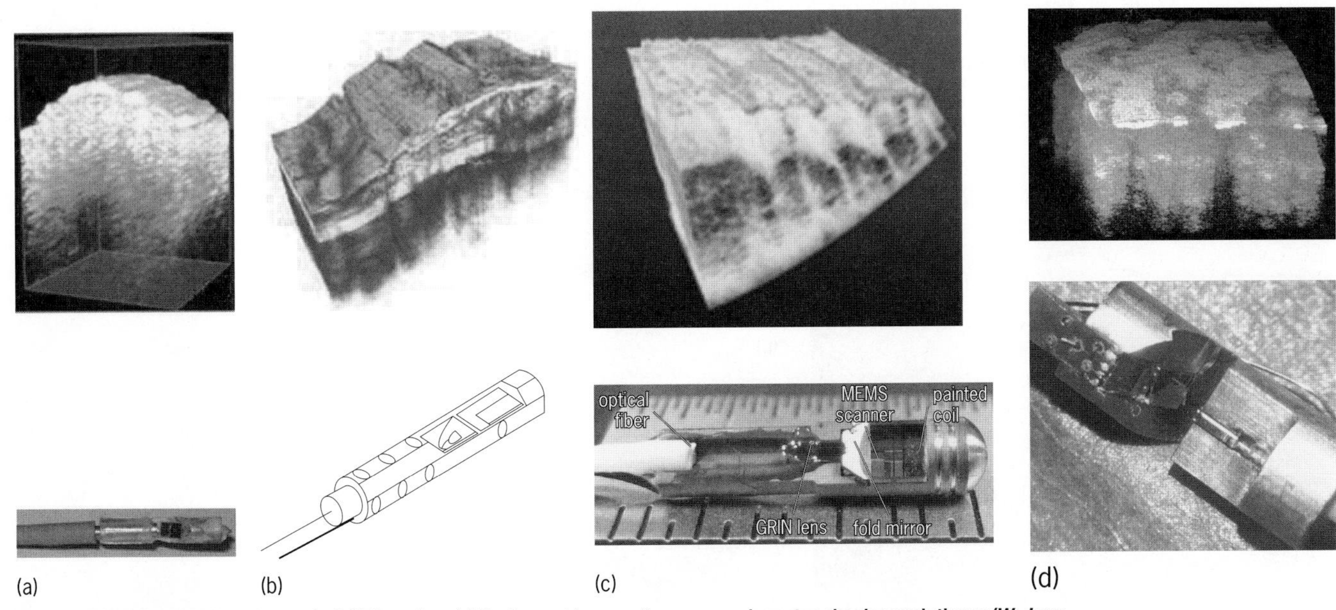

Fig. 2. MEMS-based 3D endoscopic OCT imaging. (*a*) Probe and image of cancerous hamster cheek pouch tissue (*W. Jung et al., Three-dimensional endoscopic optical coherence tomography by use of a two-axis mircorelectromechanical scanning mirror, Appl. Phys. Lett., 88:163901, 2006*). (*b*) Probe and *in vitro* image of hamster cheek pouch (*A. D. Aguirre et al., Two-axis MEMS scanning catheter for ultrahigh resolution three-dimensional and en face imaging, Optics Express, 15:2445–2453, 2007*). (c) Probe and in vivo image of a finger tip. GRIN = graded-index (*K. Kim et al., Two-axis magnetically-driven MEMS scanning catheter for endoscopic high-speed optical coherence tomography, Optics Express, 15:18130–18140, 2007*). (*d*) Probe and in vivo image of a human finger (*S. Guo et al., Three-dimensional optical coherence tomography based on a high-fill-factor MEMS mirror, paper NTuB3, Proc. OSA Spring Optics Photonics Congress, Vancouver, BC, Canada, April 2009*).

(a)

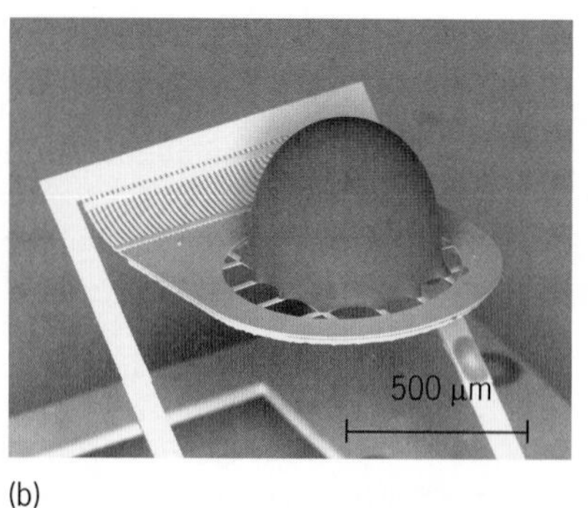

(b)

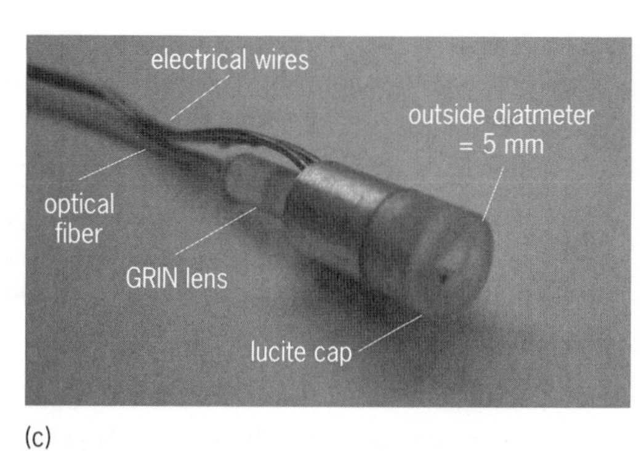

(c)

Fig. 3. Confocal endoscopic imaging. (*a*) Magnetically actuated tunable microlens. PDMS = polydimethylsiloxane (*C. Siu, H. Zeng, and M. Chiao, Magnetically actuated MEMS microlens scanner for in vivo medical imaging, Optics Express, 15:11154–11166, 2007*). (*b*) Electrothermal tunable microlens. (*c*) Confocal microscopic probe with the microlens shown in (*b*) installed. GRIN = graded-index (*A. Jain and H. Xie, Microendoscopic confocal imaging probe based on a LVD microlens scanner, IEEE J. Sel. Top. Quant. Electron., 14:1329–1338, © 2005, IEEE*).

using electrostatic, electrothermal, or electromagnetic MEMS mirrors. The outer diameters of these MEMS-based endoscopic OCT probes range from 2.8 to 8 mm.

A few such miniature probes and corresponding 3D OCT images are shown in **Fig. 2**. Figure 2*a* shows a 4-mm-diameter probe and a 3D OCT image of cancerous hamster cheek pouch tissue acquired at 3 frames per second. An electrostatic MEMS mirror similar to the one shown in Fig. 1*a* was used in this probe. The volume of the OCT image was 1 × 1 × 1.4 mm^3. Figure 2*b* shows a 5-mm-diameter probe and an in vitro 3D OCT imaging of hamster cheek pouch acquired at 4 frames per second. The employed MEMS mirror is similar to the one shown in Fig. 1*b*. The 3D OCT image was 1.8 × 1.0 × 1.3 mm^3 and the axial resolution was approximately 4 μm in tissue. Figure 2*c* shows a 2.8-mm-diameter probe and an in vivo 3D OCT image of a finger tip. The electromagnetic micromirror shown in Fig. 1*c* was used in this probe. Figure 2*d* shows a picture of a 5.8-mm-diameter probe (the outer tube is not shown) and an in vivo 3D OCT image of a human finger acquired at 2.5 frames per second. The electrothermal MEMS mirror shown in Fig. 1*d* was used. The volume of the OCT image is 2.3 × 2.3 × 1.6 mm^3.

Confocal microendoscopic imaging. Confocal microscopy uses a pinhole at the confocal point to achieve submicrometer resolution. By changing the focal point of the light beam in the sample, high-resolution cross-sectional imaging can be achieved. The imaging depth of confocal microscopy is about 300 μm in most biological tissues, corresponding to a focal point tuning range of about 400 μm in air. Conventionally, this depth scan is realized by placing samples on a motorized microscope stage. This requires biopsy. For in vivo endoscopic imaging, tunable microlenses must be used. Tunable microlenses with more than 100-μm tuning ranges have been developed. **Figure 3** shows two examples, a magnetically actuated MEMS microlens scanner (Fig. 3*a*) and an electrothermally actuated microlens scanner (Fig. 3*b*). The magnetic microlens consists of a microfabricated ferromagnetic nickel platform and

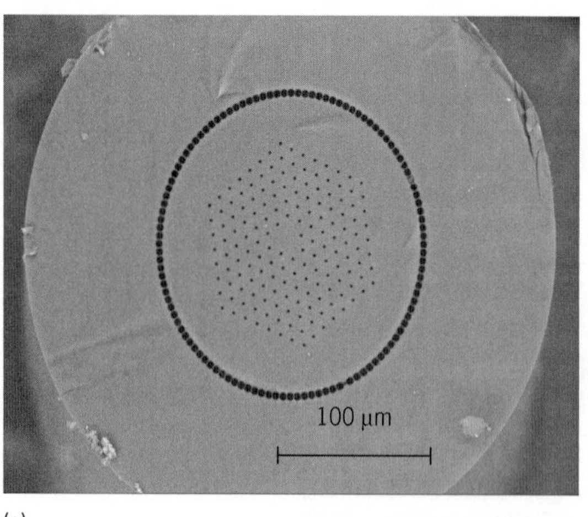

(a)

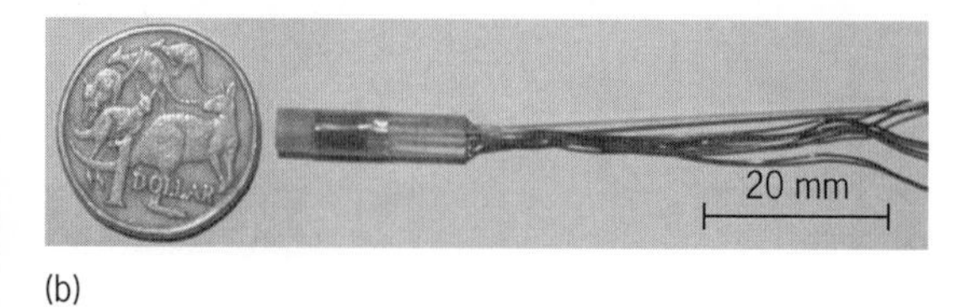

(b)

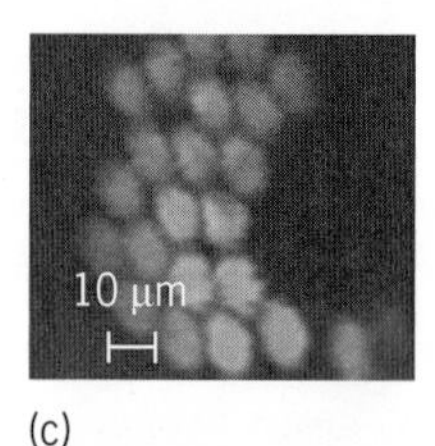

(c)

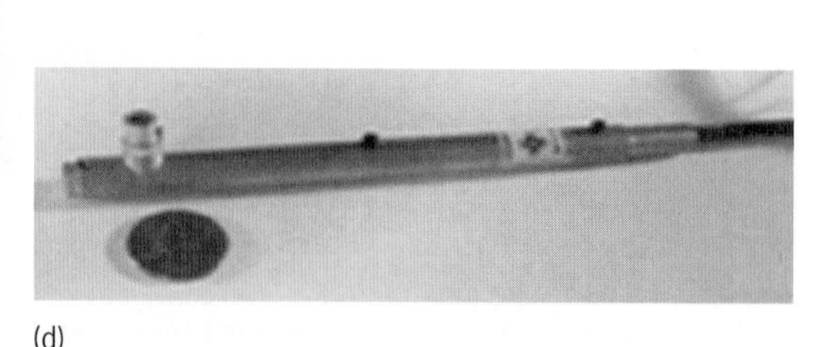

(d)

10 μm

(e)

Fig. 4. Nonlinear optical endoscopic imaging. (*a*) Cross section of a DCPCF. (*b*) Endoscopic NLOM probe using an electrothermal MEMS mirror. (*c*) NLOM image acquired by the probe in *b* (*D. Morrish et al., Nonlinear imaging by an endoscope probe incorporating a tip-tilt-piston microelectromechanical system mirror, paper NWD1, Proc. OSA Spring Optics Photonics Congress, Vancouver, BC, Canada, April 2009*). (*d*) Endoscopic NLOM probe using an electrostatic MEMS mirror. (e) NLOM image acquired by the probe in *d* (*W.Y. Jung et al., Miniaturized probe based on a microelectromechanical system mirror for multiphoton microscopy, Optics Lett., 33:1324–1326, 2008*).

a planoconvex polydimethylsiloxane (PDMS) microlens. The device area is 2 mm × 3 mm. The diameter and numerical aperture (NA) of the microlens are 1 mm and 0.2, respectively. The tunable range is 125 μm. The electrothermal microlens has a tuning range of 700 μm. The device footprint is only 2 mm × 3 mm, and such a microlens has been packaged into a small (5-mm-outer-diameter) probe (Fig. 3*c*). Confocal microscopy with a fixed lens but a MEMS mirror for lateral scanning has also been reported. In this case, the depth scan is obtained by precisely moving a stage or the entire probe. Lateral resolutions of about 0.9 μm have been achieved.

Nonlinear optical endoscopic imaging. Nonlinear optical microscopy (NLOM) uses nonlinear optical effects such as multiphoton absorption (MPA), second-harmonic generation (SHG), and third-harmonic generation (THG) to achieve submicrometer resolution. Since the nonlinear effects are detectable only when the light intensity is very high, only the signal from the focal point is collected. Different from confocal microscopy, NLOM can filter out the excitation light to achieve a very high signal-to-noise ratio, as the nonlinear effects generate doubled or tripled frequencies. In other words, the nonlinear signal has a different color compared to the input light. Conventional bench-top NLOM systems can easily handle the two colors in free space. For endoscopic applications, due to the size limitation, optical fibers must be used, but the coupling efficiency and output beam profile of optical fibers are wavelength dependent, imposing challenges to accommodate two colors which are separated far from each other in terms of wavelength. Recently, double-clad photonic crystal fibers (DCPCFs) have been developed. A DCPCF consists of a large core surrounded by both an inner and an outer cladding (**Fig.** 4*a*). The excitation light propagates through the single-mode core, and the induced, wavelength-shifted light signal is collected by the inner cladding, which has large numerical aperture that leads to high coupling efficiency. A typical DCPCF has a 16-μm single-mode core and a 163-μm inner-cladding diameter. A titanium:sapphire (Ti:sapphire) femtosecond laser is needed to generate the high-power laser pulses.

Endoscopic NLOM has been demonstrated by combining MEMS and DCPCF. Figure 4*b,d* shows a forward-viewing 5-mm-diameter and a side-viewing 10-mm-diameter NLOM probe, respectively, and Fig. 4*c,e* shows their corresponding NLOM images of fluorescence beads with 6-μm axial resolution and 1-μm lateral resolution. An electrothermal and an electrostatic micromirror were respectively employed.

Prospects. Among these optical biopsy techniques, endoscopic OCT is most mature. In the next few years, confocal microscopy or NLOM may be integrated with OCT to provide both a few millimeters imaging depth and submicrometer resolution. Among the MEMS actuation mechanisms, electrothermal and electromagnetic actuation provide better overall performance for the endoscopic applications. More investigation is still needed to increase the fill factor of MEMS mirrors and improve the packaging design. With codevelopment of MEMS, biophotonics, and probe assembling, in vivo 3D microendoscopy for early cancer detection and cancer surgery will soon be clinically available. This advance holds the promise of saving millions of lives worldwide and improving the quality of life of survivors.

For background information *see* CONFOCAL MICROSCOPY; MEDICAL IMAGING; MICRO-ELECTRO-MECHANICAL SYSTEMS (MEMS); MICRO-OPTO-ELECTRO-MECHANICAL SYSTEMS (MOEMS); NONLINEAR OPTICS; OPTICAL COHERENCE TOMOGRAPHY in the McGraw-Hill Encyclopedia of Science & Technology.

Huikai Xie

Bibliography. A. D. Aguirre et al., Two-axis MEMS scanning catheter for ultrahigh resolution three-dimensional and *en face* imaging, *Optics Express*, 15:2445–2453, 2007; L. Fu et al., Nonlinear optical endoscopy based on a double-clad photonic crystal fiber and a MEMS mirror, *Optics Express*, 14:1027–1032, 2006; A. Jain and H. Xie, Microendoscopic confocal imaging probe based on a LVD microlens scanner, *IEEE J. Sel. Top. Quant. Electron.*, 13:228–234, 2007; K. Kim et al., Two-axis magnetically-driven MEMS scanning catheter for endoscopic high-speed optical coherence tomography, *Optics Express*, 15:18130–18140, 2007; Y. Pan, H. Xie, and G.K. Fedder, Endoscopic optical coherence tomography based on a microelectromechanical mirror, *Optics Lett.*, 26:1966–1968, 2001.

Kinetic typography

Typography, unlike the spoken word, dance, music, or film, is not inherently kinetic or dynamic. The letters that make up most alphabets in most languages were designed to be read flat, frontal, and upright. But letters can be animated. And in the process of becoming kinetic, typography can take on the intonations and the voice of the spoken word, the emotional characteristics of dance or music, or the narrative qualities of film. This has resulted in new ways of reading, viewing, and accessing texts on Web sites. Visual communication thrives at the crossroads of technology and culture. The advent of television, film, video, and the computer have influenced new paradigms of visual aesthetics even though the molecular components of communication—letters and text—have remained the same for thousands of years.

Pictograms and ideograms—visual markings that represent ideas and thoughts—were used in the first attempts at recording information for what can be termed time-independent communication. The modern Western alphabet evolved in this manner as a simple, modular symbol system through which thoughts and ideas could be externalized and fixed onto a flat surface by the human hand. This system had to be learned, and many people did not have the resources, training, or opportunity to

Fig. 1. In the 1951 film *The Thing from Another World*, a darkened backdrop burns away and reveals streaking rays of vibrating light that pierce through the letterforms of the title. (*RKO Radio Pictures Inc.*)

benefit from this powerful communication tool until the fifteenth century when the German goldsmith Johannes Gutenberg began producing 42-line Bibles and other printed matter using a moveable type printing press. Gutenberg translated handwritten and calligraphic letterforms found in medieval illuminated manuscripts into simplified typographic characters cast as physical, three-dimensional objects that could be used and reused to print multiple copies of the same text efficiently and expediently.

The packaging and distribution of portable texts in the form of printed and bound books allowed ideas, information, and knowledge to enter the vast arena of public consciousness. The advent of the printing press gave thousands access to the power of the written word and spread literacy throughout the world. The printed word became a sacred body with intrinsic meaning. The discipline known as typography has evolved since Gutenberg as a practice that embodies the design, production, and application of interchangeable letterforms in order to convey a message to an audience.

Fig. 2. Designer Norman McLaren used cut-out letterforms in a stop-motion animation sequence for the Canadian Board of Tourism that was shown in Times Square, New York, on the Sony Epok Animated Electric Screen, a 720-ft^2 signboard made up of thousands of light bulbs activated by film. (*National Film Board of Canada*)

Role of the motion picture and animation. Typography evolved alongside printing technologies over the past 600 years. But it was the advent of the motion picture and animation that played an important role in the short history of kinetic typography. In 1839, Louis Jacques Mande Daguerre introduced the first successful photographic process, the Daguerreotype, for capturing still images on silver plates using a camera. Paralleling the development of photographic technology was a simple toy known as a thaumatrope, from the Greek thauma, which means wonder, magic, or miracle. The toy consists of a string that suspends a disc containing two images, one on each of its sides. When the disc is spun, the images are superimposed over one another and appear as a single image. This "perceptual" phenomenon is known as persistence of vision. The human eye holds on to images for a split second longer than they are actually projected, so that a series of quick flashes is perceived as one continuous image.

The thaumatrope, along with the photographic experiments in the 1870s of Etienne Jules Marey in France and Edward Muybridge in the United States, paved the way for the development of the motion picture camera. The actual development of moving-image technology occurred simultaneously in England, France, Russia, and the United States in the late nineteenth century. The first practical machine for projecting successive images onto a screen was the praxinoscope, invented by Emile Reynaud in 1877. Subsequently, George Eastman developed the first commercial roll camera, which replaced the cumbersome plate-image photographic technology with flexible, light-sensitive film. And in 1899, William Friese-Greene patented the first camera capable of capturing motion on a strip of light-sensitive film.

Frame-by-frame animation is the process of rendering each frame of film by hand to create a sequence of images the viewer sees as one continuous motion. Most of the early frame-by-frame animations were created as entertainment using illustrated characters rather than live action. Animators used hand-rendered, representational subjects in an attempt to provide the viewer with a simulated link to reality. They were usually domestic and farm animals—cats,

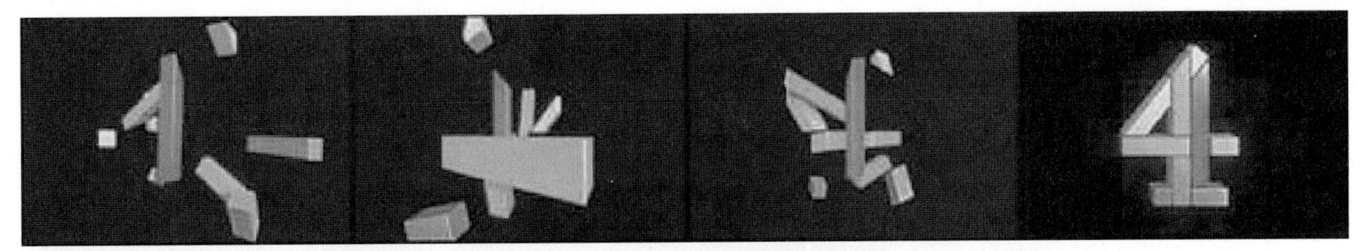

Fig. 3. Channel 4 was the first new British television network to be established since BBC2 in 1964. The Channel 4 identity was the first corporate identity specifically devised to exploit the medium of television to the fullest and to provide effective branding devices that could be developed as the character of the network evolved. The modular symbol was designed to represent the diversity of Channel 4's program sources, all of which originate from outside the company itself. In this sequence, the various parts converge in space to assemble the numeral 4. Produced in 1982, this was the first time pure computer animation techniques had been used for a U.K. television company symbol. *(Lambie-Nairn)*

dogs, mice, and ducks—whose characterization and anthropomorphism meant a human audience could relate to them.

Early examples of kinetic typography. The makers of horror and monster films in the early to mid-twentieth century used title-card technology to establish the premise of their stories and evoke an emotional response, which heightened the experience of the theater audience. In the classic 1933 film, *King Kong*, massive jungle leaves provide a slow, wiping transition between the title cards printed with monumental letterforms to imply large-scale movement through a dense, mysterious environment. In the 1951 film, *The Thing from Another World*, a darkened backdrop literally burns away and reveals streaking rays of vibrating light that pierce through the letterforms of the title and into a suspense-filled audience.

The use of text in film was originally limited to title cards—two-dimensional surfaces on which dialogue, announcements, warnings, news items, and credits were handwritten or printed. The pioneer filmmaker, George Melies, whose 1902 film, *A Trip to the Moon*, remains a masterpiece of early motion-picture technology, experimented with animated letterforms in advertising films as early as 1899. Another pioneer, D. W. Griffith, incorporated title cards in his films, *The Birth of a Nation* (1915) and *Intolerance* (1916), both of which were considered narrative and structural breakthroughs in the history of the motion picture as an art form. In *Intolerance*, Griffith used a series of title cards as significant components of the film. Each card contained a composition of static letterforms superimposed over a photographic background that introduced the historic time period about to be presented.

In the 1920s and 1930s, Walt Disney led the technological advancements in animation by rendering smoother frame actions, synchronizing music, sound, and dialogue, and incorporating color. But these efforts toward the perfection of realism caused a revolt among animators and designers who wanted to break creative boundaries. The advent of television technology and the exponential growth of the advertising industry in the early 1930s created a prime arena in which to experiment with film and animation. Instead of presenting static postcard realism in the background of a composition, commercial animators began incorporating symbolic and iconographic elements to create messages. Most specifically, they explored the use of letterforms. In the 1950s and 1960s, innovators such as Norman McLaren, Saul Bass, and Pablo Ferro used animation techniques to hand-render seamless compositions—and interactions—of type and image on film for television commercials, short films, and film titles (**Fig. 1**).

Initially, animated letterforms were restricted to supporting roles, such as announcing the beginning of films, identifying television stations, or providing information in commercials. In 1961, the animator Norman McLaren brought the animated letterform to center stage in an advertisement commissioned by the Canadian Board of Tourism. McLaren used cutout letterforms in a stop-motion animation sequence that was shown in Times Square, New York, on the Sony Epok Animated Electric Screen, a 720-ft^2 signboard made up of thousands of light bulbs activated by film (**Fig. 2**). Type historian Beatrice Warde described it as follows:

"Do you wonder that I was late for the theater when I tell you that I saw two Egyptian A's . . . walking off arm in arm with the unmistakable swagger of a music-hall comedy team? I saw base serifs pulled together as if by ballet shoes, so that the letters tripped off literally sur les pointes . . . I saw the word changing its mind about how it should look (caps? lower case? italic?) even more swiftly than a woman before her milliner's mirror . . . after forty centuries of the necessarily static Alphabet, I saw what its members could do in the fourth dimension of Time, 'flux,' movement." *(From R. S. Hutchings, ed., Alphabet 1964: International Annual of Letterforms, Kynoch Press, Birmingham, U.K., 1964)*

The role of digital technologies. The desktop computer has transformed a predominantly passive communication experience into an interactive one and increased the accessibility and control of information. Information has grown more complex and is now distributed by an overwhelming array of means. The activity of reading has diversified with such formats as e-mail, web sites, smart phones, and e-book readers.

Consequently, the historical evolution of two-dimensional, static letterforms arranged and fixed in a horizontal line is shifting course. Type is no

Fig. 4. A collage of overlapping letterforms in various sizes to convey an explosion of type from the swinging bat of a capital letter T. (*Ned Drew*)

longer restricted to the characteristics found in the medium of print such as typeface, point size, weight, leading, and kerning. Letterforms have behavioral, anthropomorphic, and otherwise kinetic characteristics, including text that liquefies and flows, and three-dimensional structures held together by lines, planes, and volumes of text through which a reader may travel. These are only a few examples of the impact digital technology is having on the once simple and humble letterform (**Fig. 3**).

The page-frame as a compositional structure, with the conventional hierarchies of headings, columns, and margins is also expanding. In contrast to the materiality and permanence of the print medium, readers now navigate through bodies of text in a digital environment such as a web site or mobile telephone. Pointing and clicking has become a commonplace action to access multiple pages in a Web site. Smart-phone navigation systems allow the reader to touch a screen and slide through pages of text. All of this activity falls under the category of kinetic typography.

Production of kinetic typography. Kinetic typography is generally produced with mass-market software applications, such as Adobe® Flash® and Adobe® After Effects®; Apple® Motion; programming languages such as Processing (http://www.processing.org); and an array of freeware and other custom software.

The montage and collage are two types of structures found in kinetic formats. Montage structures involve the planning, assembling, and editing of a multiframe sequence to create a narrative—a cause-and-effect event or continuous series of events usually with a beginning, middle, and an end. Here, letterforms serve as both a verbal and visual function. Freezing a montage sequence captures a single moment in a larger continuous motion.

A collage is a more impressionistic—almost printlike—assembly of individual frames into a sequence without a fixed order of events. Here, letterforms are not always intended to be read but most always serve as components to a visually arresting typographic texture. Freezing a collage sequence reveals an individual, "stand-alone" composition, much like a poster or painting (**Fig. 4**).

For background information *see* CINEMATOGRAPHY; COMPUTER GRAPHICS; MULTIMEDIA TECHNOLOGY; PRINTING; TYPE (PRINTING); WORLD WIDE WEB in the Encyclopedia of Science & Technology.

Matthew Woolman; Jeff Bellantoni

Bibliography. J. Bellantoni and M. Woolman, *Type in Motion: Innovations in Digital Graphics*, Thames and Hudson, London, and Rizzoli International Publications, New York, 1999; R. Carter, B. Day, and P. Meggs, *Typographic Design: Form and Communication*, 3d ed., John Wiley & Sons, Hoboken, NJ, 2002; P. B. Meggs and A. Purvis, *Megg's History of Graphic Design*, 4th ed., John Wiley & Sons, Hoboken, NJ, 2006; M. Woolman, *Motion Design: Moving Graphics for Television, Music Video, Cinema, and Digital Interfaces*, RotoVision, SA, Crans, Switzerland, 2004; M. Woolman and J. Bellantoni, *Moving Type: Designing for Time and Space*, RotoVision, SA, Crans, Switzerland, 2000.

Laser cooling of solids

The term "laser cooling" most often is associated with cooling dilute gases of atoms or ions to extremely low temperatures by employing the Doppler effect to reduce their thermal translational energy.

This area of science has progressed immensely. However, it is not widely know that in 1929, some 46 years before Doppler cooling of atoms was even contemplated, the physicist Peter Pringsheim suggested the possibility of cooling solids by optical means. In the solid phase, atoms do not possess relative translational motion; their thermal energy is largely contained in the vibrational modes of the lattice. Laser cooling of solids (or optical refrigeration) is similar to atom cooling: Light quanta in the red tail of the absorption spectrum are absorbed from a monochromatic source followed by spontaneous emission of more energetic (blue-shifted) photons, a process known as fluorescence up-conversion. In the case of solids, the extra energy is extracted from lattice phonons, the quanta of vibrational energy in which heat is contained. The removal of these phonons cools the solid.

Laser cooling of solids can be exploited to achieve an all-solid-state cryocooler. Intensive research has been motivated by the advantages of compactness; the absence of vibrations, moving parts, or fluids; high reliability; and of having no need for cryogenic fluids. Space-borne infrared sensors are likely to be the first beneficiaries, with other applications requiring compact cryocooling reaping the benefits as the technology progresses. For low-power, space-borne operations, ytterbium-based optical refrigeration could outperform conventional thermoelectric and mechanical coolers in the temperature range between 80 and 170 K (−193 and −103°C).

Cooling with light. The process of optical refrigeration is somewhat counterintuitive: Shining a high-power laser at some solid objects makes them cooler, rather than hotter. This surprising phenomenon occurs only in special high-purity materials that have appropriately spaced energy levels and emit light with high quantum efficiency. To date, optical refrigeration research has been confined to glasses and crystals doped with rare-earth elements and to direct-band-gap semiconductors such as gallium arsenide, in which radiative recombination between free electrons and holes is allowed. Although laser cooling of rare-earth-doped solids has been successfully demonstrated, observation of net cooling in semiconductors has remained elusive to date, due primarily to material purity issues.

Figure 1 schematically depicts the optical refrigeration processes for a two-level system with a group of closely spaced ground-state energy levels and a similar collection of excited-state levels. Photons from a laser with energy $h\nu$ excite atoms from the top of the ground state to the bottom of the excited state; ν is the frequency of the light and h is Planck's constant. The excited atoms reach quasi-equilibrium with the lattice by absorbing vibrational energy from the lattice or phonons. Spontaneous emission (fluorescence) follows with a mean photon energy $h\nu_f$ that is higher than that of the absorbed photon. This process has also been called anti-Stokes fluorescence.

Cooling by anti-Stokes fluorescence can be described as reaction (1), where q is the heat removed

$$h\nu + q \rightarrow h\nu_f \tag{1}$$

from the solid. The cooling efficiency or fractional cooling energy for each photon absorbed is given by Eq. (2), where $\lambda = c/\nu$ is the wavelength of the laser

$$\eta_c = \frac{q}{h\nu} = \frac{h\nu_f - h\nu}{h\nu} = \frac{\lambda - \lambda_f}{\lambda_f} \tag{2}$$

light, and $\lambda_f = c/\nu_f$ is the mean wavelength of the fluorescence.

Cooling with rare-earth-doped solids. The invention of the laser prompted many attempts to observe the optical refrigeration experimentally, but the usual heating generated by the absorption of laser light frustrated these efforts. In 1995, scientists first achieved net laser cooling in solids. These experiments overcame two technical challenges; they used materials in which the vast majority of optical excitations recombine radiatively (that is, materials with a high quantum efficiency) and for which there is a minimal amount of parasitic heating due to unwanted impurities. Both of these critical engineering issues are ignored in the idealized situation described by the above equations, but are key to experimental success.

The material used for laser cooling was an ytterbium-doped fluoride glass. Ytterbium is one of the rare-earth elements. For many years rare earths have been used in lasers and optical amplifiers largely because of their high quantum efficiency. The key optical transitions in rare-earth-doped ions involve 4*f* electrons, which have smaller electronic orbitals than the filled 5*s* and 5*p* outer shells. This shielding limits interactions with the surrounding atoms and lessens the probability of excited rare-earth ions producing heat rather than emitting light photons.

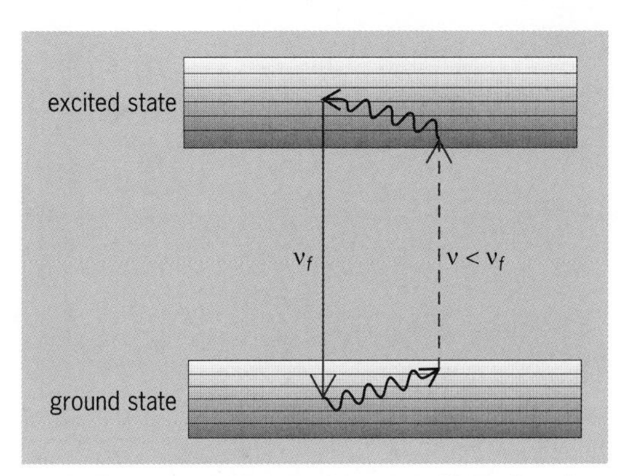

Fig. 1. Energy-level diagram of an atom with two broad levels, which is embedded in an otherwise transparent solid. Laser photons of energy $h\nu$ (h = Planck's constant, ν = frequency) excite atoms near the top of the ground-state level to the bottom of the excited state. Radiative decays occurring after thermalization emit photons with average energy $h\nu_f > h\nu$.

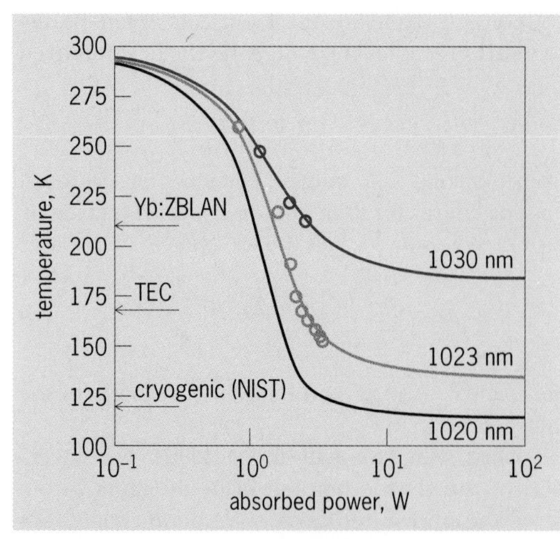

Fig. 2. Data (circles) and model fits (lines) using Eq. (3) for temperature versus absorbed power in an ytterbium-doped YLF crystal at different wavelengths. A temperature of ~115 K (−158°C) is predicted at the peak of the E4-E5 excitation (1020 nm). Arrows indicate the previous cooling record in an ytterbium-doped fluoride glass (Yb:ZBLAN), the lowest temperature accessible by a standard thermoelectric cooler (TEC), and the NIST-defined cryogenic temperature (123 K or −150°C).

The initial proof-of-principle experiments achieved cooling to only 0.3 K below the ambient temperature. Subsequent improvements led to cooling by 85 K in similar ytterbium-doped fluoride glasses, to a temperature of 208 K (−65°C). More recently, cooling from room temperature to 155 K (−118°C, $\Delta T \sim 145$ K) was demonstrated in an ytterbium-doped yttrium lithium fluoride (YLF) crystal at a laser wavelength of 1023 nm (**Fig. 2**). Optical refrigeration has therefore entered the so-called cryogenic regime by surpassing the performance of standard thermoelectric coolers, which exhibit vanishing efficiency at temperatures below 170 K (−103°C).

This record operation in a YLF crystal was made possible by tuning the laser light to the sharp resonance between levels in the ground-state and excited-state levels (centered at 1020 nm) in the crystalline host, and by the relatively high doping density (5%) that led to a superior absorption efficiency compared to that of a fluoride glass host. This becomes clearer from examination of Eq. (3), a

$$\eta_c = p(\nu)\frac{h\nu_f}{h\nu} - 1 \qquad (3)$$

more realistic expression than Eq. (2) for the cooling efficiency, because it includes various nonradiative and parasitic losses.

Here Eq. (2) has been modified by introducing a new efficiency parameter $p(\nu) < 1$, which denotes the probability that an absorbed photon at frequency ν will lead to a fluorescence photon outside the material. In the ideal case represented by Eq. (2), this probability was assumed to be unity. In real materials there is always a competition that determines whether the laser photon is absorbed on the dopant ion (as shown in Fig. 1) or on some contaminant atom or other impurity

Maximizing the cooling effect. For a given material, the lowest temperature at which η_c changes sign is essentially its minimum achievable temperature (T_m) by laser cooling. This occurs at a particular frequency ν_m (or wavelength λ_m) such that $\eta_c(T_m,\nu_m) = 0$. A suitable cooling material, therefore, will have

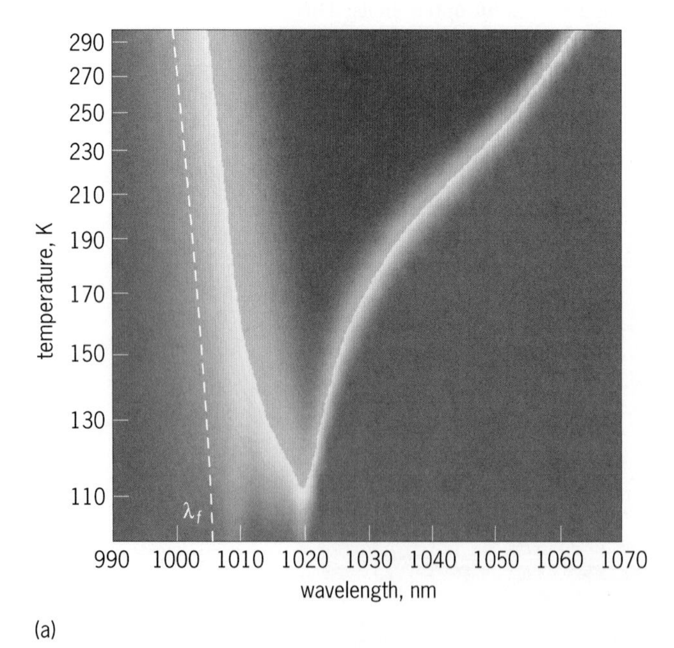

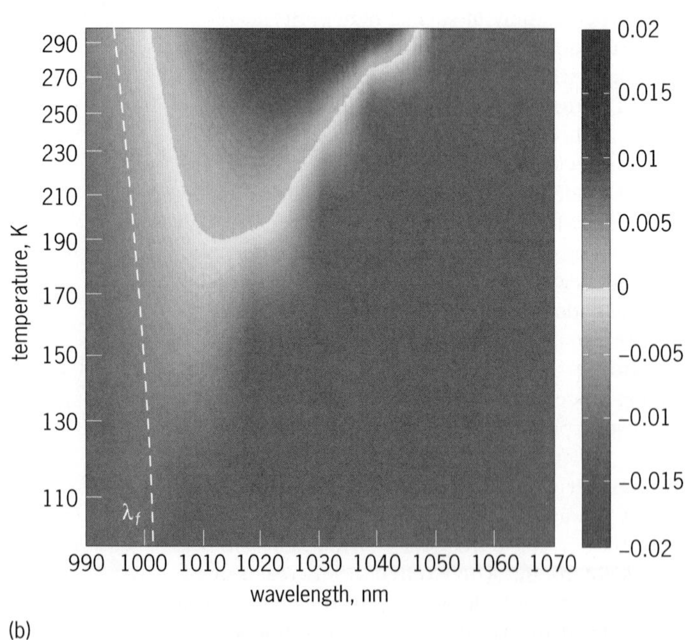

Fig. 3. Comparison of maps of cooling efficiency versus temperature and excitation wavelength for (*a*) 5%-doped Yb:YLF crystal and (*b*) 1%-doped Yb:ZBLAN glass. The minimum achievable temperature is $T_m \sim 190$ K (−83°C) in Yb:ZBLAN as compared to $T_m = 110$ K (−163°C) in Yb:YLF. Superior performance of YLF is attributed to higher doping concentration and to sharper resonance in the absorption spectrum at $\lambda \sim 1020$ nm. The broken (white) line corresponds to mean fluorescence wavelength (λ_f) as a function of temperature.

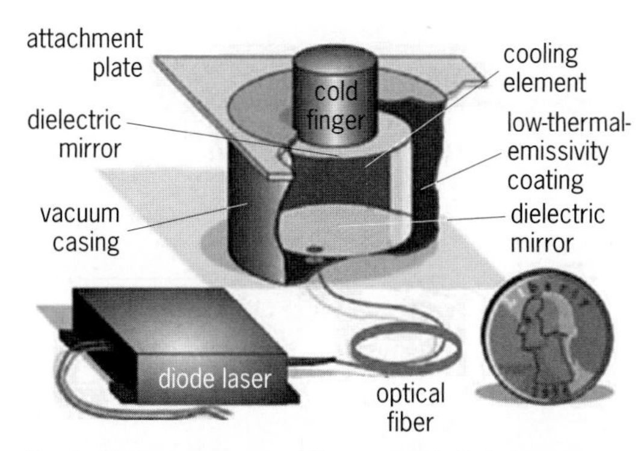

Fig. 4. Optical cryocooler. Fiber-coupled diode laser excites the cooling element, which is housed in a vacuum chamber and suspended so that it does not touch the warm chamber walls. Dielectric mirrors trap laser radiation for maximum absorption. The cold finger is attached to the cooling element and is shielded from fluorescence by a dielectric mirror. The object that one wants to cool is connected to the cold finger.

large absorption by the cooling ion, small background absorption (high purity), and high quantum efficiency. For comparison, thermo-spectral maps [contour maps of $\eta_c(T,\nu)$] for Yb:ZBLAN glass and Yb:YLF (crystal) are shown side by side in **Fig. 3**.

From the above expressions for the efficiency, one can see that the efficiency is inversely proportional to the energy of the laser photon. The numerator, $h\nu_f - h\nu$, is typically limited by the characteristic energy of thermal vibrations in the solid, kT, where k is Boltzmann's constant and T is the absolute temperature. The cooling efficiency therefore increases when lower-energy, longer-wavelength laser light can be used. For ytterbium-based optical refrigerator coolers, the pump laser light has to have wavelengths near 1 μm. In 2000, optical refrigeration in a glass doped with thulium, which used light near 1.9 μm, was shown. These experiments verified the scaling of efficiency with pump wavelength, demonstrating nearly a factor-of-2 enhancement in the cooling efficiency over ytterbium-doped systems.

Optical refrigeration has advanced from basic principles to a promising technology. Cooling of rare-earth-based materials has now approached cryogenic operation. In the next few years, optical refrigeration should be useful in applications, such as satellite instrumentation, where compactness, ruggedness, and the lack of vibrations are important. A sketch of cryocooler device based on optical refrigeration is shown in **Fig. 4**. Many more applications would be possible if the basic efficiency of optical refrigerators could be improved. If the fluorescent photons can be recycled by means of photovoltaic devices then the efficiency can be improved. Optical refrigerators clearly have the potential to make a significant impact on the electronics and photonics of the future.

For background information *see* FLUORESCENCE; INFRARED ASTRONOMY; LASER; LASER COOLING; PHONON; RARE-EARTH ELEMENTS; YTTERBIUM in the McGraw-Hill Encyclopedia of Science & Technology.

Richard I. Epstein; Mansoor Sheik-Bahae

Bibliography. R. I. Epstein et al., Observation of laser-induced fluorescent cooling of a solid, *Nature*, 377:500–503, 1995.; R. I. Epstein and M. Sheik-Bahae, *Optical Refrigeration*, Wiley-VCH, 2009; C. E. Mungan, M. I. Buchwald, and G. L. Mills, All-solid-state optical coolers: history, status, and potential, in S. D. Miller and R. G. Ross Jr. (eds.), *Cryocoolers 14*, pp. 539–548, ICC Press, Boulder, CO, 2007; Optical refrigeration sets solid-state cooling record, *Phys. Today*, 63(4):14, April 2010; D. V. Seletskiy et al., Laser cooling of solids to cryogenic temperatures, *Nat. Photonics*, 4:161–164, 2010; M. Sheik-Bahae and R. I. Epstein, Optical refrigeration, *Nat. Photonics*, 12:693–699, 2007.

Lithic use-wear

Lithic use-wear or microwear analysis is the examination of surfaces on experimental or archeological stone tools to determine tool use. Primarily used to determine the functions of chipped stone tools, in particular those produced on silicates such as chert, flint, quartz, and obsidian, it also has been used on ground stone artifacts as well. It typically involves the use of equipment to magnify surfaces so that features such as edge chipping, striations or scratches, and abrasion or polish can be more easily observed, interpreted, and recorded. Traditionally, archeologists have relied on optical microscopes to analyze stone tools. However, new technology, in the form of digital imaging of surfaces and measurement of surface microtopography using lasers, is changing the manner in which lithic use-wear analysis is done.

History of functional analysis. Scientific study of stone tools generally began in the mid-nineteenth century. Around this time, the speculative functional approach was employed as a way to ascertain how stone tools may have been used. This approach was based essentially on the notion that tool form can be used to determine function. More or less, the morphologies of archeological stone tools were compared to other tools with known uses. If their shapes or designs were similar, then it was inferred that their uses must also be similar. Comparisons were made between stone tools and metal and wooden implements whose functions were known. Not surprisingly, this approach also included comparisons between archeological tools and those used by indigenous peoples, who still relied on stone tool technology for many of their daily tasks.

Experimental studies to determine stone tool use began around the same time as the speculative functional approach. Such experimental studies were typically of two types: efficiency studies and direct verification. In efficiency studies, a stone tool was used typically for a task that the experimenters already believed it should be able to perform to test how well it could be done. For the most part, efficiency studies did not rely on examination of the wear on the tested tools. Direct verification relied on experimental use to prove or disprove that a tool could be used in the successful completion of a suspected function. Direct verification incorporated the comparison of the wear traces on the experimental tool with those on the prehistoric artifact, but was usually accomplished using the naked eye or very low levels of magnification, involving hand lenses. Overall, this work was very unscientific and lacked procedural standardization or objective recording of results.

Work by Sergei Semenov. Lithic use-wear analysis, as a systematic and reliable method to determine stone tool use, received a tremendous boost with the dissemination of the results of experiments and analyses conducted by the Russian archeologist Sergei A. Semenov (1898–1978). Semenov revolutionized functional analysis of stone tools by demonstrating a method to examine stone tools under low magnification using an optical microscope and a comparative experimental collection of tools. The English translation of his book, *Prehistoric Technology*, in 1964 exposed Western archeologists to this new technique and it received much attention. Semenov primarily concentrated on striations, but noted many other important aspects associated with use-wear analysis, including edge chipping, polishes, tool position during use, and duration of use.

Low-power approach. Ruth Tringham, following a visit to Semenov's lab in 1966–1967, and her students at Harvard University concentrated on the examination of the edges of experimental stone tools at low magnification in the late 1960s and throughout the 1970s. The resulting development of what came to be known as "low-power" use-wear analysis proved quite successful as a method for analyzing stone tools to reconstruct past human behavior. The low-power approach relies on magnifications generally lower than 100-fold using stereoscopic microscopes and dark-field illumination (angled light) to specifically view edge damage (microchipping), edge rounding, and scratches or striations (**Fig. 1**). These attributes can determine the location of use on the stone tool, the direction of use (that is, longitudinal, transverse, or rotary), and the general hardness of the worked material (that is, soft, medium, or hard). For example, soft materials typically include meat, fresh hide, and some plants; medium hardness materials include wood; and hard materials include antler, bone, stone, and shell.

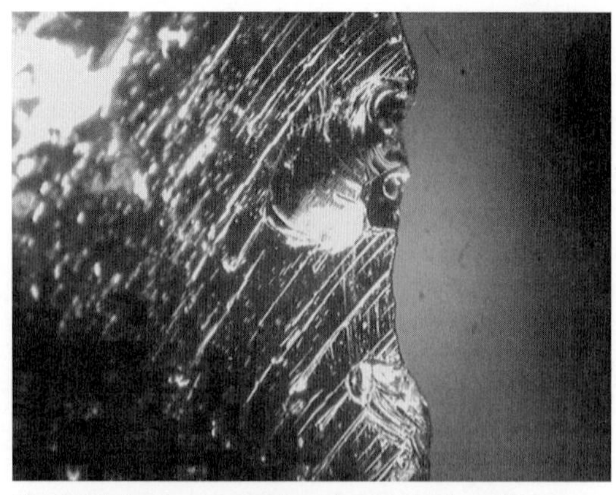

Fig. 1. Photomicrograph of an obsidian tool used to cut wood taken using a stereoscopic microscope and dark-field illumination (magnification: 40×). Note the edge chipping and striations used to identify the hardness of the worked material and the direction of use.

Fig. 2. Photomicrograph of a flint tool used to saw shells taken using a compound/metallographic microscope and bright-field illumination (magnification: 200×). Note the micropolish and the fine striations in the polish zone to identify the motion and the worked material.

High-power approach. Following on the heels of the low-power technique was the development of another method that allowed archeologists to examine stone tool surfaces at much higher magnifications. What was eventually called "high-power" use-wear analysis was based on the work of Lawrence H. Keeley at Oxford University in the 1970s. This technique focused on the polishes produced on worn stone tools as a way to identify tool function. Following the publication of Keeley's seminal work on high-power use-wear analysis, entitled *Experimental Determination of Stone Tool Uses*, in 1980, many archeologists adopted this method as a way to reconstruct stone tool use. The high-power approach involves magnifications at significantly higher levels (typically greater than 100- to 500-fold) and relies upon the ability to detect micropolishes on stone tool surfaces using bright-field illumination (incident or vertical light transmitted through the microscope lenses) and compound/metallographic microscopes. Polishes are described in terms of their brightness, volume, degree of linkage, and extent (**Fig. 2**). Topographic features, such as striations and, to a lesser degree, edge chipping, are also used to determine the location of use on the tool, the type of action (that is, scraping, whittling, cutting, sawing, or drilling), and the type of worked material (that is, wood, dry hide, meat, or bone). The main advantage of high-power use-wear analysis is the ability to assign wear to more specific worked materials; however, disadvantages include the increased amount of time required to analyze a tool and the loss of detail in surface topography when using a metallographic microscope at high magnifications.

Scanning electron microscopy. A third method used to determine stone tool function based on use-wear relies on scanning electron microscopy (SEM). It usually involves very high magnification (greater than 500- to 10,000-fold) and provides excellent depth of field. Although this method provides very detailed images of stone tool surfaces, such as the morphology of the interior surfaces of striations, it presents a number of potential issues not found with the two other methods discussed thus far. The main drawbacks of using the SEM include the time and cost associated with sample preparation (that is, some samples need to be coated in metal to increase conductivity in the vacuum chamber) and the small number of samples from an artifact assemblage that can be reasonably examined.

However, one further advantage of using SEM on stone tools is the ability to observe residues of worked materials, although, strictly speaking, this does not constitute the observation and interpretation of surface wear on tool surfaces. The type of worked material may be identified, possibly down to the level of genus or taxon, based on the presence of primarily inorganic residues preserved on the surface, sometimes within the stone itself, or in the microcracks of the stone tool. Residues adhering to the stone tool surfaces may be matched to organic tissues or inorganic substances from known sources. Archeologists have variously identified residues from plants [for example, starch grains, pollen grains, phytoliths (small diagnostic silica structures that form in the veins of plants), and raphides (long, needle-shaped crystals, usually consisting of calcium oxalate, occurring as a metabolic by-product in certain plant cells)] and animals (for example, keratin, erythrocytes, calcium carbonate, and aragonite).

Quantification of use-wear. One of the main criticisms of the optical microscopic techniques, though, is their reliance on human observation and qualitative descriptions of use-wear characteristics such as edge chipping, striation types, and polish types. They also have been criticized because descriptions of use-related damage are qualitative assessments that have not been standardized within the discipline. Primarily because of the concern over the subjectivity of the visual techniques, some archeologists have been working to develop more objective, quantitative methods for determining the functions of stone tools. So far, most of this work has focused on surface wear, but archeologists also have turned to documenting the function of stone tools based on chemical analysis to identify worked material types. Attempts to quantitatively record, describe, and discriminate use-wear have involved the use of techniques based on various forms of computer-assisted surface texture analysis using digital images, including expert systems, and types of surface metrology techniques for measuring microsurfaces. Texture analysis is based on recording the reflectivity of surfaces to produce images of surface features on a microscopic scale. Metrology uses a variety of techniques, including laser profilometry, atomic force microscopy (AFM), and laser scanning confocal microscopy (LSCM), to measure changes in surface structure on a micrometric scale. Some surface analyses incorporate fractal geometry (**Fig. 3**). The measurement of the elemental composition of stone tool surfaces is used to determine what specific materials were contacted by the tools and has included measurement of various elements, including calcium, manganese, phosphorus, and sulfur. This is based on the notion that the elements from the worked material adhere to or become trapped in the tool surface and can be detected long after the tool was used. Techniques used to measure elemental composition of stone tool surfaces include ion beam analysis (IBA), laser ablation inductively coupled plasma mass spectrometry (LA-ICP-MS), proton-induced x-ray emission spectroscopy (PIXE), and Rutherford backscattering (RBS). Most of the methods to quantitatively document use-wear are still in the experimental stages of development, although minimal testing on stone tools recovered from archeological excavations has been done with some success.

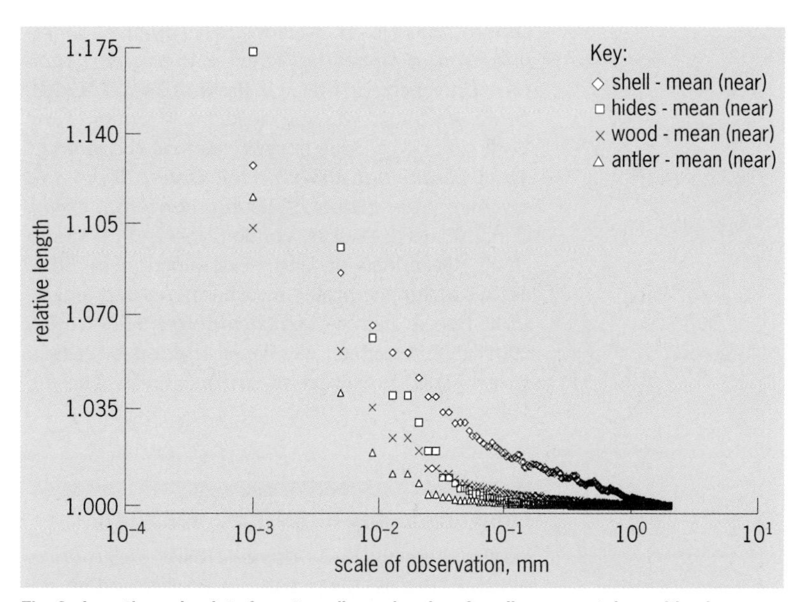

Fig. 3. Length-scale plots from two-dimensional surface line scans taken with a laser profilometer. The measurements were taken on the used ("near") surfaces of experimental flint flakes used on four different worked materials. The graph represents the conversion of the line scans to mean measures of relative length, which allows for the mathematical characterization and discrimination of the used tool surfaces. In this graph, tools used on different contact materials can be mathematically discriminated from each other based on their different relative lengths at the finest scales of measurement, beginning at about 0.1–0.001 mm.

For background information *see* ANTHROPOLOGY; ARCHEOLOGY; CONFOCAL MICROSCOPY; OPTICAL MICROSCOPE; PALEOLITHIC; PHYSICAL ANTHROPOLOGY; PREHISTORIC TECHNOLOGY; SCANNING ELECTRON MICROSCOPE in the McGraw-Hill Encyclopedia of Science & Technology. James Stemp

Bibliography. A. A. Evans and R. E. Donahue, Laser scanning confocal microscopy: A potential technique for the study of lithic microwear, *J. Archaeol. Sci.*, 35:2223–2230, 2008; R. L. K. Fullagar, Residues and usewear, pp. 207–234, in J. Balme and A. Paterson (eds.), *Archaeology in Practice: A Student Guide to Archaeological Analysis*, Blackwell,

Oxford, 2006; L. H. Keeley, *Experimental Determination of Stone Tool Uses: A Microwear Analysis*, University of Chicago Press, 1980; G. Odell, *Lithic Analysis*, Kluwer Academic/Plenum, New York, 2004; S. A. Semenov, *Prehistoric Technology: An Experimental Study of the Oldest Tools and Artefacts from Traces of Manufacture and Wear*, Cory, Adams & Mackay, London, 1964; W. J. Stemp et al., Quantification and discrimination of lithic use-wear: Surface profile measurements and length-scale fractal analysis, *Archaeometry*, 51:366–382, 2009; P. C. Vaughan, *Use-Wear Analysis of Flaked Stone Tools*, University of Arizona Press, Tucson, 1985.

Long noncoding RNAs (long ncRNAs)

Long noncoding RNAs (long ncRNAs), a group of ncRNAs (that is, RNAs without protein-coding potential), are emerging as important regulatory components in the cell. Many ncRNAs can be clearly classified on the basis of established features into ribosomal RNA (rRNA), transfer RNA (tRNA), small nuclear RNA (snRNA), small nucleolar RNA (snoRNA), microRNA (miRNA), and so forth. These groups have distinct features of their own, and the members of each group are thought to share similar function. However, substantial numbers of ncRNAs are unclassified in either of these two groups. The dominant ones among them are relatively long (more than 200 nucleotides up to 100,000 nucleotides or more), so they are collectively called long ncRNAs or macroRNAs. The term mRNA-like ncRNAs is also applicable to many of them, which share several features with protein-coding messenger RNA (mRNA), including posttranscriptional modifications through 5′ capping, splicing, and polyadenylation tailing (addition of adenine nucleotides to the 3′ end of mRNA molecules). However, unlike the other ncRNA groups, there is no common feature among the long ncRNAs that suggests their function.

Long ncRNAs are thus a collection of miscellaneous ncRNAs, and many are still enigmatic. However, they attract intense scientific interest, because they are the dominant RNA group in humans and mice, and well-studied examples have been shown to exert essential functions.

Abundance and localization in the genome. Protein-coding regions account for only 2% in the human genome. Previously, the remaining 98% was thought to be "junk spacer" between important regions. However, recent technical advances in RNA analysis have identified unexpected numbers and amounts of long RNAs that lack protein-coding potential. It is now known that as much as 70% of the mouse genome (including noncoding regions) is transcribed and yields ncRNAs (including long ncRNAs) as well as protein-coding mRNAs. In terms of amount, approximately two-thirds of long polyadenylated RNAs (more than 200 nucleotides in length), which comprise mRNAs and mRNA-like ncRNAs, are estimated as ncRNAs in humans. This means that long ncRNA is dominant over protein-coding mRNA in humans and mice.

In addition to noncoding regions, transcription of many long ncRNAs extends to and overlaps with protein-coding regions in either a sense or antisense manner, forming a complex intercalation with protein-coding genes and sometimes other ncRNAs. Some long ncRNAs are transcribed from the ultraconserved regions (UCRs), where genomic sequence is exceptionally conserved among species. This suggests a critical function of these regions or the ncRNAs (or both), which are conserved beyond evolution.

Mechanism of action and function. The established mechanism by which long ncRNA exerts its function typically occurs in two ways ([A] and [B]; see **illustration**). In the first mechanism [A], long ncRNA binds with some partner (or partners). This binding either affects the function of the partner (partners) or recruits/tethers the partner (partners) to the specific genomic region (where the ncRNA is transcribed in some cases, but it may be away from the site of transcription in other instances) and sometimes away from the site of transcription. In contrast, in the other mechanism [B], the ncRNA molecule is not important for its function. It is the transcription per se through the region, rather than the RNA molecule, that contributes to the function.

Among a small number of long ncRNAs (compared to the total number) that have been studied intensely, many are involved in the regulation of gene expression, typically by altering chromatin structure or chromatin modification (epigenetic regulation) or by controlling transcription factors or machinery. In fact, 38% of the long ncRNAs bind with well-characterized chromatin-modifying protein complexes [including PRC2 (polycomb repressive complex 2), which is responsible for an essential repressive chromatin modification] and regulate gene expression through them. In contrast, several emerging reports now suggest that some long ncRNAs have a different function (and mechanism of action) from the aforementioned ones, as shown at the bottom of the illustration.

Long ncRNA in epigenetic regulation. HOX transcription factors are essential regulators for specifying positional identities of cells during development. In humans, *HOX* genes are clustered on four different genomic regions that are designated *HOXA*, *HOXB*, *HOXC*, and *HOXD*. The long ncRNA *HOTAIR* is transcribed from the *HOXC* cluster, binds with and recruits PRC2 to the *HOXD* locus on a different chromosome, and consequently regulates *HOXD* gene expression. Recently, an extensive role of *HOTAIR* at the genome-wide level (as well as playing a role in disease) has been reported.

Radiation-induced DNA damage stops the cell cycle by suppressing the *cyclin D1* gene, and long ncRNAs transcribed from the promoter region of the *cyclin D1* gene are involved in this process. These ncRNAs bind with and tether a protein, TLS, to the

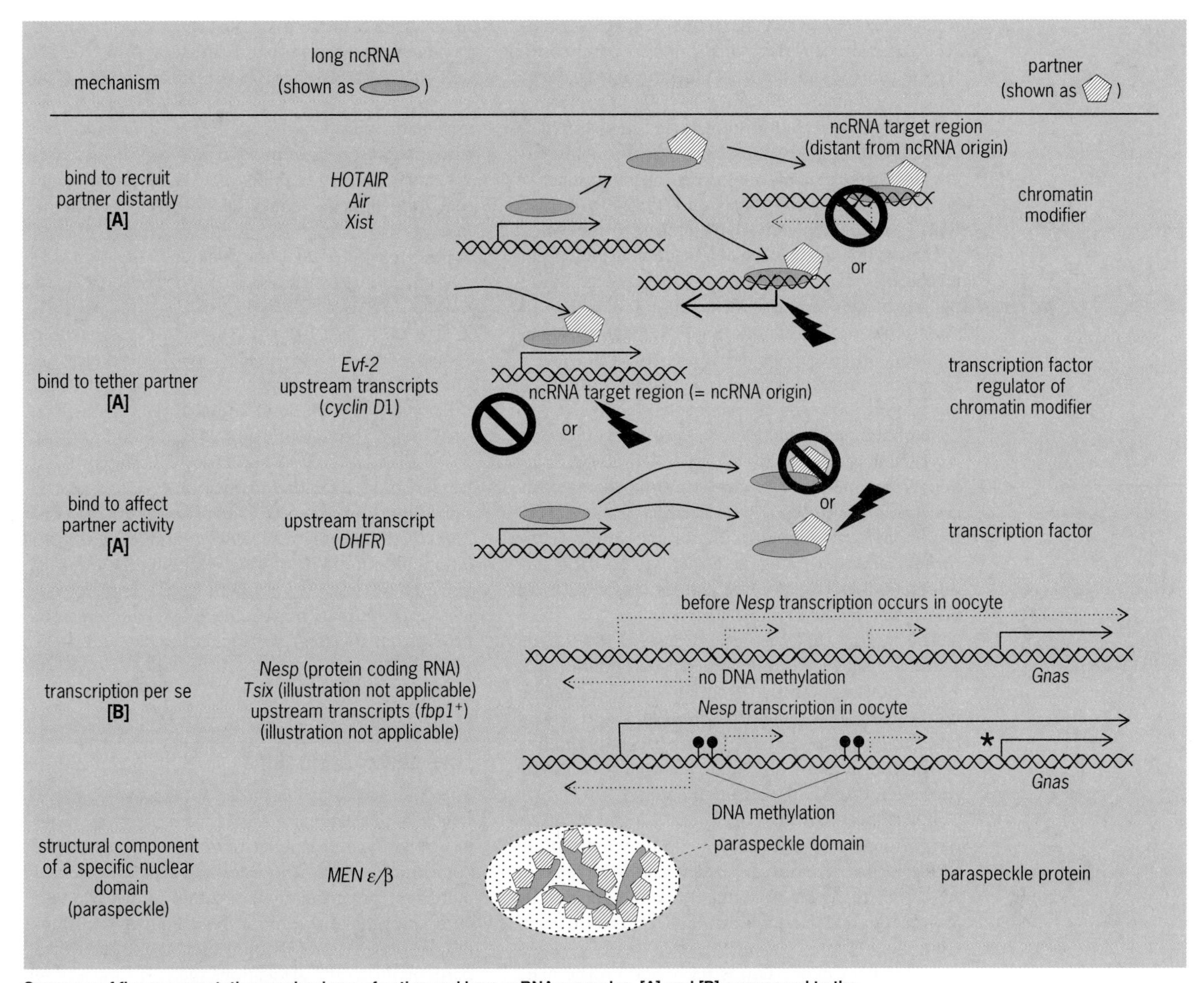

Summary of five representative mechanisms of action and long ncRNA examples. [A] and [B] correspond to the classifications in the text. Gray oval = long ncRNA; hatched pentagon = binding partner; double spiral = genomic DNA; arrow along DNA = active transcription of ncRNA or mRNA; dotted arrow = repressed transcription; prohibition mark = repression; zigzag (bolt) = activation. *_Gnas_ promoter is protected from DNA methylation, presumably as a result of its active chromatin modification. Note that the illustrations are not applicable to some ncRNA examples listed.

cyclin D1 promoter region, where TLS inhibits histone acetyltransferase (a chromatin modifier) to suppress promoter activity.

Long ncRNA in genomic imprinting. This is a special case of epigenetic gene regulation. Genomic imprinting refers to differential DNA methylation marks between oocytes and sperm at specific sites, which are established during germ-cell development. For example, the methylated gene promoter in the oocyte remains inactive after fertilization (as a maternal allele), whereas the unmethylated paternal allele is active, which results in paternal allele-specific expression of this imprinted gene.

Many loci with genomic imprinting express long ncRNAs in an allele-specific manner. Some of them are known to regulate allele-specific expression of other genes in *cis*. For example, the long ncRNA *Air* (also referred to as *Airn*) is expressed only from the paternal allele, and *Air* binds with and recruits the chromatin modification enzyme G9a to the paternal *Slc22a3* gene promoter to repress the gene. *Air* ncRNA is also responsible for repression of the paternal *Igf2r* gene via a different (G9a-independent) mechanism. Thus, a single ncRNA can employ multiple mechanisms.

Moreover, in order to establish several DNA methylation imprints in developing oocytes, transcription per se of overlapping protein-coding genes is important. Although the RNAs involved are not ncRNAs, the effect and mechanism to put DNA methylation are likely independent from the coding potential of the RNAs involved are likely independent from their protein-coding potential. For example, two promoters in the mouse *Gnas* locus are methylated as imprinting marks in developing oocytes, and this is dependent on transcription of a protein-coding *Nesp* gene from far upstream through the promoters (see illustration).

Long ncRNA in X-chromosome inactivation. This is another special example of epigenetic gene regulation. Female cells have two X chromosomes, whereas male cells have only one. To compensate for the gene dosage on the X chromosome between females and males, one of the two X chromosomes is entirely inactivated in human and mouse female cells. The long ncRNA *Xist* is essential for this process. *Xist* RNA spreads out to coat the entire inactive X chromosome, and it is assumed that PRC2 binds with *Xist* and is delivered to the entire chromosome.

Long ncRNA in regulation of transcription factors and machinery. Several long ncRNAs are involved in regulation of gene expression by interacting directly with transcription factors. Sonic hedgehog, a key signaling protein in organogenesis, induces *Dlx-5/6* homeobox gene expression in a specific brain region during mouse development. This contributes to development and function of a specific type of neuron, which is achieved by a long ncRNA *Evf-2*. Sonic hedgehog induces *Evf-2* transcription from an ultraconserved region known as the *Dlx-5/6* enhancer region. *Evf-2* then binds with a transcription factor protein Dlx-2 and enhances its activity by stabilizing interactions between Dlx-2 and its target (*Dlx-5/6* enhancer), where *Evf-2* originates.

The human *DHFR* gene is repressed in a condition that makes cells quiescent. In that condition, a long ncRNA transcribed from an upstream region binds with both the *DHFR* promoter and a general transcription factor IIB. This prevents binding of the transcription initiation machinery with the promoter, thereby repressing the gene.

Transcription per se. For several long noncoding (and protein-coding) RNAs, it is known that transcription per se is the important factor for certain functionality rather than the RNA molecule. The aforementioned example with *Nesp* is one such example.

Xist ncRNA expression is repressed on the male and female active X chromosomes. This process is regulated by another long ncRNA *Tsix*, which is transcribed through the *Xist* transcription unit in an antisense direction. Evidence suggests that *Tsix* transcription through the *Xist* promoter makes the *Xist* promoter epigenetically inactive.

During *fbp1*$^+$ gene induction in fission yeast, transcription of long ncRNAs from an upstream region occurs in the sense direction of the gene. This transcription makes local chromatin epigenetically permissive to *fbp1*$^+$ transcription.

Long ncRNA in disease. Several examples are known in which a nucleotide variation of long ncRNA increases the risk for a certain disease, suggesting a possible (yet unidentified) role of the ncRNA [for example, myocardial infarction-associated transcript (*MIAT*) ncRNA].

There are a few long ncRNA examples whose role in disease is understood in more detail. For example, a long ncRNA that is antisense to tumor-suppressor gene *p15* is expressed in leukemia cells. This ncRNA renders a negative chromatin modification at the *p15* gene. Because tumor-suppressor genes are frequently inactivated in cancer, such a mechanism could be generalized. On the other hand, a recent report shows that expression of the *HOTAIR* ncRNA (described above) correlates with breast cancer prognosis. Overexpression of *HOTAIR* changes chromatin modification at the genome-wide level through PRC2 retargeting, which results in increased invasiveness of the tumor cells. These facts suggest that long ncRNAs could be potential therapeutic targets.

For background information *see* CANCER (MEDICINE); CHROMOSOME; GENE; GENETICS; GENOMICS; HUMAN GENOME; NUCLEOTIDE; PROTEIN; RIBONUCLEIC ACID (RNA); TRANSCRIPTION; TUMOR SUPPRESSOR GENES in the McGraw-Hill Encyclopedia of Science & Technology. Takashi Nagano

Bibliography. C. S. Bond and A. H. Fox, Paraspeckles: Nuclear bodies built on long noncoding RNA, *J. Cell Biol.*, 186:637–644, 2009; T. R. Mercer, M. E. Dinger, and J. S. Mattick, Long non-coding RNAs: Insights into functions, *Nat. Rev. Genet.*, 10:155–159, 2009; C. P. Ponting, P. L. Oliver, and W. Reik, Evolution and functions of long noncoding RNAs, *Cell*, 136:629–641, 2009; J. E. Wilusz, H. Sunwoo, and D. L. Spector, Long noncoding RNAs: Functional surprises from the RNA world, *Genes Dev.*, 23:1494–1504, 2009.

Low-power radio links

The communication of information between places where a permanent infrastructure, such as wires, power grids, and cellular towers, is absent has become increasingly important. Recent advances in medicine, environmental sensing, and low-power computing have created a need for ways to transmit and receive data without such an infrastructure, particularly in critical locations such as in and around the human body, throughout the fuselage of an airplane, or from the inside of a collapsed building. These types of applications put a heavy burden on the radio links to operate with very little support and to be very small while remaining robust. For instance, a radio intended for implantation in the human body should either run for many years on a single battery or be able to harvest power from the environment, since battery replacement is likely to require surgery. Similar limits exist on other applications, where the cost of a battery in weight and size can be large compared to the cost of electronics. Conventional radio links, such as Bluetooth, require large batteries and often have lifetimes measured in hours or days rather than months or years. Overcoming these limitations requires changing the way wireless information is transmitted and received, especially for applications in which the information communicated may be sparse.

Radio architecture. A basic radio consists of a transmitter and a receiver (**Fig. 1**). In a conventional radio, data are sent wirelessly using an electromagnetic wave called a carrier at a single frequency. At the transmitter, the incoming data stream is encoded

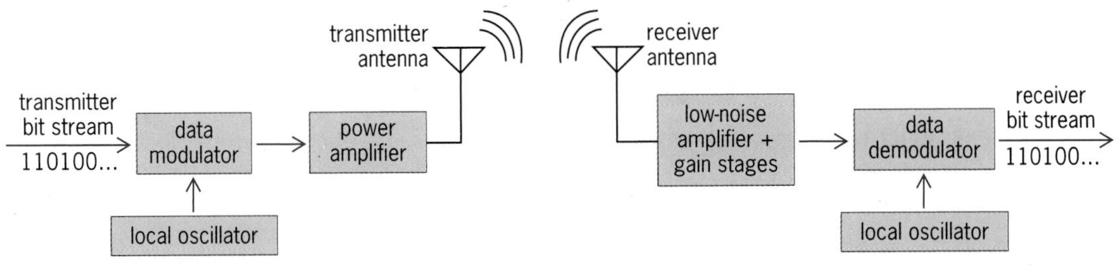

Fig. 1. The transmitter and receiver blocks in a wireless link.

by modulating the amplitude, frequency, or phase of the output of an oscillator circuit that generates this carrier. The modulated information is then radiated out by the antenna using a power amplifier. In the United States, the power radiated into the environment at a given carrier frequency is regulated and restricted based on standards issued by the Federal Communications Commission (FCC) to limit interference. The radiated signal energy spreads into the environment, and a very small part of it is captured by the receiver antenna. A low-noise amplifier (LNA) at the receiver amplifies this captured signal while adding as little noise as possible. The signal is then demodulated at the receiver by a tuned oscillator circuit (local oscillator), which decodes the transmitted data stream. The data are usually represented as ones and zeros. This basic architecture works well for long transmission distances and continuous high-data-rate bit streams.

Saving power in wireless links. The power amplifier, local oscillators, low-noise amplifier, and gain stages, as well as the modulators and demodulators, each consume a significant amount of power in a conventional radio system. In conventional systems, all the circuit blocks are always "on," thereby consuming power even when they are not communicating. For low-data-rate applications, where there is less information to communicate, the easiest way to save power in a wireless link is to turn on the most power-hungry blocks (the power amplifier and receiver) only when they are needed. These blocks of the radio, which are responsible for sending power into the environment through the antenna and detecting it from the antenna, consume more power than most of the other blocks. The trick to this strategy is in knowing when to turn these blocks on and off. To do this effectively, the transmitter and receiver blocks must decide together when to put data into the air and when to look for the data. If, for instance, a transmitter has information to send, but the receiver it is trying to contact is off, the data will be lost. If the receiver is on when the transmitter has no information to transmit, power will be wasted. Optimizing this communication cycle can be a challenge. There are several approaches to solving this problem.

Sleep-wake radios. One such approach is based on regular, periodic waking of the radio nodes. In this type of radio link, all the nodes in the network wake up periodically for a short period of time and go back to sleep mode if no activity is found. In the case of communication between two nodes, a node that wants to communicate with another node sends an RTS (request to send) packet and waits for a clear signal (CTS; clear to send) from the receiving node. If it does not receive a response from the other node, it keeps sending the RTS signal followed by a wait duration. At some point, the receiving node wakes up and sends a clear signal, after which the actual data transmission proceeds between the two nodes. After the data transfer, the nodes go back to their normal sleep-wake cycle (**Fig. 2**). Depending upon how often the node wakes up, the power consumption required for useful transmission of information may still be high with this method. This method is best suited for applications where nodes need to communicate very infrequently (hourly, daily, or even less often).

An alternative version of a wake-up receiver utilizes two types of radio transmitters and receivers in

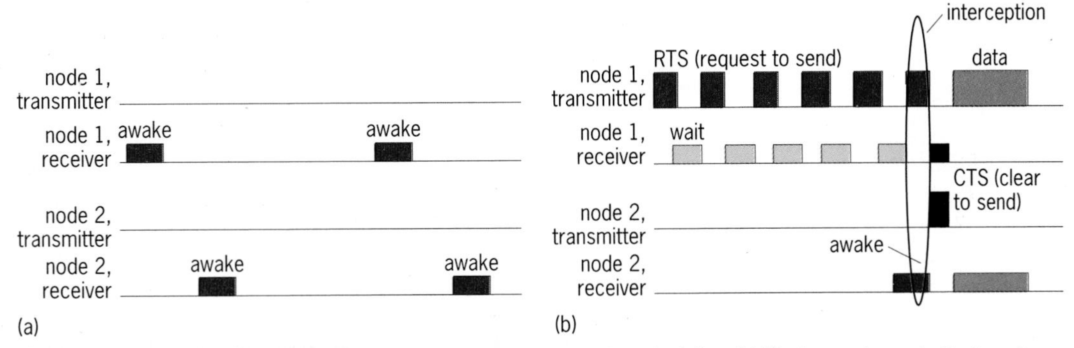

Fig. 2. Timing illustrations of activity in a network based on sleep-wake scheduling. (*a*) The two nodes periodically wake up for a short period of time and go back to sleep because there is no activity in the channel. (*b*) When a receiving node intercepts an RTS signal, it stays awake for the data exchange.

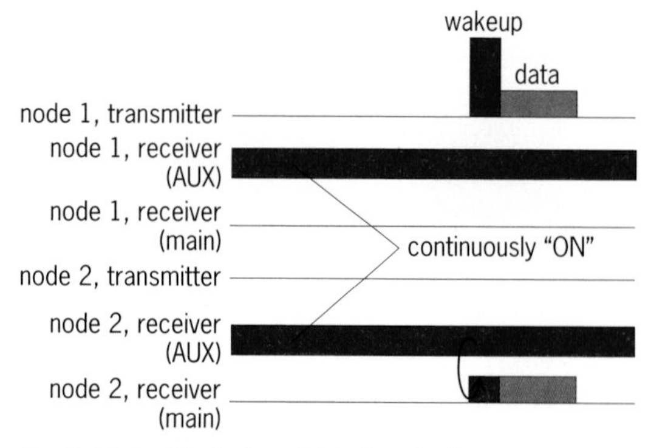

Fig. 3. Timing illustration of data transfer in an idle-listening-based wake-up radio network.

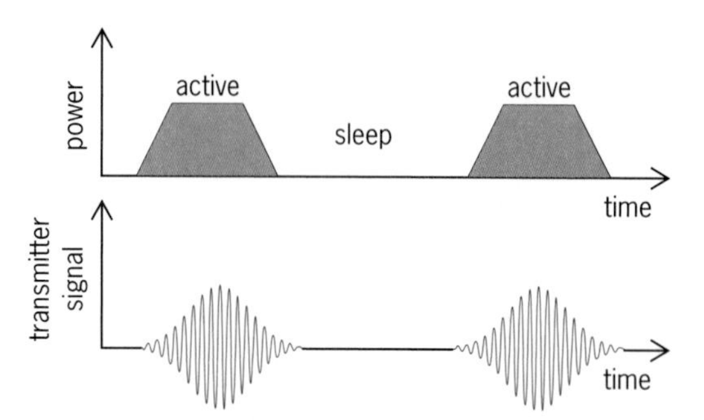

Fig. 4. Sleep-wake scheduling in an impulse-based radio.

each node. One type of wake-up receiver remains on during the idle phase, but is designed for extremely low power and consequently weak amplification of incoming signals. The low-power wake-up receiver can detect only a very strong signal. In this kind of system, when a node wants to communicate with other nodes in the system, a strong transmitter sends a very strong signal, which is detected by the wake-up receiver. The wake-up receiver then wakes up the higher-power main receiver to facilitate data communication between the two nodes through a stronger receiver and a weaker transmitter (**Fig. 3**). In this scheme, the need for a very strong signal for the wake-up receiver means that significant power is required each time the radio wakes up. Consequently, this scheme is also best for cases where there is very infrequent communication (hourly, daily, or even less often). Both sleep-wake schemes are subject to regular FCC standards based upon their carrier frequencies.

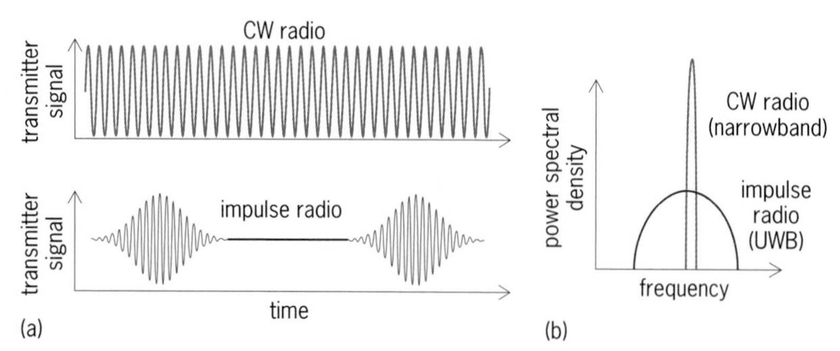

Fig. 5. Comparison of impulse radio with continuous-wave (CW) radio. (*a*) Waveforms. (*b*) Corresponding spectra.

Impulse radios. Sleep-wake radios work well in situations where there are very few events, separated by minutes, hours, or longer periods, to be communicated. For example, in detecting the presence of a toxic gas or some other dangerous event, sleep-wake radios are effective. However, this type of radio is not as effective when a constant level of communication is required, such as in a medical monitoring application. In this case, one would like to transmit data at a steady low rate, switching off the transmitter and receiver at regular intervals between transmissions to save data. In order to do this, a different type of radio is required. Rather than transmitting a continuous wave, as described above, where data are encoded in the phase or frequency of that wave, the system can encode data as the presence or absence of energy within a frequency range at a set time. In this way, rather than transmitting many cycles of a wave, one can transmit short bursts of power, called wavelets, that are only nanoseconds in duration. By transmitting or not transmitting these wavelets at particular times (**Fig. 4**), digital bits (ones and zeros) are communicated wirelessly from one radio to another. These pulsed radios are known as impulse radios, owing to the short duration of their transmitted signals. Alternatively, they are also called ultra-wideband (UWB) radios, because the short pulses have a wide spectrum (**Fig. 5**). Since the bursts are short in time, they do not require the transmitter's power amplifier to be on for very long, thus saving power at the transmitter. In order to receive these bursts, however, the receiver must either be fast and continuously on, consuming large amounts of power, or know to look for the information at approximately the right time, turning off at all other times. The latter can be accomplished if the two (or more) communicating radio transceivers are synchronized to a common clock. Once this is achieved, communication can proceed on a common time scale.

Transmitter-receiver synchronization. Synchronization of transmitters and receivers can be accomplished in a variety of ways. Each radio can contain an accurate oscillator that is well matched to those of other radios. Radios can be synchronized by a master radio sending out a clock signal that other radios use to phase-match their local oscillators, thereby making a local copy of the common clock within the radio. Alternatively, local oscillators can be driven to a common phase, frequency, and overall time scale by a process of nonlinear synchronization via pulse-coupled oscillators. The idea of pulse-coupled oscillators is modeled on a natural distributed phenomenon observed in Southeast Asian fireflies, which are thought to use local information to influence one another and ultimately light up in unison. Such phenomena have been replicated in silicon electronics and have been shown to effectively establish a global clock by which radios can be timed. This type of synchronization has the additional advantage of not requiring an explicit master node to transmit a clock signal to the entire system, but rather

Illustrative performance of low-power radio designs for low-data-rate applications

Radio type	Receiver power	Receiver sensitivity	Transmitter power	Radiated power	Data rate	Bandwidth	Carrier frequency
Continuous-wave (CW) radios*	330 μW	—	700 μW	140 μW	300 kbps	1 MHz	2.4 GHz
Idle listening radios†	52 μW	−72 dBm	N/A	N/A	100 kbps	100 MHz	2 GHz
Impulse radios‡	12 μW	−87 dBm	8 μW	2 μW	100 kbps	500 MHz	3.5–4.5 GHz

* SOURCE: B. W. Cook et al., Low-power 2.4-GHz transceiver with passive RX front-end and 400-mV supply, *IEEE J. Solid State Circ.*, 41:2757–2766, 2006.
† SOURCE: N. M. Pletcher, Gambini, and J. Rabaey, A 52μW wake-up receiver with -72 dBm sensitivity using an uncertain-IF architecture, *IEEE J. Solid State Circ.*, 44:269–280, 2009.
‡ SOURCE: R. Dokania et al., An ultra-low-power dual-band UWB impulse radio, *IEEE Trans. Circ. Syst. II*, 57(7), 2010.

establishing a leader node in an ad hoc fashion, enabling a more scalable network.

FCC regulations. Because transmissions from an impulse radio occupy a wider band of frequency spectrum, they are designed for a different "carrier" frequency range and FCC standard. For this type of radio, the standard in part 15 of the FCC regulations limits both the average power transmitted and the peak power of an impulse. For radio-frequency irradiation of the human body, the specific absorption rate of energy (SAR) may be of concern, although there are no specific FCC regulations that govern this level. However, joint standards of the Institute of Electrical and Electronics Engineers and the American National Standards Institute (IEEE/ANSI) establish a national consensus on a working threshold for adverse biological effects in humans at 0.08 W/kg in 1 g of tissue for 15 to 30 min with a safety margin of a factor of 50. These power margins are far in excess of the communication power required for impulse radio communication, which is of the order of hundreds of microwatts. This circumstance implies that this type of radio is well suited to biomedical applications.

Prospects. Because of the wide variety of potential applications of low-power wireless communication, there is a significant body of research in this area. While commercial radios exist that can be operated at 10 W, this level is still considered to be too high for applications within the human body and for long-lifetime environmental sensing. Most of the architectures discussed in this article are still actively being studied as candidate technologies to drive down power consumption of radios to 10 W or less. While physics imposes limits on transmission distances and data rates of such radios, these power levels do indeed seem possible, and will be the likely result of research into low-power radio links. The **table** shows performance numbers for some of the lowest-power radios whose descriptions have been published to date for each type of radio discussed in this article. These radios are the product of research and are not commercially available at this time.

For background information *see* AMPLIFIER; ANTENNA (ELECTROMAGNETISM); DATA COMMUNICATIONS; DEMODULATOR; MODULATOR; POWER AMPLIFIER; RADIO; RADIO RECEIVER; RADIO TRANSMITTER; ULTRAWIDEBAND (UWB) SYSTEMS in the McGraw-Hill Encyclopedia of Science & Technology.

Alyssa B. Apsel; Rajeev K. Dokania; Xiao Wang

Bibliography. I. F. Akyildiz et al., A survey on sensor networks, *IEEE Comm. Mag.*, 40(8):102–116, 2002; A. P. Chandrakasan et al., Low-power impulse UWB architectures and circuits, *Proc. IEEE*, 97:332–352, 2009; B. Otis and J. Rabaey, *Ultra-Low Power Wireless Technologies for Sensor Networks*, Springer, New York, 2007; O. Simeone et al., Distributed synchronization in wireless networks, *IEEE Signal Process. Mag.*, 25(5):81–97, 2008.

Marine mycology

Marine fungi are not a taxonomically or physiologically defined group of organisms; rather, they are an ecologically defined group. According to a commonly accepted definition, they are divided into *obligate marine fungi*, which grow and sporulate exclusively in the marine or estuarine environment, and *facultative marine fungi*, which may grow in marine as well as freshwater or terrestrial habitats. Consequently, they range from residents, that is, those that complete their life cycles exclusively in marine habitats and thus are never found outside of the marine environment, to transients, that is, those that occur in the sea accidentally by being washed or blown into it.

The total number of recognized species of marine fungi was established recently as 530, comprising 424 Ascomycota species, 94 species of anamorphic (asexual or imperfect) fungi, and only 12 Basidiomycota species. However, it is clear that the overall biodiversity will be much more pronounced. Currently, more than 100 species are newly described per decade, and certain geographic regions (including Africa, Australia, China, and South America) appear severely understudied.

Ecology of marine fungi. In general, marine fungi are found in virtually all nonliving or living marine habitats. The former comprises sediments, sandy bottoms, and even the deep sea, as well as artificial or human-made substrata such as wooden boats or pilings, polymeric cords, and so forth. Living substrata include driftwood, mangrove trunks and roots, and marine higher plants and algae. Increasingly, marine fungi also are observed from marine animals, including sponges, corals, tunicates, crustaceans,

Fig. 1. Cultivation approach to obtain endophytic fungi from mangrove plants for chemical screening. (*J. Kjer, P. Proksch, and R. Ebel, unpublished results*)

and fishes. Particularly diverse are mangrove forests, which have yielded at least 625 fungal species. As mentioned previously, this figure is probably not representative of the overall biodiversity, because the discipline of marine mycology is relatively new and many suitable marine habitats have so far been studied only very selectively. One striking example is the salt marsh plant, *Juncus roemerianus*, which has been investigated systematically over the course of many years and has yielded 117 different fungal species. In contrast, other plants from similar ecosystems have never been studied so far.

Methods to assess fungal biodiversity. Most of the data on fungi have been obtained by classical cultivation approaches (**Fig. 1**). In some cases, these approaches have been based on rather tedious experimental protocols, such as isolation of single spores. There is increasing evidence that these methods suffer from a cultivation bias toward fast-growing fungi that are able to grow rapidly on the media used, and specific precautions are required to prevent ubiquitous genera, including *Aspergillus*, *Penicillium*, and *Cladosporium*, from dominating the isolates. An alternative that is increasingly becoming popular uses molecular biological methods, which are based on extraction of total DNA from an environmental sample. Via the polymerase chain reaction (PCR) technique using fungal-specific primers [for example, ITS (internal transcribed spacer) or laccase genes], genes from each strain or species in the microbial assemblage are amplified. Separation is achieved by cloning or by special gel electrophoretic techniques, including denaturing gradient or temperature gradient gel electrophoresis (DGGE or TGGE, respectively). Then, individual microorganisms are identified on the basis of their respective DNA sequences, which are compared to existing sequence databases such as GenBank. However, it should be stated that these methods, even though in many cases they are considered far superior to cultivation, on the one hand rely on the availability of suitable reference sequences (which certainly is not the case for less frequently encountered taxa) and on the other hand also might introduce a different type of bias, especially during PCR.

Fungal communities in marine plants and animals. Over the last few decades, it was shown that marine algae harbor a diversity of associated microflora, and some of these associations have been termed mycophycobiosis. Using the aforementioned molecular biology–based protocols, fungal associates that very likely represent true or obligate marine strains have been detected in different algal species. In most cases, these symbionts resist currently available cultivation methods and are thus considered unculturable (or, in other words, yet to be cultured). In the last few years, similar studies have also been conducted with marine animals, most importantly sponges, and once more the results hint at the presence of a specialized associated fungal microflora (which has long been accepted for the association between sponges and bacteria).

Bioactive compounds and their potential industrial application. The chemical potential of marine-

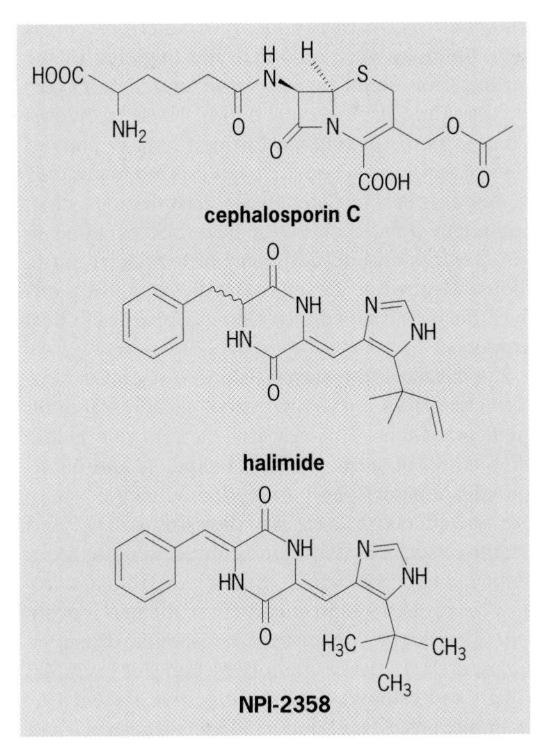

Fig. 2. Structures of clinically relevant natural products from marine-derived fungi.

derived fungi first became evident shortly after World War II, when Italian researchers discovered the antibiotic cephalosporin C (**Fig. 2**) from a culture of a *Cephalosporium* species isolated from seawater close to a sewage outlet off the coast of Sardinia. This exciting discovery was developed into one of the leading classes of clinically used antibiotics to treat bacterial infections. Still today, the chemical backbone of modern second- and third-generation cephalosporins is produced by fermentation of fungi, albeit from terrestrial strains. Since the 1990s, marine-derived fungi have been evaluated more systematically for their potential to produce pharmacologically active natural products, and they continue to deliver interesting lead structures for new drug candidates, that is, chemical models that can be structurally modified and optimized to better meet clinical needs for therapy of human diseases. Of particular interest is halimide (Fig. 2), which was previously characterized by researchers as a tubulin-depolymerizing agent and thus interferes with the dynamic remodeling of the cellular shape of actively dividing cells, most importantly cancer cells. Halimide served as a template for the closely related, but synthetically obtained analog, NPI-2358 (Fig. 2). Currently, NPI-2358 is being evaluated clinically for its effectiveness in treating certain forms of cancer, including advanced non–small-cell lung cancer, for which only a few alternative drugs are available.

The chemical potential of marine fungi, especially of "creative" genera such as *Aspergillus, Penicillium, Fusarium*, and, *Acremonium*, is widely acknowledged. As a result of their inherent aptitude to large-scale fermentation, they are of great interest for the industrial production of pharmacologically active natural products. An emerging trend in natural-products chemistry is the identification and sequencing of biogenetic gene clusters that are responsible for the production of clinically relevant secondary metabolites, with the ultimate goal of their heterologous expression in suitable standardized host strains. In this context, it is relevant to note that genes involved in the biosynthesis of β-lactam antibiotics (such as the aforementioned cephalosporin C) have also been detected in marine fungi.

A different potential biotechnological application of marine fungi is microbial biotransformation, a technology that is established mainly in the field of steroids and in which the enzymatic capabilities of terrestrial fungi are used on an industrial scale to provide drugs relevant for various therapeutic indications, including anticontraceptives, antiasthmatics, or anti-inflammatory drugs.

Marine fungal biotechnology. In addition to the pharmaceutical sector, marine fungal enzymes are currently being investigated for a variety of biotechnological applications. Of particular interest are extracellular degradative enzymes such as cellulases, xylanases, and the lignin-degrading enzymes, lignin peroxidase, manganese peroxidase, and laccase. Several strains of marine fungi have been shown to produce these enzymes, rendering them potentially useful for bioremediation of waste that is generated, for example, by paper, pulp, or textile mills, or by alcohol distilleries. Even though only a few deep-sea fungi have been described, recent studies indicate their enormous potential. Because of their adaptation to the ecological conditions of their habitat, they might represent a unique source for low-temperature-active, salt-tolerant, or pressure-tolerant enzymes, which would be particularly well suited for a broad range of industrial processes.

Future outlook. Marine mycology is still a young scientific discipline, and so far the true biodiversity of marine fungi can only be estimated. Exciting developments still lie ahead, especially when currently understudied geographic regions or entire ecosystems (such as the deep sea) are investigated more thoroughly. Currently, molecular biology–based studies are being undertaken, and they indicate the existence of truly specialized fungal communities associated with algae, mangrove plants, and marine invertebrates, displaying individual adaptations to their respective hosts. Similar to endophytic or insect-associated fungi in terrestrial habitats, these may represent an enormous source of hidden biodiversity.

Moreover, there has been a steady rise in interest in marine-derived fungal strains on the part of the natural-products community, in both academia and industry. In fact, in China, systematic evaluation of the metabolic diversity of marine-derived fungal strains has begun on a very broad scale. A finer understanding of the molecular basis of natural-product biosynthesis and its mechanisms of regulation will contribute to making better use of the enormous chemical potential of marine-derived fungi, for example, in optimizing culture conditions or, ultimately, in expressing entire biogenetic pathways in host systems that are more suited for industrial-scale fermentation. Marine-derived fungi thus represent a valuable resource for the discovery of pharmacologically active novel secondary metabolites, as well as specially adapted enzymes, with a variety of biotechnological applications. Given the discrepancy between the actual number of cultivated strains and the estimated biodiversity of fungi in marine habitats, marine-derived fungi are a heavily underexplored source of industrially relevant small molecules and enzymes. Therefore, it can be expected that this interesting group of organisms will continue to yield relevant discoveries in the future.

For background information *see* ANTIBIOTIC; BIODIVERSITY; ENZYME; FUNGAL BIOTECHNOLOGY; FUNGAL ECOLOGY; FUNGI; MARINE ECOLOGY; MICROBIAL ECOLOGY; MYCOLOGY; PHARMACY in the McGraw-Hill Encyclopedia of Science & Technology.

Rainer Ebel

Bibliography. R. Ebel, Natural product diversity of marine fungi, pp. 223–262, in *Structural Diversity II*, vol. 2 of L. Mander and H. W. Liu (eds.), *Comprehensive Natural Products II: Chemistry and Biology*, *Chemistry and Biology*, Elsevier, Oxford, 2010; E. B. G. Jones et al., Classification of marine Ascomycota, anamorphic taxa and Basidiomycota, *Fungal Divers.*, 35:1–187, 2009; C. Raghukumar, Marine fungal biotechnology: An ecological perspective, *Fungal Divers.*, 31:19–35, 2008; C. A. Shearer et al., Fungal biodiversity in aquatic habitats, *Biodivers. Conserv.*, 16:49–67, 2007; A. Zuccaro et al., Detection and identification of fungi intimately associated with the brown seaweed *Fucus serratus*, *Appl. Environ. Microbiol.*, 74:931–941, 2008.

Medicinal fungal research in China

Medicinal fungi are those that are used mainly for medical purposes. These fungi are an important part of traditional Chinese medicine (TCM). Generally, medicinal fungi are macrofungi that produce a fruit body or sclerotium. Most of these macrofungi are species of Basidiomycetes, whereas others are species of Ascomycetes. These fungi are commonly known as medicinal mushrooms. However, there are other medicinal fungi, including some molds and yeasts, that belong to the microfungi. For example, *shenqu* (massa medicata fermentata), a well-known TCM to treat maldigestion that has been in use in China for thousands of years, was prepared through the natural fermentation of several kinds of microorganisms, including yeasts. (It is now made mainly by using a single species of yeast.) Similarly, molds grown on tofu (bean curd) have been utilized to cure furuncles (small cutaneous abscesses; also known as boils). In fact, many medicinal herbs, including *Ganoderma lucidum*, *Wolfiporia extensa*, *Polyporus umbellatus*, *Cordyceps sobolifera*, and *Ophiocordyceps sinensis*, were recorded and depicted in the earliest Chinese medical literature and in other succeeding Chinese medicinal books. These traditional Chinese herbs have gone through long periods of clinical verification and are even utilized in contemporary hospitals in China. In the past decades, modern science and technology have been applied in the research and development of medicinal mushrooms. Meanwhile, dozens of medicinal mushrooms have been developed into Chinese pharmaceutical products.

Pharmacological research. Chinese scientists have paid great attention to pharmacological research on medicinal fungi. This research field covers nearly all pharmacological aspects, including antitumor, anti-inflammatory, and immunomodulating activities, and effects on diseases of the cardiovascular and cerebrovascular system (for example, hypertension, diabetes, hyperlipidemia, cardiac arrhythmia, coronary heart disease, and leukopenia), the nervous system (for example, neurasthenia, insomnia, dizziness, and neuralgia), the respiratory system (for example, cough and asthma), and the digestive system (for example, gastric and duodenal ulcers, gastritis, and hepatitis). Among the medicinal fungi, *G. lucidum* has been the most thoroughly and comprehensively studied. In fact, this fungus displays great versatility with regard to pharmacological functions. Moreover, no side effects have so far been reported, even when the fungus has been taken for long periods at high dosages. Among the activities of this fungus, the mechanism of its antitumor effect has been elucidated. This antitumor activity is mediated by the immune system. Fungal polysaccharides act on mononuclear macrophages and lymphocytes, promoting messenger RNA (mRNA) expression and protein biosynthesis of tumor necrosis factor-α (TNF-α) and interferon-γ (IFN-γ). The polysaccharides also increase the formation of interleukin-1 and -6 (IL-1 and IL-6) of the macrophages. All of these cytokines may kill tumor cells by inhibiting their proliferation and inducing apoptosis (programmed cell death).

Cordycepin (3′-deoxyadenosine) is produced mainly by *C. militaris*. This compound has broad antibiotic functions, including antibacterial and antitumor effects [for example, its antileukemic activity against terminal deoxynucleotidyl transferase-positive (TdT$^+$) leukemic cells under the protection of adenosine deaminase inhibitors]. In addition, cordycepin has been found to have a remarkable hypolipidemic function. As a nucleoside, its hypolipidemic mechanism is different from that of statins [inhibitors of 3-hydroxy-3-methyl-glutaryl-CoA reductase (HMGR)] or fibrates [agonists of peroxisome proliferator-activated receptor α (PPAR-α)]. The responsive target of cordycepin is likely to be adenosine monophosphate (AMP)–activated protein kinase (AMPK). Thus, this work may lay the foundation for developing novel hypolipidemic drugs.

However, it should be realized that, except for the polysaccharides, many medicinal mushrooms

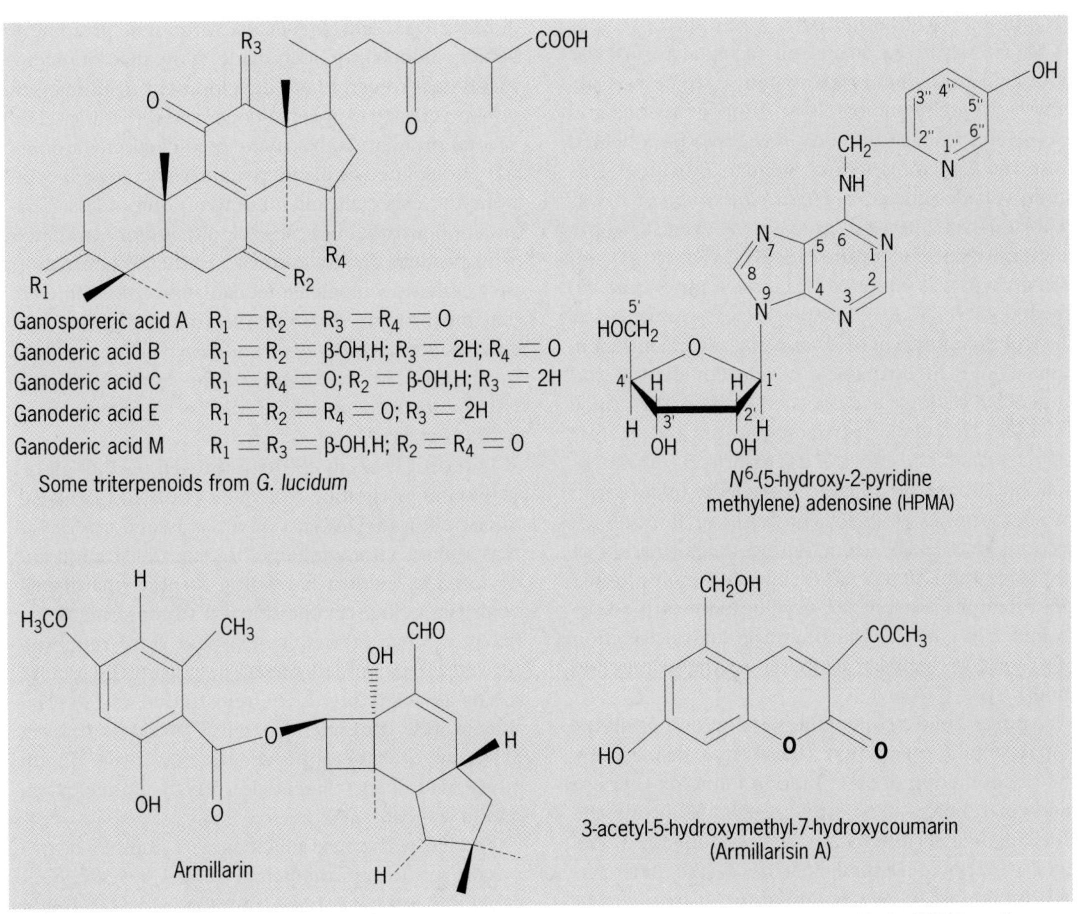

Novel compounds isolated from medicinal fungi, including some triterpenoids from *G. lucidum*, armillarin, HPMA, and armillarisin A.

lack definite active components. For this reason, the pharmacological mechanisms of these medicinal fungi are still unclear.

Chemical research. As previously stated, *G. lucidum* is one of the medicinal mushrooms whose chemical components have been thoroughly investigated. Many compounds, including triterpenoids (see **illustration**), nucleosides, steroids, alkaloids, and polysaccharides, have been isolated from this medicinal fungus.

Armillaria mellea, which has a symbiotic relationship with the TCM *Gastrodia elata*, is another extensively studied medicinal mushroom. Chinese scientists have isolated 17 novel sesquiterpenes (terpenes with the formula $C_{15}H_{24}$ and their derivatives), including armillarin (see illustration), armillaridin, and armillaricin, from cultured *A. mellea* mycelia. Moreover, 23 novel sesquiterpenes have been reported by foreign researchers. In addition, Chinese scientists have isolated eight purines from this fungus, including two novel compounds [one of which is N^6-(5-hydroxy-2-pyridine methylene) adenosine, or HPMA (see illustration)] that have shown prominent cerebrum-protective and lipid-lowering effects.

Chinese scientists have also studied the chemical components of *O. sinensis*. However, except for the sterols, nucleosides, and polysaccharides, no special components with distinct pharmacological functions have yet been identified in this fungus.

Material from *A. tabescens* has been used as a folk remedy in southern Jiangsu Province for curing diseases of the digestive system, including cholecystitis (inflammation of the gallbladder) and gastritis. In the 1970s, Chinese scientists isolated the active compound, armillarisin A (see illustration), from this mushroom and synthesized it using chemical methods. It has become a common drug in China.

In chemical terms, though, the active components of many macrofungi remain unclear, except for the polysaccharides frequently isolated from some of these fungi (including *Lentinula edodes*, *P. umbellatus*, *W. extensa*, *Tremella fuciformis*, *Hericium erinaceus*, and *Trametes versicolor*). Many medicinal mushrooms have historically been used in TCM, but the active components with novel or distinctive structures, definitive therapeutic effects, and development potential have not yet been identified.

Other applications. In recent decades, Chinese scientists have applied molecular biology and biotechnology to medicinal fungal research, including the application of DNA molecular markers in phylogenetic analysis, target-gene cloning and expression, bioconversion, and novel solid fermentation techniques. Some examples are described here.

Ophiocordyceps sinensis is a rare and precious TCM. However, its anamorph (asexual form) was uncertain until the 1980s, when a strain was obtained through multibatch isolation of tissues and ascospores from a fruit body of *O. sinensis* collected from the Kangding area of Sichuan Province. This strain was designated as *Hirsutella sinensis*. It was reported that *H. sinensis* grew slowly even at its optimum growth temperature [15–20°C (59–68°F)] and that growth was inhibited when the temperature exceeded 25°C (77°F). The species was confirmed as the true anamorph of *O. sinensis* by several independent studies involving microcycle conidiation and molecular biological evidence. Incidentally, since TCM uses the whole "herb" (sclerotium and stroma) of *O. sinensis*, the effect of its anamorph *H. sinensis* may not always be equal to that of the teleomorph (sexual form) *O. sinensis*. Furthermore, multiple infections may easily occur during the formation of the sclerotium, and it is possible that multiple fungal infections assist in the development of the sclerotium and stroma. The pharmacological function of *O. sinensis* may be attributed to these infecting fungi.

A novel solid fermentation technique [bidirectional solid fermentation (BSF)] was inspired by the phenomenon of deterioration in moldy Chinese medicinal herbs. The principle is to grow the true medicinal fungus on the comminuted (reduced to powder) medicinal herb (the herb has two functions: to be a solid culture matrix and to provide the substrates for fungal bioconversion). Because of fungal metabolism, the fermented matrix acquires increased pharmacological activities compared to the original herb or fungus itself. Sometimes, this matrix may keep the herb's activity while reducing the herb's toxicity. This technique thus may provide a new method for preparing novel TCM products. Indeed, a new preparation, "Huai Qi fungal substance," which was produced by using the fungus *Trametes robiniophila* (*Huai-er* in Chinese) to ferment the medicinal herb *Astragalus membranaceus* (*Huang-qi* in Chinese), has shown stronger immune-promoting, antiviral, and liver-protecting activities compared to *T. robiniophila*. Another example is the toxicity-attenuating research using the medicinal plant *Tripterygium wilfordii*. In this case, the plant's toxicity was reduced while its immunosuppressive activity was preserved.

Ergot (*Claviceps* species, particularly *C. purpurea*) extract has been used as an oxytocic (labor-inducing) and postpartum hemostatic drug for maternity patients. The active components are ergometrine, ergotamine, and ergotoxines, of which ergometrine is the most active. However, wild ergot resources are restricted, and ergot fungi are plant pathogens. In the 1950s and 1960s, Chinese scientists fulfilled ergometrine production requirements first by solid fermentation and then by liquid fermentation of certain strains of *C. microcephala*. Since the 1990s, ergocryptine production has been realized by liquid fermentation of *C. purpurea*.

Challenges and prospects. Although prominent achievements have been made in the field of medicinal fungal research and development in China, scientists still face many challenges and difficulties. The pivotal problem is that many medicinal mushrooms have no defined or distinctive pharmacological components, especially novel active compounds. This situation greatly increases the difficulties associated with product quality control and the determination of the pharmacological mechanisms of these medicinal mushrooms. Most of the fungal pharmaceutical products (approved as traditional Chinese patent medicines) can be sold only on the domestic market. Others are sold merely as dietary supplements in markets abroad.

In recent years, with the increased pace of global economic integration, the State Food and Drug Administration (SFDA) of China has issued a series of new rules and regulations for drug registration and declaration. Standards relating to new pharmaceuticals (including preparations of Chinese medicinal herbs and natural medicines) have been much improved and expanded, greatly increasing the requirements associated with the registration and declaration of new drugs. Thus, fungal products that the SFDA might have approved as new drugs in the 1990s will need to meet the recent higher examination standards.

To deal with these challenges, a number of proposals for future medicinal fungal research and development have been put forward: (1) follow the contemporary pharmacological concept, comply with the new SFDA rules and regulations, and strengthen active component-oriented research and development; (2) pay more attention to undeveloped medicinal mushrooms, such as *Phellinus* species, *Taiwanofungus camphoratus*, *Inonotus obliquus*, and the toxic mushrooms; (3) broaden the research field and promote the discovery of drug lead compounds from fungi occupying special environments, such as plant endophytes, marine fungi, and mushroom endophytes; and (4) consolidate biotechnological applications in medicinal fungal research and development, including gene cloning and expression, synthetic biology, enzyme engineering, and biocatalysis. In addition, if some medicinal mushrooms possess some pharmacological effects but have no definite active components, they may go to market via the nutraceutical road, as long as they have been demonstrated to be safe.

For background information *see* FUNGAL BIOTECHNOLOGY; FUNGAL ECOLOGY; FUNGI; MEDICINE; MUSHROOM; MYCOLOGY; PHARMACEUTICAL CHEMISTRY; PHARMACEUTICALS TESTING; PHARMACOLOGY; YEAST in the McGraw-Hill Encyclopedia of Science & Technology.

Ping Zhu; Hua-An Wen

Bibliography. R. Y. Chen and D. Q. Yu, The chemical components of *Ganoderma lucidum*, pp. 157–218, in Z. B. Lin (ed.), *Modern Research on Ganoderma lucidum*, 2d ed., Beijing Medical University Publishing House, Beijing, China, 2001; R. Y. Chen, J. G. Yu, and D. Q. Yu, The chemistry of

Ganoderma lucidum, pp. 311–318, in Z. Y. Song (ed.), *Modern Research on Chinese Medicinal Herbs*, vol. 2, Beijing Medical University and Peking Union Medical College United Publishing House, Beijing, China, 1996; Y. Q. Chen et al., Determination of the anamorph of *Cordyceps sinensis* inferred from the analysis of the ribosomal DNA internal transcribed spacers and 5.8 rDNA, *Biochem. Systemat. Ecol.*, 29:597–607, 2001; Q. C. Fang and D. C. Yue, Ergot, p. 183, in Z. Y. Song (ed.), *Modern Research on Chinese Medicinal Herbs*, vol. 2, Beijing Medical University and Peking Union Medical College United Publishing House, Beijing, China, 1996; J. Hu et al., Research on the microorganism in the traditional Chinese medicine "Shenqu," *J. Mudanjiang Med. College*, 25(2):19–20, 2004; E. N. Kodama et al., Antileukemic activity and mechanism of action of cordycepin against terminal deoxynucleotidyl transferase-positive (TdT^+) leukemic cells, *Biochem. Pharmacol.*, 59(3):273–281, 2000; Z. Q. Liang, Anamorphs of *Cordyceps* and their determination, *Southwest China Journal of Agricultural Sciences*, 4(4):1–8, 1991; Z. B. Lin, Pharmacological effects of *Ganoderma lucidum*, pp. 219–283, in Z. B. Lin (ed.), *Modern Research on Ganoderma lucidum*, 2d ed., Beijing Medical University Publishing House, Beijing, China, 2001; G. T. Liu, Pharmacological effects of *Ganoderma lucidum*, pp. 326–331, in Z. Y. Song (ed.), *Modern Research on Chinese Medicinal Herbs*, vol. 2, Beijing Medical University and Peking Union Medical College United Publishing House, Beijing, China, 1996; X. J. Liu et al., Isolation and identification of the anamorphic state of *Cordyceps sinensis*, *Acta Mycol. Sin.*, 8:35–40, 1989; Z. Y. Liu et al., Molecular evidence for the anamorph-teleomorph connection in *Cordyceps sinensis*, *Mycol. Res.*, 105:827–832, 2001; J. Y. Wang, Brief introduction of *Cordyceps sinensis* research, *Bull. Biol.*, 32:45, 1997; J. S. Yang, The chemistry of *Armillaria mellea*, pp. 56–67, in Z. Y. Song (ed.), *Modern Research on Chinese Medicinal Herbs*, vol. 2, Beijing Medical University and Peking Union Medical College United Publishing House, Beijing, China, 1996; P. Z. Zhang, X. M. Xie, and R. G. Shu, Study on the chemical constituents of Linglei fungal substance, pp. 120–124, in *National Symposium on Medicinal Fungi*, Jiangxi, China, 2008; H. B. Zhu, Studies on the hypolipidemic effects of cordycepin and its possible target, p. 210, in *National Symposium on Medicinal Fungi*, Jiangxi, China, 2008; P. Zhu et al., Utilization of $3'$-deoxyadenosine in the preparation of hypolipidemic drugs, Chinese Patent ZL 200310101650.7, 2003; Y. Zhuang et al., Preparation of medicinal fungal new type bidirectional solid fermentation engineering and Huai Qi fungal substance, *Chin. Pharmaceut. J.*, 39(3):175–178, 2004.

Medicinal mushrooms and breast cancer

The fungus kingdom contains an estimated 1.5 million species, of which fewer than 5% have been classified. These species have provided antibiotics (penicillin), immunosuppressants (cyclosporins), and anticholesterol drugs (statins). However, the exploration of the fungal kingdom for pharmaceutically active compounds, whether for cancer or other diseases, lags behind the efforts dedicated to the plant kingdom and the marine environment. As the complexities of biostimulants derived from medical mushrooms are unraveled, there has been increasing interest in their application within cancer therapy. This nascent field, led by the early pharmaceuticalization of these compounds in East Asia, is broad and complex. Major issues of quality and reproducibility remain, especially in terms of manufacturing for preclinical and clinical studies. However, recent advances in fermentation technology in controlling the production of the complex proteoglycans and branching polysaccharides that are involved, as well as better controlled preclinical studies, have opened up this area to renewed interest. Breast cancer remains one of the most critical cancers, affecting populations in developed, and increasingly in developing, countries. Although major gains have been made in cure and control, particularly in early presentation, metastatic disease has remained a critical cause of mortality. The use of biostimulants, including those derived from medicinal mushrooms, in conjunction with current regimens, has strong potential for improving outcomes in this common cancer.

Medicinal mushrooms in diet and as nutraceuticals and pharmaceuticals. Mushrooms have long been valued as highly tasty and nutritional foods by many societies throughout the world. Early civilizations, by trial and error, built up a practical knowledge of those mushrooms that are suitable to eat and those to be avoided (including those that are poisonous or psychotropic). In many parts of the world, especially Europe, wild mushrooms are regularly collected and used directly as a main source of food or added to soups, stews, and teas. Mushrooms are considered to be a good source of digestible proteins, with a protein content that is greater than that of most vegetables and somewhat lower than that of most meats and milk. Mushrooms contain all the essential amino acids, but they can be limited in the sulfur-containing amino acids, cystine and methionine. Mushrooms contain every mineral present in their growth substrate, including substantial quantities of phosphorus and potassium, and lesser amounts of calcium and iron. They also appear to be an excellent source of vitamins, especially thiamine (B_1), riboflavin (B_2), niacin, biotin, and ascorbic acid (vitamin C). Vitamins A and D are relatively uncommon, although several species contain detectable amounts of β-carotene, which is converted to vitamin A in the liver, and ergosterol, which is converted to active vitamin D when exposed to ultraviolet irradiation. Although crude fat in mushrooms contains all the main classes of lipid compounds, including free fatty acids, monoglycerides, diglycerides, triglycerides, sterols, sterol esters, and phospholipids, levels are typically low, around 2–8% of dry weight. In China, the term yakuzen is used generally for medicinal food dishes of mushrooms.

There has been a recognition that many edible and certain nonedible mushrooms can have valuable health benefits. The edible mushrooms that demonstrate medicinal or functional properties include species of *Lentinus* (*Lentinula*), *Auricularia*, *Hericium*, *Grifola*, *Flammulina*, *Pleurotus*, and *Tremella*; others are known only for their medicinal properties, for example, *Ganoderma* and *Trametes* (*Coriolus*), which are definitely nonedible as a result of their coarse and hard texture or bitter taste. The historical evolution of usage of these essentially scarce, forest-obtained medicinal mushrooms would most certainly not have been as whole mushrooms but as hot water extracts, concentrates, liquors, or powders, and they have been used in health tonics, tinctures, teas, soups, and herbal formulae. Presently, almost all of the important medicinal mushrooms have been subjected to large-scale artificial cultivation. Also, many of the edible species of medicinal mushrooms are gaining worldwide popularity because of their unique flavors, textures, and amenability to culinary inclusion. Indeed, most people in the West who enjoy the unique organoleptic features (that is, features that have an effect or make an impression on the sense organs) of the shiitake mushroom (*Lentinus edodes*) are singularly unaware of its possible health benefits.

When used as nutraceuticals, medicinal mushrooms are consumed normally as powdered concentrates or extracts in hot water, with the extract being concentrated and used as a drink or freeze-dried or spray-dried to form granular powders that allow easier handling, transportation, and consumption. Mushroom nutraceuticals are usually crude mixtures and should not be confused with pharmaceuticals, which are almost invariably a defined chemical preparation, the specifications for which are listed in pharmacopoeia. Regular intake of these concentrates is believed to enhance the immune responses of the human body, thereby increasing resistance to disease and in some cases causing regression of the disease state.

There is a huge range of interesting pharmaceutical compounds that can be derived from medicinal mushrooms. Among the mushroom immune modulators that have been investigated, bioactive polymers from *Lentinus edodes* have been studied extensively for interesting biological effects. Moreover, *L. edodes* is the source of two preparations with well-studied pharmacological effects—*L. edodes* mycelium (LEM) extract and lentinan. Lentinan (a cell wall constituent extracted from fruiting bodies or mycelia) is a highly purified, high-molecular-weight polysaccharide in a triple helix structure that contains only glucose (Glc) molecules, mostly with β-(1→3)Glc linkages in the regularly branched backbone, and β-(1→6)Glc side chains. *Ganoderma lucidum* has been used extensively as a "mushroom of immortality" in China and other Asian countries

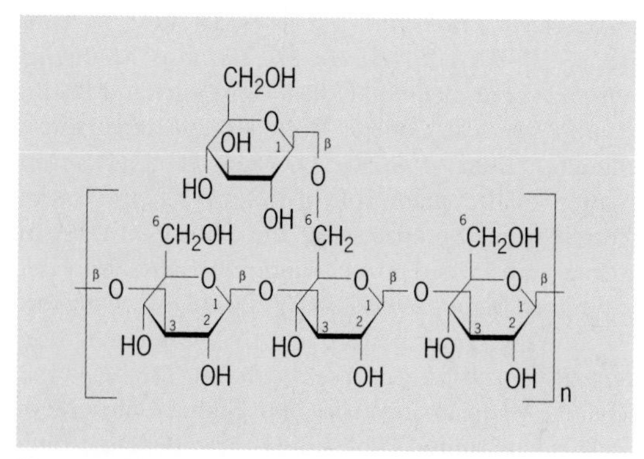

Fig. 1. Primary molecular diagram of mushroom β-D-glucan.

for 2000 years. Several major substances with potent immunomodulating action have been isolated from this mushroom, including polysaccharides (in particular, β-D-glucan; **Fig. 1**), proteins, and triterpenoids. More than 100 types of polysaccharides have been isolated from *G. lucidum*. The β-D-glucans (polysaccharides that produce D-glucose by acid hydrolysis) have been shown to be biologically active. Thus, these two species, *L. edodes* and *G. lucidum*, give some idea of the variety (and complexity) of the actual and potential pharmaceutical compounds derivable from the fungal kingdom (see **table**).

Application in breast cancer of biological response modifiers from medicinal mushrooms. The substantial literature on the use of chemotherapeutic compounds derived from medicinal mushrooms covers many solid and hematological cancers. A number of pharmaceutical grade compounds have reached late-stage clinical trials. These compounds include active hexose correlated compound (AHCC), maitake-D fraction, PSK, and PSP [these last two are proteoglycans derived from the CM-101 and COV-1 strains of *Trametes* (*Coriolus*) *versicolor* (**Fig. 2**), respectively]. PSK was reported as being active against HLA B40-positive breast cancer. However, this trial finding has never been replicated in a Western population, which points to one of the fundamental issues with these fascinating compounds—the lack

Fig. 2. *Trametes* (*Coriolus*) *versicolor* growing naturally on fallen timber.

Cross index of medically active higher Basidiomycetes mushrooms and their medicinal properties

	Antifungal	Anti-inflammatory	Antitumor	Antiviral (for example, anti-HIV)	Antibacterial and antiparasitic	Blood pressure regulation	Cardiovascular disorders	Hypercholesterolemia	Antidiabetic	Immunomodulating	Kidney tonic	Hepatoprotective	Nerve tonic	Sexual potentiator	Chronic bronchitis
Auriculariales			+			+	X	X							X
Auricularia auricula-judae															
Tremellales		+	+					+	+	+		+			X
Tremella fuciformis						+									+
Tremella mesenterica															
Polyporales															
Schizophyllum commune		X	X		X					X	X	X			
Dendropolyporus umbellatus			X							X		X			X
Grifola frondosa	+		X	X	X	X			X	X		+			+
Fomes fomentarius			+		+										
Fomitopsis pinicola		+	+		+							+			
Trametes versicolor			X	X	X						X	X			
Piptoporus betulinus	+		+		+										
Hericium erinaceus			+							X			X		X
Inonotus obliquus		X	X							X		X			
Lenzites betulina			+				+								
Laetiporus sulphureus	+		+												
Ganodermatales															
Ganoderma lucidum		X	X	X	X	X	X			X	X	X	X	X	X
Ganoderma applanatum			+	+	+					+					
Agaricomycetideae															
Agaricales sensu lato															
Pleurotaceae															
Lentinus edodes		X	X	X	X	X		X	X	X	X	X		X	
Pleurotus ostreatus			+	+	+			+					+		
Pleurotus pulmonarius	+		+					+							
Tricholomataceae															
Flammulina velutipes	+	X	X	+						X					
Oudemansiella mucida	X														
Armillariella mellea	+					X	X						X		
Hypsizygus marmoreus			X												
Marasmius androsaceus		X											X		
Agaricaceae															
Agaricus subrufescens			X												
Agaricus bisporus			+							X	X				
Pluteaceae															
Volvariella volvacea			+	+	+			+							
Bolbitiaceae															
Agrocybe aegerita	+		+					+					+		

Terms: X, commercially developed mushroom product (drug or dietary supplement); +, noncommercially developed mushroom product.

Data derived from S. P. Wasser and A. L. Weis, Medicinal properties of substances occurring in higher Basidiomycete mushrooms: Current perspectives, *Int. J. Med. Mushrooms*, 1:31–62, 1999.

of well-designed clinical studies in Western populations. To address this, major comprehensive cancer centers have begun to investigate these types of compounds in cancer patients. An early phase I/II dose-escalation clinical trial in postmenopausal breast cancer patients found a complex immunological picture when using an extract from *Grifola frondosa* (maitake mushroom), with some parameters being stimulated and others inhibited. What this means for a therapeutic effect is the subject of future research, but it gives some idea of how more needs to be learned with regard to understanding both the mechanisms of action and the optimal therapeutic strategies.

A variety of novel compounds derived from *Agaricus bisporus* (portabella mushroom), *Flammulina velutipes*, and *Pleurotus ostreatus* (oyster mushroom), as well as the aforementioned *L. edodes* (shiitake mushroom), have been isolated with potent antiproliferative effects on a variety of breast cancer cell lines. Work undertaken to understand the mechanism of action has uncovered a variety of complex effects on multiple pathways within the cancer cell lines. Thus, these compounds, in addition to their effect on the immune system, also appear to have direct effects on cancer cells. Such richness in their mechanism of action is both a blessing and a curse, in the sense that there may be no generic mechanism, only a compound-by-compound effect. Finally, there have been some fascinating studies that suggest that phytochemicals in some mushrooms (for example, *A. bisporus*) may function as weak anti-aromatase compounds and reduce the emergence and/or time course of breast cancer, thereby acting as natural chemopreventive agents. Early evidence from both mechanistic studies and epidemiological studies has given tantalizing indications that a diet rich in certain mushrooms may reduce the risk of breast cancer. However, further research, including larger-scale epidemiological studies and the testing of purified extracts (nutraceuticals) that can be given as chemopreventive supplements, is needed before any recommendations can be made. The research field for medicinal mushrooms and breast cancer remains wide open with numerous fascinating avenues to explore.

For background information *see* BREAST; BREAST DISORDERS; CANCER (MEDICINE); CHEMOTHERAPY AND OTHER ANTINEOPLASTIC DRUGS; FUNGAL BIOTECHNOLOGY; FUNGI; MUSHROOM; MYCOLOGY; ONCOLOGY; PUBLIC HEALTH in the McGraw-Hill Encyclopedia of Science & Technology.

Richard Sullivan

Bibliography. G. Deng et al., A phase I/II trial of a polysaccharide extract from *Grifola frondosa* (maitake mushroom) in breast cancer patients: Immunological effects, *J. Cancer Res. Clin. Oncol.*, 135(9):1215–1221, 2009; M. Fisher and L. X. Yang, Anticancer effects and mechanisms of polysaccharide K (PSK), *Anticancer Res.*, 22(3):1737–1754, 2002; R. Sullivan, J. Smith, and N. Rowan, Medicinal mushrooms and cancer therapy: Translating a traditional practice into Western medicine, *Perspect. Biol. Med.*, 49(2):159–170, 2006; S. P. Wasser and A. L. Weis, Medicinal properties of substances occurring in higher Basidiomycete mushrooms: Current perspectives, *Int. J. Med. Mushrooms*, 1:31–62, 1999.

Metal-oxide resistive-switching RAM technology

The emergence of portable electronics, such as cell phones, MP3 players, digital cameras, and netbooks, over the past 20 years has led to skyrocketing demand for nonvolatile (retaining information after power is removed) flash memory because of its small cell size and low power consumption. However, the further scaling of flash memory beyond 15-nanometers (feature dimension) technology is highly problematic because of the fundamental limit of the cell structure. The unit cell of flash memory is very similar to the conventional metal-oxide-semiconductor field-effect transistor (MOSFET), except for the additional floating gate to store electric charges. It is unclear whether MOSFET technology can be scaled beyond the channel length of 15 nm, since the physical constraint of quantum-mechanical-tunneling current may dominate at such short channel lengths. Meanwhile, there has been very active research into alternative nonvolatile memories to replace flash. Among the most mature are ferroelectric random access memory (FRAM), magnetoresistive random access memory (MRAM), and phase-change random access memory (PCRAM). Although MOSFET technology is not necessary to store data in these technologies, adding it into the unit cells is inevitable in most cases to ensure the correct read and write operations in the memory array, where the disturbance from neighboring cells is a serious concern. Therefore, the scalability of these technologies beyond current flash memory is implausible. We will revisit this point later in the article. In addition, because their unit cells (made of a MOSFET plus a memory element) are larger than those of flash, the cost would be higher. As a result, they are unlikely to compete directly with flash memory in the mainstream market. Indeed, the current applications of FRAM, MRAM, and PCRAM are limited to niche areas, such as aerospace systems.

Emergence of RRAM. Recently, a new class of nonvolatile memory has been actively researched, namely resistive-switching random access memory (RRAM), based on the resistive-switching (RS) phenomenon of transition-metal oxides (TMOs), such as NiO_x, TiO_x, CuO_x, HfO_x, and ZrO_x. A typical memory element consists of a layer of transition-metal oxide sandwiched between two metal electrodes, similar to a metal-insulator-metal (MIM) capacitor. The resistance across the metal-insulator-metal structure is switched between a bistable low-resistance state and a high-resistance state, depending on the bias condition. Such a resistive-switching phenomenon is not new; it was reported as early as 1962 by T. W. Hickmott, followed by a series of studies in the 1960s and 1970s. However, without any practical application in sight (the floating-gate nonvolatile memory was invented in 1967, and the demand never took off until the late 1990s), the research focused mostly on the material properties and physical understanding. The interest in transition-metal oxides quickly faded away because of the dominance of semiconductors, especially silicon, in microelectronics. It was not until 2000 that the work by A. Beck and colleagues on the resistive-switching properties of perovskite oxides stirred up the new interest in a potential alternative to nonvolatile flash memory. Active research in recent years has shown that RRAM has many superior properties. Among them are low-power operation below 3 V and 100 μA, sub-20-ns high-speed switching, and prolonged retention.

Mechanism of resistive switching. Typical resistive switching is classified into two categories: unipolar resistive switching and bipolar resistive switching (**Fig. 1**). For resistive switching to take place in RRAM, a one-time-only initialization step, namely forming, is often necessary. Forming is accomplished by applying a large forming voltage (V_{form}) across the metal-insulator-metal structure to induce oxide breakdown. By setting the appropriate compliance current (I_{comp}), the breakdown can be carefully controlled so that the damage is not permanent but reversible. After forming, the cell is in its low-resistance state and can be switched to its high-resistance state by applying a voltage larger than the reset voltage (V_{reset}). This transition from the low-resistance state to the high-resistance state is called RESET. At V_{reset}, the reset current (I_{reset}) is typically the highest of the entire resistive-switching cycle. When the cell is in its high-resistance state, a voltage larger than the set voltage (V_{set}) can once again switch it back to its low-resistance state. This transition from the high-resistance state to the low-resistance state is called SET. Typically V_{set} is smaller than V_{form}. Unipolar resistive switching is defined by V_{set} and V_{reset} having the same polarity, such as positive V_{set} and V_{reset} or negative V_{set} and V_{reset} (Fig. 1*a*). Bipolar resistive switching is defined by V_{set} and V_{reset} having the opposite polarity, such as positive V_{set} and negative V_{reset} (Fig. 1*b*). If both unipolar and bipolar resistive switching coexist, it is sometimes called nonpolar resistive switching.

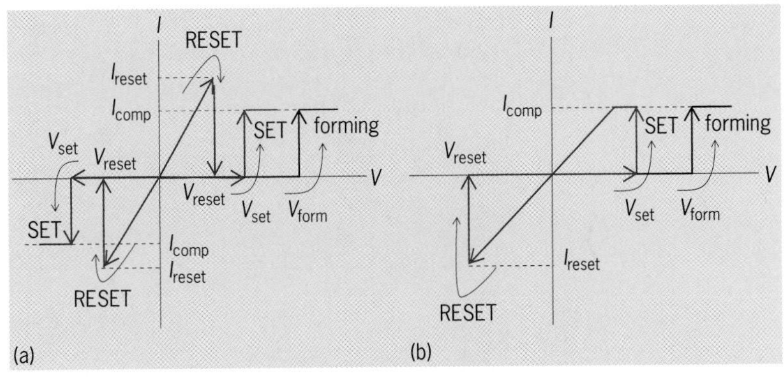

Fig. 1. Typical (*a*) unipolar and (*b*) bipolar resistive switching under dc voltage sweep.

Figure 2 illustrates the first-order filament model of the resistive-switching mechanism in RRAM. After oxide breakdown is triggered by electrical forming, one or several nanoscale conducting paths, called filaments, are formed in the otherwise insulating transition-metal oxide layer by either cation or anion migration at high electric field. Cation migration is most common in the devices with electrochemically active electrodes, for example silver (Ag) or copper (Cu). Migration of metal cations (for example, Ag^{+}) from the anode, followed by metal precipitation (reduction) at the cathode, results in the formation of metal (Ag) filaments. For the other cases, anion migration is the dominant effect. The migration of oxygen anions inside the transition-metal oxide toward the anode leaves behind the reduced (metalliclike) state of the transition-metal oxide. Unlike PCRAM, where the resistive switching is prompted by the amorphous-to-polycrystalline phase transition in entire bulk chalcogenide (compounds containing group-16 elements) materials, resistive switching in RRAM is done by rupturing and reconnecting the filaments at some local spots during the RESET/SET cycle and requires considerably less energy. As a result, RRAM is superior to PCRAM in terms of power, speed, and scalability. The reconnection of ruptured filaments at SET is quite similar to the forming process except that there is a much shorter distance of ion migration at SET, as compared with the entire thickness of the transition-metal oxide at forming. This helps to explain why V_{set} is smaller than V_{form}. The rupture at RESET may be driven by the same ion migration but toward the opposite direction when the applied electric field is reversed. In addition, the thermal effect may also play a significant role in dissolving or oxidizing metallic filaments because of the induced high temperature in the transition-metal oxide that results from Joule heating by the larger I_{reset}. It is

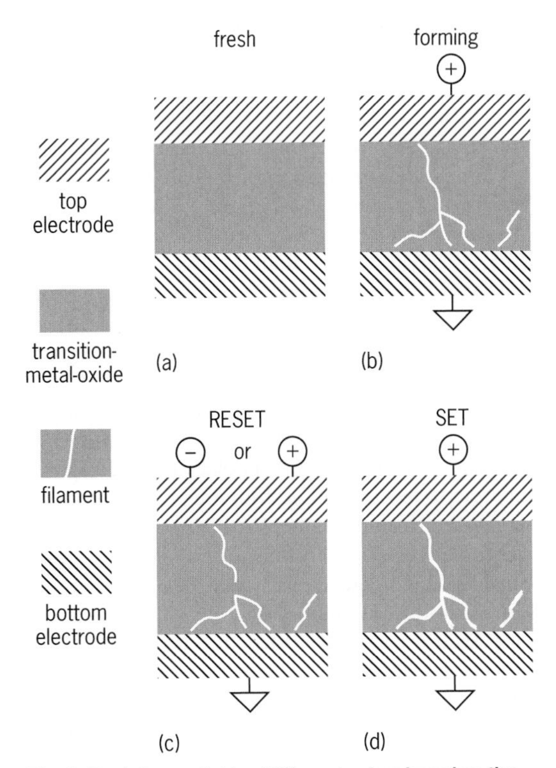

Fig. 2. Resistive-switching (RS) mechanism based on the first-order filament model. The RRAM cell is shown (*a*) before forming, (*b*) after forming, (*c*) after RESET, and (*d*) after SET.

believed that the polarity-dependent ion migration is responsible for the rupture of the bipolar RRAM, while the polarity-independent Joule heating leads to the rupture of the unipolar RRAM. Beyond the first-order model, the resistive-switching mechanism differs significantly among various transition-metal oxide systems. Many other factors, such as the composition stoichiometry, chemical interaction with the electrodes, and potential barriers at the interface, prove to be important in resistive switching. Because of its complexity, the detailed resistive-switching mechanism remains an active research topic.

RRAM cell structure. Similar to MRAM and PCRAM, where the memory states are determined by the cell resistance, RRAM is prone to read and write disturbances by neighboring cells in the memory array. Although a simple metal-insulator-metal structure is

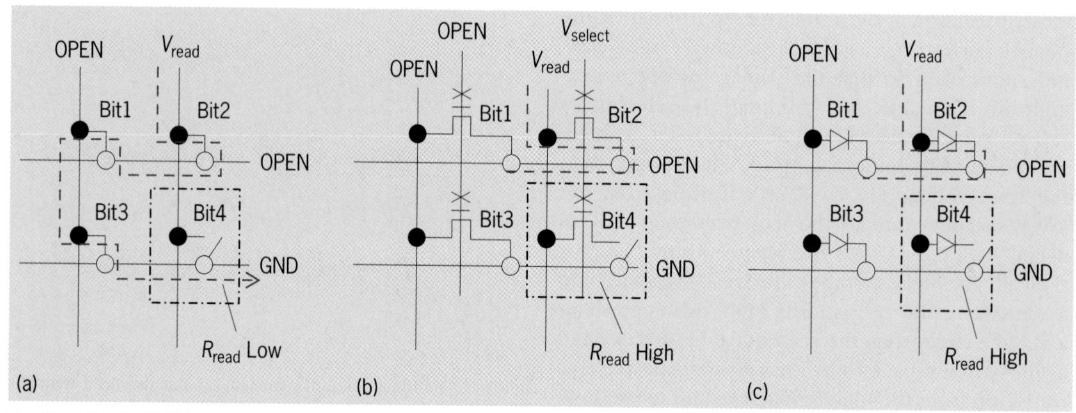

Fig. 3. A 2 × 2 RRAM array with unit cells of (*a*) 1R, (*b*) 1T1R, and (*c*) 1D1R, respectively. Bit1, Bit2, and Bit3 are all in the low-resistance state (LRS) while reading Bit4 in the high-resistance state (HRS). The parasitic path by the dashed line results in the false readout of LRS in (*a*), but not in (*b*) and (*c*) when additional access devices are employed.

very attractive in terms of its cell size and manufacturability, the unit cell of RRAM inevitably requires an extra diode or transistor to alleviate the problem of disturbances. In **Fig. 3**, we read the high-resistance-state (HRS) Bit4 in a 2 × 2 array, while Bit1, Bit2, and Bit3 are all in the low-resistance state (LRS). In Fig. 3*a*, each cell consists of only a metal-insulator-metal resistor, called 1R. The parasitic path through the neighboring Bit1, Bit2, and Bit3 renders the false readout of LRS. To cut off the parasitic path, an access transistor or an access diode may be added to yield the correct readout of HRS in Fig. 3*b* and Fig. 3*c*, respectively. Unipolar RRAM is required in order to be compatible with the unipolar conduction of diodes, while either bipolar or unipolar RRAM can be integrated with transistors. Considering the smaller size of diodes and the finite scalability of the transistors, as mentioned earlier, the one diode-one resistor (1D1R) cell structure is preferred over the one transistor–one resistor (1T1R) cell structure. Furthermore, the simple structure and fabrication process of diodes allow resistors to be stacked on top of diodes. The very compact cell and the transistor-free structure make 1D1R RRAM a strong candidate for extending nonvolatile memory beyond the scaling limit of flash memory.

Challenges of RRAM development. Before RRAM can realize its full potential as a legitimate replacement for flash memory, many technological obstacles remain to be overcome.

First, further reduction of I_{reset} is crucial for scaling RRAM below 15 nm. In the 1T1R or 1D1R cell structure, I_{reset} has to be comparable to the maximum current-drive capability of the access device, which is below 50 μA for a 15-nm transistor and below 5 μA for a 15-nm diode, according to the most optimistic estimate. Meanwhile, the best value of I_{reset} achieved by the unipolar RRAM so far is 50 μA in an NiO device, which is adequate for the 1T1R but still has a ways to go for the preferable 1D1R. Further studies on I_{reset} reduction and boosting the forward current of diodes are necessary.

Second, the cycling stability of RRAM is still not up to the standard of commercial memory, for which 10^5 switching cycles for each bit across the entire array at a working temperature of 125°C (257°F) are guaranteed. This objective is especially difficult to achieve for the unipolar RRAM, where the margin between V_{set} and V_{reset} at the same polarity is smaller. Success may require an in-depth understanding of the resistive-switching mechanism in order to tailor the device structures and materials of RRAM for improved stability.

Overall, RRAM, especially the 1D1R structure, holds particular promise for future nonvolatile memory applications because of its small cell size and superior scalability. However, considerable time and development effort may be required before it overtakes the very competitive flash memory technology.

For background information *see* COMPUTER STORAGE TECHNOLOGY; DIODE; INTEGRATED CIRCUITS; JOULE'S LAW, MAGNETORESISTANCE; PEROVSKITE; SEMICONDUCTOR MEMORIES; TRANSISTOR in the McGraw-Hill Encyclopedia of Science & Technology.

Tuo-Hung Hou

Bibliography. A. Beck et al., Reproducible switching effect in thin oxide films for memory applications, *Appl. Phys. Lett.*, 77:139–141, 2000; T. W. Hickmott, Low-frequency negative resistance in thin anode oxide films, *J. Appl. Phys.*, 33:2669–2682, 1962; T.-H. Hou et al., Flash memory scaling: from material selection to performance improvement, *2008 MRS Spring Meeting*, San Francisco, CA, March 24–28, 2008; R. Waser et al., Redox-based resistive switching memories—nanoionic mechanism, prospects, and challenges, *Adv. Mater.*, 21:2632–2663, 2009.

Metatranscriptomics

Metatranscriptomics, also known as environmental transcriptomics, is a culture-independent method by which transcriptomes (the collective RNA transcripts) of an entire community of microorganisms are analyzed. Microbial communities include bacteria, archaea, and small eukaryotes. Examples of these communities include soil, marine environments, and the human gut. It is estimated that greater than 80% of microorganisms cannot be cultured using conventional methods. Therefore, metatranscriptomics techniques are employed to circumvent the need for culturing. Metatranscriptome studies of soil and marine microbial communities have revealed niche-specific transcript expression patterns and novel genes, and they have provided insight into the ecological functions of microbes within the microbial community.

Transcriptomes. Transcriptomes are the collective RNA transcripts produced by an individual organism. These RNAs include the most well-known RNAs involved in translation: messenger RNA (mRNA), ribosomal RNA (rRNA), transfer RNA (tRNA), and the diverse and poorly understood class of small RNA (sRNA).

Analysis of the transcriptome provides greater insight into the function of organisms than a genomic study. A genomic study, which uses DNA sequencing of the entire genome of an organism, can provide information on the evolutionary origins of organisms, as well as an inventory of potential genes that could be expressed. However, genomics cannot provide answers as to how an organism responds to a given environment. Transcriptomics allows for a real-time look at the expression of genes and thus the potential function of the organism.

Most transcriptome studies focus on mRNAs, which are the RNAs that are translated into protein. Studies of mRNA expression allow scientists to gain a better understanding of potential protein profiles. Based on mRNA transcript data, it can be inferred which proteins are present in the cell and at what levels they are expressed. Because posttranscriptional and posttranslational controls ultimately regulate protein expression and levels, the actual protein profile and metabolic function can only be accurately determined by proteomics techniques (or metaproteomics when analyzing communities). Because of the difficulty associated with protein extraction and identification in environmental samples, metatranscriptomics is a more useful tool.

Transcriptomics is powerful for several reasons: The techniques to do transcriptomics studies are well established and cost-efficient; transcriptomics allows us to peer into the inner workings of the cell to see the interplay between mRNA and inhibitory RNAs (RNAi), part of the class of small RNAs; and transcriptomics can identify novel noncoding regulatory RNAs, which include RNAi and other small RNAs. Metatranscriptomics is even more powerful because of the ability to analyze a community of organisms in real-time, providing a snapshot of expression at a particular time under specific environmental conditions. Given that these communities function collectively in processes such as carbon or nitrogen cycling, it is important to view the gene expression of the community as a whole to evaluate the impact of perturbations to the community. Loss of a single organism in the community could gravely disrupt the community's collective function.

Methods to study metatranscriptomics. Initial studies to analyze the transcriptomes of environmental samples used real-time polymerase chain reaction (RT-PCR, a genetic amplification technique) and real-time quantitative PCR (RT-qPCR), which can detect the presence of known RNA transcripts and quantitate transcript levels, respectively. PCR-based approaches rely on known gene sequences because primers must be designed. Given that there is great sequence variation within functional genes from organism to organism, it is difficult to design primers to capture gene orthologs. Therefore, these approaches can target only a fraction of the total gene population. Microarrays also have been employed and proved a method capable of measuring expression levels of a large number of genes at one time. Again, the issue of diversity between orthologs prevents a complete profiling of the transcriptomes.

Metatranscriptomics uses high-throughput sequencing technology called pyrosequencing. Pyrosequencing is a DNA sequencing-by-synthesis approach that detects the release of pyrophosphate when a nucleotide is incorporated. This differs from the conventional Sanger sequencing method, where chain termination is detected with dideoxynucleotides. No specific primer is required; therefore, all complementary DNA (cDNA) can be sequenced regardless of whether the sequence or an ortholog is known.

Microbial RNA is extracted directly from an environmental sample, allowing researchers to circumvent cultivation of individual microbes. For water samples, large volumes of water can be filtered. Soil and gut samples can be taken directly from the source. Total RNA is extracted using one of a variety of commercially available kits. The greatest percentage of RNA extracted is rRNA, which can be used to determine microbe diversity (community structure). Typically, the 16S rRNA is used to determine phylogenetic diversity in a sample. If mRNA and sRNA are sought for study, there are several techniques and kits available that can selectively remove rRNA from the total RNA sample. [It should be noted that separation of mRNA from other RNAs is quite difficult in microbes because mRNAs of bacteria and archaea lack distinctive polyadenylation (poly-A) tails, which are present in eukaryotic mRNA.] RNA then is reverse-transcribed to cDNA typically by using random primers. If target RNA is thought to be at low levels, cDNA can be amplified through a variety of techniques. The cDNA is directly pyrosequenced. Several bioinformatics databases and programs can be employed to determine the species of origin, as well as the type of RNA sequence: mRNA, sRNA, or rRNA (see **illustration**).

Metatranscriptomics to identify novel small RNAs. Small (short) noncoding RNAs constitute a relatively new field of research. Unlike mRNA, they do not code for protein. Instead, they possess cellular functions, including modification of the structure and function of mRNA, tRNA, and rRNA. Microbial small RNAs are typically located in intergenic regions of

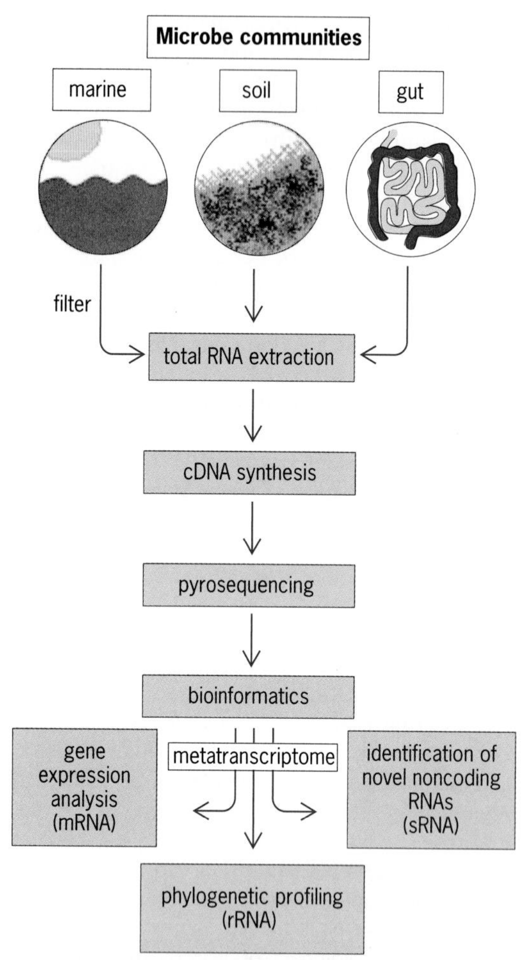

Schematic of metatranscriptomics. Total RNA is extracted directly from environmental samples, including marine, soil, and gut microbial communities. RNA is randomly primed and reverse-transcribed to generate cDNA. The cDNA is directly sequenced using high-throughput pyrosequencing. Sequence data then are analyzed using bioinformatics tools, which can determine the species of origin and type of RNA. If the RNA is novel, then origin is determined by comparison to known genome sequences; function is predicted by known orthologs. Two optional steps are employed sometimes to optimize results. If a particular type of RNA is desired, such as only mRNAs, then the rRNA and sRNA can be removed selectively from the total RNA sample. If levels of target RNAs are low, then cDNA can be amplified through one of several types of techniques. (*Courtesy of Travis W. Wheeler*)

the microbial genomes. Small RNA transcripts range in size from 50 to 500 nucleotides in length. Small RNAs are pivotal in bacterial stress responses and in bacterial virulence. Many small RNAs mediate their effects by silencing mRNAs, thus inhibiting their translation by a process known as RNA interference (RNAi). RNAi was first identified in eukaryotes (referred to as microRNA in eukaryotes) and later confirmed to be present also in prokaryotes. Although prokaryotes and eukaryotes use different enzymatic machinery for RNAi, the general mechanism is the same. Inhibitory RNA precursors are transcribed and processed into small RNAs, which are able to bind to a complementary target mRNA. This prevents the translation of mRNA into protein by either targeting the mRNA for degradation or interfering with the ability of the translational machinery to interact with the mRNA. In both cases, RNAi prevents the production of protein from its target mRNA. The field of RNAi is still in its infancy. Many of the enzymes involved in the processing of RNAi have been identified, but we have yet to fully understand the regulation, interplay, and outcome of inhibitory RNA expression.

A recent investigation demonstrated that metatranscriptomics could be employed to reveal unique microbial small RNAs from the ocean's water column. Roughly 25% of the RNA detected was either sRNA or putative sRNA (psRNA). Fine variations in psRNA of similar taxonomic groups were able to be demonstrated, suggesting a potential role for psRNA in niche adaptation. Given that small RNAs can alter whether mRNAs are translated into proteins, small RNAs are an important piece of the puzzle when trying to interpret mRNA expression data. Analyzing both mRNA and sRNA will allow us to better infer protein expression.

Soil microbial communities. Soil microbes are essential for global cycles of carbon, nitrogen, and sulfur. A single gram of soil has thousands to millions of microbes. As such, studies have been undertaken to determine the structure and function of a soil microbial community. In one study, a sandy lawn in an environmentally protected area was sampled. In total, 6 grams of soil were processed to extract total RNA. Using metatranscriptomics, soil Crenarchaeota (archaea) were able to be characterized, which cannot be cultivated in a laboratory. Using sequencing data and bioinformatics tools, mRNAs for enzymes involved in ammonia oxidation and carbon fixation were linked to Crenarchaeota, thus establishing the function of this microbe in soil. In addition, the relative abundances of bacteria, archaea, and simple eukaryotes at varying trophic levels were determined.

Marine communities. The use of metatranscriptomics to analyze marine microbial communities has provided a great deal of insight into the diversity and ecology of the oceans. A recent study compared pyrosequenced cDNA with genomic DNA sequences of a microbial assemblage from the oligotrophic (nutrient-poor) Pacific Ocean. In addition to genes coding for photosynthesis, carbon fixation, and nitrogen acquisition, several genes encoding hypothetical proteins were present. Analysis revealed previously undetected gene categories. Another study compared the metatranscriptomes and metagenomes of mid- and post-phytoplankton bloom environments in eutrophic (nutrient-rich) coastal waters, demonstrating biological differences in community composition between the two time points. Additionally, this data set was compared with that of the oligotrophic study. Analysis demonstrates some overlap between the two marine samples, but also highlights the immense diversity between two distinct marine habitats.

Human gut microbes. The human gut possesses a complex microbial community. As such, there is a medical interest in the microbial composition of the gut, most notably the study of inflammatory bowel disease (Crohn's disease and ulcerative colitis), as well as some recent reports suggesting that the microbe content of the human gut could influence overall weight and health. Profiling the microbe content of the gut is not sufficient to understanding the role of microbes in human health. It is vital to understand the functional role of these microbes, which can be studied via metatranscriptomics. Current studies have focused on transcriptomes of particular microbes, including the response to probiotics using microarray technology. These studies should be furthered with metagenomics studies.

Like soil and marine microbe communities, the gut consists of bacteria, archaea, and small eukaryotes. In addition, approximately 80% of these microbes cannot be cultured in a laboratory. Metagenomics analyses of gut mucosa and feces have detected 395 bacterial phylotypes. The predominant microbes of the gut are Firmicutes and Bacteroidetes, whereas Actinomycetes, methogens, and fungi are present at lower abundance. Although metagenomics provides insight into microbial diversity, it does not elucidate the function of these microbes within the gut.

Conclusions. Metatranscriptomics is the global study of gene expression of microbial communities at the RNA level. It provides insight into community structure in environmental samples with regard to taxonomy. It also provides an avenue by which to assign ecological function for a given taxonomic group. It enables scientists to monitor community structural and functional changes in response to environmental changes. Finally, metatranscriptomics facilitates the discovery of novel genes and small RNAs, which will further our understanding of the function of microbes in these complex communities.

For background information *see* ARCHAEA; BACTERIA; DEOXYRIBONUCLEIC ACID (DNA); ECOLOGICAL COMMUNITIES; EUKARYOTAE; INFLAMMATORY BOWEL DISEASE; MARINE MICROBIOLOGY; MICROBIOLOGY; RIBONUCLEIC ACID (RNA); RNA INTERFERENCE; SMALL INTERFERING RNA (SIRNA); SOIL MICROBIOLOGY in the McGraw-Hill Encyclopedia of Science & Technology. Rebekah L. Waikel; Patricia A. Waikel

Bibliography. J. A. Gilbert et al., Detection of large numbers of novel sequences in the metatranscriptomes of complex marine microbial communities, *PLoS One*, 8:1–13, 2008; M. A. Moran, Metatranscriptomics: Eavesdropping on complex microbial communities, *Microbe*, 4:329–335, 2009; C. Reiff and D. Kelly, Inflammatory bowel disease, gut bacteria and probiotic therapy, *Int. J. Med. Microbiol.*, 300:25–33, 2010; Y. Shi, G. W. Tyson, and E. F. DeLong, Metatranscriptomics reveals unique microbial small RNAs in the ocean's water column, *Nature*, 459:266–269, 2009; T. Urich et al., Simultaneous assessment of soil microbial community structure and function through analysis of the metatranscriptome, *PLoS One*, 3:1–13, 2008.

Microbial fuel cells

All forms of life are powered by the flow of electrons from a donor to an acceptor. In animals, this is manifested as the oxidation of organic materials (donor) and the reduction of oxygen (acceptor), coupled to the synthesis of adenosine triphosphate (ATP), the chemical currency utilized by cellular processes (**Fig. 1**). Oxygen is an ideal acceptor because it is plentiful in most environments where animals live, and its high oxidation-reduction potential (+0.82 V), a measure of its tendency to accept electrons, enables a large amount of energy to be harvested from its reduction. Unlike animals, bacteria are not limited to the use of oxygen as an acceptor and can use a wide array of alternative electron acceptors, ranging from soluble compounds such as sulfate or nitrate to insoluble minerals such as oxides of iron (III) and manganese (IV). Microbial fuel cells (MFCs) are an innovation that takes advantage of this uniquely microbial capability in order to explore the physiology of alternative electron-accepting processes, as well as to provide inspiration for applied research in bioremediation and energy production.

Fundamentals of MFC design and operation. An MFC is a device that is capable of converting chemical energy into electrical energy with the use of whole bacterial cells as the catalyst. This generally involves the oxidation of organic compounds and the transfer of electrons to an anode as a synthetic surrogate for the organism's natural electron acceptor (Fig. 1). Electrons will then flow from the low-potential anode through a wire to the high-potential cathode in a separate chamber, generating a current that can be captured or measured. At the cathode, electrons are chemically passed to an electron acceptor such as oxygen. What has transpired is essentially a compartmentalized metabolism; organic compounds are

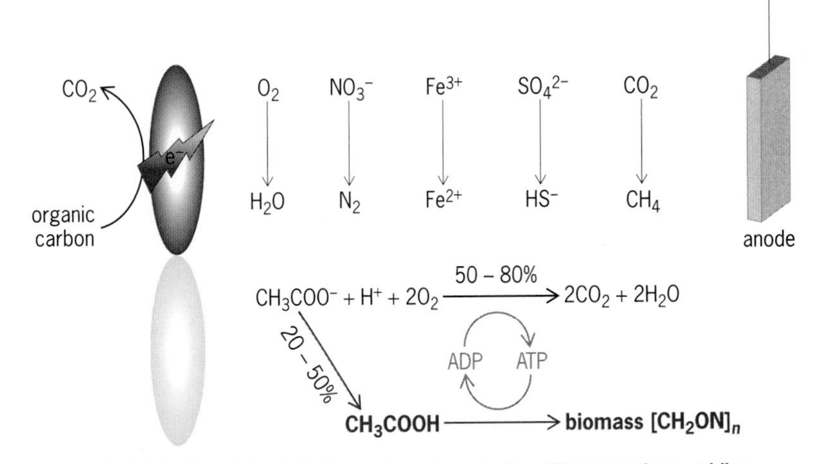

Fig. 1. Model of microbial metabolism and anode reduction. Microorganisms oxidize electron donors, either organic or inorganic, and pass the electrons through the cell energy-generating machinery to an electron acceptor (oxygen, nitrate, iron, or CO_2). Energy generated through this electron flow is captured and stored as ATP, which is subsequently used to fix carbon and produce biomass $(CH_2ON)_n$. Some microorganisms can alternatively pass electrons to an anode surface as a surrogate electron acceptor. Microbial fuel cells are based on the ability of microorganisms to pass electrons to the surfaces of electrodes during catabolic respiration.

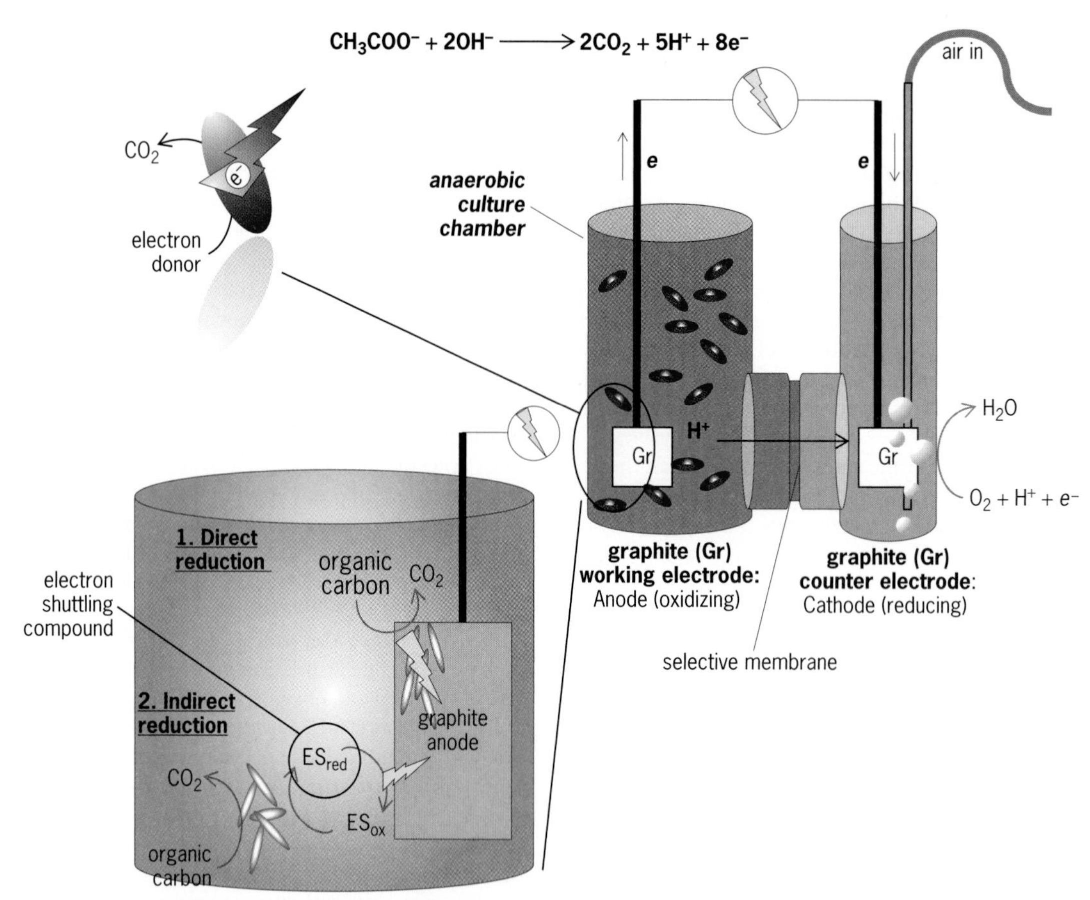

Fig. 2. Schematic of a two-chambered microbial fuel cell showing both direct and indirect reduction at the anode. In this example, acetate (CH_3COO^-) is oxidized to carbon dioxide (CO_2) and protons (H^+) with the aid of a bacterial cell. The electrons are passed to the anode and flow to the cathode, generating a current that can be captured. At the aerobic cathode, electrons meet with protons diffusing from the anode chamber to reduce molecular oxygen (O_2) to water.

oxidized at the anode, and the electron acceptor is terminally reduced at the cathode. Rather than coupling these reactions solely to synthesis of chemical energy in the cell, a portion of the free energy associated with the electron flow is captured as an electric current.

In practice, MFCs are often set up in an H-cell configuration, with the anode and the cathode located in separate chambers (**Fig. 2**). The anode chamber of the fuel cell is usually kept free of oxygen and other electron acceptors in order to generate current, as many bacteria will not transfer electrons to the anode if more favorable terminal electron acceptors are present. Between the two chambers is a membrane that impedes the movement of oxygen or negatively charged ions between the chambers, while allowing the free diffusion of positively charged protons from the anode to the cathode, thus maintaining a charge separation across the electrodes. The inputs and outputs to the system are essentially the same as those in simple respiration, but a portion of the energy that would otherwise be lost after the bacteria have harvested their energy for growth and maintenance is conserved as electrical energy. As a result, there is often a trade-off between power production (the product of electron flow per unit time or current measured in amperes and potential difference across the electrodes measured in volts) and microbial growth in an MFC. At anodes with a high (electropositive) poised potential, bacteria can extract a great deal of energy per electron transferred and grow more quickly, converting low-potential electrons into biomass. Conversely, a lower-potential anode requires a bacterium to transfer more electrons to yield a similar amount of energy, and there are fewer electrons available for growth and synthesis of biomass. Functionally, this complex relationship between potential, current, and microbial growth means that optimization of fuel cell conditions varies, based on the parameters of the system as well as on the criteria for optimization.

Mechanisms of electron transfer. The anode chamber of an MFC is an artificial environment, but it must simulate natural conditions that certain bacteria recognize. The reduction of insoluble trivalent ferric iron [Fe(III)] minerals by bacteria is the natural analog that is most similar to the anode reduction. In some environments, such as those at acidic pH values (pH < 5), ferric iron may be soluble, enabling it to interact with the microbial electron transport chain at the cell membrane, as oxygen does in aerobic respirations. However, in general, ferric iron is predominantly found in the form of aggregated insoluble hydroxides, which cannot directly interact with the bacterial cell membrane. To access these minerals, bacteria have evolved several distinct physiological strategies, the study of which has been accelerated by the experimental use of MFCs.

The three basic mechanisms of bacteria use to reduce insoluble electron acceptors can be described as direct, mediated, and chelation (**Fig. 3**). The chelation (or binding) mechanism involves the biochemical solubilization of insoluble ferric iron and is not effective in respiration of the completely insoluble carbon anode. Mediated mechanisms of electron transfer feature an electron shuttle that can accept electrons from the microbial electron transport chain, diffuse to the anode surface, and abiotically transfer electrons to the anode. Alternatively, some bacteria are capable of transferring electrons directly to mineral or anode surfaces through a contact-dependent mechanism. This can be accomplished by having electrically active proteins on the surface of the cell or possibly even via electrically conductive appendages, called pili, protruding from the cell.

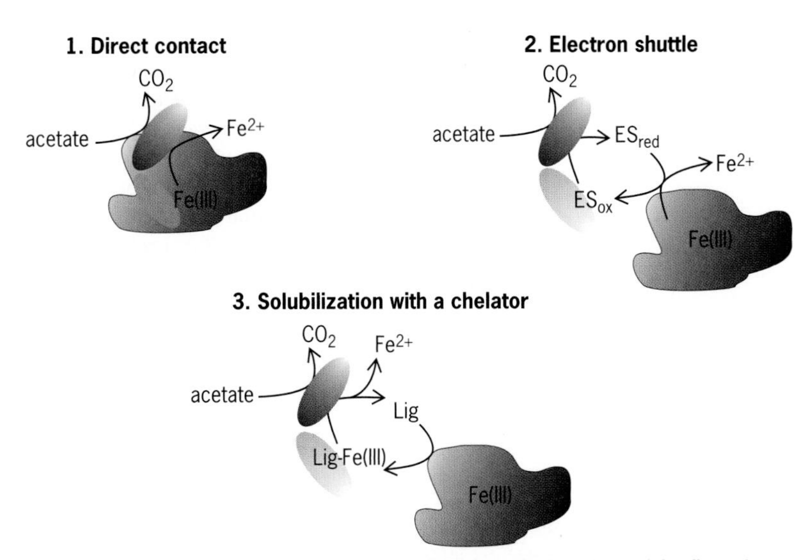

Fig. 3. Diagram of the main mechanisms of electron transfer from bacterial cells to the anode of a microbial fuel cell. The chelation mechanism is not applicable to MFCs using a carbon electrode, as the anode cannot be dissolved. Both the direct and mediated mechanisms can occur simultaneously in a microbial community, and recent research indicates that some bacteria species can switch from one mechanism to another, based on external conditions. ES = electron shuttling compound; Lig = ligand or chelator.

MFCs as a tool for exploring microbial physiology. MFCs are prominent in much of the contemporary research contributing to the understanding of the mechanisms of extracellular electron transfer. Conducting experiments in an MFC can be advantageous to the researcher for several reasons. The potential at the anode can be manipulated directly to optimize growth, and the current flowing to the cathode can be measured, which enables the monitoring of electron transfer in real time, contributing to the understanding of the process. Proteomic analysis performed on cells grown on an anode has aided in the identification of several gene products involved in the process of electron transfer. In many species, cells grown on an anode have the tendency to form conglomerates called biofilms, which can be imaged and analyzed. This allows a look at how populations of these bacteria might be organized in the natural environment. Finally, analysis of spent media in an anode is a simple way to identify microbially produced mediating compounds that allow indirect electron transfer to the anode.

Microbial ecology of MFCs. Rather than bacteria of one type being grown in an artificial pure culture, MFCs can involve communities of various bacterial species metabolizing at the anode. In the environment, various species of bacteria coexist in complex assemblages that respond to changing conditions. This is reflected in studies that aim to elucidate the population dynamics of fuel cell communities. The ability to transfer electrons to an anode was previously thought to be restricted to a few well-characterized species that are capable of iron reduction. However, studies that look at complex anode communities suggest that there are many more types of bacteria associated with this process that remain to be discovered. Environmental conditions also play a factor in shaping the electricity-generating communities of MFCs. Research performed on MFCs operated at high temperatures [55–60°C (131–140°F)] indicates that communities in these anodes are dominated by a different cohort of bacteria from those operated at more moderate temperatures [20–35°C (68–95°F)]. There is also evidence that interactions between bacteria at the molecular level drive the degradation of complex organic carbon sources at the anode. In this line of research, MFCs provide an innovative mechanism for looking at anaerobic population dynamics, as well as carbon flow in complex ecosystems.

Applications of MFC technology. Recently, the interest in alternative energy sources has driven applied research in MFCs. As this system converts organic carbon into electricity, MFC-derived power is considered "carbon-neutral" with regard to greenhouse gas emissions. The coulombic efficiency of MFCs can also be very high, with certain anodic pure cultures converting more than 90% of electrons from the input source into current. The efficiency is less than perfect because of competing electron-accepting processes (that is, hydrogen production or sulfate reduction) or because electrons are sequestered into microbial biomass. These characteristics make MFCs an attractive technology for energy production, particularly in situations where there is a waste stream that is rich in organic carbon. However, although MFCs are highly coulombic efficient, the power densities achievable by an MFC are still at least an order of magnitude below those required to make this a viable technology for commercialization. In addition, the important variables for efficient scaling of these reactors have yet to be reliably determined. To this end, for example, pilot-scale MFCs that have been implemented to treat organic waste produced at a number of breweries have fallen short of their original design goals. Although MFCs could potentially be used to treat sewage at waste treatment facilities, offering the dual benefits of remediating waste for release into the environment and alleviating the energy demanded by the process, recent studies have indicated that traditional biomethanogenesis is a more reliable technology for converting waste organics into energy. Other more interesting uses for MFCs include hydrogen production, biofuel synthesis, remediation of countless chemicals, and environmental biosensors. Alternatively, bioelectrical reactors (BERs), a relative of the MFC, use microbes to catalyze reduction of an acceptor at the cathode. Use of this technology has been shown to be capable of remediating many contaminants (including toxic heavy metals, perchlorate, and poisonous organic compounds) at the bench scale, and current efforts are focused on identification of the important scaling variables and pilot-plant installation. The applications and possibilities of microbial fuel cell research are limited only by the repertoire of microbial metabolisms, the known breadth of which is constantly expanding.

For background information *see* BACTERIAL PHYSIOLOGY AND METABOLISM; BIOCHEMICAL ENGINEERING; BIOFILM; BIOLOGICAL OXIDATION; BIOTECHNOLOGY; ELECTROCHEMICAL PROCESS; FUEL CELL; INDUSTRIAL MICROBIOLOGY; MICROBIAL ECOLOGY; MICROBIOLOGY; OXIDATION-REDUCTION in the McGraw-Hill Encyclopedia of Science & Technology.

Ryan A. Melnyk; John D. Coates

Bibliography. B. E. Logan, *Microbial Fuel Cells*, John Wiley & Sons, Hoboken, NJ, 2008; M. T. Madigan et al., *Brock Biology of Microorganisms*, 12th ed., Pearson/Benjamin Cummings, San Francisco, 2009; E. L. Madsen, *Environmental Microbiology: From Genomes to Biochemistry*, Wiley Blackwell, Malden, MA, 2008; J. C. Thrash and J. D. Coates, Review: Direct and indirect electrical stimulation of microbial metabolism, *Environ. Sci. Tech.*, 42:3921–3931, 2008; K. C. Wrighton and J. D. Coates, Microbial fuel cells: Plug-in and power-on microbiology, *Microbe*, 4:281–287, 2009.

Mobile WiMAX

Mobile WiMAX refers to a new mobile broadband radio access technology, based on IEEE802.16 standards, which has gained significant attention and momentum over the past few years. Motivated by the success of the Internet and WiFi, the data communications industry created Mobile WiMAX to promote a paradigm shift in traditional telecommunications cellular service offerings and in the ecosystem. The new paradigm is based on low-cost and Internet-friendly solutions for wide-area networks to enable innovative applications, services, and business models.

While the initial WiMAX specifications and products addressed the fixed broadband access market such as DSL (digital subscriber line) replacement and wireless backhaul, the industry quickly turned its focus to enable mobility to significantly expand the market opportunities and deployment models. The first release of mobile WiMAX was formally approved by the International Telecommunication Union (ITU) in 2008 as International Mobile Telecommunications-2000 (IMT-2000) technology, that is, a global standard for deployments of third-generation (3G) wireless networks. There have been large-scale deployments of first-generation mobile WiMAX systems in many countries and regions including the United States, Japan, Russia, and India. Altogether, there have been more than 400 deployments in more than 150 countries. Meanwhile standardization of the next generation of mobile WiMAX is ongoing as a promising contender for IMT-Advanced, or the so-called 4G systems.

A WiMAX network can be deployed as a green-field (that is, where there are no previously existing networks) and stand-alone data-voice network or as an overlay to existing fixed or mobile legacy access networks such as 2.5G or 3G cellular systems or cable and DSL networks, while supporting different levels of interworking and service continuity. The same WiMAX network architecture can accommodate a variety of usage models such as wireless backhaul to WiFi hot spots, fixed and nomadic access to customer premises equipment (CPEs) and residential gateways (RGs), and for mobile access to notebooks, smart phones, and next-generation WiMAX-embedded ultramobile devices (**Fig. 1**).

Depending on the antenna height and transmission power, the range of WiMAX cell coverage may vary, with a typical range of around 0.5–2 km (0.3–1.2 mi). From a mobility perspective, mobile WiMAX can support mobile user speeds up to 150 km/h (90 mi/h), with graceful degradation at the high end of the speed range, which is typical of all 3G and 4G technologies.

The WiMAX architecture also enables open access to Web-based applications and enhanced Internet services as well as traditional "wall-gardened" (that is, operator-managed) services in the same network. This flexibility allows operators and content providers to explore creative and mutually beneficial service offerings and Internet-friendly business models.

Standardization and certification. WiMAX end-to-end specifications and products are based on cooperative and complementary standards development efforts by the WiMAX Forum™ and the IEEE 802.16 Working Group. In fact the name "mobile WiMAX" was created by the WiMAX Forum, which was formed in June 2001 as a worldwide consortium focusing on the global adoption of WiMAX technology. The WiMAX Forum is an industry-led nonprofit organization chartered to promote and enable end-to-end standards-based WiMAX solutions and their interoperability. To achieve these goals, the WiMAX Forum has created various working groups that address requirements, end-to-end technical specifications, testing and certification, marketing, global roaming, application, and regulatory issues (**Fig. 2**).

The IEEE 802.16 Working Group, established by the IEEE Standards Board in 1999, has developed and published several versions of air interface standards for wireless metropolitan area networks (WirelessMAN), with focus on specifications for the media access control (MAC) and physical (PHY) layers. While the initial versions of the standard, 802.16/a/d, focused on fixed access, the later versions, that is, 802.16e-2005, 802.16-2009, and 802.16m, which is being developed as a basis for the next generation of mobile WiMAX systems, include many advanced features and functionalities needed to support flexible bandwidths and spectrum, enhanced spectrum efficiency, and improved security and performance at high mobility speeds.

The IEEE 802.16 standards define the structure of the physical-layer and link-layer operations that occur between mobile subscriber stations and base stations. However, the over-the-air upper-layer signaling as well as network architecture and protocols behind the base stations that are required for an end-to-end specification are outside the scope of the IEEE standard. In addition, 802.16 MAC and PHY specifications, defined with a focus on flexibility, leave open many options and allow various implementations of base station and mobile station features that, unless coordinated for interoperability, can result in incompatible products.

To address this issue, the WiMAX Forum defines minimum product compliance (interoperability and conformance) specification and required testing specifications to be used as a basis for industrywide certification of devices and base stations. In addition to developing compliance specifications and testing methodologies, the WiMAX Forum has also established certification laboratories and manages conformance and interoperability testing to ensure that all WiMAX-certified products across different implementations work seamlessly with one another.

To extend the scope of interoperability beyond the air interface and based on the service provider's requirements, the WiMAX Forum also defines end-to-end network architectures and protocols based on, and consistent with, IEEE 802.16 and WiMAX System Profile air interface definitions. The network specification efforts by the WiMAX Forum involve interactions with other standards organizations such as the Internet Engineering Task Force (IETF), the Third-Generation Partnership Project (3GPP), 3GPP2, the DSL Forum, and the Open Mobile Alliance (OMA).

Technical features and system architecture. The mobile WiMAX air interface is based on orthogonal frequency-division multiple access (OFDMA) and advanced multiantenna technologies, which are accepted by the industry at large to be the key building blocks for next-generation wireless access technologies. **Figure** 3*a* shows conceptually the structure of an OFDMA signal, where a set of narrow-band but orthogonal subcarriers ($S_1, S_2, \ldots$) or tones are used to carry modulated data in a wide-band channel. The OFDMA-based PHY structure helps with multipath effects and delay spread while also allowing a flexible framework for time and frequency scheduling of radio resources in a scalable system bandwidth.

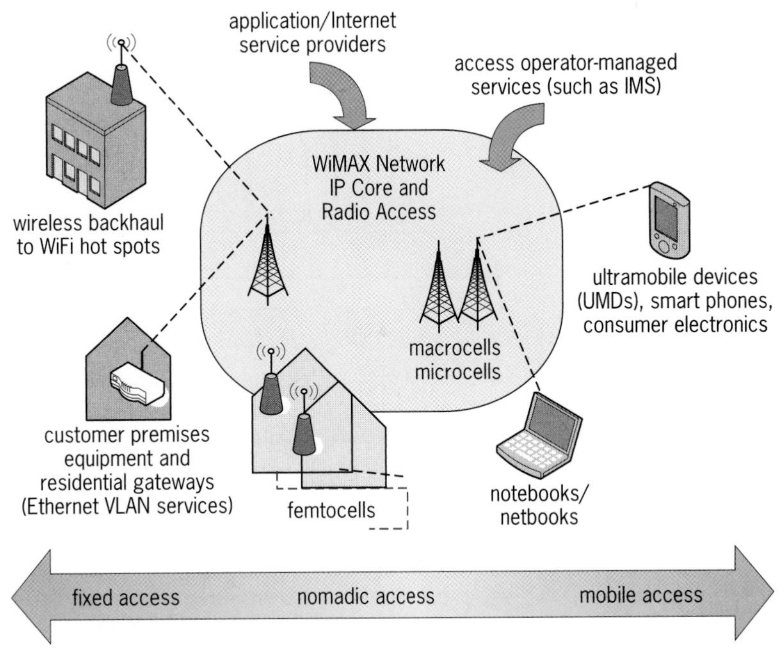

Fig. 1. Mobile WiMAX, enabling a variety of usage models in the same network. IMS = IP Multimedia Subsystem, VLAN = Virtual Local-Area Network.

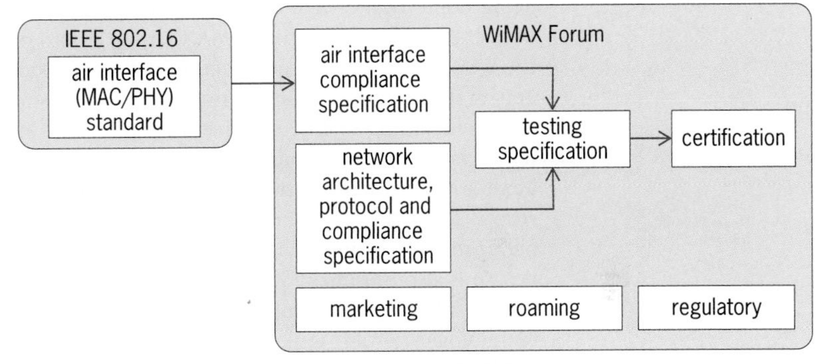

Fig. 2. WiMAX standardization in IEEE and the WiMAX forum.

Figure 3*b* shows the basic idea of multiple-input, multiple-output (MIMO) antenna systems, which effectively create and utilize multiple spatial channels between multiple transmitter and receiver antennas. The MIMO structure combined with advanced signal processing and precoding techniques provides spatial multiplexing, diversity, and beam forming, which improve system coverage and capacity and increase achievable user data rates at both high and low mobility.

The OFDMA- and MIMO-based air interface combined with advanced time and frequency scheduling and fractional frequency reuse provides very high spectrum efficiency and user data rates exceeding what 3G or enhanced 3G systems can offer, for example, data rates almost three times higher than high-speed packet access (HSPA) systems. Most early mobile WiMAX systems are deployed using 5- and 10-MHz channels in the 2.3-, 2.6-, and 3.5-GHz bands in TDD mode in which the spectrum is flexibly time shared between asymmetric downstream and upsteam data traffic.

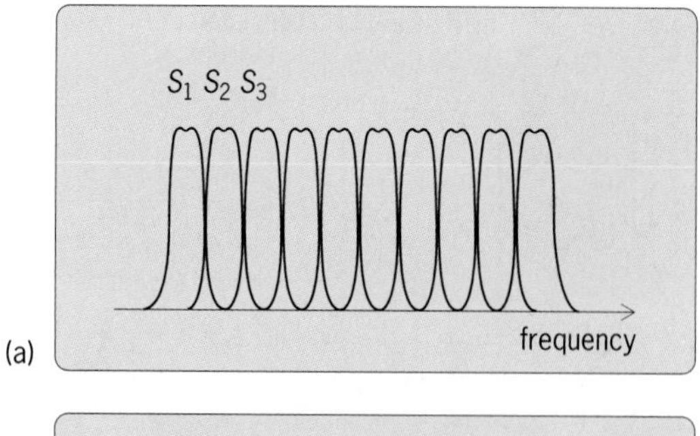

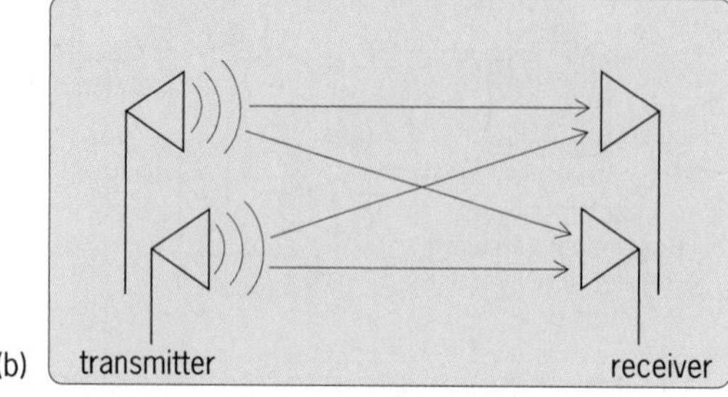

Fig. 3. Concepts of (*a*) OFDMA (orthogonal frequency-division multiple access) and (*b*) MIMO (multiple-input, multiple-output) used in the WiMAX air interface.

Work on the evolution of WiMAX networks is also being done by the IEEE 802.16 Working Group as Project 802.16m and by the WiMAX Forum as part of the WiMAX Release 2 targeting certification of products in the 2012 time frame. The key focus areas of next-generation WiMAX systems are optimized interworking with WiFi and 3G systems, support for flexible deployments using multicarrier aggregation, relays, femtocells, advanced interference management, and self-organizing networks. The air interface design is also optimized for 20-MHz channels using the latest PHY layer techniques and advanced MIMO technologies.

Network design. The WiMAX network architecture is designed to meet system requirements while maximizing the use of open standards and IETF protocols in a simple all-IP (Internet Protocol) architecture. Among the design requirements are supports for fixed and mobile access deployments as well as unbundling of access, connectivity, and application services to allow access infrastructure sharing and multiple access infrastructure aggregation. The IP network architecture in WiMAX is designed with a focus on simplicity and reuse of Internet-based open protocols in line with the new trends and requirements among operators for next-generation networks. While the WiMAX technology is primarily focused on IP protocols and solutions, it also supports Ethernet as an option that has limited usage but has importance for some fixed-access deployments. The evolution of the WiMAX network also includes voice over IP (VoIP), location-based services, and multicast and broadcast services, which are also supported by evolved 3G systems.

The WiMAX network reference architecture (**Fig. 4**) consists of several logical network entities, namely the mobile station (MS), the access service network (ASN), and the connectivity service network (CSN). This architecture differentiates a network access provider (NAP) as a business entity that provides WiMAX radio access infrastructure from a network service provider (NSP) that provides IP connectivity and WiMAX services to subscribers. The network architecture allows one NSP to have a relationship with multiple NAPs in one or different geographical locations. It also enables NAP sharing by multiple NSPs. In some cases the NSP may be the same business entity as the NAP.

In addition to focusing on OFDMA, MIMO, and IP-based technologies and despite some technical and business challenges, mobile WiMAX also aims at enabling a retail device distribution model and end-to-end infrastructure "plug and play" interoperability to lower costs of products and networks. In the retail distribution model, embedded WiMAX devices including consumer electronics are sold through conventional consumer electronics stores, and no longer require operator-specific validation and sales distribution. Following and extending the success of the WiFi Model into wide-area networks, the retail device distribution model also helps in reducing the cost of customer acquisition by removing the subsidy of devices that most operators pay in today's wireless business models.

For background information *see* INTERNET; MOBILE COMMUNICATIONS; MULTIPLEXING AND MULTIPLE ACCESS; VOICE OVER IP; WIDE-AREA NETWORKS; WIRELESS FIDELITY (WI-FI) in the McGraw-Hill Encyclopedia of Science & Technology. Kamran Etemad

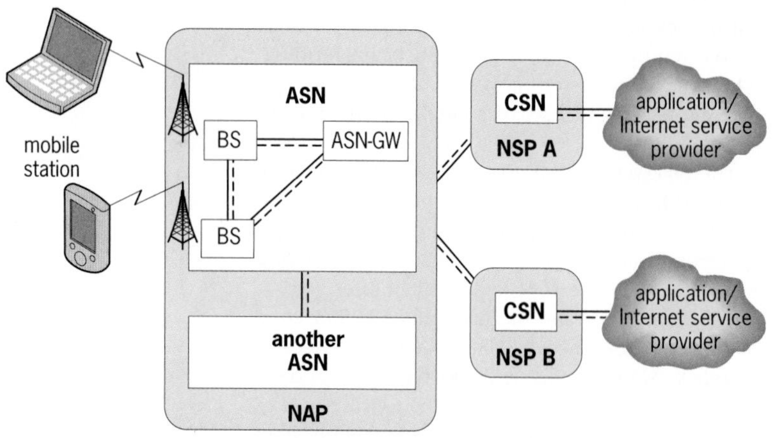

Fig. 4. WiMAX network architecture. ASN = access service network, BS = base station, ASN-GW = ASN gateway, CSN = connectivity service network, NAP = network access provider, NSP = network service provider.

Bibliography. S. Ahmadi, An overview of next-generation mobile WiMAX technology, *IEEE Comm. Mag.*, 47(6):84–98, 2009; K. Etemad, Overview of mobile WiMAX technology and evolution, *IEEE Comm. Mag.*, 46(10):31–40, 2008; P. Iyer et al., All-IP network architecture for mobile WiMAX, in *IEEE Mobile WiMAX Symposium*, IEEE, pp. 54–59, 2007; Q. Li et al., Advancement of MIMO technology in WiMAX: From IEEE 802.16d/e/j to 802.16m, *IEEE Comm. Mag.*, 47(6):100–107, 2009; F. Wang et al., Mobile WiMAX systems: performance and evolution, *IEEE Comm. Mag.*, 46(10):41–49, 2008.

Modeling effects of climate change on allergic illness

Allergic illness is responsible for a substantial proportion of health care costs in the United States, and the prevalence of allergic airway disease (AAD) has increased over the last 30–40 years. AAD includes disorders such as asthma, sinusitis, and allergic rhinoconjunctivitis. Various environmental factors are suspected to play a role in AAD, including diet and exposures to specific air pollutants (such as environmental tobacco smoke and photochemical smog) and pollen-associated allergens. Typically, pollen-related allergies in the spring are caused by pollen from trees such as birch and oak, while those in the late summer and fall are attributable to ragweed. Exposure to air pollutants can increase the sensitivity to pollen. Asthmatics who are exposed to high levels of atmospheric pollutants, such as ozone and particulate air matter, have a higher risk of developing symptoms when they subsequently are exposed to pollen. Because co-exposures to pollen and air pollutants, such as ozone, can have synergistic adverse health effects, there is a need to study co-exposures to gaseous, particulate-phase pollutants, and bioaerosols in a consistent manner. **Figure 1** shows the occurrence of ozone and ragweed pollen levels across the continental United States.

Even though AAD is a complex problem, the increase in AAD appears to be linked to various climate factors such as rising temperature, increased precipitation, and rising carbon dioxide (CO_2) concentrations. In the Northern Hemisphere, the balance of evidence strongly suggests that a significant impact of climate change is already discernible in animal and plant populations, as well as communities and ecosystems. The Intergovernmental Panel on Climate Change (IPCC) concluded that there was "good evidence" that general warming affects the timing of the onset of allergenic pollen production, which may influence pollen abundance and/or potency.

Climate change. Climate change is expected to alter the types and amounts of pollen emissions, their allergenic potency, as well as the time period when the peak emissions occur. Warming climate in the latter part of the twentieth century resulted in earlier pollination of trees, compared to grasses and weeds, and longer pollination seasons (early flowering by 6 days and delay of autumn events by 5 days, compared with the early 1960s). Controlled growth-chamber studies, where temperatures and CO_2

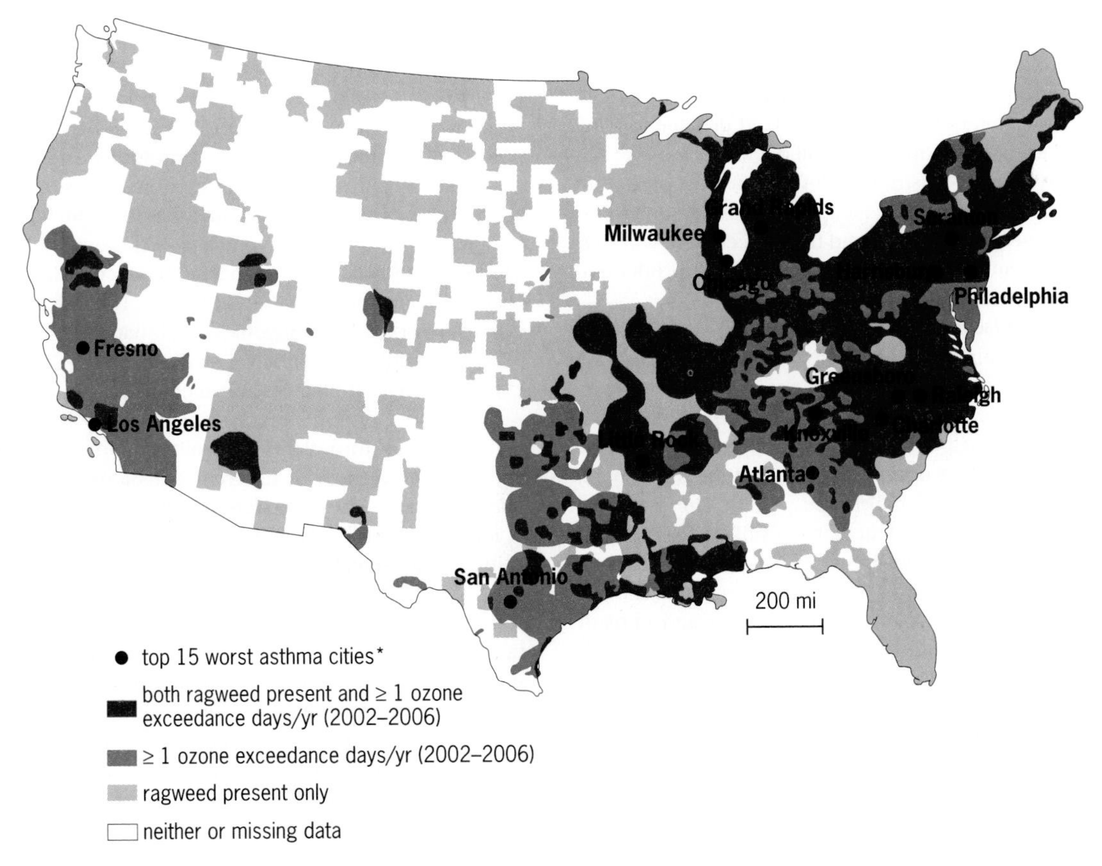

Fig. 1. Co-occurrence of high ragweed pollen levels and high ozone levels across the continental United States. Ragweed data as of 2007. Ozone data based on annual average of monitors with valid 8-h data for at least 75% of required monitoring days in each year (2002–2006). Limited to areas with at least one monitor within 100 km. (*Source: Knowlton et al., 2007, based on ozone monitoring data from U.S. EPA, U.S. Department of Agriculture, and Asthma and Allergy Foundation of America*)

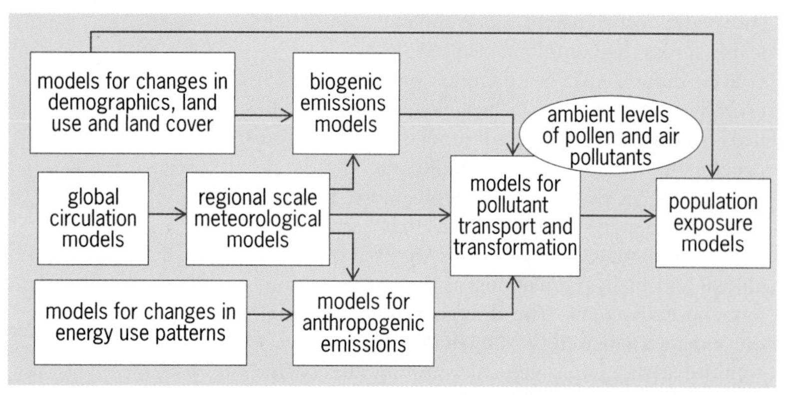

Fig. 2. Interrelationships between various computational models for assessing the impact of climate change on allergenic airway diseases.

levels can be programmed, have shown that increased CO_2 concentrations and ambient temperatures result in increased pollen production and increased allergenic potency. Weather patterns, particularly wind and precipitation, responsible for the transport and removal from the air of pollen and other pollutants, are significantly affected by climate change. For some air pollutants, additional factors that govern their photochemical transformation and precursors will also be altered so that concentrations of allergens in the air will change even with constant emissions.

General warming may even alter the distribution of plant species, which in turn modifies the locations of allergen emission. Significant changes in population demographics and land use due to natural growth of population will occur, which will not only affect the distribution of human "receptors" but also the emissions of allergens. Likewise, emissions of other air pollutants and their precursors will shift due to anthropogenic factors such as the production of electricity and vehicle emissions.

Major factors governing exposures. Assessing exposures of humans to pollen involves studying the wide range of factors that affect the spatially and temporally varying emissions profiles of pollen and the transport of pollen. Additional factors include the corresponding population distribution and contact of individuals with airborne pollen. The main factors that affect the emissions of pollen are the areal coverage by various trees, grasses, and weeds; pollination period for each plant species; temperature, humidity, soil moisture, solar radiation; and ambient levels of CO_2 and ozone. Pollen spores can be transported over tens of kilometers of distance, depending on the height of release, pollen type, and prevailing meteorological conditions. Although a very small amount of pollen released from an individual location will reach a "receptor location" far from the source, the overall contribution from pollen released from large areas can be substantial. Major factors affecting the transport of pollen include temperature, wind speed/direction, terrain, precipitation, and solar radiation.

Population density and distribution affect the potential and actual exposure to pollen, while the time-location and time-activity patterns of different individuals affect the degree of contact of these people with aeroallergens. The physiological attributes of the target population, such as breathing rates, determine the inhaled amount of aeroallergens. Similar factors are important in assessing exposures of humans to gaseous pollutants and particles, and are routinely studied in human exposure modeling. The exposure information, coupled with information on susceptibility and health effects can provide an insight into population level risks of allergic illness.

Predictive modeling. Currently, pollen forecasts and information on current conditions are provided based on pollen counts measured at about 100 locations in the United States. Even though these maps provide a coarse resolution, increasing the resolution poses very high resource demands, thus necessitating the application of computer simulation models. Because the problem of assessing human exposures is complex, involving variations across different geographical regions, times, and populations, computer models are needed that systematically follow the sequence of the steps from emissions, to transport, and finally to human contact, by using the underlying principles of physics and chemistry, along with available databases on populations, land use, and land cover. Computer models and databases can help estimate pollen emissions from trees, grasses, and weeds over a large geographical area of interest, based on the information on prevalent vegetation, CO_2 concentration, sunlight, temperature, and precipitation. These models can predict the resulting pollen concentrations using a set of mathematical equations describing the short- and long-range transport of pollen based on meteorological information. The outputs of these models can be used for developing maps of concentrations of aeroallergens and other air pollutants. These maps can help susceptible individuals by allowing them to plan their outdoor activities in order to reduce their exposures to air pollutants and allergens.

Predictive computer models are currently being widely used for forecasting air quality and for assessing the effectiveness of different emissions reduction strategies. These models describe the region of interest (usually a large geographical area) using a three-dimensional grid, typically of a resolution of 4×4 km (2.5×2.5 mi) horizontal resolution and 20–100 m (66–330 ft) in the vertical resolution. Within each cell of this grid, mass-balance equations are applied for each pollutant accounting for the emissions, flow in, flow out, deposition, and chemical transformation (formation or depletion). The solutions to these mass-balance equations at each grid cell at each time interval provide estimates of pollutant concentrations. These models can be used with historical data for model evaluation and refinement, and can be also used for studying possible future scenarios and for real-time air pollution forecasting.

To study the impact of various climate change scenarios on allergic airway diseases, multiple computer models covering different aspects of this prob-

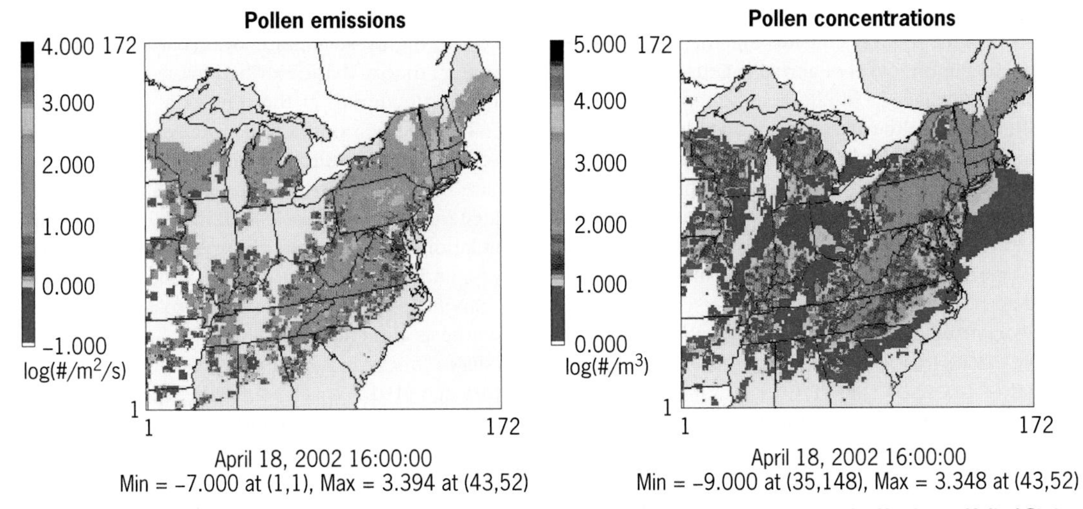

Fig. 3. Examples of estimation of birch pollen emissions and its airborne concentrations across the Northeast United States using a sequence of meteorological and contaminant fate and transport models. (*Adapted from C. Efsthathiou, 2009, Ph.D., Rutgers University*)

lem are needed. They include (1) estimating the potential changes in climate factors (CO_2 and temperature) in the future, (2) characterizing changes in demographics, land use, and energy use, (3) estimating changes in land cover and the prevalence of different pollen species under a changing climate, (4) estimating anthropogenic emissions of air pollutants and their precursors, (5) simulating atmospheric fate and transport of pollen and air pollutants, and (6) simulating the contact of humans with these chemicals (human exposure modeling accounting for indoor and outdoor dynamics or pollutants, as well as time-location and time-activity information for individuals). Results from these models need to be combined with models that quantify the changes in the allergenic potency of pollen from different allergenic plants due to changes in atmospheric CO_2 levels in order to obtain estimates of risks and costs associated with AAD in the future. **Figure 2** shows the different types of computer models, information flows through these models, and the interrelationships among them.

Computer models, tools, and databases. Global general circulation models for the current and future climate are available from the IPCC data archive, which can drive local and regional weather models such as the Weather Research and Forecasting (WRF) model. Macro-level models exist for predicting probable changes in the location and density of industrial and residential areas across the country based on recent and projected economic and land use planning patterns, available land supply, changes in the socioeconomic makeup of regional populations, and potential impacts of climate change and other environmental factors. The corresponding models for describing the fate and transport of different pollutants include the U.S. Environmental Protection Agency's (EPA's) Community Multiscale Air Quality (CMAQ) model, which takes as input the meteorological information from models, such as WRF, and calculates the spread of the pollutants over local and regional scales. In practice, these models are first evaluated and calibrated using "base case" scenarios, for which both the inputs (such as meteorological factors) and outputs (such as pollen levels) are available. **Figure 3** shows an example application where birch pollen emissions and airborne concentrations were estimated across the Northeast United States for 2002, for which detailed information on meteorology is available. These models can also be used to study "future-year" scenarios or for scenarios corresponding to specific hypotheses. Supporting information on the distribution of land use and land cover is available from the United States Geological Survey (USGS) and satellite imagery.

Even though these models cannot precisely and accurately predict the exposures in relation to specific individuals in the future, they can be used to statistically characterize exposures to selected populations by utilizing information on current or projected population distributions and related population attributes. This is pursued through randomly generating a large number of statistically representative "virtual" individuals and simulating the activities of each virtual individual in space and time. The time-location-activity information, in conjunction with information on ambient levels of pollutants, provides estimates of exposure for the particular virtual individual. When these estimates are aggregated for a large number of virtual individuals, they provide information on population exposures, which are statistically representative of the exposure to the population under consideration. These metrics can be used for comparatively analyzing different potential scenarios and for testing various hypotheses, including the impact of different degrees and patterns of warming, the impact of changing land use and land cover, as well as the interrelationships between levels of different air pollutants and aeroallergens. Even when the underlying approximations include systematic biases and uncertainties, when used comparatively these models will be appropriate for assessing the relative impacts of different hypothetical scenarios. Thus, computer models considering the impact of climate change on levels of aeroallergens and other air pollutants will help in characterizing exposures and estimating health-care costs associated with the corresponding changes in allergenic airway disease. These can help assess the effectiveness of different regulatory strategies as well as provide information that helps individuals plan their activities in order to reduce their overall exposures to aeroallergens and related air pollutants.

For background information *see* AIR POLLUTION; AIR POLLUTION, INDOOR; ALLERGY; ASTHMA; CLIMATE MODELING; GLOBAL CLIMATE CHANGE; MODEL THEORY; POLLEN; SIMULATION in the McGraw-Hill Encyclopedia of Science & Technology.

Sastry S. Isukapalli; Leonard Bielory; and Panos G. Georgopoulos

Bibliography. J. L. Gamble et al., *Review of the Impacts of Climate Variability and Change on Aeroallergens and Their Associated Effects*, U.S. Environmental Protection Agency, Washington, D.C., 2008; K. Knowlton, M. Rotkin-Ellman, and G. Solomon, *Sneezing and Wheezing: How Global Warming Could Increase Ragweed Allergies, Air Pollution, and Asthma,* Natural Resources Defense Council, New York, 2007; A. J. McMichael, R.E. Woodruff, and S. Hales, Climate change and human health: present and future risks, *Lancet*, 367(9513):859-869, 2006; J. H. Seinfeld and S. N. Pandis, *Atmospheric Chemistry and Physics: From Air Pollution to Climate Change*, 2d ed., Wiley Interscience, Malden, MA, 2006; S. Solomon et al., *Climate Change 2007: The Physical Science Basis. Contribution of Working Group I to the Fourth Assessment Report of the Intergovernmental Panel on Climate Change*, Intergovernmental Panel on Climate Change, Geneva, Switzerland, 2007.

Molecule-based superconductors

The phenomenon of superconductivity, the complete loss of electrical resistivity in a material, holds tremendous potential for applications including magnetically levitated trains, highly efficient electrical cables and generators, environmentally friendly transformers, energy storage devices, and sophisticated electronic components. Many applications already use superconductors, including magnetic resonance imaging (MRI), sensitive magnetometers based on superconducting quantum interference device (SQUID) technology, and particle accelerators.

Although the future for superconductivity is bright, many scientific and technical hurdles still need to be addressed. The superconducting transition temperature (T_c), the temperature below which a material has zero resistivity, is still well below room temperature. One of the challenges for superconductivity research is to understand the mechanism by which this remarkable phenomenon occurs. In 1957, J. Bardeen, L. Cooper, and R. Schrieffer proposed the highly successful BCS (Bardeen-Cooper-Schrieffer) theory that described many of the superconducting materials at the time. The discovery of several new superconducting systems over the past half century has resulted in unexpectedly high T_cs that are not fully described by the BCS mechanism. Understanding these exotic systems will require the formulation of a universal theory. The design, discovery, and development of new superconductors will benefit from a broad-based approach to materials synthesis and characterization.

History. Superconductivity was first discovered a century ago (1911) by H. Kamerlingh Onnes, while measuring the resistivity of mercury at temperatures near the boiling point of liquid helium. Until the 1980s, the highest T_cs were found in a class of materials called A15 compounds, with T_cs up to 23 K (−250°C) in the Nb_3Ge alloy. In 1986, the unexpected discovery by J. G. Bednorz and K. A. Müller of superconductivity in a new class of oxide materials, called perovskites, revitalized research in this field and led to the development of materials with T_cs above the boiling point of nitrogen (77 K or −196°C). Because liquid nitrogen is much cheaper than liquid helium, it is technologically important for T_cs to be above 77 K for many applications. This class of materials still holds the record, with materials such as TlBaCaCuO, having T_cs above 125 K (−148°C).

From a developmental perspective, it is important to search for superconductivity in new, and sometimes unexpected, material classes. The 2001 discovery of superconductivity in MgB_2 at 39 K (−234°C) occurred nearly 50 years after the material was first synthesized. Although considered to be a conventional (BCS-type) superconductor, its electronic structure is composed of two significantly different electronic bands near its Fermi surface. In 2008, a new family of iron-pnictide superconductors was discovered with T_cs exceeding 50 K (−223°C).

Molecule-based systems. Although most organic materials are insulators, or poor conductors, it was shown in the 1970s that conducting polymers, such as doped polyacetylene, can have conductivities that are higher than copper. Superconductivity in molecule-based materials, the focus of this article, was first reported in 1980. The discovery of fullerenes in the early 1990s enabled the T_cs in this class of materials to approach 40 K (−233°C). Molecule-based systems are ideally suited for developing structure-property correlations because their structures can be rationally tuned through chemical modification. The phase diagram (**Fig. 1**) of molecular superconductors is very similar to other superconducting families, including the cuprates, pnictides, and heavy fermion materials. The lower energy scale of soft molecular materials enables a more comprehensive understanding of fundamental physics within the confines of laboratory accessible temperatures, fields, and pressures.

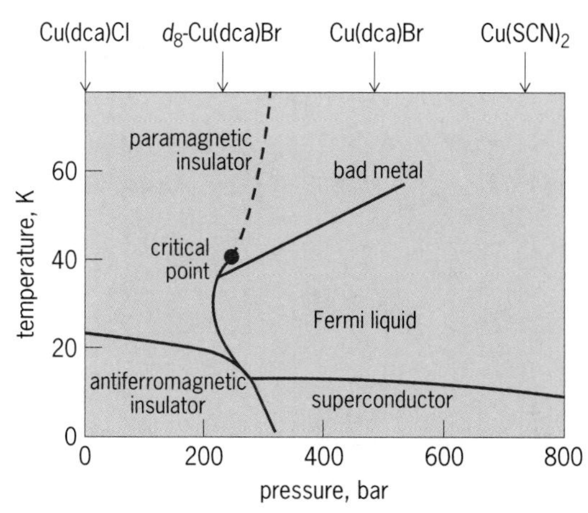

Fig. 1. The phase diagram of κ-(BEDT-TTF)$_2$X salts is similar to other classes of superconductors, including cuprates, pnictides, fullerenes, and heavy fermion materials. The superconducting regime borders an antiferromagnetic insulating state. In the case of molecule-based superconductors, the electronic state can be controlled through external pressure or chemical composition. For example, the κ-(BEDT-TTF)$_2$Cu(SCN)$_2$ salt is a superconductor at ambient pressure, whereas κ-(BEDT-TTF)$_2$Cu(dca)Cl requires 300 bar (0.03 GPa) of pressure to drive the system into a superconducting state. In the case of κ-(BEDT-TTF)$_2$Cu(dca)Br, even the slight difference in "chemical pressure" derived from deuterium substitution is sufficient to change the ground state of the material.

To date, more than 100 molecule-based superconductors have been reported in the literature. These can be divided into two classes depending on whether the molecular component primarily responsible for conductivity is an electron-donor or electron-acceptor molecule. As illustrated in **Fig. 2**, 23 electron-donor molecules have been found in superconducting salts, of which all but 2,5-bis(1,3-dithian-2-ylidene)-1,3,4,6-tetrathiapentalene (BDA-TTP) contain a tetrathiafulvalene (TTF) (or selenium-substituted) derivative. In addition, three electron-acceptor molecules have been found to form superconducting salts, of which the salts of buckminsterfullerene (C_{60}) have the highest T_cs and are the most studied.

In their neutral form, molecular solids of these species exhibit low conductivity. To form conductive (and superconductive) solids, these systems must be doped to a partially oxidized (or reduced) state. This process can be achieved through chemical oxidation (or reduction), but this process often leads to polycrystalline solids with inferior properties. High-quality crystals can be obtained through an electrocrystallization technique, whereby the crystal growth rate and quality can be controlled.

1D (TMTSF-based) systems. The first molecule-based superconductor, $(TMTSF)_2(PF_6)$ [TMTSF = tetramethyltetraselenafulvalene], was discovered in 1980 and required a pressure of 12 kbar (1.2 GPa) to stabilize a superconducting state at 0.9 K (−272.25°C). Through replacement of the PF_6^- anion with ClO_4^-, an ambient pressure superconductor with T_c = 1.4 K (−271.75°C) was crystallized. As illustrated in **Fig. 3**, the crystal structure of these cation-radical salts is characterized by one-dimensional (1D) stacks of TMTSF molecules separated by charge-compensating anions. TMTTF is a molecular analog of TMTSF in which the selenium atoms have been replaced with sulfur. Shortly after the discovery of superconductivity in TMTSF salts, it was found that at ambient pressure, the $(TMTTF)_2X$ (X = PF_6^-, AsF_6^-, SbF_6^-) salts have antiferromagnetic ground states. Advances in instrumentation have recently enabled the study of these systems under pressures as high as 10 GPa (100 kbar). Although pressure is usually detrimental to superconductivity, under these extreme conditions, it is stabilized in these TMTTF salts at temperatures of about 2.5 K (−270.65°C) and pressures of approximately 5 GPa (50 kbar).

2D (BEDT-TTF-based) systems. A milestone in the history of molecule-based superconductors occurred in 1978 when BEDT-TTF [bis(ethylenedithio) tetrathiafulvalene] was first synthesized. This electron-donor molecule has six-member sulfur-containing rings at the periphery of the TTF molecule. These sulfur atoms extend further from the long axis of the molecule allowing for increased S—S interactions between adjacent molecules. These intermolecular interactions increase the electronic dimensionality of the system from 1D to two-dimensional (2D). This 2D character is similar to that present in the inorganic cuprate and pnictide superconductors that exhibit high T_cs. Each BEDT-TTF molecule has eight peripheral sulfur atoms. During the electrocrystallization process, there is significant competition for intermolecular S—S interactions with similar energies. Partially for this reason, a wide variety of 2D packing motifs is possible for BEDT-TTF salts.

Among the BEDT-TTF salts, superconductivity was first discovered in the β-(BEDT-TTF)$_2$X (X = I_3^-, IBr_2^- and AuI_2^-) series in the mid-1980s. Here β designates a motif in which stacks of parallel BEDT-TTF molecules pack in 2D layers. Use of diamagnetic coordination polymers as the charge-compensating anions led to the discovery of κ-(BEDT-TTF)$_2$X (X = $Cu(NCS)_2^-$, $Cu[N(CN)_2]Br^-$), which have the highest T_cs at ambient pressure [10.4 and 11.6 K (−262.75 and −261.55°C), respectively] in the cation-radical family. Slightly higher T_cs can be achieved through application of pressure, as can be seen for κ-(BEDT-TTF)$_2$Cu[N(CN)$_2$]Cl [T_c = 12.8 K (−260.35°C) at 0.3 kbar (0.03 GPa)] and β′-(BEDT-TTF)$_2$ICl$_2$ [T_c = 14.2 K (−258.95°C) at 82 kbar (8.2 GPa)].

Many hundred salts of BEDT-TTF have been reported by changing the anion. One challenge in the field is to choose anions that will stabilize structures with desired properties. An important consideration is controlling the electronic coupling between BEDT-TTF layers. It is widely recognized that hydrogen-bonding interactions between the ethylene end groups of the BEDT-TTF molecules and the charge-compensating anions are important for stabilization of superconductivity. To maximize these

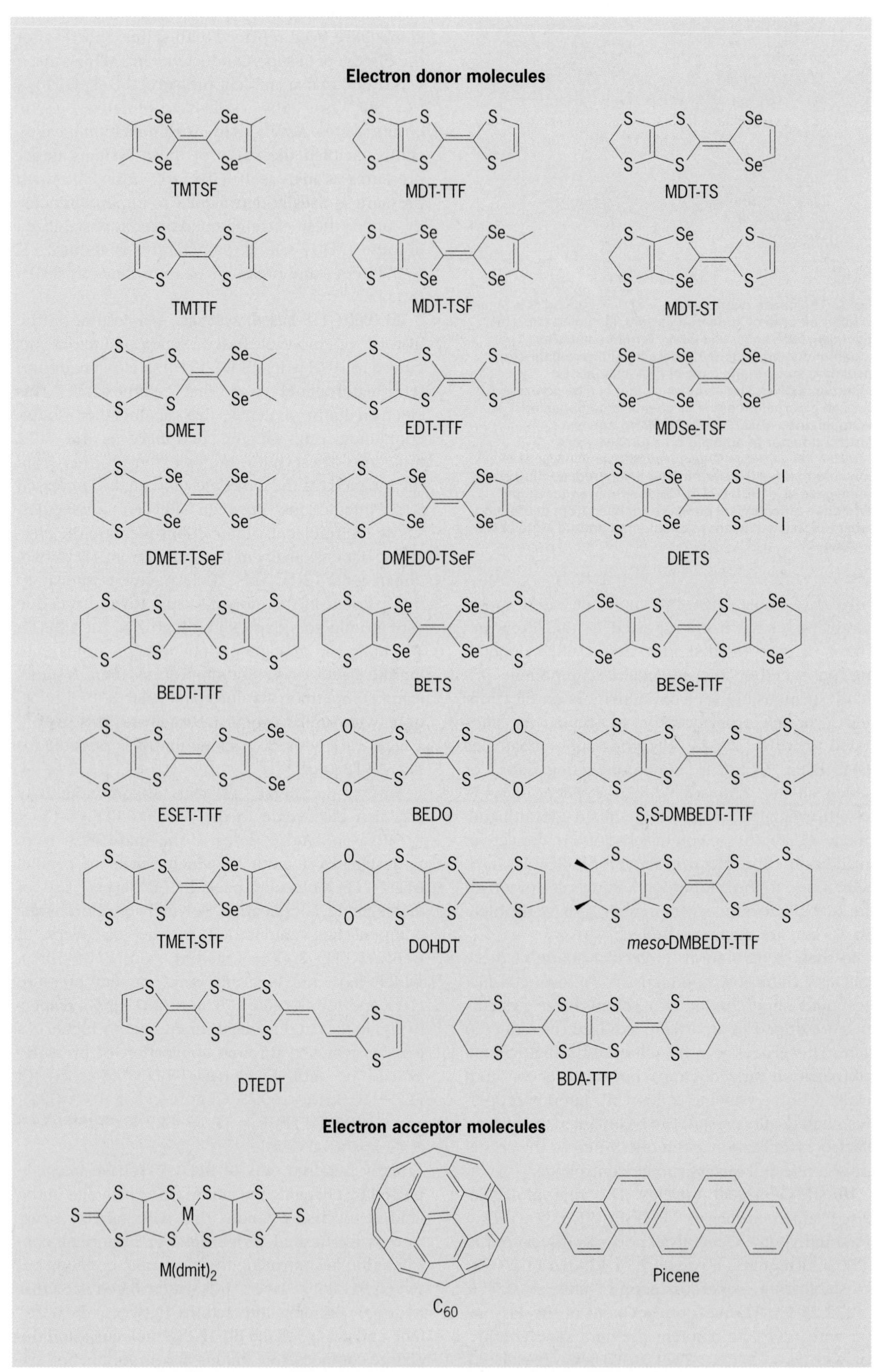

Fig. 2. The molecular building blocks that form the foundation of molecule-based superconductors.

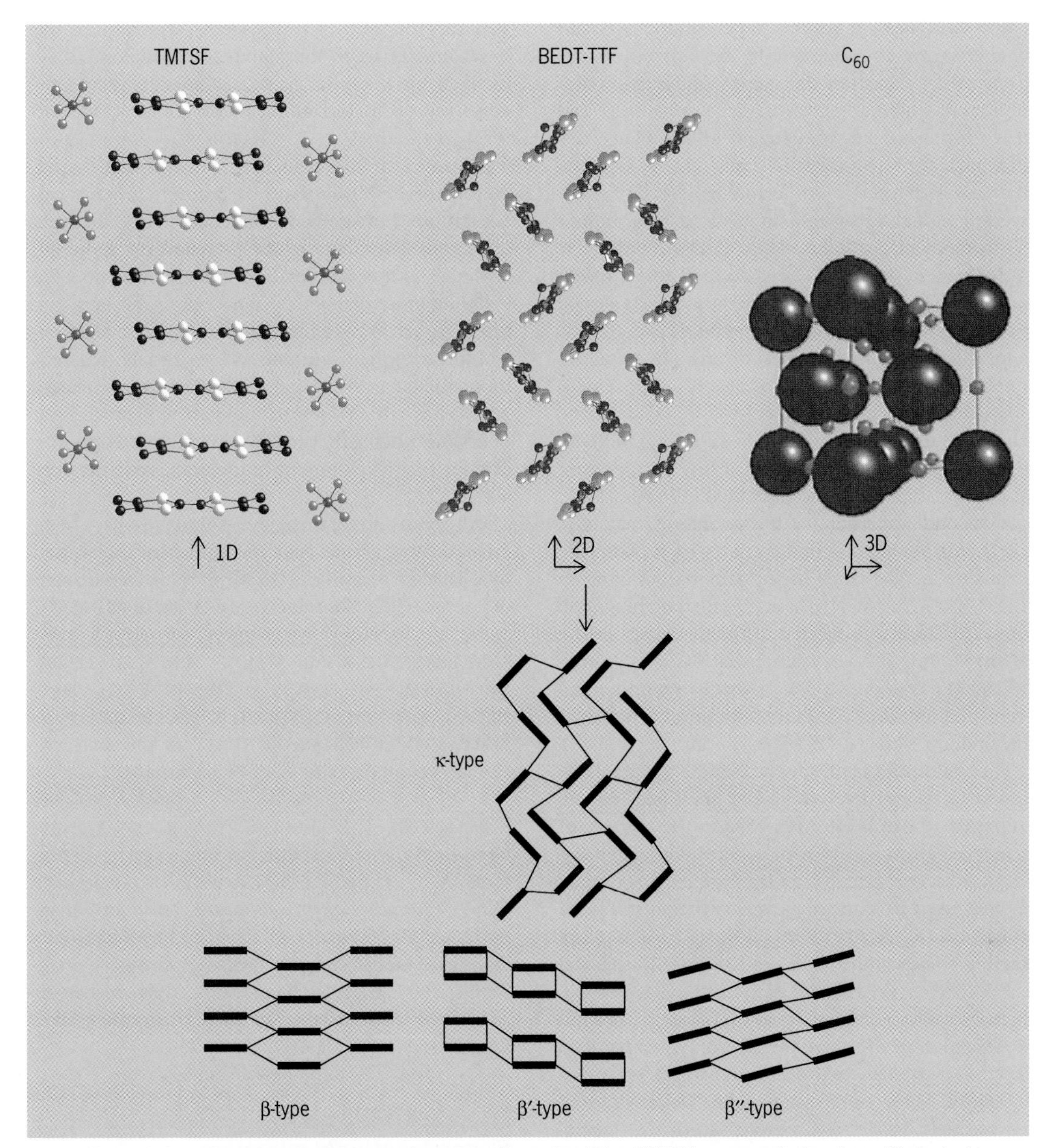

Fig. 3. Molecule-based superconductors come in various dimensionalities. TMTSF superconductors have 1D stacked structures with maximum T_c of approximately 1 K. BEDT-TTF superconductors have 2D layered structures with maximum T_c of approximately 14 K. Fullerene superconductors have 3D structures with T_cs as high as 38 K. For clarity, hydrogen atoms are not drawn. The lower half of the drawing depicts schematic representations of various 2D packing motifs commonly observed in BEDT-TTF structures. The thick lines represent the end-on view of BEDT-TTF molecules and the thin lines represent S···S intermolecular interactions.

interactions, anions were chosen with highly electronegative fluoride or oxygen atoms. Through use of an organic anion, the first completely organic superconductor, β''(BEDT-TTF)$_2$SF$_5$CH$_2$CF$_2$SO$_3$, was crystallized, which has a T_c of 5.2 K (−267.95°C). For this reason, the highly fluorinated organometallic anions, M(CF)$_3{}_4^-$ (M = Cu, Ag, Au) were chosen for study, and the highly tunable ternary family of superconductors, κ-(BEDT-TTF)$_2$M(CF$_3$)$_4$(1,1,2-trihaloethane), were crystallized. This family of superconductors can be tuned through chemical modification of the M(CF$_3$)$_4^-$ anion, the neutral co-crystallized solvent molecule, or the structural phase (κ_L or κ_H). The 30 superconducting salts in this family, with T_cs ranging from 2.1 to 11.1 K (−271.05 to −262.05°C) provide an extensive playground for developing structure-property relationships.

3D (fullerene) systems. The 1985 discovery of buckminsterfullerene (C_{60}) opened the door to a wealth of previously unimaginable research. Six years later, superconductivity was stabilized in alkali-metal-doped salts with composition A_3C_{60} (A = alkali metal). The T_c of these cubic structures increases as a function of the unit-cell parameter, with the highest T_c (38 K or −235.15°C) occurring for the Cs_3C_{60} composition. The decomposition of these salts through reaction with atmospheric oxygen adds an additional challenge for application development.

New directions. In general, superconductivity and magnetism are considered to be incompatible properties. The realization that superconductivity in a number of systems (such as cuprates, pnictides, and heavy fermions) lies in a region of the phase diagram near antiferromagnetic states (Fig. 1) suggests that incorporation of antiferromagnetic layers between conductive sheets may be a route to higher T_cs in molecular systems. The λ-$(BETS)_2FeX_4$ [X = Cl, Br] system provides a fascinating example of the interplay between a conductive lattice with π-type electrons and a magnetic lattice with *d*-type. In zero field, this material undergoes a transition to an antiferromagnetic insulating state at 8 K (−265.15°C). However, when magnetic field is applied perfectly parallel to the conducting layers, superconductivity is induced in the sample albeit at low temperature (0.1 K or −273.05°C) and high field [17 tesla (T)].

A second approach to incorporating magnetic layers into molecule conductors is to replace the diamagnetic Cu(I) ions in the superconducting κ-$(BEDT\text{-}TTF)_2Cu[N(CN)_2]X$ salts with paramagnetic ions. This approach has led to the discovery of the κ-$(BETS)_2Mn[N(CN)_2]_3$ system that exhibits antiferromagnetic ordering at 5 K (−268.15°C) under ambient pressure, but a superconducting transition at 5 K under 0.3 kbar (0.03 GPa).

The stabilization of superconductivity in alkali-doped C_{60} at relatively high temperatures has encouraged research efforts involving the doping of related electron acceptor molecules. One such example is picene, which is a planer molecule containing five fused aromatic rings. A very recent study has shown that K_x(picene) forms at least two superconducting phases with T_cs of 7 and 18 K (−266.15 and −255.15°C). It is expected that additional superconductors will be discovered in this system through modification of the alkali metal, stoichiometry, and polycyclic aromatic hydrocarbon (acene).

Outlook. Many opportunities can be envisioned for the field in the foreseeable future. For example, the incorporation of photomechanical ligands and anions into solid-state structures will provide a means to control electronic properties with light. Coupling of antiferromagnetic and conducting systems may lead to exotic superconductors that operate at temperatures higher than currently achievable. The incorporation of ferromagnetic components has the potential to lead to novel electronic states with application for spintronic devices. The trapping of metastable phases through crystallization under extreme conditions is expected to lead to new structural types with switchable electronic properties. As is key for stabilizing high T_c superconductivity in the inorganic cuprate systems, band-filling control through doping has the potential to open a new dimension in molecule-based superconductors, but structural and magnetic defects must be minimized and controlled. For applications, ultimately it will be important to grow thin films with atomically prefect interfaces between dissimilar layers. A comprehensive understanding of the electronic and magnetic structure of these novel systems will require the development of crystal-growth methodology to selectively grow sizable crystals of specific phases for advanced characterization, including neutron scattering techniques.

Advances in the field of molecule-based superconductors will be driven by multidisciplinary collaborations between synthetic chemists, theoreticians, physicists, and materials scientists. Although advances in instrumentation and computation are enabling more detailed understandings of these systems, we are only beginning to understand and control these complex and intricate materials. Whereas inorganic chemists have a toolbox of approximately 100 atoms with which to build structures, the use of organic chemistry presents an unlimited opportunity for the development of molecule-based materials.

[All or portions of the above manuscript have been created by Argonne National Laboratory, operated by UChicago Argonne, LLC, for the U.S. Department of Energy under Contract No. DE-AC02-06CHI1357.]

For background information *see* A15 PHASES; ANTIFERROMAGNETISM; FERMI SURFACE; FERROMAGNETISM; FULLERENE; MAGNETIC FIELD; MAGNETISM; ORGANIC CONDUCTOR; PARAMAGNETISM; PEROVSKITE; SUPERCONDUCTIVITY in the McGraw-Hill Encyclopedia of Science & Technology.

John A. Schlueter

Bibliography. T. Ishiguro, K. Yamaji, and G. Saito, *Organic Superconductors*, Springer-Verlag, Berlin, 1998; A. G. Lebed (ed.), *The Physics of Organic Superconductors and Conductors*, Springer-Verlag, Berlin, 2008; N. Toyota, M. Lang, and J. Müller, *Low-Dimensional Molecular Metals*, Springer-Verlag, Berlin, 2007; J. M. Williams et al., *Organic Superconductors (Including Fullerenes)*, Prentice Hall, Englewood Cliffs, NJ, 1992.

Monarch butterfly migration

Eastern North American monarch butterflies (*Danaus plexippus*) use a time-compensated sun compass to navigate south during their spectacular fall migration. Recent experiments show that the circadian clocks that provide time compensation for the sun compass mechanism reside in the antennae and not in the brain. This surprising finding poses a novel clock-compass connection that may apply widely to insects.

Annual migratory cycle. The yearly migration of eastern North American monarch butterflies (**Fig. 1**) is the longest repetitive migration among insects. During the fall migration, millions of butterflies from the eastern United States and southern Canada make the yearly journey to overwinter in roosts in a small area in central Mexico, with some traveling distances of 4000 km (2500 mi).

Migrating monarchs are in reproductive diapause, a condition in which the butterflies suspend mating behavior and exhibit arrested reproductive develop-

Fig. 1. Migrating monarchs in flight. (*Photo by Dennis Curtin*)

ment. Migrants also have increased cold tolerance and abdominal fat stores, exhibit a marked increase in longevity, and display a strong urge to fly south. Reproductive arrest persists at the overwintering site until the early spring, when the butterflies reproduce and fly northward to lay fertilized eggs on newly emerged milkweed plants (the obligate food source of the larvae) in the southern United States. Another two generations of reproductively competent, short-lived spring and summer butterflies appear to follow the progressive emergence of milkweed northward to repopulate the range of the eastern population of monarch butterflies. In the fall, decreasing day length and other environmental events trigger the migratory generation and, once again, the long journey south commences.

Time-compensated sun compass: a major navigational tool. Migratory butterflies provide a model for biologists interested in the neural integration of information about time and space, which is an essential component of navigation for a variety of animals. The remarkable navigational abilities of monarch butterflies are part of a genetic/epigenetic program that is initiated in migrants. It is not learned, as the butterflies migrating are always on their maiden voyage, and those that make the trip south are at least two generations removed from the previous generation of migrants.

A major navigational strategy used by migrating monarchs utilizes a time-compensated sun compass. The sun itself is not a reliable source of directional information because its location in the sky changes throughout the day. To maintain a constant bearing through the day, monarchs thus need the ability to compensate for the changing location of the sun and do so with a circadian clock. That monarch butterflies use a time-compensated sun compass to help them navigate has been shown in disappearance-bearing studies of free-flying migrants and in studies using a flight simulator that allows study of flight trajectories from tethered butterflies during a sustained period of flight.

The skylight cues that the sun compass uses for directional information are the sun itself and polarized and spectral patterns in the sky that result from the scattering of sunlight. Polarized light in the ultraviolet range is sensed by the dorsal rim area of the monarch eye, whereas the sun and spectral gradients are sensed through the main retina. Flight simulator studies suggest that the sun is the most dominant signal for proper sun compass orientation, with polarized light serving as a backup system when the sun is not visible and patches of blue sky can be seen. The importance of skylight spectral gradients for directional information is not yet known.

Skylight cues sensed by the eyes appear to be transmitted through a complex circuitry to the central complex, a midline structure in the central brain, which, based on studies in the desert locust, appears to function as the site of the sun compass. Thus, the central complex integrates skylight information from both eyes, and this information is processed ultimately in the motor system, leading to directed flight behavior.

The importance of the circadian clock in regulating the time-compensated component of sun compass orientation has been shown through controlled "clock-shift" experiments, in which the timing of the daily light–dark cycle is either advanced or delayed, causing predictable alterations in the flight orientation of butterflies. Prior studies have demonstrated that such timing information in monarchs is generated by a unique clock mechanism, which is based on an intracellular transcriptional feedback loop and involves the function of two distinct cryptochrome (CRY) proteins. Monarch CRY1 is the *Drosophila* CRY homolog and functions as a blue-light circadian photoreceptor for the clockwork, whereas CRY2 is vertebrate-like and functions as the main transcriptional repressor of the clock feedback loop. Using highly specific anti-"clock" antibodies, immunostaining studies have revealed an area of the brain called the *pars lateralis* as the site of the brain clocks, which appear to connect with the *pars intercerebralis* and with the central complex. Based on these findings, it therefore was assumed that the brain housed the circadian clock used for time-compensated sun compass orientation. However, this assumption was incorrect. Instead, it was shown that the antennae are necessary for time-compensated sun compass orientation. The antennae contain circadian clocks, and these clocks are necessary for proper time compensation (**Fig. 2**).

Role of the antennae in time-compensated sun compass orientation. To determine the functional role of the antennae in the migratory flight behavior of monarchs, the flight orientation of migrant butterflies with and without antennae was analyzed in a flight simulator. Migrants with intact antennae showed a southerly orientation and were able to time-compensate their orientation appropriately after being clock-shifted. However, as a group, those without antennae showed a random flight orientation. This suggested that the antennae are necessary

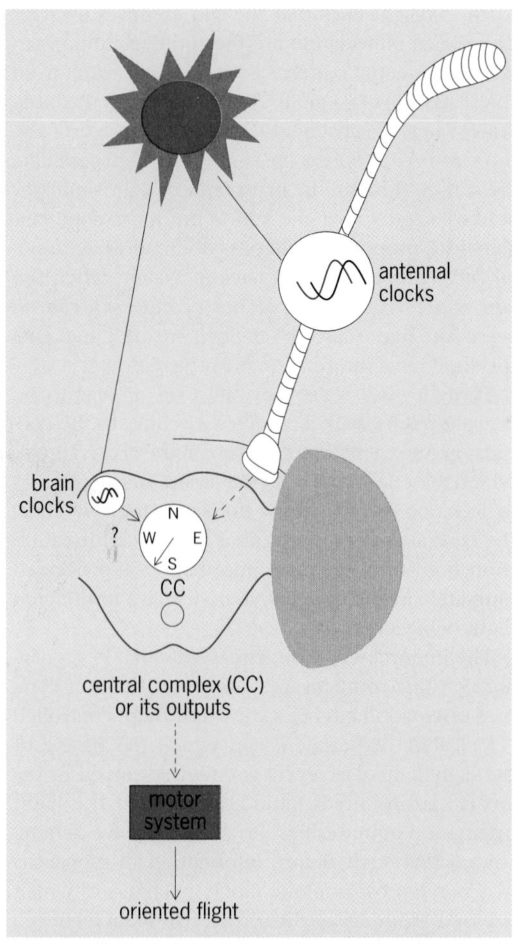

Fig. 2. The antennal clocks' time-compensated sun compass orientation in monarch butterflies. The principal timing component used for monarch flight orientation is likely to be located in the antennae. The antennal clocks communicate with the sun compass system located in the central complex (CC), where both timing and spatial information could be integrated. Ultimately, the integrated signal is transmitted to the motor system, leading to oriented flight behavior. The role of the brain clocks in this mechanism, however, cannot be ruled out (designated by the "?").

for time-compensated sun compass orientation. Additional studies showed that antennal amputation did not disrupt flight stability, a known function for insect antennae, revealing that the effect of amputation was on flight orientation. Moreover, antennal removal did not alter the molecular oscillations of the messenger RNA (mRNA) levels of the core clock genes *period* (*per*) and *timeless* (*tim*) in the brain, a measure of the functionality of the brain clocks, indicating that antennal amputation does not alter brain clock ticking.

Based on these results, it seemed possible that the antennae may contain the circadian clocks that mediate the time-compensated flight behavior. Molecular and biochemical experiments showed that the antennae indeed contain circadian clocks, whose molecular underpinnings parallel those of brain clocks. The genes coding for the main core components, *per*, *tim*, and *cry2*, as well as their corresponding proteins, oscillated in a circadian manner in the antennae of butterflies (maintained in constant conditions), in phase with those observed in the brains of the same butterflies. Importantly, antennae isolated and maintained in culture showed light sensitivity and their molecular oscillations persisted, showing that the antennae actually house light-entrainable, tissue-autonomous clocks that can function independent of the brain.

To test further the role of the antennal clocks in the sun compass orientation of migrant butterflies, light input to the clock was manipulated by covering antennal flagella with either black paint (to block light input) or clear (control) paint, and individuals were housed (thereby entrained) for several days in light–dark cycles. For butterflies with antennae painted black (maintained in virtual constant darkness), the timing of the antennal clocks "free-ran" as the molecular oscillations of *per* and *tim* gradually drifted with a decreasing amplitude. In contrast, control butterflies possessed light-entrainable clocks with unaltered molecular oscillations. Butterflies from both groups then were flown in a flight simulator. In contrast to control butterflies, which oriented appropriately, the butterflies with antennae maintained under constant dark conditions exhibited an altered sun compass orientation that was consistent with the idea that the antennae house the major clocks involved in time compensation. It remains possible that the brain clocks provide some role in time compensation, but the data overwhelmingly suggest that the antennal clocks are necessary for proper time compensation.

Studies are now directed at identifying the way in which the antennal clocks communicate with the sun compass. Is there a neural pathway involved? Does a diffusible substance provide the communication? The answers should be forthcoming.

For background information *see* BEHAVIORAL ECOLOGY; BIOLOGICAL CLOCKS; BRAIN; CRYPTOCHROME; ECOLOGY; FLIGHT; INSECT PHYSIOLOGY; LEPIDOPTERA; LIGHT; MIGRATORY BEHAVIOR in the McGraw-Hill Encyclopedia of Science & Technology.

Christine Merlin; Robert J. Gegear; Steven M. Reppert

Bibliography. C. Merlin, R. J. Gegear, and S. M. Reppert, Antennal circadian clocks coordinate sun compass orientation in migratory monarch butterflies, *Science*, 325:1700–1704, 2009; S. M. Reppert, The ancestral circadian clock of monarch butterflies: Role in time-compensated sun compass orientation, *Cold Spring Harbor Symp. Quant. Biol.*, 72:113–118, 2007.

Multimoment microphysical parameterizations in cloud models

Cloud microphysical parameterizations are techniques for representing sub-grid-scale microphysical processes using grid-scale information in atmospheric cloud-scale-resolving numerical models. A multimoment cloud microphysical parameterization is one that uses two or more of the following moments including, but not limited to, number concentration (number of hydrometeors per unit volume), characteristic diameter of hydrometeor size distribution (inverse of slope of size distribution), mixing ratio (hydrometeor density divided by density of air), and reflectivity (related to radiation power scattered back to a radar) to predict the evolution of aerosols, clouds, and precipitation. The prediction of changes in aerosols, clouds, and precipitation involves the development of parameterizations of processes such as nucleation (initiation of cloud particles with or without aerosols), vapor diffusion growth (loss or gain of hydrometeor mass owing to a phase change of water vapor), collection and breakup (one type of particle collecting another and destruction of hydrometeors by collisions), freezing (phase change of liquid water to ice), melting (phase change from ice water to liquid), and sedimentation (fallout of hydrometeors, usually at their terminal velocities). The prediction of aerosols and hydrometeors also requires the need for numerical approximations for positive-definite solutions to transport and turbulent mixing.

The need for multimoment cloud microphysical parameterizations became apparent in the 1980s with the simulations of hailstorms, became more common in the 1990s, and was considered to be essential for accurate prediction of aerosol, cloud, and precipitation processes in the 2000s. By 2010, most modern numerical models used multimoment cloud microphysical parameterizations. The reason for this sudden increase in use of multimoment cloud microphysical parameterizations owes to the resulting significant improvement in the accuracy of numerical model solutions in studies of moist atmospheres in which clouds and precipitation form. For example, in 2005 storm simulations, striking examples included better representation of sedimentation with three-moment parameterizations. In 2006, it was demonstrated that superior hailstorm simulations (compared with observations) were obtainable using three-moment parameterizations, rather than single- and double-moment parameterizations. By 2010, it was demonstrated using three-moment cloud parameterizations that evaporation rates were altered significantly enough to substantially influence tornadic storm simulation evolution and produce better simulation comparisons with observations.

Multimoment microphysics parameterizations use multiple predictive variables, each of which is related to an integral quantity of diameter to some power. Multimoment schemes have the greatest improvement over those of single-moment schemes in the accuracy of numerical solutions for simulations that employ smaller-scale grid spacing, such as in clouds models or in small-grid spacing (order of 10 to 100s m) numerical weather prediction models. However, multimoment schemes also have been found to reduce errors in larger-scale models such as those that resolve clouds in storm systems and hurricanes (grid spacing of order of 1–10 km). They also have been used in global climate models (grid spacing of order of 100 km), which do not resolve individual clouds or cloud systems. Some multimoment parameterizations preserve physical conservation laws such as preserving the number concentration of collector particles for certain collection processes, which is not the case for some processes for single-moment parameterizations. For these reasons and advantages such as those mentioned above, and because they are relatively computationally inexpensive, multimoment parameterizations have become quite popular. This is particularly true for bulk microphysical parameterizations, which use hydrometeor size distribution (spectral number density) functions to represent microphysical variables. These functions have single to two or three parameters that describe the hydrometeor size distribution's shape, and are easy to integrate.

Microphysical moments. In microphysical parameterizations, moments, M, of hydrometeor size distribution functions or spectral number density functions, $n(D)$, are often used and defined in terms of a form of a diameter, D, to some power I, multiplied by $n(D)$ and then integrated over D [Eq. (1)]. Some

$$M(I) = \int_0^\infty D^I n(D)\, dD \qquad (1)$$

of most often used moment-related variables in multimoment parameterizations for hydrometeor related processes are the zeroth-moment ($\sim D^0$), which is related to number concentration; the first-moment ($\sim D^1$), which is related to diameter; the second-moment ($\sim D^2$), which is related to surface area and parameterization of electric space charge on particles; the third-moment ($\sim D^3$), which is related to mass or mixing ratio; the fourth-moment ($\sim D^4$), which can be related to mass-weighted mean size; the fifth-moment ($\sim D^5$), which is related to the variance or the positive square root of variance (standard deviation) of the mass or volume distribution, as well as a radar reflectivity in some cases; and the sixth-moment ($\sim D^6$), which is related to radar reflectivity for Rayleigh scatterers. Higher moments can be examined for particular uses, and moments do not need to be integer values. Moments also can be described in terms of mass m to the power I, where the zeroth-moment ($\sim m^0$) is related to number concentration, the first-moment ($\sim m^1$) is related to mass or mixing ratio, and the second-moment ($\sim m^2$) is related to radar reflectivity.

The spectral number density function, $n(D)$, describes the number concentration of particles between D and $D + dD$ with the total number concentration written as N_T. The gamma distribution and log-normal distribution for bulk multimoment microphysics parameterization principles are used as examples herein. The first seven moments for gamma and log-normal distributions functions as described above are given in **Table 1**. The single-parameter gamma distribution is shown in

TABLE 1. Variables related to the zeroth-through sixth-moments

Related to the zeroth-moment $N_T = \int_0^\infty n(D)\,dD$
Related to the first-moment $\overline{D}^{N_T} = \frac{\int_0^\infty D\,n(D)dD}{\int_0^\infty n(D)dD}$
Related to the second-moment $A_T = \int_0^\infty \pi D^2 n(D)dD$
Related to the third-moment $Q = \frac{1}{\rho_a}\int_0^\infty m(D)n(D)\,dD$
Related to the fourth-moment $D_m = \frac{\int_0^\infty Dm(D)n(D)dD}{\int_0^\infty m(D)n(D)dD}$
Related to the fifth-moment $\sigma^{*2} = \frac{\int_0^\infty D^3 (D - D_m)^2 n(D)dD}{\int_0^\infty D^3 n(D)dD}$
Related to the sixth-moment $Z = \int_0^\infty D^6 n(D)\,dD$

Note that for a sphere mass is $m(D) = \pi\rho/6D^3$.

TABLE 2. Variables related to the zeroth-through sixth-moment for the one-parameter gamma distribution

Related to zeroth-moment N_T
Related to first-moment $\overline{D}^{N_T} = D_n \frac{\Gamma(v+1)}{\Gamma(v)}$
Related to second-moment $A_T = \pi D_n^2 N_T \frac{\Gamma(v+2)}{\Gamma(v)}$
Related to third-moment $Q = \rho\frac{D_n^3}{\rho_a}\frac{\pi}{6}\frac{N_T\Gamma(v+3)}{\Gamma(v)}$
Related to fourth-moment $D_m = D_n\frac{\Gamma(v+4)}{\Gamma(v+3)}$
Related to fifth-moment $\sigma^{*2} = D_n^2\left[\frac{\Gamma(v+5)}{\Gamma(v+3)} - \left(\frac{\Gamma(v+4)}{\Gamma(v+3)}\right)^2\right]$
Related to sixth-moment $Z = D_n^6\frac{N_T\Gamma(v+6)}{\Gamma(v)}$

Eq. (2). This form of the gamma function is used

$$n(D) = \frac{N_T}{\Gamma(v)}\left(\frac{D}{D_n}\right)^{v-1}\frac{1}{D_n}\exp\left(-\left[\frac{D}{D_n}\right]\right) \quad (2)$$

to derive the forms of the moment related variables that appear in **Table 2**.

The spectral number density function for the log-normal distribution may be written as Eq. (3). This is

$$n(D) = \frac{N_T}{\sqrt{2\pi}\sigma D}\exp\left(-\frac{\left[\ln\left(D/D_n\right)\right]^2}{2\sigma^2}\right) \quad (3)$$

used to derive the log-normal, moment-related variables that appear in **Table 3**. Multimoment bulk microphysics parameterization models are now more often used in cloud prediction and simulation models than a decade ago. With bulk parameterization models, a size distribution function is assumed such as the gamma or log-normal distribution. In the one-moment bulk parameterization scheme, typically the total mixing ratio, Q, for a hydrometeor species is predicted, probably as this was what was predicted in the first cloud models. In these simpler and earlier models, a specified distribution slope intercept N_o along with the prognosed Q, are used to diagnose the characteristic diameter, D_n, (which is 1 over the slope of the distribution), and the value of N_T.

In one common form of two-moment bulk-parameterization schemes, N_T and Q are predicted, and N_o and v are diagnosed or specified. Prior to the time of the emergence of the first two-moment schemes, some analytical multimoment models were designed to predict D_n, Q, and Z. However, these were limited to the application to one-dimensional models only. Later, a form of a three-moment bulk parameterization, with predictions of N_T, Q, and Z was used to diagnose v, which was found from iteration using results from the predictive equation for Z. Two-moment parameterizations with functional fits of distribution shape parameter to the mean-mass diameter and of equations relating characteristic diameter to the shape parameter from observations also have been proposed and provide very good solutions for some microphysical processes compared to three-moment parameterizations.

Bin multimoment microphysical parameterization models are often considered the type of parameterization able to most accurately represent, for example, rain distribution evolutions in rain clouds. They have bins representing the spectrum of drops from very small cloud droplet sizes (D = 2–10 μm) to larger raindrops (D = 4–8 mm) for bin parameterizations of raindrop development. Each bin is usually exponentially larger than the previous size/mass bin owing to the wide spectrum of liquid water drops that are possible, which ranges over three orders of magnitude. For liquid water drop sizes, bins often

TABLE 3. Variables related to the zeroth-through sixth moments for the log-normal distribution (The breadth parameter is given by σ = 0.5)

Related to zeroth-moment N_t
Related to first-moment $\overline{D}^{N_T} = D_n \exp\left(\frac{\sigma^2}{2}\right)$
Related to second-moment $A_T = \pi D_n^2 N_T \exp(2\sigma^2)$
Related to third-moment $Q = \rho N_T \frac{D_n^3}{\rho_a}\frac{\pi}{6}\exp\left(\frac{9\sigma^2}{2}\right)$
Related to fourth-moment $D_m = D_n \exp\left(\frac{7\sigma^2}{2}\right)$
Related to fifth-moment $\sigma^{*2} = D_n^2[\exp(8\sigma^2) - \exp(7\sigma^2)]$
Related to sixth-moment $Z = D_n^6 N_T \exp(18\sigma^2)$

will increase by 2, $2^{1/2}$, $2^{1/3}$, or $2^{1/4}$, times the previous size bin over perhaps 36, 72, or 144 bins. A shortcoming of bin models is the excessively large computation resources needed to make use of them in large three-dimensional models. At a minimum, number concentration is predicted with these schemes, although mixing ratio and reflectivity can be predicted or calculated as well. Considering number concentration with mixing ratio prediction improves the bin parameterization results against using just number concentration by limiting anomalous spreading of the distribution in analytical test problems of the process of particles collecting other particles.

Alternatively, there is the hybrid-bulk/bin microphysical parameterization. The moments predicted include some or all of the following such as number concentration, mixing ratio, and reflectivity. These moments are advected, diffused, and filtered in bulk space, but the microphysical source and sink terms are computed after a transformation to bin parameterization space. After microphysical process computations, including sedimentation, evaporation/condensation, melting /freezing, collection/breakup, and others are computed, the bins are summed back into bulk parameterization space. In describing these different microphysics schemes and moments, it is essential to use a number of mathematical relationships. One shortcoming with hybrid-bulk/bin parameterizations compared to bin parameterizations is that the bin parameterization solution may not be preserved from time step to time step, because it is replaced by the bulk parameterization solution each time step for advection, diffusion, and filtering. Nevertheless, the use of a hybrid-bulk/bin scheme appears to produce a more accurate solution than can be expected than with a bulk scheme.

Bulk predictive equations. The predictive equations for a three-moment bulk microphysical parameterizations for cloud models are given as follows, and include equations for mixing ratio, which is related to the third-moment [Eq. (4)],

$$\frac{dQ}{dt} = \frac{\partial}{\partial x_i}\left(\rho_a K_b \frac{\partial Q}{\partial x_i}\right) + \delta_{i3}\frac{1}{\rho_a}\frac{\partial\left(\rho_a \overline{V_T}^{Q} Q\right)}{\partial x_i} + SQ \tag{4}$$

where

$$\frac{d}{dt} = \frac{\partial}{\partial t} + u_j \frac{\partial}{\partial x_j}$$

which is the total or substantial derivative and includes local time-time tendency and advection tendency (transport by wind), where t is time, and u_j and x_j are the Cartesian velocities and distances, respectively (that is, $j = 1, 2, 3$ [u, v, w wind components] and $j = 1, 2, 3$ [x, y, z Cartesian directions]). The turbulent eddy-mixing coefficient is K_b in the diffusion term. The mass-weighted terminal velocity is $\overline{V_T}^{Q}$ in the vertical flux term. The microphysical source/sink term is SQ, which includes nucleation, vapor diffusion, collection of similar and other particle species, melting, and freezing among other processes. The Kronecker delta is δ_{i3}, which is 1 when $i = 3$ and 0 otherwise. The equation for prediction of number concentration, which is related to the zeroth-moment [Eq. (5)], where $\overline{V_T}^{N_T}$ is the number concentration-weighted terminal velocity, and the microphysical source/sink term is SN_T.

$$\frac{dN_T}{dt} = \frac{\partial}{\partial x_i}\left(K_b \frac{\partial N_T}{\partial x_i}\right) + \delta_{i3}\frac{\partial\left(\overline{V_T}^{N_T} N_T\right)}{\partial x_i} + SN_T \tag{5}$$

Next reflectivity, which is related to the sixth-moment, can be predicted using Eq. (6), where $\overline{V_T}^{Z}$ is the reflectivity-weighted terminal velocity, and the microphysical source/sink term is SZ.

$$\begin{aligned}\frac{dZ}{dt} &= \frac{\partial}{\partial x_i}\left(K_b \frac{\partial Z}{\partial x_i}\right) + \delta_{i3}\frac{\partial\left(\overline{V_T}^{Z} Z\right)}{\partial x_i} + SZ \\ &+ \frac{\rho_a^2}{(\pi/6)^2\rho^2}\frac{(5+\upsilon)(4+\upsilon)(3+\upsilon)}{(2+\upsilon)(1+\upsilon)(\upsilon)} \\ &\times \left[2\frac{Q}{N_T}\frac{dQ}{dt} - \left(\frac{Q}{N_T}\right)^2 \frac{dN_T}{dt}\right]\end{aligned} \tag{6}$$

This equation can be used to find the shape parameter of the distribution. In principle, predictive equations can be developed for any of the moments, and microphysical quantities, such as mixing ratio and number concentration, can be derived from these predictive equations.

Example with gravitational sedimentation. For the process of sedimentation, comparisons of the one-, two- and three-moment bulk schemes with a reference solution using a bin model is shown for the vertical profiles in time of number concentration and mixing ratio (**Figs. 1***a* and **2***a*). The diagnostic total number concentration values in the one-moment bulk-parameterization model increases with time (Fig. 1*b*). Note also that the area under the total number concentration curves increases dramatically with time until much hail falls out to the bottom of the domain. This owes to the overall lack of conservation with one-moments schemes, although all multimoment schemes can lack conservation to some degree depending on the microphysical process being represented. For the mixing ratio, the one-moment scheme maxima are somewhat in line with the bin-parameterization model although the peaks are significantly accentuated over the bin parameterization model peaks (Fig. 2*b*). In addition, total integrated mixing ratio (area under the curve) increases with time and then begins to decrease as much hail falls out to the bottom of the domain (Fig. 2*b*). The two-moment bulk-parameterization model results indicate that as precipitation falls, it progressively lags in time against the bin parameterization model results for the total number concentration, and although the mixing ratio peaks nearly are in line with

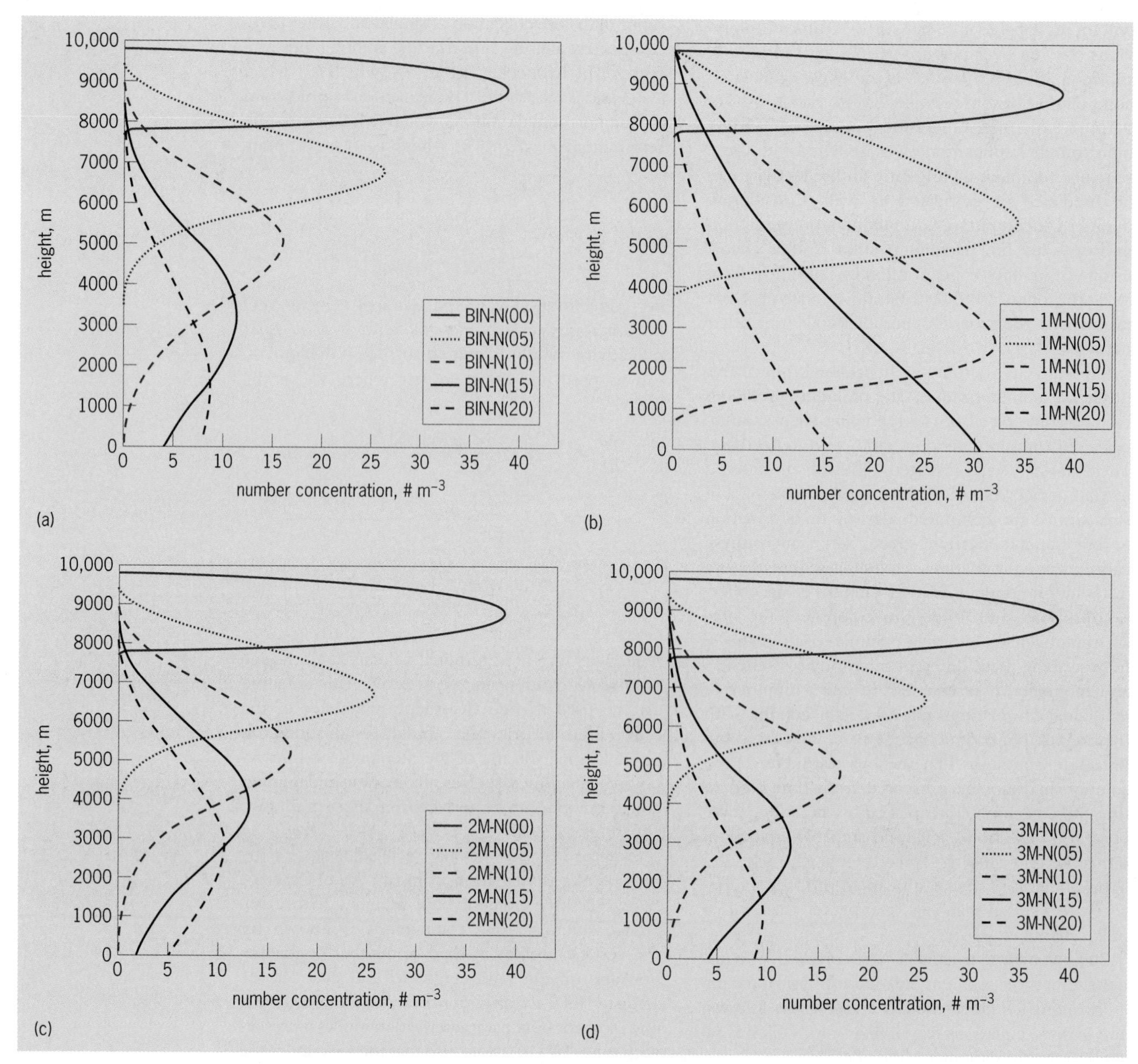

Fig. 1. Height (m) profiles of the total number concentration (number per m^{-3}) for every 5 min in the simulations: (*a*) bin model scheme; (*b*) one-moment bulk scheme; (*c*) two-moment bulk scheme; (*d*) three-moment bulk scheme.

the bin-parameterization model, the amplitudes are progressively smaller with time (Figs. 1*c* and 2*c*). Finally, it is shown that the number concentration (Fig. 1*d*) and mixing ratio (Fig. 2*d*) with the three-moment parameterization model performs nearly identical to the bin-parameterization model results (Figs. 1*a* and 2*a*). In general the more moments that are predicted, the better probability that there will be for conservation for moments near those predicted.

Outlook. The results for the one-, two-, and three-moment bulk-parameterization models presented herein for sedimentation show that lower-moment bulk-parameterizations perform less reliably than those that use three-moment bulk-parameterization schemes as compared to a bin model. Others have found that two-moment schemes with analytical expressions for the shape parameter as a function of the mass-weighted mean diameter D_m can perform as well as three-moment models for certain processes. The future seems to be directed toward widespread use of multimoment bulk and eventually hybrid-bulk/bin models. Computer limitations will limit the use of bin models in prediction models for perhaps many years to come. Only so many computational processing units (CPUs) or cores can be linked together and work most all the time for three-dimensional bin models to become practical for use in cloud modeling.

For background information *see* CLOUD; CLOUD PHYSICS; METEOROLOGICAL RADAR; METEOROLOGY; PRECIPITATION (METEOROLOGY); WEATHER FORE-

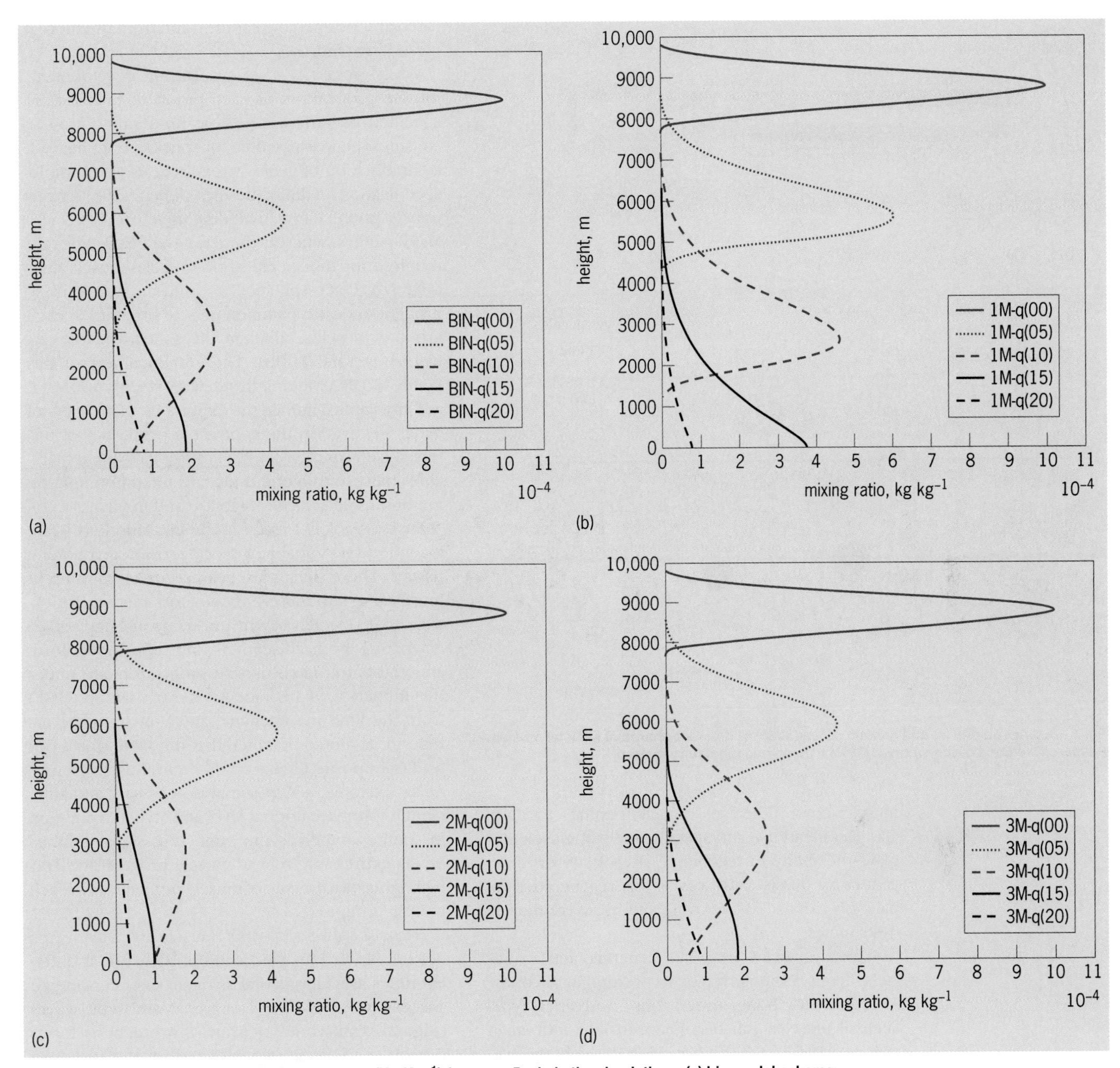

Fig. 2. Height (m) profiles of the total mixing ratio (Kg Kg^{-1}) for every 5 min in the simulations: (*a*) bin model scheme; (*b*) one-moment bulk scheme; (*c*) two-moment bulk scheme; (*d*) three-moment bulk scheme.

CASTING AND PREDICTION in the McGraw-Hill Encyclopedia of Science & Technology.

Jerry M. Straka; Katharine M. Kanak

Bibliography. B. S. Ferrier, A double-moment multiple-phase four-class bulk ice scheme, Part I: Description, *J. Atmos. Sci.*, 51:249–280, 1994; J. A. Milbrandt and M. K. Yau, A multimoment bulk microphysics parameterization, Part I: Analysis of the role of the spectral shape parameter, *J. Atmos. Sci.*, 62:3051–3064, 2005; R. E. Passarelli, Jr., An approximate analytical model of the vapor deposition and aggregation growth of snowflakes, *J. Atmos. Sci.*, 35:118–124, 1978; J. M. Straka, *Cloud and Precipitation Microphysics: Principles and Parameterizations*, Cambridge University Press, 2009; C. L. Ziegler, Retrieval of thermal and microphysical variables in observed convective storms, Part I: Model development and preliminary testing. *J. Atmos. Sci.*, 42:1487–1509, 1985.

Muscle development and regeneration

Skeletal muscle is the most voluminous tissue in humans. There are about 670 different muscles in the human body, including 170 distinct skeletal muscles in the head. The skeletal musculature serves crucial functions in vertebrates. The body muscles, including trunk and limb muscles, are responsible for posture and locomotion, whereas the head

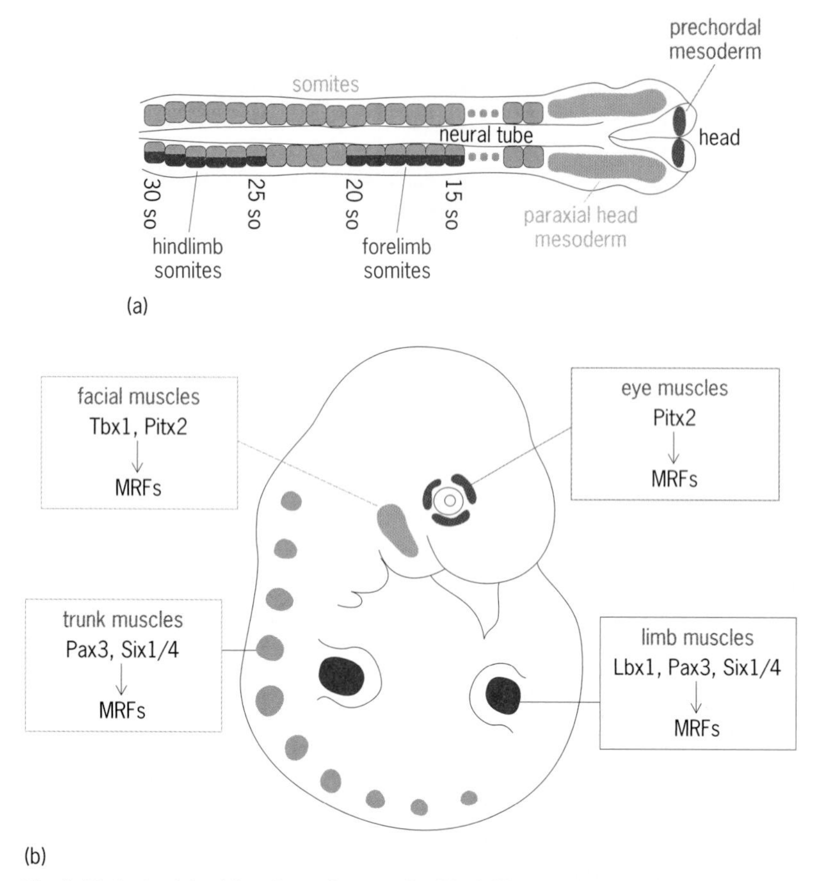

Fig. 1. Distinct origins (*a*) and genetic cascades (*b*) of different groups of skeletal muscles in different regions of the embryo. MRFs = myogenic regulatory factors.

muscles control the eyes, cranial openings, food uptake (mastication), and speech. Skeletal muscle displays the ability to regenerate after injury or in degenerative diseases. However, this regenerative capacity is exhausted over time and in severe muscle dystrophies.

Skeletal muscles are heterogeneous, and a muscle's identity is conferred by its position in the body, insertion into bone, innervation, and fiber type. Skeletal muscles can be organized into four main groups according to their position in the body: eye, face, trunk, and limbs. Different intrinsic and extrinsic regulatory pathways control muscle formation at different places and at different developmental times.

Developmental myogenesis. Myogenesis, which is the formation of muscle tissue, begins early in embyonic development.

Embryological origin. All skeletal muscle cells originate from the mesoderm (one of the three germ layers in the early embryo). Each group of muscles has a distinct mesodermic embryological origin. Myogenic cells of trunk and limb muscles derive from somites. Somites are transient segmented structures of the paraxial mesoderm that straddles the neural tube. Forelimb and hindlimb muscles originate from the lateral parts of somites 15–20 and somites 25–30, respectively. Head muscles (eye and facial muscles) originate from paraxial head mesoderm, which is located anteriorly to the somites, and from prechordal mesoderm (**Fig. 1***a*).

Intrinsic program of myogenesis. During development, muscle progenitors of each group of muscles are specified by different genetic programs (Fig. 1*b*). No single gene is essential for specification; instead, a combination of genes is required. As an example, specification of limb muscles depends on various transcription factors, including the *Lbx1*, *Pax3*, and *Six1/4* genes, whereas the *Tbx1* and *Pitx2* genes are required for the specification of facial muscle cells (Fig. 1*b*). Once specified to a muscle fate, muscle progenitors use a common muscle program in each group of muscles, which involves the myogenic regulatory factors (MRFs). The four members of this family of DNA-binding proteins (Myf5, MyoD, Mrf4, and myogenin) induce the expression of a variety of genes involved in the contractile properties of mature skeletal muscle cells. The MRFs have the remarkable ability to trigger muscle fate and differentiation in nonmuscle cells both in vitro and in vivo.

Extrinsic signals that regulate myogenesis. Muscle cell fate is acquired via signaling molecules from surrounding tissues. These signals are distinct for each group of muscles. Trunk muscle progenitors respond to extrinsic signals such as Wnt and Shh (sonic hedgehog) emanating from adjacent neural tube, notochord, and ectoderm. Limb muscle progenitors are under the influence of hepatocyte growth factor (HGF) from the limb mesenchyme and Wnt from the ectoderm. It should be noted that the same signal can lead to opposite effects on different muscle groups. As an example, a Wnt signal is necessary and sufficient to promote normal Myf5 and MyoD expression in somites, whereas the same Wnt signal inhibits MyoD expression in head muscle progenitors. This highlights the diversity of muscle progenitors in each muscle group.

Embryonic and fetal myogenesis. Muscle diversity occurs not only in various places in the body but at different times. Developmental myogenesis occurs in two successive waves: the embryonic and fetal waves (**Fig. 2**). Embryonic or primary myogenesis forms primary muscle fibers generated by the fusion of embryonic myoblasts originating from embryonic muscle progenitors. These primary fibers establish the scaffold for secondary muscle fiber formation generated from fetal or secondary progenitors. Fetal myogenesis correlates with muscle growth, innervation, and fiber type formation during development. The transcription factor Pax3 defines the embryonic muscle progenitors and is required for their formation, whereas Pax7 labels fetal muscle progenitors and is required for their formation. In the absence of both Pax3 and Pax7 activity, muscle development is arrested. Embryonic and fetal myoblasts differ in their gene expression and in their in vitro characteristics, leading to different muscle fibers in vivo and in vitro. Recently, a transcription factor, Nfix, has been identified as a major regulator of the switch between embryonic and fetal myogenesis, regulating directly the transcription of several fetal specific genes and

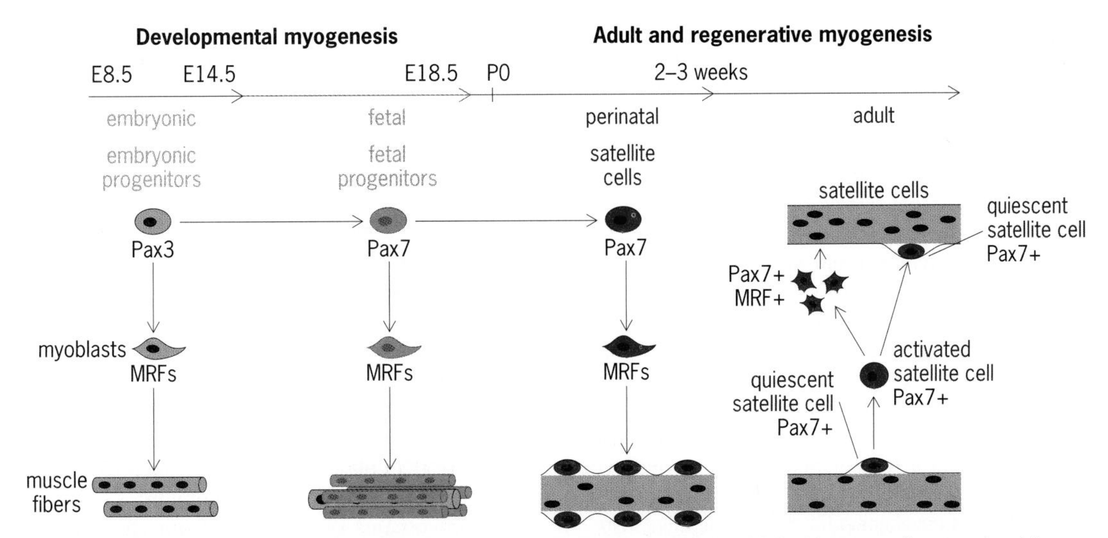

Fig. 2. Embryonic, fetal, and adult myogenesis in the mouse. Skeletal muscle is established in successive steps involving different types of muscle progenitors: embryonic/primary and fetal/secondary progenitors during development, and satellite cells in the adult. In mouse embryos, embryonic myogenesis occurs from E8.5 (embryonic day 8.5) to E14.5. From E14.5 to E18.5, fetal myogenesis allows muscle growth in the embryo. From E18.5, satellite cells appear that are responsible for perinatal growth and muscle regeneration. From 3 weeks postnatal, satellite cells are quiescent. Upon injury, quiescent satellite cells are activated, proliferate, and differentiate to reform the damaged muscle. During this process, a subset of satellite cells is used to renew the satellite cell pool.

concomitantly inhibiting embryonic specific genes. Here again, the same signaling pathway can have distinct effects on embryonic and fetal myogenesis: The Wnt pathway is not required for limb embryonic myogenesis, but it is critical for fetal myogenesis.

These different genetic requirements for various groups of muscles in the body and over time are relevant in the context of muscle dystrophies. Different groups of muscles are affected in various myopathies. Distal myopathies are characterized by progressive weakness in the muscles of the distal limbs, whereas limb girdle muscular dystrophies lead to atrophy of pelvic and shoulder girdle muscles. Oculopharyngeal muscular dystrophy causes muscle weakness mainly in eye and facial muscles. The reasons are not known why only a subset of muscles are affected, while other muscle groups escape the disease, but they are probably linked to the aforementioned muscle heterogeneity during development.

Adult and regenerative myogenesis. Adult muscle progenitors are called satellite cells. Satellite cells are found along muscle fibers underneath the basement membrane (Fig. 2). Based on this histological definition, satellite cells are first observed at 2–3 days before birth. These satellite cells ensure the perinatal growth of skeletal muscles (until 3 weeks in mice), homeostasis, and the regeneration process of damaged muscles in the adult. During perinatal growth, the satellite cells are incorporated into fibers to contribute to skeletal muscle growth. In the adult under normal conditions, satellite cells are mainly in a quiescent state. It is only upon muscle injury or in degenerative muscle diseases that quiescent satellite cells are activated to proliferate and trigger the muscle program and fuse to form nascent multinucleate myofibers. Although satellite cells are considered to be the main players in skeletal regeneration, other cell types also have a regenerative myogenic potency, including cells originating from the bone marrow or the blood vessels (vascular and hemepoietic lineages).

Molecular signature of satellite cells. The transcription factor Pax7 is the hallmark of quiescent and activated satellite cells in adult muscle. Satellite cells constitute a heterogeneous population, which can be in different states: quiescent, proliferating, or committed to differentiation. The quiescent state is important for satellite cells to retain their proliferative and differentiative potential throughout life. However, characterization of the molecular signature of quiescent satellite cells has identified various characteristic markers that are not specific to muscle tissue. Another aspect of muscle heterogeneity is that satellite cells from different muscle groups (such as head and trunk muscles) have different molecular signatures, reflecting their developmental history. Satellite cells, though, do not retain their specific developmental muscle phenotypes in heterotopic situations (that is, situations occurring in an abnormal location).

Satellite cell activation during regenerative myogenesis. In response to muscle injury, satellite cells are activated, proliferate, and then differentiate to reconstitute damaged muscle fibers. The whole process takes approximately 14 days in the mouse. The specific signals that trigger satellite cell activation and then differentiation are not well elucidated. Several signals provided by the damaged muscle fibers and surrounding cells (connective tissue or blood vessels) have been shown to be involved in activation of satellite cells during regenerative myogenesis. These signals include HGF, fibroblast growth factor (FGF), and insulin growth factor (IGF). During regenerative myogenesis, activated satellite cells have two roles.

The first is to provide cells committed to muscle differentiation that will form the regenerated fiber, and the second is to generate cells that return to the quiescent state to maintain the progenitor pool (Fig. 2). Both asymmetric division and orientation of cell division along the fiber are thought to be involved in these processes. Notch and Wnt signaling pathways have been shown to be involved in asymmetric self-renewal and symmetric expansion of satellite cells, respectively. However, it remains unclear how all these factors cooperatively regulate quiescence and activation of the satellite cells. The molecular identification of satellite cells and the actors involved in their regulation represent an important issue for exploration with regard to the use of satellite cells for transplantation in degenerative muscular dystrophies.

Link between developmental and regenerative myogenesis. Despite distinct requirements for each muscle progenitor population during developmental and regenerative myogenesis, a pool of resident progenitors is maintained in developing muscles during embryonic, fetal, perinatal, and adult myogenesis. The transcription factors Pax3 and Pax7 define this progenitor-cell population during all stages of muscle formation (Fig. 2). In addition, long-term lineage studies have shown that the same region of the embryo provides the progenitors at all stages of developmental and adult myogenesis for a muscle group. The dorsal and lateral parts of somites provide muscle progenitors for the trunk and limb, respectively, during embryonic, fetal, and adult myogenesis, whereas the head mesoderm gives rise to satellite cells of head muscles. Adult muscle regeneration is generally thought to recapitulate developmental myogenesis. During the adult regeneration process, embryonic and fetal markers are reexpressed transiently. The signals that regulate developmental and regenerative myogenesis, although not completely identified, display some similarity. Signaling pathways such as the Wnt and Notch pathways involved in fetal myogenesis are clearly also involved during the adult muscle regeneration process. Identification of the factors that regulate developmental and adult myogenesis will open new avenues to improve our knowledge about the treatment of muscle injuries and diseases.

For background information *see* CELL FATE DETERMINATION; DEVELOPMENTAL BIOLOGY; DEVELOPMENTAL GENETICS; EMBRYONIC DIFFERENTIATION; EMBRYONIC INDUCTION; GERM LAYERS; MORPHOGENESIS; MUSCLE; MUSCULAR SYSTEM; MUSCULAR SYSTEM DISORDERS; REGENERATIVE BIOLOGY in the McGraw-Hill Encyclopedia of Science & Technology

Delphine Duprez

Bibliography. S. Biressi, M. Molinaro, and G. Cossu, Cellular heterogeneity during vertebrate skeletal muscle development, *Dev. Biol.*, 308:281–293, 2007; S. Tajbakhsh, Skeletal muscle stem cells in developmental versus regenerative myogenesis, *J. Intern. Med.*, 266:372-389, 2009; F. S. Tedesco et al., Repairing skeletal muscle: Regenerative potential of skeletal muscle stem cells, *J. Clin. Invest.*, 120:11–19, 2010; E. Tzahor, Heart and craniofacial muscle development: A new developmental theme of distinct myogenic fields, *Dev. Biol.*, 327:273–279, 2009.

National Academy of Sciences report on forensic science

On February 18, 2009, the National Academy of Sciences (NAS) released the report *Strengthening Forensic Science in the United States: A Path Forward*. The report was the culmination of more than two years of work by a committee selected by the NAS. More than 80 witnesses were called to testify during the deliberations of the committee. The final report contains 13 recommendations for actions that would, if implemented, greatly strengthen the practice and reliability of forensic science by setting new standards, providing increased funding for research and education, developing and transferring best practices, and by creating a new oversight body at the federal level. Since this report came out, the U.S. Congress has held hearings and legislation is being prepared. The reaction of the forensic science and justice communities has been unprecedented.

Genesis of the committee. In 2004, the Consortium of Forensic Science Organizations went to the U.S. Congress and requested that a forensic science commission be formed that would look into the status and needs of forensic science. For one reason or another, this request, though granted by Congress, was never carried out. Finally, in 2006, a subcommittee of the Senate Judiciary Committee appropriated funds to the National Academy of Sciences to form a committee to study forensic science. This Committee was given eight charges to:

1. Assess the present and future resource needs of the forensic science community, to include state and local crime labs, medical examiners, and coroners;
2. Make recommendations for maximizing the use of forensic technologies and techniques to solve crimes, investigate deaths, and protect the public;
3. Identify potential scientific advances that may assist law enforcement in using forensic technologies and techniques to protect the public;
4. Make recommendations for programs that will increase the number of qualified forensic scientists and medical examiners available to work in public crime laboratories;
5. Disseminate best practices and guidelines for the collection and analysis of forensic evidence to help ensure quality and consistency in the use of forensic technologies and techniques to solve crimes, investigate deaths, and protect the public;
6. Examine the role of the forensic community in the homeland security mission;
7. Examine interoperability of automated fingerprint information systems;
8. Examine additional issues pertaining to forensic science as determined by the Committee.

The Committee was made up of 16 members. The chair was a federal judge. There were five practicing forensic scientists, a crime lab system manager, several attorneys, and representatives from various academic institutions. A total of eight meetings were held over a period of more than 2 years. A draft of the report was crafted and then reviewed by an independent, 21-member committee of experts that was anonymous to the NAS Committee.

Report. The report of the NAS Committee was released in February 2009. It contains 13 recommendations. The recommendations range from the formation of a new forensic science oversight agency to the relationship of forensic science to homeland security, and from coroners and medical examiners to standard language and reports. Many areas of forensic science were studied and cited in the report.

In the course of their study, the Committee found that the field of forensic science faces many challenges. These include great disparities in the operation of federal, state, and local laboratories, a general lack of mandatory standardization, certification, and accreditation, differing needs because the range of disciplines within forensic science is so broad, problems relating to the interpretation of forensic evidence, the need for research to establish limits and measures of performance, the admission of scientific evidence in court, and the political realities of the criminal justice system.

Recommendations. Below is a list of the 13 recommendations with some comments about each one. The recommendations are heavily paraphrased, as some of them are quite lengthy.

Recommendation 1. The first recommendation is pivotal. It calls for the formation of an independent federal agency, The National Institute of Forensic Sciences (NIFS), which would oversee the implementation of the other recommendations and act as the coordinating agency for research funding. Before recommending the formation of an independent agency, the Committee studied the existing agencies, such as the National Institute of Justice, the National Science Foundation, and the National Institute of Standards and Technology. None of them had the mission or skill set to effectively coordinate the national forensic science enterprise. Not surprisingly, there has been a lot of resistance from people in forensic science and criminal justice who feel that such an agency would usurp their influence.

Recommendation 2. The second recommendation calls for the adoption of a standard terminology to describe the association of evidence from a crime scene with a known person or object. This includes terms such as "match" and "consistent with." The meanings of such terms are not agreed upon by the forensic science community, and thus when they are used in court, the jury is often not sure of the meaning and can become confused. The other part of this recommendation calls for the development of model laboratory reports that all laboratories would use. This came about because many laboratories use abbreviated lab reports that present little more than the demographics and identifying information about the suspect, crime, and submitting department, along with a brief description of the evidence and the final conclusion. No information is given about the tests that were done, the results of each test, how these contribute to the final conclusion, estimates of error rates with each test, or other cautions. The reason often given for this type of report is that judges, juries, and attorneys do not have the scientific background to understand the report. However, proponents of this recommendation argue that forensic science will never be regarded as a true science unless such data is included in reports. The committee suggested that an abbreviated report can be furnished as a type of "executive summary" but that all relevant information must be included in the report itself. Use of a standard report would be tied to the accreditation of the laboratory to enforce its adoption.

Recommendation 3. The third recommendation addresses the need for more research in the areas of accuracy, reliability, and validity in forensic science. Although this is specifically aimed at the areas of pattern evidence, such as fingerprints, there are many areas of forensic science that could use more research into the scientific bases for their analysis and interpretation. It was generally agreed by the Committee that there needs to be more science in forensic science and research in these areas would address this.

Recommendation 4. The fourth recommendation is one of the most controversial and misunderstood of all. It would remove public crime laboratories from the budgetary and managerial control of law enforcement agencies or prosecutor's offices. There is nothing in the recommendation that calls for physical removal of the laboratories or for complete independence, although the committee considered that to be a worthy goal. Instead, it would put scientists at the helm of crime labs and would give them a complete budget line of their own. There are several reasons for this recommendation. First, crime labs should not have to compete with police departments for their budgets. The police department would come out on top every time. Second, law-enforcement managers often do not have the education, training, or culture to appreciate the needs of scientists, such as continuing professional development, training, and the opportunity to attend forensic science meetings and conduct research. In addition, there is a significant public perception that public forensic science laboratories associated with law enforcement agencies are part of the law enforcement and prosecution team, and therefore they, like the police, are biased. This can be destructive to forensic science laboratories that see themselves as unbiased scientists who speak for the evidence.

Recommendation 5. The fifth recommendation concerns what is commonly known as contextual or conformational bias. Forensic scientists in a public crime laboratory commonly receive evidence from criminal investigators who are seeking evidence against a suspect in a crime. The evidence that is submitted often includes materials generated by the crime such as fingerprints, bullets, drugs, hairs, and so on. The scientist is also furnished with control samples from the suspect. The scientist is normally also given the circumstances surrounding the case, including the background of the suspect's alleged involvement in the case. The scientist, armed with this information and these samples, will normally characterize and compare the evidence. This focus on the suspect, and no one else, can easily cause bias on the part of the examiner that is very difficult to minimize or account for. On the other hand, when witnesses to a crime are asked to identify the suspect, lineups are used so that the witness will not be forced to focus on the one person the police are investigating. Even when cases are verified by other examiners, they are often given the results of the first examination, thus subjecting them to bias. The report calls for research into conformational bias and the development of strategies to minimize it.

Recommendation 6. The sixth recommendation calls for the NIFS, in conjunction with the National Institute of Standards and Technologies and other relevant organizations, to develop standards for forensic science, including proficiency testing, methods, interpretation of evidence, and reliability. These standards should reflect best practices. When they are developed, the NIFS will develop strategies to disseminate these strategies and standards to the whole field.

Recommendation 7. The seventh recommendation sets up guidelines for the development of mandatory accreditation of forensic science laboratories and mandatory certification of forensic scientists. At this time, there are good mechanisms and processes for accrediting crime laboratories, chiefly by the American Society of Crime Laboratory Directors–Laboratory Accreditation Board. More than 80% of U.S. public crime labs are accredited. However, this process is not mandatory. There are no consequences for not being accredited and most private labs or police identification units need not be accredited. Likewise, there are several organizations that certify various classes of forensic scientists, but very few are mandatory. Again, there are no consequences for not being certified. This is a serious problem that must be addressed if forensic science is to gain credibility as a scientific pursuit.

Recommendation 8. The eighth recommendation concerns quality assurance and control. Although these procedures are mandated by the American Society of Crime Laboratory Directors (ASCLD) in the accreditation of laboratories, the non-accredited laboratories often do not have stringent procedures for quality assurance/acceptance criteria (QA/AC) . The emphasis here is on ensuring accuracy in forensic analyses.

Recommendation 9. There are many forensic science organizations in the United States and the rest of the world that have codes of ethics. There are wide differences in the scope, rigor, and thoroughness of these codes and they all carry some hierarchy of sanctions up to and including expulsion of a member from the organization. The fact is that even this sanction does not prohibit the violator from practicing forensic science or even testifying in court. Thus, serious violators are not removed from the system and may repeat their unethical behavior without fear of losing their livelihood. The ninth recommendation calls for the development of an enforceable consensus national code of ethics for forensic scientists. The report suggests that a way be sought to tie the ethical code to certification, so that a serious violator could lose certification and be prohibited from practicing forensic science or testifying in court as an expert.

Recommendation 10. The tenth recommendation addresses the problems inherent in attracting excellent students to the study of forensic science and of attracting research faculty from prestigious universities to the field to help tackle the most pressing scientific issues. It calls for the Congress to make funds available for scholarships for students and research funds for researchers. Currently, very little is spent, and only a few colleges and universities have an interest in forensic science research. There is a great deal of untapped potential in academia that could make a real difference in how forensic science is practiced.

Recommendation 11. The medicolegal death investigation system, or systems, are in critical shape in the United States. These are state, rather than federal issues, and there is a patchwork of different systems in each state. Some states have coroners, who are generally elected and are most often not physicians, although they usually make use of medical doctors to conduct postmortem investigations. Other states have medical examiner systems in which the medical examiner is appointed by the county or other level of government and is a medical doctor, but in most cases does not have to be a forensic pathologist. The eleventh recommendation would provide funding to the NIFS to distribute to states with the stated purpose of replacing and eventually abolishing coroner systems in favor of medical examiners. The recommendation also includes a provision for enticing more medical students to enter pathology and become certified in forensic pathology through mechanisms such as loan forgiveness. Funding is also provided for the education of coroners, medical examiners, and pathologists in the processes and procedures of medicolegal death investigations.

Recommendation 12. The twelfth recommendation deals with issues concerning local, state, and national fingerprint databases. Most large cities, all states, and the Federal Bureau of Investigation (FBI) maintain extensive databases of sets of fingerprints obtained from people in a variety of criminal, civil, security, employment, and other situations. A major problem is that the local and state databases do not, in many cases, communicate with other state or federal databases. This makes it more difficult to search for people by their fingerprints. This recommendation would provide funding to develop or migrate to a common computer platform for storing fingerprint data.

Recommendation 13. The thirteenth recommendation of the NAS report concerns the relationship between forensic science and homeland security. There is a great deal of overlap in the interests, procedures, and research questions of the forensic science and security communities. There is much to be learned and exchanged between the two communities. Funds would be provided to develop the maximum intelligence possible from forensic evidence so it can be used for security purposes.

Recent developments. Since the NAS report came out in 2009, there have been three hearings in the Congress: one in the House, and the other two in the Senate. There have also been numerous discussions among Congressional staffs and the forensic science community. Legislation was expected to come out of the Senate Judiciary Committee in the second quarter of 2010. Although little is known about the contents of the legislation, it is clear that the formation of the NIFS (*Recommendation* 1) will not take place, at least not in the near future. There is little sentiment in the Congress in favor of creating another independent agency. (The last one was the Department of Homeland Security.) Instead, there will apparently be an office created in the Department of Justice to facilitate the adoption of the other twelve recommendations to the extent this is possible.

Outlook. The staff of the National Academy of Sciences who oversaw the Committee indicated that they have never seen so much publicity and discussion following the issuance of a report. This is a testament to the importance of forensic science and the large number of stakeholders in this field. Everyone wants to see a robust, well funded, and reliable forensic science system. This report has made a start.

For background information *see* CRIMINALISTICS; FINGERPRINT; FORENSIC EVIDENCE in the McGraw-Hill Encyclopedia of Science & Technology.

Jay A. Siegel

Bibliography. Committee on Identifying the Needs of the Forensic Science Community, *Strengthening Forensic Science in the United States: A Path Forward*, National Academies Press, Washington, D.C., 2009.

Neuropsychology of memory revisited

Localized brain lesions with observable behavioral consequences have historically produced "natural" experiments in which neuropsychologists could use ingenious tests to probe into different processes that underlie human psychological functions, including memory. This approach was inaugurated in the late nineteenth century by the French neurologist Paul Broca, who localized the "speech center" of the brain in a specific convolution of the frontal lobe. Since then, single case studies of cerebral lesions and the ensuing dissociations between damaged and spared functions have provided unquestionable evidence that multiple separate processes are behind fundamental and seemingly unitary experiences, such as memory. Neuropsychological data acquired during the convalescence of these patients, and in some cases throughout the rest of their lives, acquire true meaning once the neurological basis of the syndrome is described. In the past, postmortem examination was limited to gross autopsy findings and narrow local pathological examinations. With the introduction of high-resolution neuroimaging methods and novel digital technologies, it is now possible to create detailed maps of the whole brain of those neurological patients who have turned their adverse conditions into crucial contributions to the study of the human brain.

Memory has classically been the subject of philosophical or psychoanalytic introspection. It eluded rigorous scientific investigation until experimentalists in the mid-twentieth century focused their empirical method on learning. The behaviorists, however, were not concerned with the brain or mind processes that operated between the stimulus (the "input") and the measurable reaction (the "output"), so it was actually a handful of amnesic patients who laid down the foundations of modern memory research. One of them in particular, patient H.M. (Henry Gustav Molaison, 1926–2008), showed that pervasive and persistent damage to specific memory functions could occur in the absence of any other detectable change in cognitive or intellectual capacities.

Patient H.M. A long escalation of convulsive episodes, attributable to a bicycle accident that patient H.M. sustained at the age of seven, led to such incapacitation that H.M. and his parents accepted the risks of a drastic experimental operation. Surgeon William B. Scoville had performed hundreds of frontal lobectomies at Hartford Hospital in Connecticut. Moreover, in 30 severely psychotic patients, the operation was extended further into the middle of the brain in the belief that this would give it stronger therapeutic impact. This well-rehearsed surgical approach seemed suitable to access and disrupt the structures in H.M.'s brain that were notoriously implicated with epileptic activity. H.M.'s surgery was carried out on August 25, 1953, after which H.M. continued to have seizures, but less frequently and far less incapacitating than before. However, the surgery produced a striking and totally unexpected result: when H.M. recovered from the operation, the young patient could not recognize members of the hospital staff (with the exception of Dr. Scoville, whom H.M. had known for several years), he was disoriented, and he could not remember any day-to-day events. When a proper psychological examination was performed 19 months later, H.M. thought it was March 1953.

Multiple memory systems. Brenda Milner was the experimentalist who first properly tested patient H.M.'s memory against other scales of psychological performance, including his intelligence quotient

(I.Q.). In fact, his I.Q. score was higher after the surgery; it had risen from 104 to 112, presumably because his seizures had become less frequent. She recorded that H.M. could hold the number 584 in memory for up to 15 min if he rehearsed it continuously out loud. In contrast, his memory of nonverbal material, such as complex visual stimuli (which are difficult to rehearse mentally), decayed very quickly. A formal demonstration of this effect came later when Milner used a "delayed pair comparison" paradigm that consisted of presenting two stimuli in succession separated by a variable time interval, sometimes adding distracting elements between trials. The fact that H.M. could register sensory information normally, but it escaped him in less than a minute, demonstrated a new and crucial distinction between immediate and long-term memory. H.M.'s difficulties with learning were thereafter documented consistently. A notable exception to these findings (and perhaps the most significant functional dissociation described in H.M.) was his ability to improve on a visual-motor skill task over subsequent trials without remembering any of the sessions. In what is now considered an iconic experiment, he was asked to draw between two outlines of a star, which would otherwise be trivial were it not for the fact that he had to perform the task using the image reflected in a mirror (**Fig. 1**). Demonstration of intact motor skill learning in H.M. without his recollection of the tasks was the catalyst for a large and long-lasting body of work that eventually "dissected" multiple memory systems in the brain. The concept of "declarative memory" (that is, the conscious recall of facts and events), dependent on the medial temporal lobes (MTL) and the hippocampus, which were impaired in H.M., became separated from other memory systems, such as motor and perceptual learning, which were spared in H.M.

Anatomy of memory. The word hippocampus is an anatomical term dating back to the sixteenth century and was used in early neuropsychological literature to refer to the elongated bulge on the floors of the brain's lateral ventricles, a pair of elegantly laid fluid-filled canals inside the hemispheres. This gross structure actually contains a complex three-dimensional architecture of separate cytoarchitectonic (pertaining to the cellular arrangement within a tissue or structure) fields, one of which is also called the hippocampus. The interrelation between different MTL structures changes, depending on their position along the temporal lobe. Accordingly, the entorhinal cortex, which represents the main neocortical relay for the hippocampus, transcends into the parahippocampal cortex, which was spared in H.M. It is evident, therefore, that the use of the term hippocampal lesion can create confusion from a modern, functional perspective. However, neurochemical and connectivity studies in humans and other animals, many of them inspired by the watershed case of patient H.M., have gradually provided a more reliable map of the structures contained within the MTL. Thus, individual patterns of memory impairment could be better explained on the basis of the different topography of localized lesions along this region of the brain.

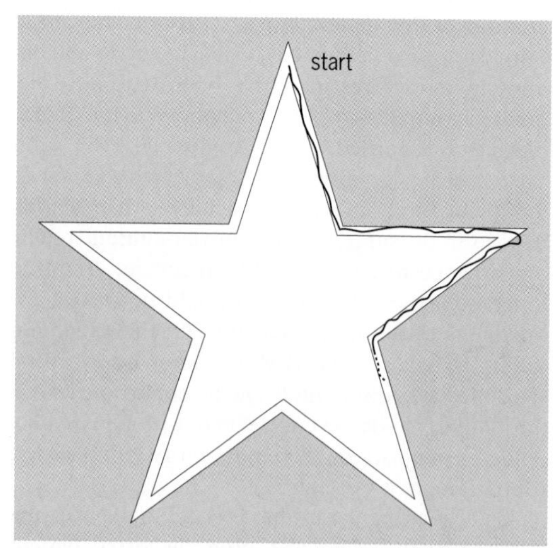

Fig. 1. Patient H.M. showed improvement in this mirror drawing task over a few days without recollection of ever having done the task.

The fact that some hippocampal tissue and adjoining parahippocampal cortex had been spared by the surgery in 1953 became evident the first time researchers used magnetic resonance imaging (MRI) to look at H.M.'s brain. This first view of the brain, even though it was not clear enough to visualize finer anatomical detail, demonstrated that Scoville had originally overestimated the length of the lesion and also provided some comparison with other amnesic patients, including E.P. who had a memory impairment with a different etiology and greater severity than H.M.'s and whose lesion in fact extended to the posterior parahippocampal cortex. Remote memories were intact in patients H.M. and E.P., even though the span of retrograde memory loss was different in each case. In each patient, however, it appeared that autobiographical memories from early life could be retrieved. Consolidation of memories is a gradual process; it involves the hippocampus and a structure called the amygdala, which is situated at the tip of the temporal lobes and which was removed or damaged completely in these patients. Nevertheless, in the premorbid period memory could have become independent of the MTL and supported by neocortical areas in the lateral temporal or frontal lobes.

Brain of patient H.M. MRI protocols can localize macroscopic neurological events in the brain, and it is possible to delineate the anatomical boundaries of the lesions noninvasively while the patient is still alive and undergoing neuropsychological examination. However, despite significant recent advances, MRI is severely limited in terms of resolution. Thus, it is impossible to see cellular-level disease and finer structural borders, especially in anatomically complex regions such as the MTL and hippocampus. The opportunity to provide direct anatomical valida-

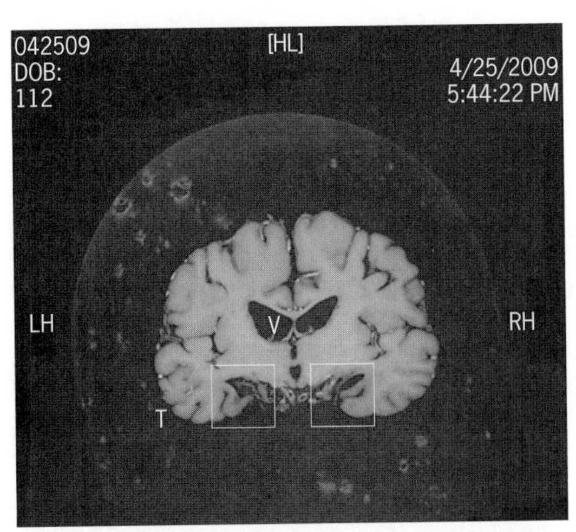

Fig. 2. High-resolution MRI of the formalin-fixed brain of patient H.M. Boxes show the lesions in both hemispheres of the medial temporal lobes. T: temporal lobe; V: ventricle.

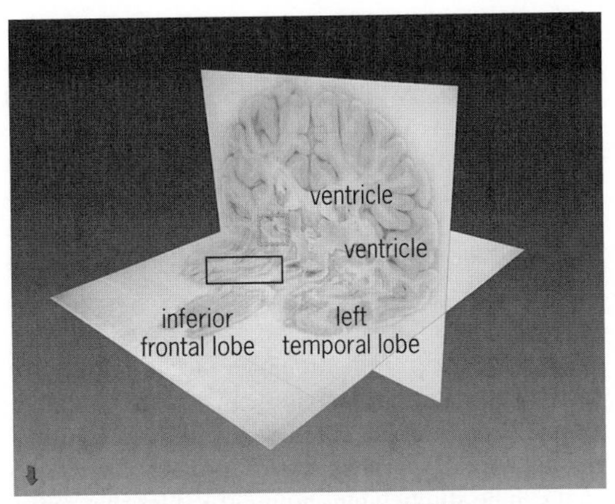

Fig. 3. Two orthogonal views through the three-dimensional digital reconstruction of the whole brain of patient H.M. The rectangular box delimits the surgical lesion in the right medial temporal lobe. The square box delimits spared tissue of the hippocampus and parahippocampal gyrus.

tion for five decades of neuropsychological research on patient H.M. occurred when he passed away at age 82 on December 2, 2008. Because consent had been given for the postmortem examination of his brain, researchers were able to initiate a series of experiments designed to illustrate the effects of the lesion and the pathological processes that occurred as a consequence of the surgery. Naturally, it was expected that many of the effects of the original lesion would be masked by the natural and morbid consequences of aging. These, in fact, were already evident from the most recent MRI scans conducted on H.M. just a few years before his death.

The neurological examination of the brain of the most studied amnesic patient in the history of neuroscience was conceived to go far beyond typical autopsy protocols. Researchers at the University of California, San Diego, proposed using newly designed computer-controlled microscopy techniques and Web-based digital tools to create a map of the entire brain of patient H.M., which then could be shared with researchers around the world. The autopsy was performed on the morning after H.M.'s death at Massachusetts General Hospital in Boston. The brain was chemically hardened and fixed in formalin for several weeks until it was ready to be transported to California. Multiple high-resolution MRI scans of the brain specimen were acquired in Boston and in San Diego (**Fig. 2**) before the sample was prepared for microscopic analysis.

Since the invention of the microscope, histological methods have been developed to allow the observation of biological tissue through transmitted illumination at high magnification. Specimens are typically only a couple of centimeters wide; these are sectioned and embedded in wax or ideally cut frozen, and the slices are stained with multiple agents to reveal different microscopic features. Researchers wanted to avoid sampling the brain of H.M. into many small blocks for routine processing, as was done for another notable brain, that of Albert Einstein, because this method leads to the loss of considerable data. Instead, on December 2, 2009, the team of The Brain Observatory at the University of California, San Diego, began performing an unprecedented dissection procedure that lasted 53 hours and resulted in a complete catalogue of 2401 "giant" frozen tissue slices through the whole brain. Although very challenging from a technical standpoint, this method was a crucial step toward the three-dimensional reconstruction of the brain. During the procedure, a photographic image of the cut surface of the brain was taken before each histological slice. The result of this imaging process was a full digital model of the whole brain at a resolution that is thirty times higher than that of a typical MRI (**Fig. 3**).

Digital library for the brain. This data set is ideal to inspect the complex three-dimensional geography of the surgical excision from multiple angles, but more resolution is necessary to map the anatomical borders and the effects of the lesion in detail. Therefore, a large-scale histological pipeline was established in order to stain tissue sections that are progressively being mounted on oversized glass slides (**Fig. 4**). This work is expected to continue for several years, adding more and more layers of information to the final map of the brain. Researchers also

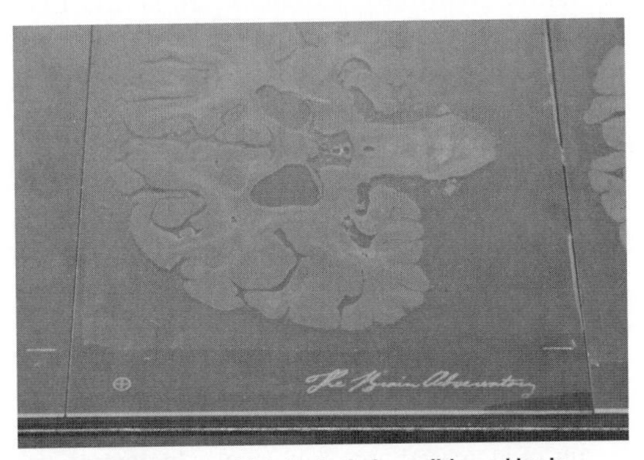

Fig. 4. Large-format histological glass slide and brain section, unstained.

wanted to enable remote access to the collection of slides digitally via the Web. Often, and notably in the case of H.M., multiple investigators conducted experiments with the same medical patient. The literature on these cases is rich in hypotheses that cite structural correlates of specific neuropsychological results; therefore, it is opportune to provide researchers with the means for revisiting their studies based on the explicit anatomical images relative to the case. The solution became available only very recently with the introduction of computer-controlled microscopes and digital pathology. Each stained histological slide can be scanned at cellular resolution, producing tens of thousands of microscopic images that can be stitched into monolithic files that represent the whole section. These images can be converted into a format that is viewable via the Web without requiring large bandwidth or high download speeds. In practice, anyone with access to the Web can inspect any region of the brain, visualizing cell assemblies and even single neurons.

Outlook. The human brain is unique on account of its multilevel structural and functional interconnectivity. A comprehensive model of a patient's brain must be able to account for neurological evidence that might be distributed across multiple structures and that can be evident at different levels of resolution. MRI protocols, if conducted longitudinally while the patient is alive and postmortem at higher resolution, can help describe global changes that occur in the brain after the onset of a syndrome. These data, combined with digital multiresolution maps of the brain, represent the modern blueprint for the neurological validation of neuropsychological cases that continue to shape our knowledge of human cognitive function.

For background information *see* AMNESIA; BRAIN; COGNITION; INFORMATION PROCESSING (PSYCHOLOGY); LEARNING MECHANISMS; MAGNETIC RESONANCE; MEDICAL IMAGING; MEMORY; NEUROBIOLOGY; PERCEPTION; PSYCHOLOGY; SEIZURE DISORDERS in the McGraw-Hill Encyclopedia of Science & Technology. Jacopo Annese

Bibliography. S. Corkin, What's new with the amnesic patient H.M.?, *Nat. Rev. Neurosci.*, 3:153–160, 2002; W. B. Scoville and B. Milner, Loss of recent memory after bilateral hippocampal lesions, *J. Neurol. Neurosurg. Psychiatr.*, 20:11–21, 1957; L. R. Squire, The legacy of patient H.M. for neuroscience, *Neuron*, 61:6–9, 2009.

New vertebrate species

Despite the extreme level of human-induced habitat destruction on Earth, which has resulted in more than 16,000 species worldwide being threatened with extinction, the last decade has seen a dramatic boom in the discovery of new vertebrate species. Such unprecedented taxonomic discoveries are the result of increased use of molecular biology in taxonomic studies and, most importantly, of increased numbers of scientific expeditions in regions (such as the Amazon and the Congo) that have been rarely, or never, surveyed previously. Although many of the newly described species are small and cryptic (appearing identical but genetically quite distinct), some very large and unlikely animals, such as petrels, cats, elephants, pandas, and whales, are being newly described. In the present article, the state-of-the-knowledge in species and subspecies discoveries is discussed, and information is compiled from published taxonomic revisions—essentially a reinterpretation of the taxonomy of known animals—and from new field discoveries of animals. In particular, special attention is paid to tetrapod vertebrates, that is, amphibians, reptiles, mammals, and birds.

Taxonomic revision. In many instances, taxonomic revisions are motivated by discoveries of new species in the field. In order to accommodate a newly discovered animal, taxonomists are often induced to modify known taxonomic relationships. Sometimes, in the process, they propose new species status to animals that were deposited, but unnoticed for decades, in museum collections worldwide. For instance, the recent discovery of Ayres black uakari monkey (*Cacajao ayresi*) along the Aracá River in northern Amazonia prompted a taxonomic revision of black uakaris, resulting in the description of a second species, the Neblina uakari (*Cacajao hosomi*), a known animal but thought previously to be the same animal described in the 1800s by Alexander von Humboldt. Thus, one field discovery resulted in the elevation of black uakaris from 1 to 3 species.

The exploration of areas never before sampled accounts for about 40% of the new mammal species described from 1993 to 2007. In many cases, these species are restricted to regions (for example, Brazil and Madagascar) that are seriously threatened by human occupation and with a high degree of endemism. (Endemic species are those with restricted ranges; thus, if these species were lost from a single region, they would be globally extinct.) In addition, a significant contribution to global biodiversity has emerged with the application of molecular genetic techniques, as has been employed, for example, in the analysis of newly described species of *Microcebus* (mouse lemurs) in Madagascar. Another example is the separation of the 2 previously known orangutan subspecies into 2 different species, *Pongo abelii* in Sumatra and *Pongo pygmaeus* in Borneo, with the latter being subdivided further into 3 new subspecies. The same happened with Central African elephants, where distinct known populations have gained full species status.

For the period from 2000 to 2010, a vast number of new species have been described: 3878 species of fish, 1193 species of amphibians, 990 species of reptiles (including turtles, geckos, and monitor lizards), 423 species of mammals, and 86 species of birds (including New World Psittacidae, some owls, a Brazilian falcon, and one petrel from the Pacific island of Vanuatu). Among the mammals, some of the most notable new species are two new whales, an elephant, a leopard, a panda, various monkeys (from Asia, South America, and Africa), 40 species of lemurs, and the giant Laotian rock rat (belonging to the Diatomyidae family thought to be extinct for 11 million years). The rate of discovery among vertebrates from 2000 to 2010 was almost 600 species per year, with more than a third of them (excluding fish) from South America and about 63% from the combination of South America and Asia (**Figs. 1** and **2**).

The contribution of developing countries to the list of new species is striking. As the frontiers of deforestation advance toward still intact or little-explored remote regions, access becomes easier for scientists. Discoveries are also associated with timber extraction projects and infrastructure work. Prior to the installation of large impact projects, such as construction of roads, dams, and so on, or logging concessions, companies are obliged to finance environmental impact assessments, which in many cases result in numerous new discoveries (for example, many new species in Papua New Guinea, Madagascar, Brazil, and Peru). Unfortunately, the same newly discovered animals sometimes become immediately threatened by the very infrastructure projects that helped finance their discovery. This has been the case for *Saguinus fuscicollis mura* (Mura's saddleback tamarin) in northwestern Brazil, Malagasy lemurs, and orangutans from Borneo and Sumatra.

The distribution of the discoveries around the world has been found to peak in the tropical countries of South America, Asia, Africa, and Oceania (Figs. 1 and 2). New vertebrates, however, have been found even in developed countries, including France, Switzerland, Germany, and the United States, where investment in scientific research is greater.

Amphibians and reptiles. Amphibians are the group of tetrapods with the most newly described species over the last decade. A total of 1193 new species have been described, representing approximately 18% of all known amphibians. The annual rate of discoveries for this taxon between 2000 and 2009 was about 115 new species per year. The number of newly described reptiles, 990 species (10.7% of all known reptiles), almost equaled that of amphibians. Moreover, in addition to more refined morphological and behavioral analyses (for example, taxonomic studies based on frog vocalizations), the advent of molecular techniques has resulted in the taxonomic partitioning of several species complexes.

Some findings are special for their uniqueness, such as the description of the purple frog, the last remnant of a new family of amphibians, Nasikabatrachidae, from the Western Ghats of India. This species is considered a living fossil (a living species belonging to an ancient stock otherwise known only as fossils), with its closest relatives being found only in the Seychelles, more than 3000 km (1864 mi) away. The time of separation between the purple frog and its Seychelles relatives is estimated at 130 million years ago. Among reptiles, the family Varanidae (monitor lizards) comprises the largest lizards in the world, including the Komodo dragon. Eleven new species have been described in the genus *Varanus*, including a giant, secretive, and forest-dwelling Philippines monitor lizard (*Varanus bitatawa*), which is more than 2 m (6.6 ft) in body length. Of the 21 species of boas (family Boidae)

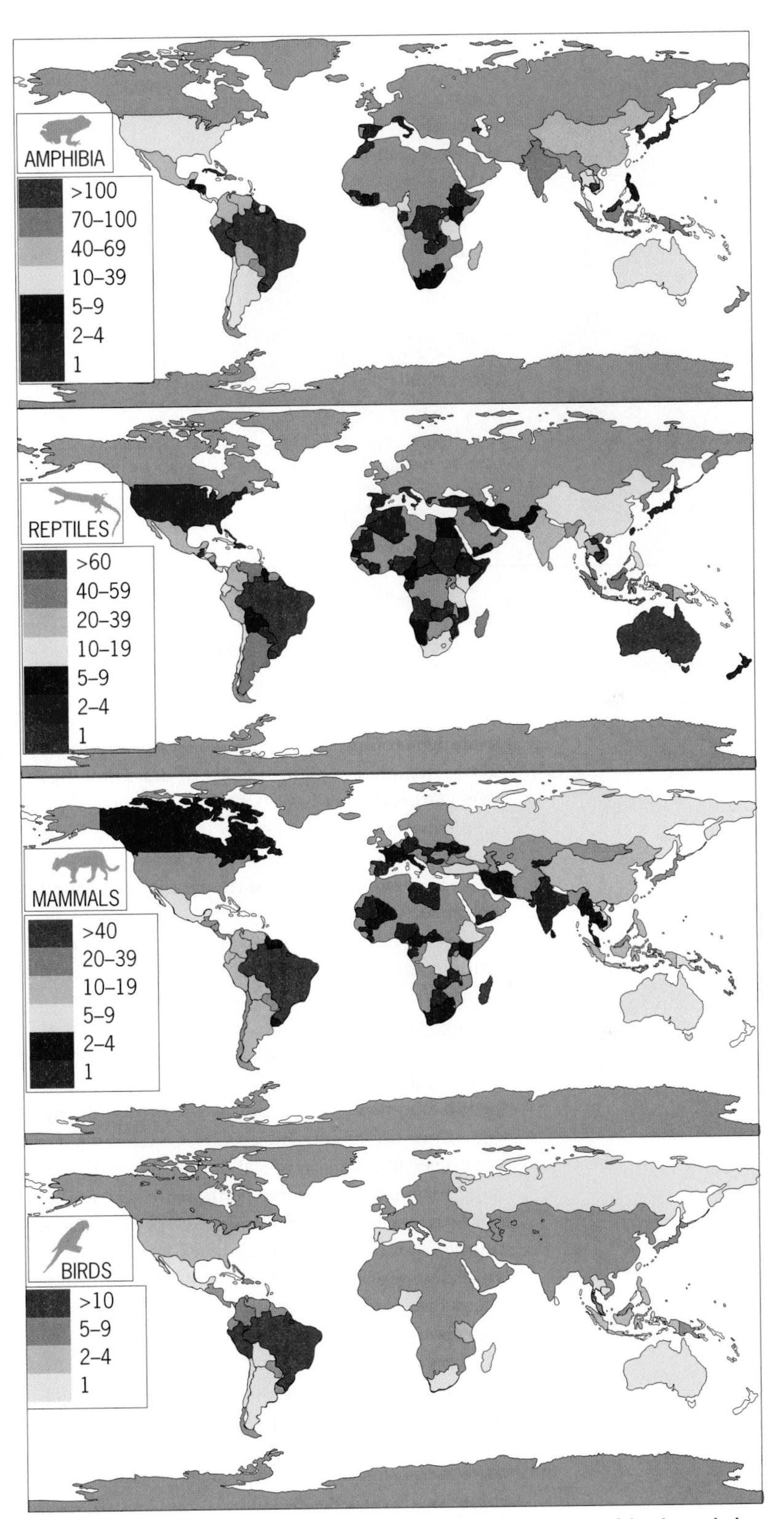

Fig. 1. World distribution of the discoveries of vertebrates (except fishes) for the period 2000–2010.

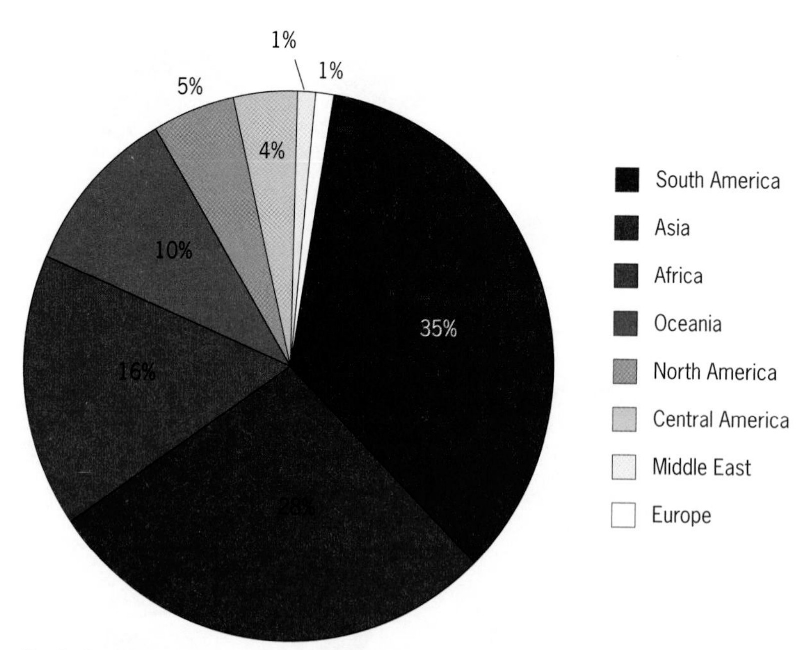

Fig. 2. Percentage of new vertebrate discoveries based on locations around the world for the period 2000–2010.

described to date, 7 (33.3%) were described since 2000, including a new species of anaconda in Bolivia (*Eunectes beniensis*). One of the most diverse families of reptiles, the geckos (family Gekkonidae), had the most newly described species since 2000, with 235 species (24.1% of the reptiles described in this period).

In general, the highest numbers of new species of amphibians and reptiles come from South America, Southeast Asia, and Oceania (see **table**). Brazil is the leading country, totaling 241 new species of reptiles and amphibians. In Africa, the country that stood out was Madagascar. Contrary to expectations, other African countries showed only modest contributions to both groups. Overall, most discoveries came from tropical countries, which have large expanses of tropical rainforests. As expected, countries containing deserts contributed more new reptile species than amphibians. Sahara Desert and Middle East countries (including Mauritania, Algeria, Egypt, Iraq, Iran, and Afghanistan) had significant numbers of reptiles described, but no frogs. This is clearly related to the natural history of these organisms, since reptiles are much more adaptable to arid climates than amphibians and tend to diversify in these environments.

Mammals. A review of new mammal species found between 1992 and 2006 showed a strong taxonomic and geographical bias in the new discoveries. New discoveries were above the expected rate for Afrotherian insectivores, primates, rodents, bats, marsupials, and monotremes, whereas discoveries were below the expected rate for eulipotyphlan insectivores, carnivores, ungulates, and cetaceans. A geographical bias is apparent since 71% of the discoveries come from continents and only 28% from islands. On continents, 62% of the findings come from the New World, in particular Brazil and Andean countries. In the Old World, new discoveries are concentrated in the continental regions of sub-Saharan Africa, tropical and temperate Asia, Australia, and Europe. A rate of 24 new descriptions of mammals per year has been estimated. In addition, some 60% of the newly described mammal species are cryptic species, but 40% are large and distinctive.

Birds. In stark contrast to the above data, new species discoveries for birds were much less abundant. The South American countries have made the greatest contribution, although birds from Africa, Asia, and Europe have been described. Most of the discoveries were made in tropical forests, revealing distinct and conspicuous species such as parrots and parakeets (Psittacidae), for example, *Pionopsitta aurantiocephala* and *Aratinga pintoi*, as well as some hummingbirds (*Eriocnemis*, *Topaza*, and *Coeligena*). A new species of kiwi (*Apteryx rowi*) from New Zealand and the Vanuatu petrel (*Pterodroma occulta*) are examples of large and distinctive birds that have been discovered. On the other hand, the Cryptic forest-falcon (*Micrastur mintoni*) was first recognized by its distinctive voice. Some nocturnal birds, such as owls and nightjars (*Asio*, *Glaucidium*, *Ninox*, *Otus*, *Athene*, and *Caprimulgus*), that have secretive habits also were discovered in the last decade. As a further example, 4 new species of the genus *Scytalopus* (family Rhinocryptidae) were recently recognized. These passerine birds are very similar morphologically and, in many instances, may be distinguished only by their vocalizations and molecular data. For birds, the largest contribution to the new species pool comes from taxonomic revisions rather than field discoveries in new areas.

Discussion. The findings analyzed here involve both the specific and subspecific level. Despite widespread discussions on the topic of species concepts, subspecies level has not proven effective for conservation planning, since often a single species can harbor dozens of subspecies (mainly in mammals and birds). In order to fulfill the goal of preserving the existing biodiversity, the diversity of organisms must be considered, including the diversity of geographic forms and the genetic diversity of organisms. To this end, it is necessary for conservation planning to consider the most refined taxonomic level as possible.

In spite of its importance for biodiversity conservation, efforts for the discovery of new species are still rather small. In recent years, the discovery of species has been progressive and continuous, justifying an investment in the training of professionals in taxonomy and molecular biology as well as more funding for expeditions to search for new species in areas not yet explored. Unfortunately, although the rate of new discoveries is high and encouraging, the rate of species loss as a result of habitat destruction is even higher. Recent extinction rates are 100 to 1000 times their prehuman levels in groups taxonomically

Ranking of countries in terms of discoveries of amphibians, reptiles, mammals, and birds for the period 2000–2010

Rank	Amphibians		Reptiles		Mammals		Birds	
1°	Brazil	153	Brazil	88	Madagascar	57	Brazil	18
2°	Peru	110	Australia	63	Brazil	47	Peru	14
3°	Papua New Guinea	86	Argentina	55	Tanzania	18	Colombia	6
4°	India	72	Vietnam	54	Ecuador	17	Ecuador	6
5°	Indonesia	72	Indonesia	53	Peru	16	Cuba	4

diverse and well known, from widely different environments.

For background information *see* BIODIVERSITY; CONSERVATION OF RESOURCES; ECOLOGICAL COMMUNITIES; ENDANGERED SPECIES; LIVING FOSSILS; POPULATION ECOLOGY; RAINFOREST; SPECIATION; SPECIES CONCEPT; TAXONOMY; ZOOGEOGRAPHY in the McGraw-Hill Encyclopedia of Science & Technology.

Fabio Röhe; Sergio Marques Souza; Claudia Regina Silva; Jean Philippe Boubli

Bibliography. S. D. Biju and F. Bossuyt, New frog family from India reveals an ancient biogeographical link with the Seychelles, *Nature*, 425:711–714, 2003; J. P. Boubli et al., A taxonomic reassessment of black uakari monkeys, *Cacajao melanocephalus* group, Humboldt (1811), with the description of two new species, *Int. J. Primatol.*, 29:723–749, 2008; G. Ceballos and P. R. Ehrlich, Discoveries of new mammal species and their implications for conservation and ecosystem services, *Proc. Natl. Acad. Sci. USA*, 106:3841–3846, 2009; C. P. Groves, *Primate Taxonomy*, Smithsonian Institution Press, Washington, D.C., 2001; J. Köhler et al., New amphibians and global conservation: A boost in species discoveries in a highly endangered vertebrate group, *BioScience*, 55:693–696, 2005; A. P. Paglia et al., A luta pela proteção dos vertebrados terrestres, *Sci. Amer. Brasil*, Edição Especial, no. 39, 2010; D. A. M. Reeder, K. M. Helgen, and D. E. Wilson, Global trends and biases in new mammal species discoveries, *Occas. Pap. Mus. Texas Tech. Univ.*, 269:1–35, 2007; F. Röhe et al., A new subspecies of saddleback tamarin, *Saguinus fuscicollis* (Primates, Callitrichidae), *Int. J. Primatol.*, 30:533–551, 2009; P. Uetz, The original descriptions of reptiles, *Zootaxa*, 2334:59–68, 2010; L. J. Welton et al., A spectacular new Philippine monitor lizard reveals a hidden biogeographic boundary and a novel flagship species for conservation, *Biol. Lett.*, in press, 2010.

Optogenetics

Optogenetics is a field of bioengineering research that centers on the development and use of tools that enable the use of light to control specific biological processes in order to understand how those processes causally contribute to complex cellular and organismal functions. By turning off a specific biological process, it is possible to understand what complex or emergent functions that process is necessary for. On the other hand, by stimulating a specific biological process, it may be possible to see what complex functions that process is sufficient to initiate. As might be inferred from the name, the most common optogenetic strategy is to express specific foreign genes that encode for light-receptive proteins in cells, in vitro or in vivo, and then to illuminate the cells to activate the resulting light-receptive protein, thus driving the specific signaling pathway that is downstream of the protein. These genes are taken from cells of other organisms and introduced using transgenic vectors, such as viruses.

Molecular tools. One of the most widely used sets of optogenetic tools is a group of genes that encode for light-driven proteins that, when expressed in neurons, allow neuronal activity to be turned on and off with pulses of light. This is of great importance because the brain is made up of hundreds of cell types that are very densely wired to form a connected circuit, and it is difficult to resolve how these different cell types and pathways are connected to drive computation and behavior. Additionally, how these circuit elements go awry in brain disorder states is unclear.

Neurons can be thought of as electrical devices that signal by changing the differential distribution of charged particles, or ions, across their cellular membranes. Accordingly, neural optogenetic tools work by translocating charge from one side of the membrane to the other in response to pulses of light. An early molecular strategy was to express in mammalian neurons the genes that encode for a three-protein invertebrate phototransduction cascade found in the eyes of flies. While such neurons could be electrically activated by pulses of light, the time scale over which neural activity could be modulated on and off was slow, perhaps in part because of the multigene nature of the system, which means that signaling must take place between proteins that may be distributed in different parts of a cell.

In 2005, the single gene that encoded for the protein channelrhodopsin-2 (ChR2) was expressed in neurons. This is a light-gated ion channel (a channel opened by light that controls the passage of ions through the cell membrane) that a species of green algae uses to navigate in water for optimal light exposure. This protein resembles in three-dimensional structure and in some biochemical aspects the kinds of mammalian rhodopsins that reside in the human retina and enable vision. ChR2-expressing neurons, when illuminated by brief pulses of blue light, underwent very rapid depolarization, that is, reduction in the electrical potential difference across the cell membrane, and could even fire just one millisecond-time-scale "spike" or "action potential," a basic transmissible unit of neural information, in response to just one brief pulse of light (**Fig. 1**).

Other optogenetic tools soon followed, including, in 2007, a light-driven chloride pump from an archaebacterium (Halo/NpHR) that, when expressed in a neuron and illuminated, resulted in hyperpolarization (increase in the potential difference across the cell membrane, making the interior more electronegative), hence inhibiting spike generation (**Fig. 2**); and, in 2010, a new set of silencers from archaebacteria and fungi (including Arch, eNpHR3.0, Mac, and others; **Fig. 3**) that enabled multicolor, powerful shutdown of neuron spiking in the awake-behaving brain (**Fig. 4**). Innovation in the field is continuing, as mining of genomic data is leading to many tools with novel and improved powers being continually discovered.

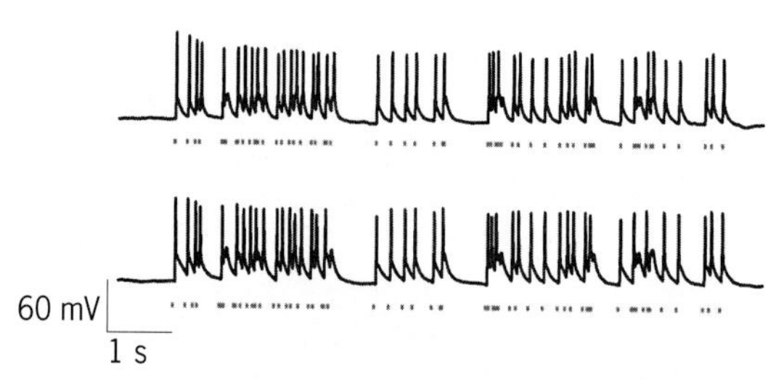

Fig. 1. Spikes (action potentials) elicited by pulses of blue light (represented by dashes beneath the spikes) in two different neurons expressing the optogenetic molecule ChR2. The train comprises blue-light pulses that are distributed in time pseudorandomly according to Poisson statistics, a distribution of times akin to the times between clicks heard on a Geiger counter when a radioactive substance is nearby, which is not a bad model of naturalistic neural firing. (*Adapted from E. S. Boyden et al., Millisecond-timescale, genetically targeted optical control of neural activity, Nat. Neurosci., 8:1263–1268, 2005*)

Applications to basic and clinical science. These neural optogenetic tools have found many uses in neuroscience, thanks to the increasingly common presence of lasers, LEDs (light-emitting diodes), and other bright light sources in neuroscience laboratories (driven in part by synergistic, increasing interest

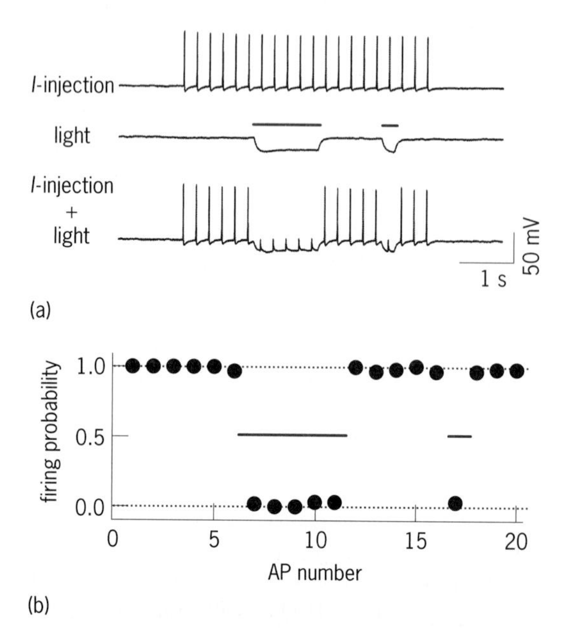

Fig. 2. Light-driven spike blockades demonstrated for neurons expressing Halo. (*a*) Blockade demonstrated for a representative neuron in the brain region known as the hippocampus expressing Halo. *I*-injection = neuronal spike firing induced by injection of current into the neuron using a glass electrode; 20 spikes are induced in a rhythmic fashion; light = neuronal hyperpolarization induced by periods of yellow light (represented by the bars above the recordings); *I*-injection+light = yellow light drives Halo to block neuron spiking, leaving spikes elicited during periods of darkness intact. (*b*) Blockade averaged across a population of neurons expressing Halo (data from seven neurons). Spike firing probability is plotted versus action potential (AP) number. Yellow light again drives Halo to block neuron spiking, reducing the spike firing probability to zero. (*Adapted from X. Han and E. S. Boyden, Multiple-color optical activation, silencing, and desynchronization of neural activity, with single-spike temporal resolution, PLoS ONE, 2(3):e299, 2007*)

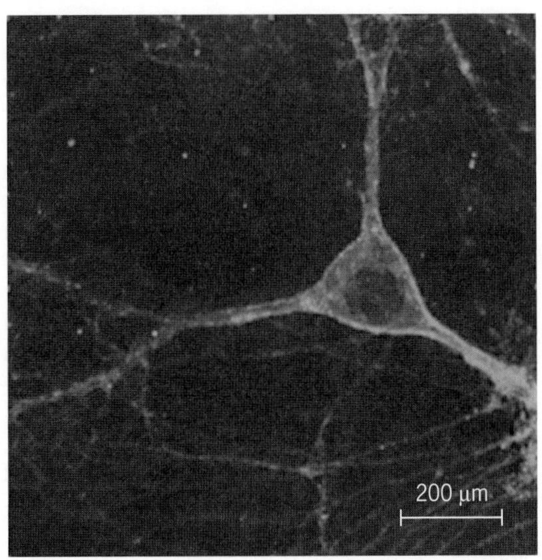

Fig. 3. Neuron expressing the optogenetic molecule Arch, here fused to green fluorescent protein (GFP), which allows it to be seen under a fluorescence microscope. ***(Adapted from B. Y. Chow et al., High-performance genetically targetable optical neural silencing by light-driven proton pumps, Nature, 463:98–102, 2010)***

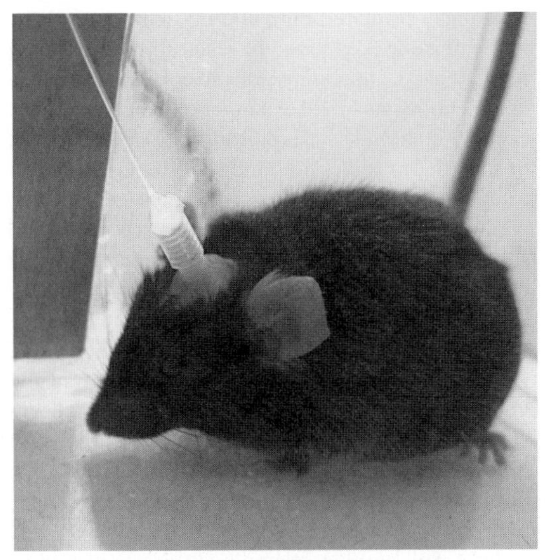

Fig. 5. A mouse whose neurons have been made light-activatable through viral expression of ChR2 in specific target neurons, with a blue laser-coupled fiber inserted into a cannula to deliver light into its brain.

in microscopy). ChR2 is commonly used to drive a pathway in order to see what its power to modulate downstream circuits or behavior is. Neural silencers can be used to quiet the activity in a defined set of cells or circuits, in order to see what behaviors use those cells or circuits and when they make their

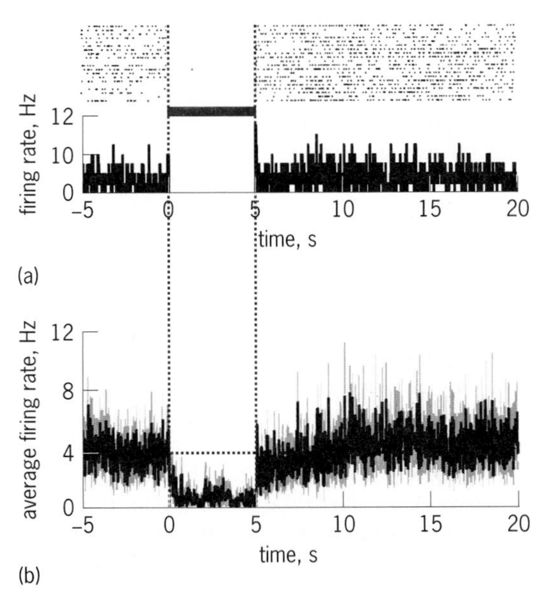

Fig. 4. Neural activity in neurons expressing Arch, before, during, and after 5 s of illumination with yellow light. (*a*) Activity in a representative neuron expressing Arch, shown as a spike raster plot—that is, a plot in which each spike is represented by a dot—displayed above a histogram of spike firing rate, which shows spike firing rate for each 20-ms-duration period throughout the light delivery process, averaged across trials. (*b*) Population average of instantaneous firing rate in a population of neurons expressing Arch. Black lines show mean; gray lines show mean ± standard error; data from 13 neurons. ***(Adapted from B. Y. Chow et al., High-performance genetically targetable optical neural silencing by light-driven proton pumps, Nature, 463:98–102, 2010)***

contribution. For example, in 2008, different numbers of neurons in different mouse brains were made photosensitive with ChR2, and then illuminated to see if the mouse would respond (**Fig. 5**). By doing this, it was possible to count the minimum number of neurons required to signal a sensation to the brain in a way that allowed for decision making. The same group of researchers expressed ChR2 in long-range projection neurons, whose output projections (axons) traveled long distances from their cell bodies, and then photostimulated the distant axon terminals in different brain regions to see what their local targets were. In 2007, a group expressed ChR2 in mice, targeting a class of hypothalamic neurons (lying in a ventral part of the brain, and involved in the control of a number of important physiological functions) that are compromised in narcolepsy patients, and showed that stimulating these neurons with light could cause the mice to awaken. A group in 2006 took blind mice that lacked their normal photoreceptors and delivered ChR2 to spared neurons in the retina by means of virus vectors, thus enabling the retina to respond to light again and to send signals to the brain.

This last example highlights the translational, that is, medical, potential of neural optogenetic tools: they can be used to deliver information into the brain very precisely to yield specific potential therapeutic benefits. Many hundreds of thousands of people have implantable electrical stimulators in their body to modulate their nervous systems, for example, cochlear implants for deaf people and deep brain stimulators for Parkinsonian patients. Used in conjunction with implantable lasers or LEDs, optogenetic reagents may enable ultra-precise optical manipulations of defined cell types in the brain, thus resulting in powerful treatments with minimal side effects. In 2009, ChR2 was expressed in the nonhuman primate brain, and no cell death or immune reaction was found to result from expression of this algal protein in neurons that are similar to human neurons. As of 2010, a number of academic and corporate entities are exploring translational use of opsins (the protein component of the visual receptor pigment rhodopsin) as a potential gene-device combination therapy for human disease.

Other optogenetic strategies. Optogenetics can be used in cells other than neurons, and used to drive signaling pathways other than just ionic localization. For example, a small protein from the protist Euglena can be used to drive the production of the important signaling molecule cyclic AMP (cAMP) in response to light. Another example is the use of a blue-light-sensitive plant protein called FKF1, which normally binds another protein in response to light, to drive the expression of a specific gene in response to light. Still another example is the use of a blue-light-sensitive plant protein coupled to a cytoskeletal remodeling protein, so that when it is expressed in a cell and illuminated, the cell cytoskeleton changes and the cell moves. These tools can be used to trigger very precise events in cells in response to pulses of light, enabling scientists to study the downstream effects in a spatially and temporally precise way. For example, triggering a precise cascade in a cancer cell would allow scientists to see what happens downstream—for example, does it result in cell death?—and then, if the consequences are interesting, further analysis of the cascade may yield mechanistic insight into how to develop a new cancer therapy, perhaps resulting in discovery of novel drug targets.

Chemically augmented optogenetics. Some optogenetic strategies require chemical cofactors. For example, a strategy that has been used in the fruit fly *Drosophila* is the use of ultraviolet light-labile compounds conjugated to small chemicals, in conjunction with transgenically expressed receptors for those chemicals. When ultraviolet light hits such compounds, bonds break and the chemical is released and is free to bind to the receptor, thus driving signaling downstream. As another example, a photosensitive molecule that changes its geometry in response to light can be bound to a so-called agonist, a molecule that is then tethered to a receptor; when light hits the photosensitive molecule, the agonist will flip around and bind to the receptor, activating it. This strategy can be very versatile, as almost any receptor is amenable to such manipulation, but the requirement for chemicals makes the experiments somewhat more complex than those using gene-only methods.

Optogenetics offers an unprecedented ability to enter information into and to control biological systems, enabling scientists to understand how those biological systems work. The ability to use optogenetic tools to reveal principles of normal and abnormal biological system function will enable new treatments for brain disorders and other areas of human health need.

For background information, *see* BIOPOTENTIALS AND IONIC CURRENTS; BIOTECHNOLOGY; CELL (BIOLOGY); LIGHT; NERVOUS SYSTEM (VERTEBRATE); NEURON; SOMATIC CELL GENETICS; VISION in the McGraw-Hill Encyclopedia of Science & Technology.

Edward S. Boyden

Bibliography. E. S. Boyden et al., Millisecond-timescale, genetically targeted optical control of neural activity, *Nat. Neurosci.*, 8:1263–1268, 2005; B. Y. Chow et al., High-performance genetically targetable optical neural silencing by light-driven proton pumps, *Nature*, 463:98–102, 2010; X. Han et al., Millisecond-timescale optical control of neural dynamics in the nonhuman primate brain, *Neuron*, 62:191–198, 2009; X. Han and E. S. Boyden, Multiple-color optical activation, silencing, and desynchronization of neural activity, with single-spike temporal resolution, *PLoS ONE*, 2(3):e299, 2007; F. Zhang et al., Multimodal fast optical interrogation of neural circuitry, *Nature*, 446:633–639, 2007.

Parallax and the brain

Vision serves many roles. One of the most important roles is the ability to quickly and accurately perceive the relative positions of objects as we move among them. Perhaps it is for this reason that the visual system employs a variety of different methods to perceive the relative depth of objects in a scene, often making use of parallax (the change in the apparent relative orientations of objects when viewed from different positions). Indeed, the perception of depth, that is, the ability to judge spatial relationships in three dimensions (3D), is an interesting neural processing problem because the visual system must recover the 3D from a flat, two-dimensional retinal image.

Binocular stereopsis. The best known of the processes for perceiving the relative depth of objects in a scene is binocular stereopsis, which is the ability of the visual system to detect the slight differences in object position as viewed from the two laterally displaced eyes (**Fig. 1**). While the observer perceives only a single "cyclopean" view centered on the point of fixation, each of the two eyes has a slightly different view. The static positional parallax between the two views allows the visual system to recover the relative depth of objects in the view. Mimicking these differing views to the two eyes through the use of 3D glasses allows a person to perceive vivid 3D images on an objectively flat movie or television screen. The neural processing underlying stereopsis relies on neurons in the visual system that are sensitive to the parallax differences, or disparities, between the two eyes. This means that these neurons are most responsive when an object's image falls in different locations on the two retinas. Some of these cells are sensitive to the parallax of an object nearer than the fixation point, meaning that the two images fall on relatively more lateral retinal locations. Other

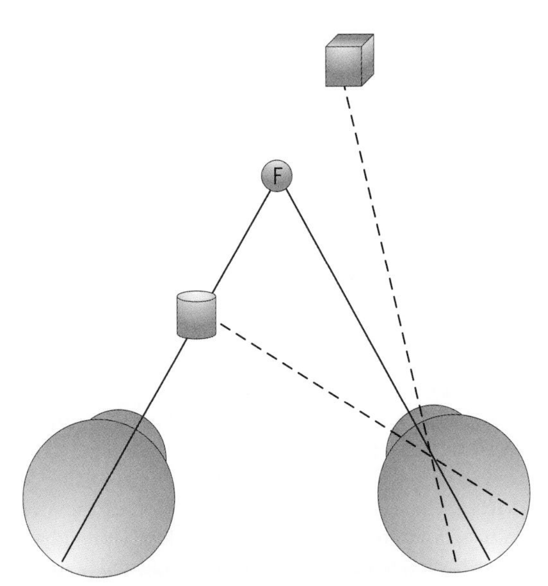

Fig. 1. Illustration of the positional parallax used for binocular stereopsis. The two eyes foveate (focus on) the same point, here indicated by F. While the retinal images of all three objects fall on the same point in the left eye, these same three objects fall on different retinal locations in the right eye.

neurons are sensitive to objects farther than the fixation point, meaning that the two retinal images fall on comparatively more nasal retinal locations.

Motion parallax. While stereopsis has been a focus of vision research for decades, many investigators, including Charles Wheatstone, who first described binocular stereopsis in the mid-1800s, recognized that movement provides another equally potent source of depth information. Even Hermann von Helmholtz, a famous physiologist writing in the late 1800s, noted that complex scenes appeared to have vivid depth when viewed by a translating observer. However, this dynamic cue to depth was difficult to study until the development of computerized displays. In fact, only recently have we begun to understand how the visual system uses parallax created by movement of the observer—motion parallax—to perceive relative depth in a scene.

Eye movements provide an important step in understanding how the brain processes motion parallax. The brain has two general types of eye movements: fast and slow. The fast, or saccadic, eye-movement system is used to change the point of gaze quickly. This system directs the foveal region of the retina, the region with the highest visual acuity, toward a point of interest in the visual scene. In contrast, as an observer translates, the slow eye-movement system generates a compensatory slow eye movement in the opposite direction to keep the fovea directed at that point of interest. The function of this slow eye-movement system is to maintain stable fixation, thereby improving visual acuity or sharpness of vision for that point. If the eyes did not move during the observer translation, the visual scene would be a blur and visual acuity would be very poor. As the observer continues to move, the saccadic eye-movement system jerks the eyes ahead to fixate on a new point, with the fixation again maintained by the slow eye-movement system.

Therefore, as a translating observer views a scene, the observer's eyes move to keep their gaze fixed on a particular point. As a result of these compensatory eye movements, the image of this fixation point remains stationary on the observer's retina; however, other objects, both nearer and farther, move on the observer's retina (**Fig. 2**). This retinal image motion is a dynamic form of parallax called motion parallax. The magnitude of the motion parallax is proportional to the object's distance from the fixation point and is in opposite directions for objects nearer and farther than the fixation point.

Extra-retinal information and pursuit signal. By itself, this retinal image motion is ambiguous with regard to whether an object is nearer or farther than the fixation point. To resolve this ambiguity, the visual system needs additional extra-retinal information, such as knowing the direction of observer translation. While a variety of possible extra-retinal signals (vestibular, kinesthetic, and proprioceptive) have been suggested, the visual system appears to use the direction of eye movement, in the form of an extra-retinal pursuit eye-movement signal, to determine which direction of retinal image motion is being generated by objects in near and far depth. Therefore, to recover the unambiguous perception of depth from motion parallax, the visual system uses this extra-retinal eye-movement signal to disambiguate the retinal image motion and determine which direction of the retinal image motion comes from objects nearer than fixation and which direction comes from objects farther than fixation.

Neurons in the primate middle temporal (MT) area, a cortical area associated with visual motion processing, appear to be the site where the retinal image motion is combined with the pursuit eye-movement information. Normally, MT neurons respond maximally to a particular direction of retinal image motion and typically do not respond to pursuit eye movements. However, the response of some MT neurons to retinal motion appears to be modulated

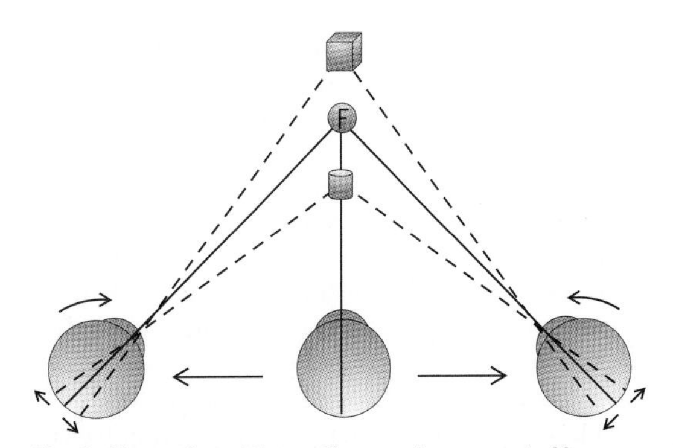

Fig. 2. Illustration of the motion parallax generated by observer translation to the left or to the right. Observer translation in opposite directions results in eye movements in opposite directions. The retinal images of objects nearer and farther than the fixation point (F) move in opposite directions, and reverse with opposite directions of observer translation.

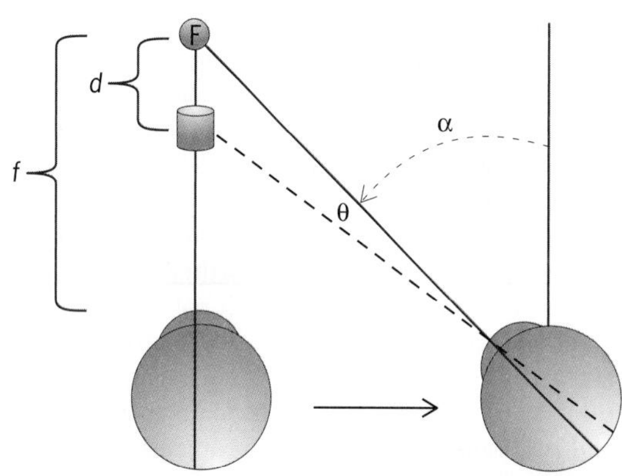

Fig. 3. Illustration of the stimulus parameters for motion parallax generated by observer translation to the right. The viewing distance to the fixation point (F) is given by the value f; α denotes the angle of the eye; angle θ is the retinal position of the object; d is the relative depth.

by the direction of the pursuit signal, with one direction of pursuit increasing the response to retinal motion and the other direction of pursuit decreasing the response. This response signals a depth-sign selectivity for motion parallax that is based on an extra-retinal pursuit signal.

The visual system uses both the retinal image motion and the pursuit signal to generate an estimate of perceived depth magnitude in a rather simple way (**Fig. 3**). The viewing distance to the fixation point (F) is given by the value f. The symbol α denotes the angle of the eye, and, more important, $d\alpha/dt$ gives the change in the angle, or the measure of the eye movement. The angle θ is the retinal position of an object, and $d\theta/dt$ gives the change in relative position, or the retinal image motion of the object. To determine the relative depth (d) of an object displaced from the fixation point, the visual system appears to use the formula

$$d/f \approx d\theta/d\alpha \quad \text{or} \quad d \approx (d\theta/d\alpha)f$$

where $d\theta/d\alpha$ provides a ratio used to scale the viewing distance (f) to provide an estimate for relative depth (d). This is similar to binocular stereopsis, where viewing distance is also used to scale relative depth from retinal disparity.

Underlying neural mechanisms. The similarities between motion parallax and binocular stereopsis are also found in the underlying neural mechanisms. The mechanisms that serve motion parallax and binocular stereopsis appear to interact; if an observer adapts to one type of stimulus, aftereffects of this adaptation are observed when viewing another type. This interaction suggests that the two systems share, at some level, a common neural substrate. Research in infants suggests that motion parallax and binocular stereopsis develop at roughly similar times (14–16 weeks of age), with motion parallax developing a bit earlier than binocular stereopsis, but after the development of motion perception and pursuit eye movements. Perhaps the development of motion perception, pursuit eye movements, and motion parallax provides the neural foundation for these parallax-type depth cues, thus guiding the development of binocular stereopsis. This is an interesting developmental course because all of these abilities are affected by a developmental visual disorder known as esotropia (convergent strabismus, or cross-eye, which occurs when one eye fixes upon an object and the other deviates inward).

In addition, our understanding of the neural mechanisms that serve motion parallax could help explain some of the dysfunction that is tragically found with drunk driving. Alcohol intoxication is well known to affect the slow, pursuit eye-movement system, which is the basis for one type of field sobriety test. This disruption of pursuit by alcohol intoxication affects the perception of depth from motion parallax. Considering the importance of motion parallax in driving, this alcohol-induced disruption of the ability to accurately perceive and make depth judgments probably has some role in drunk driving and is an important direction for further study.

For background information *see* BRAIN; EYE (VERTEBRATE); NEUROBIOLOGY; OPTICS; PERCEPTION; PSYCHOLOGY; STEREOSCOPY; VISION in the McGraw-Hill Encyclopedia of Science & Technology.

Mark Nawrot

Bibliography. I. P. Howard, *Seeing in Depth*, vol. 1, Porteous, Toronto, 2002; I. P. Howard and B. J. Rogers, *Seeing in Depth*, Vol. 2, Porteous, Toronto, 2002; J. W. Nadler et al., MT neurons combine visual motion with a smooth eye movement signal to code depth sign from motion parallax, *Neuron*, 63:523–532, 2009; E. Nawrot, S. Mayo, and M. Nawrot, Development of motion parallax in infancy, *Atten. Percept. Psychophys.*, 71:194–199, 2009; M. Nawrot and L. Joyce, The pursuit theory of motion parallax, *Vis. Res.*, 46:4709–4725, 2006; M. Nawrot and K. Stroyan, The motion/pursuit law for visual depth perception from motion parallax, *Vis. Res.*, 49:1969–1978, 2009.

Pedestrian protection

Pedestrian protection has been overlooked for a long time, as motor vehicles have been favored over other road users. Vehicle safety was synonymous with occupant protection, and separating pedestrians from vehicle traffic was the only approach to protect vulnerable road users. However, the situation has changed, and pedestrian protection has gained more importance because of legislative and consumer testing requirements. The car occupant often is the custumer, and in most cases does not consider the role of a vulnerable road user when buying a car. Based on these considerations, including the relevance of pedestrians in accidents, the focus on pedestrian protection for passenger vehicles is clearly driven by public authorities.

Accident statistics show the need for measures to protect vulnerable road users, especially pedestrians, who account for approximately 15% of all road fatalities in Europe and 35% in Japan. Since the

European Directive and the Japanese Regulation on Pedestrian Protection became effective in 2005, new vehicle models have to fulfill mandatory pedestrian protection requirements. Consumer testing in Euro New Car Assessment Programme (Euro NCAP) and Japan New Car Assessment Program (JNCAP) had already considered pedestrian protection tests before the legislation was introduced. Efforts toward harmonization have led to a global technical regulation, which is the basis for worldwide approval regarding pedestrian protection requirements for future cars. The testing protocols prescribe the use of subsystem tests based on the developments of the European Enhanced Vehicle-safety Committee (EEVC) Working Group 17, with free-motion impactors representing the human head as well as the upper and lower leg. This work originated from the studies done in the 1980s.

Innovative active safety systems help to reduce the impact velocity and prevent accidents. Nevertheless, passive safety systems are required for protecting pedestrians to mitigate injuries in the case of an unavoidable accident.

In a vehicle to vulnerable road user accident scenario, the primary impact, which consists of the interaction of the human with the vehicle, is followed by the secondary impact on the ground. In the primary impact, there are three characteristic impact areas on the car's front. In a typical collision with a pedestrian, the legs first hit the bumper and then the body starts wrapping around the hood's leading edge, and finally the head impacts the hood or the windshield (windscreen) area (**Fig. 1**).

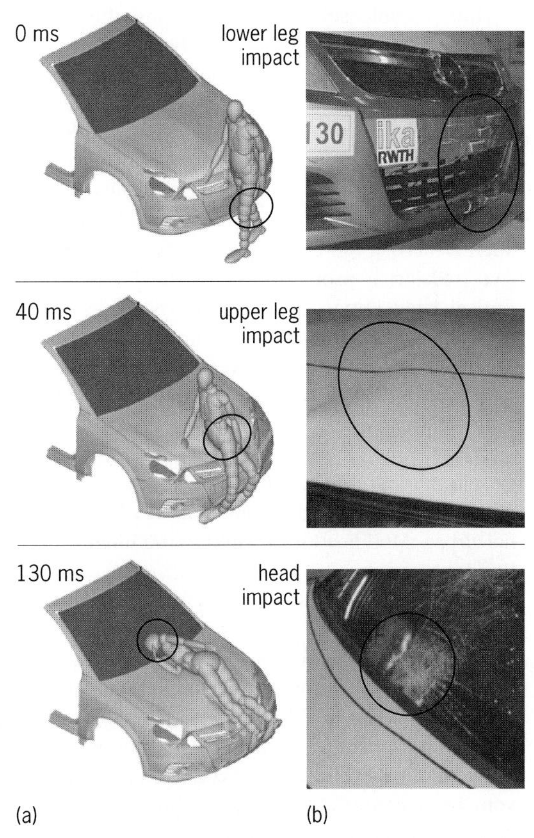

Fig. 1. Vehicle and vulnerable road user impact scenario in a real-world accident. (*a*) Simulation. (*b*) Hardware test.

Providing protection zones for the relevant body parts requires intelligent solutions because of the demanding targets related to the human biomechanical limits. In general, these solutions for mitigating accident consequences focus on providing more crush space between the vehicle's outer skin and the hard components, such as the engine block. To attain this goal, there are two approaches: passive and deployable systems. Passive-only systems include adjusting the stiffness of the hood and surrounding parts in the relevant pedestrian impact area and lowering clearance of the hard engine components. Deployable systems then are applied if the passive-only solutions are not sufficient to fulfill the protection requirements. Most likely, a combination of deployable systems and passive measures will be needed for effective protection.

The development of deployable protection systems, such as pop-up hoods or external airbags, will gain more importance because of the tightening demands of legislation and consumer testing within the coming years. This will mean higher investments for both developing the systems and implementing them in vehicles, including the repair costs in the case of a false deployment.

Advanced vehicle systems. Designing efficient vehicle systems for vulnerable road user protection is a challenge, with conflicting requirements. Active safety systems, focusing on the accident prevention, make use of pedestrian recognition, driver warning systems, or brake assistance. However, these systems should be supported by passive safety systems.

There has been a significant improvement in passive safety measures during the last few years. Purely passive solutions imply material or geometry optimizations, which are preferred in terms of simplicity and cost. Deployable systems such as active bumpers add a level of protection for the legs, and pop-up hoods are designed to protect the torso and the head. In 2005, pop-up hoods were first implemented in European vehicles such as the Citroën C6 and Jaguar XK. The portion of vehicles equipped with such systems is increasing. However, benefits of these systems should be questioned carefully because of the possibility of new design problems being introduced, such as the exposed sharp rear edge of the hood and activation timing.

In an accident, the emphasis for improving the survival of the victim is on the critical head impact. From the resultant acceleration of the head, the head injury criterion (HIC) value is calculated, which determines the severity of the impact. The HIC considers the average acceleration as well as acceleration peaks to the 2.5th power multiplied by the corresponding impact time. An HIC value below defined threshold values with a certain safety margin and minimum deformation space is the goal for original equipment manufacturers (OEMs). To achieve this, an optimum acceleration pulse is necessary.

One possible solution for reducing the impact severity could be a U-shaped airbag system, combined with a pop-up hood function to increase the deformation space. Such a system has the advantage of protecting various body parts. Both the additional deformation space in the hood area and coverage of the hard structure, especially the A-pillars (vertical supports for a car's windshield), can be realized. Such an airbag system is able to offer comprehensive pedestrian protection, with the focus on the critical head impact region. The system can even be optimized by integrating an impact plate, forming a smooth transition between the hood and the windscreen, and absorbing the head impact energy as desired.

The overall influence of these advanced systems is shown by upgrading an average-shaped sedan vehicle with a pop-up hood in combination with an external airbag. Modifications related to the hinges in the hood rear area, the latch in the front, and the integration of the airbag module are necessary. When fully deployed, the airbag volume is approximately 90 L (23.8 gal). High-speed sequences of dummy tests at the Institute for Automotive Engineering [(Institut für Kraftfahzeuge RWTH Aachen University (IKA)], at Aachen University, Germany, with an unmodified and a modified vehicle using the airbag system at a velocity of 40 km/h (25 mph) are compared in **Fig. 2**. Head and neck loads could be reduced by approximately 50% because of the larger deceleration space and lower impact acceleration level of the airbag system.

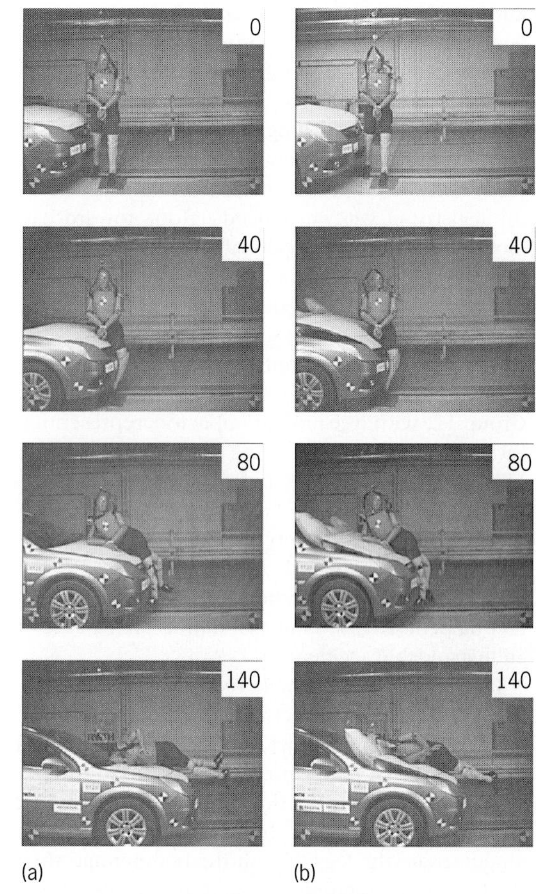

Fig. 2. Polar-II Dummy tests at IKA with series and modified vehicle, *t* in microseconds. (*a*) Unmodified vehicle. (*b*) Modified vehicle.

Current and future testing methods. Presently, the progress of vehicle systems for vulnerable road user protection is proceeding faster than the development of their corresponding test procedures. Current testing procedures are more than 20 years old, without major adjustments for the changes in vehicle fronts, and the strong competition in the safety area among the OEMs are reasons for this. Testing procedures need be updated to reflect the technological improvements in the protection solutions and vehicle fronts. Future vehicles will be designed based on the testing procedures, so that the challenge of such a testing is to bring the real-world scenarios into the test lab. Testing has to be robust, repeatable, and feasible, including costs, too.

Current hardware testing is done by using linear head and leg impactor tests according to the existing procedures (**Fig. 3*a***). Injury criteria that are specific to pedestrian body parts are used to compare the protection potentials of the vehicles.

Today's approach, as a form of advanced testing, is a combination of hardware and virtual testing. The next step is to take into account the different front shapes of the vehicles, which lead to different impact kinematics. For virtual testing, usually the impact scenario of the full-scale pedestrian is simulated. In multibody or in coupled vehicle-pedestrian simulations that analyze the kinematics (Fig. 1*a*), all relevant values can be determined concerning the impact velocity, angle, timing, and location of the body parts relative to the vehicle. Afterward, hardware investigations are done using the previously determined boundary conditions (Fig. 3*b*).

New testing procedures and subsystem tests are being investigated to improve the biofidelity. For higher front vehicles, such as SUVs, an upper-body mass is added to a flexible legform impactor (Flex-PLI) to improve the kinematics. For head motion, a rotational trajectory or an eccentric neck mass is used to better reflect the dependency of the head on the body motion as well as the head-neck interaction (Fig. 3*c*). Additional dummy tests with the Polar-II Pedestrian Dummy can be conducted to test real-world system behavior (Fig. 2). This average (50th percentile) male dummy is the most advanced physical representation of a pedestrian worldwide and may be very beneficial for research work.

Outlook. Pedestrian protection constitutes a design problem with challenging targets because of significant goal conflicts with other requirements, such as occupant protection, when providing a deformation zone in the windscreen region. Certain front-end design constraints that are fundamental to everyday driving performance as well as easy maintenance and repair are demanded for passenger cars (for example, hood stiffness). Considering these requirements, intelligent systems, including accident prevention, structural measures, and deployable

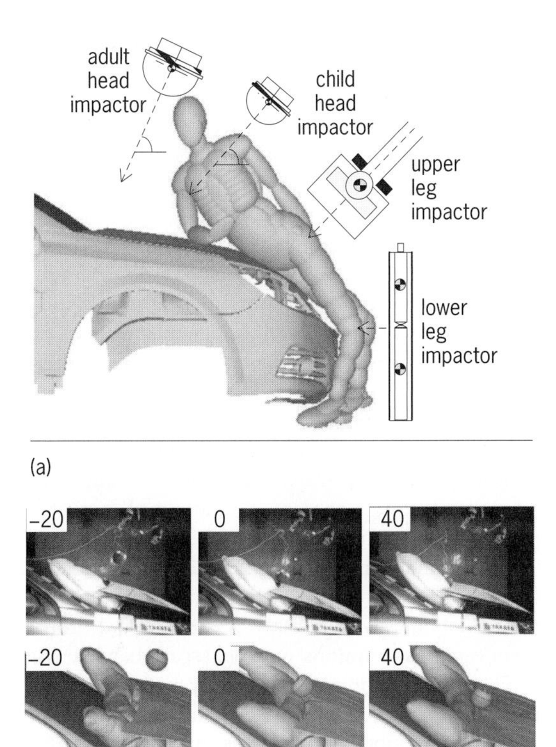

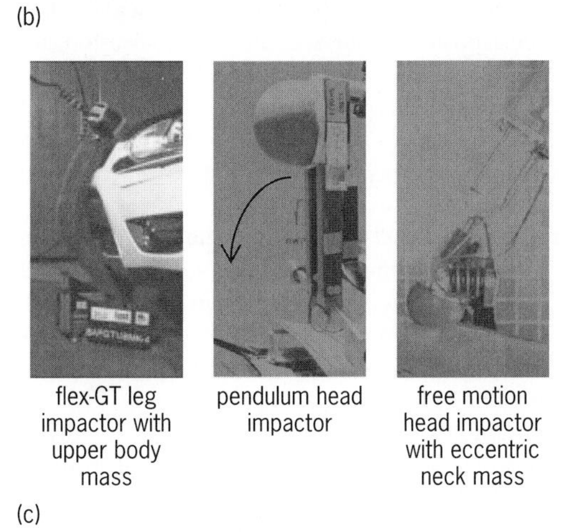

Fig. 3. Current and future subsystem tests. (*a*) Current test methods. (*b*) Advanced testing. (c) Future testing.

deformation space-increasing methods (active bumpers, pop-up hoods, and airbags) are currently being studied. Some of these solutions have already been applied to production vehicles. From the customer's point of view, today's vehicles offer a considerable level of pedestrian protection and are much safer than were vehicles in the past. However, continuous improvement in this area can only be achieved by forcing legislation and developing the necessary test procedures related to these laws. In the future, the efficiency and feasibility of these measures have to be monitored and proven, for example, by accident reconstruction and statistics.

For background information *see* AUTOMOBILE; BIOMECHANICS; HUMAN-FACTORS ENGINEERING; KINEMATICS; VIRTUAL MANUFACTURING in the McGraw-Hill Encyclopedia of Science & Technology.

Jens Bovenkerk

Bibliography. J. Bovenkerk et al., *Benefits of New Testing Methods for Deployable Pedestrian Protection Systems*, Runner-Up Award Winner of 3rd ESV Student Safety Technology Design Competition, Stuttgart, Germany, 2009; J. Bovenkerk, *New Modular Assessment Methods for Pedestrian Protection in the Event of Head Impacts in the Windscreen Area*, Paper no. 09-0159 of 21st ESV Conference Proceedings, Stuttgart, Germany, 2009; EEVC Working Group 17 Report, *Improved Test Methods to Evaluate Pedestrian Protection Afforded by Passenger Cars*, 1998, update 2002; I. Kalliske and J. Bovenkerk, *New and Improved Test Methods to Address Head Impacts, Deliverable D333C* (Rep. no. AP-SP33-021R) of the European FP6 Research Project on Advanced Protection Systems (APROSYS), 2009; T. Kinsky, *Pedestrian Protection as Challenge for Modern Automotive Engineering*, 18th Aachener Kolloquium Fahrzeug-und Motorentechnik, Aachen, Germany, 2009; OECD, International Road Traffic and Accident Database, 2005; United Nations Economic Commission for Europe (UNECE) Global Registry, *Agreement Concerning the Establishing of Global Technical Regulations for Wheeled Vehicles, Equipment and Parts Which Can Be Fitted and/or Be Used on Wheeled Vehicles*(ECE/TRANS/132 and Corr.1)—Addendum: Global technical regulation no. 9 Pedestrian Safety ECE/TRANS/180/Add.9 and /Add./Appendix 1, 26 January 2009; O. Zander et al., *Evaluation of a Flexible Pedestrian Legform Impactor (Flex-PLI) for the Implementation within Legislation on Pedestrian Protection*, Paper No. 09-0277 of 21st ESV Conference Proceedings, Stuttgart, Germany, 2009.

Pelagic ecosystem recovery after end-Permian mass extinction

During the last 540 million years, the history of life on Earth has been punctuated by severe mass extinctions, brief periods when biodiversity collapsed. At the boundary between the Paleozoic and Mesozoic eras (approximately 252 million years ago), the end-Permian mass extinction was the most devastating global-scale event ever recorded, resulting in the loss of more than 90% of marine species. During long and intense debate about the potential causes of this major crisis, considerable effort has been devoted to investigating the unusual ecological patterns that arose during its aftermath. Until recently, several studies suggested that the biosphere took between 5 and 30 million years to reach the levels of biodiversity seen before the extinction. However, recent findings on pelagic organisms such as ammonoids and conodonts indicate a rapid and explosive recovery in less than 2 million years. This duration contrasts radically with patterns observed for benthic biota (living on or within the sea floor), which are based mainly on fragmentary data. It also suggests that recovery

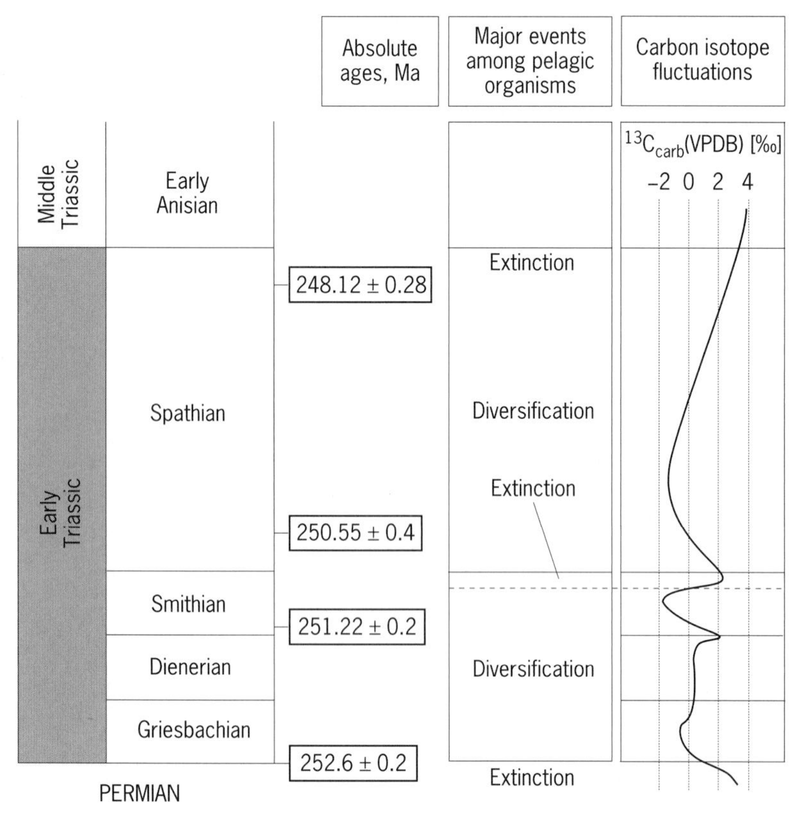

Fig. 1. Chronostratigraphic subdivisions of the Early Triassic with absolute ages and simplified trends for pelagic recovery and geochemical fluctuations during this time interval. The Early Triassic is commonly subdivided into four substages (Griesbachian, Dienerian, Smithian, and Spathian), with well-defined boundaries in terms of ammonoid and conodont events. For instance, the beginning of the Smithian corresponds to the appearance of several new ammonoid families, and the end-Smithian coincides with a marked extinction event. Ma = million years ago.

rates for numerous taxa need to be reevaluated and that ecosystem reorganization may occur rapidly after a mass extinction.

Early Triassic aftermath. All mass extinctions are followed by phases during which the surviving species recover and diversify. It has usually been assumed that the end-Permian mass extinction affected ecological assemblages so deeply that the postcrisis recovery spanned at least the entire Early Triassic (about 5 million years; **Fig. 1** shows the chronostratigraphic subdivisions of this period and their respective durations). This time interval is notorious for its low biodiversity and the unusual proliferation of opportunistic and generalist taxa. The recent discovery of successive carbon isotope anomalies suggests severe fluctuations of the global carbon cycle throughout the Early Triassic. In addition, it appears that some of these geochemical events may have been concomitant with marked climate shifts, for example, those seen during the end of the Smithian substage. Consequently, although it has not been totally proved, potential recurrent or prolonged harsh environmental conditions, including, for instance, anoxia (oxygen deficiency), euxinia (an environment of restricted circulation and stagnant or anaerobic conditions with high levels of hydrogen sulfide), collapse of food webs, and productivity decline, are generally admitted and proposed for the Early Triassic.

Fig. 2. Ammonoid (*a*) and conodont (*b*) fossils from the Early Triassic. (*Conodont three-dimensional scan; courtesy of Nicolas Goudemand, Universität Zürich, Zürich, Switzerland*)

During the Early Triassic, benthic marine taxa often exhibited unusual abundance patterns or reduced body size, possibly related to the aforementioned assumptions of global-scale harsh environmental conditions. This is why benthic organisms, comprising mainly brachiopods and mollusks such as gastropods and bivalves, have largely influenced the classical model of a delayed recovery after the end-Permian mass extinction. However, data on their postcrisis survival and diversification are sparse, frequently truncated, and based on imprecise dating. Furthermore, until recently, a persistent problem in studies of extinction selectivity and recovery patterns was that no absolute ages were used to calibrate these events.

Explosive rebound of pelagic organisms. Taxa living in the water column of the open ocean are called pelagic and are very different from benthic organisms in both appearance and way of life. Ammonoids and conodonts predominated among Early Triassic open-marine biotas (**Fig. 2**).

Ammonoids represent an abundant, highly diversified, and geographically widespread fossil group of marine cephalopods with an external shell (Fig. 2*a*). Present-day coleoids, including octopus, squids, and cuttlefishes, are closely related to this now totally extinct group. Ammonoids were abundant throughout the Permian, but they were nearly wiped out during the end-Permian mass extinction. Only two or three species survived, and a single species may have been the stem group of their postcrisis diversification. Significant progress has been made recently by researchers using diversity analyses combined with absolute age calibrations, which show that Triassic ammonoids diversified rapidly in the first million years after the extinction (**Fig. 3**).

Obviously, times for recovery varied widely across marine groups: ammonoids rediversified long before the oceanic realm returned to a steady state. Based on newly available absolute ages, ammonoids reached diversity values equal to, if not higher than, Permian values during the Smithian, only 1–2 million years after the mass extinction (Fig. 3). Still higher diversity values followed during the Spathian substage.

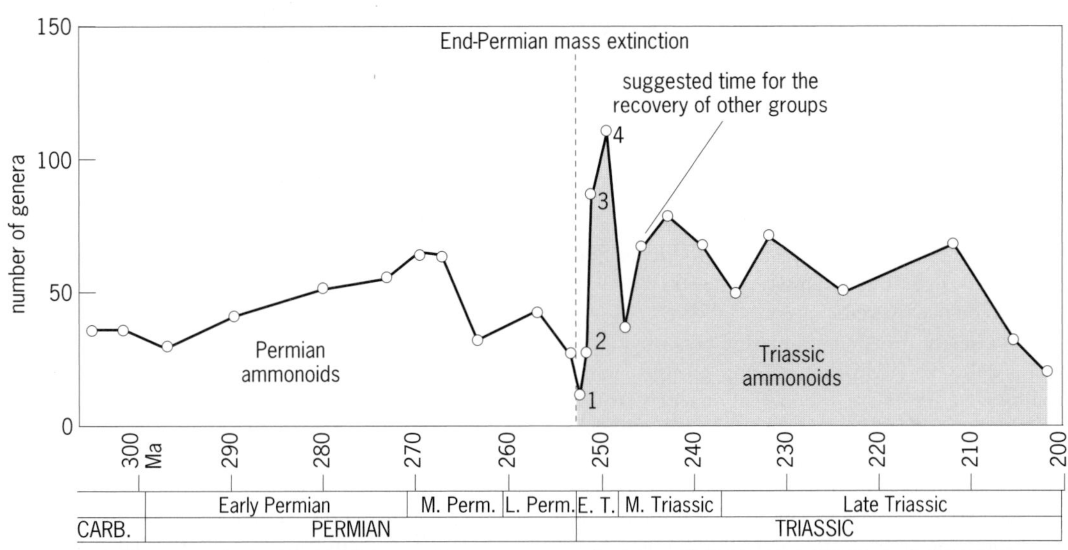

Fig. 3. Total number of ammonoid genera for the Permian and Triassic (Carb. = Carboniferous; M. Perm. = Middle Permian; L. Perm. = Late Permian; E. T. = Early Triassic; M. Triassic = Middle Triassic; 1 = Griesbachian; 2 = Dienerian; 3 = Smithian; 4 = Spathian). Note the peak diversity value during the short Early Triassic time interval. The end-Smithian extinction event cannot be shown at this temporal scale.

They were unsurpassed during Middle and Late Triassic times, when diversity oscillated around the previous Permian maximum value. Thus, by producing more than 200 genera in less than 5 million years, this explosive Early Triassic ammonoid diversification clearly contrasts with the hypothetical delayed recovery suggested for benthic groups. This extraordinary diversification also was accompanied by the formation of a marked latitudinal diversity gradient during most of the Smithian and Spathian substages, as well as by a progressive change from cosmopolitan to latitudinally restricted distributions of taxa. Ammonoids were highly sensitive to temperature, and shifts of their geographical distributions mainly reflect fluctuations in temperature belts. Therefore, these data are interpreted as an increase in the latitudinal temperature gradient during the Early Triassic. However, this trend was neither a continuous nor a gradual process: It was interrupted at least once during the end-Smithian. During this brief event, ammonoid diversity value fell and distributions became once more cosmopolitan, coinciding with a major short-term disturbance in the global carbon cycle. Remarkably, this event did not delay Early Triassic ammonoid diversification.

Conodonts are also thought to have been a major component of Early Triassic pelagic ecosystems. As with ammonoids, they are now totally extinct. Conodonts are often considered to have been very small jawless vertebrates that resemble modern eels. They are known mostly in the fossil record from teeth-like elements (Fig. 2*b*). They may have been less affected by the end-Permian mass extinction, showing only a gradual decline in families and genera. Interestingly, a recent study indicated that their diversity and high evolutionary dynamics tend to parallel the ammonoid rebound and probably recorded the same global environmental changes. This study also established that conodonts reached a first maximum diversity value during the Smithian, followed by a marked extinction at the end of that geological substage. They then diversified during the Spathian, but their diversity gradually declined from the late Spathian. They disappeared finally during the end-Triassic mass extinction, around 201 million years ago.

Early Triassic pelagic recovery: lessons and prospects. The explosive recovery of pelagic organisms prompts further scientific investigations. One of the major causes speculated for the end-Permian mass extinction is volcanic degassing by large continental flood basalts (high-temperature basaltic lava flows from fissure eruptions that accumulate to form a plateau) called traps, presently located in Siberia. It has been argued that a late eruptive phase of the Siberian traps probably also triggered the marked event at the end of the Smithian substage; this brief time interval corresponds to major ammonoid and conodont extinctions and marked changes in the global carbon cycle and climate. Consequently, a multiphased geological scenario with successive marked radiations and extinctions can be proposed for the Early Triassic recovery.

From a biological point of view, a major question remains: How did pelagic organisms, unlike benthic taxa, flourish in the presumably unstable and harsh environmental conditions prevailing during the Early Triassic? Even if ammonoid and conodont modes of life are still poorly known, they were so diverse both morphologically and taxonomically that it is probable that they occupied many ecological niches and exploited a great variety of food resources. Moreover, numerous benthic and pelagic taxa known in the Permian seem to have disappeared during the mass extinction, only to reappear in the fossil record during the Smithian. Paleontologists frequently refer to these taxa as Lazarus taxa. Coupled with the nondelayed diversification of pelagic organisms, this suggests that Early Triassic trophic webs based on abundant and diversified primary producers, such as algae, were far from devastated and were already functioning less than 2 million years after the mass extinction. Additionally, it opens up the possibility that organisms other than ammonoids and conodonts also recovered rapidly, even in the benthic realm, which is only fragmentarily known. Unfortunately, the Early Triassic primary producers that would have made this explosive recovery possible are yet to be discovered.

Recoveries of pelagic and benthic taxa obviously exhibit different environmental and group-specific dynamics. Nevertheless, the indications are that the duration of the recovery is probably overestimated, at least for some marine taxa. The modern biosphere is also probably heading toward a new mass extinction, and these discoveries remind us that the recovery of surviving species after an extinction is a very long process at the human scale, taking several tens of thousands of generations at the very least, even for pelagic species.

For background information *see* BIODIVERSITY; CEPHALOPODA; CONODONT; ECOSYSTEM; EXTINCTION (BIOLOGY); FOSSIL; GEOLOGIC TIME SCALE; MARINE ECOLOGY; PALEOECOLOGY; PERMIAN; POPULATION ECOLOGY; TRIASSIC in the McGraw-Hill Encyclopedia of Science & Technology.

Arnaud Brayard

Bibliography. A. Brayard et al., Good genes and good luck: Ammonoid diversity and the end-Permian mass extinction, *Science*, 325:1118–1121, 2009; D. H. Erwin, *Extinction: How Life on Earth Nearly Ended 250 Million Years Ago*, Princeton University Press, Princeton, NJ, 2006; T. Galfetti et al., Timing of the Early Triassic carbon cycle perturbations inferred from new U-Pb ages and ammonoid biochronozones, *Earth Planet. Sci. Lett.*, 258:593–604, 2007; M. J. Orchard, Conodont diversity and evolution through the latest Permian and Early Triassic upheavals, *Palaeogeogr. Palaeoclim. Palaeoecol.*, 252:93–117, 2007; J. L. Payne et al., Large perturbations of the carbon cycle during recovery from the end-Permian extinction, *Science*, 305:506–509, 2004.

Permanent-magnet synchronous machines

Tremendous advances in the development of permanent-magnet synchronous machines have taken place over the past two decades, particularly in high-efficiency interior-permanent-magnet (IPM) motors and associated drive technology. A close examination reveals that different technological advancements and market forces have combined, sometimes in fortuitous ways, to accelerate the development of permanent-magnet synchronous machines. In a conventional synchronous machine, an electromagnet is produced by direct current passing through windings placed on the rotor of the machine. In permanent-magnet (PM) machines, the electromagnet is replaced by modern hard permanent magnets with good magnetic properties, such as neodymium boron iron (NdBFe). The PM synchronous machines are available as both PM generators and PM motors. Because of their use in modern line-start as well as soft-start inverter-fed ac (alternating-current) motor drives in intensive energy-consuming devices such as air conditioners and electric and hybrid-electric vehicles, as well as heavy-duty-cycle ventilation fans and pumps, IPM synchronous motors can readily benefit an energy-hungry world.

Magnet placement. Permanent magnets may be placed in the rotor in one of three ways: surface-mounted, inset in the rotor, or embedded inside the rotor. The last type is called an interior-permanent-magnet (IPM) synchronous machine. Each type has its advantages and disadvantages in terms of design, application, and efficiency. Since the relative permeability of the surface-mounted type is almost equal to that of the air, it behaves like a round-rotor synchronous machine. It produces electrical voltage and power when the machine is acting as a generator, or torque when the machine is acting as a motor. On the other hand, a rotor of either of the other two types behaves as if it has distinct poles (that is, as a salient-pole synchronous machine). These other two types produce voltage and two components of power (or torque) by varying the inductances of the machine in the air gap. One component of the developed torque is called the magnet torque, and results from the permanent-magnet excitation. The second component of torque is called the reluctance torque, and results from the variation of the air-gap reluctances. The round-rotor machine produces only magnet torque or power.

IPM motor design. At starting, motors draw large currents when connected directly across the normal voltage supply. Motors are usually started using a reduced voltage; otherwise a motor started directly across the supply is referred to as a line-start motor. **Figure 1** shows the rotor of a three-phase,

Fig. 1. Rotor of an interior-permanent-magnet (IPM) machine. (*K. Kurihara and M. A. Rahman, High-efficiency line-start interior permanent-magnet motors, IEEE Trans. Ind. Appl., 40(3):789–796, 2004*)

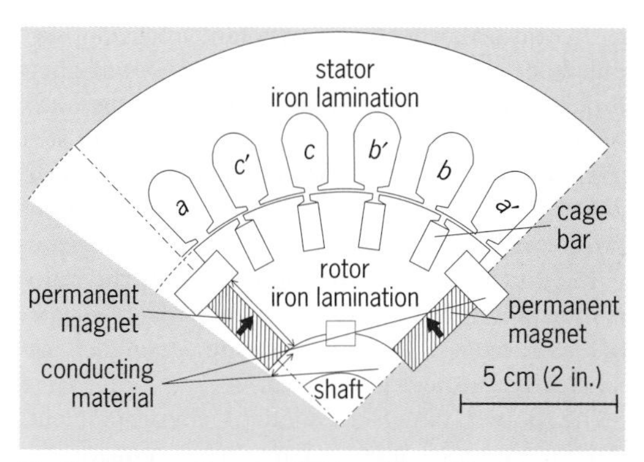

Fig. 2. Cross section of an IPM machine with magnets in V formation. Arrows on permanent magnets indicate the direction of magnetization. Axial length of the machine is 70 mm (2.8 in.). (*K. Kurihara and M. A. Rahman, High-efficiency line-start interior permanent-magnet motors, IEEE Trans. Ind. Appl., 40(3):789–796, 2004*)

four-pole, line-start, IPM motor. A one-quadrant cross section of the IPM rotor of Fig. 1, with a set of paired neodymium boron iron permanent magnets in V formation, is shown in **Fig. 2**. The top and bottom conducting materials reduce intermagnet leakage flux. This V-shaped IPM machine develops two components of torque. Three-phase distributed winding coils are inserted in slots in the stator. Thus, a represents the beginning of the stator distributed coil in phase a, and a' is the end of that coil; similar notation is used for phases b and c. The rotor of an IPM motor is designed with permanent magnets embedded below the conduction cage bars. The magnet polarity is oriented in a partial V shape within the IPM rotor such that it creates variations of machine inductance along the direct (d) and quadrature (q) axes of the machine. Thus, the machine behaves like a permanent-magnet-excited, salient-pole synchronous machine, without the physical variation of the air gaps that is characteristic of a conventional salient synchronous machine.

Applications of PM synchronous generators. Applications of large PM synchronous generators in electric utility systems are still limited, but these generators are widely used for emergency, standby, and isolated power supply sources. There is a recent trend to use surface-mounted PM rotors in large, low-speed wind generators. The advantages of surface-mounted synchronous PM machines include easy mounting of the permanent magnets on the rotor surface, where they can be held in place by an adhesive retainer or a stainless-steel sleeve. Permanent-magnet ac generators are used as the starter/alternator in standard (gasoline-powered) automobiles.

In recent years, IPM generators have been used to charge the on-board battery modules in electric and hybrid-electric vehicles, and as a 60- or 50-Hz ac power source for various electrical loads for modern cars. **Figure 3** illustrates the arrangement of the IPM motor and IPM generator in hybrid-electric vehicles. The IPM motor is paralleled with the internal-combustion engine via a hybrid transmission system. The IPM generator is also mounted on the same front axle to charge the on-board car battery, and for other auxiliary purposes.

IPM synchronous motor applications and performance. In contrast to the limited scope of IPM synchronous generators, IPM synchronous motors are utilized in energy-efficient ac motor drives. (A motor with hardware control accessories is called a motor drive.) These IPM motors are becoming widely accepted in applications that require high performance in order to succeed in the competitive motor drive market for quality products and improved services. This is because of their advantageous features, such as high torque-to-current ratio, high power-to-weight ratio, high efficiency with high power factor, low noise, and robustness. These motors are popular because they overcome the limitations of both the ac induction motor and conventional synchronous motors. The control of an IPM synchronous motor is relatively simple. Furthermore, it meets almost all the requirements for modern high-performance industrial motor drives. Although there are a few challenges in designing and operating an IPM synchronous motor, they have been successfully met over the past 25 years, resulting in the emergence of modern IPM motors.

The control and operation of modern IPM synchronous motor drive systems lie at the core of high-performance and efficient IPM motor technology for contemporary large-scale industrial applications. This is because the IPM synchronous motor is a hybrid machine, in which the total motor-developed torque consists of the permanent-magnet torque component and the reluctance torque component. It is fundamentally a salient-pole-type PM

Fig. 3. IPM motor and IPM generator of hybrid-electric vehicle. (*M. Kamiya, Development of traction drive motors for Toyota hybrid system, Proceedings of International Power Electronics Conference (IPEC 2005), Niigata, Japan, April 4–8, 2005*)

Performance comparisons of IPM motor and induction motor		
Motor type	IPM motor	Induction motor
Input voltage (V_i), V	130	200
Input current (I_i), A	3.11	3.43
Input power (W_i), W	687	818
Rotor speed (n), rpm	1500	1434
Torque (T), N · m	3.82	4.00
Efficiency (η), %	87.3	73.3
Power factor (pf), %	98.1	68.8
Output power (P_o), W	600	600
Efficiency × power factor product, %	85.6	50.4
Maximum output (P_{max}), W	960	1240

synchronous motor, and hence it produces more power. This leads to more output torque. The performance of the line-start IPM synchronous motor of Figs. 1 and 2 is presented in the **table**. A comparison of the performances of the 4-pole, 3-phase, 600-W-output IPM synchronous motor and a standard 4-pole, 3-phase, squirrel-cage induction motor with the same 600-W output shows that the IPM synchronous motor yields results superior to those of the induction motor in each category except developed torque and maximum output power. The efficiency and efficiency-power-factor product of the IPM synchronous motor are 14% and 35% higher, respectively, than those of the induction motor. Furthermore, the 30% improvement in power factor of the IPM motor is quite beneficial for the ratings of the inverter used for soft-starting purposes. One of the most successful applications of IPM synchronous motors lies in their use in the compressor pumps of modern energy-efficient air conditioners.

Figure 3 illustrates the application of the IPM machines in commercial parallel hybrid-electric vehicles as both the traction motor and the recharging generator. **Figure** 4 depicts the special feature of the hybrid IPM synchronous machine in developing reluctance torque in addition to permanent-magnet torque. The reluctance torque was 54% of the total traction torque required per unit vehicle weight at a maximum speed of 6000 rpm for the 2000 model of a hybrid-electric vehicle, and 60% of the total traction torque per unit vehicle weight at a maximum speed of 13,500 rpm for the 2010 model. Thus, the ratio of reluctance torque to permanent-magnet torque increased from 1.17 to 1.50 for newer models of parallel hybrid-electric vehicles.

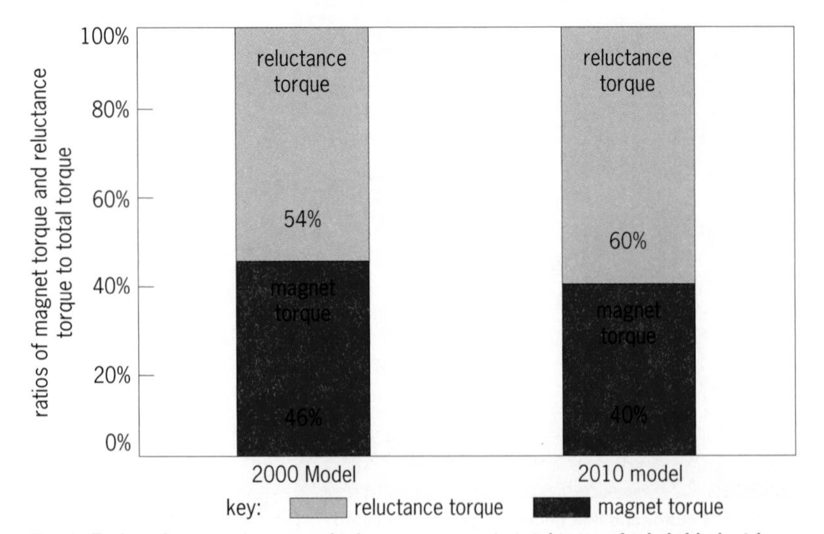

Fig. 4. Ratios of magnet torque and reluctance torque to total torque for hybrid-electric vehicle (Toyota Prius). (*M. Kamiya, Development of traction drive motors for Toyota hybrid system, Proceedings of International Power Electronics Conference (IPEC 2005), Niigata, Japan, April 4–8, 2005 for 2000 model*)

The power ratings of modern energy-efficient IPM synchronous motors have been dramatically expanded by three orders of magnitude during the past 2 decades. IPM motor technology encompasses not only more powerful hybrid IPM synchronous motors, but also the combination of intelligent power electronics modules, variable power-factor controls, direct torque control, indirect vector control, maximum-torque-per-ampere control, minimization of torque ripples, field-weakening control using new intelligent techniques, and optimization of reluctance and magnet torques with minimum losses over wide speed ranges for high-performance ac synchronous motor drives. There are many specific applications that require advanced research and development in modern permanent-magnet synchronous machine systems.

For background information *see* ALTERNATING-CURRENT GENERATOR; ALTERNATING-CURRENT MOTOR; ELECTRIC ROTATING MACHINERY; ELECTRIC VEHICLES; GENERATOR; INDUCTION MOTOR; MAGNET; MAGNETIC MATERIALS; MOTOR; SYNCHRONOUS CONVERTER; SYNCHRONOUS MOTOR in the McGraw-Hill Encyclopedia of Science & Technology.

M. Aziz Rahman

Bibliography. M. Kamiya, Development of traction drive motors for Toyota hybrid system, *Proceedings of International Power Electronics Conference* (IPEC 2005), Niigata, Japan, April 4–8, 2005; K. Kurihara and M. A. Rahman, High-efficiency line-start interior permanent-magnet motors, *IEEE Trans. Ind. Appl.*, 40(3):789–796, 2004; M. A. Rahman and G. R. Slemon, Promising applications of neodymium boron iron magnets in electrical machines, *IEEE Trans. Magn.*, 21(5):1712–1716, 1985.

Phylogeography and biogeography of fungi

The geographical distribution of organisms is shaped by a number of factors, including ecological requirements, competition with other species, and historical events. However, the historical factors that shaped the current spatial occurrence of fungi have traditionally been understudied. The scarcity of the fossil record in fungi has made the evaluation of the importance of historical events in these organisms difficult. Before molecular techniques [such as polymerase chain reaction (PCR) and cycle sequencing] dramatically increased the availability of DNA sequence data, historical factors were only rarely studied. These studies were largely restricted to a few fungi that share distribution patterns with flowering plants. Molecular data, in tandem with new statistical tools, provided an avenue for analyzing the geographical distribution of genealogical lineages in a phylogeographical framework. The term phylogeography describes the analysis of the historical processes responsible for the contemporary geographic distributions of individuals. This is achieved by studying the geographic distribution of individuals in light of the patterns associated with a gene genealogy.

Fungi. The kingdom Fungi includes a variety of heterotrophic organisms that have chitin cell walls. Although traditionally a number of unrelated groups of organisms were referred to as fungi, molecular studies have supported that several of these were unrelated, such as slime molds (Myxomycota), which are a basal group of protists, and water molds (Oomycota), which belong to Stramenopiles and are actually closer to brown algae and diatoms than to fungi. The majority of species of true fungi belong to the two crown groups, Ascomycota (sac-fungi) and Basidiomycota (mushrooms and relatives), and numerous more basal groups. Most fungi are terrestrial, but some basal groups are aquatic (some crown group species have secondarily evolved an aquatic lifestyle). Fungi can live on dead material (saprobes), parasitize other organisms, or be involved in several types of mutualistic symbioses (for example, mycorrhiza and lichens).

Traditionally, scientists believed that the "everything is everywhere" paradigm for microorganisms applied to fungi, which usually have small propagules (structures, such as spores, that aid in dispersal and propagation). This has led to a widespread notion that the distribution of fungi is shaped primarily by ecological conditions rather than being explained by historical events. However, recent studies of distribution in tandem with molecular approaches to phylogeny have made substantial advances related to our understanding of the diversity and biogeography of fungi. These studies revealed that fungi, like plants and animals, have discrete distribution patterns and that the current distribution of fungi is shaped by a complex combination of numerous factors.

Molecular data revolutionized species delimitation. In addition to the scarcity of fossils, another major reason for the misconception of widely distributed species is a severe underestimation of species diversity by morphology-based concepts of species recognition. Up to 1.5 million species of fungi are estimated to exist in nature. Since there are approximately 100,000 described species, only a small fraction is currently known. Molecular data helped to revolutionize the species delimitation in fungi. These organisms have only a limited number of morphological characters, and several studies have shown that similar structures evolved independently in these fungi, causing a high level of homoplasy (correspondence between organs or structures in different organisms acquired as a result of evolutionary convergence or of parallel evolution) of phenotypic characters among fungi. This makes it difficult, if not almost impossible, to distinguish between taxonomically informative and noninformative characters without using molecular data in a phylogenetic framework. However, understanding the species circumscription of organisms is pivotal for our understanding of biodiversity.

Distribution patterns of fungi. Indeed, the number of widely distributed species is generally higher in fungi than in plants or animals. This includes cosmopolitan species, which basically occur worldwide where suitable habitat exists. Among cosmopolitan species, a group of fungi that have an antitropical (often referred to as bipolar) distribution are often distinguished. These are species that occur in the polar regions of the Northern and Southern Hemispheres and at high altitudes in tropical mountain systems, such as the high Andes. Molecular data support that these widely disjunct (geographically discontinuous) populations belong to the same lineages. Molecular studies provided evidence that these distribution patterns, which span the whole globe, are caused by long-distance dispersal.

The impact of glaciations on European lichen flora as the explanation of eastern North American–eastern Asian disjunctions in several genera was studied in the premolecular era. However, although molecular studies indicate that there is a relatively close relationship between eastern North American and eastern Asian fungi, the relationships are not as close as suggested by morphological data.

A further group of fungi with large geographical ranges is pantropical species that occur in all tropical regions of the world. This group of fungi is less well studied, and recent studies on selected tropical fungi suggest that several of these morphologically circumscribed species represent different species. Currently, there are not enough data available to estimate the number of truly pantropical species, but it can be expected to be much lower than that suggested by the current species circumscription. Among tropical species, there are also species that, similar to distribution patterns found in animals and plants, occur only in the tropics of the Old World (paleotropical) or the Americas (neotropical).

In circumpolar fungi, species occur in polar or boreal regions throughout the Northern Hemisphere. Recent studies have supported the monophyly (development from a single common ancestral form) of at least some of the species with this distribution type, which is explained as a result of long-distance dispersal. Ongoing long-distance gene flow has been demonstrated in *Porpidia flavicunda*, a lichen occurring in the Arctic and boreal zones of the Northern Hemisphere. The lack of fixed alleles and the wide sharing of haplotypes in this species suggest recurrent long-distance gene flow of propagules.

In general, long-distance dispersal has been identified as being common to explain distribution patterns in fungi. However, our view has changed from that of traditional fungal biogeography. Although traditional views suggested that long-distance dispersal was common and explained the large geographical ranges of species, molecular data have reshaped our understanding of the impact of long-distance dispersal. Recent studies have showed that long-distance dispersal events were much rarer. Instead, the general picture emerging from numerous studies is that long-distance dispersal was often followed by speciation and diversification events on different continents and that transoceanic dispersal was limited. Hence, multiple events of dispersal and vicariance [fragmentation or splitting of one contiguous area into spatially separate (disjunct) areas by the appearance of a new barrier] are responsible for the observed patterns of geographical ranges.

A large number of studies have demonstrated that several species that were believed to have cosmopolitan distribution based on a morphological circumscription represent distinct lineages on each continent or geographical region. For example, *Parmelina quercina*, a foliose (leaflike) lichen that occurs worldwide in areas with Mediterranean climate (such as the Mediterranean region, southern California, and southwestern Australia), has been shown to represent distinct lineages on each continent. In fact, these morphologically similar species are often not closely related. In a study on the white-rotting fungus *Hyphoderma setigerum*, only one of nine cryptic species (that is, species that are basically morphologically undistinguishable, but that are genetically distinct) has a geographic range spanning more than one continent. These results underline the problematic nature of a morphological species recognition in fungi with relatively few and highly plastic characters. In other cases, molecular studies have provided evidence that the diversity within certain ecosystems is dramatically underestimated by morphology-based species delimitations. In several groups of tropical fungi, molecular data hint at a much higher diversity, showing that currently accepted species belong to a number of distinct lineages.

Human activity also has a severe impact on the distribution of fungi, especially those associated with crops or tree species. For example, at least 200 ectomycorrhizal fungi are known that have been moved from native ranges to novel habitats. The majority of recorded introductions are associated with plantations of eucalypts or pine trees in the Southern Hemisphere. In another study, some Italian populations of *Heterobasidium annosum*, the root-rot pathogen of conifers, were shown to have been introduced recently from North America. Intercontinental movement from Europe to North America has been shown for the deadly poisonous *Amanita phalloides*, which is an invasive species that is rapidly expanding its range on the west coast of North America.

Inference of ancestral areas. Molecular data, coupled with newly developed statistical tools to address ancestral ranges, allow testing of hypotheses related to the origins of taxa and past and recent migration events. These studies include a variety of different fungi. For example, the mammal pathogen *Histoplasma capsulatum* was shown to have originated about 3–13 million years ago in Latin America, and the foliose lichen genus *Remototrachyna* has been shown to have its ancestral range on the Indian subcontinent. Furthermore, the destructive dry rot fungus *Serpula lacrymans* was shown to have an Asian origin, with rapid subsequent global spread. In the case of the fly agaric (*A. muscaria*), molecular data demonstrated that it actually consists of three cryptic species. The ancestral population of the complex probably evolved in the Siberian-Beringian region and underwent fragmentation. These fragmented populations subsequently evolved into different species and expanded their range in North America and Eurasia.

Although a number of studies suggest that taxa evolved in those areas where they have their current center of diversity, it has been shown that, in some groups, the migration into a different geographical area and climate has allowed a shift in diversification. This includes the ectomycorrhizal Inocybaceae, a fungal family that is currently common and widespread in north temperate regions, but has been found to have originated in paleotropical settings, with several relictual (remnant) lineages persisting in the tropics. Parmelioid lichens, in contrast, have been suggested to have originated in temperate regions and subsequently diversified into one of the most diverse clades of ascomycetes, after migrating into tropical regions.

For background information *see* BIODIVERSITY; BIOGEOGRAPHY; FUNGAL ECOLOGY; FUNGAL GENOMICS; FUNGI; LICHENS; MUSHROOM; MYCOLOGY; PALEOECOLOGY; PHYLOGENY; POPULATION DISPERSAL; SPECIATION in the McGraw-Hill Encyclopedia of Science & Technology. H. Thorsten Lumbsch

Bibliography. J. Geml et al., Beringian origins and cryptic speciation events in the fly agaric (*Amanita muscaria*), *Mol. Ecol.*, 15:225–239, 2006; P. R. Johnston, Causes and consequences of changes to New Zealand's fungal biota, *New Zeal. J. Ecol.*, 34:175–184, 2010; H. T. Lumbsch et al. (eds.), Special issue: Phylogeography and biogeography of fungi, *Mycol. Res.*, 112:423–484, 2008; P. B. Matheny et al., Out of the Palaeotropics?: Historical biogeography and diversification of the cosmopolitan ectomycorrhizal mushroom family Inocybaceae, *J. Biogeogr.*, 36:577–592, 2009; J. Walker and K. A. Pirozynski, Pacific mycogeography: A preliminary approach, *Aust. J. Bot. Suppl. Ser.*, 10:1–172, 1983.

Picoprojectors

A picoprojector is a small projector that fits easily in the palm of a hand or a pocket, as shown in **Fig. 1**. Picoprojectors typically produce between 8 and 15 lumens (lm). Ten lumens, for example, produces a usable 12-in. (30.5-cm) or so image in a room with moderate ambient light or a 40-in. (102-cm)

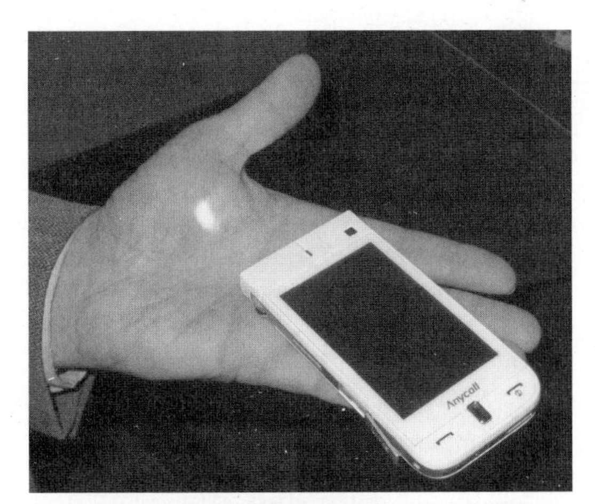

Fig. 1. Samsung Anycall W9600 projector phone at the Consumer Electronics Show 2010, now shipping in Korea. (*Photo credit: Matthew Brennesholtz*)

or larger image in near-total darkness. Many, but not all, picoprojectors are battery-powered. As projector efficiency and battery capacity increase, more are expected to be battery-powered. With the power consumption and battery capacity typical of today, a picoprojector can run for 1–2 h on an internal battery before recharging is needed.

Some picoprojectors are pure projectors and require an external source to supply a video signal. These pure projectors are becoming less common, however, as picoprojectors are embedded into other systems. To date, picoprojector modules have been embedded into media players, digital still cameras, and cell phones. Because of the very low cost, small size, and low power consumption of media player circuitry, media players have been embedded into many mobile projectors, not just picoprojectors. This allows the projector to be used with preloaded content and without the hassle of a video connection at the point of use.

In digital still cameras with embedded picoprojectors, the camera can be used to take photos or video and then immediately show these images to a larger group. Picoprojectors embedded into digital still cameras typically do not have an embedded media player but normally do accept external video input.

Some cell-phone handsets are available with embedded picoprojectors. All of these handsets contain a camera and media player as well as the picoprojector, and some can also accept external video. These handsets also typically have a 2-2.5-in. (5-6-cm) or larger touch-screen liquid-crystal display (LCD) for control, Web browsing, and video. With current picoprojector technology, these handsets tend to be relatively large.

It is expected that picoprojectors will be embedded into additional product types in the future, including digital video cameras, laptop and netbook computers, mobile television receivers, and handheld games. Virtually all mobile electronic products have been proposed as targets for embedded picoprojectors.

Picoprojectors were initially proposed in 1997, and prototypes were shown privately for several years before they were first demonstrated for the public at the Consumer Electronics Show (CES) in January 2008. The first commercial picoprojectors appeared shortly thereafter, in March 2008. The first commercial picoprojector embedded into a cell-phone handset appeared, along with many other new picoprojector models, at CES in 2009. CES 2010 showed a continuing expansion of available picoprojector products.

Picoprojector markets. Picoprojectors are being marketed toward both the professional and consumer markets. In the professional market, it is suggested that picoprojectors can be used for spur-of-the-moment presentations. In addition, depending on the picoprojector connectivity, streaming video material or other Internet content can be shown.

The marketing orientation for consumers is almost pure entertainment. In addition to watching preloaded content, Internet downloads, and Web browsing, consumers can use picoprojectors for capturing (with the built-in camera) and displaying images and video.

To date, sales of picoprojectors have been modest, and it is not really well understood how consumers or professionals will use them. Most picoprojectors sold so far have gone to early adopters.

The cell-phone handset market is seen as very promising for picoprojectors. Like the cell-phone camera market, the picoprojector handset market will start with a price premium and bulky handsets. Now cameras embedded into cell phones are nearly universal and have no noticeable price premium or handset size penalty. This low price premium occurs because the price to a handset manufacturer of a basic cell-phone camera module has been reduced to as little as $1 through innovative design, wafer-scale assembly techniques, and very high volumes. Although no one expects the near-100% penetration of picoprojectors into the handset market enjoyed by cameras, some do foresee 10% penetration levels. With a 10% penetration level, picoprojectors would sell at a rate of more than 100 million units per year. Even the most optimistic observers of the market do not expect this penetration level before 2015.

There are additional low-volume and high-value markets for picoprojector technology. For example, two companies make systems that detect subcutaneous veins by infrared imaging and project an image of the veins on the patient's skin. This allows medical personnel to find veins for needles with less trial and error.

Picoprojector specifications. The **table** indicates the approximate range of picoprojector specifications from 2010 to 2014.

The requirements of one major handset manufacturer that must be met before it would consider embedding picoprojectors are (1) a module cost of less than $30, (2) a module height of less than 7 mm, (3) a module volume of less than 5 cm^3, (4) a module output of 9–10 lm, and (5) power consumption of

Key specifications of a picoprojector module

Specification	Typical 2010 value	Expected 2014 value	Justification
Lumens	10–12 lm	12–50 lm	Higher lumens allow larger images to be shown in areas with higher ambient light
Power consumption	2–15 W	0.5–10 W	Maximize battery lifetime in hand-held systems
Lumens per watt	2–5 lm/W	8–20 lm/W	Maximize light output in power-limited designs
Module height	10–12 mm	5–7 mm	Allow slim picoprojector designs
Size	4–10 cm³	2.5–7 cm³	Allow compact handsets and picoprojectors
Resolution	HVGA–VGA	VGA–WVGA–720p	Higher resolution needed for Web browsing and other text-based applications
Aspect ratio	4:3 and other narrow formats	16:9 and other wide formats	Wide aspect ratio needed for entertainment; also simplifies many text-based applications such as Web browsing

SOURCE: Updated values from Insight Media 2009 Picoprojector Report. Used by Permission of Insight Media.

1.0–1.5 W. Note that the height is a separate specification from the module volume. This height specification would allow the manufacturer to produce the slim handsets desired by consumers.

The 1-W preferred and 1.5-W maximum power consumption are the figures given by virtually all handset manufacturers. This power limitation is based on available battery capacity. Manufacturers do not want to put larger batteries in their handsets, because that would add both weight and cost.

Picoprojector technology. All picoprojectors use a single imaging device to modulate the red, green, and blue colors needed to produce a full-color image. Although some early prototypes were monochrome, all current picoprojectors produce full-color images. Desktop and larger projectors can use either one imager or three imagers to make the three primary colors, but price and module size restrictions limit picoprojectors to single-panel designs. In addition, all picoprojectors use solid-state light sources. Most use LEDs, but a few are laser-based.

Color-sequential technology is currently the most popular approach to picoprojector design. In this approach, a single microdisplay makes first the red, then the green, and then the blue image. This sequence is repeated so rapidly that the eye sees the three images as a full-color image. The minimum acceptable color-field rate is 180 Hz, and up to 2 kHz or beyond is used in some designs.

Modules. Handset makers and others who desire to embed picoprojectors into their products want to buy a complete picoprojector module, with the internal details of the module not really of concern to them. The typical internal layout of a color-sequential digital light processing [DLP™, Texas Instruments (TI)] or liquid-crystal-on-silicon (LCoS) picoprojector module is shown in **Fig. 2**. These modules are currently available from a number of vendors.

A picoprojector module typically takes in power and digital video and puts out a projected image. The module includes a light source, a microdisplay or other imaging component, projection optics, and a mechanical structure to hold all these components in the correct positions. A drive ASIC (application-specific integrated circuit) for the imager and the light source may be either embedded into the module or supplied as separate components, so that the product manufacturer can incorporate them on the motherboard.

In addition to the specifications for the input digital video and power, picoprojector modules typically have specified mechanical and thermal interfaces. A company with little or no understanding of the optics and electronics internal to a module

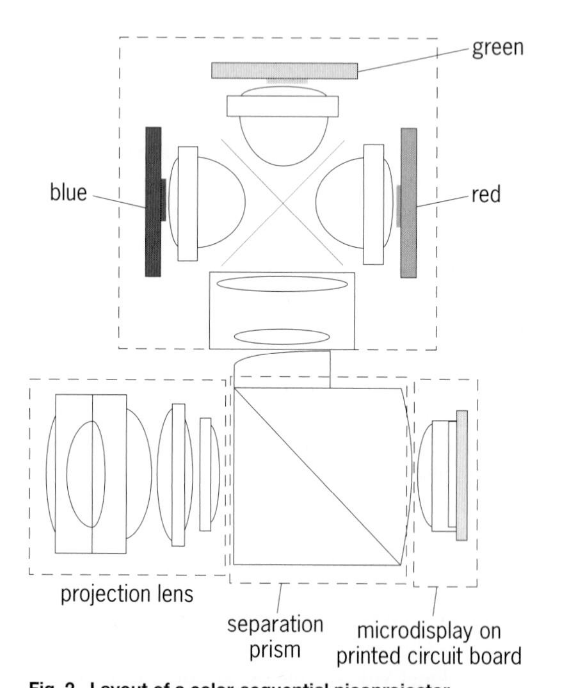

Fig. 2. Layout of a color-sequential picoprojector illumination module with red, green, and blue LEDs on heat sinks, collection optics, and dichroic filters. In a module with a DLP microdisplay, the separation prism is a total-internal-reflection (TIR) prism; in a module with a LCoS microdisplay, it is a polarizing beam splitter.

could design a product based purely on these external specifications.

Imaging technology. Four different imaging technologies are used in commercially available technologies: (1) DLP technology from TI, (2) color filter array liquid-crystal-on-silicon (CFA-LCoS) technology from Himax, (3) color-sequential liquid-crystal-on-silicon (CS-LCoS) technology from Micron Technology and Syndiant, and scanning laser systems using a single micro-electro-mechanical systems (MEMS) scanning mirror from Microvision.

DLP imagers have one small mirror per pixel. Currently, the smallest mirrors used give a pixel pitch of about 7.4 μm, although smaller mirrors are theoretically possible. Each mirror can tilt between an "on" and an "off" state, depending on the input video signal. Because each pixel in the system can produce only a bright or dark state at any given instant, time-division multiplexing using a bit-plane scheme is used to produce gray scale.

DLP projectors are available at all levels of light output from 10-lm picoprojectors through 35,000-lm projectors for large-venue applications. These and other large projectors are based on three DLP imagers, one for each primary color. In smaller projectors, including DLP picoprojectors, a single DLP imager and color sequential technology are used. Larger projectors, including digital cinema projectors, are based on three DLP images, one for each primary color.

The up-front design and development costs of a digital micromirror MEMS device are very high, and Texas Instruments can spread these development costs over its broad range of projectors. Although start-up companies have tried to develop a competitive chip, so far none has produced a commercial product.

CFA-LCoS consists of a liquid-crystal layer atop a complementary metal-oxide semiconductor (CMOS) silicon backplane with a patterned layer of red, green, and blue filters. In principle, it works like an ordinary small liquid-crystal display except that it is reflective instead of transmissive. These imagers are relatively easy to manufacture and can accept the same digital video signal used for the handset display. Because three independent subpixels are required for each full-color pixel, the devices are relatively large and have trouble meeting the 7-mm height requirement. They are also very inefficient because the color-filter layer absorbs most of the light from the white LED that is normally used with the system.

In 2008, DLP and CFA-LCoS each had about a 50% market share. Although DLP is maintaining its market share, the market share for CFA-LCoS has been declining, as the more efficient CS-LCoS market share grows.

The financial and intellectual-property barriers to entry into the CS-LCoS market are not as high as the barriers to entry into the MEMS market in competition with DLP. Therefore, it is possible that additional CS-LCoS companies may appear at any time.

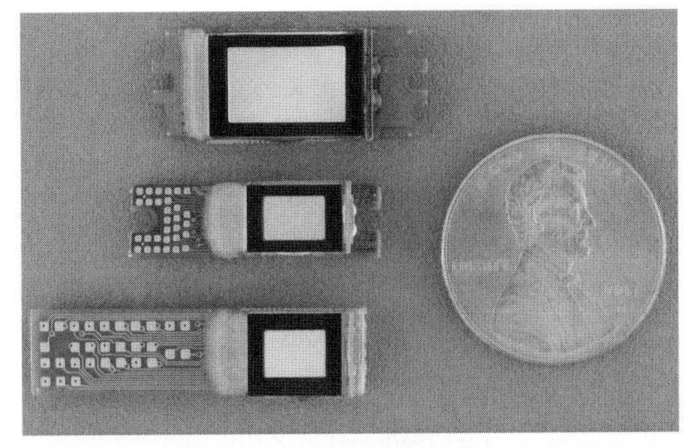

Fig. 3. Three color-sequential LCoS imagers: (*top*) SYL2061 with 1024 × 600 pixel array; (*middle*) SYL2030 with 854 × 480 pixel array; (*bottom*) SYL2010 with 854 × 600 pixel array. (*Photo courtesy of Syndiant, Inc.*)

CS-LCoS pixels can be quite small; for example, Syndiant currently uses a 5.4-μm pixel in its picoprojector products to produce an 854 × 480 pixel array in a 0.21-in. (6.35-mm) diagonal imager. In comparison, a DLP imager with a 0.21-in. diagonal has a 480 × 320 pixel array. In applications in which resolution is an important property, including Web browsing, this gives CS-LCoS an advantage over both DLP and CFA-LCoS microdisplays. **Figure 3** shows three color-sequential LCoS microdisplays. The DLP has an appearance very similar to these except for variations in the package design.

Scanning laser projectors operate by scanning a laser beam in a raster pattern, much like a cathode-ray tube (CRT). The laser beam is then modulated at the pixel rate, which can be 50 MHz or more. The scanning is normally done with a single biaxial moving mirror made with MEMS technology to scan both horizontally and vertically, or by two MEMS mirrors, one for the horizontal and the other for the vertical scan. **Figure 4** shows a biaxial scanning mirror from Microvision. Currently, the only commercial picoprojector available that uses the scanning laser technology is produced by Microvision, which uses a single-mirror system.

Scanning laser projectors are not a new technology and were proposed almost as soon as the laser was invented. The earliest serious proposal for a scanning laser projector was detailed in a 1966 Texas

Fig. 4. MEMS mirror scanner. The mirror is approximately 1 mm in diameter and is mounted for two-axis tilting. (*Photo courtesy of Microvision, Inc.*)

Instruments internal memo. Scanning systems without lasers are even older and date back to the 1930s. Therefore, any basic patents on scanning laser technology have expired and the development of one or two suitable MEMS mirrors is not a serious problem. Entry into the scanning laser picoprojector market is limited mostly by the cost and availability of suitable lasers and the performance limitations of scanning laser technology compared to pixel-based picoprojectors.

Light sources. Three types of light sources are currently used in picoprojectors: white light-emitting diodes (LEDs), red, green, and blue (RGB) LED sets, and RGB laser sets.

White LEDs are extremely efficient, with LEDs producing 80 lm/W and more. However, white LEDs are currently usable only with CFA-LCoS microdisplays, which have very low efficiency. Therefore, the efficiency in lumens per watt of current systems using white LEDs is relatively poor. Newer, more efficient designs of picoprojectors are expected to emerge in the future to take advantage of the very high efficiency of white LEDs.

RGB LED sets are typically used with color-sequential systems such as DLP or CS-LCoS. Although the efficiency of an RGB LED set is not as good as that of a white LED, the light can be used much more efficiently by the color-sequential microdisplay. The most power-efficient picoprojectors currently available are based on RGB LED sets in color-sequential designs.

RGB laser sets can be further subdivided into two subcategories called "fast" and "slow" lasers. Slow lasers are modulated at the color field rate of a color-sequential microdisplay-based picoprojector. Typically, this means modulation in the range of 2–20 kHz. Fast lasers must be modulated at the pixel rate in mirror projector designs. This modulation frequency is typically of the order of 50 MHz.

Red and blue diode lasers suited for use in picoprojectors are available, and all these lasers are capable of being modulated at 50 MHz or higher. Obviously, if these lasers are intended for use in a color-sequential picoprojector requiring only 20 kHz modulation, they can be put in much simpler and therefore less expensive packages. These red and blue diode lasers are available from multiple vendors.

Currently, no green diode lasers suited for use in picoprojectors are available. All available green lasers are frequency-doubled infrared (IR) lasers. For example, a 1064-nm IR laser can be frequency-doubled to 532 nm, which produces a very good green color for displays; 1047-nm IR lasers, frequency doubled to 523 nm, also can be used.

The IR laser itself can be either a direct-emission IR laser or a diode-pumped solid-state (DPSS) laser. In a DPSS design, an IR laser, typically 808 nm, pumps a laser crystal that emits the 1064- or 1047-nm light, which is then frequency doubled to green. This DPSS design is relatively low-cost and efficient and is used, for example, in green laser pointers. Because of persistence in the laser crystal, this design cannot be modulated faster than about 20 kHz. This makes the design suited for color-sequential systems, such as the DLP or CS-LCoS, but it is not usable in scanning laser systems. For a scanning laser system, it is necessary to use a direct-emission 1064- or 1047-nm IR diode and then frequency-double the IR to green. Typically, these lasers are more complex, less efficient, and more expensive than DPSS designs. Green lasers suited for scanning laser projectors are available from both Corning and Osram.

Currently, lasers of any design are very expensive compared to RGB LED sets or white LEDs. Therefore, LEDs provide the light source for the vast majority of picoprojectors sold today. Laser prices are coming down rapidly and performance is increasing, but this is true of LEDs as well. Only time will tell whether lasers will significantly penetrate the picoprojector market.

For background information *see* LASER; LIGHT-EMITTING DIODE; LIQUID CRYSTALS; MICRO-ELECTRO-MECHANICAL SYSTEMS (MEMS); OPTICAL PROJECTION SYSTEMS; PHOTOMETRY in the McGraw-Hill Encyclopedia of Science & Technology. Matt Brennesholtz

Bibliography. V. Bhatia et al., Compact and efficient green lasers for mobile projector applications, *J. Soc. Inform. Display*, 17(1):47–52, 2009; M. S. Brennesholtz and E. H. Stupp, *Projection Displays*, 2d ed., John Wiley & Sons, Chichester, U.K., 2008; M. S. Brennesholtz, New-technology light sources for projection displays (invited paper), *Society for Information Display 2008 International Symposium Digest of Technical Papers*, vol. XXXIX, paper 56.4, Los Angeles, CA, May 18–23, 2008; K. Guttag, High-resolution microdisplays for pico-projectors (invited paper), *Society for Information Display 2010 International Symposium Digest of Technical Papers*, vol. XLI, paper 71.1, Seattle, WA, May 23–28, 2010.

Pollination mutualisms by insects before the evolution of flowers

A preeminent association between flowering plants and insects is pollination. Pollination is a mutualism in which two interactors reciprocally benefit: a host plant receives the service of insect pollination in return for a reward provided for its insect pollinator. Typically, the reward is nectar or pollen, but occasionally the provision can be a mating site, resin for nest construction, floral aroma, or even the attraction of plant-generated heat. Evidence from the fossil record and from the inferred ecological and phylogenetic relationships between flowering plants (angiosperms) and their insect pollinators indicates that these types of associations initially were launched during the Early Cretaceous, 125–90 million years ago. It was from this interval of time that flowering plants experienced their initial radiation, as did major groups of insects, especially Thysanoptera (thrips), Coleoptera (beetles),

Diptera (flies), Lepidoptera (moths and butterflies), and Hymenoptera (sawflies, wasps, ants, and bees). However, until recently, very little was known about more ancient modes of insect pollination, those that predated the appearance of flowering plants or that occurred before angiosperms became dominant in terrestrial ecosystems.

Types of evidence. There are several dietary precursors to the consumption of pollen and nectar as a reward for pollination. As far back as the Early Devonian, 416–397 million years ago, small, mandibulate (chewing) arthropods were consuming the spores of well-known, early land plants. Evidence for this feeding style consists of the sporangial contents of single species of host plants found as identifiable spores in coprolites (preserved fecal pellets) that were defecated by small arthropods. Later evidence from the Early Permian, 299–271 million years ago, includes the prepollen and pollen of seed plants found in the coprolites of insects such as booklice (**illus.** *a*). Of course, pollen consumption by insects is a pattern that continues today. However, it is especially demonstrated by the gut contents extracted from extinct, mid-Mesozoic sawflies (illus. *b*). These sawflies typically consumed gymnosperm pollen provided by gnetaleans (illus. *c*), which are represented currently by *Ephedra* (Mormon tea), *Gnetum* (a tropical shrub or vine), and *Welwitschia* (an unusual plant of the southern African Namib Desert). Other extinct lineages that provided pollen for insect consumption included seed ferns and, notably, cheirolepidiaceous conifers, whose distinctive pollen has been found as gut contents and clumps plastered to the heads and mouthparts of insects (illus. *d*). In contrast, fluids (such as the nectarlike pollination drops of gymnosperms) are not preserved in dispersed coprolites or the guts of insects. Thus, insect feeding on fluids is inferred by distinctive, siphonlike mouthparts, such as those from extinct species of flies (illus. *e*) or scorpionflies (illus. *f*).

Several features of the ovulate and pollen organs of many gymnospermous plants directly implicate fluid feeding by insects bearing siphonate mouthparts. These features included elongate tubular micropyles, minute openings in the integument at the tip of an ovule through which the pollen tube commonly enters and only occurs in gymnosperms; illus. *g*): deep channels through ovulate-covering tissue that allowed access to pollination-drop rewards secreted by deep-seated ovules (illus. *h*), and channels formed by the partial closure of a bivalved ovule, allowing proboscis-bearing insects to imbibe inner pollination drops (illus. *i*). This varied evidence, including coprolites and gut contents from mandibulate insects that fed on pollen grains, the distinctive long-proboscate mouthparts of insects that fed on pollination drops, and plant structures that received the size-matched proboscises of insects, collectively indicates the presence of pollination during the preangiospermous Mesozoic. Evidently, there were two major modes of insect pollination: (1) long-proboscid, fluid-feeding insects that fed on

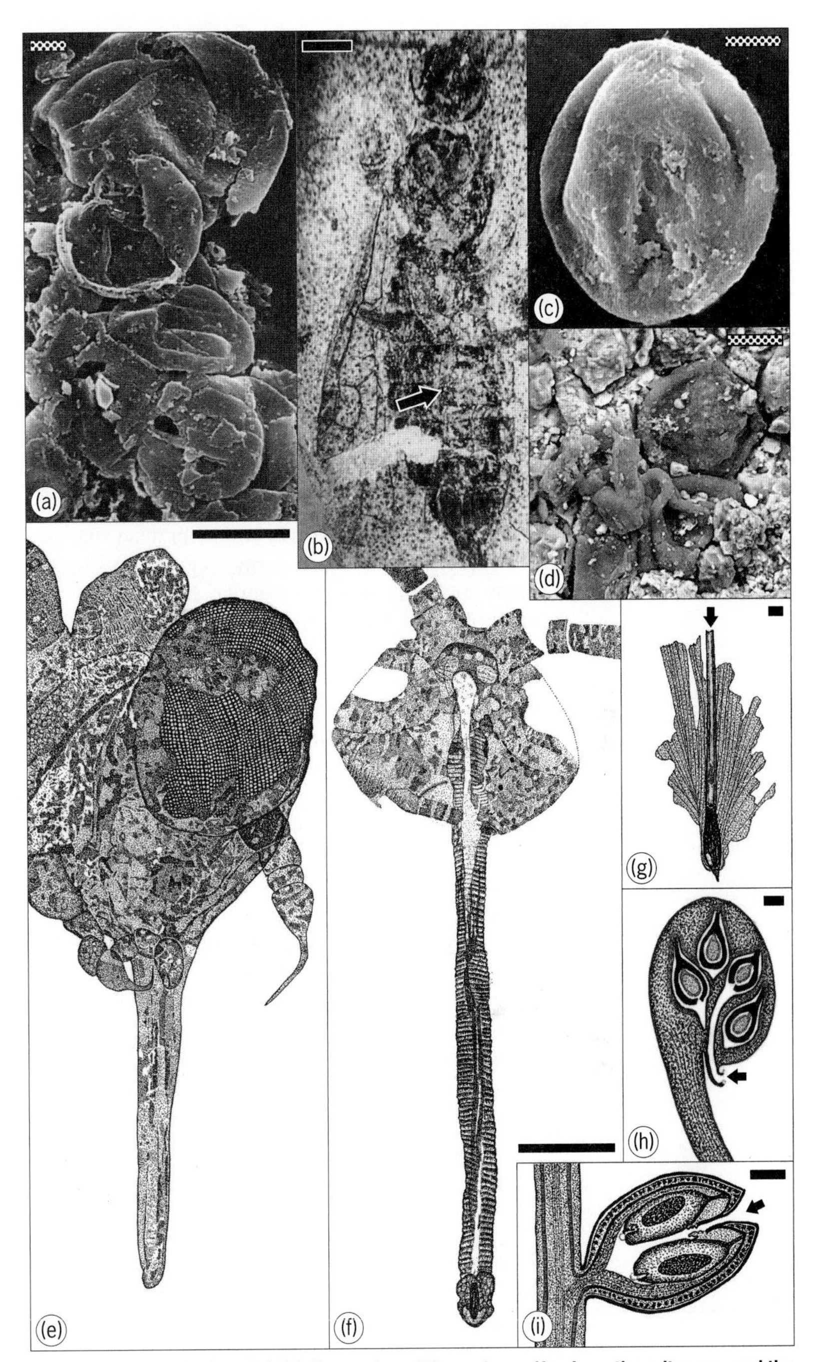

Plant mutualism displayed. (*a*) Pollen grains of the early conifer, *Lunatisporites* sp., and the glossopterid, *Protohaploxypinus perfecta*, from the gut of the booklouse, *Parapsocidium uralicum*, from the Early Permian Chekarda site of Russia. (*b*) The sawfly, *Ceroxyela dolichocera*, from the Lower Cretaceous of Baissa, Transbaikalia, Russia, with *Eucommiidites* type pollen (gnetalean?) in its midgut (indicated by an arrow). (*c*) The *Eucommiidites* type pollen extracted from the midgut of the sawfly in panel *b*. (*d*) Pollen of the cheirolepidiaceous conifer, *Classopollis* sp., found smeared on the head of a small brachyceran fly, from the same locality as panel *b*. (*e*) Head, proboscis, and associated mouthparts of the nemestrinid fly, *Florinemestrius pulcherrimus*, from the mid–Early Cretaceous of Liaoning, China. (*f*) The head and proboscis of the aneuretopsychid scorpionfly, *Jeholopsyche liaoningensis*, from the same formation as panel *e*. (*g*) *Problematospermum ovale*, an enigmatic seed with a 12-mm (0.48-in.) micropyle, from the late Middle Jurassic of Inner Mongolia, China. (*h*) *Caytonia sewardi*, with a 4-mm (0.16-in.) integumental tube, from the mid-Jurassic deposits at Yorkshire, United Kingdom. (*i*) *Leptostrobus cancer*, with an intervalve, 4.5-mm (0.18-in.) channel allowing access to concealed pollination drops, from the same site as panel *h*.

gymnosperm reproductive organs, in which the reward was principally nectar; and (2) mandibulate, chewing insects, whose adults consumed mostly pollen and whose larvae tunneled through vegetative tissue and ovules of plant hosts, including extinct bennettitaleans and the extant cycads.

Major players. The spectrum of mandibulate insect pollinators that occurred before the advent of angiosperms, during the Permian, included the Psocoptera (booklice) and Grylloblattodea (rock crawlers), as well as extinct groups related to orthopteroids (currently, grasshoppers and relatives) and hemipteroids (currently, cicadas, aphids, true bugs, and relatives). After the end-Permian extinction and during the Mesozoic, this spectrum of insects was replaced by various lineages of Coleoptera, whose life cycles were associated intimately with bennettitaleans and cycads, and some taxa that became affiliated with the large, showy flowers of basal angiosperms (for example, members of the magnolia and laurel families). In contrast, an array of small, fluid-feeding, long-proboscate insects included early forms of Trichoptera (stem-group caddisflies), Mecoptera (aneuretopsychine scorpionflies), and Neuroptera (butterfly-like kalligrammatid lacewings), as well as three or four major lineages of brachyceran Diptera (tanglevein flies, horseflies, and flower-loving flies). With the exception of the brachyceran Diptera, this gymnosperm-pollinating assemblage was replaced, during the diversification of angiosperms in the Early Cretaceous, by the Lepidoptera (glossate moths), more advanced groups of Diptera (hover flies and bee flies), many lineages of Hymenoptera (wasps and bees), and probably blister beetles with siphonate proboscises.

Based on the structure of ovulate and pollen organs consistent with a pollinator function, gymnospermous plant hosts for these lineages were predominantly cycads, bennettitaleans, seed ferns, cheirolepidiaceous conifers, and gnetaleans. Cycads and bennettitaleans typically were composed of fleshy, thick stems that were predominantly beetle-pollinated and involved sacrifice of internal tissues to larvae for completion of their life cycle. Various seed fern groups included caytonialeans, corystosperms, and lepidostrobaleans that bore ovulate organs (illus. *g–i*) and secreted pollination drops for imbibation by insects, and produced pollen that may have been consumed as particles by species with accommodating siphons. Gnetaleans were considerably more diverse during the Mesozoic and undoubtedly were insect-pollinated, based on a variety of reproductive features, including pollen consistent with insect vectoring. A plant host with the most distinctive and bizarre type of insect pollination was the cheirolepidiaceous conifer, *Alvinia bohemica*. This ovulate cone bore a trichome- and nectary-lined funnel that was directed outward and connected by a "pipe" to an inward-placed ovule that secreted a pollination drop. The deep funnel structure was surrounded by various conspicuous appendages that may have served as a lure to small insects or to large insects with elongate mouthpart siphons.

Evolutionary biology. The earliest significant evidence for insect pollination is from the Permian, even though earlier occurrences extending to the Early Devonian indicate the presence of spore and pollen consumption by mandibulate insects. Permian pollination was mediated overwhelmingly by mandibulate insects, although there is some evidence for small, caddisfly-like insects with extended proboscises designed for fluid uptake. After the demise of this assemblage at the end-Permian mass extinction, new lineages of both mandibulate chewing and long-proboscid fluid-feeding insects originated by the mid-Mesozoic. Although mandibulate beetles were associated predominantly with bennettitaleans and cycads, the former hosts became extinct during the Cretaceous, whereas the latter survive today, probably with similar life cycles. A distinctively different evolutionary trajectory was followed by the long-proboscate lineages of scorpionflies, lacewings, true flies, and others that diversified during the Middle Jurassic and lasted approximately 60 million years as pollinators, principally on a variety of seed ferns, cheirolepidiaceous conifers, and gnetalean hosts. However, this assemblage of gymnosperm pollinators came to a close with the global turnover of gymnosperm to angiosperm floras that commenced during the mid–Early Cretaceous and was mostly completed during the early Late Cretaceous. Their replacements, on angiosperms, are long-proboscid lineages that are familiar to us today: glossate moths, bee flies and hover flies, wasps and bees, and a broad range of other holometabolous insects that independently evolved siphonate mouthparts.

For background information *see* DIPTERA; FLOWER; INSECTA; MAGNOLIOPHYTA; MUTUALISM; PALEOBOTANY; PALYNOLOGY; PLANT-ANIMAL INTERACTIONS; POLLEN; POLLINATION in the McGraw-Hill Encyclopedia of Science & Technology.

Conrad C. Labandeira

Bibliography. R. Gorelick, Did insect pollination increase seed plant diversity?, *Biol. J. Linn. Soc.*, 74:407–427, 2001; V. A. Krassilov, A. P. Rasnitsyn, and S. A. Afonin, Pollen eaters and pollen morphology: Co-evolution through the Permian and Mesozoic, *Afr. Invert.*, 48:3–11, 2007; H. W. Krenn, J. D. Plant, and N. U. Szucsich, Mouthparts of flower-visiting insects, *Arthropod Struct. Devel.*, 34:1–40, 2005; C. C. Labandeira, The pollination of mid Mesozoic seed plants and the early history of long-proboscid insects, *Ann. Missouri Bot. Gard.*, vol. 96, 2010; C. C. Labandeira, J. Kvaček, and M. B. Mostovski, Pollination drops, pollen, and insect pollination of Mesozoic gymnosperms, *Taxon*, 56:663–695, 2007; M. P. Proctor, P. Yeo, and A. Lack, *The Natural History of Pollination*, 2d ed., Timber Press, Portland, OR, 1996; D. Ren et al., A probable pollination mode before angiosperms: Eurasian, long-proboscid scorpionflies, *Science*, 326:840–847, 2009.

Positive train control

From the time of the steam engine in the 1830s, innovation has been a hallmark of the U.S. rail industry. However, the train control technology used on U.S. railroads has remained virtually unchanged for several decades, aside from design upgrades using microprocessors and modern electronics. These twentieth-century systems still rely heavily on locomotive crews to comply with standard operating rules, follow verbal commands from dispatchers, and obey trackside rail signals.

However, accidents and train collisions can sometimes occur because of inattention or distraction, misinterpretation of wayside signal indications, or incapacitation of the crew. In a few selected areas, automatic train control (ATC) and automatic train stop (ATS) systems are used to intervene in train operations by warning the crew or causing the trains to stop if they are not being operated safely. However, these old systems using mechanical relays and extensive track circuits are expensive to install and maintain, and, therefore, have not been employed extensively.

With the advent of technologies such as computer networking, wireless communication, and global positioning system (GPS) tracking, the railroad industry has been developing the next generation of train management, called positive train control (PTC), a system that is more affordable than the conventional ATC systems and will virtually eliminate human errors and keep trains properly separated.

Positive train control systems. Positive train control systems are integrated command, control, communications, and information systems for controlling train movements with safety, security, precision, and efficiency. A typical configuration of PTC usually consists of four components: the office system, the wayside devices (signals and interface units), the locomotive onboard systems, and the communication network (**Fig. 1**).

The communication network connects these different segments through commands and messages sent using predefined protocols and a data encryption scheme. The most important message flow conveys the movement permission (authorities) from the office system to the onboard locomotive systems. These movement authorities are generated by vital logic processors to ensure that there is no conflict in train movements. Even without PTC, if the locomotive engineers follow these movement authorities, there will be safe separation of trains, and collisions will not occur. However, accidents can still occur as a result of human errors. Through GPS or other train-tracking devices, the positions of the locomotives hauling the trains can be tracked. When PTC detects that a train may be about to violate a movement authority or relevant mandatory directive from the dispatching system, the locomotive engineer receives a warning message to bring the train back within safety limits. If the engineer fails to do so, the onboard system will impose automatic braking to stop the train before this violation can occur. The primary objective of PTC is to eliminate human errors that potentially can lead to collisions but it can provide other protections too. The core functions of PTC, as defined by federal regulation 49CFR236 Subpart I, include preventing train-to-train collisions, preventing over-speed derailments, preventing train incursions into work zones, and preventing movement of a train through a switch left in the "wrong position."

When the PTC is activated, an onboard database keeps track of terrain information, such as the grades, degrees of curves, speed limits, signal locations, and switch locations. The PTC system developed in the United States for freight systems uses GPS to track train locations. By knowing the location, the onboard system can compare the speed of the train with that stored in the database for that location. If the speed of the train is found to be above the allowable limit, penalty (enforcement) braking will be imposed automatically to stop it. Penalty braking is initiated by activating a valve, called P2A, within the locomotive air-brake system. The braking force during penalty braking is the maximum, unless emergency braking is used. This is the general surveillance method used to prevent over-speed derailments.

Collision avoidance. The wayside communication network sends movement authorities to the trains equipped with the PTC system on the lead locomotive. In signal territories, these authorities are generated by the safety interlocking logic built within the signal systems. In nonsignal territories, these authorities are given by the dispatchers, but are verified by the safety logic of the safety server prior to delivery to the lead locomotive of a train. The safety logic, either in the signal system or in the safety server, ensures that trains are kept at a safe distance and will never collide with each other, as long as the authorities are not violated. The movement authorities require the trains not to overrun a target point, usually a signal location, a switch location, or just a location with a milepost. The PTC monitors the position and the speed of the trains by GPS and sensors

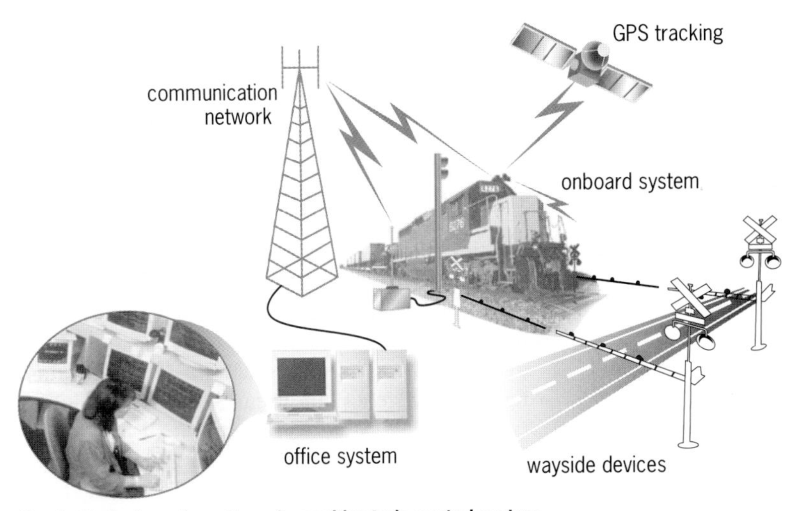

Fig. 1. Typical configuration of a positive train control system.

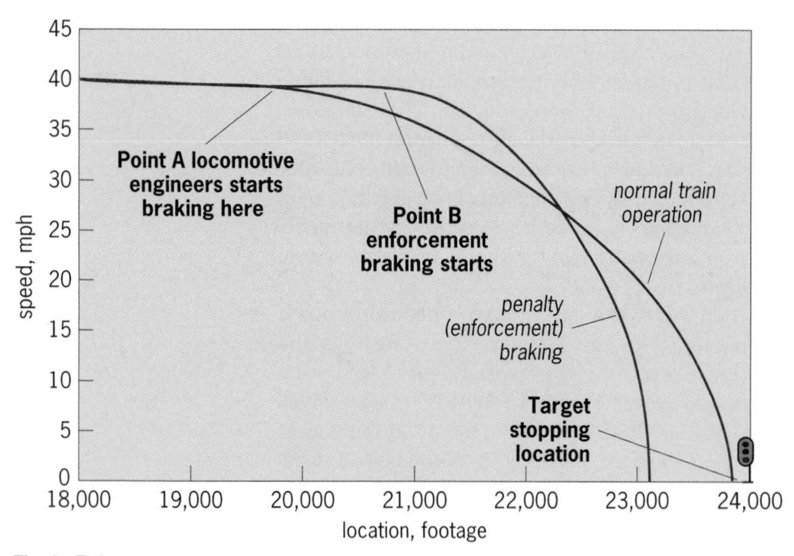

Fig. 2. Enforcement braking when the locomotive engineer ignores a red signal shown.

and continuously calculates the distance required to stop the train through an enforcement braking action. Knowing this distance, the PTC can determine the last moment at which it should activate the enforcement to prevent the train from overrunning the target. If engineers slow down the trains properly, the calculations will show that such enforcement activation is not necessary. In the event that enforcement is required, the PTC will send a signal to the locomotive air-brake control system. When the signal is received, the P2A valve within the air-brake control system will induce a full-service braking, normally by a 26-psi (179-kPa) reduction in the brake pipe pressure to propagate braking action through the train. Such continuous surveillance and readiness for enforcement will prevent any possible human error or negligence from violating the movement authorities. The trains are guaranteed to be kept within their allowed movement authorities, and therefore train collisions will not occur.

Figure 2 illustrates how PTC technology provides collision avoidance protection by enforcing a stop signal, as an example. At around location 24,000 ft, a wayside signal shows an indication of red, meaning that the train needs to stop in front of it. Because a freight train is long and heavy, it takes a long distance for it to stop. Sometimes the stopping distance can be as long as 2 mi (3.2 km). Thus, the locomotive engineer needs to start initiating the air-brake system of the train at point A, around 19,500 ft, in order to stop the train at the signal. During the operation of PTC, the GPS tracking module informs the onboard system where the train is. The wayside device interfacing with the signal system sends the signal indication, in this case a red stop signal, to the onboard system through the communication network. Currently, the trend within the industry is to adopt a 220-MHz VHF (very high frequency) radio network for communication. With the terrain information provided by the database and the train information provided by the dispatching system, a braking algorithm residing in the onboard computer system continuously calculates the braking distance at which penalty enforcement braking is required. At point B, around 20,700 ft, the system determines that without enforcement, the train will overrun the red stop signal and activates the P2A valve to stop the train automatically. The red signal normally indicates that another train is in the next signal block (space between the signals). Should the train be allowed to continue and overrun the red signal, there is a possibility that it would collide with another train in the next signal block.

Note that in Fig. 2, when the train stops through penalty braking, it may be some distance from the stop signal. A safety offset is usually included in the braking distance calculation, as some of the information, especially the train information, may not be too precise. As PTC is considered a safety-critical system, overrunning a target such as a red signal is not acceptable; therefore, a safety margin is added to ensure that there is no underestimation of the braking distance that can lead to a target overrun. There is an ongoing effort to develop a more accurate braking algorithm so that the train can stop as close to the target as possible with negligible risk of overrunning it.

There are other possible functions and business applications that can be applied in the PTC platform, once the computer and communication networks are in place. Examples are stopping the trains because of other alarms, such as those caused by high winds, landslides, or flooding. Another example is forcing the train to pace its trip properly to save fuel. However, these supplemental functions are not too far along in the development cycle yet.

Development projects. PTC has gone through several generations of development, starting as Advanced Train Control System (ATCS) in the mid-1980s. Since then, the railroads have attempted limited pilot projects based on the ATCS concept. Burlington Northern Santa Fe (BNSF) tested its Advanced Railroad Electronic System (ARES) in the Iron Range Region (northeast Minnesota), with some success. BNSF and Union Pacific Railroad (UPRR) cooperated to develop Positive Train Separation (PTS), another form of train control system based on the ATCS concept, in the Northeast Pacific Region. Subsequently, Amtrak developed and deployed its Advanced Civil Speed Enforcement System (ACSES) on the Northeast Corridor (NEC), which runs between Boston, Massachusetts, and Washington, D.C. This system is based on the use of track transponders, not GPS, to track train positions.

The Advanced Civil Speed Enforcement System was not considered suitable for freight operation, so the Federal Railroad Administration (FRA) teamed up with the Association of American Railroads (AAR) and the freight railroads in 1995 to initiate the North American Joint PTC project (NAJPTC), with the objective of developing a PTC system that would be suitable for mixed-traffic operation (that is, freight and passenger rail) and could be deployed

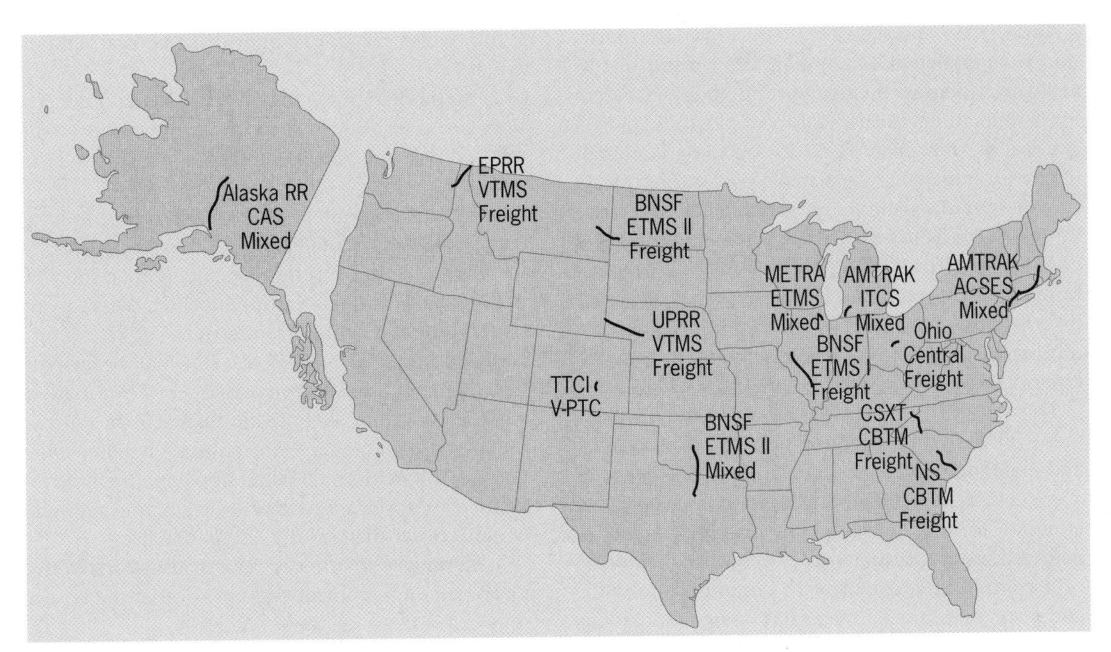

Fig. 3. Various pilot projects as of 2007.

nationwide with advanced features, such as moving block operation and protection for PTC-unequipped trains. The project did not lead to the actual deployment of PTC, but some of the PTC-related technologies resulting from this project were used in PTC systems subsequently developed by others.

PTC categories. Since the late 1990s, the FRA has sponsored several demonstration projects to explore various approaches for implementing PTC. Some railroads have also developed their own versions of PTC. As of the end of 2007, various developments with and without government funding were attempted (**Fig. 3**). Generally, as defined in 49CFR236 Subpart I in the Code of Federal Regulations, PTC can be classified into three categories depending on the implementation: nonvital overlay, vital overlay, and stand-alone.

Nonvital overlay. A nonvital overlay PTC system is an overlay on the existing method of operation, not built in accordance with the safety assurance principles generally recognized by the industry and contained in Appendix C of federal regulation 49CFR236. Examples of existing methods of operation that generate movement authorities for trains are a centralized traffic control (CTC) signal system, track warrant control, and direct traffic control. The safety assurance principles include fail-safe operation and closed-loop designs, among others.

Vital overlay. The vital overlay system is an overlay on the existing method of operation and is built in accordance with the safety assurance principles. For example, some vital overlay systems may use multiple processors in the locomotive onboard system for redundancy to ensure that the failure of a single microprocessor will not lead to disaster.

Stand-alone. A stand-alone system built on a newly constructed track or existing track performs the PTC functions without relying on another train control system, such as the signal system. A stand-alone system can also be built on an existing track as a replacement for an existing signal or train-control system. For some advanced features, such as moving block operation, a stand-alone PTC system without any physical control blocks rather than an overlay system needs to be employed. Almost all the systems shown in Fig. 3 are either nonvital overlay or vital overlay systems.

Deployment. Before 2008, the only PTC developments were pilot projects installed on about 100-mi (160-km) lines. The total length of these pilot projects was a little more than 2200 route miles (3540 km). The industry was struggling to perfect the technology in these pilot programs. Debugging, redesigning, and retrofitting to correct or enhance the systems occurred frequently and intensively in some of these developments. The railroads also found it difficult to justify the wide-scale deployment of PTC systems economically. Therefore, wide-scale deployment did not occur. At the same time, systems performing functions similar to PTC had already been deployed in Europe, China, Japan, and other countries to enhance operational safety. Safety was the paramount factor in their decision to deploy PTC-like systems. These overseas systems might use a different method of train tracking, such as track transponders instead of GPS, or a different method of wireless communication, such as a GSM cellular network instead of a 220-MHz VHF network. Nonetheless, they performed the main function of PTC to prevent train collisions.

On September 12, 2008, a collision between a Metrolink commuter train and a Pacific Union freight train in Chatsworth, California, resulting in 25 fatalities and hundreds of injuries, changed the landscape of PTC development completely. The engineer on the commuter train was not aware of, and overran, a stop signal controlled by a California Transportation Commission (CTC) system. As a result of this accident, Congress included in the 2008 Rail Safety Improvement Act the mandatory deployment of PTC systems by December 31, 2015, on Class 1 railroad lines with annual traffic of 5 million gross tons (MGT) or with hazardous shipments classified as poisonous by inhalation (PIH) material, and on intercity and commuter railroad lines. This basically calls for a nationwide deployment of PTC. By one estimate, about 69,000 route miles (111,000 km) and 20,000 locomotives are required by this federal mandate to be equipped with PTC systems.

This statutory requirement hastened the development of PTC in the rail industry. Instead of piecemeal and unconnected developments, the industry made a concerted effort to develop technology in areas common to all, especially when interoperability is involved. Interoperability in PTC means the ability of a controlling locomotive to communicate with, and respond to, any railroad's PTC system, including uninterrupted movement over property boundaries. In the United States, locomotives owned by a railroad and equipped with onboard equipment unique to that railroad can haul a freight train across the boundary between one railroad and another, known as run-through operation. Therefore, it is important to have an interoperability standard to enable the onboard system to communicate and function properly, no matter what territory the train traverses.

The Class 1 railroads formed the Interoperable Train Control (ITC) Committee to develop this interoperability standard, which is scheduled to be completed in late 2010. Interfaces, messages, and protocols are important aspects of this standard; however, the standard will also specify other aspects, such as the system architecture, braking algorithm, communication networking, locomotive platform, and wayside equipment. It is important not to underestimate the significance of this work, as the successful implementation of the interoperability standard is a prerequisite for nationwide PTC deployment.

Completing the interoperability standard, finalizing the PTC designs, and deploying the PTC integrated systems nationwide by 2015 are challenging tasks. The railroad industry, however, is committed to meeting the deadline. In addition to the interoperability standard, priorities have been placed on developing the new 220-MHz data radio network with sufficient bandwidth and throughput to handle PTC, and developing the improved braking algorithm, as was discussed earlier. However, once the nationwide initial rollout of the PTC system is completed, work will continue on enhancing the technology and leveraging the platform to improve its operating efficiency and line capacity.

Developments that are underway include:

1. Higher-accuracy train-tracking technology to provide high-integrity measurements within 10–20-cm (4–8-in.) resolution, so that the PTC system can accurately distinguish, on a rail line with multiple tracks, the track the train is traversing even if the tracks' center-to-center distance is around 4–4.6 m (13–14 ft).

2. Methodology for data encryption and encryption key distribution, so that the operation of the railroad will not be affected by adding this layer of security.

3. Advanced digital radio technologies, such as using computer modeling and neural networks, to increase the data throughput and reduce interference, especially in a metropolitan area where many railroads operate concurrently.

Outlook. It is envisioned that PTC will be an ever-evolving technology. Once it is deployed nationwide, there will be a wholesale change in the railroad operating technology. As the railroads become more comfortable with the new technology, it will be expanded to applications that will give the railroads benefits other than safety. These include:

1. Merging with other systems, such as locomotive diagnostics and control systems, so that the cost of onboard systems can be reduced.

2. Merging with locomotive operation guidance systems. These separate systems continuously make evaluations of operating parameters based on current traffic conditions, terrain, and locomotive performance and recommend to the locomotive engineers how to control the trains in real time to improve train handling, fuel efficiency, and equipment reliability and durability.

3. Implementing moving-block or flexible-block operation instead of the current fixed-block operation. Trains are currently separated by blocks with fixed distances, determined by the longest possible stopping distances for freight trains. Short and light trains are unnecessarily separated by large gaps because of the constraints of fixed blocks. A conceivable operation with PTC is to have no block designation, but to separate trains using real-time calculation of their actual stopping distances. Thus, the headway of trains can be shortened and the capacity of a rail line can be increased.

The possible free flow of real-time data and operating parameters among the central office, trains, wayside devices, and rail workers means that rail companies can improve their business practices, as well as their safety records. These business applications provide better maintenance planning, optimize tactical decisions on everyday operations, afford better on-time performance, and improve asset management, all of which boost customer satisfaction, thereby increasing freight volumes and ridership. Potentially, PTC can be a strategic platform to elevate the U.S. railroad industry to a new level of competitiveness based on service, performance, and safety.

For background information *see* AIR BRAKE; ALGORITHM; CRYPTOGRAPHY; LOCOMOTIVE; NEURAL NETWORK; RADIO BROADCASTING; RAILROAD CONTROL SYSTEMS; RAILROAD ENGINEERING; SATELLITE NAVIGATION SYSTEMS in the McGraw-Hill Encyclopedia of Science & Technology Terry Tse

Bibliography. P. J. Detmold, New concepts in the control of train movement, *Journal of the Trans-*

portation Research Board, No. 1029, Transportation Research Board of the National Academies, Washington, D.C., pp. 43–47, 1985; Federal Railroad Administration (FRA), *Implementation of Positive Train Control Systems*, FRA, U.S. Department of Transportation, 1999; Federal Railroad Administration (FRA), *Positive Train Control Systems Economic Analysis*, Docket No. FRA-2006-0132, Notice No. 1 RIN 2130-AC03, FRA, U.S. Department of Transportation, 2009; Federal Railroad Administration (FRA), *Positive Train Control: Final Rule*, Docket No. FRA-2008-0132, Notice No. 3 RIN 2130-AC03, FRA, U.S. Department of Transportation, January 15, 2010; *Rail Safety Improvement Act of 2008*, 110th Cong., 2nd sess., H.R. 2095; The Railroad Association of Canada and the Association of American Railroads, *Advanced Train Control Systems Operating Requirements*, 1984.

Protein crystallization

Proteins are biological macromolecules that carry out functions crucial for living organisms. Proteins make thermodynamically possible and kinetically achievable all biochemical processes and in this way regulate and coordinate the birth, growth, nutrition, excretion, division, reproduction, and, eventually, death of the organism. Proteins are fundamental elements in the replication and transfer of the genetic information stored in DNA and RNA molecules, thus ensuring the survival of the species. Proteins are the main structural element supporting the shapes and the morphological complexity of the living world. Proteins transduct the chemical and neuron signals needed for the adjustment of the organisms to their ever-changing environments.

Protein molecules are heteropolymers of about 20 amino acids linked by peptide bonds into linear chains (**Fig. 1**). One protein chain may contain up to several hundred amino acid residues. In these amino acids, the amine group is attached at the α-position and is separated by one methylene group from the carboxyl residue. The amino acids differ in the side chain attached to the same α-position: the side chains vary from a single hydrogen atom in glycine to the bulky arginine, in which a three-carbon aliphatic chain is capped by a complex guanidinium group. Counting the amine group, the side chain, the carboxyl group, and remaining hydrogen atom, the α-carbon atom has four different ligands, and it is chiral. In all proteins in nature, this carbon atom is in the L-conformation, so that all naturally existing amino acids are α-L-amino acids. The amino acid residues can be classified according to their polarity as nonpolar or polar, and as positively or negatively charged. The nonpolar residues consist of straight or branched aliphatic chains and may contain aromatic rings; the polar ones contain alcohol or amide groups. The positive side chains contain one or more extra amino groups, which associate with a positive hydrogen ion in acidic solutions. The negative amino acids contain an extra carboxylic group, which may dissociate and form an anion in basic solutions.

In their native environments, the protein chains do not exist as random coils but are folded into compact structures. These structures are unique and representative for every protein. The folded chain is not fixed in place, and considerable dynamics exist. These dynamics are important for understanding the many aspects of protein behavior and function, with two examples being binding to substrates and nucleic acids and resistance to aggregation and proteolysis. In folded protein molecules, the hydrophobic nonpolar residues are tucked inside and are mostly in contact with other such residues from the same molecule. The charged and polar residues are mostly on the outside and ensure favorable interactions with water and therefore protein solubility. The amino and carboxyl groups in the side chain are weak bases or acids and have pK_a (acid dissociation constant) values in a broad range, from 4.0 for glutamic acid to 12.48 for arginine. Hence, at a set pH value, some of the surface residues are charged and others are neutral. In this way, the pH of the environment in which a protein exists and functions regulates its charge.

The proteins are divided into two general classes: globular and transmembrane. The globular proteins are water-soluble and are present in the fluid part of the cytoplasm (cytosol) of live cells and in the intracellular and body fluids such as blood, stomach juice, and so on. The transmembrane proteins are embedded in the cellular membranes and serve to transmit signals between the cellular environment and the cytosol, to transfer nutrition into cells and waste out, and to strengthen and control the membrane. To be able to perform these and other functions, the transmembrane proteins are structured as a hydrophobic midriff capped with two hydrophilic plates: one facing the cytosol and the other facing the cellular environment.

Crystals of proteins. In the early days of biology and biochemistry, it was assumed that crystals such as those of diamond, quartz, or calcite existed only in the mineral world, and that no living organism would contain them. The first protein crystals, of hemoglobin, were discovered accidentally in 1840, in dried earthworm blood. By 1871, crystals of hemoglobin from nearly 50 species had been reported. By the end of the nineteenth century, crystals of numerous other proteins of both plant and animal origin had been obtained. Crystal formation was considered a criterion for purity of the protein preparation. In the further discussions of the nature of proteins, their ability to form crystals was one of the major arguments supporting their molecular nature and refuting the hypothesis, common at that time, that proteins were colloid particles of disordered matter (**Fig. 2**).

Protein crystals are of great interest in science and technology because their formation underlies a number of human pathological conditions. An example is

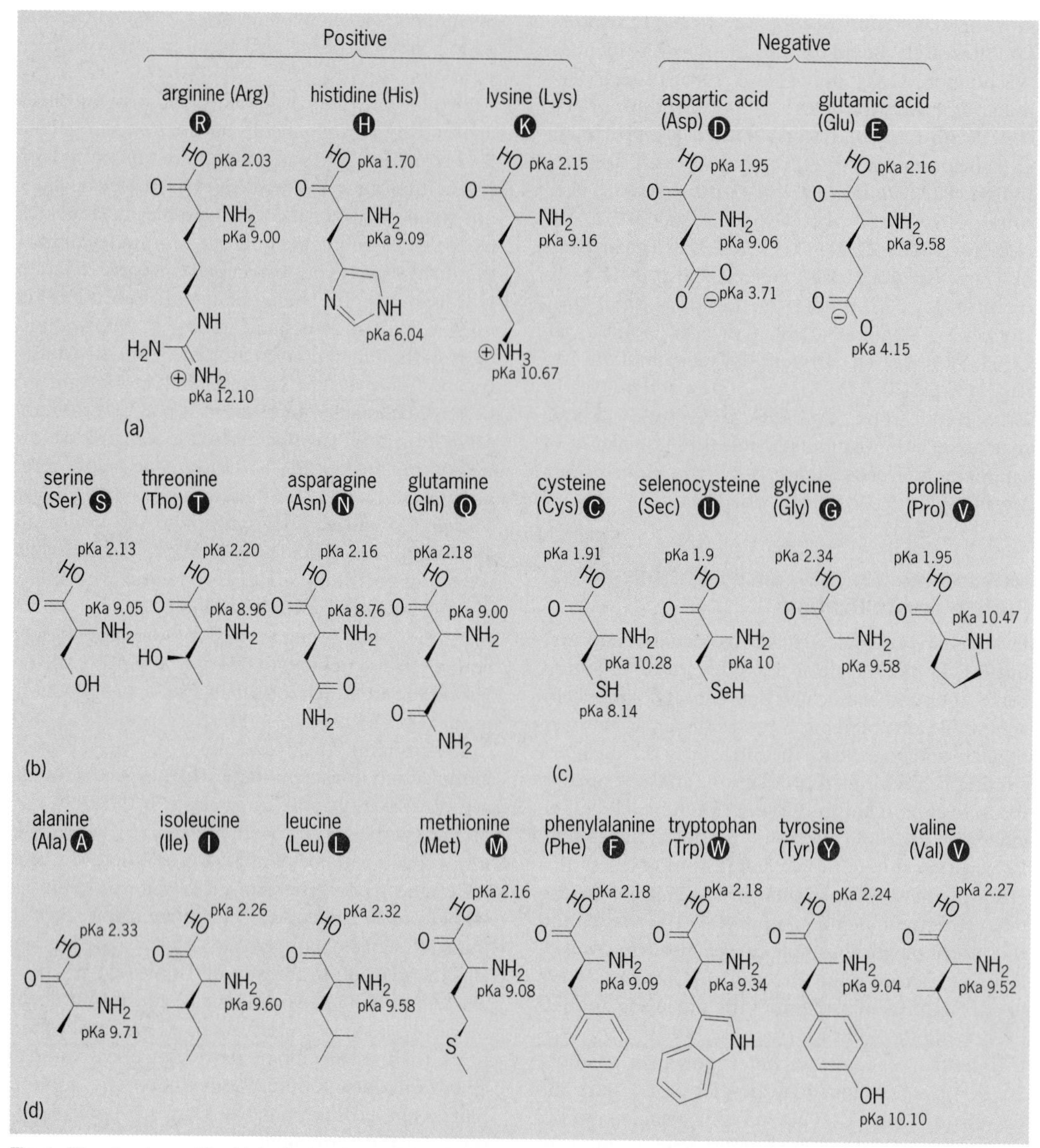

Fig. 1. **The structure of the biological amino acids and selenocysteine. (*a*) Amino acids with electrically charged side chains. (*b*) Amino acids with polar uncharged side chains. (*c*) Special cases. (*d*) Amino acids with hydrophobic side chains. (*pKa data from CRC Handbook of Chemistry, by Dan Cojocari, available under a Creative Commons Attribution—Noncommercial license*)**

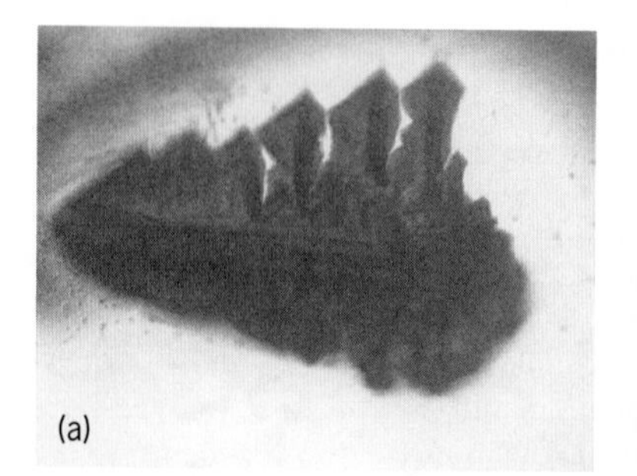

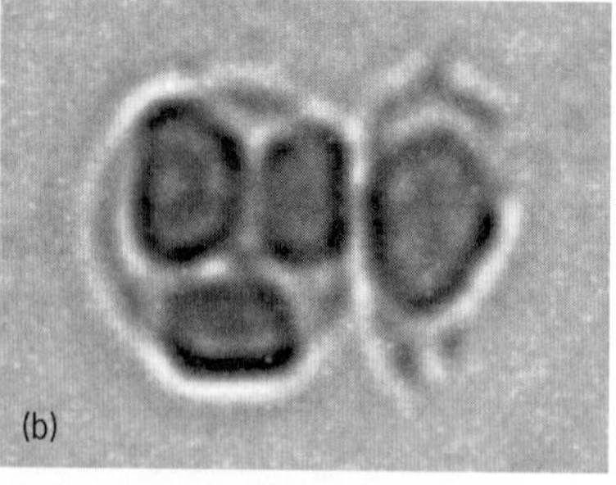

Fig. 2. **Crystals of human hemoglobin C, a relatively rare mutant associated with CC disease. (*a*) A crystal of about 300 μm in width that was grown in the laboratory and imaged with bright-field microcopy. (*b*) Two red blood cells containing crystals of hemoglobin C; the diameter of the cell on the right is about 7 μm; imaging by differential-interference contrast microscopy.**

the crystallization of hemoglobin C and the polymerization of hemoglobin S, which cause, respectively, hemoglobin-CC and sickle-cell diseases. The formation of crystals and other protein condensed phases of the crystalline bodies in the eye lens underlies the pathology of cataract formation. A unique example of benign protein crystallization in humans and other mammals is the formation of rhombohedral crystals of insulin in the islets of Langerhans in the pancreas. The suggested function of crystal formation is to protect the insulin from the proteases that are present in the islets of Langerhans and to increase the degree of conversion of the soluble proinsulin.

Another field that relies on protein crystals and dense liquid droplets is pharmacy, in which the slow crystal dissolution rate is used to achieve the sustained release of medications, such as insulin,

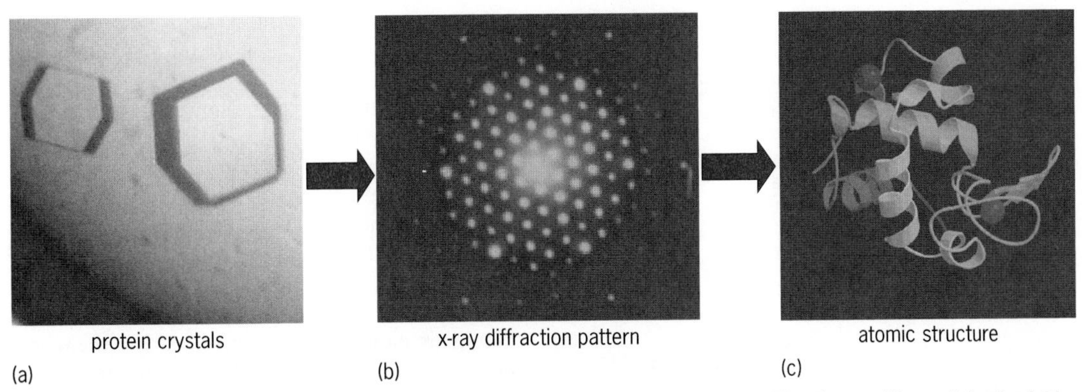

Fig. 3. Crystallography is the main route to structure determination. The preparation of diffraction-quality crystals of soluble and membrane proteins is the bottleneck in this process. (*a*) Protein crystals. (*b*) X-ray diffraction pattern. (*c*) Atomic structure.

interferon-α and human growth hormone. Also underway is work on the crystallization of other therapeutically active proteins, for example, antibodies against foreign proteins, which can be dispensed as microcrystalline preparations. If the administered dose consists of a few equidimensional crystallites, steady medication release rates can be maintained for longer periods than for doses comprised of many smaller crystallites.

Traditionally, protein crystals have been used in the determination of the atomic structure of protein molecules by x-ray crystallography. This method contributes about 87% of all protein structures solved, with the majority of the other determinations done by nuclear magnetic resonance (NMR) spectroscopy (**Fig. 3**).

Methods of protein crystallization. The laboratory techniques for protein crystallization account for the fact that the expression, extraction, and purification of a protein are often time-consuming procedures. Hence, wide use is being made of various micro methods of protein crystallization, employing microdialysis and microdiffusion cells, allowing one set of crystallization experiments in protein solution droplets of 5–20 μl containing only 0.1–0.5 μg of the protein (**Fig. 4**).

There are numerous commercial cells for microdialysis. The basic arrangement is a membrane or a gel plug, impenetrable for the protein, that lets through the low-molecular components of the solution. The protein solution and a buffer is held on one side of the membrane, while the precipitant is on the other. Crystallization occurs when the precipitant, which lowers the protein solubility, diffuses through the membrane into the protein-containing cell.

In the simplest hanging-drop cell, a drop of protein solution containing a small amount of precipitant is placed on a glass plate. Next, the plate with the drop is turned over and placed on top of a cup containing a precipitant solution in concentration sufficient to precipitate the protein. After several days of (mainly water) vapor diffusion, the precipitant and protein concentration in the drop rises and the protein starts to crystallize. The sitting-drop technique is similar, with the drop sitting on top of a pedestal rising above the solution reservoir. In the sandwiched-drop technique, the protein solution is held between two surfaces. Recently, a microbatch technique was proposed, in which the protein and precipitant solutions are mixed together and held between two layers of silicone, perfluorosilicone, or lipid oil. The water has extremely low solubility in any of these oils, so its escape from the solution droplet is very slow. This provides for a gentle increase in supersaturation and enhances the chances of crystal formation.

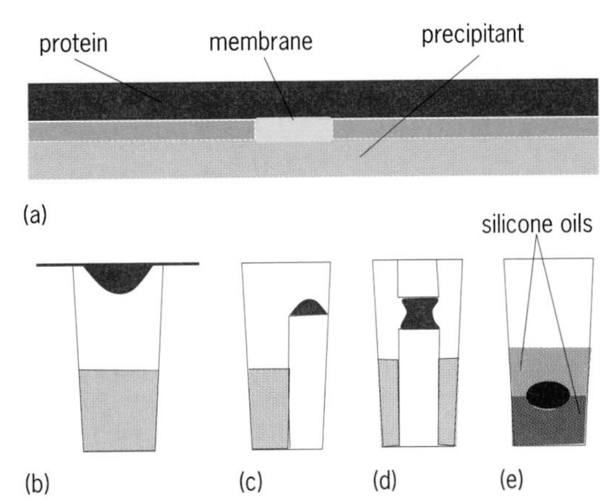

Fig. 4. Laboratory methods of protein crystallization. (*a*) The microdialysis method, in which the volumes holding the protein and the precipitant are separated by a semipermeable membrane and the crystals form in the protein volume. (*b*) The hanging-drop method. (*c*) The sitting-drop method. (*d*) The sandwiched-drop method. (*e*) The microbatch method.

The crystallization conditions of protein, precipitant and buffer, type and concentration of additives, pH, heterogeneous "nucleants," and others are usually found by trial and error, which involves screening tens or even hundreds of drops. At present, special computer-controlled installations (robots) are used to screen thousands of protein crystallization conditions. Complete automation of the liquid handling increases the speed, precision, and reproducibility of the experiment. Recent advances of these methods, based on microfluidics, have reduced the size of the solution droplets for

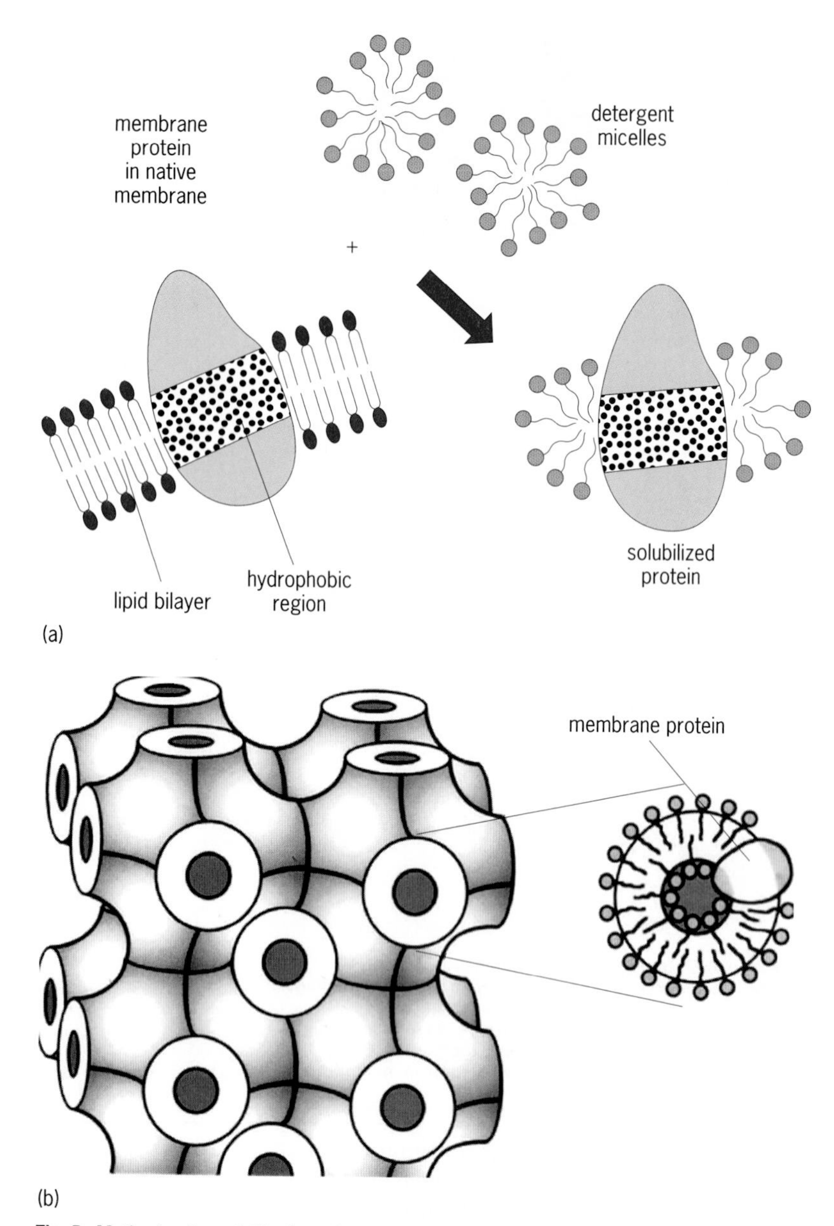

Fig. 5. Methods of crystallization of membrane proteins. (*a*) Solubilization by coating the hydrophobic midriff of the molecules with detergent. (*b*) The membrane protein molecules are embedded in lipidic cubic structures, like the one shown on the left, in which the protein molecule is in an environment resembling that in the cell membrane.

crystallization down to the nanoliter scale, thus reducing the amount of protein required.

To somewhat rationalize the trial-and-error approach, the influence of various ions on the protein solubility, dependence of growth rates of different crystal faces on the protein concentration in the solution, convection and flow rates close to the growing crystal, and epitaxy of the protein crystals have all been investigated. Relatively weak interactions between the large protein molecules in solution (per unit molecular area), peculiarities of protein structure, and the multicomponent composition of the solutions make the crystallization process extremely sensitive to physical and chemical conditions. The same protein can crystallize in several crystalline forms, depending on the solution composition and pH. For instance, lysozyme crystallizes in tetragonal, monoclinic, triclinic, and orthorhombic polymorph modifications.

Crystallization of membrane proteins. In their native environments, membrane proteins are embedded in the phospholipid bilayer of the cell membrane, where they are held by their hydrophobic midriffs. Two approaches to extract the membrane proteins from the membrane and to make them mobile have been designed. In the first approach, the membrane proteins are solubilized by coating the midriff with detergent molecules, akin to enclosing them in a detergent micelle. A search for crystallization conditions of the solubilized protein is then done as for the water-soluble proteins. The second approach relies on the existence of three-dimensional (3D) structures in high-concentration solutions of some lipids. These structures consist of a lipid bilayer warped into a frame and interlaced with aqueous channels. Both the lipid and the aqueous subphases in these structures are continuous, and this allows free diffusion of the protein (embedded in the lipid bilayer) and precipitant (in the aqueous channels), leading to crystallization (**Fig. 5**).

Nucleation of protein crystals. The formation of protein crystals is a first-order phase transition. Accordingly, it is characterized with nonzero latent heat. More significant for the crystallization kinetics is the second feature of first-order phase transitions: the discontinuity in the protein concentration at the phase boundary. As a result of this discontinuity, the solution–crystal interface possesses nonzero surface free energy. If a small piece of a crystal forms in a supersaturated solution, the surface free energy of the emerging phase boundary makes this process unfavorable. Thus, a very limited number of crystal embryos appear as a result of the few fluctuations that overcome the free-energy barrier. The first step in crystal formation, in which the kinetics of the phase transformation are determined by this barrier, is called nucleation (**Figs. 6** and **7**).

Nucleation for protein crystals is the crucial part of the search for protein crystallization conditions in the field of structural biology, as typically most of the crystallization trials yield no crystals. In cases in which protein crystallization underlies a pathology, suppression of nucleation is needed. The use of protein crystals in industry and pharmacy also requires precise control of nucleation, which can only be achieved with understanding of the nucleation mechanisms.

Determinations of the rates of nucleation of protein crystals have revealed that they are about 10 orders of magnitude (that is, by a factor of 10 billion) slower than predicted. This finding may be the principal explanation for the difficulty of protein crystallization, and hence it was studied in detail. It was found that the reason for these extremely slow nucleation rates is the fact that protein crystal nucleation does not follow the mechanism envisioned by this classical theory, according to which molecules that comprise the solution assemble in well-ordered

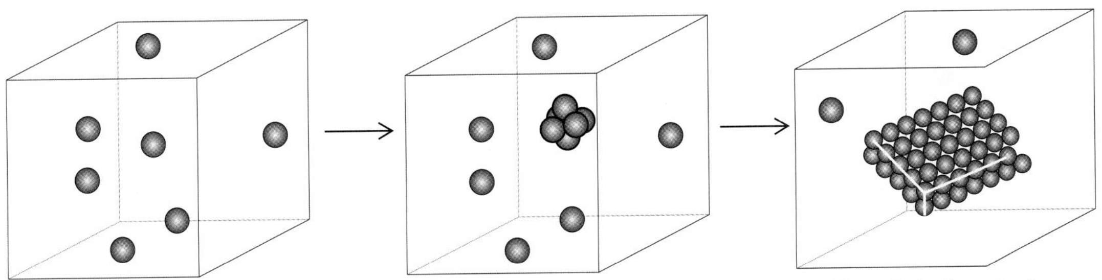

Fig. 6. Crystallization is the formation of a 3D translationally symmetric array of molecules, as illustrated schematically on the right, from a disordered and dilute state in the solution, as illustrated on the left. In contrast to what is shown, a crystal often consists of billions of parallel layers. Because of the surface free energy of the interface between the crystal and the solution, which emerges during crystallization, the formation of the first crystalline embryo, called the nucleus and shown in the middle, has to overcome a barrier. This first step of crystallization is called nucleation and is followed by the second step, growth of the nucleus to macroscopic crystals, called the nucleation-and-growth scenario.

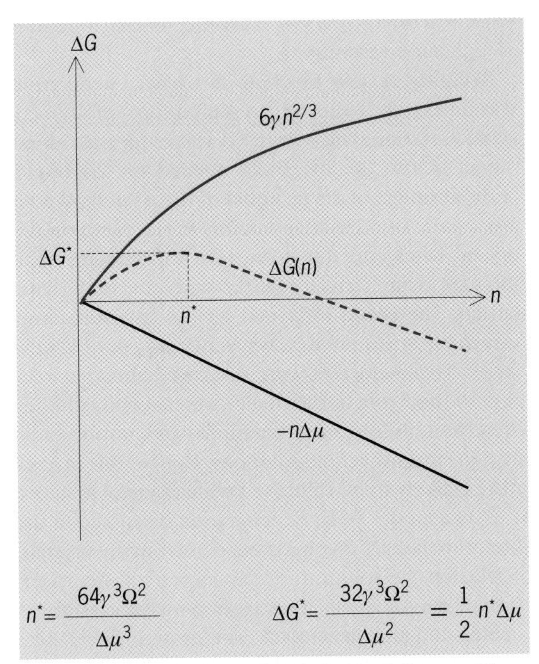

Fig. 7. Free-energy pathway for nucleation. The transition of n molecules from a solution supersaturated by $\Delta\mu$ with respect to a crystal to this crystal is accompanies by a free-energy gain of $-n\Delta\mu$. The formation of the interface between the new crystal and the solution leads to a free-energy loss of $6\gamma n^{2/3}$. The sum of these two contributions is the free energy along the nucleation pathway, $\Delta G(n)$, which has a maximum at n^*. This n^* is the nucleus size, and $\Delta G^* = \Delta G(n^*)$ is the nucleation barrier. The values of n^* and ΔG^* emerging from the classical nucleation theory for cubic nuclei are shown.

arrays. A two-step mechanism of protein crystallization was proposed. In step one, a cluster of about 1 million protein molecules in a dense liquid forms in the solution. In step two, a crystal nucleus of about 10 protein molecules forms inside this cluster. The reason for the slow overall nucleation rate of the crystal is that the clusters occupy a very low volume fraction, about 10^{-8}–10^{-6}, and are highly viscous (**Fig. 8**).

The two step mechanism shows that enhanced nucleation rates may be possible under conditions that favor the existence of greater numbers of low-viscosity clusters. Investigations of the properties of the dense liquid clusters, still in their infancy, have revealed that they are controlled by the hydrophilic, hydrophobic, and specific interactions among the protein molecules. Thus, by controlling these interactions, it may be possible to achieve a higher cluster volume fraction, more favorable cluster properties, and precise control of the nucleation rate.

Mechanisms of Growth

The elementary act of crystal growth is the attachment of molecules from the solution. This attachment occurs at sites called kinks, in which an incoming molecule has half the number of neighbors that it would have in the crystal bulk. The kinks are defined as special sites for growth because of two specificities of attachment there: the kinks are retained after the attachment, and the attachment does not alter the surface free energy of the crystal. The rate constant of growth of a crystal is determined by two factors: the density of kinks on the interface with the growth medium, and the barriers, both entropic and enthalpic, for incorporating a molecule into a kink.

Protein crystals typically exhibit smooth interfaces because of high surface free energy between the crystal and the growth medium. The smooth interfaces between crystals and solution are usually crystal planes with a high density of molecules, designated with low Miller indices. This is because these planes are the slowest to grow; the faster-growing planes taper out because of geometry and disappear from the crystal faceting. Thus, the notion that a macroscopic crystal is faceted by planes that minimize its surface free energy is a misconception, albeit a common one. In fact, the anisotropy of the surface free energy only affects the shape of crystals of near-equilibrium size—that is, the size of the nucleus.

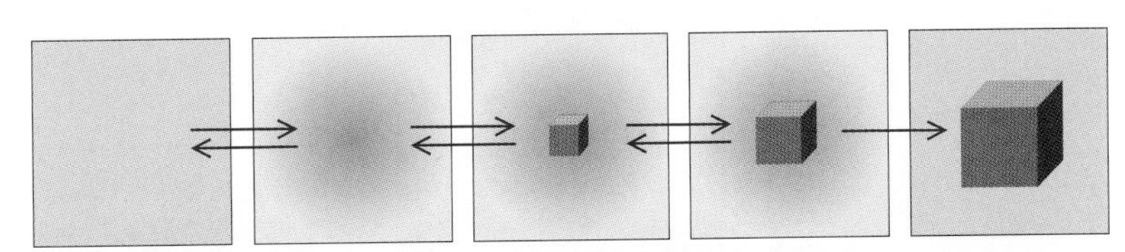

Fig. 8. The two-step mechanism of nucleation. In a dilute solution, depicted in the leftmost panel, a droplet of a dense liquid forms, as depicted in the second panel, and this is the first step of the mechanism. A crystalline nucleus forms within this droplet, as depicted in the third panel, and this is the second step of the mechanism. The nucleus then grows to a macroscopic crystal, as in the last two panels.

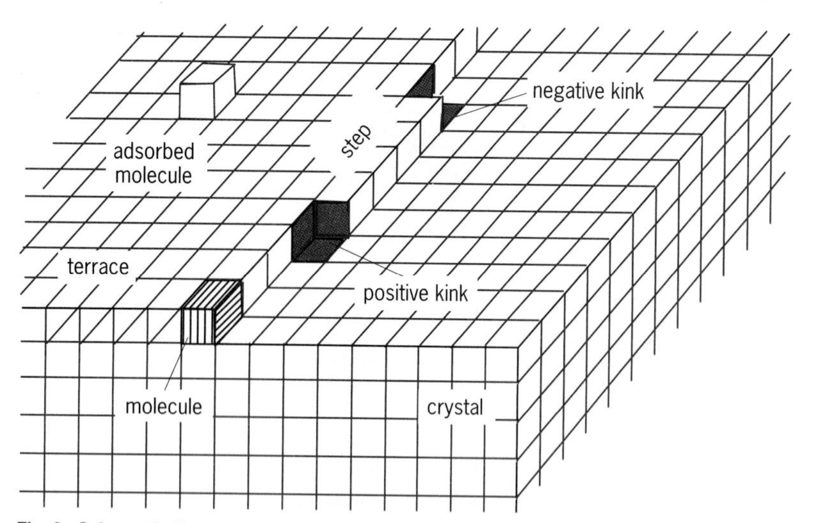

Fig. 9. Schematic illustration of the structure of the surface of a faceted crystal. The edges of the incomplete layers are called steps. The flat terraces between the steps are the singular crystal planes. The kinks on faceted crystals are located at the steps.

From the above discussion, we see that crystals in contact with the solution are typically faceted and follow the layer growth mode. In this mode, a new lattice layer, typically one lattice spacing high, is deposited on the smooth surface of the previous lattice layer. The edges of the incomplete layers are called steps. The flat terraces between the steps are the singular crystal planes. The kinks on faceted crystals are located at the steps (**Fig. 9**).

Experimental determinations of the step density on the surfaces of growing protein crystals have yielded numbers of the order 10^{-2}–10^{-3}. The density of the kinks along the steps is from 10^{-2} to 1. This relatively low density of kinks on the surface of a growing protein crystal highlights the significance of three issues for the regulation of the crystal growth rate: the mechanism of generation of kink, the pathway of the molecules from the solution into the sparse kinks, and the kinetics of incorporation of a molecule into a kink. In turn, the issue of kink density subdivides into generation of steps and generation of kinks along the steps.

Layer generation. During the growth of faceted crystals, new layers are generated and spread to cover the whole face. New layers are generated, by several mechanisms, only in supersaturated solutions. A common layer-generation mechanism is by screw dislocations, piercing the growing facet.

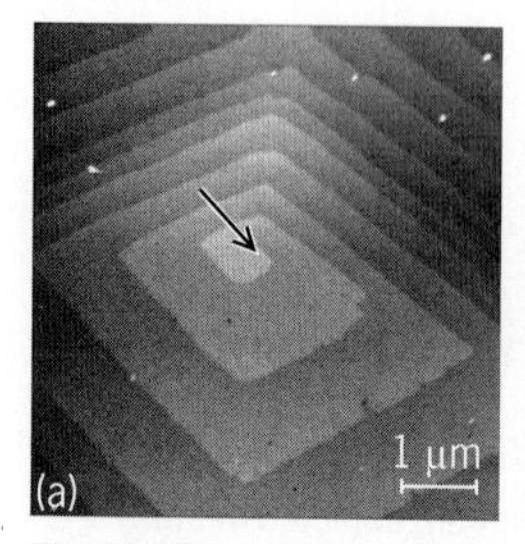

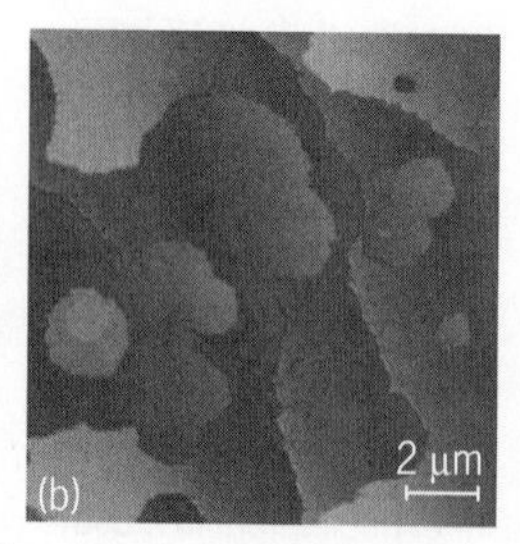

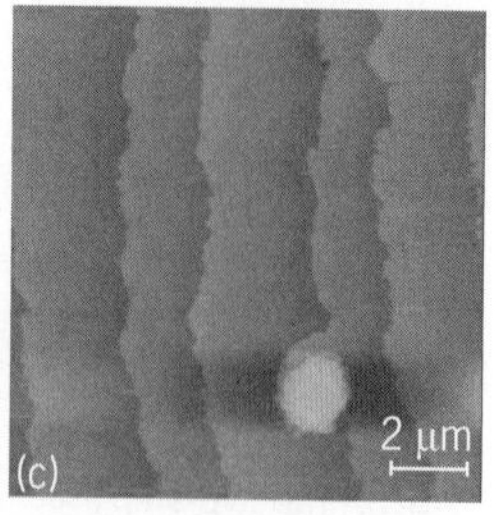

Fig. 10. Three mechanisms of generation of crystalline layers. (*a*) Screw dislocation outcropping on the face, showing a (100) insulin face. (*b*) 2D nucleation, showing a (111) ferritin face. (*c*) Landing and subsequent crystallization of metastable clusters of dense liquid, showing a (0001) face of a lumazine synthase crystal.

The dislocation produces a step on the facet, which terminates and is pinned at the point where the dislocation outcrops on the surface. The step grows in a supersaturated solution and, because of the pinned end, twists into a spiral around the dislocations. If step motion is isotropic, the spiral will be circular; if the velocity of step propagation is faster in some directions and slower in others, the spiral will develop edges at the faster directions and become polygonized. A typical example of this mechanism can be seen during insulin crystallization (**Fig. 10**).

Another common layer-generation mechanism is by two-dimensional (2D) nucleation of islands of new layers. This is the original layer-generation mechanism, put forth by I. N. Stranski and R. Kaischew in the 1930s. This mechanism operates at high supersaturations.

Recently, a new mechanism of layer generation was discovered during crystallizations of the enzyme lumazine synthase. In a certain supersaturation range, below the threshold needed for 2D nucleation, droplets of dense liquid of the protein, which are several hundred nanometers in size, land on the crystal facet and transform into a crystalline matter that is in perfect registry with the underlying lattice. The island of layers, several lattice parameters thick, spreads sideways, generating several new steps. These clusters were discussed above in relation to their role in the nucleation mechanism; they are common in protein solutions, and, importantly, also in small-molecule solutions. Hence, this mechanism is likely to be valid for a wide range of systems.

Other modes of layer generation discussed in the literature mostly involve gross defects in the crystals: occlusions of solution, or the imperfect incorporation of microcrystals into larger growing crystals.

Step and kink generation. The growth rate R of a faceted crystal, measured in a direction perpendicular to the growing facet, is the product of the step velocity and the mean step density. The step density is determined by the rate of step generation by one of the three mechanisms discussed above. The step velocity is determined by three factors: the density of kinks along the step; the barrier for incorporation of molecule into a kink; and the concentration of the molecules in the immediate vicinity of the kink, determined by the pathway taken by the molecules from the solution into the kinks. These three factors are discussed below.

In general, kinks along the steps are generated by one of three mechanisms: thermal fluctuations of the step edge, "one-dimensional (1D) nucleation" of new molecular rows, and the association of 2D clusters diffusing on the terraces between the steps.

J. W. Gibbs suggested that the edges of the unfinished layers (the steps) fluctuate and in this way create kinks. It was recently shown that the same mechanism applies in supersaturated solutions and determines the kink density during the growth of a step (**Fig. 11**).

Several other cases of steps with low kink density due to high kink energy have been studied, and it

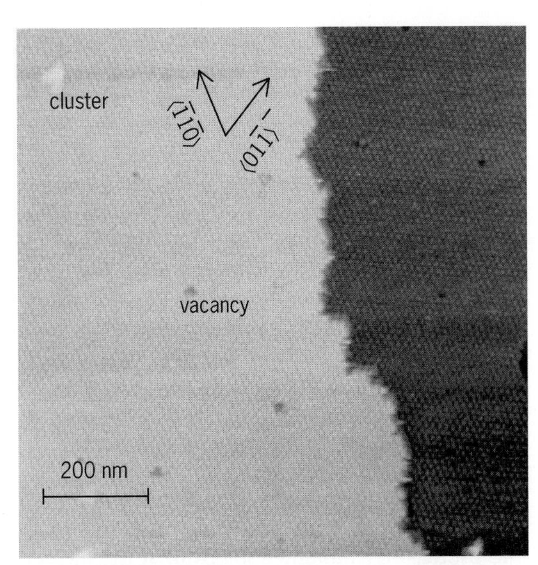

Fig. 11. Molecular structure of a growth step on an apoferritin crystal. The dark color is the lower layer and the light color is the advancing upper layer. Adsorbed impurity clusters and surface vacancies are indicated. The kinks are the ends of the unfinished rows of molecules at the step edge.

was found that instead of growing with correspondingly low velocity, the steps use additional mechanisms of kink generation: by "1D nucleation" of new molecular rows or by the association to the steps of clusters on the terraces. In contrast to 3D and 2D nuclei, a 1D nucleus cannot be defined thermodynamically. However, a 1D nucleus can be defined kinetically as a molecular row of length such that its probability of growing is equal to its probability of dissolving (**Fig. 12**).

A novel mechanism of kink generation was demonstrated for the crystallization of insulin, in which 2D clusters of several insulin molecules on the terraces between steps associate with the steps. This mechanism operates at moderate and high supersaturations, whereas, as discussed previously, at low supersaturations only kinks generated by thermal fluctuations exist. The mobility of clusters of several molecules is not surprising. They may be ordered or disordered, akin to a 2D liquid formed in the pool of insulin hexamers adsorbed on the terraces. A 3D analog of this process is layer generation by the landing of dense liquid droplets on the surface of an existing crystal.

Investigations of the barrier for incorporating molecules into kinks revealed that, as in the thermodynamics of protein crystallization, water structuring is the main determinant of the kinetics of crystal growth. It was shown that the rate of the elementary step of crystallization, that is, the attachment of a molecule from solution to an existing growth site on the crystal surface, is determined not by the rate of decay of a hypothetical transition state, but by the kinetics of destruction of the "shell" of structured water via a "diffusion-limited" kinetics scenario. Hence, the free-energy barrier that must be overcome by a protein molecule on its way toward

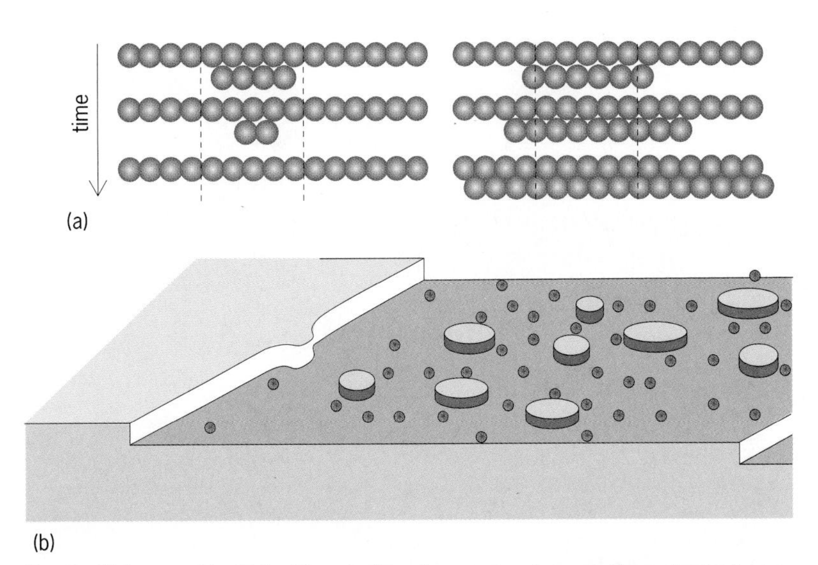

Fig. 12. Kink generation (*a*) by 1D nucleation of new molecular rows. Rows shorter than a critical length, denoted with vertical dashed lines, dissolve, as shown at left. Rows that are longer than this critical length grow. The association of (*b*) 2D clusters preformed on the terraces to the steps yields protrusions (one such protrusion is shown) that are rich in kinks. The protrusions spread sideways and promote the step forward.

its incorporation sites is determined by the strength of this water shell—that is, by the free energy of the hydrogen bonds within the layer of structured water. It was shown that the characteristic length of

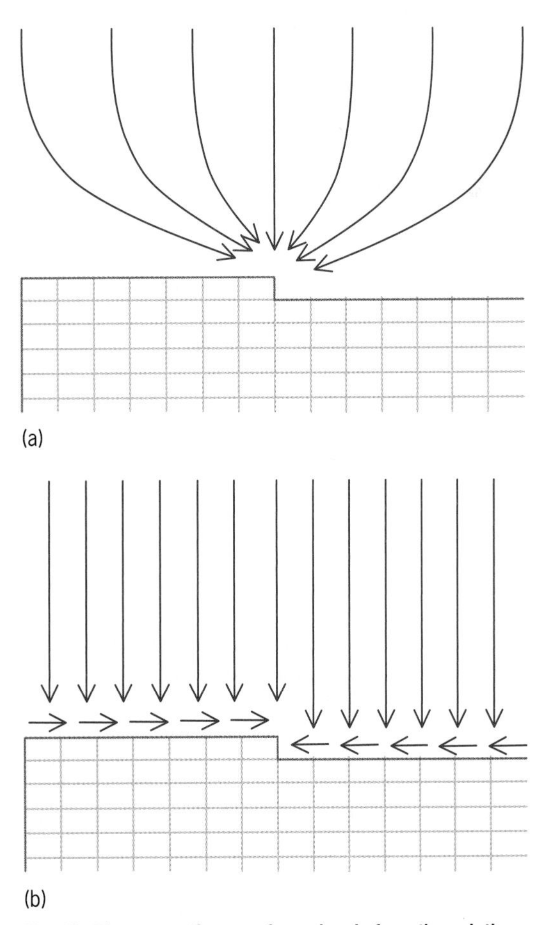

Fig. 13. The two pathways of a molecule from the solution into a step. (*a*) Direct incorporation from the solution. (*b*) Adsorption on the surface followed by surface diffusion and incorporation into a step.

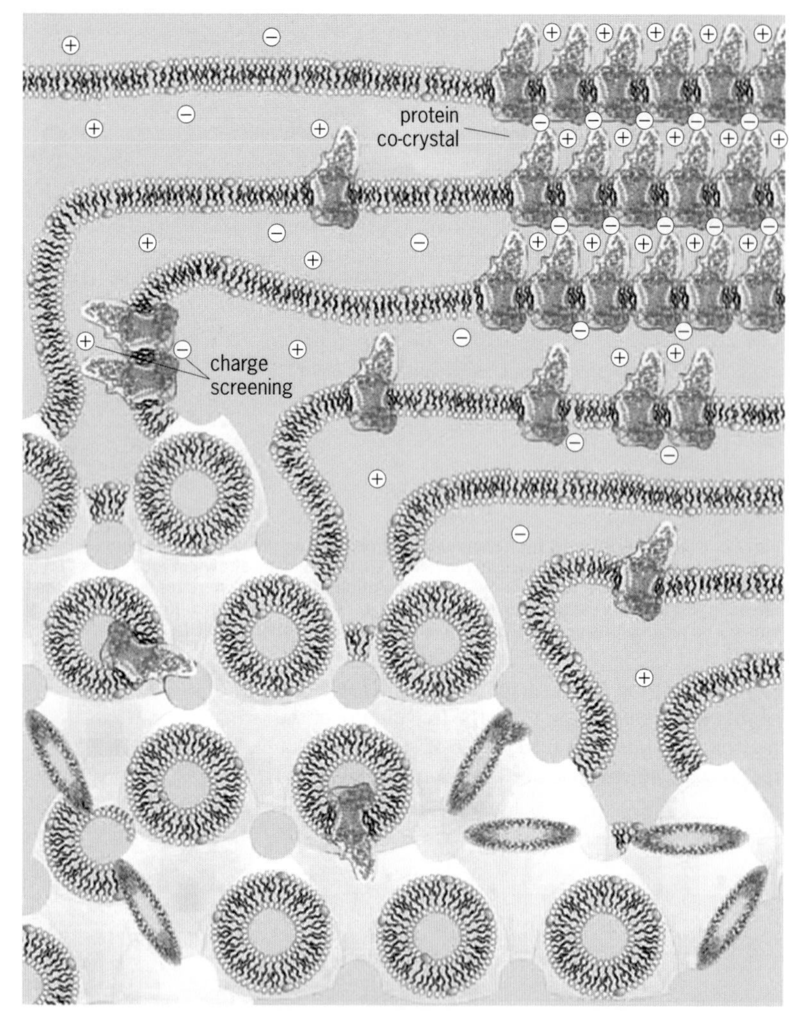

Fig. 14. Mechanism of the growth of membrane protein crystals in lipid cubic phases. The crystal is integrated into a lamellar subregion of the lipid structure. A disordered region surrounds the crystal and serves as an interface for the transfer of molecules between the lipid cubic phase and the crystal. (*From M. Caffrey, Membrane protein crystallization, J. Struct. Biol., 142(1):108–132, 2003*)

this free-energy barrier will be equal to the combined thickness of the solvation layers of the incoming molecule and the surface—that is, about 10–15 Å. Experiments with insulin revealed that when acetone was used to remove the structured water shell, the kinetic constant for the incorporation of molecules into a crystal increased by an order of magnitude.

Growth from solution. During crystal growth from solution, the solute molecules have two possible pathways between the solution and the kinks: they can be incorporated directly, or they can first adsorb on the terraces between the steps, diffuse along them, and then reach the steps (**Fig. 13**).

If a crystal grows by the direct incorporation mechanism, the competition for supply between adjacent steps is mild. On the contrary, competition for supply confined to the adsorption phase is acute: it retards step propagation and acts as a strong effective attraction between the steps. This dramatically affects the stability of the step train, the appearance and evolution of step bunches, and, ultimately, the crystal quality and utility.

The two mechanisms can be discerned by monitoring the adsorbed solute molecules on the crystal surface. Direct imaging of the diffusion of fluorescently labeled lysozyme molecules along the surface of a growing crystalline faces was monitored in elaborate recent experiments. Electron microscopy of flash-frozen samples has in several cases revealed the presence of adsorbed solute molecules on the crystal surface.

Indirect evidence for the growth mechanism of several systems has been sought by comparing the velocities of isolated steps to those of closely spaced steps. Slower growth of dense step segments was interpreted in favor of the surface diffusion mechanism for the proteins lysozyme and canavalin.

A study of the growth processes of crystals of the proteins ferritin and apoferritin at the molecular level compared the fluxes of molecules entering the steps to those leaving the steps. This and other pieces of evidence for this system allowed the conclusion that during the growth of ferritin and apoferritin crystals, the molecules from the solution enter the steps via a state of adsorption on the terraces between steps.

Growth mechanism of membrane protein crystals. The mechanisms of growth of membrane protein crystals are relatively less studied. It is generally assumed that solubilized membrane proteins follow growth mechanisms similar or identical to those of globular proteins. An interesting mechanism was found to operate during the growth of membrane protein crystals embedded in lipid cubic phases: crystal nucleation is accompanied by transformation of the lipid cubic structure into a lamellar one. The crystals are integrated into the lamellar structure, and they both grow as molecules associated to the crystal from the lipid cubic phase. A region of relatively disordered lipid bilayers surrounds the crystal and serves as an interface for the transfer of molecules from the lipid phase into the crystals during growth (**Fig. 14**).

For background information *see* AMINO ACIDS; CELL MEMBRANES; CRYSTAL DEFECTS; CRYSTAL GROWTH; CRYSTALLIZATION; CRYSTALLOGRAPHY; FREE ENERGY; HEMOGLOBIN; MICROFLUIDICS; NUCLEATION; PEPTIDE; PROTEIN; SUPERSATURATION in the McGraw-Hill Encyclopedia of Science & Technology.

Peter G. Vekilov

Bibliography. T. Bergfors, *Protein Crystallization*, 2d ed., International University Line, La Jolla, CA, 2009; A. McPherson, *Crystallization of Biological Macromolecules*, Cold Spring Harbor Laboratory Press, Cold Spring Harbor, NY, 1999; P. G. Vekilov, What determines the rate of growth of crystals from solution?, *Crystal Growth and Design*, 7(12):2796–2810, 2007; P. G. Vekilov and A. A. Chernov, The physics of protein crystallization, in *Solid State Physics*, vol. 57, H. Ehrenreich and F. Spaepen (eds.), Academic Press, New York, 2002.

Pulsar navigation

Throughout history, celestial sources have been utilized for vehicle navigation. Great ships have successfully sailed Earth's oceans utilizing only these celestial aides. Many vehicles operating in the space environment also navigate utilizing celestial sources for their missions. However, most space-vehicle operations rely heavily on Earth-based navigation solutions to complete their tasks. As the cost of vehicle operations continues to increase, spacecraft navigation is evolving away from Earth-based solutions toward increasingly autonomous methods. For vehicles operating near Earth, the current Global Positioning System (GPS), and similar human-developed systems, can provide a complete navigation solution comprised of referenced time, position, and attitude. However, these human-developed systems have limited scope for operations of vehicles relatively far from Earth. Thus there is a need for an autonomous celestial-based system that can be used to provide a complete navigation solution for spacecraft missions.

Recently discovered celestial sources may provide answers to navigating throughout the solar system and beyond. Neutron stars, also referred to as pulsars, spin at exceptionally fast rates, with immense magnetic fields that create a stable, predictable, unique radiation signature. These celestial sources provide periodic pulsed radiation that can be utilized in a navigation system for spacecraft. Given their large distances from Earth, pulsars provide good signal coverage for operations within the solar system, and, conceivably, the Galaxy. Issues with these sources exist that make their use complicated. However, further algorithmic and experimental study may address these complications for deep-space navigation missions.

Prediction and discovery of pulsars. During the 1930s physicists began to develop the theory of general relativity and stellar structure. The theoretical existence of black holes arose, predicted to be small superdense objects with immense gravitational fields, so large that even light could not escape. These theories also predicted that upon their collapse, stars with insufficient mass to create a black hole would produce several types of ultradense, compact objects. One such proposed object is a neutron star. A neutron star is the remnant of a massive star that has exhausted its nuclear fuel and undergone a supernova explosion. Young neutron stars typically rotate with periods of the order of tens of milliseconds, while older neutron stars eventually slow down to periods of the order of several seconds. A unique aspect of this rotation is that it can be extremely stable and predictable. In addition to maintaining stable rotation periods, neutron stars produce immense magnetic fields. Under the influence of these strong fields, charged particles are accelerated along the field lines to very high energies. As these charged particles move in the pulsar's strong magnetic field, powerful beams of electromagnetic waves are radiated out from the magnetic poles of the star. If the neutron star's spin axis is not aligned with its magnetic field axis, then an observer will sense a pulse of high-energy photons as the magnetic pole sweeps across the observer's line of sight to the star (**Fig. 1**). Neutron stars that exhibit this behavior are referred to as pulsars. Since no two neutron stars are formed in exactly the same manner, the pulse frequency and shape produces a unique identifying signature for each pulsar. Because of their unique pulses, pulsars would act as natural beacons, or celestial lighthouses, on an intergalactic scale.

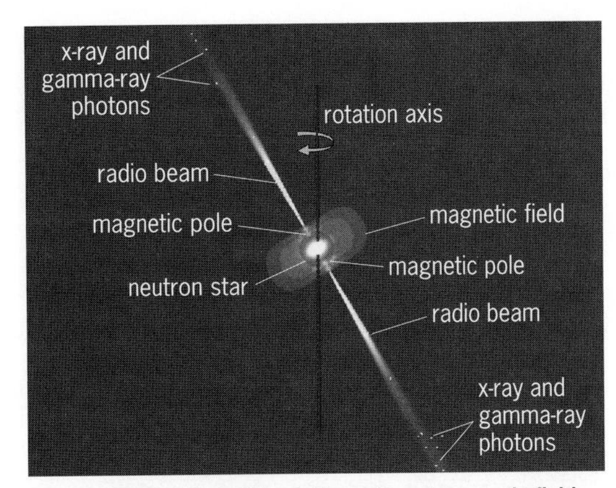

Fig. 1. Neutron star with rotation axis and magnetic field.

In 1967, Anthony Hewish and Jocelyn Bell discovered radio pulsations during a survey of scintillation phenomena due to interplanetary plasma at the radio frequency of 81.5 MHz. Among the expected random noise, signals emerged timed at regularly spaced intervals, having a period of about 1.337 s and constant to better than one part in 10^7. Since their discovery, pulsars have been found to emit at the radio, infrared, visible (optical), ultraviolet, x-ray, and gamma-ray energies of the electromagnetic spectrum. (**Fig. 2**).

History and principles. In 1974, G. S. Downs presented a method of navigation for orbiting spacecraft based upon radio signals from a pulsar. However, both the radio and optical signatures from pulsars have limitations that reduce their effectiveness for spacecraft missions. Optical pulsar-based navigation systems would require a large aperture to collect sufficient photons, since few pulsars exhibit bright optical pulsations. The large number of visible sources requires precise pointing and significant processing to detect pulsars in the presence of bright neighboring objects. This is not attractive for many spacecraft-vehicle designs. At the radio frequencies that pulsars emit (from around 100 MHz to few gigahertz) and with their faint emissions, radio-based systems would require large antennas of the order of 25 m (80 ft) in diameter or larger to detect sources, which would be impractical for most spacecraft. Also, neighboring celestial objects including the Sun, Moon, close stars, and Jupiter, as well as distant objects such as supernova remnants, radio

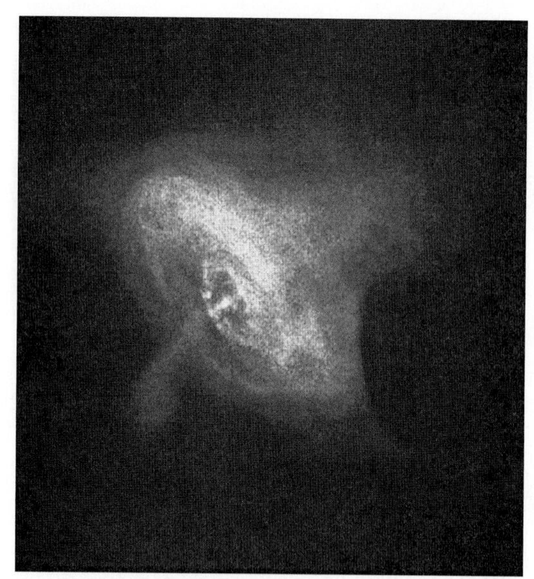

Fig. 2. Crab Nebula and pulsar in the x-ray band, imaged by NASA's *Chandra X-ray Observatory.* (*NASA, CXC, SAO*)

galaxies, quasars, and the galactic diffuse emissions are broadband radio sources that could obscure weak pulsar signals. Additionally, the low signal intensity from radio pulsars would require long signal integration times to raise weak received signals to an acceptable signal-to-noise ratio.

During the 1970s, astronomical observations within the x-ray band of 1–20 keV (2.4×10^{17}–4.8×10^{18}Hz) yielded pulsars with x-ray signatures. T. J. Chester and S. A. Butman proposed the use of pulsars emitting in the x-ray band as an improved option for navigation. Sensors potentially with areas of the order of 0.1 m^2(1 ft^2) could be used for x-ray detection, which is a much more reasonable size for a spacecraft than large optical telescopes or radio antennas. Also, there are fewer x-ray sources to contend with and many pulsars have unique x-ray signatures, which are not obscured by closer celestial objects.

Using x-ray detectors to capture photons emitted from a collection of spatially diverse pulsars, a spacecraft's position can be estimated from relative time-of-arrival measurements, which determine the offset of the time that a pulsar signal arrives at a spacecraft compared to its arrival at the solar system barycenter. To first order the spacecraft position vector $\mathbf{r}$ is related to the arrival time difference given by the equation below, where t_b is the time of pulse arrival

$$t_b - t_{\text{obs}} = \Delta t = \mathbf{n} \cdot \mathbf{r}/c$$

at the solar system barycenter, t_{obs} is the time of observation at the spacecraft, $\mathbf{n}$ is the unit direction from the origin to the pulsar, and c is the speed of light. The right-hand side of the equation is the dot product of the vectors $\mathbf{n}$ and $\mathbf{r}/c$, which is the product of the magnitudes of these two vectors and the cosine of the angle between them. Since the pulsars are so distant from the Earth, the unit direction to the pulsars can be considered constant throughout the solar system. Three or more time-difference measurements to different pulsars are needed to solve for $\mathbf{r}$. The time used in this equation is referred to as coordinate time, or the time measured by a standard clock at rest in the inertial frame whose origin is the solar system barycenter. A spacecraft's clock, unless actually at rest (zero velocity) with respect to the solar system barycenter and not within any gravity field, does not measure coordinate time. A spacecraft's clock measures proper time, or the time a standard clock measures as it travels along a four-dimensional space-time path. The goal of a pulsar-based navigation system would be in part to provide accurate position information of the spacecraft. This could be accomplished only by accurately timing pulsar signals and then correctly transferring this time to the solar system barycenter. If a performance goal of such a navigation system is to provide position information with accuracy of the order of less than 300 m (1000 ft), then the system must accurately time pulses to at least less than 1 μs ($\approx 300/c$). To achieve these types of accuracies, general and special relativistic effects on a clock in motion relative to an inertial frame and within a gravitational potential field must be considered.

During the late 1990s, the Naval Research Laboratory's Unconventional Stellar Aspect (USA) experiment, in which an x-ray detector was mounted on the *Advanced Research and Global Observation Satellite* (*ARGOS*), provided an observation platform for conducting pulsar-based spacecraft navigation testing (**Fig. 3**). Actual proof of proposed position and timing navigation using the Crab pulsar was developed by S. I. Sheikh and D. J. Pines, who corrected the timing of pulsars for proper motion and relativistic effects. In this work they processed the x-ray data from the USA experiment

Fig. 3. *ARGOS* satellite with the USA experiment.

TABLE 1. Position offsets from Crab pulsar observations by USA detector observations*

Observation date	Duration, s	Observed pulse cycles	Time-of-arrival difference, 10^{-6} s	Position offset, km
December 21, 1999	446.7	13,332	53.75 (5.8)	16.1 (1.8)
December 24, 1999	695.9	20,770	−31.02 (5.2)	−9.3 (1.6)
December 26, 1999	421.7	12,586	−37.16 (6.3)	−11.1 (1.9)

*Numbers in parentheses are the standard deviations of the measurement errors.

and computed the position accuracies displayed in **Table 1**.

While these position inaccuracies were not small, they demonstrated that position estimates could be obtained using data from the USA detector with photons received from the Crab pulsar. Further investigations with higher-fidelity detectors and spacecraft dynamic models yielded enhanced position accuracies using a recursive Kalman filter estimation approach for a variety of candidate orbits and pulsars.

Navigation system comparison. Although the intent of a pulsar-based navigation system is to complement human-made navigation systems for use in regions where human-made systems are unable to reach, comparisons between the two types of systems are useful. **Table 2** provides a list of various aspects of both types of systems, including information about the number of sources or satellites, as well as the timing and spherical error probable (SEP) position accuracies for spacecraft traveling in low Earth orbit (LEO), medium Earth orbit (MEO), or interstellar space. SEP represents the radius of a sphere that contains 50% of all likely position fixes that are made from "noisy" measurements. Many algorithmic techniques used by GPS could potentially be implemented within the pulsar-based system, such as differential or relative positioning, and integer wavelength ambiguity-resolution algorithms.

Additional applications. The use of x-ray pulsars, and other variable x-ray sources, is not limited to single spacecraft navigation. Other applications are available from this technology.

Differential or relative position. An orbiting base station may be used to detect pulsar signals and to transmit pulse arrival times and signal errors to other spacecraft. This base station can also be used to monitor and update pulsar ephemeris information. Ideal locations for these base stations include geosynchronous orbits, and Sun-Earth and Earth-Moon Lagrange points.

GPS system navigation complement. A pulsar-based system can be used to complement, or provide an alternative to, spacecraft that rely on Earth-based navigation aids, or when GPS is obscured or unavailable.

Radio-based systems on Earth. Systems based on x-ray sources cannot be used for navigation of air, land, and sea vehicles because of the absorption of x-ray signals in the atmosphere, but systems based upon radio signals have been suggested for this purpose. Using these sources as celestial clocks or navigation aids would still be a viable alternative to other methods, as long as an application could support large antennas, such as on large naval vessels.

Planetary rovers. With rovers suggested for future planetary missions, a pulsar-based navigation system could provide a navigation system for these exploratory vehicles. Upon landing, the rover's base station could monitor pulsar signals and provide a relative positioning system for rovers that navigate over the surface terrain. This may work well on

TABLE 2. Navigation system comparison

Characteristic	Global Positioning System (GPS)	Global Navigation Satellite System (GLONASS)	X-ray pulsar and sources
Number of sources available per spacecraft	24	24	>1000
Visible sources (at antenna)	12	12	Field-of-view (several)
Signal wavelength	L1: 0.1903 m L2: 0.2442 m	L1: 0.185–0.187 m L2: 0.144 – 0.146 m	10^{-11}–10^{-8} m
Cycle/pulse period	L1: 6.35×10^{-10} s L2: 8.15×10^{-10} s	L1: 6.24×10^{-10} s L2: 8.03×10^{-10} s	~0.001–10^{6} s
Solution accuracy			
Time	30 ns	15 ns (1-sigma)	10–100 ns
Position	<30 m (SEP, SPS*)	<30 m (SEP)	30–300 m
Usable signal	LEO - MEO	LEO - MEO	Interstellar
Issues	Atmospheric effects Multipath High-power signal Human-controlled	Atmospheric effects Multipath High-power signal Human-controlled	Above atmosphere only Line-of-sight only Many faint objects Universe-controlled Subject to star quakes Long detector dwell times

* SPS = (GPS) Standard Positioning Service. This is the unencrypted service available to civil GPS users.

planetary bodies that have a thin or negligible atmosphere, including Earth's Moon.

For background information *see* CALCULUS OF VECTORS; CELESTIAL NAVIGATION; CRAB NEBULA; ELECTRONIC NAVIGATION SYSTEMS; ESTIMATION THEORY; NEUTRON STAR; PULSAR; RELATIVITY; SATELLITE NAVIGATION SYSTEMS; SPACECRAFT NAVIGATION AND GUIDANCE; X-RAY ASTRONOMY in the McGraw-Hill Encyclopedia of Science & Technology.

Darryll J. Pines; Suneel I. Sheikh

Bibliography. T. J. Chester and S. A. Butman, Navigation using x-ray pulsars, NASA Tech. Rep. no. N81-27129, pp. 22–25, June 15, 1981; G. S. Downs, Interplanetary navigation using pulsating radio sources, NASA Tech. Rep. no. N74-34150, pp. 1–12, October 1, 1974; A. Hewish, et al., Observation of a rapidly pulsating radio source, *Nature*, 217:709–713, 1968; W. G. Melbourne, Navigation between the planets, *Sci. Amer.*, 234(6):58–74, June 1976; S. I. Sheikh, et al., Spacecraft navigation using x-ray pulsars, *J. Guid. Contr. Dynam.*, 29:49–63, 2006.

Quantum coherence in photosynthesis

Photosynthetic organisms include trees, plants, grasses, ferns, mosses, algae, seaweeds, and certain special bacteria. What they share in common is the ability to capture energy from sunlight and use it to drive biochemical reactions that enable the organism to grow. In other words, photosynthesis uses water, carbon dioxide, and the energy of the Sun to produce biomass. That biomass forms the base of food chains all over Earth and is produced on an astounding scale: Every hour, about 12 million tons of biomass are produced through photosynthesis. Photosynthesis does not occur just on land; half of the photosynthetic biomass worldwide is produced in oceans and lakes. Moreover, because oxygen is a by-product of most photosynthetic organisms, the impact of photosynthesis on our environment is of immense importance. Indeed, all the oxygen in our atmosphere derives from this incredible machinery that has evolved to make food from carbon dioxide and water.

Photosynthetic process. An overall (unbalanced) chemical equation to summarize photosynthetic production of sugars, which is taught in school, can be written as

$$CO_2 + H_2O + \text{sunlight} \rightarrow \text{sugar} + O_2$$

A deeper issue is how this transformation takes place. It turns out that it is very complicated and involves many proteins cooperating to capture the Sun's energy and then convert it to forms that ultimately can be used to promote chemical reactions, such as the splitting of water into oxygen. Photosynthesis is not a single reaction; it is a network of many reactions.

The central players in the photosynthetic process are reaction centers (or photosystems). These protein complexes function in a manner similar to solar cells. They store energy captured from the Sun as a pair of positive and negative charges separated in space. Usually, two different photosystems work in tandem and wire this electrical potential in series, as we do batteries in a radio, to generate enough electrical energy (or voltage) to power the photosynthetic machinery. The "wires" in this case are not really physical connections; they involve still more proteins shuttling electrons and protons. The reaction centers are machines fueled by the Sun's energy, and they need to run smoothly and continuously regardless of whether the sky is cloudy or clear. A key companion to reaction centers evolved to orchestrate the capture, distribution, and regulation of the solar energy: proteins called light-harvesting complexes.

Half the chlorophyll (the green pigment) visible outside one's window is bound into light-harvesting complexes in higher plants. Those chlorophyll molecules are recruited, first, to absorb the energy from the blue and red parts of the Sun's spectrum, and, second, to transfer the energy with great efficiency to reaction centers (**Fig. 1**). The proteins sit in membranes in the chloroplast (thylakoid membranes), with light-harvesting complexes surrounding the reaction centers. The Sun's energy is absorbed by a molecule in one of the light-harvesting complexes and then transferred to the reaction center, where it is stored as a pair of separated charges.

Researchers have studied for many years how this light-harvesting process occurs, and as a result, the basic scheme as just outlined is well understood. Nonetheless, "tricks" used to move energy efficiently continue to be discovered, and these insights will ultimately help in designing new technologies. Most recently, research has exposed a role played by quantum coherence in moving the energy within light-harvesting complexes isolated from marine algae. This observation is fascinating scientists now because it suggests that a biological system can make use of the strange laws of quantum mechanics for its function. To understand what this means, it is necessary first to discuss how molecules absorb energy from light.

Molecules capture energy from light. A light-absorbing molecule can be thought of as a framework holding electrons that are made to dance when they are hit by a light wave. An electron does not leave the molecule; it just oscillates like a tuning fork that has been tapped against a surface. Similar to the tuning fork, which resonates at a particular frequency (a musical note), the molecule resonates at a specific frequency. To "listen" to that resonance we measure an absorption spectrum that shows a peak at the molecule's resonance frequency. For example, the broken line in **Fig. 2** shows the absorption resonance (or absorption band) of the naphthalene-type molecule labeled N-2. This absorption band is centered at around 44,000 cm^{-1}. (Here, cm^{-1} is an energy unit that can be converted to frequency by multiplying by the speed of light; 1 eV = 8066 cm^{-1}.)

Once a photon of light is absorbed at the molecule's absorption frequency, its energy is held captive by the dancing electrons for a rather short time, about 1 nanosecond (10^{-9} s). It is therefore important that photosynthetic systems quickly move this energy to molecules in the reaction center, where it can precipitate the formation of long-lived separated charges. Otherwise the energy will be lost as emitted light or heat.

Energy transfer from molecule to molecule. The photophysical process that transfers energy (excitation energy) from one molecule after it absorbs light to another molecule that has not absorbed light is known as electronic energy transfer or resonance energy transfer. Energy transfer is important in a variety of applications and instances, from degradation of plastics to the development of sophisticated sensors, and therefore it has been studied extensively. The most common form of energy transfer is understood to be caused by a weak interaction between molecules. When one molecule is excited by light, its dancing electron initiates a dance of the electrons on a nearby molecule. At some point the electron on the first molecule stops oscillating and all its energy is transferred to the second molecule, which is now excited just as if it had absorbed the photon in the first place. The electrons do not move from one molecule to another; they just transfer their dancing motions. By a series of such hops, energy can be transported among the light-harvesting proteins in photosynthesis. However, although this mode of energy transfer is prevalent, it turns out not to be the sole mechanism in photosynthetic light harvesting.

Quantum mechanics makes molecules share excitation. When we add a second naphthalene unit to N-2 to make DN-2 (Fig. 2), we anticipate that the absorption band would look the same but would be twice as intense because there are now two light-absorbing units per molecule. However, we actually observe a striking change in the absorption spectrum of DN-2 compared to N-2: Instead of a single absorption band, there are two bands, one at an energy of around 40,000 cm^{-1} and the other at around 47,000 cm^{-1}, as shown by the solid line in Fig. 2. This splitting of the absorption band comes from the strong electronic interaction between the two naphthalene units and is explained by the quantum-mechanical superposition principle. An electronic excitation on the left-hand naphthalene is indistinguishable from one on the right-hand naphthalene. The quantum-mechanical states of the system are formulated as linear combinations of these possibilities, just like Schrödinger's famous cat that is both dead and alive. Excitation is therefore shared quantum mechanically between the two naphthalene units when the electronic interaction between the molecules is significantly larger than it is in the case described in the previous section, in which it is sufficient only to cause the energy to hop, not resonate. The phenomenon can be imagined as a synchronous dance of electrons on each of the two naphthalene molecules

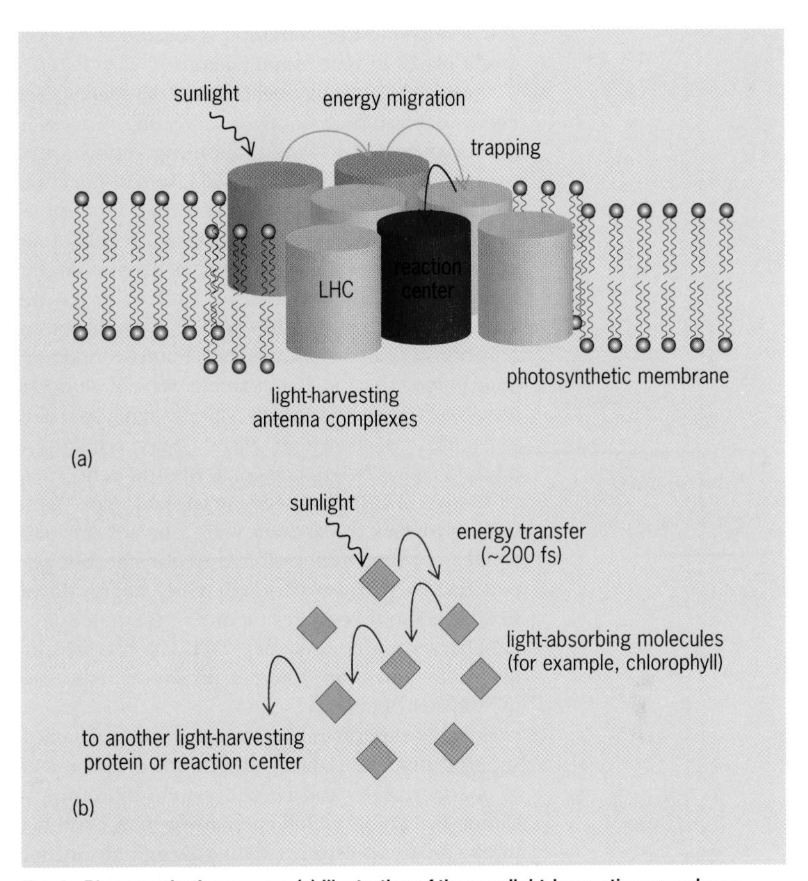

Fig. 1. Photosynthetic process. *(a)* Illustration of the way light-harvesting complexes (LHCs) are arranged in the photosynthetic membrane to surround a reaction-center protein. *(b)* Diagram of the molecules (for example, chlorophyll) that are held in each light-harvesting complex. Energy is transferred from one molecule to another within a complex before it is transferred to another light-harvesting complex en route to the reaction center.

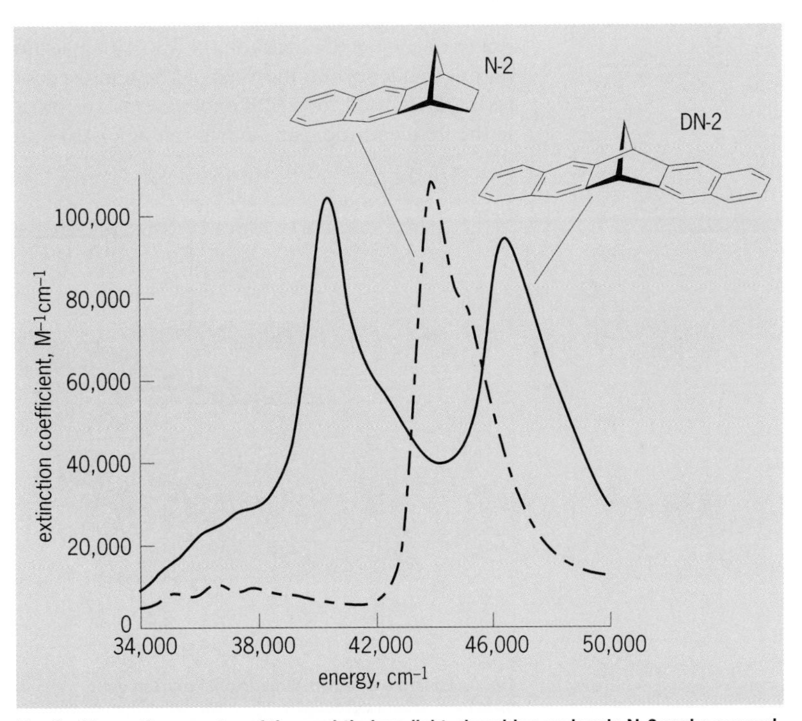

Fig. 2. Absorption spectra of the naphthalene light-absorbing molecule N-2 and a second molecule that consists of a framework holding two of these light-absorbing groups, DN-2.

and, as a result, the excitation is located at two different places in space simultaneously.

"Imaging" of quantum-coherent energy sharing. Scientists have studied the light-harvesting process in photosynthesis for many years using sophisticated experiments that employ pulsed lasers to inject excitation energy into a protein and then time subsequent events against a clock set as the delay time between the laser pulse that excites the system (the "pump") and one that is used later to probe the state of the system. In this way the dynamics of energy transfer can be followed on a femtosecond time scale (1 fs $= 10^{-15}$ s). For example, we can work out how fast the energy jumps among light-absorbing molecules, as shown in Fig. 1. Very recently, a method called two-dimensional photon echo spectroscopy (2DPE) has been developed. This technique provides researchers with a means not only to measure time scales of energy transfer, but also to detect the pathways through which energy flows. In addition to this clearer picture of the function of light-harvesting complexes, 2DPE has revealed evidence for quantum-coherent processes that assist the transfer of energy.

How the experiment detects quantum-mechanical superposition states that evolve during the processes of energy transfer can be understood by returning to our analogy in which each absorption band is a tuning fork. Impulsively exciting a single absorption band is like hitting one tuning fork; it rings, and the sound gradually decays because of frictional forces in the environment. In the 2DPE this ringing is detected as long-lived bands on the diagonal position in a two-dimensional plot of the frequencies excited by the "pump" laser and those detected as the signal; the frequencies are labeled (ω_1, ω_1) and (ω_2, ω_2) in **Fig. 3**. These coordinates mean that the input frequency of the light (on the x axis) equals the output frequency (on the y axis). The femtosecond laser pulses used for 2DPE experiments are broad in the frequency domain, so two tuning forks of different frequencies, ω_1 and ω_2, can be struck in concert. In the case of real tuning forks we would hear the frequencies of each tuning fork as well as a low-frequency beating at the difference frequency, $\omega_1 - \omega_2$, because of interference of the sound waves. If the molecular absorption bands are able to form quantum-mechanical superposition states, even very weakly, then signal amplitude beats can similarly be detected in 2DPE as a function of the pump-probe waiting time. In the quantum case we actually detect two kinds of beats in the two-dimensional spectra: one in the lower off-diagonal region and one in the upper off-diagonal region—that is, at the coordinates (ω_1, ω_2) and (ω_2, ω_1). These features, shown in Fig. 3, are measured to determine their rise and fall in amplitude out of phase with each other as a function of the time delay between pump and detection, like two pistons in an engine. This helps us detect these signatures of quantum coherence.

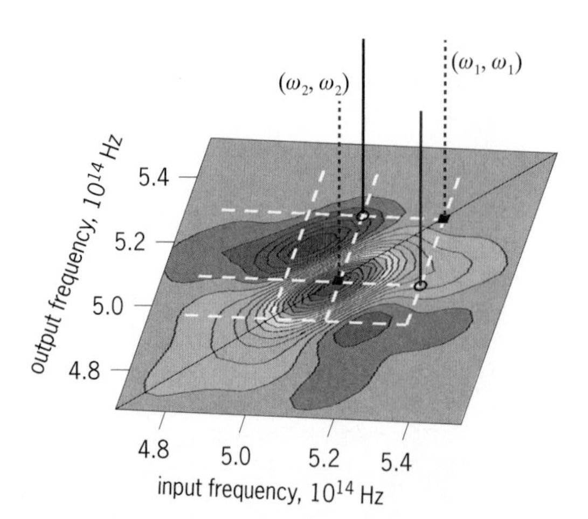

Fig. 3. Example of a two-dimensional photon echo spectrum. Beats detected in the lower off-diagonal and upper off-diagonal regions, at the coordinates (ω_1, ω_2) and (ω_2, ω_1), are shown as black sticks.

Prospects. There are several reasons why observing quantum-mechanical processes that aid the transfer of excitation energy in photosynthetic light-harvesting proteins is interesting. First, researchers can now study how quantum probability laws are harnessed for this function and ask how they can be beneficial over classical probability laws (those familiar from chemical kinetics). Second, these 2DPE experiments provide experimental evidence that nontrivial quantum-mechanical phenomena can be used by a biological system. This result is fascinating because it shows that quantum coherence can survive even in complex systems at normal temperatures, which was not expected based on previous experiments or theory. Third, the work has stimulated a field of research in which it is asked: Just how "quantum" are these processes in biology? This topic, however, which is related to the field of quantum information, is beyond the scope of the present article.

For background information *see* CHLOROPHYLL; COHERENCE; NONRELATIVISTIC QUANTUM THEORY; PHOTOCHEMISTRY; PHOTOSYNTHESIS; QUANTUM MECHANICS; RESONANCE (QUANTUM MECHANICS); SUPERPOSITION PRINCIPLE; ULTRAFAST MOLECULAR PROCESSES in the McGraw-Hill Encyclopedia of Science & Technology.

Gregory D. Scholes

Bibliography. R. E. Blankenship, *Molecular Mechanisms of Photosynthesis*, Blackwell Science, Oxford, UK, 2002; Y. C. Cheng and G. R. Fleming, Dynamics of light harvesting in photosynthesis, *Annu. Rev. Phys. Chem.*, 60:241–262, 2009; E. Collini et al., Coherently wired light-harvesting in photosynthetic marine algae at ambient temperature, *Nature*, 463:644–648, 2010; G. S. Engel et al., Evidence for wavelike energy transfer through quantum coherence in photosynthetic systems, *Nature*, 446:782–786, 2007; B. R. Green and W. W. Parson, *Light-Harvesting Antennas in Photosynthesis*, Kluwer, Dordrecht, The Netherlands, 2003; D. M. Jonas, Optical analogs of 2D NMR, *Science*, 300:1515–1517, 2003; G. D. Scholes, Long-range resonance energy transfer in molecular systems, *Annu.*

Rev. Phys. Chem. 54:57–87, 2003; B. W. Van Der Meer, G. Coker, and S.-Y. S. Chen, *Resonance Energy Transfer: Theory and Data*, VCH, New York, 1994.

Quantum electronic properties of matter

The quantum-mechanical description of matter aims to explain how diverse properties of materials arise from the mutual interactions of many atomic nuclei and their associated electrons. While a more or less complete theory of ordinary electrical conduction in metals was developed in the early part of the twentieth century, newer phenomena such as high-temperature superconductivity and ballistic conduction in carbon-based electronic devices challenge our basic understanding of condensed-matter physics. It is an important goal to understand the subtle interactions leading to these phenomena, and if possible to control them. Can these inherently quantum-mechanical behaviors be extended to room temperature? If so, remarkable applications could be envisioned.

Electronic self-energy. Quantum theory explains that subatomic particles such as electrons behave as waves whose energies are determined by their environment. For the single electron in a hydrogen atom, the environment consists of the nearby, oppositely charged nucleus. The electron's motion is determined by a few simple parameters such as the strength of the attractive electrostatic interaction and the masses of the particles, leading (in an elementary treatment) to a discrete set of energy levels of the electron. This is analogous to the properties of vibrating strings, which carry sound waves of definite frequencies depending on their fundamental properties such as length, density, cohesive forces, and tension.

This simple description of an atom is incomplete because it ignores the detailed nature of the vacuum. Even in a volume free of all matter, quantum mechanics predicts the existence of electric forces that pass into and out of existence; these ephemeral forces are called vacuum fluctuations. These forces can distort the shape of the electron's orbits and alter their energies in the same way that an electric voltage accelerates the electrons through a wire to conduct electricity. In quantum mechanics, an oscillating electric field is represented by photons, swiftly traveling particles that comprise light, and so the vacuum fluctuations can be viewed as a constantly changing population of photons. In this view, we think of the actual electron as a "bare" electron (the fictitious entity that exists in the absence of anything, even the vacuum) dressed with a fluctuating population of "virtual" photons, which the electron continuously emits and reabsorbs on an infinitesimally short time scale. These processes alter the electronic energies, but more importantly, they also cause electrons in higher-energy orbits to decay to their lowest energy level by emitting a real photon. This occurs when one of the virtually emitted photons escapes the electron, carrying off the excess energy associated with the decay as a bit of light.

Despite this complicated picture, the original, simple description of the hydrogen atom gives a remarkably good estimate of the energy levels. The interactions with the vacuum do not qualitatively alter the electronic spectrum of the atom, but the electronic binding energy is changed in ways that can be calculated if we take the vacuum fluctuations into account. This difference between the bare and the altered energies is called the electronic self-energy.

Electrons in condensed matter. We can view electrons in a solid in a similar way, by considering a single bare electron to be dressed not only by photons but also other background entities. Three particular entities commonly dress the electrons in condensed matter. The first two, phonons and plasmons, are particles that exist in exact analogy to photons. Whereas photons are particles that represent traveling oscillations of the electric field, phonons and plasmons represent oscillations of the density of the atomic nuclei and valence electrons, respectively. (Phonons of very long wavelengths are familiar as the sound waves generated in air or along a plucked string.) A third, more complicated entity is formed when a background electron is promoted to a formerly unoccupied state, leaving a positively charged "hole" behind. (This hole is not fictitious but has a real physical meaning: for all intents and purposes the holes act as actual positive particles, carrying electrical current under the influence of an applied voltage for instance.) The newly excited electron and its partner hole are called an electron-hole pair and it is these pairs which dress the electron.

Like the vacuum fluctuations of the number of photons, these background entities can be either virtual—appearing and disappearing quickly around an electron—or real, with the ability to carry away energy associated with the decays of electrons from higher to lower energy states. And to make things more complicated, each of these entities (and their sub-parts) is further dressed by their own virtual particles altering their interactions, and so on. For the original electron in a metal, the cumulative effect of all of these interactions can be very strong indeed, altering the energy levels to a great extent, or even transforming the nature of the material itself, as in the case of superconductors.

Graphene. Graphene, a single, atomically flat layer of carbon atoms arranged in a honeycomb lattice, is the building block of carbon nanotubes, graphite, and C_{60} molecules, and has received a great deal of attention because of its novel electronic properties (**Fig. 1***a*). Measurements of its electrical conduction and band structure have shown that it exemplifies the importance of the interactions discussed above.

As in other crystalline solids, in graphene each bare electron is bound to all the atoms, passing from one to the next with a well-defined momentum for a given energy. The important energy levels for electrical conduction in graphene are distributed in two

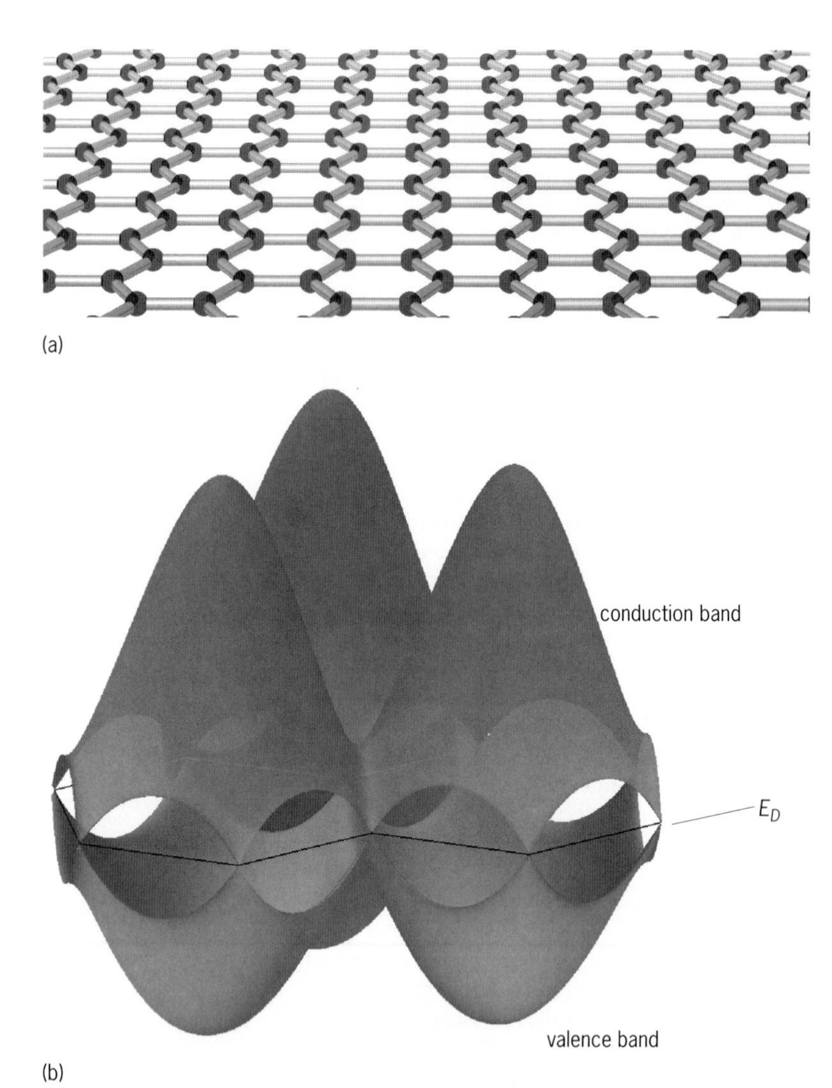

Fig. 1. Structure of graphene. *(a)* **Two-dimensional layer of carbon atoms arranged in a honeycomb lattice.** *(b)* **Bare electronic band structure (in the absence of many-body interactions). The vertical direction represents the energy of the conduction and valence bands as a function of momentum. The Dirac crossing energy E_D is indicated.**

energy bands having the unusual property of touching for only a very few particular momenta. The energy at which these bands contact each other is called the Dirac energy E_D. The symmetry of the atomic lattice protects these crossing points from sustaining energy gaps as occur in semiconductors (Fig. 1*b*). Thus, graphene is a zero-gap semiconductor, or a zero-overlap semimetal.

Graphene is exceptional because of the ease of altering its carrier density. Neutral graphene (with exactly four electrons in its valence band) has its highest occupied energy level exactly at the Dirac energy. Perturbing the material by applying potentials on the order of ± 1 electronvolt is sufficient to sweep the electron density through a range of $\pm 10^{14}$ electrons/cm^2. Such potentials are easily achieved by applying gate voltages in device geometries, by chemical modification, or by confinement in nanometer-scale wires or dots. This explains why graphene's properties are tunable and why it is useful in many possible device schemes.

What is even more unusual is that, near the Dirac energy, the electrons propagate as effectively massless particles. This means that their bare energy is proportional to their momentum, as is the case for massless photons. Other properties of photons, such as having a constant velocity in any reference frame, are also obeyed by the electrons in graphene, and many of the novel properties of graphene follow from this massless behavior. But there are also significant differences between graphene electrons and photons, chief among which are the slower speed of the electrons (around 300 times slower than the speed of light *c*) and the fact that, as charged particles, electrons may interact strongly with each other through their electric fields. *See* NOBEL PRIZES FOR 2010.

ARPES measurements of graphene's self-energy. The premier experimental tool for determining the self-energy corrections in solids is angle-resolved photoemission spectroscopy (ARPES), in which a photon with energy E_p promotes an electron, originally bound to a solid, into a free electron that can be counted by a detector, as first described by Albert Einstein in 1905. If the energy of the outgoing electron E_{out} is measured, then the initial electron's binding energy E_B can be inferred from the simple relationship $E_B = E_p - E_{\text{out}}$ (**Fig. 2**). If the angle of the outgoing electron is measured, we can similarly infer the momentum of the electron when it was in the solid. And if instead of a single well-defined energy E_B, we find a distribution of energies, we can infer the lifetime of the electrons from the width of this distribution.

A useful and equivalent way to look at this process is not the removal of a negatively charged electron,

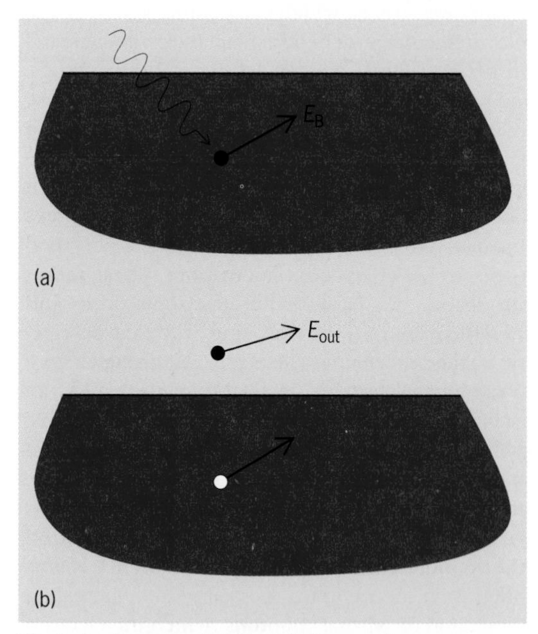

Fig. 2. Angle-resolved photoemission spectroscopy (ARPES) measurement. *(a)* **An incoming photon (wiggly line) of energy E_p about to excite an electron bound to the solid with binding energy E_B.** *(b)* **The initially bound electron is promoted to a free electron state with energy E_{out}. A hole, said to have a binding energy $E_B = E_p - E_{\text{out}}$, is left behind in the solid.**

but rather the creation of a positively charged hole in the solid (Fig. 2*b*). This picture is useful because for sufficiently large photon energies, typically in the soft x-ray spectrum, the outgoing electron leaves the solid too quickly for any strong interactions, and so the measured photoelectron spectrum reflects the energy, momentum, and lifetime of the hole left behind in the solid.

Recently, great strides have been made in the ARPES technique because of improvements in the energy and momentum resolution and in detection efficiency. These arose both by advances in synchrotron-based x-ray sources, which provide sufficient numbers of photons of very well-defined energy, and detectors which can efficiently discriminate the energies and momenta of the electrons with very high resolution.

Using such a synchrotron-based apparatus, Aaron Bostwick and his colleagues directly measured the electronic band structure of graphene grown on a silicon carbide substrate near the Dirac crossing (**Fig. 3***a*). The bare band structure in this region should look like the adjoining schematic: two cones meeting at a single point. Since the experiment probes only the band structure in a single momentum direction, the data look like a planar cut through the conical bands, so an "x" shape is observed.

The width, or fuzziness, of the features in the data is inversely proportional to the lifetime of the hole created at each energy. The data are sharpest near zero binding energy because these least-bound carriers have few states to decay to and hence the longest carrier lifetime. But for holes at larger (negative) binding energies, the features are broader, reflecting a short lifetime due to decay processes. Apart from a small contribution due to electron-phonon scattering, the short lifetime of the carriers can be ascribed mainly to electron-hole pair generation and electron-plasmon scattering. The lifetime where the bands cross appears to be particularly short, attributed to enhanced scattering of the hole states there by plasmons.

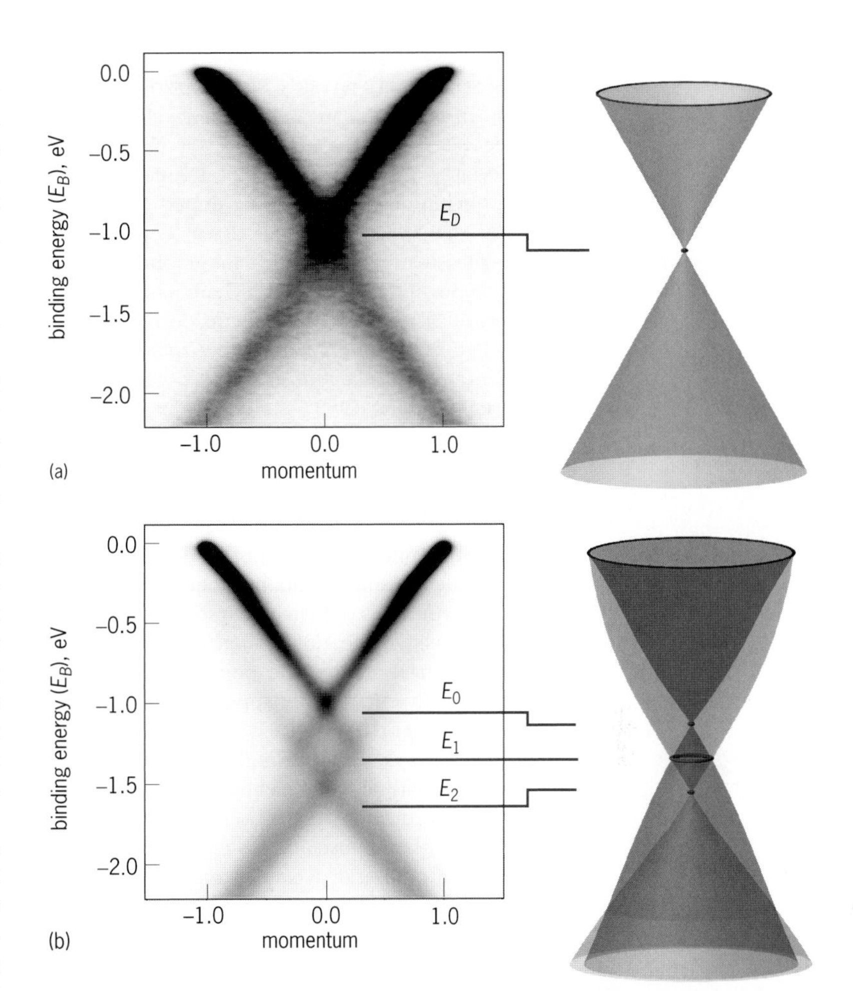

Fig. 3. Electronic band structures of graphene layers grown on silicon carbide. (*a*) Band structure of slightly electron-doped layer, left, gives the structure near crossing of the bands in graphene. The bands observed are similar to the expected conical band structure, right. Momentum is expressed in arbitrary units (*adapted from A. Bostwick et al., Quasiparticle dynamics in graphene, Nat. Phys., 3:36-40, 2007*). (*b*) Band structure of graphene on hydrogen-passivated silicon carbide. In the presence of enhanced electrostatic forces, more bands are observed, which cross at three different energies. The uppermost crossing is between ordinary hole bands; the middle, ring-shaped crossing is between hole and plasmaron bands; the lower crossing is between two plasmaron bands (*adapted from A. Bostwick et al., Observation of plasmarons in quasi-free-standing doped graphene. Science, 328(5981):999–1002, 2010*).

If the graphene is grown on a silicon carbide crystal which has been passivated by hydrogen atoms, the mutual interaction between graphene's electrons is greatly enhanced, because the substrate dielectric strength (which diminishes the strength of the Coulomb interaction) becomes much weaker. In such samples, there is a dramatic effect on the energy spectrum (Fig. 3*b*). Instead of two bands crossing at a single energy, four bands are observed to cross at three different energies, as indicated in the schematic model. This shows that the actual band structure of doped graphene is far more complicated than the simple conical bare bands predicted within a single-particle model.

The apparent splitting of the bands can be attributed to the formation of new composite particles called plasmarons, consisting of a hole bound to a plasmon, the two propagating together with the same velocity and strongly interacting with each other. In this picture, the photoemission process leads to creation of either an ordinary hole or a plasmaron with about equal probability. That such a new propagating charged entity emerges is a testament to the strong self-energy effects in graphene, which depend on both the energy and momentum of the bare electrons. These data are the first observation of "plasmaronic" band structure not only for graphene, but for any material. Because plasmons, already known to couple strongly to photons, are now demonstrated to couple strongly to the charge carriers in graphene, it suggests that graphene is likely to play a role in plasmonics, the merging of electronic and photonic technology at the nanoscale.

For background information *see* BAND THEORY OF SOLIDS; BRILLOUIN ZONE; CONDUCTION BAND; CRYSTAL STRUCTURE; FERMI SURFACE; HOLE STATES IN SOLIDS; PHONON; PHOTOEMISSION; PHOTON; PLASMON; QUANTUM ELECTRODYNAMICS; QUANTUM

THEORY OF MATTER; RELAXATION TIME OF ELECTRONS; SOLID-STATE PHYSICS; SUPERCONDUCTIVITY; VALENCE BAND in the McGraw-Hill Encyclopedia of Science & Technology. Eli Rotenberg

Bibliography. A. Bostwick et al., Observation of plasmarons in quasi-free-standing doped graphene. *Science*, 328(5981):999–1002, 2010; A. Bostwick et al., Quasiparticle dynamics in graphene, *Nat. Phys.*, 3:36-40, 2007; A. K. Geim and K. S. Novoselov, The rise of graphene, *Nat. Mater.*, 6:183–191, 2007; R. A. Mattuck, *A Guide to Feynman Diagrams in the Many-Body Problem*, 2d ed, Dover Publications, 1982.

Quantum signatures of chaos

From weather phenomena, to population dynamics, to the orbits of Jupiter's moons, chaos appears everywhere around us. All chaotic systems exhibit an extreme sensitivity to tiny changes in initial conditions that has come to be popularly illustrated by the butterfly effect: The flapping of a butterfly's wings can trigger the eventual development of a tornado half a world away. A majority of real-world phenomena, natural or human-made, are governed by nonlinear equations of motion that can lead to chaos, and this circumstance has inspired a systematic study of nonlinear dynamics and chaos theory over many decades. Recent advances in optics, solid-state physics, and nanoscience are now making it possible to control systems at the level of individual atoms and photons. At this microscopic level governed by the laws of quantum mechanics, the concept of nonlinear dynamics and chaos becomes ambiguous and difficult to characterize. The problem arises from the uncertainty principle, which forbids exact knowledge of initial conditions, and from the Schrödinger equation, which appears to forbid nonlinear dynamics. The study of how chaos manifests itself at the quantum level is therefore crucial for understanding the fundamental connections between the quantum world and our macroscopic classical world, and also for developing a toolbox of useful quantum control techniques. Quantum control of nonlinear systems is important for a variety of future applications, including control of molecular processes for chemical and biological purposes, controlling electron transport in solid-state devices, and performing high-precision measurements. Furthermore, due to the ever-increasing miniaturization of transistors on a computer chip, quantum chaos has gained new relevance in the field of quantum computation.

Classical versus quantum. Whereas Newton's laws describe classical dynamics, quantum evolution obeys the Schrödinger equation. A classical system is described by a set of variables, for example the position and momentum of a particle. The set of all possible values of these variables is the phase space that the system moves in. Each combination of allowed values of the variables defines a point in phase space. As the system evolves from point to point, it traces out a trajectory in phase space. Quantum states, by contrast, are described by wave functions (distributions of probability amplitudes), which, due to the uncertainty principle, cannot be localized to a point in phase space. In a classical system, enough constraints on the motion (for example a limit on the total energy) will produce regular (nonchaotic) trajectories; insufficient constraints allow trajectories to explore phase space in a complex manner, and initially close trajectories can diverge at an exponential rate, which is the definition of chaos. At first glance, there appears to be no corresponding exponential sensitivity to initial conditions in a quantum system described by the Schrödinger equation. In fact, under certain conditions, quantum dynamics can suppress chaos and limit spreading of the quantum probability distribution, a phenomenon known as Anderson localization. Such problems have led to questions regarding the very existence of chaos in the quantum regime, and whether it is possible to find quantum-classical correspondence in chaotic systems. *See* ANDERSON LOCALIZATION.

Signature of chaos in static properties. In classical dynamics, constraints on motion arise from fundamental symmetries of the system (for example, translational symmetry leads to momentum conservation), and these symmetries lead to constraints also on the analogous quantum system. Conversely, the lack of symmetries or constraints that lead to classical chaos should also affect properties of the quantum behavior. Martin Gutzwiller showed the first such correspondence by deriving a formula to calculate the quantized energy levels of a quantum system from the orbits of the corresponding classical chaotic system. An important quantum signature of chaos can be found in the distributions of the energy levels of regular versus chaotic quantum systems. In regular systems, they tend to be closely spaced, whereas chaotic systems exhibit level repulsion; strong interactions lead to larger average spacing between the energy levels. Hence, a static property of the quantum system, the energy-level distribution, can be used to infer whether the corresponding classical system is chaotic or not. Furthermore, quantum wave functions can have increased amplitude (scars) along unstable classical orbits. This scarring effect has been observed in various chaotic systems.

Dynamical signatures of chaos. The key characteristic of classical chaos, sensitivity to small perturbations, can be identified at the quantum level if one considers perturbations to the system parameters (such as slight changes in external forces) rather than perturbations of the initial conditions. In chaotic systems, an exponential decay in overlap (closeness) of wave functions can occur for quantum systems that evolve with slightly different system parameters. In some cases, the rate of exponential decay is related to the Lyapunov exponent, which

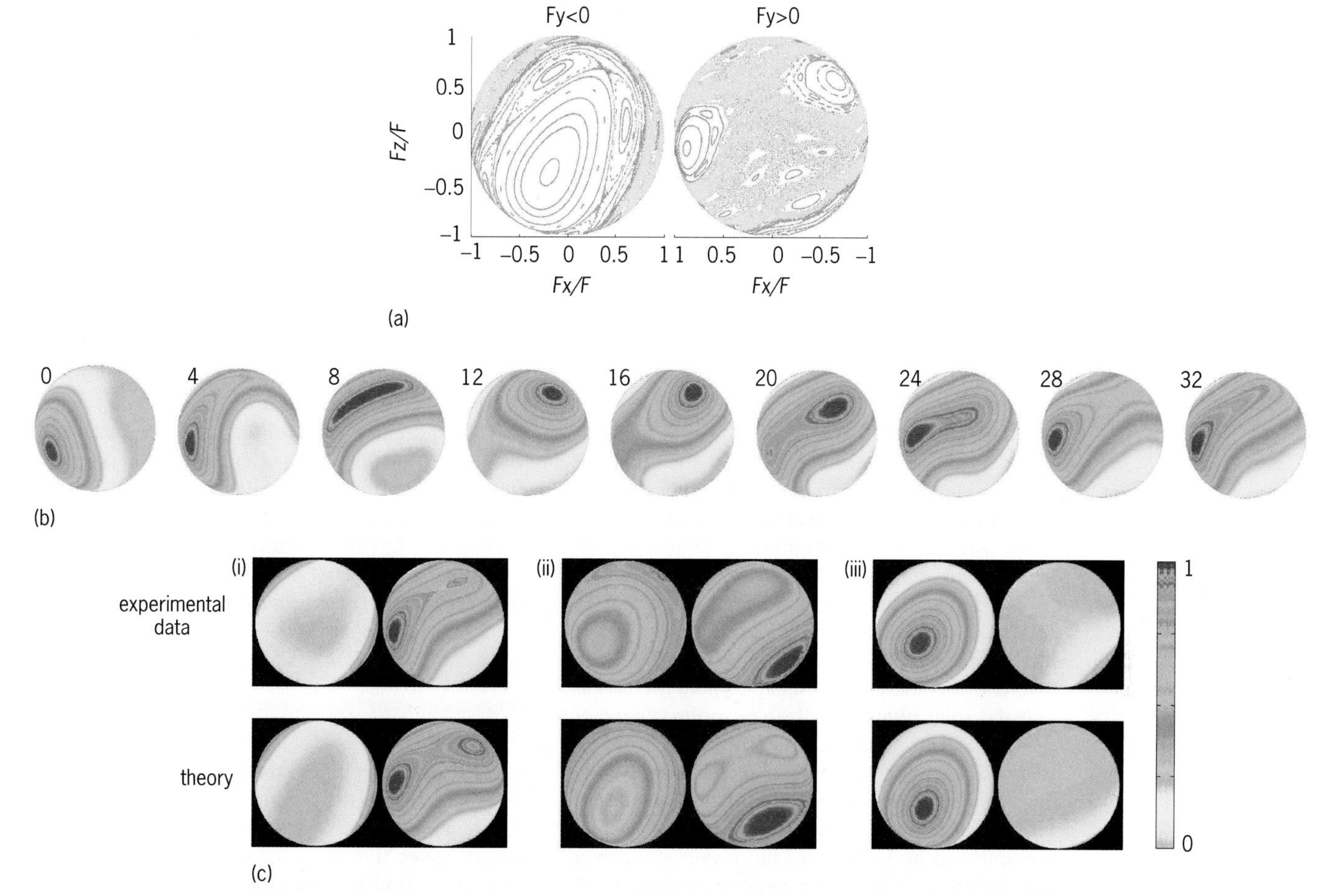

Chaos in a kicked top. (*a*) **The orientation of the spin** *F* **of the top is plotted on the surface of a sphere of fixed radius. The plot shows the classical trajectories of a large number of initial points, forming islands of periodic orbits (solid and broken lines) interspersed with a sea of chaotic trajectories (shading). Both hemispheres are shown separately.** (*b*) **In experiments using cesium atoms to realize a quantum kicked top, snapshots of the measured quantum probability distribution at different times show dynamical tunneling between islands of periodic orbits through a classically forbidden sea of chaos. The kick number (time step) is indicated at the top left of each image.** (*c*) **Time averages of the quantum distributions for three different initial conditions clearly reflect the classical phase-space structures in part** *a***. (i) Initial state is centered on an island in the right hemisphere shown in part** *a***. (ii) Initial state is centered in the chaotic sea. (iii) Initial state is centered on the island in the left hemisphere. Both hemispheres are shown separately. The experimental data (top) agree well with theoretical predictions (bottom).**

characterizes the rate of exponential divergence of classical trajectories. Interestingly, signatures of chaos can also be observed in purely quantum phenomena such as tunneling and entanglement. Tunneling refers to the ability of quantum systems to move through an energy barrier, something forbidden in classical dynamics. In chaotic systems, there are classically forbidden regions in phase space even without an energy barrier. Sharp barriers exist between regular and chaotic regimes, and a classical trajectory cannot move from a regime of regular motion to one of chaos. However, quantum systems can tunnel through these dynamical barriers, a phenomenon known as dynamical tunneling. The tunneling rate is very sensitive to system parameters and can be enhanced by chaos because the system can exist in a combination (superposition) of regular and chaotic energy levels that are not closely spaced. Since the tunneling rate depends on this energy spacing, it can be greatly increased. Such behavior is labeled chaos-assisted tunneling.

Of particular relevance for the design of future quantum computers is the possibility of rapid generation of entanglement in chaotic systems. Entanglement refers to stronger-than-classical correlations that are a crucial ingredient in quantum computation and communication. The connection between chaos and entanglement generation between parts of the system, or between the system and its environment, is an active area of research. Recent work has also sought to identify classical chaos at the level of individual trajectories by analyzing a quantum system coupled to a measurement device. Under certain circumstances, the measurement records can trace out trajectories consistent with classical chaos.

Experimental studies. Most studies of quantum chaos thus far have been theoretical due to the significant experimental challenges involved in initializing, manipulating, and measuring quantum systems. Nevertheless, a growing number of experiments have begun to observe quantum fingerprints of chaos. Early work involved studies of ionization, that is, the removal of an electron from highly energetic states of a hydrogen atom through excitation by a microwave field. At low frequencies, the observed strong dependence of the ionization on the field intensity agreed well with a classical chaotic model of a nonlinear oscillator driven by the field. At higher frequencies, Anderson localization was observed. Experiments in microwave cavities have simulated quantum chaotic billiards and led to the observation of "scars" in the probability distributions.

The motion of cold atoms in modulated standing waves of light provides an ideal system for studies of quantum chaos. This has allowed the experimental observation of Anderson localization, dynamical tunneling, and chaos-assisted tunneling. Working with the magnetic moments of the electrons and nuclei of cold atoms gives access to an even more powerful suite of experimental tools to precisely prepare, manipulate, and measure the atoms using magnetic fields and laser light. A recent experiment used laser-cooled cesium atoms to realize a quantum version of the "kicked top," a classically chaotic system often used as a paradigm for quantum chaos. This system consists of an angular momentum (spin) associated with, for example, a spinning top or a single atom. The spin can be mathematically represented by a vector F and evolves under an alternating series of "kicks" and "twists," where a kick is a fast rotation by a fixed angle around the y axis, and the twist is a rotation around the x axis by an angle proportional to the x component of the spin. These kicks and twists of the atomic spin are implemented by pulses of magnetic fields and laser light, respectively. Since rotations change the direction but preserve the magnitude of the spin, the state of the classical spin can be represented in phase space as a point on the surface of a sphere of fixed radius proportional to the magnitude of the spin. Trajectories are visualized by marking the location (orientation) of the spin on the sphere once per kick/twist cycle. **Illustration *a*** is such a stroboscopic plot for experimentally accessible parameters, showing trajectories resulting from many different initial conditions. Islands of regular trajectories embedded in a sea of chaotic trajectories are clearly visible. In the experiment, quantum probability distributions were prepared in different parts of phase space and their subsequent evolution observed. Dynamical tunneling between regular islands was observed, and the time-averaged quantum probabilities closely matched the classical phase-space structures (illus. *b*, *c*). Sensitivity to perturbation was higher in chaotic than in regular regions. Furthermore, signatures of chaos in the time evolution of entanglement were observed for the first time, showing that the time-averaged entanglement between the electronic and nuclear components of the atomic spin was larger for initial states localized in chaotic versus regular regions. Future studies could potentially use the collective dynamics of many-atom ensembles to observe the transition from quantum to classical behavior in macroscopic quantum systems subject to measurement.

For background information *see* CHAOS; NONLINEAR PHYSICS; NONRELATIVISTIC QUANTUM THEORY; QUANTUM COMPUTATION; QUANTUM MECHANICS; QUANTUM TELEPORTATION; SCHRÖDINGER'S WAVE EQUATION; TUNNELING IN SOLIDS; UNCERTAINTY PRINCIPLE in the McGraw-Hill Encyclopedia of Science & Technology. Shohini Ghose; Poul S. Jessen

Bibliography. R. Blumel and W. P. Reinhardt, *Chaos in Atomic Physics*, Cambridge University Press, Cambridge, U.K., 1997; S. Chaudhury et al., Quantum signatures of chaos in a kicked top, *Nature*, 461:768–771, 2009; W. K. Hensinger et al., Experimental tests of quantum nonlinear dynamics in atom optics, *J. Opt. B: Quantum Semiclass. Opt.*, 5:R83–R120, 2003; E. Ott, *Chaos in Dynamical Systems*, Cambridge University Press, Cambridge, U.K., 2002; H.-J. Stockmann, *Quantum Chaos, An Introduction*, Cambridge University Press, Cambridge, U.K., 2007.

Ramjet engines

The ramjet engine relies on forward velocity to compress inlet air for the combustion process. In order to operate efficiently, a ramjet must be traveling at several times the speed of sound. At the present time, there is little demand for commercial or military transport in this flight regime, so the primary application for ramjets is in military single-use missiles. There are a number of systems around the world in various stages of development ranging from exploratory research to production. The scramjet engine is a close cousin, with the differentiating factor being the combustion process speed—subsonic for a ramjet and supersonic for a scramjet.

The field of ramjet engines, a class of propulsion systems, includes hardware design, ground testing, fuel composition, and flight performance. Flow analysis is performed to understand how ramjet engines work and aerodynamic principles are used to enhance operation.

Analysis method. Since, in theory, a ramjet engine contains no moving parts, it essentially becomes a mechanism for efficiently managing inlet and combustion flows to produce thrust. An inlet is used to decelerate and compress inlet air, and to introduce the air flow to a combustion chamber. Fuel and air are then mixed to promote complete combustion without introducing undue flow losses associated with most flame-holding devices. An exit nozzle is used to provide a sonic throat to control the combustion pressure and combustion chamber Mach number, and then reaccelerate the combustion products through a supersonic expansion process.

For convenience in design and analysis, station numbers are assigned to the ramjet engine cycle **(Fig. 1)**. The freestream condition is described as station 0. Station C corresponds to the inlet entrance, which is also known as the "capture" station. Station 2 is the location where air is introduced to the combustion chamber, which is also known as the "dump" station. This may be an axial dump or a side dump from one or several locations. Station 3 is located at the upstream end of the combustion chamber, following the dump station but before the air and fuel is mixed, and no combustion has occurred. Station 4 is the entrance to the nozzle following combustion, station 5 is the nozzle sonic point, and station 6 is the exit plane.

Of primary interest is the net jet thrust produced by the engine. This is determined by taking the difference between the exit stream thrust and the inlet stream thrust (Fig. 1). Embedded within this calculation are the air management process, fuel delivery rate, sources of pressure losses, and the combustion process.

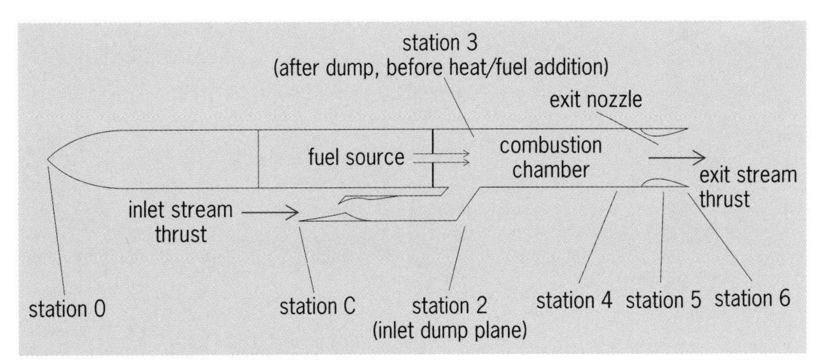

Fig. 1. Ramjet analysis station nomenclature.

Supersonic inlet. Before describing the flow analysis method, it is essential to understand the basic principles of supersonic inlet design. A ramjet missile inlet is used to decelerate the freestream flow (and increase pressure) and deliver it to the combustion chamber with minimum losses. Typically this is done with a series of shock waves to transition the flow from the freestream supersonic flow conditions to subsonic flow conditions. The inlet is modeled by defining a theoretical area that corresponds to the area of a stream tube of air captured at freestream conditions. This area is expressed as a ratio to the capture area, which is the projected geometric area of the inlet face at zero angle of attack. Because of nonideal flow in an actual application, this air capture ratio, A_0/A_C, is typically less than unity. A common inlet design practice to improve performance is to create air bleed holes or slots that further reduce the "captured" air. However, for nonsymmetric inlet configurations this ratio can be greater than one at angles of attack other than zero.

A supersonic inlet will typically have a "design point," also known as the design Mach number. When the flight Mach number, M_0, reaches the design Mach number, the leading edge shock impinges on the cowl (**Fig. 2***a*). When the flight Mach number, M_0, is lower than the design Mach number, the leading edge shock falls outside the cowl and a portion of the air is not captured, with a corresponding reduction in the calculated air capture ratio, A_0/A_C (Fig. 2*b*). At the design Mach number or above (Fig. 2*c*), the maximum amount of air is captured (Fig. 2*d*).

The next important factor in determining the inlet operating mode is the back pressure from the combustion chamber. This pressure determines the position of the terminal shock, or location where the flow transitions from supersonic flow to subsonic flow. When the combustion chamber pressure is low, the terminal shock will be strong, and be positioned well inside the inlet diffuser at an area greater than the minimum throat area. This is referred to as supercritical inlet operation (**Fig. 3***a*). As the combustion pressure is increased by combustion of fuel or exit area reduction, the terminal shock is pushed closer to the critical point, where the shock is positioned near the inlet minimum area (Fig. 3*b*). Typically the lowest pressure losses are seen near the critical point. Here, the pressure loss may be expressed in terms of the ratio of total pressure (P_T), also referred to as stagnation pressure, which is the value that results when the flow is brought to rest without energy loss, or isentropically) at station 2 to the freestream total pressure (P_{T_2}/P_{T_0}); this ratio is known as the inlet total pressure recovery. As the back pressure is raised further, the shock is pushed farther out the inlet opening and air is spilled around the cowl (Fig. 3*c*). The result is a decrease in the air capture ratio (A_0/A_C), but this occurs at nearly constant pressure recovery. The resulting curve of pressure recovery versus capture ratio is known as a cane curve, owing to its shape (Fig. 3*d*). As pressure continues to rise, pressure oscillations and flow instabilities result, called inlet buzz, until the last stable operating point is reached. This is obviously an undesirable operating point, so margins are defined for both supercritical and subcritical operation to help ensure proper engine operation at a range of flight conditions and fuel settings.

Cycle analysis procedure. The technique employed in ramjet modeling essentially attempts to balance engine operation at the nozzle sonic point with the inlet operating conditions. As discussed, changes in

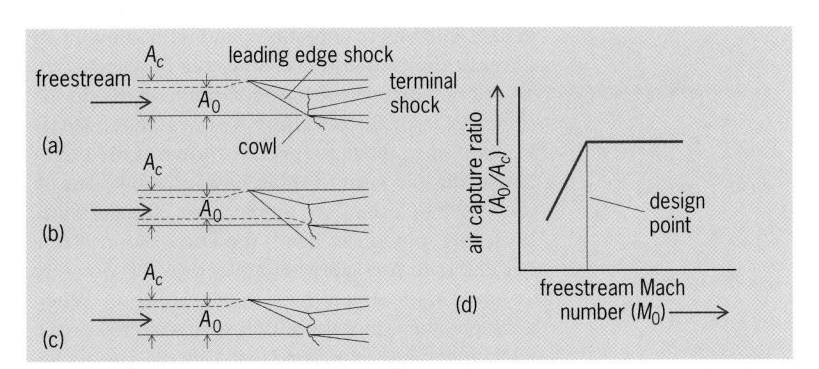

Fig. 2. Inlet design point. Diagrams show the entrance to the supersonic inlet when the flight Mach number, M_0, is (*a*) equal to the design Mach number, (*b*) below the design Mach number, and (c) above the design Mach number. (*d*) Dependence of the air capture ratio, A_0/A_C, on the freestream or flight Mach number, M_0.

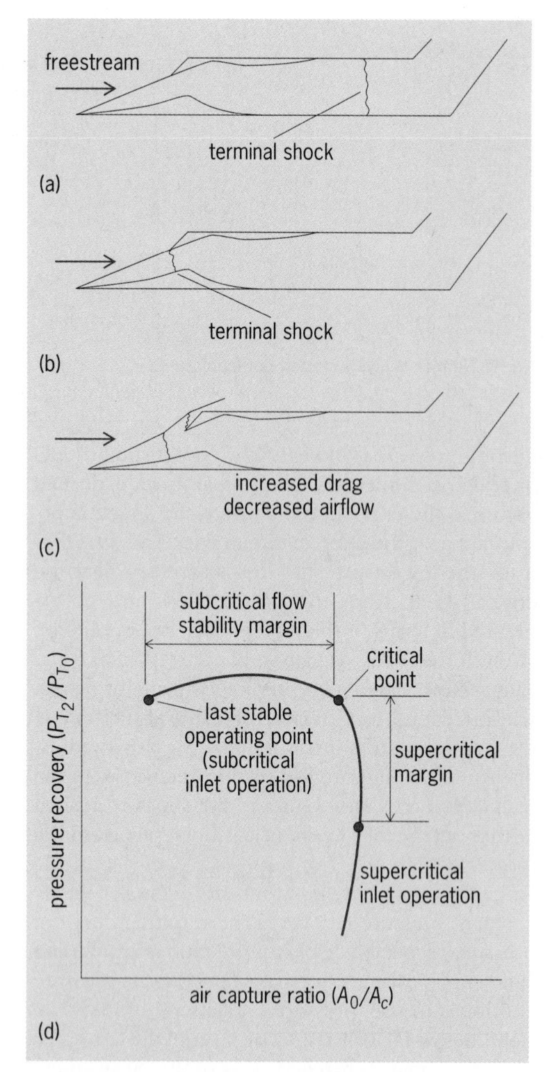

Fig. 3. Inlet operating points. Diagrams show the supersonic inlet for (*a*) supercritical inlet operation, (*b*) operation at the critical point, and (*c*) subcritical inlet operation. (*d*) Graph of inlet total pressure recovery versus air capture ratio (a cane curve), with the operating points of parts (*a*), (*b*), and (*c*) identified.

combustion conditions affect the inlet back pressure, which in turn affects the inlet air capture, leading to an iterative solution process. To begin the process, the inlet is assumed to be operating supercritically, at a point defined on the cane curve. This defines inlet dump conditions, and with known fuel flow rate the combustion analysis can be carried out. This in turn defines the flow conditions at the nozzle sonic point, or throat. This is a convenient analysis location since the flow speed is known at the throat and isentropic compressible flow relations can be used to relate mass flow, temperature, and pressure. With this solution in hand, the combustion chamber entrance pressure is compared to the pressure assumed in the supercritical inlet calculation. When a disagreement in station 2 pressures is shown, the inlet operating point is adjusted in the proper direction with resulting changes in air capture and pressure recovery, and a new combustion solution is calculated. This process is continued until the inlet operation matches the combustion conditions. Once a converged solution is reached and engine mass flow rate is known, the exit stream thrust and inlet stream thrust can be calculated to determine net engine thrust.

Combustion process. The most important parameter to be determined in the combustion process is the resulting total temperature (the temperature of the fluid brought to rest without energy losses) from specified flow rates of air and fuel. The total temperature is first estimated through an equilibrium thermochemistry calculation using the known air temperature, combustion pressure estimate, and fuel-to-air mass flow ratio, f/a. The resulting calculation yields the predicted combustion exhaust composition, thermodynamic properties, and total temperature (T_T). Since these estimates are based on ideal equilibrium thermochemistry, the actual combustion temperature is typically seen to be a lower value due to heat losses and incomplete combustion.

To model this temperature deficit, the combustion efficiency is defined. The two most prevalent definitions used for combustion efficiency involve temperature rise and fuel/air ratios. Temperature-rise combustion efficiency relates the actual rise in total temperature determined from testing to the theoretical rise as a percentage. Thus, this efficiency may be defined by the equation below.

$$\text{combustion } T_T = \text{air } T_{T_0} + (\text{efficiency} \cdot T_T \text{ theoretical rise})$$

A slightly more complicated but more intuitive method yields the fuel/air combustion efficiency. For this definition, a curve of combustion total temperature versus f/a is drawn (**Fig. 4**), usually with a maximum near the stoichiometric point (where all the fuel is predicted to react with the air). Below the stoichiometric point, the mixture is lean, and above this point the mixture is fuel-rich. In order to calculate combustion efficiency, the predicted total temperature is calculated for the known f/a ratio.

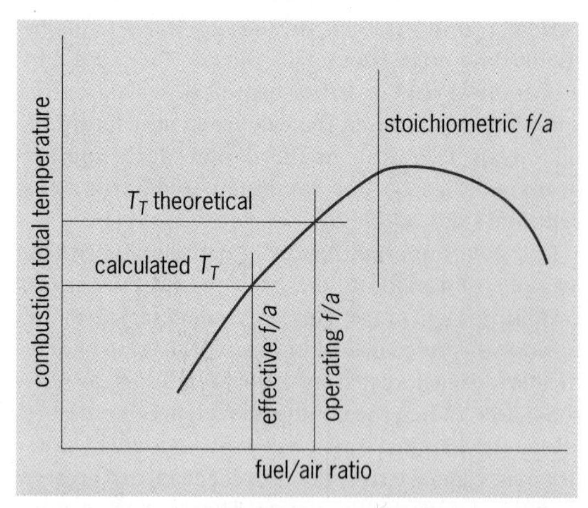

Fig. 4. Plot of combustion total temperature versus fuel/air ratio, used in the definition of fuel/air combustion efficiency.

Test data will yield a lower temperature value due to the reasons mentioned above, which would correspond to a lower effective value of *f/a*. The ratio of these two *f/a* values as a percentage forms the fuel/air combustion efficiency definition. This efficiency can be viewed in an abstract sense as a measure of how much fuel is actually reacting in the combustion process.

The other important variable in ramjet combustion chamber flows is the "burner pressure recovery." This is normally characterized in ground testing by measuring pressures at the dump station and the nozzle entrance. The relationship is expressed as a function of combustion chamber operating conditions, and can be used in the cycle analysis thrust predictions. Alternatively, the individual station pressure losses can be calculated on a theoretical or empirical basis considering component losses due to the inlet dump and any flameholders, in addition to the air/fuel mixing process and heat addition from combustion.

For background information *see* AERONAUTICAL ENGINEERING; AIRCRAFT PROPULSION; COMPRESSIBLE FLOW; HYPERSONIC FLIGHT; MACH NUMBER; NOZZLE; PRESSURE; SCRAMJET; SHOCK WAVE in the McGraw-Hill Encyclopedia of Science & Technology.

Patrick W. Hewitt

Bibliography. P. W. Hewitt, Ramjets, *Advanced Hypersonic Propulsion Short Course, Hypersonic Educational Initiative*, National Institute of Aerospace, Hampton, VA, October 24, 2007; P. G. Hill and C. R. Peterson, *Mechanics and Thermodynamics of Propulsion*, 2d ed., Addison Wesley, 1992; J. J. Mahoney, *Inlets for Tactical Missiles*, AIAA Education Series, American Institute of Aeronautics and Astronautics, Reston, VA, 1991.

Reappraisal of early dinosaur radiation

The origin and early diversification of dinosaurs occurred during the Late Triassic Period (235–201.3 million years ago). By the beginning of the Early Jurassic (201.3–176 million years ago), dinosaurs were a dominant part of terrestrial ecosystems across the globe in terms of both diversity and abundance. Vertebrate paleontologists by the mid-1990s thought that the origin and rise of dinosaurs was fairly well understood. However, the discovery of new specimens and reevaluation of previous hypotheses during the past few years have resulted in a much better understanding of the early diversification of dinosaurs and their close relatives. Dinosaurs are defined as the most recent common ancestor of *Triceratops* and birds (specifically, the house sparrow, *Passer domesticus*), and all their descendants. Thus, by definition, dinosaurs consist of two major lineages: the Ornithischia and the Saurischia (**Fig. 1**). Familiar ornithischian dinosaurs include a variety of forms, such as the armored thyreophorans (stegosaurs, ankylosaurs, and their relatives), the dome-headed pachycephalosaurs, horned ceratopsians (including *Triceratops*), and the "duck-billed" hadrosaurs. All known ornithischians were herbivorous or omnivorous. Saurischian dinosaurs consist of two major lineages: the sauropodomorphs, which include the gigantic long-necked quadrupedal sauropods and their relatives; and the theropods, a bipedal, largely carnivorous group that includes birds (Fig. 1). Studies on the origin and initial diversification of dinosaurs focus on the earliest members of the ornithischian, sauropodomorph, and theropod groups. Renewed study of the earliest members of these groups and the closest relatives of dinosaurs (here called "basal dinosauromorphs") has revealed new insights into the patterns and processes related to the rise of dinosaurs (Figs. 1 and **2**).

Age of the earliest dinosaurs. A significant part of the reevaluation of the early diversification of dinosaurs has come not from new fossil discoveries, but from new information about the age of the sedimentary deposits in which early dinosaur fossils are found. The oldest dinosaurs known come from the Ischigualasto Formation of northwestern Argentina; these fossils are radioisotopically dated to approximately 230 million years ago. Early representatives of all three major dinosaur lineages are found in the Ischigualasto Formation, including the ornithischian *Pisanosaurus*, the sauropodomorph *Panphagia*, and the theropods *Herrerasaurus* and *Eoraptor*. Thus, the earliest dinosaur fossils demonstrate that the group had already split into the three major groups that persisted for approximately 170 million years. Previously, it was thought that some dinosaur-bearing vertebrate assemblages from the rich fossil record of western North America were approximately the same age as the Ischigualasto Formation. However, new uranium-lead (U-Pb) radioisotopic dates from the Chinle Formation of the western United States indicate that the oldest dinosaurs from western North America are 10–15 million years younger than those from Argentina. Also, with recent recalibration of the Late Triassic timescale, it is now clear that these North American assemblages

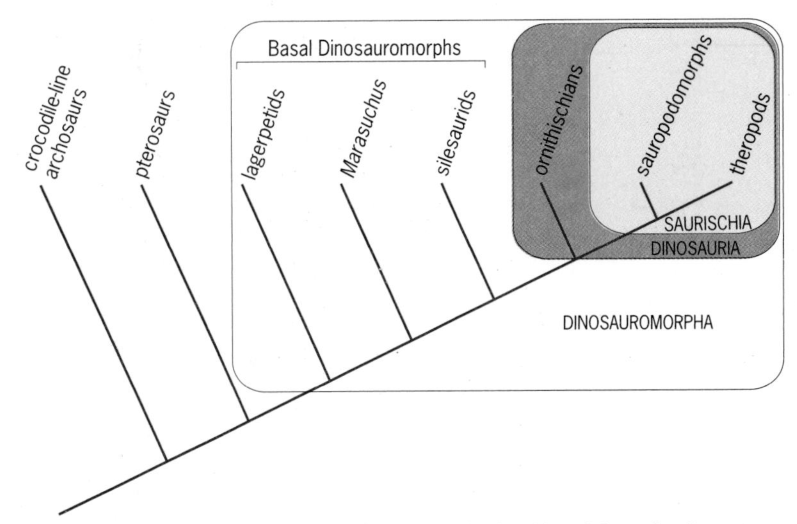

Fig. 1. Evolutionary tree (cladogram) displaying the relationships of the major dinosaur groups and their relatives.

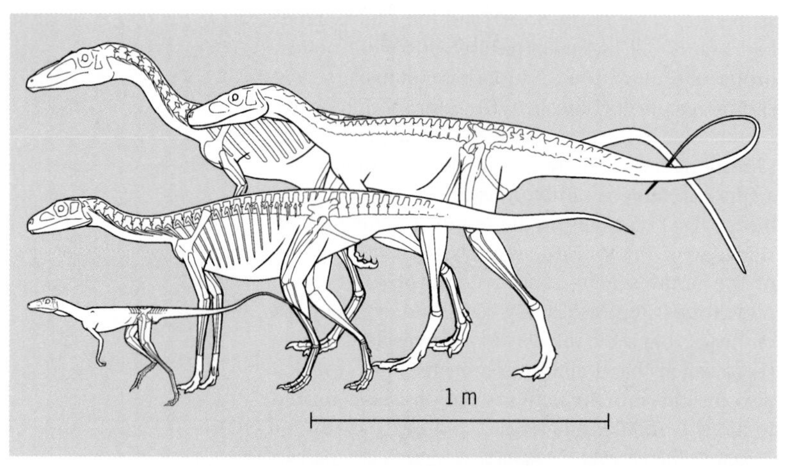

Fig. 2. Early dinosaurs and basal dinosauromorphs that coexisted during the Late Triassic in New Mexico. Front to back: the lagerpetid *Dromomeron*, the silesaurid *Eucoelophysis*, the theropod dinosaur *Chindesaurus*, and the theropod dinosaur *Coelophysis*. (*Artwork by Donna Braginetz*)

(as well as others across the globe) are from the Norian stage rather than the older Carnian stage that they were originally assigned to.

Early dinosaur diversity. Redating of early dinosaur-bearing rocks has changed the understanding of dinosaur diversity during the Triassic Period. Prior to these new data, it was generally thought that dinosaurs were rare during the Carnian, and diversified quickly after a mass extinction of other terrestrial vertebrates at the Carnian-Norian boundary. With the redating of many supposed "Carnian" dinosaur assemblages as Norian, it is now clear that there was no Carnian-Norian extinction event and that dinosaurs did not diversify across this boundary. In some places such as North America, dinosaurs remained rare (in terms of both number of species and relative abundance) throughout the Late Triassic and did not dominate their respective ecosystems until the Early Jurassic, immediately after the end-Triassic extinction event. In contrast, assemblages from South America (Argentina and Brazil), southern Africa, and Europe indicate that dinosaurs were relatively species-rich and abundant in these regions soon after their appearance. The diversity of these assemblages mainly consists of sauropodomorph dinosaurs, but sauropodomorphs are curiously absent from the Triassic of North America. In contrast, Triassic ornithischians were extremely rare and restricted to the southern continents. Only three Triassic ornithischian dinosaur specimens have been discovered: *Pisanosaurus* and an unnamed heterodontosaurid from Argentina, and *Eocursor* from South Africa.

Patterns of early dinosaur evolution. The biogeographic differences in early dinosaur assemblages described above have led some paleontologists to suggest that early dinosaur distribution and diversity were controlled by latitudinal gradients because North American assemblages were at low paleolatitudes during the Late Triassic, whereas South American, southern African, Indian, and European assemblages were at mid- to high paleolatitudes. Modern terrestrial and marine biotas show strong latitudinal gradients, generally thought to be controlled by climate. Thus, it could be that early dinosaur distribution and diversity were controlled in part by paleoenvironment, particularly for sauropodomorphs.

Another new discovery is that early dinosaurs coexisted with their close relatives for a protracted period of time. Basal dinosauromorphs represent several lineages that are closely related to, but just outside of, Dinosauria. Sometimes called "dinosaur precursors" or "protodinosaurs," these lineages were thought until recently to have gone extinct prior to the appearance of the first dinosaurs. Instead, new discoveries from North America, Europe, and South America demonstrate that basal dinosauromorphs coexisted with dinosaurs for at least 15 million years. Groups that coexisted with early dinosaurs include the lagerpetids (for example, *Dromomeron*) and silesaurids (for example, *Eucoelophysis* and *Silesaurus*) [Figs. 1 and 2]. This long-term coexistence indicates that dinosaurs did not outcompete their close relatives, and that basal dinosauromorphs were diverse and widespread during the Late Triassic in both the southern and northern portions of Pangaea.

A final new contribution to the understanding of early dinosaur evolutionary patterns is the investigation of rates of evolution and morphological disparity (variance of the phenotype; for example, shape and size) among early dinosaurs and other contemporary reptiles. These data indicate that dinosaurian disparity and rates of evolution during the Late Triassic were not appreciably different from those of the equally diverse and more abundant crocodile-line archosaurs. Only during the Early Jurassic, after the end-Triassic extinction, did dinosaur disparity increase relative to that of the crocodile lineage. This suggests that there was nothing intrinsically different about early dinosaurs compared to their nondinosaurian contemporaries. The researchers of the original studies interpreted this to mean that dinosaurs were not evolutionarily superior to other lineages at the time, and therefore that it was unlikely they outcompeted non-dinosaurian reptiles that were extinct by the end of the Triassic Period. However, these analyses should be regarded as preliminary because they are particularly sensitive to the type of species and morphological characters that are chosen for the analysis.

Opportunism, competition, and adaptation. Taking all of these new insights on early dinosaur evolution together, what processes allowed dinosaurs to become successful components of terrestrial ecosystems for such a long period of time? Three main ideas have been proposed: (1) dinosaurs radiated opportunistically after the extinction of other Triassic reptiles, after either a Carnian-Norian or end-Triassic extinction event; (2) dinosaurs outcompeted other reptiles; and (3) dinosaurs possessed one or more adaptations that gave them an evolutionary advantage. An opportunistic radiation of dinosaurs at the Carnian-Norian boundary is no longer supported because there is no longer much evidence for a terrestrial extinction event during this time. A radiation after the end-Triassic extinction is still plausible and consistent with available evidence, but it does not address the ultimate cause for dinosaurian success. Even if the success of dinosaurs was opportunistic, there are still other reasons why dinosaurs radiated prolifically in the wake of a mass extinction, relative to other reptilian groups that also survived. The second and third hypotheses outlined above are not mutually exclusive, and it is generally inferred that some sort of adaptation led to dinosaurs outcompeting other groups. There is a general consensus that there is little positive evidence for competitive scenarios, but it can be very difficult to detect competition in the fossil record. Just because competition cannot be detected in data patterns does not mean that it did not occur. There is some evidence that a sustained elevated growth rate is a dinosaurian trait, and this feature may have contributed to early dinosaur success. However, high growth rates do not fully explain the complex biogeographic and diversity patterns observed in early dinosaur evolutionary history.

Ultimately, it is likely that the rise of dinosaurs was driven by multiple processes at different spatial and temporal scales. To disentangle the story, research will need to focus on developing well-dated, high-resolution records of faunal change from many areas at a local to regional scale. This focus, along with continued analysis and interpretation of new fossil discoveries, should help elucidate the causes for the rise of dinosaurs.

For background information *see* DATING METHODS; DINOSAURIA; ECOLOGICAL COMPETITION; ECOLOGICAL SUCCESSION; EXTINCTION (BIOLOGY); FOSSIL; JURASSIC; ORNITHISCHIA; PALEONTOLOGY; REPTILIA; SAURISCHIA; TRIASSIC in the McGraw-Hill Encyclopedia of Science & Technology.

Randall B. Irmis; Sterling J. Nesbitt; William G. Parker

Bibliography. M. J. Benton, Origin and relationships of Dinosauria, in D. B. Weishampel, P. Dodson, and H. Osmolska (eds.), *The Dinosauria*, 2d ed., pp. 7–19, University of California Press, Berkeley, 2004; S. L. Brusatte et al., Superiority, competition, and opportunism in the evolutionary radiation of dinosaurs, *Science*, 321:1485–1488, 2008; S. Furin et al., High-precision U-Pb zircon age from the Triassic of Italy: Implications for the Triassic time scale and the Carnian origin of calcareous nannoplankton and dinosaurs, *Geology*, 34:1009–1012, 2006; R. B. Irmis et al., A Late Triassic dinosauromorph assemblage from New Mexico and the rise of dinosaurs, *Science*, 317:358–361, 2007; S. J. Nesbitt et al., A complete skeleton of a Late Triassic saurischian and the early evolution of dinosaurs, *Science*, 326:1530–1533, 2009; S.J. Nesbitt et al., Ecologically distinct dinosaurian sister group shows early diversification of Ornithodira, *Nature*, 464:95–98, 2010; S. J. Nesbitt, R. B. Irmis, and W. G. Parker, A critical re-evaluation of the Late Triassic dinosaur taxa of North America, *J. Syst. Palaeontol.*, 5:209–243, 2007.

Recent advances in tornado observations

Tornadoes are violently rotating columns of air in contact with the ground, underneath cumuliform clouds, that are capable of inflicting tremendous damage. In order to learn safely how the airflow in tornadoes varies in space and time, and how and why tornadoes form (tornadogenesis), it is necessary to probe them through remote-sensing techniques and by leaving instrument packages in their path. Ground-based mobile Doppler radars, mounted on trucks, have been especially useful in mapping the wind field near the ground in tornadoes. In the springs of 2009 and 2010, a major field experiment in the Plains of the United States known as VORTEX2 (Verification of the Origin of Rotation in Tornadoes Experiment), involved many mobile radar systems and in situ probes.

Tornado structure. It is well known from laboratory and numerical simulations of vortical columns of air in contact with the ground that the spatial variations of the wind (both the horizontal component and the vertical component) in a tornado are controlled by the relative amount of rotational flow (the vertical component of vorticity) to the amount of upward forcing of air above the vortex (caused by buoyancy in the cloud overhead), the "swirl ratio," and surface friction caused by turbulent eddies when the air is slowed down as it rubs against the ground. Above the friction layer, the air is in cyclostrophic balance, as an inward-directed pressure gradient force is balanced by an outward-directed centrifugal force (which is proportional to the square of the rotational wind speed). In the friction layer, however, the air speed is slowed down, so that the inward-directed pressure gradient force overwhelms the outward-directed centrifugal force, and air flows radially inward to the center of the vortex and then turns upward. As vorticity at the ground increases, the central pressure lowers, and air is forced downward. Eventually, as the swirl ratio is increased, the air may be forced all the way down to the surface, and the structure of the tornado markedly changes; that is, air flows into the tornado, but is diverted upward well away from the center, and air flows downward in the center. The highest rotational wind speeds are found near where the air flows upward, leading to a zone of high lateral wind shear inside the radius of maximum wind. This shear zone may become barotropically unstable, and the tornado then breaks up into smaller, satellite vortices (becoming a "multiple-vortex" tornado) that rotate around the center of the broader vortex.

Mobile radars have been used to verify observationally what is expected from theory. While satellite vortices in tornadoes have been seen frequently, they have also been detected on rare occasions by ground-based, mobile Doppler radars. The tornado center appears to be mostly devoid of scatterers and therefore hollow (**Fig. 1**), as a result of the radially outward centrifuging of hydrometeors and

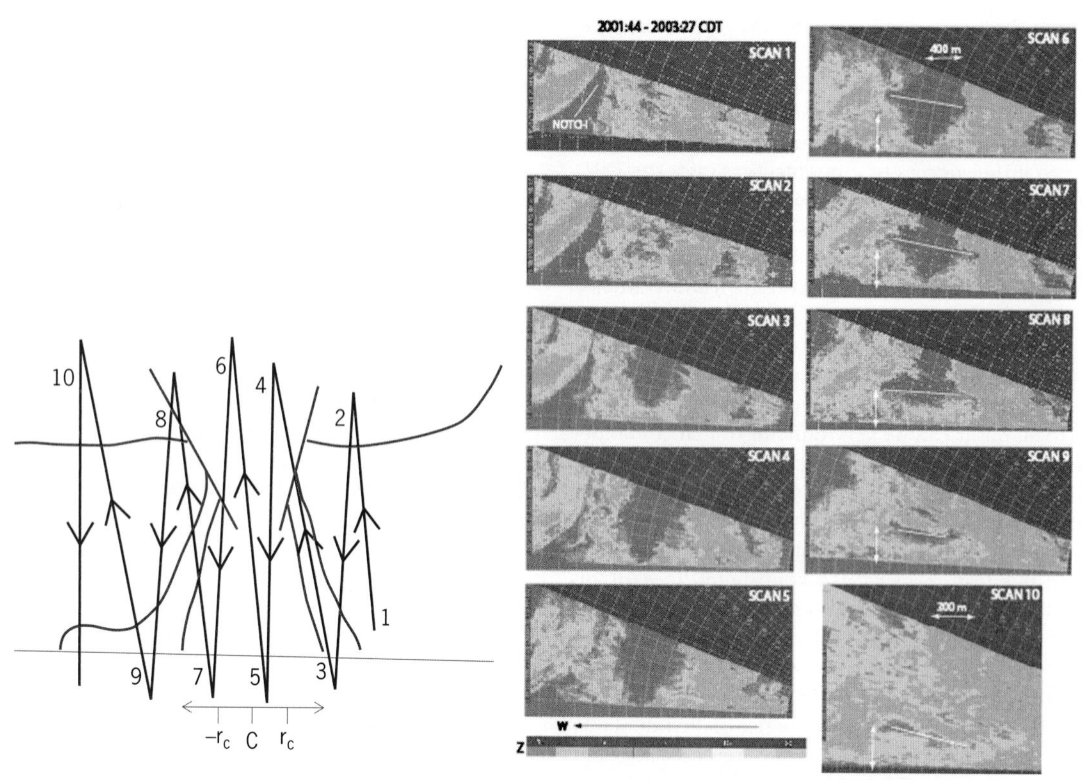

Fig. 1. Vertical cross sections of radar reflectivity through a tornado in south-central Kansas on May 12, 2004, from data collected by the University of Massachusetts mobile W-band (3-mm wavelength) Doppler radar. (*a*) Scans through a tornado relative to the cloud base, condensation funnel, and debris cloud, as determined from a boresighted video camera; "C" is the center of the tornado, "± r_c" marks the right and left-hand parts of the core, where the RMW (radius of maximum wind) is found. (*b*) Radar reflectivity scans; scan 5 cuts through the center of the tornado and reveals a hollow inner core. *(Bluestein et al., The structure of tornadoes near Attica, Kansas, on 12 May 2004: high-resolution, mobile, Doppler radar observations, Mon. Weather Rev., 135:475–506, 2007, courtesy of the American Meteorological Society)*

other scatterers. The hollow center is referred to as a "weak echo hole." In some of the largest and strongest tornadoes, it extends almost to the top of the parent storm (**Fig. 2**). The strongest rotational wind speeds in tornadoes at about 30 m above the ground have been measured by Doppler radar as high as 135 m s^{-1}, in rough accord with damage-based wind-speed estimates. From single Doppler wind data using a technique known as "ground-based velocity track display" (GBVTD), the estimated rotational winds vary linearly from zero at the center to a maximum at the "radius of maximum wind (RMW)" as a solid body would, but then decrease much more rapidly at greater distances from the center (**Fig. 3**). It has been very difficult to measure the wind speed in the lowest 10 m above the ground, where friction should cause a large vertical variation in wind speed, using Doppler radar because of the spreading of the radar beam with distance from the radar and the scattering of radiation off the ground from the antenna sidelobes, leading to "ground clutter" contamination. To resolve the actual variation in wind in the lowest 10 m, surface probes have been placed in or near tornadoes, and a mobile Doppler radar has been used to estimate the wind above the ground at the same time a surface probe made a measurement. High-resolution and higher-frequency Doppler radars have also been used to get closer to the ground (within 50 m), and more recently a pulsed Doppler lidar with a 10-cm-wide collimated beam has been used.

Pseudo-RHI of UMass X-Pol data, 0229-0230 UTC, 13.0 deg
reflectivity

height, m ASL: 0, 5000, 10000, 15000, 20000
range, m: 0, 10000, 20000, 30000, 40000, 50000, 60000
dBZ: 0, 5, 10, 15, 20, 25, 30, 35, 40

Fig. 2. Vertical cross section of radar reflectivity factor through a tornadic supercell in southwest Kansas during the evening of May 4, 2007, as seen by the University of Massachusetts X-Pol mobile Doppler radar. The EF-5 tornado hit the town of Greensburg, Kansas. The arrow points to a weak-echo hole that extends from near the ground up to at least 12 km. *(Robin Tanamachi, University of Oklahoma, Norman, and collaborators)*

The winds inferred from the extent of damage at the surface have been compared with mobile Doppler radar measurements 20–40 m above the ground in a very strong tornado. It was found that the radar-based estimates of damage overestimated the damage-survey-based estimates, for a variety of possible reasons. In 2007, the Enhanced Fujita (EF) scale replaced the Fujita (F) scale of tornado

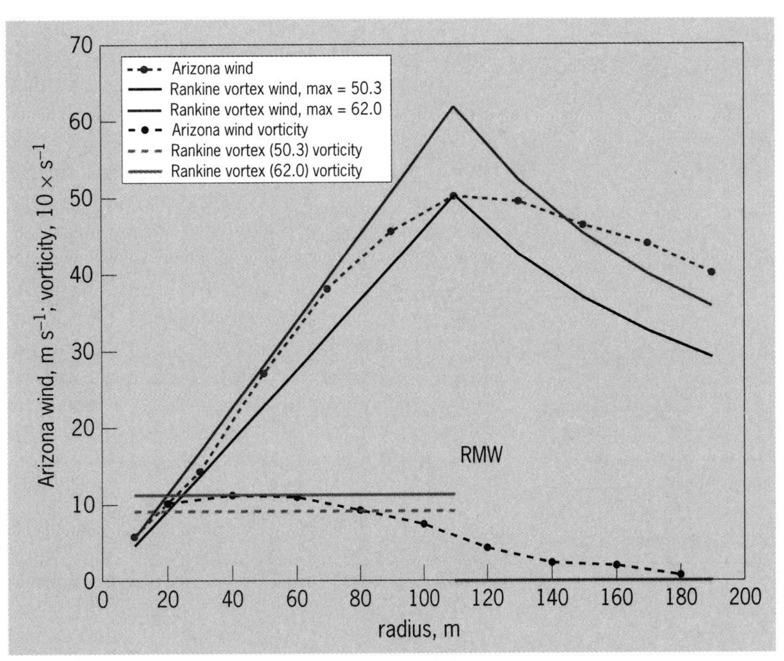

Fig. 3. Azimuthal (rotational) wind component and vorticity as a function of distance from the center (radius) of a tornado in south-central Kansas on May 12, 2004, as determined by the ground-based velocity-track-display (GBVTD) method using data collected by the University of Massachusetts W-band (3-mm wavelength) mobile Doppler radar. The data are fitted to the profile of a Rankine vortex [solid body rotation up to the radius of maximum wind (RMW) and potential flow, that is, having no vorticity, outside]. ***(Robin Tanamachi, University of Oklahoma, Norman, and collaborators)***

Fig. 4. The MWR-05 XP (Meteorological Weather Radar-2005, X-band, Phased-array), mobile phased-array Doppler radar probing a tornado in southeastern Wyoming on June 5, 2009, during year 1 of VORTEX2. ***(Chad Baldi, ProSensing, Inc.)***

intensity to take into account variations in structural integrity.

Most tornadoes rotate cyclonically (counterclockwise in the Northern Hemisphere), but some have been observed on rare occasions to rotate anticyclonically. Doppler radars usually detect a "hook echo" associated with the parent cyclone that spawns tornadoes in supercells, the most prolific of tornado-producing storms and those that produce the strongest tornadoes. Mobile Doppler radars have detected hook echoes that (unlike most, which are curved in a counterclockwise manner) are curved in a clockwise direction, coincident with anticyclonic rotation. (The discussion here is limited to the Northern Hemisphere; in the Southern Hemisphere, cyclonic rotation is clockwise.) In some instances, an anticyclonic tornado was observed. When anticyclonic tornadoes are observed in supercells, they usually occur along the flanking line associated with the rear-flank gust front, away from a companion cyclonic tornado or mesocyclone.

Instrumented probes have been successfully placed in the path of tornadoes. Some of these instruments, such as Hardened In-Situ Tornado Pressure Recorder (HITPR) probes, are cone-shaped and are dropped alongside a road that a tornado is expected to cross. Others are instruments mounted on tripods (sticknet) or small masts (tornado pods). Measurements have been made with HITPR in which the pressure has fallen about 100 hectopascasls (hPa), in rough accord with what would be expected for cyclostrophic balance, hydrostatic balance, and observed wind speeds. However, the hydrostatic assumption breaks down near the ground in tornadoes. Video recordings have been photogrammetrically analyzed to yield estimates of the speed with which debris flies by. An Unmanned Aircraft System (UAS) was used for the first time during the second year of VORTEX2 to make measurements in supercells where it is not otherwise possible to do so. No measurements have yet been made in or near tornadoes.

A digital infrared camera has been used in an attempt to estimate the temperature variations across the cloud base before a tornado forms to see how much baroclinicity (temperature difference on a constant-pressure surface) is present. This technique proved difficult to implement when, as is often the case, there is intervening precipitation. However, it was found that the temperature along the edge of a tornado condensation funnel decreased with height at the moist-adiabatic lapse rate, as would be expected in a saturated updraft.

Tornadogenesis. The source of rotation in cyclonic tornadoes in supercells has been thought to originate in horizontal vorticity associated with environmental vertical wind shear and that generated along the edges of evaporatively cooled "pools" of air in convective storms produced when rain falls into unsaturated air. The vorticity must be enhanced by convergence, after it has been tilted into the vertical, along the edge of an updraft or downdraft or along the interface between a downdraft and an updraft. The details of how this process happens sometimes, but not all the time, are not yet understood.

To gain insight into tornadogenesis, there have been efforts to catch a tornado in the act of forming by probing supercells with two (or more) mobile Doppler radars and then using single-Doppler velocity (line-of-sight) data from pairs of mobile Doppler radars (at different viewing angles) separated along a "baseline." The three-dimensional wind field is inferred from the "dual-Doppler" analysis, and its evolution is thus determined. However, it is difficult to position the mobile Doppler radars at the right place and at the right time to capture tornadogenesis. In a limited number of instances, mobile radars have been located advantageously, and dual-Doppler analyses documenting parent storm and tornado evolution have been obtained, augmenting what had already been learned from earlier airborne Doppler radar observations. These earlier observations did not include the lowest several hundred meters above the ground and were made at relatively long intervals of about 5 min. Earlier fixed-site, ground-based radar observations were made on scales too large to resolve the tornado, owing to the relatively long range from the tornado, and were made at relatively long intervals of approximately 1–2 min. It has been found that the "rear-flank gust front" seems to play an important role and that multiple surges of outflow, sometimes relatively cool, but sometimes not, may also be significant. Many more cases to be analyzed were collected during VORTEX2, during which simultaneous thermodynamic measurements were made using instrumented cars that traversed selected critical regions of supercells. The results are forthcoming. What happens in supercells when tornadogenesis occurs will be contrasted with what happens in supercells when tornadogenesis does not occur; there are more cases of the latter than of the former, since tornadogenesis is a rare event.

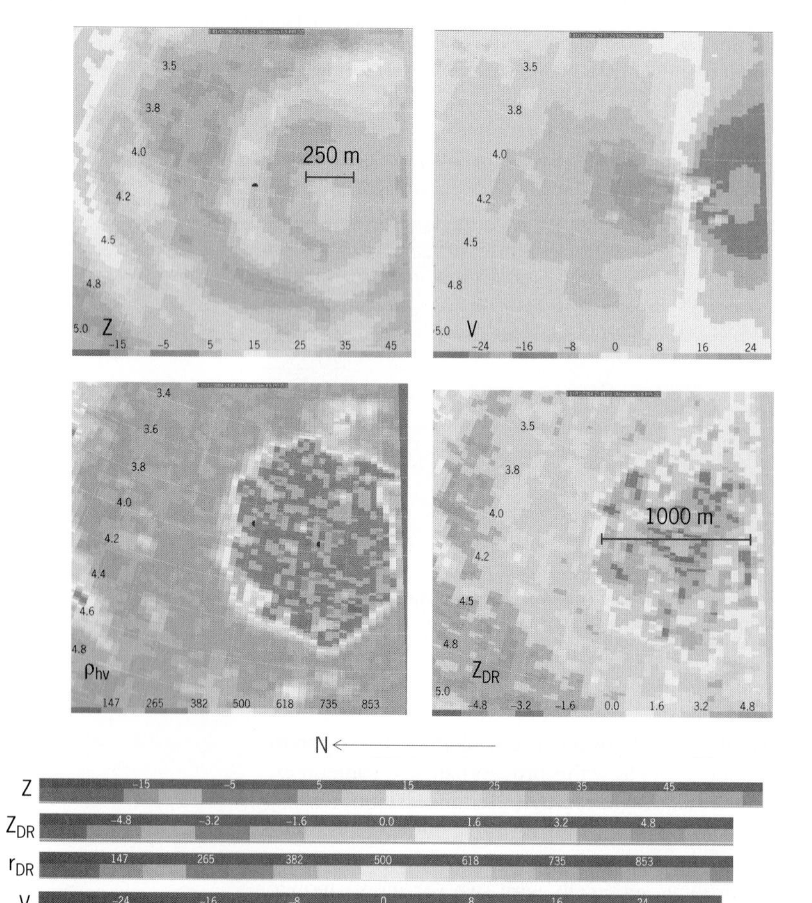

Fig. 5. Radar reflectivity factor in dBZ, Doppler velocity (the scales to the right indicate motion away from the radar, those to the left indicate motion toward the radar) in m s^{-1}, cross correlation coefficient (×1000) ρ_{hv}, and differential reflectivity Z_{DR} in dB at a constant elevation angle through a tornado (near the ground) in south-central Kansas on May 12, 2004, as seen by the University of Massachusetts W-band (3-mm wavelength) mobile Doppler radar. ***(Bluestein et al., The structure of tornadoes near Attica, Kansas, on 12 May 2004: high-resolution, mobile, Doppler radar observations, Mon. Weather Rev., 135:475–506, 2007, courtesy of the American Meteorological Society)***

Since wind speeds in tornadoes are typically on the order of 50–100 m s^{-1}, it takes air about 5–10 s to circulate around completely at a typically observed radius of 100 m. This time period is called the "advective time scale." In order to observe tornadogenesis, one must therefore make observations of the wind field every 5–10 s or so, and thus observations made every minute or more miss the complete evolution of tornadogenesis. For this reason, "rapid-scan" radars have been developed and used in the past several years. These radars make use of electronic scanning, rather than slower mechanically scanning antennas. The electronic scanning technique employs phased-array technology, in which a radar beam is deflected by changing the phase systematically along a line of antenna elements or by changing the frequency systematically along a slotted waveguide. The National Severe Storms Laboratory (NSSL) located in Norman, Oklahoma, has a fixed-site, phased-array Doppler radar. There are two mobile Doppler radars that make use of both mechanical and electronic scanning. One was developed at the National Center for Atmospheric Research (NCAR) in Boulder, Colorado, and the other was adapted at the Naval Postgraduate School in Monterey, California, for meteorological use from radars used for military applications (**Fig. 4**).

Previous observations indicate that in some instances, the tornado, seen in Doppler radar data as a Tornado Vortex Signature (TVS), forms aloft and propagates downward to the ground, while in others it forms everywhere in a column simultaneously. In the latter case, observations were made roughly every 5 min, so that a rapid-scan radar may be able to see if the tornado forms aloft and builds downward on time scales of less than a minute. Several cases of tornadogenesis were collected during VORTEX2, but data analysis is still in progress, so the final results are forthcoming. Preliminary analysis, however, does suggest that in at least one case, a cyclonic tornado formed aloft and descended to the ground very quickly in just tens of seconds, while in another case an anticyclonic tornado formed near the ground and built upward. The latter process is like that of a nonsupercell tornado or "landspout." The source of vorticity in an anticyclonic tornado has been hypothesized to be horizontal vorticity generated at the leading edge of the rear-flank downdraft, and then tilted by an updraft within the flanking line; at the southern (northern) end of the flanking line, anticyclonic (cyclonic) vorticity is produced.

Tornado detection. Tornadoes have been detected as a region of very strong lateral shear in Doppler velocity having vertical and temporal continuity (TVS). When a TVS propagates downward to the ground over a period of 10–20 min, advance warning can be given based on the detection of the TVS.

However, when the radars are relatively far from a storm (>20 km or 12.4 mi), the lowest beam is well above the ground, owing to the curvature of the Earth. It has been proposed that networks of low-powered Doppler radars be implemented, using towers with relatively dense spacing of the radars, so that the lowest beam from the closest radar to a storm is below the cloud base. Such a test-bed network has been set up in central Oklahoma called "CASA," for the Center for Collaborative Adaptive Sensing of the Atmosphere.

Even when a Doppler radar detects a TVS, however, in the absence of simultaneous visual observations, there is no way of knowing for sure if there is a tornado. Polarimetric Doppler radars have been used to distinguish between airborne debris and hydrometeors. Raindrops are flattened as they fall, whereas airborne debris tumbles, so that there is no preferred orientation. It turns out that Z_{DR}, the differential reflectivity, a measure of the relative amount of horizontal to vertical backscattered radiation, and ρ_{HV}, a measure of how well the horizontally and vertically polarized backscattered radiation are correlated with each other, are relatively low when there is debris and high when there is rain. Thus these quantities aid in tornado detection (**Fig. 5**). In addition, research has been done and is ongoing to determine whether polarimetric signatures can be found that are precursors to tornadogenesis.

For background information *see* BAROCLINIC FIELD; BAROTROPIC FIELD; CLOUD; CYCLONE; DOPPLER RADAR; DYNAMIC METEOROLOGY; STORM; STORM DETECTION; TORNADO; VORTEX in the McGraw-Hill Encyclopedia of Science & Technology.

Howard B. Bluestein

Bibliography. H. B. Bluestein et al., The structure of tornadoes near Attica, Kansas, on 12 May 2004: high-resolution, mobile, Doppler radar observations, *Mon. Weather Rev.*, 135:475-506, 2007; H. B. Bluestein et al., Close-range observations of tornadoes in supercells made with a dual-polarization, X-band, mobile Doppler radar, *Mon. Weather Rev.*, 135:1522-1543, 2007; H. B. Bluestein et al., A mobile, phase-array Doppler radar for the study of severe convective storms, *Bull. Am. Meteorol. Soc.*, 91:579-600, 2010; D. C. Dowell et al., Centrifuging of hydrometeors and debris in tornadoes: radar-reflectivity patterns and wind-measurement errors, *Mon. Weather Rev.*, 133:1501-1524, 2005; C. D. Karstens et al., Near-ground pressure and wind measurements in tornadoes, *Mon. Weather Rev.* (in press), 2010; D. McLaughlin et al., Short-wavelength technology and the potential for distributed networks of small radar systems, *Bull. Am. Meteorol. Soc.*, 90:1797-1817, 2009; J. C. Snyder et al., Attenuation correction and hydrometeor classification of high-resolution, X-band, dual-polarized mobile radar measurements in severe convective storms, *J. Atmos. Ocean. Tech.* (in press), 2010; R. L. Tanamachi et al., Infrared thermal imagery of cloud base in tornadic supercells, *J. Atmos. Ocean. Tech.*, 23:1445-1461, 2006; J. Wurman, The multiple-vortex structure of a tornado, *Weather Forecast.*, 17:473-505, 2002; J. Wurman and C. R. Alexander, The 30 May 1998 Spencer, South Dakota, storm. Part II: Comparison of observed damage and radar-derived winds in the tornadoes, *Mon. Weather Rev.*, 133:97-119, 2005; J. Wurman et al., Low-level winds in tornadoes and potential catastrophic tornado impacts in urban areas, *Bull. Am. Meteorol. Soc.*, 88:31-46, 2007; J. Wurman et al., Dual-Doppler analysis of winds and vorticity budget terms near a tornado, *Mon. Weather Rev.*, 135:2393-2405, 2007.

Reference printing conditions

Reference printing conditions provide the interface between design, printing data preparation, proofing, and printing and define the intended relationship between data in a computer and color on a printed sheet, regardless of the printing process used. Symbolically, reference printing conditions represent the transition of the printing industry from a craft-based and nonstandardized film environment in the late 1970s to today's digital-data-based and standardized manufacturing process. This has been truly a revolution in which computer capability, digital data storage capability, standards, and cooperation among industry associations have worked together to change the basic philosophy of printing and publishing.

Background. Historically, within the printing and publishing industry the goal of design and prepress was to prepare printing masters specific for the printing process and equipment to be used. Both industry trade associations and standards groups focused on defining printing aims for different applications and classes of equipment. Initially, it was to define the densities of the cyan, magenta, yellow, and black (CMYK) solids and two-color overprints (blue = C + M, red = M + Y, green = Y + C), along with the single-color tone reproduction curves. These aims were associated with specific inks, papers, and printing processes and were unique to a particular printer or to a printing application or industry trade association. The Specifications for Web Offset Publications, or SWOP specifications, are one of the earliest U.S. efforts by a trade association to define a printing process for a particular printing application, in this case, a magazine publication. Most importantly, they were based on the common practice of exchanging content data (for printing) as film halftone separations.

The writers of early graphic arts standards followed this same approach, although they did use colorimetric instead of densitometric definitions of the solids. Thus, the principle ISO (International Organization for Standardization) standard in this area (ISO 12647, *Graphic technology—Process control for the production of half-tone colour separations, proof and production prints*) has different parts for different printing processes, and within each part addresses a variety of papers suitable for that printing process.

Once the graphic arts standards groups decided to use colorimetry instead of densitometry to define printing aims, it became necessary to develop a standard, based on the work of the CIE (Commission Internationale de l'Eclairage or International Commission on Illumination), but unique to the graphic arts application that defined the specific options to be used for colorimetric measurements and computations. The first version of this document was ANSI/CGATS.5 (American National Standards Institute/ Committee for Graphic Arts Technologies Standards), which was replaced by ISO 13655:2009, *Graphic technology—Spectral measurement and colorimetric computation for graphic arts images*.

Digital revolution. In 1980, the growth in computer speeds and data storage capability allowed the printing industry, for the first time, to consider digital manipulation and storage of content. Because this digital capability was expensive, it was slow to penetrate the workflow. And although it had a significant impact in prepress, its impact on printing, and printing and proofing standards was not really felt until the early 1990s.

The first step toward implementing the concept of reference printing conditions was the use of test targets with patches having many combinations of CMYK values that could be printed or proofed using the traditionally defined printing aims. Using an agreed-upon combination of target patches and making colorimetric measurements of the printed results allowed the printing and proofing systems to be compared quantitatively by using reflection spectrophotometers and computing CIELAB data. (CIELAB is a CIE color scale for measuring color independent of how, or on what device, the color was produced.) This also allowed printing by different systems, done to similar process control aims, to be compared. The first of these targets, published in 1993, was called the ANSI IT8.7/3 data set and had 928 target patches.

Such mappings of CMYK values of target patches to the color measured on the printed sheet for a specific set of printing aims (such as ink, paper, color of the solids, and tone reproduction) soon became the reference for digital data preparation (the relationship of CMYK to color). When such mappings were done by, and endorsed by, the trade group that had traditionally defined the printing aims they became the reference characterization data or reference printing condition for that process.

The first such data set was the result of a cooperative effort between ANSI CGATS (Committee for Graphic Arts Standardization) and SWOP and was published in 1995 as CGATS TR 001. When it came time to give final approval to CGATS TR 001 many of the representatives of both SWOP and CGATS were apprehensive about saying that this represented SWOP. In fact, the final recommendations from the SWOP Committee said, "The data of Annex A provide the first publicly available colorimetric characterization of sheetfed offset proofing (printing) that meets the aims defined in CGATS.6 for Type 1 printing (derived from and complimentary to SWOP)."

The characterization data based on a data set, such as that contained in IT8.7/3 or the newer IT8.7/4 target with 1617 patches, both maps the outer color gamut (range of colors) of the printing condition and also maps the color area internal to the gamut shell. The bulk of the IT8.7/3 data points represented two-, three-, and four-color overprints of tints within the gamut shell. While the gamut shell is largely determined by the paper, ink color, and transparency, the internal colors are affected by ink trap, ink transparency, the tone reproduction of the individual colors, and so on. Thus, characterization data maps the behavior and interaction of the complete printing process. However, such characterization data applies only to the specific ink, paper, and press used to create the test form at a specific point in time.

Generalizing characterization data to make it more broadly representative is part of the reference printing condition concept. The simple steps that were taken with TR 001 were to control the printing to be close to the specified SWOP aims. Sheets were then selected from the pressrun using weighted averages of solid and midtone density to produce a sample set that represented the SWOP aims. These were then measured spectrophotometrically in several labs and the results averaged to produce the final data set. Today, such data are created using real reference data for the outer points of the gamut shell and computer modeled within-gamut interactions to achieve tone reproduction, overprinting, and smooth data transitions, as well as data spacing that facilitate profile building in color management systems.

Evolving digital printing world. In 1999, the publication of ANSI CGATS.12/1, *Graphic technology—Prepress digital data exchange—Use of PDF for composite data—Part 1: Complete exchange (PDF/X-1)*, which later became ISO 15930-1:2001, firmly established characterization data and reference printing conditions as an integral part of the transition from an analog (film) to a digital workflow in the printing industry. That standard said that all data "shall be color corrected and adjusted for a single characterized printing condition prior to exchange." The groups that created this standard realized that characterization data, such as TR001, were not perfect but were a useable reference for comparing prepress, proofing, and printing. They became the digital data definition of the process with which they were associated, that is, the reference printing conditions.

The advent of the PDF/X standard for prepress data interchange and the increasing availability of digital data aided the development of printing plates and platesetters that could bypass the film-exposure step and create plates directly from digital data. Some color-proofing systems were already working from digital data, but once direct-plate exposure was available, the all-digital workflow became both a reality and a goal. Again, reference printing conditions became the data reference to which both printing and proofing were aimed.

This opened the door for color management systems. The reference printing conditions represented the average data from a carefully managed series of tests that match as closely as possible the printing aims established by industry associations or standards groups. Actual printed matter, even if it matches the same solids, solid overprints, and tone reproduction curves, usually does not exactly match printing on another press or a reference printing condition. Proofing, with a whole different set of interactions and colorants certainly cannot be made to match simply based on the solids and tone reproduction. Because electronic data can be manipulated, and because the observer sees only the color produced not the magnitude of the individual CMYK or RGB components of that color, a color management system can adjust the data within gamut to make the output of one system have the same appearance as the output of another system.

This evolved into a workflow in which data used by both printing and proofing were adjusted electronically to allow both to match the reference characterization data (the reference printing condition). Today, this reference data is manipulated, smoothed, and otherwise adjusted so that it represents what the printing was intended to be, not what any particular press or proofer produced. It has become the characterization data for an idealized virtual printing system.

The next logical step was to use the same reference printing condition for printing intended for the same application but using different processes. One of the earliest efforts was the U.S. publication printing market. Some publications, because of their run length, are printed using gravure instead of offset. For example, advertisers expect their ads to look the same irrespective of the printing process used. Because the gamut of gravure can be larger than that of offset, and because gravure electronic cylinder engraving requires that the data be manipulated to properly drive the engraving system, the gravure printers have been able to match the same characterization data—the same reference printing condition—that was used for the offset version of the publication.

Over the past few years, characterization data sets have been created for all of the various printing process and paper types defined in the ISO 12647 printing standards. The International Color Consortium (ICC) has established a registry for publically available characterization data. The full PDF/X International Standards require inclusion of, or pointers to, the characterization data for which the content data has been prepared.

Where to next? More recently, data manipulation and data smoothing have been used to create characterization data that more closely represents the aims of a particular process than can be achieved in actual printing. This is truly creating a reference printing condition for a virtual (ideal) press, albeit based on practical testing to determine the primary interactions experienced in real printing.

There have also been three recent activities that are helping to move the reference printing concept forward.

First, in 2009 the ISO Technical Committee 130 (TC130) published ISO/TS 10128, *Graphic technology—Methods of adjustment of the color reproduction of a printing system to match a set of characterization data*. This technical specification assumes that the printing process can match the outer gamut specified and describes three ways that incoming data can be adjusted to allow the within-gamut data to also match the reference printing condition data when printed.

Second, ISO 12647-7 2007, *Graphic technology—Process control for the production of half-tone color separations, proof and production prints—Part 7: Proofing processes working directly from digital data*, is based on the premise that all proofing from digital data is based on a set of characterization data (a reference printing condition) which the proof is expected to match colorimetrically.

Third, ISO TC130 (ISO Technical Committee 130) at its fall 2009 meeting authorized a Preliminary Work Item with the title ISO 15339, *Graphic technology—Printing of digital data—Part 1: Basic principles.* This has become known as the "process-agnostic specification" because one of its basic assumptions is that the aims for most printing can be described as a series of gamuts that range from typical newsprint up through and beyond Type 1 (high-quality coated) sheetfed paper. Each gamut (hopefully no more than six or seven) will have an associated characterization data set that will have tone reproduction characteristics based on a common principle. Additional parts will describe compromises that may be associated with different printing processes and methods or additional definitions to help users of those processes match the reference printing conditions described in Part 1.

The goal is that all printing—flexographic, offset, gravure, electrophotographic, ink-jet, and even new processes—can use these gamuts and reference printing conditions as their primary reference, particularly for copy preparation aims and data exchange. This is an ideal goal, but one that it will take a while to reach. In many ways, when completed ISO15339 will represent the embodiment of the reference printing condition concept.

For background information *see* COLOR; COLOR VISION; COMPUTER GRAPHICS; ELECTRONIC DISPLAY; INK; PHOTOGRAPHY; PRINTING in the McGraw-Hill Encyclopedia of Science & Technology.

David Q. McDowell

Bibliography. D. Q. McDowell, Reference printing conditions, what are they & why are they important? *Prepress Bull.*, 88(6):42–44, 1999; D. Q. McDowell, Reference printing conditions: Where can it lead us?, *TAGA 2002 Proceedings*, Ashville, NC, pp. 353–366, 2002.

Regulation of leaf shape

The arrangement and shape of leaves of flowering plants are some of the most diverse characteristics in plant biology. The striking morphological differences in these above-ground structures can range from simple leaves, with each leaf consisting of a one-blade unit, to compound leaves, in which a blade is subdivided into two or more separate parts called leaflets. Examples of such leaf-shape variation include simple leaves of *Arabidopsis thaliana*, peltate leaves (which have the petiole attached to the lower surface instead of the base) of *Tropaeolum majus* (Nasturtium), pinnately compound leaves of tomato (pinnate leaves have their parts arranged like a feather, branching from a central axis), and palmately compound leaves of *Pachira aquatica* (money tree; palmate leaves have lobes that radiate from a common point) [**Fig. 1**]. Within these different leaf shapes, the blade of the leaf may have smooth margins, toothlike serrations, or lobes, each of which may be elaborated to various degrees (**Fig. 2***a,b*). Despite the great morphological diversity in leaves across seed plants, the common trait of all leaves is that they function as the main photosynthetic organ of the plant. Leaves also share certain defining traits that determine their early development independent of final leaf form. For instance, all leaves arise from the flanks of an organ called the shoot apical meristem. The shoot apical meristem is a dome-shaped structure composed of a population of stem cells, and it is these cells that regulate the patterns of cell division activity that lead to leaf initiation. Once leaves initiate from the flanks of the shoot apical meristem, their lateral position inherently polarizes the leaves along their dorsiventral axis. This dorsiventral polarity is also specified in both external and internal tissues. The common internal tissues of leaves differentiate during this stage of leaf development, during which structures such as the vascular system are formed. Despite these commonalities of early development in simple and compound leaves, evolutionary studies suggest that compound leaves may have arisen more than once in the flowering plant lineage, with frequent reversions back to the simple leaf form. With the advent of molecular biology, classical theories that suggest or refute homology between simple and compound leaves can be revisited.

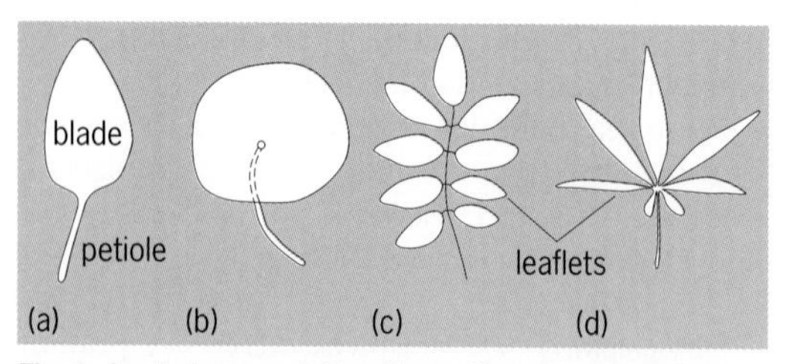

Fig. 1. Leaf-shape variation: Illustration showing various leaf shapes. (*a*) Simple leaf with a single blade and petiole. (*b*) Peltate leaf. The insertion point of the petiole (dashed line) occurs on the underside of the leaf. (*c*) Compound leaf. A compound leaf has multiple blade units called leaflets. (*d*) Palmately compound leaf. Each leaf blade radiates out from the petiole.

Fig. 2. Regulation of leaf shape: (*a*) simple-leafed tobacco species (*Nicotiana benthamiana*); (*b*) compound-leafed tomato species (*Solanum lycopersicum*); (*c*) *Lanceolate* tomato-leaf mutant; (*d*) *bipinnata* tomato-leaf mutant; (*e*) *pPIN1:PIN1:GFP* tomato shoot showing the presence of auxin-transporting and auxin-concentrating PIN proteins [as determined by the expression of GFP (green fluorescent protein)] correlated with the initiation and development of leaflets (arrowheads); (*f*) *entire* tomato-leaf mutant.

Acquisition of leaf shape. Acquisition of leaf shape occurs very early in leaf development, shortly after emergence from the shoot apical meristem. During this stage of leaf development, that is, the primary morphogenesis stage, leaf shape can be acquired either through limited cell division along the leaf margin or through prolonged phases of division (and outgrowth) separated by regions of growth suppression. This creates various leaf shapes. Leaf shape also can be enhanced and modified at a later stage, called secondary morphogenesis, through differential expansion of the blade. In simple leaves, cell division activity is generally uniform along the leaf margin and is short-lived, leading to a simple leaf shape consisting of one blade unit that attaches to the leaf base by the petiole. Compound leaves are subdivided into smaller units—leaflets—and require both enhancement and suppression of cell division activity in discrete regions along the leaf margin. This prolonged cell division activity, together with unique patterns of enhancement and suppression along the leaf margin in compound leaves, is one of the defining key characteristics altering leaf shape. Although the exact molecular mechanism that regulates organogenic activity along the leaf margin is currently unknown, evidence suggests that class I *KNOTTED-LIKE HOMEOBOX* (*KNOX1*) genes may play a distinct role in generating leaf shape. *KNOX1* genes are expressed throughout the shoot apical meristem and are required for meristem formation and maintenance. In simple leafed species such as *Arabidopsis*, *KNOX1* genes are no longer expressed at sites of leaf initiation. These *KNOX1*-absent sites are correlated with leaf primordium formation. In most simple-leafed species, *KNOX1* genes are permanently excluded from developing leaf primordia. In contrast, *KNOX1* gene expression is reestablished in leaf primordia in compound-leafed species, such as tomato, suggesting that *KNOX1* genes play a role in maintaining an indeterminate environment during leaflet formation. An exception to this patterning process is found in a subset of legumes, including pea and alfalfa, which have both simple- and compound-leafed species. These species of legumes are an interesting group in which to study leaf shape because they do not show the expression pattern of *KNOX1* genes typical of most compound-leafed species. Rather, this group uses the floral gene *FLORICAULA* (*FLO*, first named in snapdragon) or *LEAFY* (first named in *Arabidopsis*) to generate compound leaves. Recently, other species with dissected leaves, including *Cardamine hirsuta*, a close relative of *Arabidopsis*, have been analyzed in great detail and have contributed further information about the genetic regulation of leaf shape.

Leaf shape mutants. The study of leaf-shape mutants has been fundamental to our understanding of leaf developmental processes. The model plant system, *A. thaliana*, has been used widely and has been the source of many genetic discoveries about leaf development. However, because it is a simple-leafed species, it is of limited value in studies to understand how complexity in leaf shape is generated. Therefore, other species, such as tomato and *C. hirsuta*, have come to the research forefront and have added to our knowledge of the genetic regulation of leaf shape. Tomato is an excellent species in which to study leaf-shape regulation because of the abundance of leaf forms in cultivated and wild tomato species, as well as the existence of many tomato leaf-shape mutants. One particular tomato-leaf mutant, *Lanceolate* (*La*), has a simple leaf shape consisting of a single blade and petiole (Fig. 2*c*). Although this mutant has been studied for decades, the mutation that causes the phenotype has only been identified recently. It was found to be a gain-of-function mutation in a transcription factor [TCP (TEOSINTE BRANCHED1, CYCLOIDEA, PCF)] that promotes differentiation during leaf development. Thus, in *La* tomato mutants, leaves differentiate prematurely to produce a simple leaf shape. Another recently identified classical tomato-leaf mutant, *bipinnata* (*bip*), has been shown to have a loss-of-function mutation. Regulation of the *KNOX* genes is thought to be controlled in a dosage-sensitive manner through interaction with *BIP*, and loss of *BIP* causes the mutant plants to have increased leaf dissection and marginal serrations (Fig. 2*d*). Currently, novel approaches that utilize natural variation in leaf shape, such as that seen in the tomato species complex, are being explored and will increase our knowledge of the genetic mechanisms and environmental factors that regulate leaf shape.

Auxin and leaf shape. Plant hormones play a critical role in the development of the plant. In particular, the phytohormone auxin plays a role in nearly every developmental process. Auxin is a small signaling molecule that is transported from cell to cell in a directional manner, creating auxin gradients (low to high concentrations) that regulate both spatial and temporal developmental processes. During the vegetative stage of plant development, auxin plays many roles, including positioning leaves on the shoot apical meristem and establishing the vascular system in leaves and stems. More recently, auxin has been identified as a key player in regulating leaf shape. The PIN-FORMED (PIN) family of auxin efflux carriers is one of the most well studied auxin transporters. Asymmetric PIN1 orientation in cells causes auxin efflux and leads to convergence points or points of auxin maxima during leaf development. In compound-leafed species such as tomato, these regions of high auxin concentration have been located at sites of leaflet initiation (Fig. 2*e*, arrowheads). In tomato, the simple-leafed loss-of-function *entire* (*e*) mutant is shown to be caused by a lesion in an auxin-response gene *LeIAA9* (Fig. 2*f*). Other genes (*AUX/IAA*) repress the transcription of auxin-responsive genes until they are degraded upon auxin perception. *ENTIRE* has been shown to play a role in marginal dissection by repressing blade outgrowth between serrations in simple leaves and between leaflets and lobes in compound leaves. Although *ENTIRE* plays a repressive role during blade outgrowth, additional factors, including the *NAM/CUC* (*NO APICAL MERISTEM/CUP-SHAPED COTYLEDON*) genes, are also a requirement. In *Arabidopsis*, leaf serrations are generated by the repression of growth between leaf serrations via *CUC2*. In compound leaves, the *NAM/CUC* genes play a role in organ boundary establishment and separation. Together, both auxin and boundary genes are a requirement in determining leaf shape. Although it appears that auxin and other genes that regulate domains of enhancement and repression along the leaf margin are a requirement during leaf-shape generation, it is still unclear how the very early patterns are established and how these signals then target other histogenic processes such as cell division and vascular development. However, all recent studies on leaf shape do suggest that development of leaf shape is a complex process that requires many genetic players. Currently, with the utilization of new approaches, including computer modeling, studies on natural variation on the microevolutionary time scale, and genomics, our understanding of leaf-shape evolution will be greatly facilitated.

For background information *see* APICAL MERISTEM; AUXIN; GENE; LEAF; LEGUME; MUTATION; PLANT DEVELOPMENT; PLANT GROWTH; PLANT ORGANS; PLANT PHYSIOLOGY; TOMATO; TRANSCRIPTION in the McGraw-Hill Encyclopedia of Science & Technology.

Julie Kang; Neelima R. Sinha

Bibliography. M. Barkoulas et al., From genes to shape: Regulatory interactions in leaf development, *Curr. Opin. Plant Biol.*, 10:660-666, 2007; T. Blein, A. Hasson, and P. Laufs, Leaf development: What it needs to be complex, *Curr. Opin. Plant Biol.*, 13:75-82, 2010; C. Champagne and N. Sinha, Compound leaves: Equal to the sum of their parts?, *Development*, 131:4401-4412, 2004; D. P. Koenig and N. R. Sinha, Genetic control of leaf shape, pp. 433-441, in K. Roberts (ed.), *Handbook of Plant Science*, Wiley, Chichester, U.K., 2007; N. Uchida et al., Coordination of leaf development via regulation of *KNOX1* genes, *J. Plant Res.*, 123:7-14, 2010.

Relations among structure, properties, and function in biological materials

Biological materials have exceptional mechanical properties in comparison to synthetic or engineered materials. Wood has a strength (the ability to carry a load) per unit weight that is comparable to that of the strongest steels. Seashell, bone, antler, and keratin have a toughness (the ability to resist fracture) more than 10 times greater than engineering ceramics. And mature bamboo stalks or culms have slenderness ratios (the ratio of length to diameter) that are remarkable even by standards of modern engineering. Natural materials are optimized to fulfill the complex biological and mechanical requirements posed by the way plants and animals function. Frequently, these requirements are mechanical in nature. For example, tree trunks and branches as well as animal legs and wings support or propel the organism. In doing so, their structural materials (wood and bone) are highly stressed. Beyond this, living natural materials are unique in that they can self-repair and adapt to changes during their lifetime (bone) or during the cycle of a year (wood), modifying their properties according to the requirements placed on them.

A hierarchical material architecture, spanning multiple length scales from the nanometer to the millimeter and beyond, in which material structure, interfaces, and function are optimally adapted at each level of the hierarchy, is thought to give rise to the unique combination of mechanical properties of biological materials. Correlations among microstructural features, principles of function, and performance optimization at all length scales in biological materials are currently attracting much research attention in the hope that these can be emulated in new and improved engineering materials.

Efficiency. In *On The Origin of Species* (1859), Charles Darwin suggested, "Natural selection is continually trying to economize in every part of the organization. . . . For it will profit the individual not to have its nutriment wasted in building up a useless structure." This idea of efficiency, by which we mean that biological materials function using as little matter and energy as possible, has been explored and developed in science and engineering since his day. The remarkable efficiency of biological materials, which manifests not only in the living organism but also in the "harvested" or dead state, is an important reason why biological materials have played a key role in the development of civilization and technology for tens of thousands of years, if not longer. Today, the field of biomimetics, or biomimicry, the study of the principles of function and optimization in biological materials in order to deploy these in the development of novel materials and design, is a thriving research area.

Structural hierarchy. How do biological materials achieve such mechanical and overall efficiency, which makes them so attractive, often for multiple functions? Why can they still compare so favorably with synthetic materials? Are they made from particularly strong and tough components that are superior to those used in the synthesis of engineering materials? Investigating biological materials from a materials point of view, we find that they consist primarily of a relatively small number of soft polymeric and brittle ceramic building blocks (see **table**). Examples include polymers such as cellulose and chitin [polysaccharides (sugars)]; proteins that can either be relatively stiff, such as collagen and keratin, or rubbery, such as elastin, resilin, and silk; and ceramics (calcium salts such as calcite, aragonite, hydroxyapatite, and silica). Occasionally, the materials are enriched with metal ions such as zinc, iron, or manganese. Generally, and very importantly, biological materials contain water, which functions as a plasticizer. From this limited list of ingredients, nature fabricates a remarkable range of composites with a hierarchical structure from the nanometer to the macro scale. For example, solid wood, bamboo, and palm are fiber composites composed of cellulose fibers (bundles of stiff cellulose fibrils) embedded in a lignin-hemicellulose matrix. Multiple, plywoodlike layers of this fiber composite form highly elongated, hollow cells of varying wall thickness. Glued together, these cells form a cellular solid, which in the first order may be approximated as a honeycomblike structure. The exoskeleton (cuticle) of insects consists of a composite in which chitin fibers are embedded in a protein matrix. Layers of this composite form a plywoodlike structure that encases the entire organism and has multiple functions, ranging from mechanical to optical ones. In regions where particularly high abrasion or wear resistance is required, the chitin composite frequently contains metal ions (for example, zinc). Hair, nail, horn, hoof, wool, bird feathers, and reptilian scales are all made of keratin fibers. The dominant ingredient of the mollusk shell is the stiff mineral calcium carbonate (in the form of calcite or aragonite), frequently in the shape of micrometer-scale platelets and needles that are glued together in a brick-and-mortar–like fashion with a few percent of "soft" protein. Enamel is composed mainly of hydroxyapatite. Collagen fibers are the basic structural element for soft and hard tissues in animals, such as tendon, ligament, skin, blood vessels, muscle and cartilage, and the cornea. Bone is formed of collagen fibrils that are mineralized with hydroxyapatite (**Fig. 1**). Bundles of fibrils combine into collagen fibers, which form the plywoodlike structure of the lamellar bone of osteons. Osteons are cylindrical structures that constitute cortical (dense) bone and contain a central canal. This Haversian canal surrounds blood vessels and nerve cells throughout the bone. It is connected to the lacunae, cavities within the dense bone that contain living cells (osteocytes) that constantly de- and reconstruct bone tissue.

Considering that cellulose and chitin fibers have a stiffness (resistance to elongation in tension or deflection in bending) that is about the same as that of nylon fishing line but much less than that of steel, and that the lignin–hemicellulose matrix into which they are embedded has properties very similar to that of epoxy resin, and knowing that the biological mineral hydroxyapatite has a fracture toughness comparable to that of ceramics, we must conclude that it must be the structure and arrangement of the components rather than the components themselves that give rise to the striking efficiency of natural materials.

Material characterization. If we wish to better understand Nature's principles of optimization and function with the intent of using them in the development of new materials, we need to investigate the details of a material's structure and the effect of different levels of the hierarchical structure on performance and failure. Much information can be gained from tensile or compressive tests, in which the material's deformation in response to either a tensile or a compressive load is measured. From these tests we can determine material properties such as stiffness, strength, and toughness. Mechanical tests are even more informative, if in parallel to them the microstructure of the material and structural changes on loading are recorded by, for example, optical or electron microscopy, x-ray tomography, or x-ray diffraction. That way, important information on the deformation mechanisms can be gained at several levels of the tissues' hierarchy. Using results from such combined experimental studies to model material performance has made it possible to understand why wood is as elastic as it is, and why wood

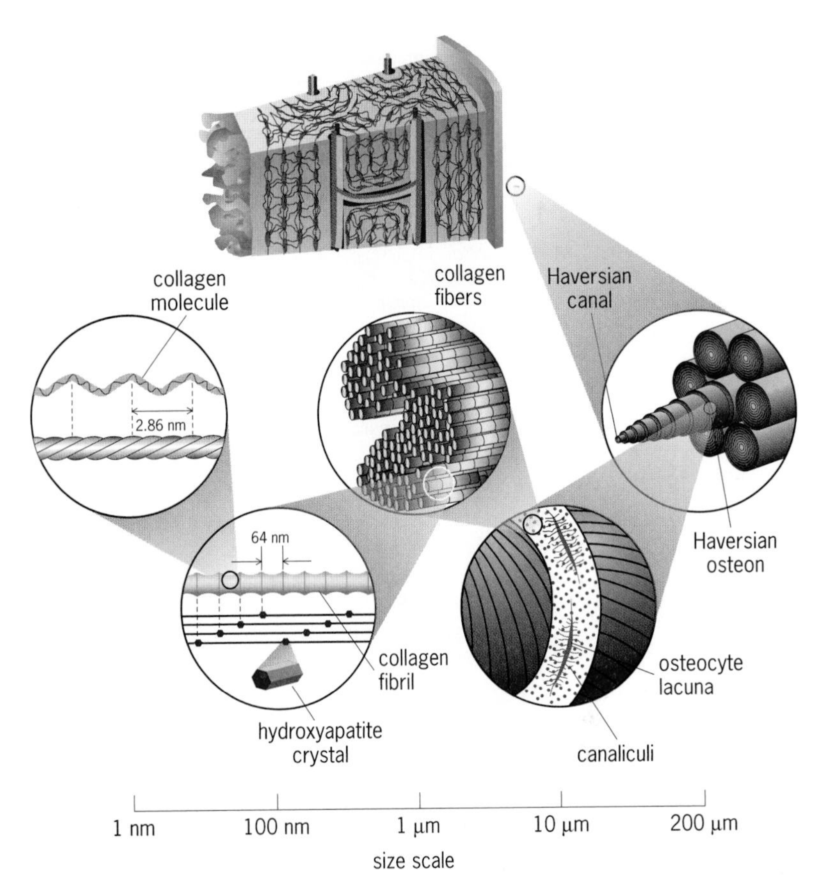

Fig. 1. The structural hierarchy of bone. (*Redrawn after P. Ball, Made to Measure: New Materials for the 21st Century, Princeton University Press, p. 198, 1999*)

cell walls, when loaded in tension, act like elastic springs. Considering that the wood cell wall contains a large amount of highly aligned and stiff cellulose fibers, such elastic performance is surprising until one realizes that the fibers are wound helically around the cell. The imaging techniques also show that the cellulose fibrils carry the load practically without deformation because, as a result of the relatively "soft" lignin–hemicellulose matrix in which they are embedded, they can shear against one another. The same functional principle, load transfer from one stiff component to the next via shear deformation within a thin layer of a significantly softer matrix material, has also been observed in bone, antler, dentin, enamel, and nacre (mother-of-pearl). In bone, the stiff fibrils transfer the load to each other via the softer matrix into which they are embedded. In nacre, stiff microscopic mineral particles effectively transfer the load via a thin polymeric glue layer that joins them.

Investigating both the structure and mechanical performance of the other biological materials mentioned above, one finds that virtually all of them are composite materials with a complex hierarchical structure in which, at the nanometer length scale, a varying percentage of stiff fibrils or ceramic platelets reinforce a polymer matrix. It is this structural principle at the nanometer length scale that causes biological materials to have combinations of properties

Biological materials and their building blocks

Building blocks	Material	Occurrence and function
POLYMERS	**Dominant elements:** C, N, O, H, etc.	
Polysaccharides (fibrous)	Cellulose $(C_6H_{10}O_5)_n$	Tensile load-bearing material in plant cell walls.
	Chitin $(C_8H_{13}O_5N)_n$	Tensile load-bearing material that, embedded in a protein matrix, forms the exoskeletons of insects (ants, locusts, etc.) and crustaceans (crabs, lobsters, etc.). In lobsters, for example, it is mineralized with CaO_3.
Proteins (fibrous)	Collagen	Major component in skin, arterial wall, muscle, cartilage, tendon, ligament, bone.
	Silk	Silkworm cocoon, Spider silks (dragline, viscid, etc.)
	Keratin	Hoof, horn, wool, birds feathers, beaks, gecko skin attachment
Proteins (rubbery/elastomeric)	Elastin	Keeps arteries and other large blood vessels in shape. Provides many tissues with elasticity: e.g., lung, ligament, tendon, skin, bladder, elastic cartilage.
	Resilin	Found in wing hinges of insects, counteracting muscles to restore wing to resting position.
	Abductin	Found in shell-opening ligaments of bivalve mollusks (two-part sea shells).
Proteins (globular)	Enzymes, hormones, antibodies, etc.	Serve regulatory, maintenance and catalytic roles in living organisms.
CERAMICS	**Dominant elements:** Ca, Si, Mg, etc.	
Minerals	Hydroxyapatite $[Ca_9(PO_4)_3(OH)]$	Bone, teeth (dentin, enamel), young mollusks
	Calcite $(CaCO_3)$	Crustaceans (crabs, lobsters, etc.), mollusk shells (seashells), echinoderm (starfish, brittle stars, sea urchins, etc.), sponge spicules, some brachiopods (lampshells)
	Aragonite $(CaCO_3)$	Many mollusks (sea shells), some Foraminifera
	Dolomite $[CaMg(CO_3)]$	Echinoderm teeth
	Magnesite $(MgCO_3)$	Sponge spicules
	"Amorphous" hydrated silica ("biogenic opal") $[SiO_2(H_2O)_n]$	Sponge spicules, limpet radula (teeth)
METALS	Dominant elements: Zn, Mn, Fe, etc.	Crucial for biochemistry. Of structural importance in ionic form in, e.g., the exoskeleton of insects, where they increase the material's wear resistance.

that are contradictory to those in synthetic ones. In contrast to engineered materials, biological materials can be strong and tough simultaneously. Their toughness is also a result of the hierarchical structure, soft phases, and interfaces between the different phases, which hinder the propagation of a straight crack and absorb a large amount of energy when they fail. These combined properties allow biological materials to withstand, within limits, occasional extreme overloads, such as those that a tree in a storm or a person in a fall experiences. **Figure 2** illustrates how significant the contribution of composite design is to mechanical performance. When collagen fibers are mineralized with hydroxyapatite, both components dramatically improve their "weak" property. The stiffness of collagen increases by a factor of about 20, while the toughness of the otherwise brittle mineral, hydroxyapatite increases by a factor of over 100. Such remarkable improvements of performance are rare in synthetic materials. Recently, a biomimetic nacre-inspired material came close. It was fabricated by freeze-casting (ice-templating) alumina powder and then infiltrating the porous structure with poly(methylmethacrylate), which is the material of which Plexiglas® is made.

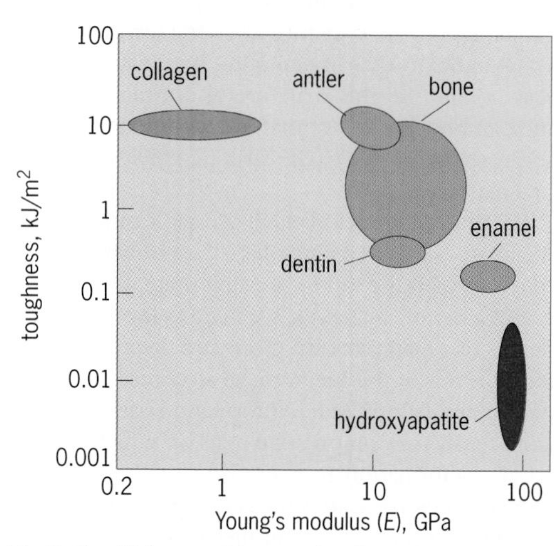

Fig. 2. Combining a soft but tough polymer (collagen) with a stiff but brittle ceramic (hydroxyapatite) results in stiff and tough composites (bone, dentin, enamel, and antler).

Formation of biological materials and synthesis of engineered materials. The formation of composite materials, which occurs primarily at the nanometer to micrometer length scales, is one of Nature's principles of optimization for structural efficiency, as it loads each of the components in a way that takes advantage of the component's best properties, such as mineral stiffness or polymer toughness. Another optimization strategy takes advantage of holes or porosity. At the micrometer and millimeter scales,

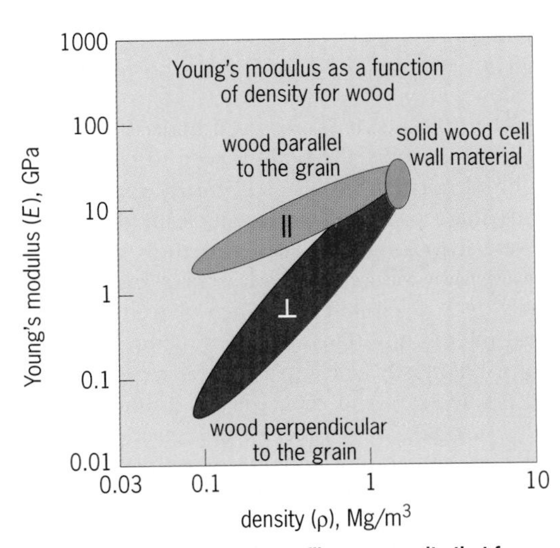

Fig. 3. Foaming the fully dense fiber composite that forms wood cell walls into the cellular material "wood" lowers both density and Young's modulus (stiffness). Parallel to the long axis of the honeycomblike wood structure or wood grain, the stiffness decreases approximately linearly with a decrease in density, perpendicular to it much more significantly with the density, to a power higher than 2.

many biological materials have a cellular structure (frequently due to their formation by cells) similar to that of foams or honeycombs. Taking advantage of the high efficiency of the composite material at a lower level of the structural hierarchy, the introduction of porosity permits a significant weight reduction with only moderate loss in performance, such as stiffness. Wood is a good example (**Fig. 3**). Its cell-wall material has a density of about 1500 kg/m^3 and a Young's modulus (stiffness) of about 35 GPa. Because preferred loading directions exist in the trunk and branches of the tree in which the wood forms, we might expect and in fact do find that wood is anisotropic, meaning that it has properties that are different along and across a tree's trunk. The wood cell-wall material forms cells with a shape and microstructural arrangement that places the material where it is mechanically most needed, resulting in a honeycomb-like cell structure in which the long cell axis is aligned along the tree trunk and branches. Figure 2 shows the Young's modulus plotted against density for wood cell-wall material and woods of different densities, ranging from ultralow-density balsa (about 100 kg/m^3) to lignumvitae (up to about 1400 kg/m^3), the wood with the highest density. It illustrates the effect of density (synonymous here to porosity, because we may assume the cell-wall material properties to vary relatively little for woods of different densities) and structure on wood's mechanical performance. Loading wood parallel to the grain, which means parallel to the long axis of the cells, the cell-wall material primarily stretches, making most efficient use of its cellulose fiber composite. Loaded in this fashion, the stiffness of the material decreases only linearly with density. If it is loaded perpendicular to this axis, the wood's stiffness decreases with the square of the density. Thus a 10-fold decrease in stiffness in one case is a 100-fold decrease in the other. This anisotropy in structure, which results in an even more marked anisotropy in mechanical properties, is the reason that wooden components are cut parallel to the grain when load-bearing ability is required and across the grain when impact resistance and energy absorption are desired.

Two important design strategies found in biological materials—the formation of fiber composites and cellular solids—are individually standard strategies used in materials science and engineering when designing efficient structures. However, combining both principles of optimization in one material is not common yet, and hierarchically structured materials are still rare. The study of biological materials and correlations among their structure, properties, and function has considerable potential for exploring additional design strategies, which may achieve in synthetic materials what we take for granted in natural ones—namely, efficiency, multifunctionality, and the ability to adapt and self-repair.

For background information *see* BONE; CELL WALLS (PLANT); CELLULOSE; COLLAGEN; COMPOSITE MATERIAL; ELASTICITY; HEMICELLULOSE; LIGNIN; MATERIALS SCIENCE AND ENGINEERING; NATURAL FIBER; POLYMER; STRENGTH OF MATERIALS; WOOD PROPERTIES; YOUNG'S MODULUS in the McGraw-Hill Encyclopedia of Science & Technology. Ulrike G. K. Wegst

Bibliography. P. Ball, *Made to Measure: New Materials for the 21st Century*, Princeton University Press, 1999; P. Fratzl and R. Weinkamer, Nature's hierarchical materials, *Prog. Mater. Sci.*, 52:1263, 2007; L. J. Gibson and M. F. Ashby, *Cellular Solids, Structure and Properties*, 2d ed., Cambridge University Press, 1999; S. A. Wainwright et al., *Mechanical Design in Organisms*, Princeton University Press, 1982; U. G. K. Wegst and M. F. Ashby, The mechanical efficiency of natural materials, *Phil. Mag.*, 84:2167–2186, 2004.

Scarab beetle iridescence

The color of various insects, such as beetles and butterflies, in the natural world has attracted the attention of scientists since at least the time of Robert Hooke (1635–1703). Sir Isaac Newton (1642–1727) understood that the colors that are produced must be a result of the presence of "thin film structures." It is now commonly recognized that the colors produced by insects and perceived by an observer are a result of the microstructures that are present on their bodies. In other words, the colors are produced by the interaction of light with the periodic structures on their bodies.

The present article deals with the iridescence of scarab beetles, particularly the color of *Chrysina gloriosa* or *Plusiotis gloriosa*, which possess a metallic green reflection that is circularly polarized. [Circularly polarized light consists of two perpendicular electromagnetic plane waves of equal amplitude but differing in phase by 90°; the resulting electric (or magnetic) field vector traces a circle as it approaches an observer.] It should be noted that linearly polarized light (that is, light in the form of a plane wave) is quite commonplace in nature, but circularly polarized light is quite rare. Among a number of species of beetles that have been examined, so far only the scarabs possess circularly polarized reflectance. Often, scarab beetles are referred to as jewel beetles, based on the fact that reproductions of them have been used as ornaments in many Asian countries. The term iridescence, as used in different fields of study, can refer to different characteristics; in this article, the term is used to mean a change in the hue of the object possessing the perceived color as the angle of vision is varied, which is a definition quite similar to what C. W. Mason used in 1927. The term metallic is often used to describe the saturation or the purity of color.

Color and reflectance. The color of beetles has been studied since the early 1900s, when Albert A. Michelson observed that some scarab beetles possessed a metallic reflection and that the reflection was circularly polarized. Michelson described the color of the beetle *P. resplendens* as follows: "[This] is a beetle whose whole covering appears as if coated with an electrolytic deposit of metal with a lustre resembling brass. Indeed, it would be difficult for even an experienced observer to distinguish between the metal and the specimen." Although *C. gloriosa* is green in color, it does possess the metallic luster to which Michelson referred. Michelson noted that the reflection was circularly polarized, but he did not specify the handedness. However, he did mention that the handedness reverses if one looked at the reflectance from the blue part of the visible spectrum to the red; that is, the polarization was found to reverse near the red end of the spectrum and was completely reversed in the extreme red. He also postulated that "the effect must therefore be due to a screw structure of ultramicroscopic, probably of molecular dimension." Although Michelson did not pursue this further, others (such as A. C. Neville and S. Caveney) investigated the origin of circularly polarized reflection in several scarab cuticles, using electron microscopy. It was found that a "helicoidal structure" is responsible for the color (selective reflection) and handedness of the circular polarization. Scarab beetle cuticles were seen to be analogous to cholesteric liquid crystals (a type of liquid-crystal material in which the elongated molecules are parallel to each other within the plane of a layer, but the direction of orientation is twisted slightly from layer to layer to form a helix through the layers).

Microstructure analysis of the exocuticle. In an effort to understand the microstructure responsible for these optical effects, examinations were undertaken to study the structure of the exocuticle of the *C. gloriosa* beetle, which selectively reflects left-circularly polarized light and possesses a brilliant metallic appearance (**illus.** *a*). If left-circularly polarized light is blocked by the use of a quarter-wave plate and a polarizer, the beetle "loses" its characteristic bright green reflection. The reflectance of the *C. gloriosa* beetle has a broad halo from 500 to 600 nm, with two peaks at 530 (green) and 580 nm (yellow).

When the beetle is observed under a reflected-light microscope, the body is seen to consist of a richly decorated mosaic of regularly spaced polygons that cover the entire cuticle of the beetle, where it looks green. Such structures have been observed by a number of researchers, but most notably by A. Pace in 1972. When viewed in bright-field microscopy (illus. *b*), the structure seems to consist of mostly hexagonal cells (approximately 8–10 μm in size), and each cell has a bright yellow core surrounded by green. The regular lattices of the cells contain not only hexagonal cells but also cells that are pentagonal and heptagonal. It was also noticed that, as the curvature of the beetle exocuticle increased, the number of heptagons decreased a bit and the number of hexagons decreased more, whereas the number of pentagons increased the most. In other words, the more curved the surface of the beetle was, the more pentagons were found, thus leading to higher disorder with increasing curvature. Although hexagonal packing affords the most efficient use of space on a plane, defects (pentagons and heptagons) are essential for tessellating (tiling) a curved surface, thus leading to higher disorder of the structures on the head, thorax, and abdomen of the beetle because of the curved nature of its body. Quantification of these structures found on the beetle cuticles has shown that it is energetically

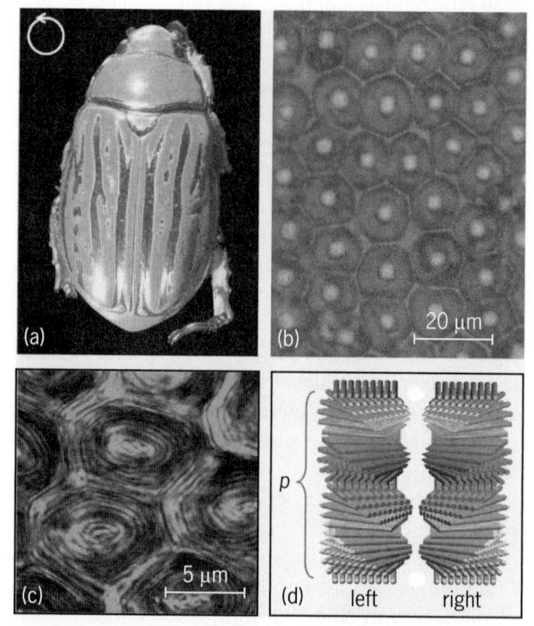

Views of scarab beetle iridescence. (*a*) Photograph of the beetle *C. gloriosa* displaying bright metallic appearance as seen in unpolarized light or with left-circularly polarized light. (*b*) An image using a reflected-light microscope of the exoskeleton of the beetle *C. gloriosa* showing the bright core surrounded by darker reflection. (*c*) An *x–y* section using a confocal light microscope showing the concentric rings resolved at high magnification and present near the free surface. (*d*) A schematic representation of the cholesteric helix for right- and left-handedness (p = pitch of the helix).

unfavorable to create a completely hexagonal packing on the exocuticle of the scarab beetles.

In an effort to better understand the structure of the beetle exocuticle, a laser scanning confocal microscope was employed to reconstruct a three-dimensional (3D) map of the underlying structure, using the autofluorescence of the beetle, where the fluorescence was excited by a 488-nm laser line of an argon-ion laser. The cells under the fluorescence microscope exhibited nearly concentric bands that are bright and dark (illus. *c*), which have been attributed to the underlying structure of the exocuticle. The 3D reconstruction suggests the existence of a nested arc that is similar to what would be observed for a cholesteric liquid crystal.

We have alluded to the fact that the color of the reflected light by the beetle is analogous to the selective reflection of cholesteric liquid crystals. Such liquid crystals possess long-range orientational order, described by a unit vector **n**, known as the director; for a cholesteric liquid crystal, the equilibrium director structure is a helix (illus. *d*). The director advances uniformly, tracing a helix of pitch *p*. Pace noticed the polygonal structures in the transmission electron micrographs from the exoskeleton of *C. gloriosa*, collectively calling them a Bouligand structure (also known as focal conic defects). Extensive studies have been carried out on the similarity of textures observed in crabs and other organisms to cholesteric liquid crystals, and their role in morphogenesis has been explored. Such structures form when a cholesteric liquid crystal has a surface exposed to air (in other words, a free surface). The structure that is found by confocal microscopy (using nondestructive imaging techniques) on the beetle exocuticle is completely analogous to the structure found on the free surface of cholesteric liquid crystals. In fact, the 3D microstructure of the beetle elytra (forewings) shows unmistakable similarity to the cholesteric focal conic texture at a free surface.

Because of its helical structure, the cholesteric phase exhibits selective reflection when the pitch of the helix is comparable to the wavelength of visible light. Of course, the reflection has the same handedness as the cholesteric helix. Hence, when unpolarized light is incident on the cholesteric helix, with the helical axis oriented normal to the surface, it reflects 100% of the light with the same handedness. Unpolarized light can be thought of as a mixture of left- and right-circularly polarized light; therefore, light of the same handedness (approximately 50%) is reflected, whereas the rest is transmitted. It should be pointed out that there is very little absorption in a cholesteric fluid; however, this may not be the case for the beetle exocuticle. For helices oriented at some other angles or for light that is incident at some oblique angle, the optical properties are a bit more complicated (and beyond the scope of this article). Thus, the patterns found on scarab beetles are surmised to be largely a consequence of the array of focal conic defects formed at the free surface of a cholesteric liquid crystal formed from chitin, and the color is a result of the selective reflection mediated by the defect array on the surface.

Explanation and purpose. Insofar as the colors and circularly polarized reflection are created by an underlying cholesteric phase, one might wonder about the purpose of such colors and the resulting polarization. Can the beetles in fact sense the circular polarization? Questions of this sort are beginning to be answered. In a recent study, the response of *C. gloriosa* toward different light stimuli was studied. It was found that these beetles exhibit flight orientation that is dependent on the polarization of the light, thereby indicating that these beetles are sensitive to circular polarization of the light. It is possible that this sensitivity to circular polarization allows *C. gloriosa* to communicate in some fashion, because the signals are independent of their orientation.

A number of other beetles and butterflies also produce iridescent colors as a result of the periodic structures on their wing scales. The beetle *Calidea panaethiopica* exhibits a complex color pattern that contains blue-green iridescent stripes. In this case, the color is produced by a multilayer structure that has tiny cups, with the cups producing two different colors that are color-mixed to provide the perception of a single color. Such is the case with the butterfly *Papilio palinurus*; each wing scale is about 120 μm in length, with 5–10-μm-diameter bowls, and each is lined with a multilayer stack of alternating layers of chitin and air. The distinct green color of the wing results from an additive color mixing of yellow and blue reflections. The yellow-colored reflection is from the bottom of the bowl, and the blue results from two reflections at 45° at the edge of the bowl. It is remarkable that the natural world has a number of different but elegant solutions to producing iridescent colors for a variety of purposes.

[Acknowledgment: MS acknowledges support from the National Science Foundation under the grant DMR-0907529. MS also acknowledges support from WCU (World Class University) program through the National Research Foundation of Korea funded by the Ministry of Education, Science and Technology (R32-2008-000-10142-0).]

For background information *see* COLEOPTERA; COLOR; CONFOCAL MICROSCOPY; INSECT PHYSIOLOGY; LIQUID CRYSTALS; OPTICS; PHYSIOLOGICAL ECOLOGY (ANIMAL); POLARIZED LIGHT; REFLECTION OF ELECTROMAGNETIC RADIATION in the McGraw-Hill Encyclopedia of Science & Technology.

Mohan Srinivasarao; Matija Crne;
Vivek Sharma; Jung Ok Park

Bibliography. P. Brady and M. Cummings, Differential response to circularly polarized light by the jewel scarab beetle *Chrysina gloriosa*, *Am. Nat.*, 175:614–620, 2010; C. W. Mason, Structural colors in insects, III, *J. Phys. Chem.*, 31:1856–1872, 1927; A. A. Michelson, On metallic colouring in birds and insects, *Philos. Mag. Ser. 6*, 21:554-567, 1911; A. C. Neville and S. Caveney, Scarabeid beetle exocuticle as an optical analogue of cholesteric liquid crystals,

Biol. Rev., 44:531–569, 1969; A. Pace, Cholesteric liquid crystal-like structure of the cuticle of *Plusiotis gloriosa*, *Science*, 176:678–680, 1972; V. Sharma et al., Structural origin of circularly polarized iridescence in jeweled beetles, *Science*, 325:449–451, 2009; M. Srinivasarao, Nano-optics in the biological world: Beetles, butterflies, birds, and moths, *Chem. Rev.*, 99:1935–1961, 1999.

Securing wireless implantable health-care devices

Implantable health-care devices are being used in diverse applications, including patient identification, continuous monitoring of patient vital signs, and automatic delivery of medications. Many implantable devices have the ability to communicate wirelessly with other devices. This capability can be very useful; for example, an implanted insulin pump might be reprogrammed to alter its dosage without its having to be removed from the patient. However, wireless communication has vulnerabilities: an attacker may be able to eavesdrop on patient information, track or spoof patients, or even cause direct physical harm to patients by maliciously sending commands that affect the operation of an implanted device. Security issues for implantable devices deserve particular attention because of the permanent nature of these devices.

A major challenge in securing implantable devices is the extreme resource constraints facing them. Many devices are completely passive, with no power or computational resources of their own. This makes it difficult to implement traditional security schemes that require performing cryptographic operations on data. Thus, any practical security scheme must be as efficient as possible.

Classification of devices. Wireless implantable health-care devices can be roughly classified into three categories, based on their primary function and communication model (**Fig. 1**).

Implantable identification devices (IIDs). These devices are used strictly for providing personal information. When it is scanned by an external reader, an IID uses the energy in the reader's signal to wirelessly emit a unique identifier. The best known of these devices is the VeriChip Corporation's RFID (radio-frequency identification) tag, which was approved for human implantation by the U.S. Food and Drug Administration in 2004. The serial number emitted by this device can be used to access a person's medical information in a database called VeriMed. This allows quick access to vital information even if the patient is unresponsive or unconscious, as in a medical emergency. Similar implantable devices can have other practical uses in health care. For example, physicians can use IIDs to gain access to sensitive hospital areas or to certain patient records. IIDs might also be used to verify a patient's identity before performing an operation or administering drug treatment.

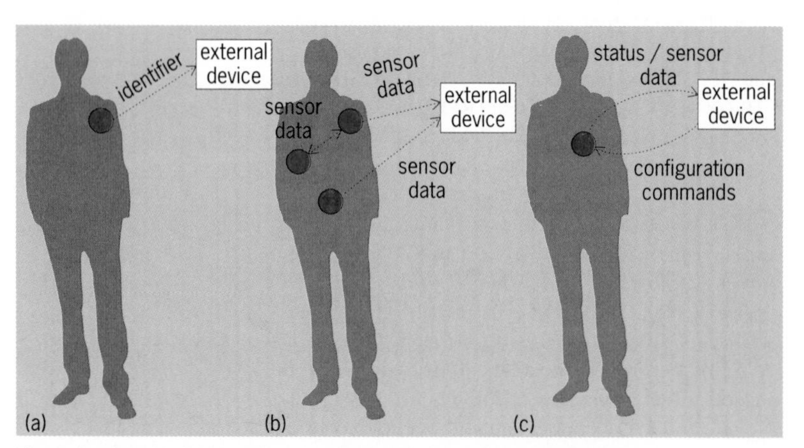

Fig. 1. Implantable (*a*) identification, (*b*) monitoring, and (*c*) control devices.

Implantable monitoring devices (IMDs). These devices are capable of measuring physiological characteristics of the patient. Examples include blood glucose sensors and electrocardiogram monitors. Monitoring devices communicate by sending their sensed physiological data to an external device. One IMD may also send its data to another such device for data aggregation purposes.

Implantable control devices (ICDs). These devices are capable of altering physiological characteristics of the patient. Some control devices also have integrated monitoring capabilities. Examples include artificial pacemakers and drug delivery devices. Control devices receive commands from an external device to allow adjustment of the settings on the implanted device. They may also send information on their current status (as well as sensed data) to an external device.

Securing identification devices. An attacker can attempt to harvest information from an IID by using his or her own reader or by eavesdropping on the communication between an IID and a valid reader. The harvested information may be used directly or to carry out further attacks, such as tracking (where an attacker with access to a large number of readers monitors people's locations as they move within range of different readers) or cloning (where an attacker replays a harvested identifier to a valid reader and attempts to pass the attacker off as someone else).

A number of countermeasures to these attacks have been proposed, and will be briefly surveyed. These countermeasures were often originally developed with general-purpose RFID tags in mind, but may be readily adapted for IIDs.

Limiting IID information content. This approach was employed in the design of the VeriChip device, which contains simply a 16-digit serial number. In order to gain more information, the attacker must access the VeriMed database. If the database is sufficiently secure, no patient information is at risk of being compromised. However, this approach alone does not prevent cloning or tracking.

Distance measurement. An IID can release its identifier only to readers within a certain range. An attacker would thus have to be in close proximity to the IID in order to harvest its identifier; this countermeasure also makes tracking difficult.

Blocking. An IID relies on a second device called a blocker to prevent it from being scanned by readers. The blocker transmits certain identifiers that collide with the blocked IID's transmission, making it impossible to obtain a clear reading. The blocker can be turned on or off at will, allowing the IID's owner to control when it can be read. Blocking effectively eliminates threats, but it also disables the IID's functionality when it is turned on.

Using shared secrets. An IID can be preconfigured to share secret information (such as a password) only with valid readers. These secrets can be used to ensure that only valid readers can identify the tag correctly. However, the IID must be capable of cryptographic operations.

Varying the identifier. Several methods of varying the identifier emitted by a device have been proposed. In minimalist cryptography, the basic idea is for an IID to have a set of identifiers and choose one from the set to send each time it receives a reader signal. A valid reader has access to the entire list of identifiers. This scheme makes it more difficult for an attacker to clone or track a tag. In reencryption schemes (**Fig. 2**), public-key cryptography is used to randomly reencrypt tag identifiers (using a nonsecret public key) whenever a tag is interrogated. A reader can uniquely identify a tag only when the reader possesses a secret private key.

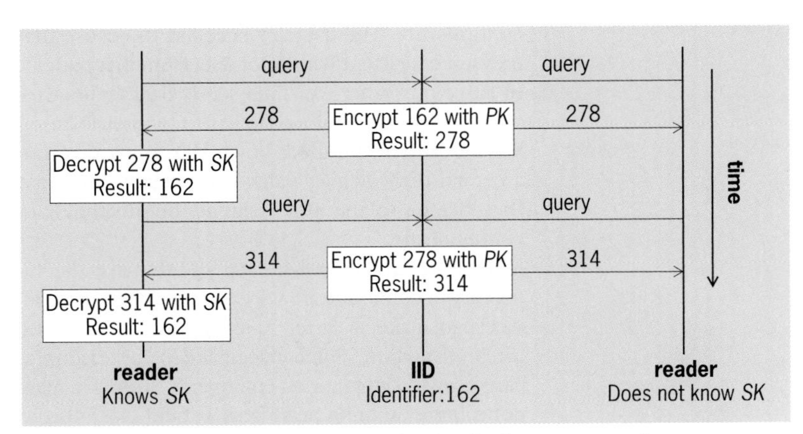

Fig. 2. Example of a reencryption scheme, showing the time sequence of messages between an implantable identification device (IID) and two different readers. The IID randomly reencrypts its identifier with a public key PK each time it receives a reader query. A reader that knows the secret private key SK can uniquely identify the tag; a reader without knowledge of SK simply sees a continuously changing identifier.

Securing monitoring devices. Since IMDs collect medical data that need to be transmitted wirelessly, they are subject to many of the same threats that face IIDs. In addition, IMDs are subject to denial-of-service attacks. An attacker might flood the wireless channel with meaningless transmissions, making the IMD unable to send data. This type of attack is difficult to defend against.

Addressing security in IMDs boils down to at least three main goals: preventing eavesdropping on patient information, preventing unauthorized devices from accessing patient information, and allowing receivers to verify the source of patient data to prevent data falsification.

Data encryption. To prevent eavesdropping, patient data from IMDs must be encrypted before wireless transmission. Two options exist for encryption: secret-key or symmetric encryption, and public-key or asymmetric encryption. While more flexible, asymmetric encryption is also much more computationally expensive. Although symmetric encryption is attractive in IMDs for its efficiency, it presents the challenge of key management: how to securely generate, store, and update secret keys between a sender and a receiver.

One approach to IMD key management is to use asymmetric cryptography to securely derive symmetric keys between communicating parties (such as an IMD and an external reader) without either one having prior shared secrets. The derived keys allow efficient symmetric encryption of the patient data.

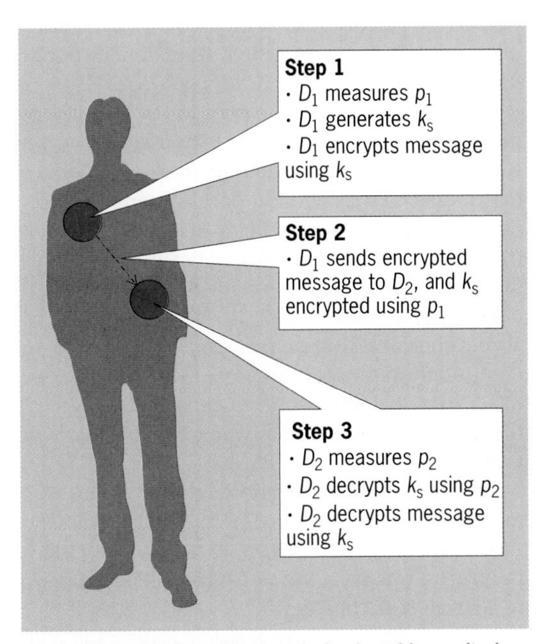

Fig. 3. Biometric-based keying. An implantable monitoring device (IMD) D_1 measures a physiological quantity p_1 and uses it to encrypt a key k_s, which is then sent to IMD D_2.

A more novel approach that is specific to IMDs is biometric-based keying (**Fig. 3**), in which two IMDs that wish to communicate measure the same physiological quantity of the patient and use it to secure an encryption key generated by the sender. For this to work, both devices must be able to measure the same quantity.

Data access control. An IMD sending data to an external device needs some way of authenticating the external device to ensure that it is valid. To do this, many of the same mechanisms outlined for IIDs could be used (for example, using shared secrets).

Data spoofing. A receiver of patient data must be able to verify that the data actually belong to the patient they claim to be from. One potential solution is using statistical or machine learning techniques such as neural networks to analyze incoming patient data continuously. The data are checked to ensure that they are consistent with prior data from that patient. In situations where an IMD sends data to another such device, a technique similar to biometric-based keying can also be used: both IMDs must measure a certain physiological value (thus establishing that they belong to the same patient) before they can communicate.

Securing control devices. Many ICDs have the capability to communicate wirelessly with external devices, to make it easier to configure the ICDs after implantation. Some pacemakers, for example, can receive configuration commands through a magnetic "programming head" that is held close to the pacemaker. Clearly, this ability poses the risk of wireless reprogramming. A malicious attacker could conceivably cause direct physical harm to a patient by altering the commands sent by an external device (or sending his or her own commands) to change the electric current emitted by a pacemaker or the dosage of an implanted drug administration device.

The goal of a countermeasure to the wireless reprogramming vulnerability in ICDs is to allow the ICD to respond only to requests from a valid external device. Currently there does not appear to be much academic attention being given to addressing this vulnerability in ICDs. Limiting the communication range of the device seems to be the primary defense. ICD security can also be approached via access control at the external level: the patient is secure if only authorized personnel can access external devices that are capable of communicating with the ICD. However, this limitation cannot always be enforced.

For background information *see* BIOELECTRONICS; BIOMEDICAL ENGINEERING; COMPUTER SECURITY; CRYPTOGRAPHY; DRUG DELIVERY DEVICES; MEDICAL CONTROL SYSTEMS in the McGraw-Hill Encyclopedia of Science & Technology.

Kriangsiri Malasri; Lan Wang

Bibliography. J. Halamka et al., The security implications of VeriChip cloning, *J. Am. Med. Informat. Assoc.*, 13(6):601–607, 2006; A. Juels, RFID security and privacy: a research survey, *IEEE J. Sel. Area. Comm.*, 24(2):381–394, 2006; K. Malasri and L. Wang, Securing wireless implantable devices for healthcare: ideas and challenges, *IEEE Comm. Mag.*, 47(7):74–80, July 2009.

Self-aware networks

Self-aware computer networks (SAN) are, by design, able to observe both their own internal behavior and the external systems with which they interact. By modifying their behavior, they can adaptively achieve certain objectives, such as discovering services for their users, improving their quality of service (QoS), reducing their energy consumption, compensating for components that fail or malfunction, detecting and reacting to intrusions, and defending themselves against external attacks. The study of such networks is part of the field of autonomic communications. However, commonly used terms, such as self-awareness, can elicit different reactions from different people, including scientists and engineers from different disciplines. We all struggle with such concepts and may not have a clear understanding, or may have different understandings, of what such terms actually mean. Thus we will first review some relevant aspects of self-awareness, and then discuss how these can be embodied in the context of computer networks.

According to the philosopher John Locke, the self "depends on consciousness, not on substance." Thus, we can identify our "self" by being conscious or aware of our past, present, and future thoughts, and through the memory of actions and the planning of actions. Any concept is necessarily juxtaposed to other opposing concepts, so that the "self" needs to be related to or separated from the "other." The notion of self also includes an integrated view of our past sensory inputs over time, as well as of our current sensory inputs. Some form of autobiographical memory should be a part of self-awareness. Sensing should also include what engineers call condition monitoring; that is, are we in good health, do we feel well, are we performing our various tasks as expected? Self-awareness allows an individual to select or drive not only actions but also thinking, by choosing, for instance, to turn to a past memory or to a future plan, rather than musing over thoughts generated by current external stimuli. When we turn our attention to ourselves, we can also compare our current and past behavior to our own standards and to the behavior of other entities external to ourselves, so that we become critics or evaluators of ourselves.

This also leads to a notion of awareness of our location in different spaces: the physical space where we observe, feed, work, move, and so on. But we may also be aware of a space of actions, values, and successes. Different spatial dimensions can describe the attributes that we are interested in. The locations in attribute space are then different states that we may perceive that we (or others) are in, while the distances can represent the time and effort, or value, of moving from one set of values for the attributes to another. This spatial paradigm, which we can consider to be a hippocampus-type brain-related representation, is present in most things we do, and certainly in many of our planning activities. For example, we are aware of where we are, of where we would like to be, and of the distance between the two, including the time and effort needed to bridge the two. In many organisms (for example, dogs), the notion of self is also accompanied by markers, such as smell, that the animal recognizes and uses to mark its physical space, to self-recognize its own presence, and to detect intrusion. Thus the sensory system and the notions of self and space are intertwined. In much simpler organisms, much of the self may actually be contained in the immediate chemical or physical environment (heat, electromagnetic radiation, or pressure). In complex living organisms such as mammals, self-awareness will use a neuronal system, so that the physical embodiment of self-awareness will include a network of interacting components and subsystems.

Desirable properties of a self-aware network. A SAN should offer a distributed internal representation of itself, coupled with the ability to discover actions that it can take, such as paths to destinations and services, and have a "motor" capability for forwarding streams of packets along selected paths. The internal representation of the past and present experience of the network, based on sensing and measurement, with proactive sensing as one of the concurrent activities undertaken by the SAN, would include performance monitoring to provide an internal evaluation of how well the network is doing its job and condition monitoring to evaluate the health of the network. The coupling of the internal representation with motor control is obviously necessary if the network is to evaluate when its information is insufficient or obsolete and then sense its environment, such as other networks and network users, or to probe itself for condition monitoring or performance evaluation. A SAN should offer a distributed representation of potential future actions and plans, and also of critical evaluations of past actions and behaviors and of potential courses of action. The internal representation should allow the system to locate its components and behavior both in physical space and in different virtual spaces that are relevant to the network. Useful virtual spaces can include aspects such as security, energy consumption, delay, loss of packets, and computational overhead. Network nodes placed in this geography could then be identified via these different attributes—for example, those nodes that are secure, those that are functioning properly, and those that are energy efficient. Thus, such attributes could be used to distinguish between different parts (nodes and links) or subareas of the network. This distributed representation should be coupled to the computer network's distributed sensor system, which probes and measures the network's behavior, and its motor control, which forwards streams of packets from node to node along paths in the network to the desired destinations. A SAN should be able to sense and evaluate threats and detect intrusions and attacks. The capability for threat evaluation should be coupled with distributed motor reactions concerning packet routing and driven by self-preservation and the need to assure quasinormal operation, even in the presence of transient breakdowns or sustained attacks.

A practical approach to self-aware networks. Since the routing and conveying of packets reliably from some input or source point S to some other point or destination D is a computer network's key function, we now focus on a practical scheme for implementing a self-aware network that offers many of the capabilities that we described, based on the cognitive packet network (CPN) protocol. Addressing the requirements of the previous section one by one, we first note that in CPN, each network node maintains an internal representation of its primary role as a forwarder of traffic through the network. Specifically, for any source and destination (S-D) pair and relevant quality of service (QoS) characteristic: either (1) a node does not know which of its neighbors is the best choice as the next hop for this S-D pair and the specified QoS, but it can discover this information using smart packets (SP), which are distinct from the payload packets that the network is forwarding as part of its useful job, or (2) it does have this information based on previous experience, and the information is stored in an oracle (prediction software) that is implemented in the form of a neural network.

This distributed representation is also coupled with the network's motor capability to forward packets. When a packet with a given S-D and QoS requirement arrives at the network, then in case 1 the packet is forwarded as prescribed, while in case 2 a stream of smart packets is forwarded by the node, starting with its immediate neighbors, in search of the path that currently has the best QoS metric for reaching D. Smart packets that reach D will send back an acknowledgement (ACK) packet, whose information content is used to update the oracles using reinforcement learning (RL) concerning this S-D at all the intermediate nodes and at S. The oracles can then be used to forward subsequent packets. When a packet that travels from S to D does reach D, an ACK containing the hop-by-hop QoS experienced by the packet returns to the source, together with time stamps documenting the packet going forward and the ACK coming back. Thus, this record of the paths that were explored, the dates, and the observed QoS creates an autobiographical memory of the SAN's experience, which includes its past performance and the conditions that were prevalent at the times when this experience was collected. The most recent data in this autobiographical memory will be stored in the primary memory of the sources and nodes for rapid retrieval and for use in decision making, while older elements may be moved to secondary memory. In addition to smart packets, payload packets that are sent out from a source and do not result in an ACK coming back can also be used to estimate the packet loss rates and failures in the network.

The QoS metrics used by the RL algorithm and by the SPs and the information brought back by ACKs can include different QoS characteristics that are either specified by network users as being relevant to themselves or ingrained and relevant to the network's own performance criteria, such as the variation in packet arrival time, jitter, delay, packet loss, security, energy consumption of nodes or links, cumulative energy consumed by packets, or composite metrics that can be used to summarize several primary metrics (**Fig. 1**). Essentially any property that can be either directly measured (such as delay or energy consumption) or deduced from other

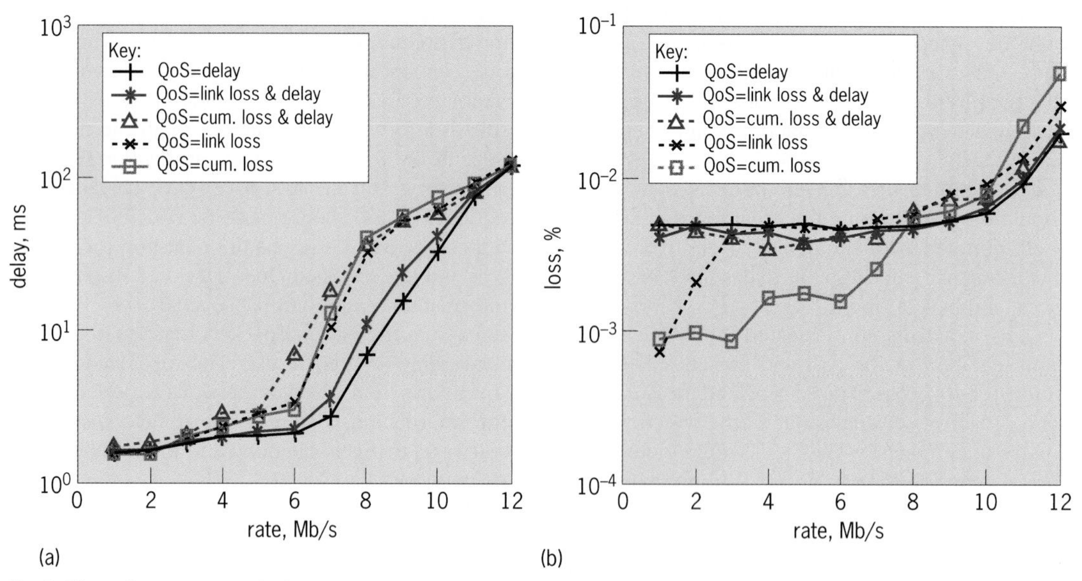

Fig. 1. The self-aware network observes its own performance and adapts itself to provide the users with the best possible QoS that the users have indicated. Thus when the users specify "delay" (QoS=delay) as their desired QoS objective, the network achieves (*a*) low delay but is unable also to offer (*b*) low loss rates. On the other hand, when low loss (QoS=cum. loss) is desired (*b*) it is indeed achieved, but (*a*) the delay may be higher.

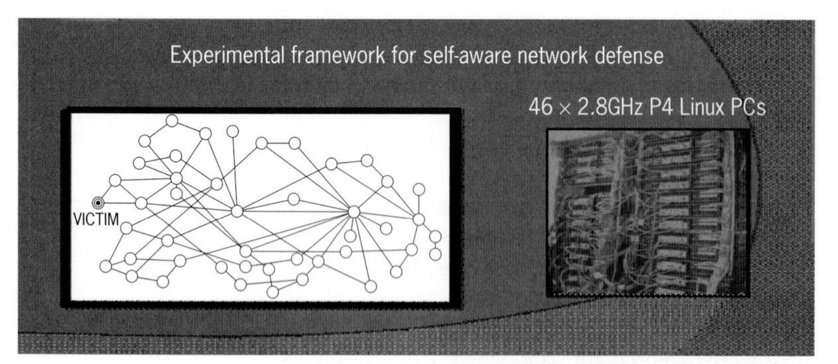

Fig. 2. Laboratory setting for experiments on the self-aware network's self-defense.

direct measurements (such as jitter) may be used in CPN.

Distributed decisions, oscillations, and "changing one's mind." When a node in a SAN makes a routing choice on the basis of its expectation that the outcome, for example, in terms of QoS will be favorable, it may not have the information concerning other nodes that might make the same choices based on similar information. Thus, independently made "best" decisions become collective "worst" decisions. There are simple ways to avoid this. In a simple case, randomization over time, the decision is made, but it is applied only after a random delay. Before it is actually applied, the node again senses the variables of interest to test whether the decision is sound. Because of the time randomization, the probability of a collision between two or more deciders is negligible. This is similar to the way Ethernet operates.

Self-awareness and network attacks. Self-observation and coupled motor reaction carry over directly to the manner in which a SAN can detect attacks and defend itself. These can be similar to attacks among living organisms. Either (Type 1) they are very discrete in trying to infiltrate "worms" or viruses that disable the network's useful activities while letting through the attack itself, which then delivers a second or third strike on the network's normal activities, or (Type 2) they can overwhelm the network's ability to handle traffic by generating an excessive number of requests that appear to be legitimate, and may also infiltrate into the network the attacking traffic of the first type (Type 1). Attacks of Type 1 can require the SAN to act by eavesdropping or testing, which may or may not be allowed depending on the type of service the SAN has agreed to provide and the agreements with the users. Attacks of Type 2 are usually detected without eavesdropping, using the observation of traffic characteristics (**Figs. 2** and **3**).

CPN offers the ability to monitor incoming traffic, to classify it as attack or nonattack traffic, then counterattack by destroying traffic that is identified as being part of an attack, and perhaps also storing it in a harmless manner for offline analysis and comparison by deviating the traffic to a decoy server (honey pot). The honey pot may also be used to restore traffic that was mistakenly identified (that is, a false alarm) as being part of an attack.

Outlook. As packet networks become extremely large and diverse, a design based on a single view of how a network should operate becomes harder to justify. Thus, each part of the network should be aware of its environment, of its users, of its resources, of its performance, and of the threats that it is facing, and should use this awareness as an operational means to adapt its behavior accordingly. This text attempts to set the stage for these ideas, and to show a simple and practical way forward.

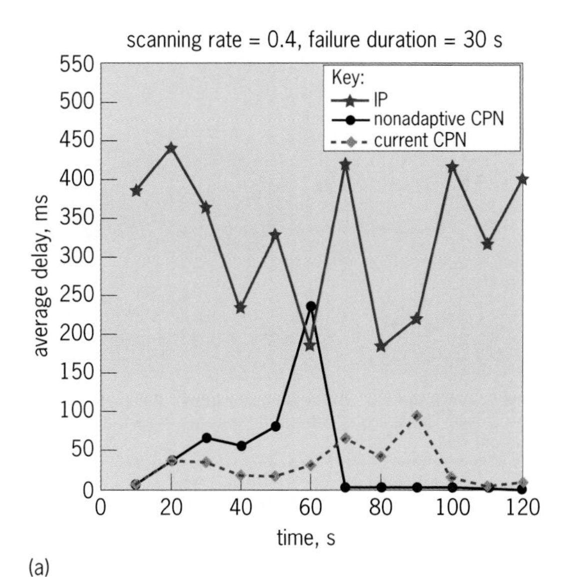

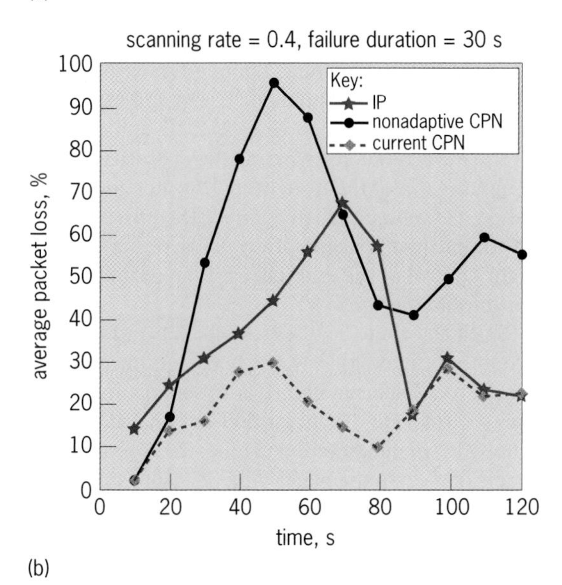

Fig. 3. During a worm-based attack, (*a*) the average packet delay and (*b*) average packet loss experienced by one network user without (IP and nonadaptive CPN) and with (current CPN) self-aware network capabilities.

For background information *see* COMPUTER SECURITY; COMPUTER-SYSTEM EVALUATION; DATA COMMUNICATIONS; NEURAL NETWORK; PACKET SWITCHING; WIDE-AREA NETWORKS in the McGraw-Hill Encyclopedia of Science & Technology. Erol Gelenbe

Bibliography. S. Dobson et al., Autonomic communications, *ACM Trans. Autonom. Adapt. Syst.*, 1(2):223–259, 2006; E. Gelenbe, Steps toward self-aware networks, *Comm. ACM*, 52(7):66–75, July 2009; E. Gelenbe, Users and services in intelligent networks, *Intell. Transport Syst.*, 153(3):213–220, 2006; E. Gelenbe, M. Gellman, and G. Loukas, An autonomic approach to denial of service defence, *Proc. International Symp. on a World of Wireless Mobile and Multimedia Networks (WoWMoM '05)*, IEEE, New York, June 2005, pp. 537–541, 2005; E. Gelenbe, R. Lent, and A. Nunez, Self-aware networks and QoS, *Proc. IEEE*, 92(9):1478–1489, 2004; E. Gelenbe, G. Sakellari, and M. d'Arienzo, Admission of QoS aware users in a smart network, *ACM Trans. Autonom. Adapt. Syst.*, 3(1):4:1–4:28, 2008.

Sensory structures in crocodilians

The Crocodilia (also Crocodylia) of today are the only living members of the great archosaurian group that ruled the world throughout the Mesozoic era [251–65 million years ago (mya)]. As such, these animals, together with birds, represent the only relatives of extinct dinosaurs, thecodonts, and pterosaurs. Biologists recognize approximately 21 extant species, divided into three major lineages: the gavials, alligators, and crocodiles. For many people, the Crocodilia are familiar amphibious reptiles that occupy most of the tropical and subtropical climates throughout the world. These animals have essentially remained unchanged since the Late Cretaceous (65.5 mya) and retain a remarkable body armor. It is interesting to think how animals that are so heavily armored are able to be sensitive to the environment. Another consideration when examining their sensory systems is the nature of their environment. Crocodilians live, mate, and hunt prey both on land and in water. Animals that thrive on both land and underwater are faced with the task of interpreting stimuli in different media.

Vision. Few studies have examined the visual characteristics of crocodilians, but an apparent lack of importance of visual cues for underwater feeding has long been noted. One study used refractive and anatomical analyses to confirm that crocodilian eyes can focus objects in air, but their ability to refocus the eye underwater is limited. In this study, the plane of focus of six species of crocodilians was examined. It was found that the eyes are generally well focused in air for distant targets, but severely defocused underwater. There are two possible ways to make an eye that is effective both in air and underwater: (1) the eye must have an extreme accommodative range (seen in turtles and diving ducks); and (2) the cornea must be flatter so that the refractive role of the air-corneal interface is reduced (such as in penguins). Crocodilians possess neither of these adaptations; thus, underwater, these animals appear to see as well as humans. These results suggest that sensory systems other than vision must play an important role in prey capture underwater. It has been found that feeding efficiency was nearly the same in clear and turbid water in free-living *Caiman sclerops*. In addition, this species can capture fish in total darkness. Therefore, the ability to hunt underwater combined with the absence of visual cues, the highly curved corneas, and the lack of any accommodative ability suggest that the crocodilian eye is adapted almost exclusively for vision in air.

Hearing. It is interesting to note that American alligators (*Alligator mississippiensis*) are highly vocal and make a variety of communication sounds, including low-frequency signals, with dominant frequencies around 100 Hz and harmonics up to 400 Hz. However, do they hear these signals well on land and underwater? Sound travels about four times faster in water depending on temperature [1484 m/s (4869 ft/s)] than in air [343 m/s (1125 ft/s)] at 20°C (68°F). The challenge of amphibious hearing is to perform the appropriate computation, depending on the environment. While the ultimate response to sound detection in both air and water is driven by the stimulation of the sensory hair cells of the inner ear, how the sound gets to the inner ear may differ between the two media. In water, the soft tissues of vertebrates are essentially acoustically transparent, so sounds travel through the body and have little or no direct impact on the sensory hair cells. Fish and amphibians solve this problem by having a dense structure in the inner ear that overlies the hair cells and moves at a different amplitude and phase from the rest of the body. In air, the problem is the matching of densities between air and the fluid-filled inner ear cavities. Terrestrial vertebrates have solved this by evolving a tympanic membrane and middle ear bones to compensate for the impedance mismatch. These two different solutions mean that moving an air-adapted animal to water, or vice versa, results in a loss in hearing sensitivity.

Crocodilians are interesting because their ear morphology and corresponding auditory brainstem structures are very similar to those of birds. Therefore, they appear to be adapted for detection of airborne sound. Yet, these animals spend significant amounts of time in water and at the air-water interface, being highly vocal and taking full advantage of the air-water interface itself for communication (**Fig. 1**). It has been noted that the hearing of the American alligator is acute in both air and water relative to birds and fish tested in the same way. In air, alligators have a frequency range similar to birds (for example, parakeets), responding to tones from 100 to 8000 Hz, with peak sensitivity around 1000 Hz. In water, alligators have a frequency range comparable to goldfish (a species considered to be hearing specialists among fish), but they are more sensitive than goldfish, responding to tones from 100 to 2000 Hz, with peak sensitivity at 800 Hz. Thus, this amphibious group has overcome the difficulties inherent in hearing in two different media in their own way. One study suggests that the pathway of sound to the ear may be different, providing a novel solution to the amphibious problem. Crocodilians possess a slit-like external auditory opening, bounded above by a large superior ear flap and bounded below and in front by a small inferior ear flap (**Fig. 2**). The anterior part of the meatus (auditory passageway) is usually open when the top of the head is out of water and closed when it is submerged; in contrast, the posterior part of the meatus is nearly always closed. Perhaps the tight control of the external ear is an adaptation for an amphibious lifestyle.

Fig. 1. Crocodilians take advantage of the air-water interface. This alligator is bellowing, creating what is called a "water dance" with his body. He is also vocalizing low-frequency sounds.

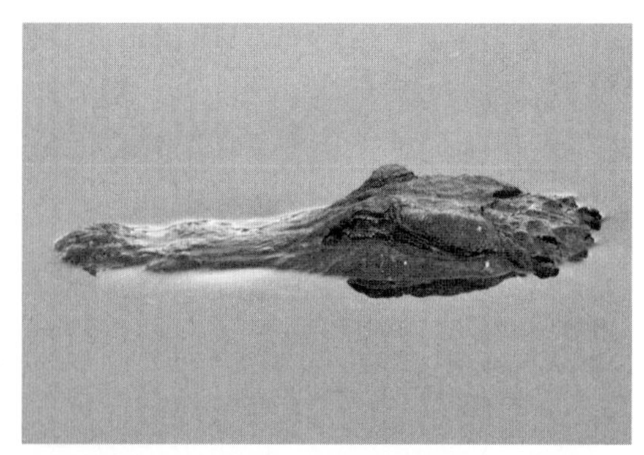

Fig. 2. Note the ears of the alligator above the surface of the water. There are slits behind the large eyes that create the ear flap.

Smell and taste. The senses of smell and taste are likely important for various behaviors in crocodilians. Crocodilians in nature or in seminatural enclosures can locate distant carrion or concealed meat, implying chemoattraction to food. In response to extracts of meat in the water, American alligators open their mouths and wave their heads. This likely introduces water-soluble chemicals to taste buds that are located on the tongue, palate, and pharyngeal walls. Behavioral and olfactometer experiments suggest that crocodilians detect both airborne and water-soluble chemicals and use their olfactory system for hunting. All species possess valvular nostrils, which are regulated by smooth muscles; paired nasal cavities, each with three conchae (bony elements); and flaps that close the rear of the buccopharyngeal passage to channel inspired air through the nasopharyngeal duct. All crocodilians enhance their olfaction by gular (throat) pumping, in which the floor of the pharynx goes up and down to pulse air through the nasal cavity. Juvenile crocodilians also increase gular pumping in response to airborne skin gland secretions. Observations of adults suggest that gular and paracloacal glands produce chemical signals that are used in mating and nesting activities, but behavioral responses to skin gland exudates are too poorly documented to ascribe pheromonal properties.

Skin. How does an armored creature maintain tactile sensitivity? Crocodilians have specialized sensory organs called dome pressure receptors (DPRs). DPRs on the crocodilian skin are stimulated by surface waves. In experiments that cover over the DPRs,

Fig. 3. Dome pressure receptors (DPRs) on the skin of crocodilians.

the animal's orientating behavior can be abolished. DPRs are round, dome-like structures that lack pores or protruding hairs (**Fig. 3**). The epidermis is 40% thinner immediately above the DPRs, whereas the keratin layer is 60% thinner and more compact. A dermal fold below each organ contains a highly branched nerve bundle. The organs are present on the faces of alligatorids, but are found all over the body of crocodiles. DPRs on the faces of alligators are innervated by the trigeminal nerve. This nerve is the largest of all the cranial nerves in the alligator. DPR responses to surface waves and tactile stimulation can be recorded from the trigeminal ganglion. Under continuous pressure, DPR responses adapt and there is no neural response when the skin between DPRs is subjected to the same pressure. Although the mechanism of transduction is unknown, it is most likely that DPRs are sensitive to pressure differences and not to particle motion. However, responses to shear have not been ruled out.

Conclusion. In summary, crocodilians are a monophyletic reptilian group that has solved the problem of being sensitive to the environment in air, underwater, and at their interfaces while maintaining a robust armor.

For background information *see* ALLIGATOR; CHEMICAL SENSES; CHEMORECEPTION; CROCODILE; CROCODYLIA; GAVIAL; HEARING (VERTEBRATE); OLFACTION; SENSATION; SENSE ORGAN; SKIN; TASTE; VISION in the McGraw-Hill Encyclopedia of Science & Technology.

Daphne Soares

Bibliography. L. J. Fleishman and A. S. Rand, *Caiman crocodilus* does not require vision for underwater prey capture, *J. Herpetol.*, 23:296, 1988; L. J. Fleishman et al., Crocodiles don't focus underwater, *J. Comp. Physiol. A*, 163:441-443, 1988; B. Fritzsch, Hearing in two worlds: Theoretical and actual adaptive changes of the aquatic and terrestrial ear for sound reception, pp. 15-42, in R. R. Fay and A. N. Popper (eds.), *Comparative Hearing: Fish and Amphibians*, Springer-Verlag, Berlin/Heidelberg/New York, 1999; L. D. Garrick and J. W. Lang, Social signals and behaviors of adult alligators and crocodiles, *Am. Zool.*, 17:225-239, 1977; O. Gleich and G. A. Manley, The hearing organ of birds and Crocodilia, pp. 70-138, in R. J. Dooling, R. R. Fay, and A. N. Popper (eds.), *Comparative Hearing: Birds and Reptiles*, Springer-Verlag, Berlin/Heidelberg/New York, 2000; D. M. Higgs et al., Amphibious auditory responses of the American alligator (*Alligator mississipiensis*), *J. Comp. Physiol. A*, 188:217-223, 2002; G. B. Schaller and P. G. Crawshaw, Fishing behavior of Paraguayan caiman (*Caiman crocodilus*), *Copeia*, 1982:66-72, 1982; D. Soares, An ancient sensory organ in crocodilians, *Nature*, 417:241-242, 2002; K. Vliet, Social displays of the American alligator (*Alligator mississippiensis*), *Am. Zool.*, 29:1019-1031, 1989; P. J. Weldona, W. Mark, and J. Ferguson, Chemoreception in crocodilians: Anatomy, natural history, and empirical results, *Brain Behav. Evol.*, 41:239-245, 1993.

Sister chromatid cohesion

A cell's genome contains instructions required by that cell to function and exist. All organisms, whether single-celled or multicellular, require a complete set of instructions contained in the form of DNA for each cell to function properly. Thus, a cell faces two challenges: (1) it must replicate its entire genome; and (2) during cell division, it must ensure that each daughter cell receives one copy of the identical sister chromatids (duplicated DNA double helices), which were produced during replication. How the eukaryotic cell orchestrates DNA replication and how it ensures correct segregation of sister chromatids are fundamental questions in cell biology. Importantly, failure to accurately segregate chromosomes during cell division carries severe consequences for daughter cells, including aneuploidy (deviation from a normal chromosome number), which is a condition thought to contribute to the development of cancer. One cellular strategy that is essential for faithful chromosome segregation is sister chromatid cohesion.

Ties that bind sisters. Cohesion is the term used to describe the physical linkages that hold sister chromatids together. Cohesion is established along the length of each set of sister chromatids as they are produced during S phase (the DNA synthesis phase of the cell cycle) and is maintained until the sister chromatids move to opposite poles of the spindle during anaphase. Because of cohesion between sister chromatids, tension results when microtubules from opposite poles attach to the sister chromatid kinetochores [small specialized structures on the surface of centromeres, which are unique regions of highly condensed chromatin (DNA plus protein)] and exert poleward forces as sisters line up on the metaphase plate (the imaginary plane located halfway between the spindle poles along which the kinetochore microtubules align their chromosomes at metaphase). This tension stabilizes bioriented sisters on the spindle apparatus until congression (coming together) of all chromosomes is complete and the cell is ready to divide. In mitosis, the

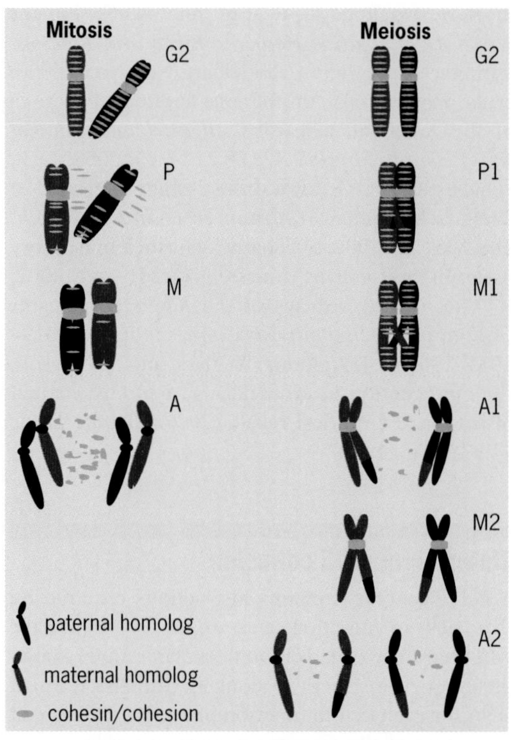

Fig. 1. Cohesion during mitosis and meiosis. Cohesion between sister chromatids is essential for proper chromosome segregation in all eukaryotic cells and relies on the cohesin complex. Although some cohesin is lost from the chromosome arms as the mitotic chromosomes condense, residual cohesin keeps the arms closely associated until anaphase. Upon anaphase onset, cleavage of centromeric and arm cohesin results in the separation of sister chromatids. In contrast, during meiosis, cleavage of cohesin occurs in a stepwise manner. Loss of arm cohesion at anaphase I allows recombinant homologs to separate and move to opposite poles. However, when arm cohesion is lost, centromeric cohesion is protected and remains intact until anaphase II, when sister chromatids segregate away from each other. Abbreviations: P: prophase; M: metaphase; A: anaphase; G: gap phase.

metaphase-to-anaphase transition occurs when cohesion along the entire length of the chromatids is released and sister chromatids are pulled to opposite spindle poles such that each daughter cell receives one and only one copy of each chromosome. In meiosis, the release of cohesion occurs in a stepwise manner during the two divisions (**Fig. 1**). At the onset of anaphase I, when homologous chromosomes segregate to opposite poles, arm cohesion is lost; however, centromeric cohesion remains intact and keeps the sister chromatids connected. Then, in the second meiotic division, loss of centromeric cohesion at the onset of anaphase II allows sister chromatids to move to opposite poles.

Cohesion complex. In both mitotic and meiotic cells, sister chromatid cohesion is mediated by an evolutionarily conserved multiprotein complex called cohesin. The cohesin complex is composed of two subunits belonging to the structural maintenance of chromosomes (SMC) family and two non-SMC subunits (**Fig. 2**). The SMC subunits (Smc1 and Smc3) are both ATPases [enzymes that hydrolyze adenosine triphosphate (ATP) into adenosine diphosphate (ADP) and phosphate]. Both have globular N- and C-termini separated by a long coiled domain, which folds back on itself at a third "hinge" domain to form a long antiparallel coiled-coil region, bringing the N- and C-termini together to create a functional ATPase. The hinge domains of Smc1 and Smc3 also interact and yield a V-shaped SMC heterodimer. One of the non-SMC cohesin subunits, which belongs to the kleisin family of proteins (kleisin is the Greek term for closure), acts to bridge the gap between the SMC ATPase domains and stimulates the ATP hydrolysis activity of Smc1. In mitotic cells, Scc1 (also known as Mcd1 or Rad21) functions as the kleisin subunit. A meiotic-specific kleisin, Rec8, is found exclusively in germ cells. The interaction of Smc1, Smc3, and Scc1/Rec8 results in the formation of a tripartite ring whose integrity is required for cohesion. The fourth cohesin subunit, Scc3 (also known as SA, STAG, or Irr1), contains multiple HEAT repeats (protein-protein interaction motifs) and physically interacts with the kleisin. Although Scc3 is essential for cohesion, its exact role is not well defined. It has been proposed that this subunit acts as an interface between cohesin and its regulators, but it also may play a structural role in holding two tripartite rings together (see handcuff model below). Two distinct models have been put forth to explain the mechanism by which an intact cohesin ring mediates cohesion (Fig. 2). The embrace model posits that the cohesin ring encircles a pair of sister chromatids. In this model, the original cohesin ring on the parent DNA molecule traps the two sister chromatids as they are formed during

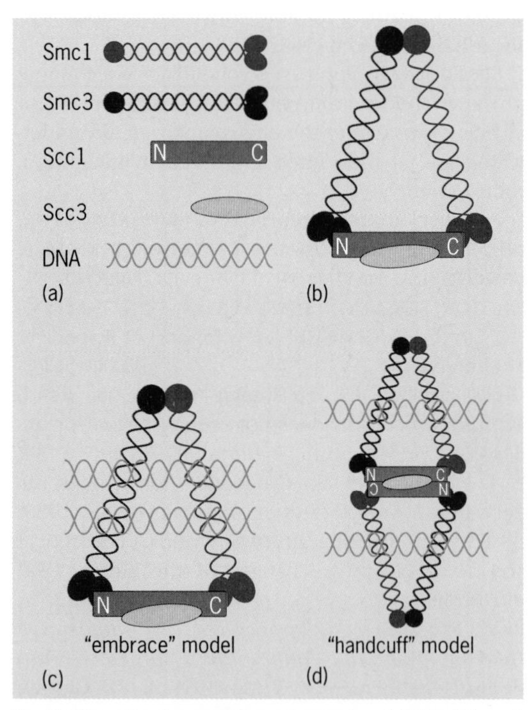

Fig. 2. The cohesin complex and cohesion. (*a*) The different cohesin subunits are shown schematically. (*b*) Cohesin subunits assemble into a functional cohesin complex. (*c, d*) The embrace and handcuff models of sister chromatid cohesion are illustrated (left and right, respectively).

DNA replication. The alternative handcuff model argues that two tripartite rings, each enclosing a single sister, hold sisters together through a ring–ring interaction that is mediated by the Scc3/SA/Irr1 subunit. Although these models differ, both stipulate that cohesin captures DNA in a topological manner instead of physically binding DNA or chromatin.

Making and breaking the cohesin-mediated linkages that hold sisters together. In order to mediate sister chromatid cohesion, the cohesin complex must first be loaded onto chromatin. The activity of a heterodimer composed of Scc2/NIPBL and Scc4/MAU-2 is necessary for recruitment of cohesin to chromatin during G1 (gap phase 1) of the cell cycle in yeast and during telophase (the final stage of mitosis or meiosis) in vertebrate cells. It is important to note that association of cohesin with chromatin is necessary but not sufficient to establish cohesin-mediated linkages between sister chromatids. An acetyltransferase (Eco1/Ctf7/Deco/ESCO1-2) is required during S phase to establish cohesion. Although the details of how cohesion is established are not completely understood, one target of this enzyme is chromatin-associated Smc3.

Once established, cohesion between sister chromatids must be maintained until it is time for sister chromatids to move to opposite poles of the spindle. Dissolution of cohesion at anaphase onset is a prerequisite for chromosome segregation and requires separase, which is a protease that cleaves the kleisin subunit of cohesin. Separase is locked in an inactive state by its regulator (securin) until all the chromosomes have attained stable bipolar attachments. A cellular surveillance mechanism, the spindle assembly checkpoint (SAC), blocks entry into anaphase by inhibiting the ubiquitination (conjugation of ubiquitin to cellular proteins) and degradation of securin. Once all chromosomes have achieved biorientation, the SAC is silenced and this triggers activation of anaphase promoting complex/cyclosome (APC/C), which is a ubiquitin ligase that targets securin for destruction. Degradation of securin liberates separase and results in cleavage of the kleisin subunit of cohesin, thereby opening the ring and releasing the cohesin-mediated linkages that hold sisters together.

Unique features of meiotic versus mitotic cohesion. In mitotic cells, activation of separase at the metaphase–anaphase transition results in loss of cohesion along the entire length of sister chromatids, and the simultaneous release of both arm and centromeric cohesion allows sister chromatids to segregate to opposite poles (Fig. 1). In contrast, separase cleavage of cohesin during meiosis occurs in a stepwise manner (Fig. 1). After meiotic crossovers occur between homologous chromosomes, cohesion along the arms of sister chromatids keeps recombinant homologs physically attached and facilitates their biorientation on the meiosis I spindle. Therefore, release of arm cohesion is necessary for homologous chromosomes to separate and segregate to opposite poles during anaphase I. Indeed, because human oocytes remain arrested at prophase I for decades, age-dependent weakening and premature loss of cohesion may be one important factor that underlies the higher incidence of meiosis I segregation errors and associated trisomic pregnancies in older women. When arm cohesin is cleaved during the first meiotic division, cohesin at the centromeric regions must be protected from separase so that sister chromatids remain associated and can achieve stable bipolar microtubule attachments on the meiosis II spindle (Fig. 1). Protection of centromeric cohesion during meiosis depends on shugoshin (Sgo)/Mei-S332 (shugoshin is the Japanese term for guardian spirit). Shugoshin localizes to the centromeres of meiotic chromosomes and prevents cleavage of centromeric Rec8 during anaphase I when arm cohesion is dissolved.

Non-cohesin-mediated cohesion. Although the cohesin complex is clearly essential for sister chromatid cohesion and proper chromosome segregation in both mitotic and meiotic cells, DNA catenation also contributes to the physical linkages that hold sister chromatids together. During the process of DNA replication, daughter duplexes become interlocked, and both biochemical and genetic experiments indicate that topoisomerase II activity is necessary during anaphase to untangle DNA linkages between sister chromatids.

Separase-independent cohesin dynamics. A number of recent studies indicate that several proteins function to modulate the association of cohesin with chromatin and to regulate the maintenance of sister chromatid cohesion until the onset of anaphase. Included among these factors are Pds5/PDS5A-B and Rad61/Wapl, which exhibit opposing activities. Both Pds5 and Wapl can independently interact directly with Rad21 (Scc1), and the presence of SA1 (Scc3) enhances these interactions, supporting the notion that Scc3 facilitates communication between cohesin and its regulators. In contrast to the pro-maintenance role of Pds5, Wapl appears to promote cohesin removal from the arms by a separase-independent mechanism. Interestingly, at least in metazoan cells and despite the persistence of cohesion, the bulk of cohesin associated with chromosome arms, but not centromeres, is removed during mitotic prophase (Fig. 1). A mitotic-specific shugoshin functions to protect centromeric cohesin from this separase-independent removal pathway by recruiting the phosphatase PP2A to the centromere and keeping Scc3 in a dephosphorylated state.

Other roles for cohesion. A *Drosophila* genetic screen for factors that facilitate communication between promoters and enhancers recovered mutations that disrupt the cohesin loader protein Nipped-B (Scc2) and implicated the cohesin complex as an important regulator of transcription and gene expression. Identification of the mutations associated with a number of human genetic disorders, now collectively known as cohesinopathies, has provided additional evidence that cohesin function extends beyond the mediation of cohesion. Cornelia de Lange syndrome (CdLS) is a dominantly inherited disorder characterized by deficits in craniofacial development, upper limb deformities, and varying degrees of mental retardation. A number of CdLS patients have heterozygous loss-of-function mutations in *NIPBL* (60% of cases) or point mutations in *SMC1* or *SMC3* (5% of cases). Roberts syndrome (RBS), on the other hand, manifests in an autosomal recessive manner and is associated with mutations that are thought to disrupt proper expression of the Smc3 acetyltransferase *ESCO2* (*Eco1*). Like CdLS, RBS patients show defects in long bone development, cleft palate, and mental retardation. Finally, SC phocomelia is also associated with mutations that affect *ESCO2* expression, but has a milder phenotype than RBS because patients tend to live longer with the former. Interestingly, sister chromatid cohesion and chromosome segregation appear to remain largely intact in patients who suffer from the aforementioned cohesinopathies, supporting the notion that cohesin is critical for activities independent of holding sister chromatids together.

For background information *see* BIOCHEMISTRY; CELL CYCLE; CELL DIVISION; CENTROSOME; CHROMOSOME; DEOXYRIBONUCLEIC ACID (DNA); GENE; GENETICS; GENOMICS; MEIOSIS; MITOSIS; MUTATION in the McGraw-Hill Encyclopedia of Science & Technology.

Sharon E. Bickel; Justin J. Gaudet

Bibliography. D. Dorsett, Cohesin, gene expression and development: Lessons from *Drosophila*, *Chromosome Res.*, 17:185–200, 2009; A. J. McNairn and J. L. Gerton, Cohesinopathies: One ring, many obligations, *Mutat. Res.*, 647:103–111, 2008; K. Nasmyth and C. H. Haering, Cohesin: Its roles and mechanisms, *Annu. Rev. Genet.*, 43:525–558, 2009; B. Xiong and J. L. Gerton, Regulators of the cohesin network, *Annu. Rev. Biochem.*, 79:131–153, 2010; N. Zhang and D. Pati, Handcuff for sisters: A new model for sister chromatid cohesion, *Cell Cycle*, 8:399–402, 2009.

Skin friction measurement

Skin friction (or wall shear) is defined as the tangential force per unit area created on a surface that is exposed to a flowing viscous fluid. This quantity is very important to the design, analysis, and understanding of all types of fluid machinery, including, as but a few examples, aircraft, ships, automobiles, piping systems, atmospheric and river flows, and circulatory and pulmonary systems. To put this in perspective, about 50% of the drag, and thus the fuel consumption, of a well-designed passenger aircraft is attributable to skin friction on the wings, fuselage, tails, and engine nacelles. Moreover, the skin friction distribution can play a very important role in identifying and correcting problem areas in fluids devices and systems. Another application for skin friction measurement is as a sensitive input signal for flow control systems. From a basic viewpoint, local skin friction values are central to all data-correlating techniques for turbulent flows, and these correlations form the basis for the development of all turbulence models used in computational fluid dynamics. Accurate measurement is critical, because computational methods still cannot provide sufficiently accurate skin friction results for complex flows. Since this quantity is so important, there is a long history of work in the area starting in the late nineteenth century.

Indirect measurement methods. Measurements can generally be divided into two broad classes of methods: indirect and direct measurements. Indirect methods rely upon some analogy or data correlation to use another measurement and imply skin friction from that other measurement. For example, surface heat flux can be measured, and skin friction can then be inferred from that, using the Reynolds analogy relating heat flux and skin friction. Such methods all presume a great deal of prior knowledge about the flow under study, so they work best for studies of flows that are well understood.

Direct measurement methods. The direct methods employ a movable surface or a small movable element of the surface surrounded by a very small gap, connected to some type of flexure. The displacement of the movable surface or surface element or the strain in the flexure is then measured to obtain the skin friction force acting on the surface or surface element directly. Measuring the friction force on a complete surface or a large area of the surface gives an average value for the skin friction, and measuring the friction force on a small surface element gives the local skin friction value. An idealized schematic of a direct measuring gage for local skin friction is shown in **Fig. 1**, where the flexure is depicted as a cantilever beam. The cantilever beam flexure is attractive because it can be made insensitive to forces that are normal to the surface, such as pressure, and sensitive to the much smaller desired skin friction forces, which are tangential to the surface. Other flexure arrangements are also used. A rigid surface element can be mounted on a subsurface layer of flexible material or on an array of micro- or nanotubes. Flexures formed of wheels with very flexible spokes have also been successfully

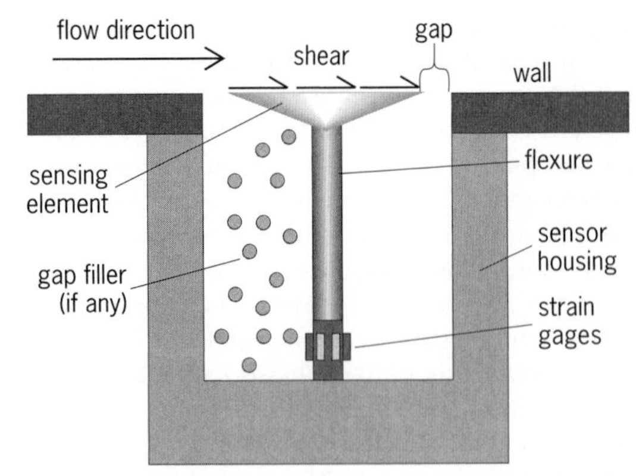

Fig. 1. Simplified schematic of a direct measuring gage for local skin friction.

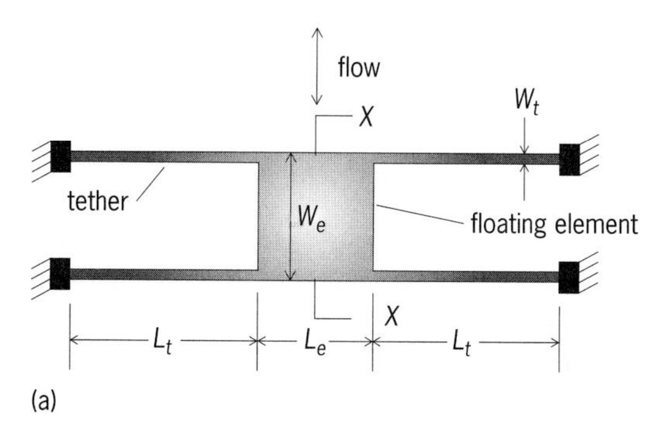

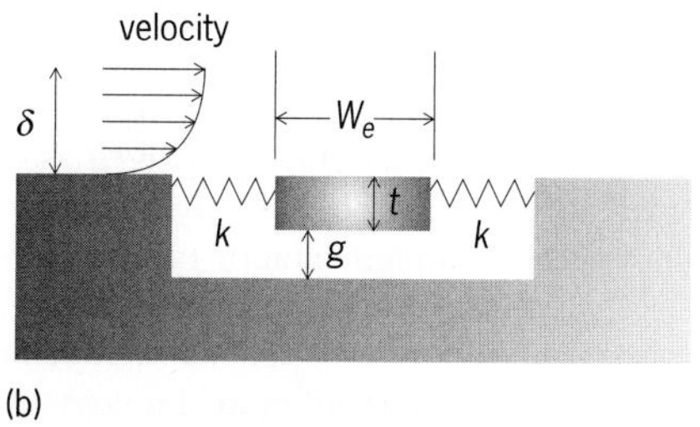

Fig. 2. Typical MEMS floating-element skin friction sensor. (*a*) Plan view. (*b*) Cross section. The spanwise length and streamwise width of the element are L_e and W_e, respectively. The element has a thickness *t* and is suspended over a recessed gap of depth *g* by silicon tethers that are represented by springs in this schematic, reflecting the fact that the tethers bend in the direction of the flow and therefore exhibit springlike behavior in that direction. The tether length and width are L_t and W_t, respectively. (*Adapted from J. W. Naughton and M. Sheplak, Modern developments in shear-stress measurement, Progr. Aero. Sci., 38:515–570, 2002*)

employed. Micro-electro-mechanical (MEMS) devices with various flexure arrangements, such as tethers, have been made (**Fig. 2**). The schematic in Fig. 1 shows strain gages mounted on the base of the flexure as the means of sensing the reaction to the load on the sensing element produced by the skin friction. It is also possible to sense the reaction to the load on the sensing element with various flexure arrangements by measuring the small displacement by means of fiber optics, capacitance change, magnetic proximity sensors, and other techniques. Some of the flexure-sensor arrangements are limited to measuring only a single component of the skin friction in the plane of the surface. General three-dimensional (3D) flows have two components of the skin friction in the plane of the surface with orientations that are unknown before the measurements are performed, so a limitation to measuring only a single component with a preset orientation can be quite restrictive.

A very viscous fluid can be used as a gap filler to provide damping, protection from fouling, and thermal stabilization. In many flow situations, damping and thermal stabilization can be very important matters, requiring careful skin friction gage design and testing. Filler leakage is often a problem.

Another issue in designing a successful skin friction gage for a particular flow situation is time response. That is obvious if the flow of interest is highly unsteady, but it can also be a significant matter in short-duration test facilities such as shock tunnels, where the total test time is measured in milliseconds. In such facilities, vibration damping is a concurrent concern.

Newer measurement concepts. The use of a flexible sheet as a flexure with a rigid surface element attached above it was mentioned previously. Recently, that concept has been generalized through the removal of the attached rigid element. Now, the response of the exposed flexible sheet itself to the shear on the surface is being measured. That has been done by measuring the distortion of the sheet with interferometry, and also by loading the sheet with a fluorescent material or nanoparticles. In the case of nanoparticles, the electrical resistance of the sheet changes as it is strained. Interferometry has also been employed to measure the time-varying thinning of an oil film applied to the surface. A relation describing that response to the shear can be written from first principles. The various sheet methods can provide the wall shear variation over a region of the surface, in addition to local values. Some of these methods require optical access to the surface, and that can be a problem in some circumstances.

Methods based on velocity measurements. The wall shear is directly related to the product of the viscosity of the fluid and the velocity gradient at the surface, so any means of measuring that gradient can be used to obtain the wall shear. There are two limitations with that approach. First, the velocity gradient can be very large, especially in turbulent flows, so velocity measurements at very small distances above the surface are often required. Second, the insertion of velocity probes into the flow is intrusive, causing disturbances downstream. Nonintrusive methods such as laser Doppler velocimetry (LDV) or particle image velocimetry (PIV) can be used, but then optical access may be an issue. A clever method is to use a miniature LDV system mounted behind the wall, with optical access through a small window in the wall.

Calibration. The matter of accurate calibration of skin friction gages is very important, as for all kinds of instrumentation. That can be a simple process for some gage concepts or a complicated one for others. In principle, the simplest cases are the measurement of the velocity gradient at the wall by laser Doppler velocimetry or particle image velocimetry, since those velocity measurements are usually presumed to be based on first principles. A gage based on the concept in Fig. 1 can also be easy to calibrate. The unit is simply turned so that the measuring surface is vertical, and the output is measured as small weights are hung from the surface with a thread. Such a procedure is greatly complicated for MEMS or other miniature gages. The next level of calibration methods employs devices that produce a well-documented flow. Examples are fully developed pipe or channel flow and that in a cone-and-plate viscometer.

For background information *see* AERODYNAMIC FORCE; BOUNDARY-LAYER FLOW; FLUID-FLOW PRINCIPLES; INTERFEROMETRY; MICRO-ELECTRO-

MECHANICAL SYSTEMS (MEMS); SHIP POWERING, MANEUVERING, AND SEAKEEPING; TURBULENT FLOW; VELOCIMETER; VISCOSITY in the McGraw-Hill Encyclopedia of Science & Technology. Joseph A. Schetz

Bibliography. J. W. Naughton and M. Sheplak, Modern developments in shear-stress measurement, *Progr. Aero. Sci.*, 38:515–570, 2002; W. Nitsche, C. Haberland, and R. Thunker, Comparative investigations of the friction drag measuring techniques in experimental aerodynamics, ICAS 84-2.4.1, *14th ICAS Congress*, Toulouse, France, International Council of the Aeronautical Sciences, 1984; J. A. Schetz, Skin friction measurements in complex turbulent flows using direct methods, in W. Rodi (ed.), *Engineering Turbulence Modelling and Experiments, 6*, pp. 421–430, Elsevier Ltd., Oxford, U.K., 2005; K. G. Winter, An outline of the techniques available for the measurement of skin friction in turbulent boundary layers, *Progr. Aero. Sci.*, 18:1–57, 1977.

Smart metering

Over the past few years, there has been keen worldwide interest in smart metering. In Europe, this interest has been strongly motivated by the European Union (EU) Energy Efficiency Directive, which among other things aims to minimize the environmental impacts of energy generation and to meet the commitments made on climate change under the Kyoto Protocol.

By putting the smart metering system in use, it is expected that consumers will reduce their electricity usage as they become aware of their actual consumption. This in turn will result in reducing the need to build more power plants and to expand existing distribution networks. Furthermore, the system will provide more accurate information to grid operators regarding how to allocate electricity use.

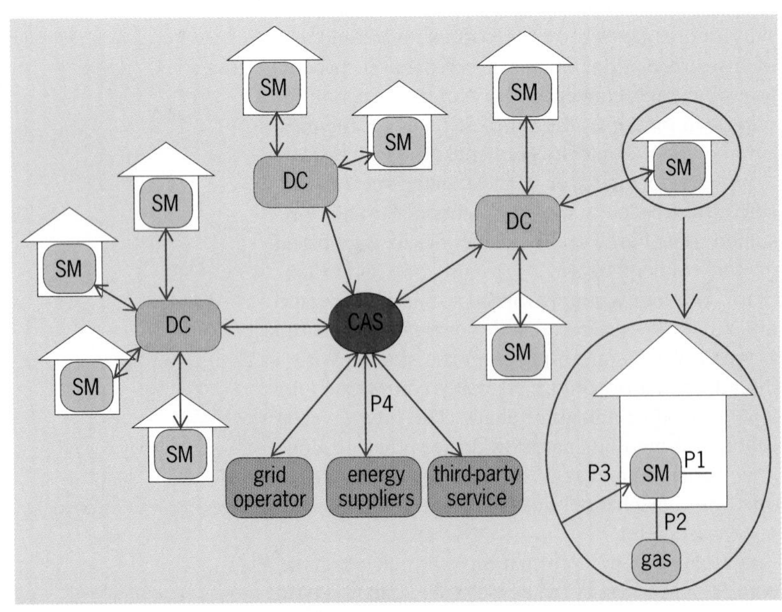

Main components of the smart metering system.

However, despite the numerous advantages gained by the smart metering system, it is still argued that the main driver behind its launch is primarily related to cost reduction.

Smart metering technology provides in principle two-way communication to upload commands to the users and to download measured data from the meters. The system extends the existing automatic meter management (AMM) approach of advanced metering. Automatic meter management basically offers the ability to remotely read electricity or gas-consumption registers without physical access to the meter, in addition to allowing remote connection and disconnection of customers. Moreover, there are many different types of meters on the market, and the system implementation choices as well as the roll-out process vary among the countries that have already embarked on implementation of the system.

System components. The smart metering system encompasses a mixture of components that belong to either energy or information technology (IT) infrastructures, and that interact with each other to carry out the system's intended services. Typically, the main components of the system (see **illustration**) are smart meter devices (SM), data concentrators (DC), central access servers (CAS), and communication ports of types P1, P2, P3, and P4.

Smart meter devices. A smart meter device is the electrical meter installed on or near the premises of a remote household. The main purpose of a smart meter is to keep track of electricity, gas, and water consumption in an advanced manner. In addition to recording metering data, a smart meter differs from a conventional one in the wide range of highly developed services that it offers, such as remote activation and deactivation of connections, and providing consumers with real-time pricing and their current amount of electricity consumption via the two-way communication medium between the meter installation and service providers. A single meter is capable of monitoring several appliances per household and switching them off and on, if required. More appliance manufacturers are producing smart appliances, which are capable of interacting with the smart meters to adapt to more efficient patterns of electricity usage.

Data concentrators. Data concentrators are located at substations. They automatically manage a portion of the smart metering system functionality, such as gathering metering data from the smart meters at remote households and sending them to control stations and the billing system, monitoring the operation of the power grid and smart meters and reporting disruptions and failures, detecting and configuring newly installed smart meters and creating repeating chains if necessary (where a meter receives a signal intended for another meter to amplify and resend it), and detecting and reporting theft or tampering attempts.

Central access servers. A central access server is a central application that takes care of the data collection, and the centralized authorization for access to the metering installation. Electricity grid operators each maintain their own cluster of servers as part of the smart metering system operation. These servers store the system's data, including metering data, in addition to the software and applications that are necessary to carry out the various operations of the system.

Communication ports. Information flows among the various components within the smart metering system are facilitated through four main types of communication ports, labeled P1, P2, P3, and P4. A port of type P1 is a read-only port that is mainly used to link the metering installation to an external device. The metering installation is linked to external devices used by the electricity grid operators through a port of type P2. A port of type P3 is a two-way communication port that is used to connect the metering installation to the central access servers through a series of intermediate nodes. A port of type P4 is located on the central access server. Through this type of port, other authorized market players, such as electricity suppliers, are given access to the metering data. A fifth type of port, P0, is located on the metering device and is used solely for configuration purposes.

Functionality. This system, which is based on a two-way communication medium, may offer diverse services other than generating and delivering real-time electricity consumption data and electricity prices. Some of these functionalities include: remotely measuring and detecting power quality; safely and remotely activating and deactivating electricity and gas connections, on either a collective or an individual basis; and monitoring distribution networks and generating disruption or fraud alerts. From the perspective of their role in an advanced intelligent distribution grid, smart meters can contribute via demand-side management to the smart-grid management for flattening or load shifting.

Demand-side management aims to achieve energy efficiency and saving, and making more electricity available on the grid, not by increasing generation, but rather by reducing demand or patterns of use while ensuring electricity quality. This activity leads not only to reducing the need to generate more electricity but also to eliminating the need to expand the distribution networks. Smart metering technology can assist reaching this goal in many different ways. The digital displays available at remote households provide consumers with real-time electricity prices and the amount of electricity consumed based on time of use, as opposed to standard tariffs given by mechanical meters. This helps to promote awareness among consumers and creates a good incentive for reducing and shifting electricity consumption since studies show that persistent feedback promotes persistent conservation behavior.

Another way in which smart metering can aid in reducing the need for more electricity generation is by being an enabler for microgeneration, which allows consumers to generate electricity and make it available on the distribution networks. The system can also help reduce the costs associated with metering companies' personnel visiting remote households to manually read the meters or perform connection or disconnection. Furthermore, the system can provide significant help in putting an end to electricity fraud or theft, which also results in financial savings.

Incentives. The launch of smart metering systems around the world has been motivated by diverse incentives, which may differ from one country to another. For example, one of the main incentives in The Netherlands was improving deregulated markets. In France it was energy efficiency and improving market competition. In Italy, improving operations and fraud detection were the main drivers behind the system. There also exist main general drivers that are common among different countries, including, to name a few, protection of the environment by reducing carbon dioxide (CO_2) emissions, enabling microgeneration and demand response, reduction of costs related to system operation and electricity consumption, elimination of manual meter reading costs, allowing for prepaid electricity, providing consumers with accurate electricity consumption bills based on time-of-use tariffs, improving management and control of the electricity grid, improving theft detection, and rapid location and restoration of interruption.

Initiatives. A number of smart metering deployment projects are taking place in countries around the world.

U.S. initiatives. Two initiatives may be mentioned in the United States. Utility Partners of America was contracted to replace 84,500 electricity meters and 33,400 water meters with smart meters in Glendale California. The program began in March 2010 and was expected to last for 14 months. In Texas, Oncor, an electricity grid operator, had deployed 300,000 meters in the Dallas area by mid-2010. The company intended to launch a massive roll out of the system that was expected to reach 3 million households by 2012. However, the roll out could be impeded by complaints from consumers that the meters are increasing their consumption bills.

European initiatives. In Europe, the roll outs of smart metering systems are regulation-driven; the deployments of these systems in EU states are effected mainly by EU energy policies. In April 2009, the EU parliament decided that 80% of all electricity consumers should have a smart meter by 2020. By mid-2010, smart metering roll outs had started, or even been completed, in a number of EU states such as Sweden, Poland, Norway, The Netherlands, France, Spain, and Italy. In July 2009, Sweden became the first country to achieve 100% penetration. Italy was the first country in Europe to witness a nearly complete massive roll out that began in 2000 and is expected to reach full penetration by 2011. Within each country, the roll out is planned to take place according to a certain time frame. For instance, the deadline for France and Spain is the end of 2010, whereas the United Kingdom aims to complete its system deployment by 2020. An example of a later deployment is a roll out of up to 1 million residential smart meters, which has been contracted by the energy supplier British Gas to the Landis+Gyr smart metering company.

In The Netherlands, despite a number of pioneering efforts in smart metering pilot projects and deployment, such as a roll out of 100,000 residential smart meters by Oxxio, the fourth largest Dutch energy supplier, the deployment of the system has been delayed at least until 2013. The delay was caused by the rejection of a bill to mandate system roll out by the Dutch senate in April 2009 due to privacy and security-related concerns. The bill then underwent a revision process in preparation for another discussion in the Dutch parliament in autumn 2010.

For background information *see* ELECTRIC DISTRIBUTION SYSTEMS; ELECTRIC POWER SYSTEMS; ELECTRICAL ENERGY MEASUREMENT; WATT-HOUR METER in the McGraw-Hill Encyclopedia of Science & Technology.

Zofia Lukszo; Layla O. AlAbdulkarim

Bibliography. L. AlAbdulkarim and Z. Lukszo, Smart metering for the future energy systems, *4th CRIS International Conference on Critical Infrastructures*, Linköping, Sweden, April 28–30, 2009; S. Darby, *The Effectiveness of Feedback on Energy Consumption*, Environmental Change Institute, University of Oxford, April 2006; G. Deconinck, Metering, intelligent enough for smart grids?, in Z. Lukszo, G. Deconinck, and M.P.C. Weijnen (eds.), *Securing Electricity Supply in the Cyber Age: Exploring the Risks of Information and Communication Technology in Tomorrow's Electricity Infrastructure*, Topics in Safety, Risk, Reliability and Quality, vol. 15, pp. 143-158, Springer, 2010; F. Koenis, Energy efficiency driving smart metering movement in Europe, *Automat. Insight*, 1(2):9–11, KEMA, March 2007; M. Togeby, *Development of AMR in Europe-National Perspectives*, European Smart Metering Alliance, 2008.

Social intelligence and the brain in the spotted hyena

It has long been known that the capacity for thought, innovation, and problem solving increases as brain size increases. Harry J. Jerison first proposed the principle of proper mass, that is, the idea that brain regions increase proportionately in size relative to the capacity for certain behaviors. For example, the relative volume of the brain devoted to the representation of the forelimb is greater in the raccoon, an animal noted for the manipulative capabilities of its forepaw, than in the dog, a species that lacks these skills. The principle of proper mass also applies to brain areas involved in cognitive skills. One explanation for the evolution of larger brains is that the ability to respond flexibly to social information requires enhanced neural processing. This increased neural requirement led to increased brain size and increased cognition.

Socially intelligent species have especially large brains, including the frontal cortex. Increased brain size has been correlated with increased social complexity and group size in a variety of species, including primates, ungulates, and some carnivores. Social cognition has been extensively studied in Old World or cercopithecine primates—for example, baboons, vervet monkeys, and macaques. These species are noted for living in hierarchically organized groups, including families based on female lineage. They exhibit a range of social awareness, including recognition of others in their groups (both relatives and nonkin), preference for higher-ranking non-kin, recognition of third-party relationships, and formation of coalitions. These characteristics are often thought of as unique to primates, but relatively few researchers have systematically evaluated social behavior in nonprimate species. Now, a body of work examining the behavior, physiology, and anatomy of the spotted hyena (*Crocuta crocuta*) [**Fig. 1**] has revealed that this species shares many similarities in social organization with Old World monkeys and also exceeds that found in other carnivore species. Since the evolutionary history of carnivores and primates diverged more than 100 million years ago, the spotted hyena offers an opportunity to determine whether social complexity and its relationship to brain organization evolved independently in carnivores.

Fig. 1. An adult female spotted hyena with her 3-month-old cub. (*Photo by Kay Holekamp*)

Social intelligence in the spotted hyena. The spotted hyena lives in large groups, called clans, containing up to 90 individuals. Members of a clan come together to cooperatively defend their territory, but typically individuals spend time alone or in small groups to hunt and feed. Although spotted hyenas do scavenge for food, about 95% of their food is derived from hunting ungulates, such as zebra, antelope, or buffalo. Hyenas often hunt alone, but success in hunting of large prey increases by about 20% with each additional clan-mate. Once a kill has been made, clan members will cooperatively defend their food from intruders, including other hyenas and lions. The social cognition necessary for such cooperation begins early in development. Young cubs are raised in communal dens, where they learn to recognize other clan members and, with the help of their mothers, learn to dominate lower-ranking individuals and refrain from attacking higher-ranking individuals. They retain this social information throughout their lives. Social information regarding the identity and rank of an individual is especially important because rank determines access to food and mate choices. Once prey has been killed, feeding is highly competitive. Dominant hyenas have first access to the kill, while less delectable options are left to lower-ranking individuals. A kill may also attract other clan-mates, but hyenas will defend their kill with the aid of kin and other allies. Spotted hyenas recognize third-party relationships and will join a fight to aid a dominant animal, even if that individual is losing the fight. They also recognize and are more likely to attack relatives of their opponents.

The clan maintains a linear dominance based on maternal rank. Offspring inherit their mother's rank. All adult females are dominant to adult males in a clan. Females are philopatric; that is, they remain with their birth or natal clan throughout their lives. Males leave the natal clan and emigrate to another clan, where they enter at the bottom of the dominance hierarchy. Immigrant males gain social status as other male clan members leave or die. Because these differences in the social life history of adult male and female hyenas are so distinctive, the brain volume differences in male and female spotted hyenas have been systematically studied. Sex differences in brain size have been extensively studied in a variety of primates, including humans. It appears that overall brain volume is greater in males than in females. However, when the larger male body size is taken into account, this difference diminishes. It is not known whether there are sex differences in total and regional brain volume in social carnivore species.

Virtual endocast. Until recently, it has been technically challenging to systematically study brain features across a wide range of species. For the majority of mammal species, intact brains are not readily available for study because this soft tissue must be immediately preserved for direct anatomical evaluation. In contrast to skeletal remains, brain tissue does not withstand exposure to the elements. Neuroscientists, anthropologists, and zoologists have studied brain morphology in a number of extinct and living species by using endocasts, which are latex or plaster impressions made from the interior surface of the skull or the endocranium. Recent advances in computerized tomography (CT) have made possible the creation of three-dimensional models of the endocasts. Using this technology, high-resolution images of multiple x-ray slices through the skull are compiled into a three-dimensional image of the endocranium, creating a virtual endocast (**Fig. 2**). Both the surface anatomy of the brain and measurements of brain volume can be obtained from these virtual endocasts. Based on CT analyses of the spotted hyena skulls, brain volume and size are highly correlated. In fact, as skull size increases, so too does brain volume. No sex difference is found in overall brain size relative to body size in adult spotted hyenas. However, analysis of regional brain volume showed that the relative volume of the anterior cerebrum, which primarily includes the frontal cortex, is significantly larger in male spotted hyenas compared to female spotted hyenas (**Fig. 3**). At the same time, regional brain analysis of the volume of the cerebellum and brain stem, which are brain regions primarily concerned with motor coordination and basic physiological functions, revealed no sex differences.

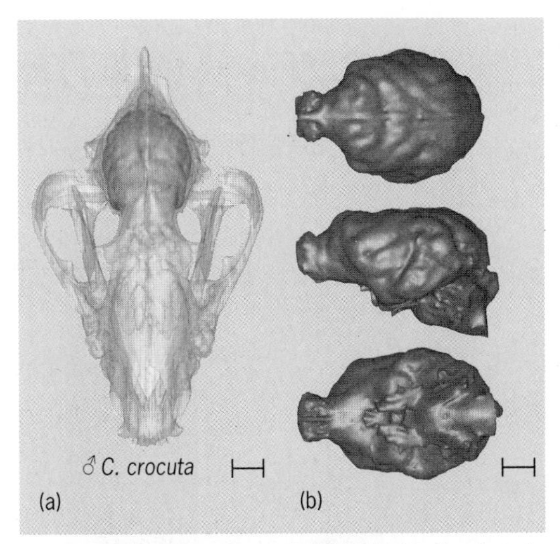

Fig. 2. A male spotted hyena skull (a) and virtual endocast (b). Details of the endocast are shown from a dorsal, lateral, and ventral view. Scale bars: 1 cm (0.4 in.).

Frontal cortex in males versus females. Social information processing demands differ between male and female adult spotted hyenas. After emigration, adult males enter new clans and must establish relationships with the higher-ranking females within the dominance hierarchy in order to survive and reproduce within the new clan. A male hyena may follow a female for weeks in the hope of establishing a relationship. Under these circumstances, it may be that immigrant males must exert more behavioral control than that required of the more aggressive and higher-ranking females. Immigrant male spotted hyenas require enhanced inhibitory control over inappropriate behaviors, such as aggression, in order

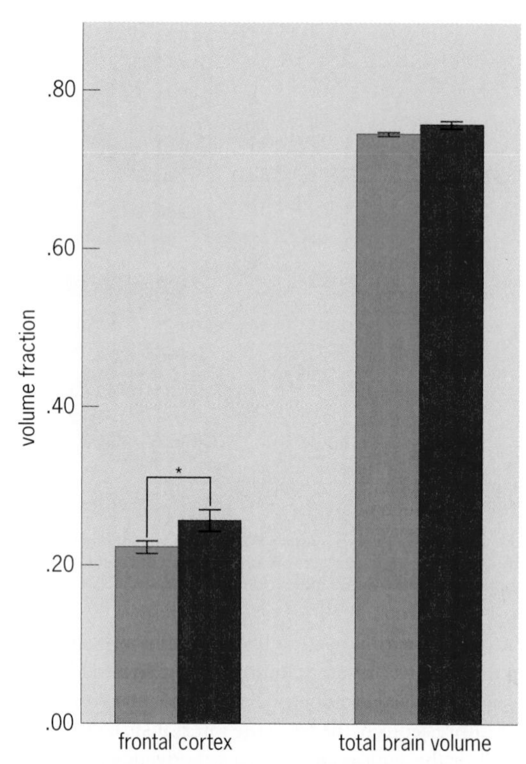

Fig. 3. Histograms comparing sex differences in the spotted hyena brain volume. Significantly greater frontal cortex volume is found in males than in females (left, indicated by the asterisk), but there is no difference between the sexes in total brain volume (right). Error bars show mean ±1 SE. Gray bars: female; colored bars: male.

to negotiate the social system in the new clan. The frontal cortex is associated with inhibitory control of inappropriate behavior. In humans, reduced frontal lobe activity is associated with impulsive acts of aggression. Disinhibition and inappropriate social responses also have been noted in patients who suffer from prefrontal cortex damage. It is not known whether the frontal cortex plays a role in the mediation of social responses in the spotted hyena, but other research data in dogs have reported disinhibition subsequent to damage of the prefrontal cortex.

Under both captive conditions and in the wild, female spotted hyenas are more aggressive than males. Captive female spotted hyenas are more likely to initiate aggression than males. In the wild, aggressive acts are committed at a higher rate and with greater intensity by females compared to males. In the field, it has been observed that wild male spotted hyenas, after being initially anesthetized, sometimes direct aggression inappropriately at females and their cubs. This behavior is never seen under any other circumstances. This observation suggests that perhaps the anesthetic, a combination of a dissociative hypnotic and anxiety-lowering drug, may influence inhibitory control of aggression. Thus, it is tempting to hypothesize that the greater frontal cortex volume found in adult male spotted hyenas in comparison to female spotted hyenas might be related to the greater inhibitory control required of males than females. If this hypothesis is correct, then future work should reveal larger frontal cortex volumes in other group-living mammals, in which individuals must inhibit inappropriate impulses in the presence of more aggressive, higher-ranking individuals.

For background information *see* ANIMAL EVOLUTION; BEHAVIORAL ECOLOGY; BRAIN; CARNIVORA; COMPUTERIZED TOMOGRAPHY; HYENA; SEXUAL DIMORPHISM; SOCIAL HIERARCHY; SOCIAL MAMMALS; SOCIOBIOLOGY; VERTEBRATE BRAIN (EVOLUTION) in the McGraw-Hill Encyclopedia of Science & Technology.

Sharleen T. Sakai

Bibliography. B. M. Arsznov et al., Sex and the frontal cortex: A developmental CT study in the spotted hyena, *Brain Behav. Evol.*, in press, 2010; R. I. M. Dunbar and S. Shultz, Evolution in the social brain, *Science*, 317:1344–1347, 2007; K. E. Holekamp, Questioning the social intelligence hypothesis, *Trends Cognit. Sci.*, 11:65–69, 2007; K. E. Holekamp, S. T. Sakai, and B. L. Lundrigan, The spotted hyena (*Crocuta crocuta*) as a model system for study of the evolution of intelligence, *J. Mammal.*, 88:545–554, 2007; J. E. Smith, S. K. Memenis, and K. E. Holekamp, Rank-related partner choice in the fission–fusion society of the spotted hyena (*Crocuta crocuta*), *Behav. Ecol. Sociobiol.*, 61:753–765, 2007.

Space flight, 2009

The year 2009 was a busy and successful one in space. The space shuttle program conducted five successful missions, including four to the *International Space Station* (*ISS*) and one to repair and refurbish the *Hubble Space Telescope*. The *ISS* was also extended to a six-person crew for part of the year. Exciting new discoveries were made in finding methane on Mars and water on the Earth's Moon.

The year's last quarter saw a successful test flight of the Ares I-X test rocket on a suborbital flight to demonstrate the first stage of the vehicle designed to replace the space shuttle. However, questions arose about the financial and schedule viability of the Ares project and its Constellation program to return U.S. astronauts to the Moon. As the year ended, the future of the program was in doubt.

Human space flight. NASA's space shuttle program had a very successful year with five successful missions. The shuttle program continued toward a retirement date of the end of fiscal year 2010 (September 30). Four missions continued the completion of the construction of the *International Space Station* project. The other shuttle mission was the last scheduled repair mission to the *Hubble Space Telescope*, expanding the observatory's capabilities and lifetime for at least 5 more years.

At 12:39 p.m. EDT on March 12, the *ISS* had a close encounter with a small piece of orbital debris (Object #25090, PAM-D debris), which passed by the station inside the warning area. Because of the late notification, beyond the timeline for maneuvering, a debris avoidance maneuver could not be performed.

As a precaution, the three crewmembers withdrew to the *Soyuz 17S* capsule at 12:35 p.m., leaving the spacecraft's hatch open (in case the *Soyuz* itself was struck). The crew returned to the *ISS* at around 12:45 p.m.

Space shuttle *Discovery* and its seven-member crew lifted off from NASA's Kennedy Space Center at 7:43 p.m. EDT on March 15. *Discovery's* STS-119 flight carried the space station's fourth and final set of solar array wings, completing the station's truss segments. The mission featured three spacewalks to help install the S6 truss segment to the starboard (right) side of the station and deploy its solar array as well as position a new Global Positioning System (GPS) navigation antenna. Air Force Col. Lee Archambault commanded the crew and Navy Cmdr. Tony Antonelli was the pilot. The mission specialists were NASA astronauts Joseph Acaba, John Phillips, Steve Swanson, Richard Arnold, and Japan Aerospace Exploration Agency astronaut Koichi Wakata. Wakata remained on the space station. The flight also replaced a failed unit for a system that converts urine to potable water. The mission ended with a landing at Kennedy Space Center on March 28, bringing the mission's final elapsed time to 12 d, 19 h, 29 min, and 33 s. Discovery traveled 8,536,186 km (5,304,140 mi) during its journey.

On March 26, the *ISS Expedition 19* crew blasted off from the Baikonur Cosmodrome, Kazakhstan, aboard a Soyuz spacecraft. NASA astronaut Michael Barratt, Russian cosmonaut Gennady Padalka, and spaceflight participant and U.S. software engineer Charles Simonyi lifted off at 6:49 a.m. CDT. Simonyi became the first private space explorer to complete a second mission in space. Previously he flew to the *ISS* in spring 2007 as Space Adventures' fifth orbital client.

NASA astronaut Mike Fincke and Russian cosmonaut Yury Lonchakov, both members of the 18th Expedition to the *ISS* and spaceflight participant Simonyi safely landed their Soyuz spacecraft in the steppes of southern Kazakhstan on April 8.

Shuttle *Atlantis* launched on May 11 on the final shuttle visit to the *Hubble Space Telescope*. Scott D. Altman commanded the final space shuttle mission to *Hubble*. Retired Navy Capt. Gregory C. Johnson served as pilot. Mission specialists were veteran spacewalkers John M. Grunsfeld and Michael J. Massimino and first-time astronauts Andrew J. Feustel, Michael T. Good, and K. Megan McArthur.

Over the course of the mission's five spacewalks, the crew added two new science instruments, repaired two others, and replaced hardware that will extend the telescope's life at least through 2014. The five spacewalks lasted 36 h and 56 min all together. There have been 23 spacewalks devoted to *Hubble*, totaling 166 h and 6 min.

With the newly installed Wide Field Camera, *Hubble* will be able to observe ultraviolet and infrared spectra as well as visible light, looking for the earliest star systems and observing planets in the solar system. The telescope's new Cosmic Origins Spectrograph will allow it to study the large-scale structure of the universe.

With ongoing bad weather in Florida, space shuttle *Atlantis* landed at Edwards Air Force Base at 11:39 a.m. EDT on May 24, completing a 13-day journey.

The launch of *Soyuz-TMA 15* on May 27 sent three new *ISS* crew members to orbit. It also marked the beginning of an era of six-person crews aboard the orbiting laboratory. Russian cosmonaut Roman Romanenko, European Space Agency astronaut Frank De Winne, and Canadian Space Agency astronaut Robert Thirsk launched aboard the Soyuz spacecraft.

Space shuttle *Endeavour* and its seven-member crew launched at 6:03 p.m. EDT on July 15 from NASA's Kennedy Space Center in Florida. The mission delivered the final segment to the Japan Aerospace Exploration Agency's *Kibo* laboratory and a new crew member to the *ISS*. Mark L. Polansky commanded, and Douglas G. Hurley served as the pilot. Mission specialists were Christopher J. Cassidy, Thomas H. Marshburn, David A. Wolf, and Julie Payette, a Canadian Space Agency astronaut. The mission also delivered Timothy L. Kopra to the station and returned Koichi Wakata to Earth.

Endeavour's 16-day mission involved five spacewalks and the installation of a platform or "porch" outside the Japanese module. The platform is permanent and will allow experiments to be directly exposed to space. On July 31, space shuttle *Endeavour* touched down at 10:48 a.m. EDT at NASA's Kennedy Space Center in Florida. STS-127 was the 127th space shuttle mission, the 23rd flight for *Endeavour*, and the 29th shuttle visit to the station.

On August 29, space shuttle *Discovery* illuminated the central Florida coast with launch at 11:59 p.m. EDT, beginning its 37th mission and a flight to deliver supplies and research facilities to the *ISS* and its crew. On board *Discovery*, STS-128, were Commander Rick Sturckow, Pilot Kevin Ford, and Mission Specialists Pat Forrester, Jose Hernandez, Danny Olivas, Nicole Stott, and Christer Fuglesang from the European Space Agency. Stott remained on the station as Expedition 20 flight engineer. Kopra returned home aboard *Discovery* as a mission specialist.

Discovery's payload included the *Leonardo* Multi-Purpose Logistics Module (MPLM), containing more than 7 tons of supplies, including life support racks and science racks. The Lightweight Multi-Purpose Experiment Support Structure carrier was also launched in *Discovery's* payload bay. Landing was at 5:53 PM PDT on September 11 at Edwards Air Force Base, ending the 14-day mission.

The Japanese *HTV/H-IIB* launched to the station on September 11. The H-II Transfer Vehicle (HTV) demonstration flight was the first flight of the Japanese resupply vehicle designed to transport cargo to the station.

Flight Engineers Jeffrey Williams and Maxim Suraev became the 21st *ISS* crew after launching in their *Soyuz TMA-16* spacecraft from the

Baikonur Cosmodrome in Kazakhstan at 3:14 a.m. EDT September 30 to begin a 6-month stay in space. Spaceflight participant Guy Laliberte also launched, flying under an agreement between the Russian Federal Space Agency and Space Adventures, Ltd.

ISS Expedition 20 Commander Padalka and Flight Engineer Barratt landed their *Soyuz TMA-14* spacecraft on the steppes of Kazakhstan October 11, wrapping up a 6-month stay. Joining them was spaceflight participant Laliberte, who spent 11 days in space.

STS-129 was launched with space shuttle *Atlantis* from Launch Pad 39A at NASA's Kennedy Space Center in Florida at 2:28 p.m. EST on November 16. Commander Charles Hobaugh led a crew that included pilot Barry Wilmore and mission specialists Robert Satcher, Michael Foreman, Randy Bresnik, and Leland Melvin.

Atlantis brought to the station about 15 tons of cargo in its payload bay and crew area, including two large carriers with heavy spare parts that were stored on the station. It also brought home a similar weight of cabin cargo from the station. Three spacewalks also placed two platforms to hold large spare parts and sustain station operations after the shuttles are retired. The crew of seven astronauts, including station crew member Stott, landed on November 27 at NASA's Kennedy Space Center in Florida at 9:44 a.m. *Atlantis'* mission lasted 11 days and traveled nearly 7.2 million km (4.5 million mi).

Canadian Space Agency Flight Engineer Thirsk, Expedition 21 Flight Engineer and Soyuz Commander Romanenko, and European Space Agency Flight Engineer De Winne landed in Kazakhstan at 1:15 p.m. local Kazakhstan time, December 1.

On December 21, NASA astronaut Timothy J. Creamer, Russian cosmonaut Oleg Kotov, and Japan Aerospace Exploration Agency astronaut Soichi Noguchi safely launched aboard a Soyuz spacecraft to the *ISS* at 3:52 p.m. CST from the Baikonur Cosmodrome in Kazakhstan.

Robotic solar system exploration. The HiRISE imager on NASA's *Mars Reconnaissance Orbiter* (*MRO*) captured enhanced-color images of Deimos, the smaller of the two moons of Mars, on February 21 (**Fig. 1**). These images show that Deimos has a smooth surface due to a blanket of fragmental rock or regolith, except for the most recent impact craters. It is a dark, reddish object, very similar to its sister moon, Phobos.

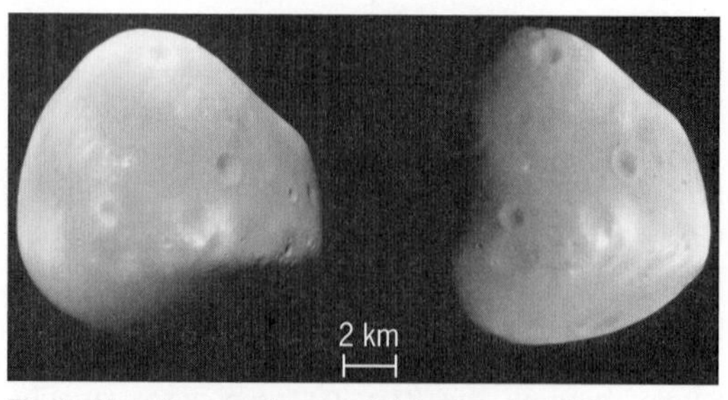

Fig. 1. *Mars Reconnaissance Orbiter* images of Mars' moon Deimos; 2 km = 1.24 mi (*NASA*).

Chang'e-1, China's first lunar probe, impacted the Moon at 4:13 p.m. Beijing Time (0813 GMT), March 1. The satellite ended its 16-month mission when it hit the lunar surface at 1.50°S latitude and 52.36°E longitude.

NASA's *Lunar Reconnaissance Orbiter* (*LRO*) launched at 5:32 p.m. EDT on June 18 aboard an Atlas V rocket from Cape Canaveral Air Force Station in Florida. The spacecraft was designed to relay more information about the lunar environment than any other previous mission to the Moon. *LRO* separated from the Atlas V rocket carrying it and a companion mission, the *Lunar Crater Observation and Sensing Satellite* (*LCROSS*), and immediately began powering up the components necessary to control the spacecraft. After a 4-1/2-day journey from the Earth, *LRO* successfully entered orbit around the Moon.

On September 24 NASA's *MRO* revealed frozen water hiding just below the surface of midlatitude Mars. The spacecraft's observations were obtained from orbit after meteorites excavated fresh craters on Mars. Instruments on the orbiter found bright ice exposed at five Martian sites with new craters that range in depth from approximately 0.5 to 2.4 m (1.5 to 8 ft). The craters did not exist in earlier images of the same sites. The ice was also observed to be evaporating in subsequent images of the area.

Also on September 24, NASA scientists announced that they had discovered water molecules in the polar regions of the Moon. Instruments aboard three separate spacecraft revealed water molecules in amounts that are greater than predicted, but still relatively small. Hydroxyl, a molecule consisting of one oxygen atom and one hydrogen atom, also was found in the lunar soil. *See* WATER ON THE MOON.

On October 9, NASA's *LCROSS* craft was intentionally crashed into Cabeus crater, a permanently shadowed region near the Moon's south pole. Nine *LCROSS* instruments successfully captured each phase of the impact sequence: the impact flash, the ejecta plume, and the creation of the Centaur crater. Preliminary data from *LCROSS* indicated the mission successfully uncovered water in the lunar crater. In November NASA confirmed the discovery of water on the Moon from the *LCROSS* impact data. This water, in the form of ice, could have come from comets that date back to the earliest days of the solar system and may be useful as a resource for future human exploration of the Moon.

In December NASA's *Cassini* spacecraft confirmed observation sunlight glinting off a liquid hydrocarbon lake in the Northern Hemisphere on the surface of Saturn's moon Titan.

Other activities. In January, a team of NASA and university scientists announced the achievement of the first definitive detection of methane in the atmosphere of Mars. This discovery indicates the planet is either biologically or geologically active. The team found methane in the Martian atmosphere by carefully observing the planet throughout several Mars years with NASA's Infrared Telescope Facility and the W.M. Keck Telescope, both at Mauna Kea, Hawaii, and the Gemini South telescope system in Chile. The team used spectrometers on the telescopes to separate the light into its component colors, as a prism separates white light into a rainbow. The team detected three spectral features (absorption lines) that together are a definitive signature of methane.

On February 2, Iran launched its first satellite. The launch of the *Omid* satellite (*Omid* meaning hope) was timed to coincide with the 30th anniversary of the Islamic revolution and United Nations' talks aimed at stopping Iran's nuclear program. The launch has highlighted international concerns that Iran will use domestically developed space technology to develop intercontinental nuclear missiles.

The launch of the *NOAA-N Prime* (*NOAA-19*) polar-orbiting weather satellite for NASA and the National Oceanic and Atmospheric Administration (NOAA), aboard a United Launch Alliance Delta II rocket, took place on February 6. This satellite was severely damaged and repaired after a mishap during processing in 2003.

On February 24, NASA's *Orbiting Carbon Observatory* satellite failed to reach orbit after its 4:55 a.m. EST liftoff from California's Vandenberg Air Force Base. The fairing on the Taurus XL launch vehicle apparently failed to separate. A mishap investigation board was immediately convened to determine the cause of the launch failure. This committee verified that the launch vehicle fairing failed to separate upon command. The launch vehicle then could not attain orbit with the added fairing mass.

A Delta II rocket carrying NASA's *Kepler* planet-hunting spacecraft lifted off on time at 10:49 p.m. EST on March 6 from Launch Complex 17-B at Cape Canaveral Air Force Station in Florida. The spectacular nighttime launch followed a smooth countdown free of technical issues or weather concerns. The *Kepler* spacecraft will watch a patch of space for 3.5 years or more for signs of Earth-sized planets moving around stars similar to the Sun. The patch that *Kepler* will watch contains about 100,000 stars. Using special detectors similar to those used in digital cameras, *Kepler* will look for a slight dimming in the stars as planets pass between the stars and *Kepler*.

Soon after launch, *Kepler* observed the atmosphere of a previously identified gas giant planet. The observation demonstrated the precision of the measurements made by the telescope, even as its calibration and data analysis software were being completed.

On March 17, the European Space Agency's *GOCE* (Gravity field and steady-state Ocean Circulation Explorer) satellite launched. It is mapping the global variations in the gravity field with extreme detail and accuracy. This will provide a model of Earth's gravity field with detail required to understand ocean circulation and sea-level change.

The European Space Agency launched *Herschel* and *Planck* on May 14 on an Ariane 5 rocket. *Herschel* has the largest mirror ever launched into space (3.5 m or 11.5 ft) and will observe the birth of stars and galaxies as well as dust clouds and planet-forming discs around stars and water in the universe. It is observing the universe in the far-infrared and submillimeter regions of the electromagnetic spectrum. *Planck* will map tiny irregularities in the cosmic background radiation left over from the big bang.

NASA's *PharmaSat* nanosatellite successfully launched at 7:55 p.m. EDT on May 19 from NASA's Wallops Flight Facility and the Mid-Atlantic Regional Spaceport located at Wallops Island, Virginia. *PharmaSat* rode to orbit aboard a four-stage Air Force Minotaur 1 rocket. Also aboard were the Air Force Research Laboratory's *TacSat-3* satellite and other NASA CubeSat Technology Demonstration experiments, which include three cube-shaped satellites, with edges measuring 10 cm (4 in.), developed by universities and industry.

On June 1 NASA announced the members of the Review of U.S. Human Space Flight Plans Committee to assist President Barak Obama determine the future direction of human spaceflight in the United States. The committee members included chair Norman Augustine, retired chairman and CEO, Lockheed Martin Corporation, and former members of the President's Council of Advisors on Science and Technology under Presidents Bill Clinton and George W. Bush, as well as current and former aerospace company executives, university professors, two astronauts, and a retired Air Force general.

The presidential panel's final report, released October 22, advised the White House that NASA is on an "unsustainable trajectory" and to preserve a "meaningful" human spaceflight program, NASA would need an additional $3 billion annually and should work closely with other countries and private companies. At the end of 2009, NASA was examining a "flexible path" of exploration that includes crewed trips to nearby asteroids. President Obama is looking to alter the course set in the previous White House, which focused on a return trip to the Moon, with the goal of eventually landing on Mars.

On July 2, the Ad Astra Rocket Company successfully demonstrated operation of its VX-200 plasma engine's first stage at full power and under superconducting conditions in tests conducted in Houston, Texas. This was the first time a superconducting plasma rocket has been operated at such a power level. The VX-200 engine is the first flight-like prototype of the VASIMR propulsion system, a new high-power plasma-based rocket, being developed privately by Ad Astra. It is hoped that VASIMR engines will enable space operations far more efficiently than today's chemical rockets, and eventually greatly speed up robotic and human transit times for missions to Mars and beyond. *See* VASIMR PROPULSION SYSTEM.

On July 8, NASA successfully demonstrated an alternative system for future astronauts to escape their launch vehicle. A simulated launch of the Max Launch Abort System (MLAS) took place at NASA's

Fig. 2. The Ares I-X test rocket launch from Kennedy Space Center, Florida, on its suborbital test flight on October 28, (*NASA/Sandra Joseph and Kevin O'Connel*)

Wallops Flight Facility in Wallops Island, Virginia. The unpiloted launch tested an alternate concept for safely propelling a future spacecraft and its crew away from danger on the launch pad or during ascent. The MLAS consists of four solid-rocket abort motors inside a composite fairing attached to a full-scale model of a crew module.

On July 15, the U.S. Senate unanimously approved ex-astronaut Charles F. Bolden, Jr. as the new NASA administrator.

In July, NASA also celebrated the tenth anniversary of the launch of the *Chandra X-ray Observatory*. During its 10 years of operations, *Chandra* has provided the strongest evidence yet that dark matter exists. It has independently confirmed the existence of dark energy and collected x-ray images of explosions produced by matter as it is swallowed by supermassive black holes. The European Space Agency (ESA) also celebrated the tenth anniversary of their *XMM-Newton* x-ray telescope. *XMM-Newton* has observed black holes, dark matter, and, closer to home, Mars' exosphere.

On October 28, the Ares 1-X 327-ft (100-m) rocket lifted off from a former space shuttle launch pad, LC-39B, at Kennedy Space Center (**Fig. 2**). This was the first time a new vehicle has launched from the complex since the first space shuttle launch in 1981. The test was a suborbital demonstration flight of the first stage of a vehicle designed to send humans into space to replace the space shuttle. The flight also provided data that can be used on other future rocket designs.

NASA's *Tracking and Data Relay Satellite-1* (*TDRS-1*) was decommissioned on October 28. The spacecraft's final traveling-wave-tube amplifier failed. As a result, all equipment that links *TDRS-1* to the ground has failed, and it can no longer relay science data and spacecraft telemetry to ground stations located at the White Sands Complex in New Mexico and on Guam. After deployment from Space Shuttle *Challenger* on April 4, 1983, a failed rocket burn stranded the first NASA science and data geostationary satellite almost 16,000 km (10,000 mi) short of its planned orbit. Small on-board hydrazine thrusters were used to place the satellite in a useful orbit after eight months.

NASA's *Wide-field Infrared Survey Explorer* (*WISE*) launched on December 14 from Vandenberg Air Force Base in California. Its mission is to map the entire sky in infrared light. A Delta II rocket placed *WISE* into a polar orbit 525 km (326 mi) above Earth.

In December, the *Hubble Space Telescope* project released the deepest image yet of the universe in near-infrared light using the new Wide Field Camera 3 that the space shuttle *Atlantis* astronaut crew placed on the telescope in May. The oldest galaxies ever identified, having formed only 600-900 million years after the big bang, were observed in this image.

Launch summary. The year 2009 topped the previous year, with 78 launch attempts (see **table**).

Russia's 29 launch attempts in 2009 were all successful, and once again Russia led the world in total launches. Ten of these were commercial and, of the other 19, 9 were devoted to the *ISS*. Of these, 4 crewless *Progress* modules were launched on Soyuz launch vehicles on *ISS* supply missions, 4 were crewed Soyuz missions, and another Soyuz launched the *Mini Research Module 2*. The module is made up of a new docking port and airlock for the *ISS*. Also known as Poisk, the module automatically docked to the space-facing port of the Zvezda service module of the *ISS*.

Russia conducted six military launches. Civilian science missions and communications satellite launches made up the rest. Launch vehicles used were Soyuz (11), Proton-M (9), Rockot (3), Soyuz 2 (2), Kosmos 3M (1), Cyclone (1), Dnepr (1), and Proton-K (1).

The United States had the second-most launches in 2009 with 25, including one international sea launch mission. This was a large increase over the 15 U.S.

Launches to Earth orbit and beyond in 2009

Country of Launch	Attempts	Successful
Russia	29	29
United States	25	24
Europe	7	7
China	6	5
Russian-Ukrainian Zenit-3SL	3	3
Japan	3	3
India	2	2
Iran	1	1
North Korea	1	0
South Korea	1	0
Total	78	74

launches in 2008. Five were commercial launches. The Delta II rocket launched eight times, the space shuttle five times, Atlas V five times, Delta IV twice, and one each for the Delta IV-heavy, Falcon I, Taurus XL, Minotaur, and Sea Launch Zenit-3SL. All were successful except for the Taurus XL launch (the *Orbiting Carbon Observatory*).

Europe saw seven successful launches of the Ariane 5 launch vehicle with five communications satellite commercial payloads and two scientific payloads. ESA's *Planck* and *Herschel* observatories were sent into orbit beyond the Moon. The other scientific Ariane 5 launch placed the French *Helios 2B* scientific satellite into a Sun-synchronous orbit. One Ariane communications satellite launched *TerreStar 1*, the heaviest geostationary communications satellite ever placed into orbit.

China dropped from 11 launches in 2008 to only six in 2009. China's only commercial launch was only a partial success with the Indonesian *Palapa D* communications satellite not getting into its proper orbit. The other missions were government related, including two military, one scientific, one commercial, and one remote sensing.

Japan had three successful launches from its Tanegashima site. Two were civilian including the HTV launch and a satellite launch. The Japanese military placed its *IGS Optical 3* reconnaissance satellite into orbit in November.

The international consortium Russian-Zenit 3SL system launched three geosynchronous communications satellites. The consortium dubbed Land Launch is a combination of the Sea Launch group and the Russian Space International Services. Launches occurred at the Baikonur Cosmodrome in Kazakhstan.

The two Indian launches were noncommercial and included the launch of a military reconnaissance satellite and a nonprofit Anusat into low-Earth orbit in the first launch. The civilian *Oceansat 2* and 6 nonprofit educational/scientific satellites attained orbit on the second launch.

On April 5, North Korea conducted its first orbital launch attempt since 1998 but the launch was a failure. The Taepodong 2 rocket failed to place its payload, the satellite *Kwangmyongsong 2*, into orbit.

South Korea attempted its first orbital launch on August 25. However, the launch was not successful and the scientific *STSAT 2A* satellite was not placed into orbit. A malfunctioning second stage failed to eject its payload fairing, preventing the rocket from achieving orbit.

For background information *see* ASTEROID; CHANDRA X-RAY OBSERVATORY; COMMUNICATIONS SATELLITE; COSMIC BACKGROUND RADIATION; EARTH, GRAVITY FIELD OF; EXTRASOLAR PLANETS; GALAXY, EXTERNAL; HUBBLE SPACE TELESCOPE; INFRARED ASTRONOMY; MARS; METEOROLOGICAL SATELLITES; MOON; PLASMA PROPULSION; REMOTE SENSING; ROCKET; SATELLITE (ASTRONOMY); SATURN; SCIENTIFIC AND APPLICATIONS SATELLITES; SPACE FLIGHT; SPACE STATION; SPACE TECHNOLOGY; SPACECRAFT PROPULSION; SUBMILLIMETER ASTRONOMY; SUN; TELESCOPE; UNIVERSE; X-RAY ASTRONOMY in the McGraw-Hill Encyclopedia of Science & Technology.

Donald Platt

Bibliography. *Aviation Week & Space Technology*, various 2009 issues; *Commercial Space Transportation: 2008 Year In Review*, Federal Aviation; Administration, January 2010 ESA Press Releases, 2009 NASA Public Affairs Office News Releases, 2009.

Star-nosed mole

The star-nosed mole, *Condylura cristata* (**Fig. 1**), is one of the true oddities of the animal kingdom. This distinction arises not only from the 22 conspicuous nasal appendages (the "star") that fan concentrically outward from its nostrils, but also from the ability of this functionally blind insectivore to successfully navigate and exploit the underwater environment for feeding on aquatic annelids and other small invertebrates. The aquatic medium poses particular challenges for terrestrial animals as they are limited to relatively brief underwater excursions because of the necessity to obtain oxygen from the atmosphere. This problem is amplified for small amphibious predators, including the star-nosed mole [40–60 g (1.4–2.1 oz)], as they carry far less oxygen stores underwater than larger aquatic animals and consume the oxygen at a much faster rate. This occurs because the oxygen requirement of each gram of cells increases disproportionately as animal size decreases, following a process called allometric scaling. Consequently, 1 g (0.035 oz) of star-nosed mole tissue consumes about 10 times more oxygen per second than does the same mass of tissue from a 500-kg (1100-lb) seal. As a result, most voluntary dives by star-nosed moles last less than 10 s in duration, with few underwater excursions exceeding 20 s. This strict limitation has led to the evolution of several remarkable specializations that have allowed

Fig. 1. The star-nosed mole in its aquatic habitat. (*Photo courtesy of Christian Artuso and Kevin Campbell*)

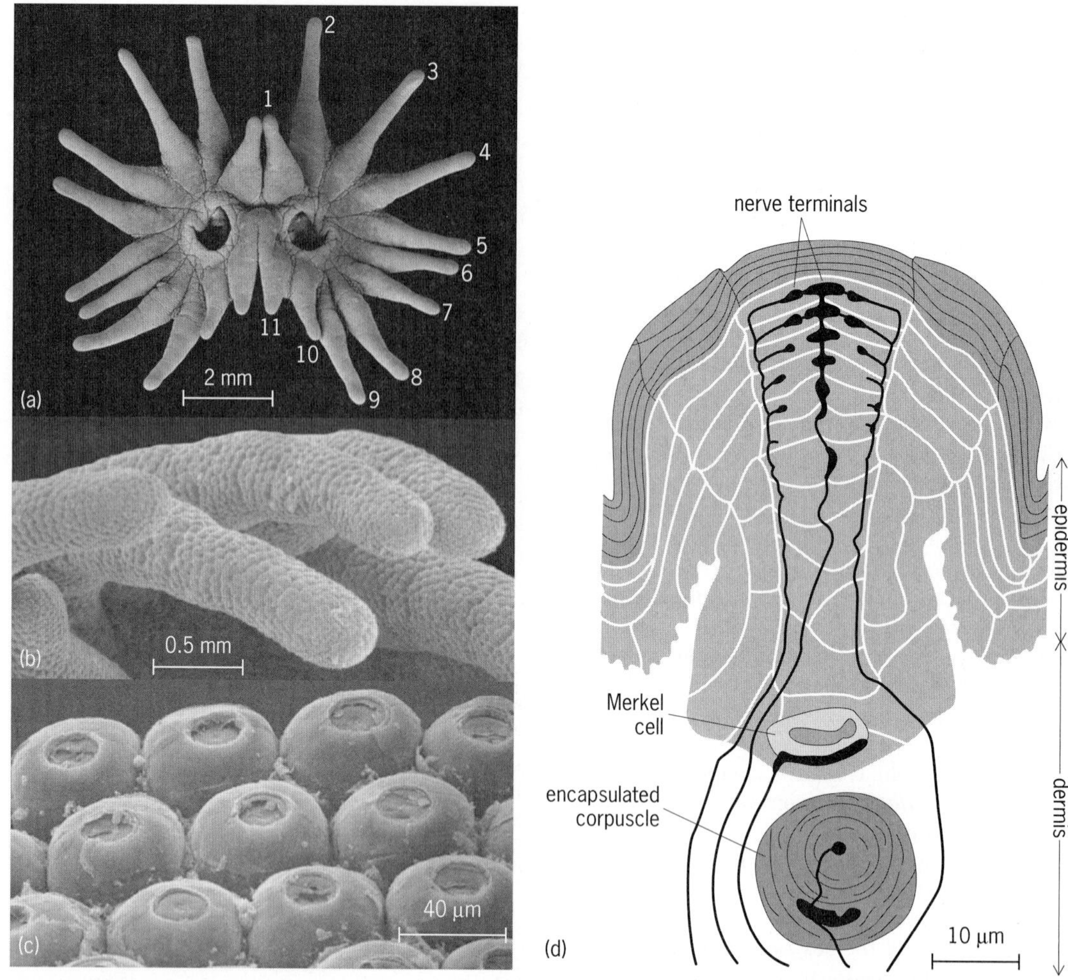

Fig. 2. Attributes of the nose of the star-nosed mole. (*a*) Low-resolution scanning electron micrograph of the nose of a star-nosed mole. Nasal ray pairs are numbered from 1 to 11. (*b*) Side view of nasal rays 6–11, illustrating the honeycomb pattern of Eimer's organs. (*c*) Close-up of the individual dome-shaped Eimer's organs. The disk shapes are the tops of the column of epidermal cells associated with sensory nerve fibers. (*d*) Schematic side view illustrating the internal structure of an individual Eimer's organ. (*Images a and d courtesy of Kenneth Catania; images b and c courtesy of Elaine Humphrey*)

this semiaquatic mole to increase the speed and efficiency with which it can locate and identify prey while submerged.

Unique nose. It is tempting to speculate that the 11 pairs of mobile projections that encircle the mole's snout are used to grasp prey. However, a closer look at the nasal surface reveals that they instead comprise one of the most sensitive organs of touch on the planet (**Fig. 2***a*). This attribute arises from the more than 25,000 small (40–50 μm in diameter) domed mechanoreceptors, termed Eimer's organs, which envelop each nasal ray in a honeycomb pattern (Fig. 2*b*). Eimer's organs are unique to moles (family *Talpidae*). Moreover, in the star-nosed mole, these sensory receptors are so tightly packed together that they can touch a pinhead at some 600 places at once. The center of each sensory dome contains a stacked column of cells that possess 5–10 free nerve endings, emanating from a triad of myelinated nerve fibers that originate in the underlying skin layer (Fig. 2*c*,*d*). The star [about 1 cm (0.4 in.) in diameter] thus contains over 100,000 touch-sensitive nerve projections; in comparison, only about 17,000 touch-sensitive nerve fibers innervate a single human hand. Consequently, the star is able to accurately detect microscopic textural surface features that differentiate the various objects and prey items that are encountered.

Despite the relatively uniform covering of Eimer's organs, not all of the mole's nasal appendages are equally touch-sensitive. Instead, the two smallest rays, which reside immediately above the mouth, act as a focus point, or fovea, and are used for highly detailed investigations (pair 11; Fig. 2*a*). The cortical brain regions that correspond to these two lowermost rays develop earlier and take up a greater area than those of the other, and larger, nasal appendages in a process termed cortical magnification. Therefore, most of the mole's nasal surface is of relatively low tactile resolution and is used primarily to scan the environment for food and other items of interest. This arrangement helps the mole efficiently deal with the immense volume of complex tactile information that it receives. (A similar division of labor is found in vertebrate eyes and brains, and is responsible for the characteristic rapid eye movements, or saccades, that accompany reading and other visual tasks.) As a result, a highly patterned behavior is observed in which the initial detection of an item is followed by a rapid jerky movement of the star for inspection by the two high-resolution appendages immediately prior to prey ingestion.

The speed with which the mole is able to scan and process information with regard to its tactile environment is astonishing. It is able to touch 10–15 different spots per second, and can recognize and consume prey items in as little as 120 ms (mean: approximately 227 ms). Significantly, the remarkable tactile acuity, extreme feeding speed, and tweezer-like front teeth of star-nosed moles have uniquely allowed them to exploit the many tiny prey items that populate their wetland environment; these food sources are too small and energetically unprofitable for shrews and other small animals to utilize. While it is clear that touch is the key source of information about the mole's external environment, recent research has indicated that olfaction (the sense of smell) also may be highly important, although in a most startling and unexpected way.

Underwater sniffing. It has long been assumed that terrestrial species may face sensory challenges while submerged because their body parts are adapted to living on land. Olfaction, or "sniffing," is one such obstacle as it involves the transport of odorants in air to the olfactory epithelium in nasal cavities, where the odorants diffuse into mucosa that contain the receptive nerve terminals. Not surprisingly, there is a complete absence of olfactory systems in some fully aquatic whale species as a result of the assumed inability to obtain olfactory signals while submerged. It thus came as a surprise that star-nosed moles are able to exploit olfaction while immersed to acquire important information about potential prey items (**Fig. 3**). This serendipitous discovery was made through the use of high-speed video recordings, in which diving star-nosed moles exhibited characteristic behavior that resembled the general pattern of sniffing [a rhythmic (6–12 times per second) exhalation and inhalation of small volumes (about 0.1 mL) of air] seen in small mammals in terrestrial environments. More telling, the sniffing behavior was not random, with the outward flow of air occurring only when the star touched down on an object, before quickly re-inhaling as the nose lifted and moved to a new location. Objects that had odorants applied to their surfaces elicited a greater number of touches and sniffs by moles than did untreated surfaces. This suggests that odorants diffuse into the air bubbles upon contact and are able to be detected at the nasal epithelium upon re-inhalation. Further experiments revealed that moles exposed to randomly laid scent trails in underwater channels were able to follow these to food rewards. In order to exclude the possibility that the extremely sensitive nerve endings of the nasal appendages were being used to detect small particles left behind during the laying of the scent trails, a fine mesh grid also was used to cover the channels. This allowed the moles only to pass air bubbles through the grid and not acquire corresponding tactile input. Under these conditions, it was found that the moles had an average accuracy of 85% (individual range: 75–100%) in locating the reward. After testing the moles with a finer grid that precluded air bubbles from passing through and contacting the scent trail, the accuracy rate decreased to chance alone.

Fig. 3. A star-nosed mole exhaling air bubbles while underwater. (*Photo courtesy of Kenneth Catania*)

Not all air emitted during underwater sniffing is re-inhaled as air bubbles are frequently dislodged and rise to the surface. In this regard, it is interesting that the lung volume of star-nosed moles is twice that of terrestrial mole species, presumably allowing them to remain submerged longer and hence extend their foraging opportunities. Star-nosed moles additionally have elevated levels of hemoglobin in their blood and myoglobin in their skeletal muscles; these proteins play important roles in oxygen transport and storage in other diving vertebrate species.

Despite previous suggestions, it is highly unlikely that star-nosed moles are able to detect weak underwater electric fields emitted by prey items, as the outer surface of their snout lacks characteristic ampullary pit type or duct gland type electroreceptors found in salamanders, fish, and platypuses. Similarly, no evidence has been found to suggest that star-nosed moles utilize ultrasonic emissions or sonar to detect prey while submerged. These results indicate that star-nosed moles employ underwater sniffing in conjunction with touch to increase the speed and efficiency with which they find prey while foraging in aquatic environments. Given that other small semiaquatic mammals are limited to relatively brief underwater excursions, they too might be expected to possess the ability to detect and exploit olfactory cues while submerged. These expectations have been borne out in American water shrews, *Sorex palustris*, which were shown to perform the same behavior and had comparable success rates (83%) in the same choice experiments as those provided to star-nosed moles. In comparison, strictly terrestrial short-tailed shrews, *Blarina brevicauda*, which were trained to retrieve food items from shallow underwater wells, were not observed to elicit aquatic sniffing behavior. This finding indicates that this ability evolved independently in the two semiaquatic lineages and is not a characteristic common to insectivores in general. Consequently, it can be speculated that the ability to smell underwater is likely a widespread phenomenon among small, predatory semiaquatic mammal species.

For background information *see* ALLOMETRY; CHEMICAL SENSES; DIVING ANIMALS; INSECTIVORE; MOLE (ZOOLOGY); NEUROBIOLOGY; NOSE; OLFACTION; PHYSIOLOGICAL ECOLOGY (ANIMAL); SENSATION; SENSE ORGAN in the McGraw-Hill Encyclopedia of Science & Technology. Kevin L. Campbell; Nadine T. Price

Bibliography. K. C. Catania, Olfaction: Underwater "sniffing" by semi-aquatic mammals, *Nature*, 444:1024–1025, 2006; K. C. Catania and F. E. Remple, Asymptotic prey profitability drives star-nosed moles to the foraging speed limit, *Nature*, 433:519–522, 2005; K. C. Catania and F. E. Remple, Tactile foveation in the star-nosed mole, *Brain Behav. Evol.*, 63:1–12, 2004; I. W. McIntyre, K. L. Campbell, and R. A. MacArthur, Body oxygen stores, aerobic dive limits and diving behaviour of the star-nosed mole (*Condylura cristata*) and comparisons with non-aquatic talpids, *J. Exp. Biol.*, 205:45–54, 2002.

STEREO mission

The Solar TErrestrial RElations Observatory (STEREO) is a two-spacecraft NASA-led mission to study the three-dimensional structure of the Sun and heliosphere. STEREO's primary scientific objectives are to understand the mechanisms by which material leaves the Sun in both large eruptions called coronal mass ejections (CMEs) and the steady flow known as the solar wind, and to determine how this material propagates through the solar system. One of the two spacecraft that comprise the mission orbits the Sun ahead of the Earth, while the other trails the Earth. The separation between the two spacecraft allows them to capture stereoscopic images of the Sun and make in situ measurements that reveal the three-dimensional structure of interplanetary CMEs and the solar wind.

The two *STEREO* spacecraft were launched together on October 26, 2006, from Cape Canaveral, Florida, and were injected into their respective orbits by gravitational slingshots around the Moon in December 2006 and January 2007. Though both spacecraft orbit the Sun at approximately the same distance as the Earth, *STEREO A* leads the Earth, while *STEREO B* trails the Earth. (The letter A refers to the fact that that spacecraft is ahead of the earth, while the letter B is for behind.) The separation between each of the two spacecraft and the Earth increases at a rate of about 22° per year. The spacecraft will be separated by exactly 180° in February 2011, which will allow observers a complete 360° view of the Sun for the first time ever.

Scientific instrumentation. The two *STEREO* spacecraft each carry four packages of scientific instruments, two that are dedicated mostly to remote observations of solar plasma and two for in situ measurements (**Fig. 1**).

Remote sensing. The first package, the Sun Earth Connection Coronal and Heliospheric Investigation (SECCHI), is composed of a set of telescopes that capture images of the Sun and heliosphere on a number of different scales. The Extreme Ultraviolet Imager (EUVI) images the Sun's atmosphere, the corona, revealing plasma with temperatures between about 10^4 and about 10^7 K, corresponding to one of four possible observation wavelengths.

The Cor1 and Cor2 instruments are both coronagraphs, special telescopes with disks that occult direct light from the Sun in order to reveal the faint extended corona. Together they cover the region from 1.3 to 15 solar radii above the Sun's surface. The Heliospheric Imager (HI) provides an even more extended view than that, allowing scientists to observe CMEs and the solar wind as they propagate deep into the solar system. Because of the *STEREO* spacecraft's locations far from Earth, the Heliospheric Imager can produce images that show interaction between the interplanetary plasma and Earth's magnetosphere directly.

The second package is STEREO/WAVES (SWAVES), which records the signature of CME liftoff and propagation using radio observations. Eruptions on the Sun can trigger bursts of radio noise when they interact with free electrons in the corona and solar wind. These bursts can be detected by radio antennae on SWAVES both as the liftoff occurs and as the disturbance passes the *STEREO* spacecraft.

In situ observations. Two instrumental packages measure the properties of the particles and fields that make up the solar wind locally. The In-situ Measurement of Particles and CME Transients (IMPACT) instrument is a suite of several detectors that measure the flux of high-energy ions and electrons in

Fig. 1. Artist's concept of one of the *STEREO* spacecraft in operation. (*NASA*)

the solar wind as well as the interplanetary magnetic fields in which these particles are embedded. The Plasma and Suprathermal Ion Composition (PLASTIC) instrument measures the velocity, density, and flow rate of particles that make up the solar wind.

From these properties scientists can determine the origin of the solar wind that arrives at the two spacecraft. CMEs often cause shocks, abrupt changes in the properties of the local solar wind, to form as they pass, and the changes that accompany these shocks can be used to determine the mechanisms that triggered and accelerated the eruption.

From the flow speed and magnetic field orientation of the plasma in the solar wind, space weather forecasters can predict how the solar wind will interact with the Earth's magnetosphere. CMEs, fast solar wind, and southward-oriented interplanetary magnetic fields are more likely to trigger geomagnetic activity than slow wind and northward-oriented magnetic fields.

Important discoveries and contributions. Though some of the instrumentation that *STEREO* carries is similar to that on other solar and heliospheric observatories in space, a *STEREO* is unique among all solar observatories in its three-dimensional view of the Sun and heliosphere. Accordingly, many of its most important contributions to solar and heliospheric physics are a direct result of this three-dimensional view.

In the case of *STEREO*'s remote-sensing instruments, its observations have allowed scientists to view and model complex, three-dimensional solar features that, previously seen only in projection, were difficult to interpret. At the same time, the in situ instruments on *STEREO* have allowed scientists to measure the properties of CMEs and the solar wind throughout interplanetary space. Previously, measurements were available only in the near-Earth environment, so *STEREO* made possible both reconstructions of interplanetary events and improved space weather forecasts that depend on these interplanetary measurements.

Three-dimensional reconstructions of solar phenomena. CMEs are one manifestation of the larger phenomenon of solar eruptions. These events often originate in dense concentrations of coronal magnetic field and heated plasma above the Sun's surface called active regions. Most scientists agree that the material that will form a CME is held in place by magnetic fields. When perturbed, a process called magnetic reconnection can break the fields that hold the CME in place and allow it to escape into interplanetary space. Often such eruptions are accompanied by an impulsive release of energy in the corona called a solar flare.

Understanding what triggers such an eruption and how reconnection proceeds once the eruption begins requires good models of the magnetic fields of erupting active regions. SECCHI observations have, for the first time, made it possible to reconstruct the magnetic field configurations of active regions directly from imagery rather than theoretical models.

At the same time, CMEs are best seen in coronagraph imagery where the bright light of the Sun is blocked out. Most CMEs that affect Earth originate near the center of the solar disk as seen in the sky and, in coronagraphs, are seen to propagate outward in all directions simultaneously—such events are known as halo CMEs.

Seen in projection from Earth, it is difficult to distinguish between CMEs that are Earth-directed and CMEs that originate on the back side of the Sun and thus travel harmlessly away from the Earth. It is equally difficult to detect the propagation speed of eruptions that do travel earthward, something space weather forecasters must know in order to predict the arrival of interplanetary disturbances at Earth. *STEREO* (sometimes used in conjunction with instrumentation near Earth) provides a multiperspective view that allows scientists to fill in the missing data (**Fig. 2**).

Many researchers have attempted to use *STEREO* to make complete three-dimensional reconstructions of CMEs and active regions. One of the greatest challenges confronting scientists who undertake such a reconstruction is that both are composed of complex, diffuse, transparent structures, so many features either change appearance or simply disappear in the perspective shift between images taken by each of the two spacecraft.

Nonetheless, researchers have successfully generated three-dimensional models of both CMEs and active regions that have helped to confirm previously existing theories about the origins and behaviors of eruptions and their structure and evolution as they pass through the heliosphere.

Solar tsunamis. That large-scale disturbances propagate outward across the Sun in the wake of eruptions has been known since observations from the *Solar and Heliospheric Observatory* (*SOHO*) spacecraft revealed these events in the 1990s. However, the exact nature of these disturbances was unknown. Because *SOHO* offered only a single perspective

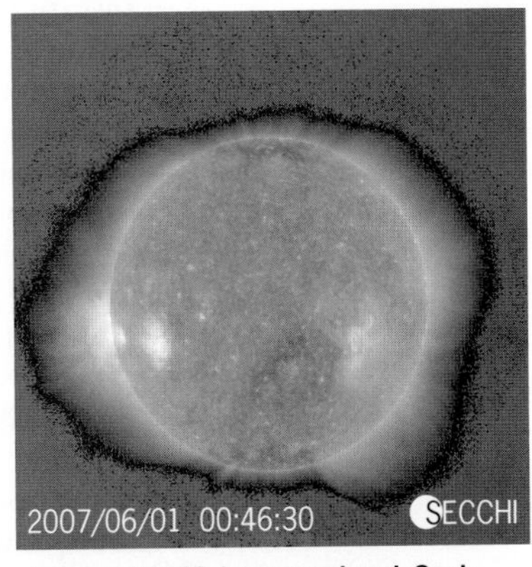

Fig. 2. Simultaneous *STEREO* views of the Sun, taken in the 28.4-nm passband. Such images can be used to construct three-dimensional views of dark coronal holes and magnetic loops in active regions. When used for 3D viewing, the *STEREO A* view, at right, is viewed by the right eye, while the *STEREO B* view, at left, is viewed by the left eye.

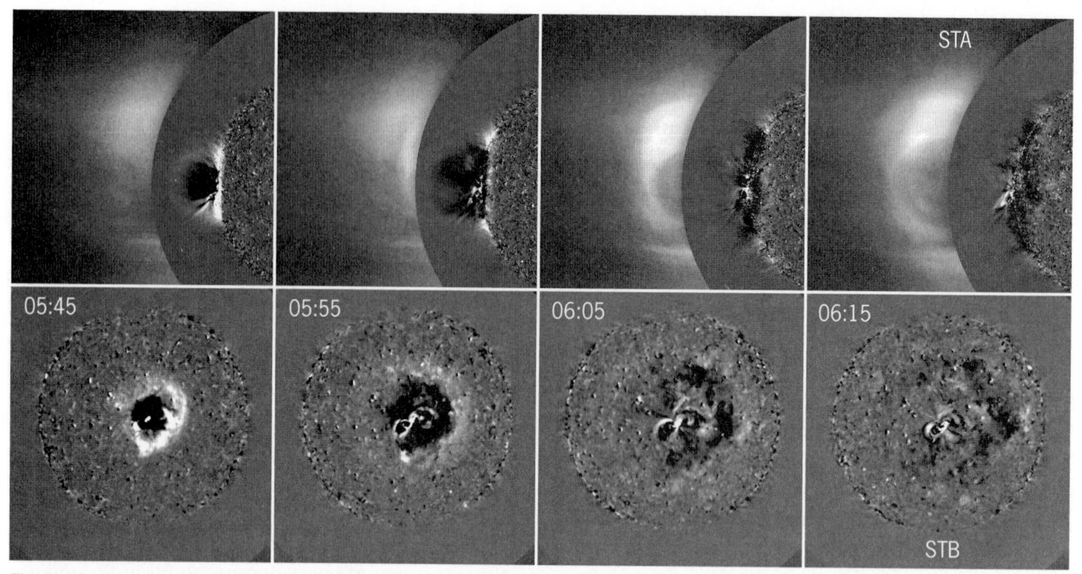

Fig. 3. Views of an extreme ultraviolet (EUV) wave from *STEREO A* (STA above) and *STEREO B* (STB below), showing how the wave expands as the event unfolds. Because the two spacecraft were separated by about 90° when the images were captured, *STEREO B* saw the event from above, while *STEREO A* saw it in profile. EUV plasma is shown in running difference images in the views from *STEREO B* and within the solar disk in the views from *STEREO A*, while coronagraph images from the Cor1 telescope are shown to the left of the solar disk (that is, from above the photosphere) in the views from *STEREO A*. (*S. Patsourakos and A. Vourlidas, "Extreme ultraviolet waves" are waves: first quadrature observations of an extreme ultraviolet wave from STEREO, Astrophys. J. Lett., 700: L182–L186, 2009, used with permission*)

on these events, it was impossible to determine if they were caused by magnetic shock waves in the corona or were simply projection effects of the erupting CME blocking light from the Sun behind it. Using observations from both *STEREO* spacecraft simultaneously, it was possible to determine that these events, now popularly referred to as solar tsunamis, were, in fact, magnetohydrodynamic waves (**Fig. 3**).

Understanding these events has implications both for our understanding of the forces that drive solar eruptions and for space weather modeling. The interaction of the wave with other magnetic structures in the corona can help scientists determine properties of the eruption itself and the surrounding solar atmosphere. Additionally, tracing the center of the outward-propagating wave allows scientists to better isolate the point of origin of CMEs and improve predictions of how such events will affect Earth.

Predicting the solar wind. Though CMEs and solar flares are the best known influences on space weather, other disturbances in the solar wind can also trigger geomagnetic activity. Some scientists have explored the possibility of placing a permanent solar wind monitor at the Earth's L5 Lagrange point, a point along the Earth's orbit around the Sun 60° behind Earth where the gravitational forces of the Earth and Sun are balanced. As the Sun rotates, areas where fast and slow solar wind originate rotate with it. As a result, regions of fast and slow solar wind reach the L5 point a few days before they reach Earth, and measurements of the solar wind there could be used to predict future conditions near Earth.

As a test of this proposal, scientists used the in situ observations made by *STEREO B* to predict conditions expected at *STEREO A* when the two spacecraft were separated by 60°. During the quiet conditions of the most recent solar minimum, these measurements were generally successful predictors of both future conditions for *STEREO A* and the arrival times of regions of particularly high-speed solar wind and magnetic field reversals.

STEREO and space weather. STEREO has been responsible for a number of improvements in the understanding of the solar corona and heliosphere, but perhaps its greatest contribution may be to the field of space weather forecasting. Since space weather events can threaten spacecraft hardware and human astronauts in orbit, as well as power grids, airliners and passengers, and global communication networks, accurate predictions are extremely important.

Improvements, like those discussed above, in the ability to determine both the structure and properties of both the solar wind and events such as CMEs and solar flares have enabled forecasters to improve the quality of their predictions over the course of the past few years. STEREO's contribution to gains in understanding of the origins of these phenomena will continue to improve our ability to model and predict solar and heliospheric activity well beyond the lifetime of the mission itself.

For background information *see* CORONAGRAPH; SOLAR CORONA; SOLAR WIND; SPACE PROBE; SUN; ULTRAVIOLET ASTRONOMY in the McGraw-Hill Encyclopedia of Science & Technology. Daniel B. Seaton

Bibliography. L. Feng et al., First stereoscopic loop reconstructions from STEREO SECCHI images, *As-*

trophys. J. Lett., 671:L205–L208, 2008; M. L. Kaiser et al., The STEREO mission: an introduction, *Space Sci. Rev.*, 136:5–16, 2008; M. Mierla et al., On the 3-D reconstruction of coronal mass ejections using coronagraph data, *Ann. Geophys.*, 28:203–215, 2010; S. Patsourakos and A. Vourlidas, "Extreme ultraviolet waves" are waves: first quadrature observations of an extreme ultraviolet wave from STEREO, *Astrophys. J. Lett.*, 700:L182–L186, 2009; K. D. C. Simunac et al., In situ observations from STEREO/PLASTIC: A test for L5 space weather monitors, *Ann. Geophys.*, 27:3805–3809, 2009; T. Wiegelmann, B. Inhester, and L. Feng, Solar stereoscopy—where we are and what developments do we require to progress?, *Ann. Geophys.*, 27:2925–2936, 2009.

STIM family of proteins

Calcium (Ca^{2+}) is a major signaling molecule that activates an array of cellular events. Accordingly, cytosolic Ca^{2+} levels are tightly controlled at approximately 100 nM. However, elevations to approximately 300 nM initiate powerful signals that control a spectrum of cell functions, ranging from short-term contractile responses to long-term events such as transcription. Thus, maintenance of cytosolic Ca^{2+} is a very important and elaborate process. Some of the regulatory elements that maintain these cytosolic Ca^{2+} levels are ion channels and exchangers located on the plasma membrane of the cell. However, the endoplasmic reticulum (ER) also plays a large role in Ca^{2+} signaling. The ER lumen stores approximately 500 μM of Ca^{2+}, which not only provides an environment conducive for proper protein folding, but also serves as a major source of Ca^{2+} for signaling.

In the 1990s, investigators found that depletion of the ER Ca^{2+} pool results in the commencement of a subsequent phase of cytosolic Ca^{2+} elevation. This phenomenon has since been carefully characterized and is termed store-operated Ca^{2+} entry (SOCE). SOCE operates under the control of stromal interacting molecule 1 (STIM1), which is the ER Ca^{2+} sensor, and Orai1, which is the store-operated Ca^{2+} channel. This signaling pathway has been identified as the major signaling pathway of intracellular Ca^{2+} increase in nonexcitable cells and regulates major physiological events, including immune cell activation and cell migration, as well as pathophysiological processes such as tumor cell metastasis.

STIM1. STIM proteins are ubiquitously expressed type IA transmembrane proteins found in species ranging from *Drosophila* to humans. Early investigations identified STIM1 as a surface-positioned protein on stromal cells with unknown function. However, while the role of STIM1 on the cell surface remains controversial, its function as a Ca^{2+} signaling protein in the ER is now well understood. Briefly, the activation of phospholipase C (PLC)-coupled surface receptors initiates complex signaling cascades, which, among many things, stimulate Ca^{2+} release from the ER (**Fig. 1**). Subsequently, STIM1 senses that ER Ca^{2+} levels have been depleted and signals a channel on the plasma membrane, Orai1, to allow Ca^{2+} into the cytoplasm from the extracellular space.

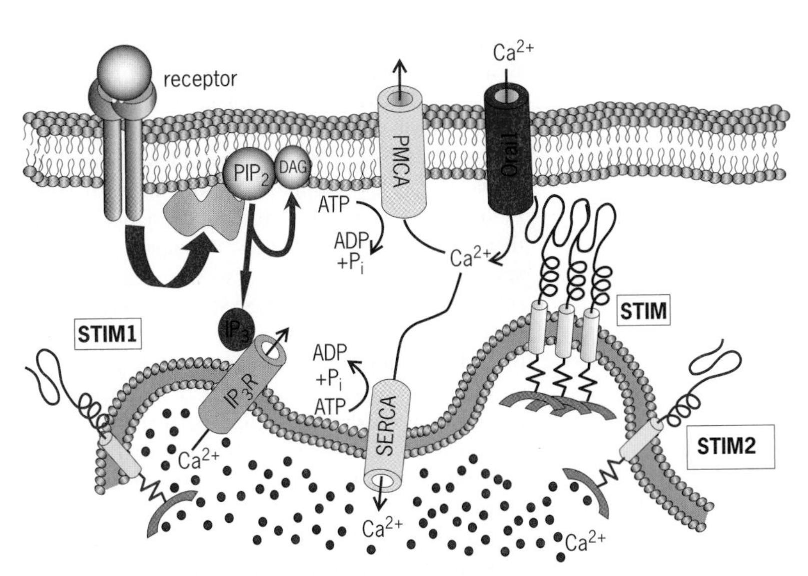

Fig. 1. The store-operated Ca^{2+} entry (SOCE) signaling pathway. Stimulation of certain plasma membrane receptors activates phospholipase C (PLC), which cleaves phosphatidylinositol 4,5-biphosphate (PIP_2) and produces inositol-3-phosphate (IP_3) and diacylglycerol (DAG). IP_3 activates a receptor (IP_3R) located in the ER membrane and results in ER Ca^{2+} pool depletion. Depletion of ER Ca^{2+} causes STIM proteins to aggregate and activate the store-operated channel Orai1. The plasma membrane Ca^{2+} ATPase (PMCA) and sarcoplasmic/endoplasmic reticulum Ca^{2+} ATPase (SERCA) use energy produced from adenosine triphosphate (ATP) hydrolysis to pump Ca^{2+} out of the cell or into the ER, respectively.

STIM1 is positioned in the ER membrane with its N-terminus facing the lumen and its C-terminus extending into the cytoplasm (Fig. 1 and **Fig. 2**). The protein contains two EF-hand motifs: one EF hand defined as a Ca^{2+}-binding canonical EF-hand motif (cEF) that pairs with an immediate atypical "hidden" non-Ca^{2+}-binding EF-hand motif (hEF). Together, this pair of EF hands mediates hydrophobic

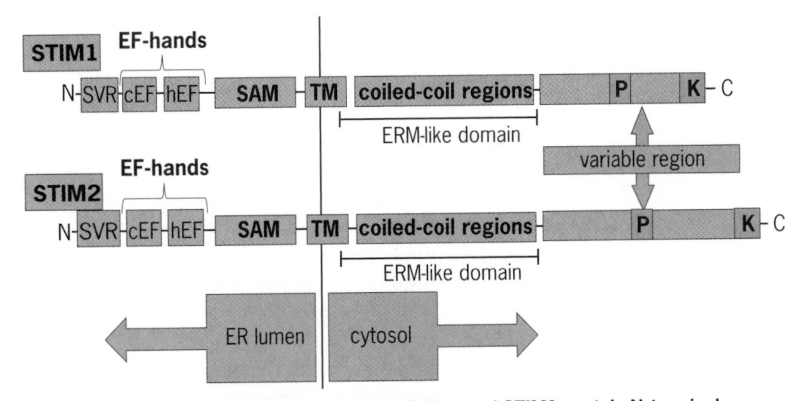

Fig. 2. Structures of STIM1 and STIM2. Both STIM1 and STIM2 contain N-terminal, canonical (cEF) and hidden (hEF) EF-hand motifs, as well as a sterile alpha motif (SAM), which extend to the ER lumen. Each protein also contains a small region of sequence variability at the N-terminus (SVR). The transmembrane domain of each protein is denoted TM. Both proteins also contain homologous C-terminal ezrin-radixin-moesin (ERM)–like domains, as well as proline (P)–rich and lysine (K)–rich regions, which extend to the cytoplasm. The structures of STIM1 and STIM2 differ significantly at their C-terminus.

interactions between the EF hand and SAM domains. Biophysical analyses show that, in the presence of Ca^{2+}, the EF hand/SAM domain is a monomeric, compact-alpha helix. However, when Ca^{2+} is depleted, the EF hand/SAM domain changes to a less alpha-helical, less compact conformation that promotes aggregation. Thus, the conformation change that occurs as a result of Ca^{2+} dissociation exposes several hydrophobic residues to water in the ER lumen. As a consequence, EF/SAM domains from adjacent STIM1 molecules aggregate in order to conceal these hydrophobic residues and form a more energetically favorable structure. It is believed that this aggregation signals STIM1 clusters to activate Orai1 channels on the plasma membrane and generate Ca^{2+} influx. Furthermore, readdition of Ca^{2+} results in disaggregation of the domain back to monomers, demonstrating complete reversibility of this Ca^{2+}-dependent conformational change.

The STIM1 C-terminus, located in the cytosol, contains two coiled-coil regions overlapping with an ezrin-radixin-moesin (ERM)-like domain followed by a serine/proline-rich region and a lysine-rich region (Fig. 2). It is speculated that the lysine-rich region provides an area of positive charge, which interacts with anionic lipids in the plasma membrane to facilitate Orai1 activation. Recent studies have described the essential ORAI-activating region within the ERM domain, termed SOAR (STIM ORAI-activating region), OASF (ORAI-activating small fragment), or CAD (CRAC-activating domain). Hence, the CAD/SOAR/OASF fragments of STIM1 are sufficient to activate Orai1, whereas the roles of the remaining portions of the cytosolic portion of STIM1 remain unclear.

STIM2. The function of STIM2 appears to be more complex than STIM1. Various studies tend to show little or no contribution of STIM2 to SOCE, depending on the cell type. However, when STIM2 is overexpressed, marked inhibition of SOCE is observed. This observation has led researchers to speculate that STIM2 may function as an endogenous inhibitor of SOCE. However, the discovery of Orai1 as the channel that mediates SOCE led to the rejection of this idea since coexpression of both STIM2 and Orai1 led to constitutive Ca^{2+} entry. Therefore, subsequent efforts have focused on defining key differences in the structures of STIM1 and STIM2 that could explain these seemingly contradictory observations and shed new light on the true cellular function of STIM2.

The structures of STIM1 and STIM2 have very high homology, with differences noted primarily in the N-terminal small variable regions (SVR) and the C-terminal variable regions (Fig. 2). Interestingly, most of the functional differences observed between STIM1 and STIM2 are conferred by sequence variations in the N-terminal or luminal regions of the proteins. Hence, dissociation of Ca^{2+} from the STIM EF/SAM domains causes rapid conformational changes that support aggregation; minor differences in the SVR domains of STIM1 and STIM2 cause the STIM2 EF/SAM domain to undergo a threefold slower conformational change and a 70-fold slower rate of aggregation upon Ca^{2+} withdrawal. Furthermore, minor differences in the STIM canonical EF-hand loop between STIM1 and STIM2 change their relative Ca^{2+} affinity, with STIM2 being active at near-resting concentrations of ER Ca^{2+}. Therefore, STIM1 is a stronger and faster activator of Orai1 than STIM2, yet greater changes in ER Ca^{2+} content are required for STIM1 activation. This has led to the hypothesis that, unlike STIM1, STIM2 functions primarily as a modulator of basal cytosolic Ca^{2+} concentration.

Critical biological functions for STIM proteins. The identification of the STIM and Orai proteins as the molecular mediators of SOCE has allowed for elegant studies that have revealed the biological role of SOCE in cell physiology and pathophysiology. It has been shown that both STIM1 and STIM2 are critical for T cell function. T cells deficient in STIM1 lack SOCE and are severely compromised in their ability to produce various cytokines, including interleukins 2 and 4 (IL-2 and IL-4) and interferon-gamma (IFN-γ), in response to stimulation of the T cell receptor. Interestingly, expression of STIM2 in these cells led to a moderate rescue in SOCE. Also, STIM2-deficient T cells had moderately less SOCE than control cells, highlighting that endogenous STIM2 does play a role in SOCE activation. However, whereas STIM2-deficient T cells only had slightly diminished SOCE, these cells were still notably impaired at cytokine production. Thus, STIM1 and STIM2 are important signaling molecules, each with distinct roles for control of Ca^{2+} influx in T cell activation.

STIM1 has also been implicated in pathophysiological events such as tumor cell metastasis. Investigators have been able to inhibit the migration of highly metastatic breast cancer cells by reducing the levels of STIM1. Furthermore, reduction of the Ca^{2+} influx that STIM1 initiates using small molecule inhibitors had a similar effect on the migration of these cells in vitro and in vivo, indicating that STIM1 and SOCE are potential cancer therapeutic targets.

For background information *see* BIOPOTENTIALS AND IONIC CURRENTS; CALCIUM; CALCIUM METABOLISM; CANCER (MEDICINE); CELL MEMBRANES; CYTOKINE; CYTOPLASM; ENDOPLASMIC RETICULUM; ION TRANSPORT; PROTEIN; SIGNAL TRANSDUCTION in the McGraw-Hill Encyclopedia of Science & Technology.

Michael F. Ritchie; Jonathan Soboloff

Bibliography. T. Hewavitharana et al., Role of STIM and Orai proteins in the store-operated calcium signaling pathway, *Cell Calcium*, 42:173–182, 2007; M. Oh-Hora et al., Dual functions for the endoplasmic reticulum calcium sensors STIM1 and STIM2 in T cell activation and tolerance, *Nat. Immunol.*, 9:432–443, 2008; J. Soboloff et al., Orai1 and STIM reconstitute store-operated calcium channel function, *J. Biol. Chem.*, 281:20661–20665, 2006; P. B. Stathopulos et al., Stored Ca^{2+} depletion-induced oligomerization of stromal interaction molecule 1 (STIM1) via the EF-SAM region, *J. Biol. Chem.*, 281:35855–35862, 2006.

Supermassive black holes

Black holes are compact massive objects from which light cannot escape. The largest of these are the supermassive black holes, which are millions to billions of times more massive than the Sun. Dormant supermassive black holes reside at the centers of nearly all present-day galaxies. Their presence has been detected through the orbital motions of stars and gas near the galactic centers. Locally the mass of a galaxy's supermassive black hole is known to be tightly related to the mass of its host galaxy. Given the very limited extent of a supermassive black hole's gravitational influence, this relation hints at a deep connection between the processes that formed the stars in the host galaxy and the processes that formed the central supermassive black hole. Astronomers would like to follow this connection back in time in order to understand the growth of supermassive black holes in galaxies, but this requires obtaining a census of all the accreting supermassive black holes in the universe.

Epoch of formation. Active galactic nuclei (AGNs) are galaxies that emit staggering amounts of energy from their nuclei. The high nuclear luminosity of an AGN can be explained only by the release of gravitational energy from gas falling toward a supermassive black hole lying at the center of the galaxy. The gas funnels into the supermassive black hole through a giant, spinning accretion disk. The accretion of material causes the mass of the supermassive black hole to grow. Quasars are the most energetic of the AGNs. When they are accreting they outshine galaxies of hundreds of billions of stars, which means that they should be easy to find, even at tremendous distances. Using optical observations, astronomers discovered that the number of quasars peaked several billion years after the big bang and then mostly disappeared. A possible explanation for their dominance at these early times (the current age of the universe is 13.7 billion years) is that galaxies in the past were closer together than they are today, so collisions between them were more common. These collisions funneled gas to the central supermassive black hole. Indeed, often the galaxies hosting quasars are observed to be interacting or merging with other galaxies. In addition, galaxies in the past had more gas that had not yet been incorporated into stars, providing fuel for the quasars. However, once the quasar consumed most of the gas in its immediate neighborhood, it likely faded with time. Since the supermassive black hole itself cannot be destroyed, its dormant state can be detected at the present time through its gravitational influence on its immediate surroundings. *See* GALAXY EVOLUTION IN THE EARLY UNIVERSE..

Role of dust. Astronomers get a very different view of the universe depending on which wavelength window they use to observe it. This can be illustrated by looking at a region of sky at two different wavelengths, the optical and the submillimeter (here 850 micrometers; **Fig. 1**). There are thousands

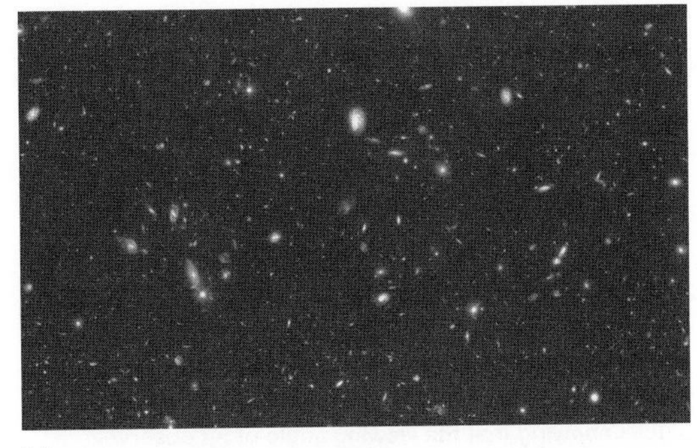

(a)

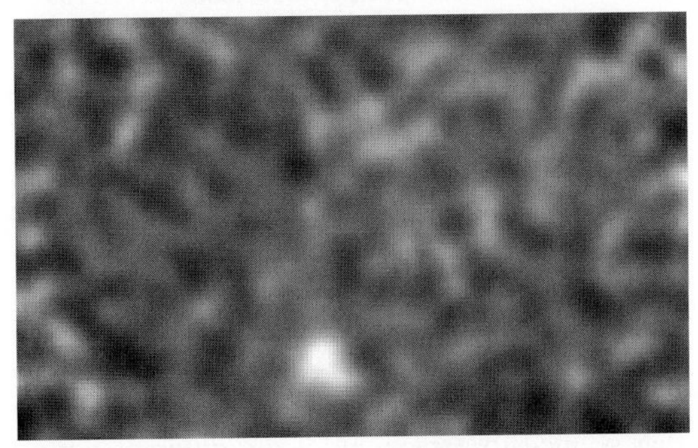

(b)

Fig. 1. Comparison of optical and submillimeter images of the same region of the sky, showing how viewing the universe at optical wavelengths gives one a very different perspective than viewing it at submillimeter wavelengths. (a) *Hubble Space Telescope* optical image (*NASA*). (b) James Clerk Maxwell Telescope submillimeter image (*Wei-Hao Wang, Academia Sinica Institute of Astronomy and Astrophysics*).

of galaxies detected in the optical and only a handful detected in the submillimeter. The submillimeter galaxies are distant, exceptionally luminous sources obscured by dust. Although they are rare, they are among the most luminous galaxies in the universe. They are forming stars at a rate that is hundreds of times greater than the rate at which present-day galaxies, including our own Milky Way, are forming stars. Similarly, if a supermassive black hole is obscured by gas and dust, then the optical light emitted by the infalling material will be absorbed and reradiated at longer wavelengths. This means that optical surveys will not find hidden supermassive black holes.

Unified model. In 1962 Riccardo Giacconi headed a rocket mission to study x-rays from the Moon. He discovered a diffuse uniform x-ray glow coming from the sky, for which he was awarded the Nobel Prize in physics in 2002. X-ray telescopes can resolve this background glow into individual sources. Early x-ray telescopes could produce images only in low-energy x-rays (0.5–2 keV), usually referred to as soft x-rays. However, even soft-x-ray surveys are unable to find all of the sources that make up the x-ray background, as they, too, can be absorbed by obscuring gas. Thus, observations are also needed at higher-energy x-rays (greater than 2 keV), usually referred to as hard x-rays, which are able to penetrate the obscuring material. It was predicted that when

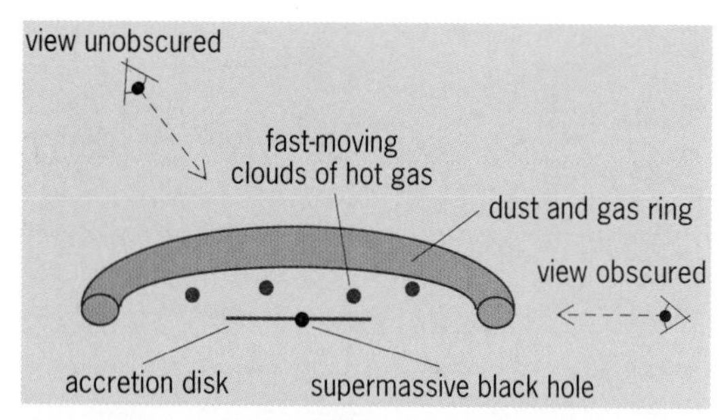

Fig. 2. Schematic of the unified model for active galactic nuclei showing that the viewing angle of an observer looking toward a supermassive black hole will determine whether or not certain optical spectral features, such as broad emission lines from fast-moving clouds of hot gas, are seen.

such observations became technically feasible, astronomers would discover a population of obscured quasars that existed several billions years after the big bang. This would be a parallel population to the unobscured quasars that had already been observed to peak at these times. The basis for this prediction was a model that astronomers had developed that seemed to explain all of the different optical emission-line features observed in the spectra of accreting supermassive black holes. In this so-called unified model, the type of AGN that one sees at optical wavelengths depends on one's line-of-sight to the supermassive black hole (**Fig. 2**). Located around the accretion disk are relatively dense, fast-moving clouds of hot gas that are responsible for the broad emission lines seen in the optical spectra of unobscured AGNs. However, if the observer's view of the AGN is obstructed by the presence of a gas and dust torus that hides the accretion disk, then only the slower-moving hot clouds farther from the supermassive black hole will be visible. There should be no luminosity or distance dependence of the obscuration, because what one observes is only a function of geometry. Thus, the expectation is that there should be equal numbers of obscured quasars and unobscured quasars at the peak of the quasar era.

Essential instrumentation. The fundamental goal of the National Aeronautics and Space Administration's *Chandra* and the European Space Agency's *XMM-Newton X-ray Observatories* was to resolve the sources that made up the hard-x-ray background. The advantages of *Chandra* for this kind of work are twofold. First, it has high positional accuracy, which minimizes misidentifications of faint optical galaxy counterparts to the x-ray sources. Second, even with the deepest observations, it does not reach the so-called confusion limit, where there are so many detected sources that it is difficult to distinguish individual sources. The deepest images of the x-ray sky have been taken with *Chandra*. These images are essential for finding hidden supermassive black holes, but they are not sufficient to fully understand the nature of the sources nor to determine the distances to the sources. Observations at many different wavelengths are required to achieve those objectives. Indeed, a comparison of a deep *Chandra* x-ray image with a deep optical image shows how modest the number of x-ray sources is compared to the number of optical sources (**Fig. 3**). This is because only a fraction of the supermassive black holes at the centers of galaxies are actively accreting at any given time. In addition, most of the sources in the optical image are star-forming galaxies, and the x-ray data are not deep enough to pick up the bulk of this population.

Hidden population. Spectroscopic observations of the optical counterparts to the *Chandra* x-ray sources have revealed that many show no evidence of supermassive black hole activity. Thus, without the x-ray observations, astronomers would not even know that these galaxies contained actively

(a)

(b)

Fig. 3. Comparison of (*a*) *Chandra X-ray Observatory* x-ray image (*NASA*) and (*b*) Subaru Telescope optical image of the Hubble Deep Field-North region. There are many more sources visible in the optical image, because star-forming galaxies are far more common in the universe than active galactic nuclei.

accreting supermassive black holes at their centers. Even these x-ray observations will miss the most obscured sources, which can block the x-ray photons from escaping. Thus, there are still more "missing" supermassive black holes to be found. The new x-ray-detected AGNs were active at much later times than the optical quasars: their numbers peak around 6 billion years (as opposed to 2–3 billion years for the quasars) after the big bang. They do not exhibit the same behavioral patterns as the distant quasars. In particular, their rates of accretion are much lower, so they radiate less intensely. However, there are so many more of them at their peak than there were quasars at their peak that they produce a substantial amount of light. The x-ray observations require a modification of the unified model: The higher x-ray luminosity sources are observed to be predominantly unobscured AGNs, while the lower x-ray luminosity sources are observed to be predominantly obscured AGNs. This means that there must be some luminosity dependence of the obscuration. The modelers' predictions for equal numbers of obscured quasars as unobscured quasars during the quasar era are not born out. There is also a distance dependence on the obscuration, since at more recent times the sources that are unobscured are less luminous than similar sources at earlier times.

Parallel history. The most likely evolutionary history to explain the earlier peak in the extremely bright quasars versus the later peak in the more moderately bright (and obscured) AGNs discovered by the x-ray observations is that the more massive supermassive black holes are being preferentially starved, while the less massive supermassive black holes continue to accrete. This "downsizing" would be akin to the downsizing observed in star-forming galaxies, where the big galaxies at earlier times, whose light mostly comes out at wavelengths longer than the optical, give way to a larger number of smaller galaxies at late times, whose star formation comes out in the ultraviolet and optical. When astronomers compare the cumulative growth of AGNs throughout cosmic time with the cumulative star formation, they find that both form most of their mass at late times, and both show downsizing effects. It is possible that whatever mechanism is quenching the star formation in the large galaxies is also switching off the AGN activity. The current thinking is that the mechanism is "feedback" from the AGN activity itself. This feedback might be strong winds from the AGN that blow the remaining gas (fuel for both star formation and supermassive black hole accretion) completely out of the galaxy.

For background information *see* BIG BANG THEORY; BLACK HOLE; CHANDRA X-RAY OBSERVATORY; COSMOLOGY; GALAXY, EXTERNAL; GALAXY FORMATION AND EVOLUTION; HUBBLE SPACE TELESCOPE; QUASAR; SUBMILLIMETER ASTRONOMY; UNIVERSE; X-RAY ASTRONOMY; X-RAY TELESCOPE in the McGraw-Hill Encyclopedia of Science & Technology.

Amy J. Barger

Bibliography. A. J. Barger (ed.), *Supermassive Black Holes in the Distant Universe*, Kluwer Academic Publishers, Dordrecht, The Netherlands, 2004; J. H. Krolik, *Active Galactic Nuclei: From the Central Black Hole to the Galactic Environment*, Princeton University Press, Princeton, NJ, 1999; B. M. Peterson, *An Introduction to Active Galactic Nuclei*, Cambridge University Press, Cambridge, United Kingdom, 1997; *Scientific American Reports, Special Edition on Astrophysics: Black Holes*, vol. 17, no. 1, 2007.

Sustainable development in steel construction

Construction as a whole accounts for approximately 13% of the gross domestic product (GDP) of the United States. Recognizing the recyclability and other characteristics of steel, this paper will focus on the leading role of the steel industry in a large number of sustainability efforts in the United States. The efficiency of steel as a primary construction material especially recognizes the changes that have taken place in steelmaking processes over the past 15–20 years. Thus, the industry has now formulated an aggressive program for energy-efficient and economical construction. Most of the mills use electric-arc furnaces, which are fed steel scrap in lieu of iron ore and coke, and continuous casting of shapes, plates, and sheet is used instead of the processing of large ingots into blooms and shapes and other products. This has led to significantly increased production efficiency and economies, with approximately 0.5 worker-hours now being used to produce 1 ton of steel. At the same time, the consumption of energy has been reduced by more than 60%, and the level of environmental pollutants and waste materials has been cut by nearly 70%.

The industry is working closely with the United States Green Building Council (USGBC), which has evolved over the past dozen years into a major force for sustainable, or "green," construction. Activities such as the Leadership in Energy and Environmental Design (LEED) program have become major factors for architects and engineers, and a large number of individuals have now been certified as LEED APs (LEED Accredited Professionals) through rigorous examinations. Although USGBC and the LEED program focus on buildings only, in all construction materials, the principles and methods of evaluation are now being applied to any number of different types of projects. The key words are energy, efficiency, economy, and environmental performance.

Structural steel supply chain. To provide a realistic and detailed assessment of the sustainability capacities of steel and steel construction, it is important to be familiar with the key segments of the industry. A representative description of the various elements of the industry focuses on the following types of companies: producers (steel mills), steel service centers, fabricators, and erectors.

For classification purposes, producers include mills that make hot-rolled steel shapes and plates and various forms of hot-rolled bars, as well as sheet and strip products. Also in this group are companies the make cold-formed steel shapes from sheet, including roll-forming and press-brake operations. The producers of preengineered buildings (metal buildings in the United States) for storage warehouses, residential construction, and rack structures are counted among the fabricators.

The American steel industry changed drastically in the early 1980s, when the construction market and the steel industry suffered enormous losses and many companies closed their doors. Integrated steel-mill companies, such as U.S. Steel and Bethlehem Steel, downsized and shuttered a number of their plants. Cities such as Pittsburgh and Bethlehem, Pennsylvania, Cleveland, Ohio, and Chicago, Illinois entered a period of depression-like conditions. This was also the time during which minimills had become productive and major contributors to the market. Minimills are steel mills that use electric-arc furnaces and steel scrap, with continuous casting in lieu of the basic oxygen furnaces of the integrated mills that use iron ore and coke and produce ingots. Minimills have become the center of activity of the American steel industry, with highly efficient production and competitive pricing for all types of products, including all of the production of hot-rolled shapes. In a word, they are no longer "mini" by any stretch of the imagination.

Effectively, the steel mills and their products fundamentally establish the sustainability characteristics of the material. The service centers, fabricators, and erectors contribute to the use of a highly recycled material, but the steel in itself is a function of the work of the mills. This is further emphasized by the Steel Recycling Institute, which was founded by the industry in the 1990s. Taken together, these efforts attest to the importance of the entire sustainability program. It also illustrates why it is critical for the industry as a whole to address the subjects jointly as well as the fact that the government and the various building-code groups have the legislative power to ensure that sustainability in all forms is properly covered.

In the United States, the design codes are prepared by private industry trade associations and professional organization groups, along with very significant input from researchers in academia. The code requirements are an integral part of the sustainability issue, primarily because efficient and economical use of the materials is a fundamental criterion for state-of-the-art codes. As a result, major contributions are provided by all interest groups, and the outcome reflects extensive use of steel along with thoughtful consideration of all aspects of sustainability.

Some current industry initiatives and results. The institutes representing the various components of the American steel production and steel construction industry are currently focusing on a number of aggressive undertakings. The fundamental principles are summarized by The Four R's program: reduce, reuse, recycle and restore.

Reduce efforts. These efforts aim at reduced energy consumption in the production of steel, as well as the reduction of greenhouse-gas emissions, primarily carbon dioxide (CO_2). The results demonstrate that energy consumption has been reduced by 60% since 1960; the reduction was 29% per ton during the 1990s alone. Energy reduction has largely been achieved through the change from basic oxygen to electric-arc furnace operations of the mills. Data on greenhouse gases are more recent, specifically, to the effect that CO_2 emissions were reduced by 47% between 1990 and 2005. In the same period, the overall industry emissions were reduced by nearly 70%. This outcome can be compared to the suggested measures of the Kyoto Protocol, which would have required reductions of only 5.2% for all U.S. industries by 2012. In this context, it is interesting to note that the United States is not a signatory to the Kyoto Protocol, and that the attitude of the American steel industry reflects significant attention to societal and world needs.

Reuse efforts. Reuse activities focus on waste reduction, and have traditionally been more important for the operations of the traditional integrated mills. The concepts certainly have worldwide applicability, since approximately two-thirds of the world's steel mills are still based on iron ore and coke. For example, blast furnace slag is reprocessed for use in road building, glass making, and other commercial applications. Similarly, the large quantities of gas that are produced in the making of coke are reprocessed and used for fuel in various applications.

Recycle efforts. Steel is recognized as the most recycled material in the world. This applies very much to today's electric-arc furnace steels, which are primarily based on the use of steel scrap. For example, the structural shapes (I-beams or wide-flange shapes) that are now produced in the United Stages average 88% recycled steel. The remainder of the steel is made up of materials such as general ferrous scrap, iron carbide (93% iron), and pig iron (which is almost pure iron), and the shapes themselves are recycled at a rate of 98%. Hollow structural sections (rectangular or round tubes) are preferred by some architects and engineers for certain projects. The recycled content of such products is 88% when produced by an electric-arc furnace mill; it is 29% when it is produced by an integrated or basic oxygen furnace mill. Many of the sheet and plate products that are used to make high-strength steel sections come from basic oxygen furnace mills; wide-flange and other hot-rolled shapes in the United States come only from electric-arc furnace mills. This means that such tubes are less green than hot-rolled shapes, a fact that is generally not known.

Restore efforts. The restoration activities of the American steel industry have focused on the redevelopment of old steel industry properties, including land. The land is sometimes referred to as "brownfields," and any such properties have the problem of ground contamination of various types. New land uses typically focus on light commercial activity. Old mill buildings may be torn down and the steel recycled. Where suitable, the buildings have been remade into offices or shopping malls.

Steel construction and green building. The United States Green Building Council (USGBC) has become a major force for sustainable construction of buildings. The well-known LEED program has evolved into an effort that provides professional certification for accredited individuals. The membership of the council of 13,500 individuals includes a broad variety of architects, engineers, contractors, representatives of material suppliers and building operation companies, and some universities. In brief, the activities of the USGBC and its LEED APs are now recognized as authoritative efforts for all sustainability aspects of building construction.

LEED standards and their applicability for buildings. The LEED standards apply to buildings in any construction material, and the various award levels are based on a point system that recognizes all aspects of sustainability. Depending on the accumulated points, a project is designated as certified, silver, gold, or platinum.

The LEED standards have evolved over the past 15 years, and all proposed criteria are voted on by the members of USGBC. The standards are now very detailed in their assessment of the categories of construction, including new construction, existing buildings, commercial interiors, building cores and shells, and building functions.

In awarding points to the individual projects, the standards are further refined to recognize the following categories, and the points that are awarded are then used to classify the project according to the USGBC award levels: (1) sustainable construction sites, (2) efficiency of water supply, (3) energy and atmosphere, (4) materials and resources, (5) indoor environment quality, and (6) innovation and design process.

Obviously the sustainability issue goes well beyond the individual construction material and how and where it is produced, which is why the LEED APs are required to have a broad background and understanding of all aspects of the built environment.

All of the above subjects are important in the context of sustainability, but for construction materials and structural systems item numbers 3, 4, and 6 are especially applicable. For steel, the current materials and particularly the products that are made through the electric-arc furnace and continuous casting processes do extremely well, as a result of the very large content of recycled materials (for example, scrap) and the nearly 100% recyclability of steel. The waste that is contributed by steel is effectively zero, since essentially all of the material is recycled.

One issue that is difficult to assess is the regional materials question, which is being addressed in the United States today. Specifically, materials that can be provided from a local or near-local plant or supplier will get a higher LEED score. This is because the environmental impact of transporting the goods from the plant to the construction site will be less than if the goods have to be shipped from a far-away location. This is one of the advantages of the largest U.S. steel producer today, the Nucor Corporation, since it has steel mills in many areas of the country, and transportation distances are therefore somewhat limited. But this also makes the assessment procedure much more complex, since each project has to be evaluated on an individual basis, and local versus larger region supply of steel scrap, for example, can be difficult to determine. Currently, the American LEED criteria award regional material points on the basis of the percentage of the total mill scrap that is supplied from within a circle of 500 mi from the project.

Current green building construction in the United States. Based on information from the McGraw-Hill Construction Analysis group's "SmartMarket Trends Report 2008" (February 2008), the value of green building construction starts at that time was larger than $12 billion. The volume was expected to increase to $60 billion by 2010. In the same vein, LEED projects in 2002 accounted for approximately 80 million ft^2 (7.4 million m^2); for 2006 it was 640 million ft^2 (59.5 million m^2).

Design of structures for deconstruction. With careful attention to constructability and deconstructability during the design and planning process, the entire life span of a structure is assessed very efficiently insofar as sustainability is concerned. In this context, it is recognized that 25–30% of the total amount of waste produced each year in the United States can be attributed to material waste from new construction, renovation, and demolition. Of the waste, 92% comes from renovation and demolition and 8% comes from new construction. Much of the waste ends up in landfills, despite the fact that significant amounts could be recycled or at least salvaged, reducing the impact on landfills.

Structural steel in all forms should be recycled, and usually is, but that depends to some extent on the locale where the projects are undertaken. The impacts on society through sustainability are serious, and a concerted education is currently underway. It is led by the USGBC. For steel, various arms of the industry are playing a lead role, such as the design-oriented groups within the industry as well as the parts of organizations such as the American Institute of Steel Construction (AISC) and the American Iron and Steel Institute (AISI) where design standards are developed. Because of the close relationship between constructability and deconstructability, designs satisfying both needs simultaneously are becoming the norm.

Constructability is aided by simple systems and details, and the principles of building in steel are critical. Thus, the analyses focus on prefabrication and the development of practical modules, and especially the types of connections and fastening systems that are used for steel facilitate simpler assembly as well as disassembly. In this respect, the standardization of connections is particularly useful, and such programs have been pursued vigorously by the steel organizations for many years. The design process therefore can be arranged to allow for considerations of reuse, which is one of the primary areas of the sustainability efforts of the U.S. steel industry. The structure should be simple to assemble and build. Doing much of the work off-site will provide higher and more reliable quality, and the structure will be completed sooner and the owner will be able to derive revenue at a much earlier stage. Sustainability is therefore achieved at all levels of the construction process.

Reversing the construction process, joints that were easy to assemble will be easier to take apart; the tools that are used are simple, the energy consumed by the deconstruction process will be much lower, and the waste materials are reduced to very small amounts. None of the waste will be structural steel.

The benefits of a total design thus will include the entire life cycle of the structure, which means that realistic construction economies are used to assess the true construction costs. These are among the most important of the benefits of carefully planned and built steel structures.

Summary. The use of steel has a long and successful history in the United States, starting with the first tall building (1883) and bridge structure (1874) using what is now referred to as structural steel. The industry has gone through a major reshaping process over the past 15–20 years, from the mills with basic oxygen furnaces to today's highly efficient mills with electric-arc furnaces and continuous casting of shapes, plates and sheets. The quality has been significantly improved, assuring users that the structures will perform as intended.

The focus of sustainability has come to the fore over the past 15 years, and the steel industry is enjoying a reputation as the one with the most sustainable construction material of all. Research, design, planning, and construction processes incorporate life-cycle considerations, and waste and related products are minimized. The industry focus continues to be on all aspects of energy, efficiency, economy, and environmental performance. It is enjoying success in all of these efforts.

For background information *see* CONSTRUCTION ENGINEERING; ELECTRIC FURNACE; RECYCLING TECHNOLOGY; STEEL; STEEL MANUFACTURE; STRUCTURAL STEEL in the McGraw-Hill Encyclopedia of Science & Technology.

Dr. Reidar Bjorhovde

Bibliography. American Iron and Steel Institute (AISI), *North American Steel: The Environmentally Preferred Material*, AISI, Washington, DC, 2008; D. Eckmann, T. Harrison, and R. Ekman, Structural steel contributions toward obtaining a LEED rating, *Modern Steel Construction*, AISC, Chicago, February, 2004; M. Pulaski et al., Design for deconstruction, *Modern Steel Construction*, AISC, Chicago, June, 2004.

The search for a supersolid

In the early part of the twentieth century, liquid helium was found to have zero viscosity at temperatures below about 2.2 K. The condition of matter displaying this property was named the superfluid state. This property was later explained as a consequence of the fact that helium-4 (^{4}He) atoms are Bose particles which, at sufficiently low temperatures and sufficiently high density, can display the quantum statistical phenomenon called Bose-Einstein condensation. This occurs when a fraction of the atoms occupy the same zero-kinetic-energy quantum state.

Speculation about a supersolid state. Like all other solids, solid helium is a crystalline state in which atoms settle into a regular three-dimensional (3D) lattice structure. Each atom is localized to a different position in space so that each atom is in a different quantum state. In the case of solid helium, however, the force between the atoms is weaker and their mass is smaller than in any other solid. Consequently, the quantum states of the atoms are spread out over a large fraction of the mean separation between them. For this reason, solid helium is referred to as a "quantum solid." This circumstance led physicists to speculate whether some fraction of the atoms somehow might be completely delocalized so that they could be in identical ground states and form a superfluid. This would be a very peculiar, "supersolid" state of matter in which quantum physics leads to a material that simultaneously exhibits an ordering of the atoms in space, on the crystal lattice, and an ordering into a single quantum state, as in the superfluid.

The first such speculation was based on the assumption that there would be a finite concentration of vacancies (sites in the 3D lattice that are not occupied by an atom) even at the absolute zero of temperature (so-called zero-point vacancies), and that these vacancies would be shared by the entire crystal or "delocalized." Theoretical papers have debated whether zero-point vacancies would be required to yield a supersolid and even whether such a state can exist with or without vacancies. Experimental measurements of the properties of solid helium have been motivated largely by a search for evidence of this state.

Experimental approaches. The experimental approaches have included measurements of macroscopic flow of mass or heat and measurements of microscopic properties to learn what are the elementary excitations of the solid. The flow experiments have searched for direct evidence of a superfluid component, by methods similar to those that revealed superfluidity in the liquid. In ordinary solids, the elementary excitations are the quantized elastic vibrations of the crystal lattice, called phonons, which are manifest on a macroscopic scale as sound waves. Liquids also transmit sound waves and therefore also have phonons as elementary excitations. However, superfluid helium has additional, unique, excitations called rotons, which are related directly to its superfluid properties. Therefore, study of the elementary excitations in the solid can also reveal the presence of an unusual quantum state.

Flow measurements. Attempts to find evidence of macroscopic superfluid mass flow have included measurement of the resistance to the motion of a sphere pulled through the solid, forced flow through small pores, and flow of liquid from one side of a solid sample to the other. Flow has been observed in some experiments, but the results have not demonstrated superflow conclusively. The most dramatic claim to have observed superflow was made for the observation of a decrease in the moment of inertia of solid helium contained in a torsion pendulum as its temperature was decreased below about 0.2 K. The period of a torsion pendulum increases with its moment of inertia. The pendulum in the experiment consisted of a hollow metal cylinder, filled with solid helium and suspended by a torsion rod attached to the center of the cylinder. If part of the solid helium becomes superfluid, it will remain at rest while the container oscillates, so the moment of inertia will decrease and the period of the pendulum will decrease. The magnitude of the effect and the temperature at which it occurs has been found to depend on the state of strain of the crystal. A substantial number of independent measurements with different torsion pendulums have been made, which indicate that the phenomenon is complicated, and it is not yet certain whether it represents a transition to a supersolid state.

Measurements of microscopic properties. Measurements that have revealed microscopic properties include nuclear magnetic resonance (NMR) on helium-3 (^{3}He) impurities, x-ray and neutron scattering, the velocity and attenuation of acoustic waves, and the propagation of heat pulses. Nuclear magnetic resonance measurements of the diffusion of small concentrations of ^{3}He atoms through solid ^{4}He provided an early look at flow on the atomic scale. It was attributed to the diffusion of the ^{3}He atoms by exchanging positions with vacancies, in this case, thermally activated vacancies rather than zero-point vacancies. At temperatures above about 1 K, the diffusion increases as the concentration of these vacancies increases. The energy required to create them was found to be of order 10 K, but it decreases with decreasing melting temperature. At lower temperatures the diffusion is independent of temperature, so the ^{3}He atoms move by the quantum tunneling process. However, it was not possible to determine from these measurements, or any others thus far, whether there are zero-point vacancies. At present there is no known relation between these excitations and possible supersolid behavior, but some of the expected consequences of their presence (for example, specific heat) have not been observed. Thus their properties are not completely known.

X-ray measurements of the lattice parameter found the lattice parameter decreasing as an exponential function of temperature, and that decrease was interpreted as arising from the decreasing concentration of vacancies with decreasing temperature. This work also found that the energy required to create a vacancy was of order 10 K.

Measurements of the velocity and attenuation of acoustic waves probe the microscopic processes that scatter phonons. Early acoustic measurements above about 1 K found that the temperature dependences were due to the presence of dislocations. Dislocations are defects in the crystal lattice where there are extra rows of atoms. A possible explanation of the torsion oscillator experiments in terms of dislocations has been proposed by the author of some of that work. Later measurements in more perfect crystals found the attenuation decreasing exponentially with decreasing temperature so that some form of excitation with minimum energy near 1 K was observed. It has not yet been identified.

Measurements of sound propagation in the solid with a few parts per million of ^{3}He impurities revealed a large attenuation peak accompanied by an increase of the velocity as the temperature was lowered below 200 mK. The anomaly was also observed in pure solids that were highly strained. A possible relationship between these observations and the subsequent measurements of changes in the moment of inertia remains to be explored.

Heat pulses can travel as waves in superfluid liquid helium and in solid helium over a narrow range of temperatures. The pulses consist of a cloud of excitations created by the heat. Depending on the temperature range, either rotons or phonons dominate the process in the liquid. Observation of heat-pulse propagation in the solid that is not carried by phonons would be evidence for the presence of a different type of excitation, analogous to rotons in the liquid. At temperatures below about 200 mK, the excitations in both liquid and solid propagate without collision at the velocity of the individual excitations. For phonons, this is the velocity of the related sound waves. However, in solid helium, at these low temperatures, heat pulses travel at a velocity that is independent of direction in the crystal and at a velocity different from the dominant phonon. The excitation carrying the heat remains to be identified.

Inelastic neutron scattering provides a direct measurement of the elementary excitations in condensed matter. The earliest measurements were all interpreted in terms of ordinary phonons, modified by the fact that solid helium crystals depart much more than other solids from the harmonic approximation. That is, the force restoring an atom to its equilibrium position is not proportional to its displacement from that position as in other solids. The only peculiarity was that the measurements in low-density crystals did not find one of the phonon modes that should be present (the transverse optic mode). At that time, no evidence was found for excitations other than the anticipated phonons.

Neutron scattering also has been used to measure the occupation of the ground state in superfluid liquid helium. Recent neutron scattering measurements attempted, but failed, to detect a finite occupation of the ground state in solid helium. Current inelastic scattering measurements are finding more evidence for excitations that are not ordinary phonons, but the results have not yet been reported in the literature.

Outlook. The experimental and theoretical study of the quantum solid ^{4}He has revealed a variety of properties that are not found in "classical" solids. As yet there is no unifying understanding of these properties and no unambiguous evidence that superfluidity coexists with the spatial order of the crystal. However, the investigation is far from complete and may yet reveal a substance as peculiar, interesting, and useful as superfluid liquid helium.

For background information, *see* BOSE-EINSTEIN CONDENSATION; CRYSTAL DEFECTS; LIQUID HELIUM; LOW-TEMPERATURE ACOUSTICS; MOMENT OF INERTIA; NUCLEAR MAGNETIC RESONANCE (NMR); QUANTUM SOLIDS; SLOW NEUTRON SPECTROSCOPY; SUPERFLUIDITY; X-RAY DIFFRACTION in the McGraw-Hill Encyclopedia of Science & Technology. John M. Goodkind

Bibliography. Proceedings of the 2009 International Conference on Quantum Fluids and Solids (QFS2009), *J. Low Temp. Phys.*, 158(1-4):1-735, 2010; J. Wilks, *The Properties of Liquid and Solid Helium*, Clarendon Press, Oxford, U.K., 1967.

Three-dimensional geographic information systems

Geographic information systems (GIS) are software tools for storing, retrieving, mapping, and analyzing geographic data. Because paper maps depict geographic phenomena in two dimensions (2D), GIS were originally designed to handle 2D data and present visual images in 2D. The digital maps created and visualized with early GIS are thus basically similar to conventional paper maps. For several reasons, there has been dramatic development in three-dimensional (3D) GIS in the past 15 years or so.

First, the world that we see around us is 3D. To represent what we see more realistically, presenting geographic data in 3D is closer to what we experience. It allows us to more quickly recognize geographic features or relations because our brain no longer needs to build a conceptual model from the symbols or colors on a 2D map. Second, representing certain geographic objects or relations in 2D has many limitations. For example, industrial contaminant released into coastal water may vary in thickness and depth at different locations. Buildings and trees have height, which affects the visual appearance and quality of a proposed urban development project. It is not possible to represent land use under bridges or inside multistory buildings on 2D maps. It is also difficult to represent the movement of objects over time using 2D maps. Third, data needed for 3D GIS is complex and considerable computing power is required to render 3D scenes. But there has been dramatic improvement in both data availability and processing power in the past 15 years. Large amounts of 3D geographic data can now be collected quickly using new technologies such as light detection and ranging (lidar). Fourth, many more software packages are now available for performing various kinds of 3D GIS functions, such as creating realistic 3D representations of undersea contaminant plumes or landscape features like buildings and trees.

Representing geographic features in 3D. To represent geographic features in 3D, a simple method is to attach a z value to the geographic coordinates (x, y) of a feature and plot this variable as height in the vertical dimension (in a sense, protruding geographic features vertically). Using this method, a point becomes a vertical line, a line becomes a vertical wall, and a polygon becomes a 3D volume. In this manner, "flat" features are protruded vertically and given a 3D appearance. In the case where a flat 2D space is treated as a continuous field fully covered by grid cells of constant size, a z value is attached to each cell to represent certain attributes (such as elevation or population density) at different locations. An example of this method is the 3D representation of terrain using digital elevation models (DEM), where the z attribute records the elevation of a geographic location (x, y). However, since geographic data, like elevation, are often irregularly distributed over space, a more efficient method called a triangulated irregular network (TIN) was developed to represent 3D surfaces. It uses triangles of variable size and density to take into account the variable density of data points instead of using a regular lattice of grid cells. But when a 3D surface is represented by these methods, it is not a true 3D representation because a z variable can take up only one value at each geographic coordinate. This kind of method is therefore known as a 2.5D representation because it cannot represent the complex shape and volume of 3D features, such as an undersea contaminant plume or underground rock layers.

There are different methods for representing true 3D features with GIS. In grid-based or volume-based representations in which a 3D space is fully covered by volumetric 3D grid cells of constant size, 3D features are represented by voxels (volumetric pixels). A voxel is the 3D equivalent of a 2D pixel (which is like an image pixel in a digital photo that can hold several variables such as color, hue, or intensity). A voxel is a volumetric element, such as a cube, which is one of a large number of orderly arranged voxels used to represent geographic features in a 3D space. Voxels, however, do not normally have their geographic coordinates explicitly encoded along with the attributes they hold. Instead, their position is inferred based on its position relative to other voxels in the 3D space. This means that an analysis based on topological relations, such as connectivity among the voxels, can be slow. Further, a high-resolution representation requires a very large number of voxels, and this in turn requires a large amount of computer memory and power to process.

These difficulties are mitigated when using vector-based representations, where geographic features

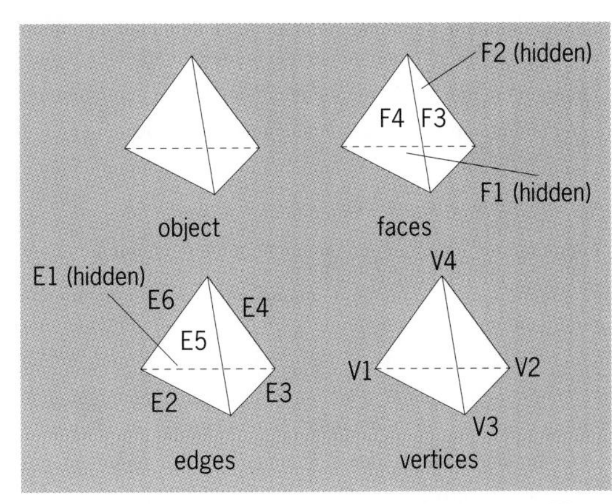

Fig. 1. Boundary representation of a triangular pyramid.

are represented as discrete 3D objects. In one method called boundary representation (or B-rep), a 3D object is represented as a collection of connected surfaces (or faces), edges, and vertices. For example, a simple triangular pyramid can be represented with four faces (F1, F2, F3, F4), six edges (E1, E2, E3, E4, E5, E6), and four vertices (V1, V2, V3, V4) [**Fig. 1**]. In this representation, an object is bounded by faces, a face is bounded by edges, and an edge is bounded by vertices. Using this method, the location and shape of a 3D feature can be explicitly represented by the coordinates of its faces, edges, and vertices. The connectivity among these geometric elements (know as topological relations) can also be recorded. In the case of the triangular pyramid, Vertex 1 is connected with Edges E1, E2, and E6, and Edge 5 is the boundary between Face 4 and Face 3 (Fig. 1). Boundary representation is a very efficient method for representing simple 3D features with empty space inside (as in the case of a house). Another vector-based representation is called the object-oriented model, which assembles the geometric elements into distinct and recognizable 3D features (such as a building or a highway) instead of treating these objects as a collection of geometric elements.

Another way to represent a geographic feature in 3D is called constructive solid geometry (CSG). This method represents a 3D feature as a combination of primitive solids, which includes cubes, cones, cylinders, and spheres. With this method, 3D objects are created by overlapping 3D objects using operations like union, intersection, and difference. This method is often used in computer-aided design (CAD) to represent features in 3D.

Use of 3D GIS. 3D GIS is used in many applications. Geographic visualization (or geovisualization) is one common use. This is the process of creating and viewing geographic images of data with the aim of discovering new knowledge of geographic patterns or relations. Because humans tend to learn more effectively with visual images than with textual information and numerical means, geovisualization is particularly suitable for exploring and analyzing large and complex data sets. 3D geovisualization can be used to analyze human movement (**Fig. 2**). In this visualization, a person's movement in time and space is represented by a 3D space-time path. This path is a continuous trajectory indicating the location of a person who moves around to visit different places within a 24-hour period. The vertical axis represents the temporal progression of such movement (from the bottom to the top), while the horizontal plane represents the extent of the geographic area.

Virtual geographic environments (VGEs) are also important applications of 3D GIS. They are used in different contexts such as urban planning, landscape evaluation, disaster response, and reconstruction of historical sites. A VGE presents a representation of the real geographic environment that looks similar to the real environment (**Fig. 3**). The representation may include photorealistic still images, animated sequences, and real-time interactive environments where users can change their viewing positions, manipulate the 3D scene, and modify elements of the VGE. It can be created with various means, including the desktop computer, the Web, and immersive virtual-reality systems. Its use may involve one or many users interacting with others, who may be remotely connected to such environments via the Web and represented as avatars. Some VGEs may be created as virtual worlds in which users can interact with simulated actors, while some may have a very high degree of realism.

One popular use of VGE is to create virtual cities or virtual landscapes that allow the visualization of proposed urban development projects. Using this application, a digital 3D representation of a city or landscape is constructed using various types of geographic data, including 2D digital maps, aerial photographs, remotely sensed images, and lidar data. Buildings or houses are often the main elements of a virtual city. When the number of buildings involved

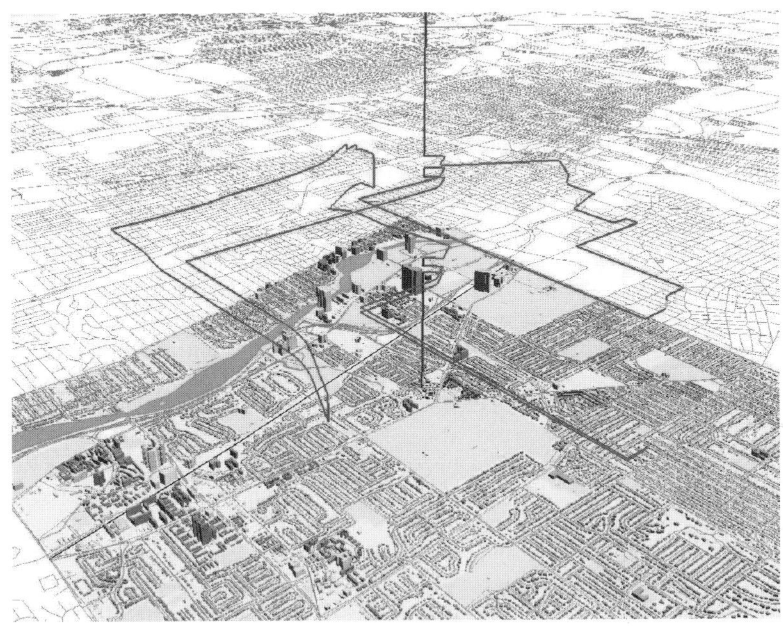

Fig. 2. Geovisualization of human movement using 3D GIS.

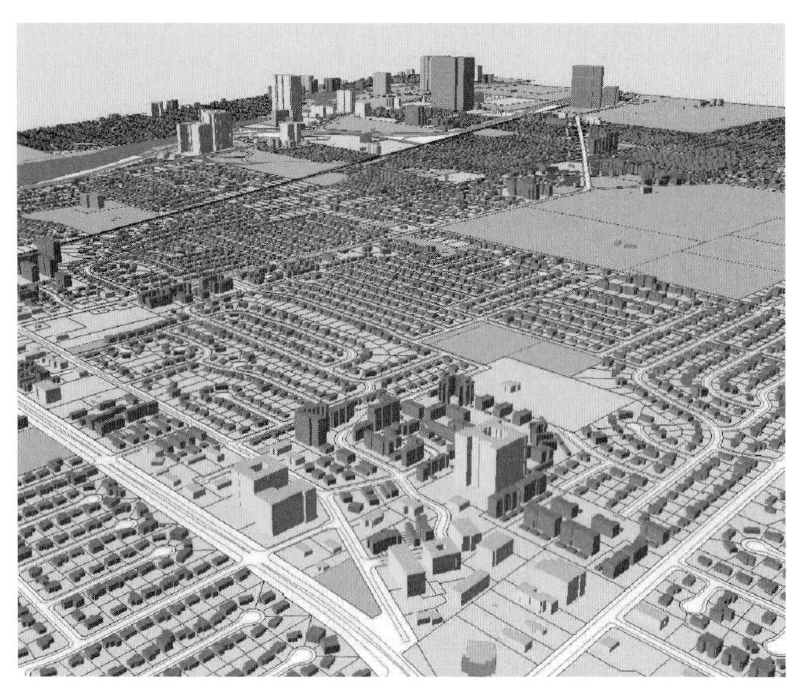

Fig. 3. An example of a virtual geographic environment.

is small, they can be constructed using 2D building footprints and wire frames. The walls and roof of a house can be textured with appropriate digital images to enhance realism. But when the number of buildings involved is large, automatic generation methods are used. The following steps can be implemented to construct a building automatically in a VGE: (1) retrieve the geographic coordinates and attributes of a building from the GIS database; (2) select the appropriate pre-constructed building model; (3) resize the building model to match the width and depth of the building lot; and (4) rotate and place the model in the correct location in the VGE. These steps are repeated until the 3D representations of all buildings in the virtual city are constructed. After a virtual city is created, it can be used to visualize how the urban landscape will change after the completion of a development project. When it is used as a participatory decision support tool it can encourage stakeholder participation and help achieve a consensus on design. It is also useful for reconstructing and visualizing historical landscapes. The virtual landscapes in some videogames, such as Microsoft's *Flight Simulator*, are also VGEs because they are constructed with real geographic data and are representations of real landscapes.

For background information *see* LIDAR; GEOGRAPHIC INFORMATION SYSTEMS; COORDINATE SYSTEMS; CARTOGRAPHY; GEOGRAPHY; REMOTE SENSING; AERIAL PHOTOGRAPHY in the McGraw-Hill Encyclopedia of Science & Technology. Mei-Po Kwan

Bibliography. M.-P. Kwan, Interactive geovisualization of activity-travel patterns using three-dimensional geographical information systems: A methodological exploration with a large data set, *Transportation Research C*, 8:185-203, 2000; J. Lee and S. Zlatanova (eds.), *3D Geo-Information Systems*, Springer 2009; H. Lin and M. Batty (eds.), *Virtual Geographic Environments*, Science Press, 2009; J. Raper, *Multidimensional Geographic Information Science*, Taylor & Francis, 2000.

Transportation's role in sustainability

Transportation plays a significant role in global energy consumption and the emission of greenhouse gases. Transportation professionals provide an effective resource for reducing transportation's impact on this energy use and resulting emissions.

It is well documented that traffic is responsible for a significant proportion of greenhouse-gas emissions and fossil-fuel consumption. Transportation engineers need to work together to ensure that greenhouse-gas emissions are reduced to a sustainable level, while meeting the traffic and transportation needs of the present generation and their economies. Transportation funding sources also need to be created to provide the resources needed to increase traffic engineering programs related to traffic operations. In particular, the U.S. government has funded priority discussions for a new transportation authorization act. Improving traffic operation efficiency and reducing congestion decreases vehicle emissions. These efforts should be carried out in conjunction with promotion of alternative energy sources and fuel economy programs. In the United States, transportation is responsible for 27% of all greenhouse-gas emissions (see **illustration**).

Many scientists agree that greenhouse-gas emissions must be reduced by 60–80% or more by 2050 to avoid serious environmental consequences. As a result, energy supply and climate change will affect all modes of transportation in every region of the world. We can also conclude that transportation infrastructure and funding challenges will require new solutions in the worldwide economy. Transportation engineers need to be ready to address these important issues.

The Institute of Transportation Engineers (ITE) has created a task force to define transportation sustainability and to identify programs that support the operation and management of transportation systems. Sustainability is not a new topic and has been defined in numerous ways throughout history. Early on, native Australian aborigines defined sustainability in their culture as taking care of the

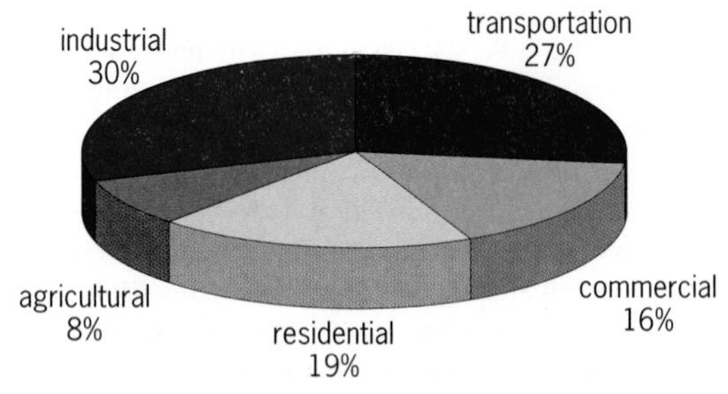

U.S. greenhouse-gas contributors. (*Source: U.S. EPA, 2002*)

Earth's resources for the next ten generations. Native American tribes had a similar definition, that they be caretakers of the Earth's resources for the next seven generations. A simple definition of transportation sustainability is the ability to meet the needs of the present generation to provide for the movement of people and goods from one location to another without compromising the ability of future generations to meet their own needs.

Goals for transportation sustainability include reducing both vehicle emissions and vehicle miles traveled. Intelligent transportation systems (ITS) and traffic engineering have the ability to improve transportation efficiency, safety, and mobility through the application of advanced information and communications technologies.

Reducing vehicle emissions improves traffic efficiency and reduces congestion. It also promotes alternative energy sources and fuel economy programs. Equally important is avoiding the use alternative fuels that can result in harmful emissions into the atmosphere.

Reducing vehicle miles traveled will require improving transit service and expanding its infrastructure as well as providing a well-connected network of streets and investing in "complete streets" for bicycles, pedestrians, and motor vehicles.

ITS and traffic engineering applications supporting sustainability. The application of intelligent transportation technologies as well as basic traffic engineering practices have been documented through comprehensive studies to reduce congestion and improve transportation safety.

ITS and advanced technologies already have a major impact on reducing transportation's greenhouse-gas emissions and fuel consumption. It is important to increase understanding by elected officials and their constituents so that they support the application, expansion, and funding of these programs. The following is a summary of numerous ITS applications and their evaluated effects, including incident management programs, traveler information systems, open-road tolling, traffic signal systems, and integrated corridor management.

Incident management. Incident management systems detect traffic roadway data and incidents using sensors, such as closed-circuit television. When a traffic incident is detected, emergency responders are dispatched and traveler information is broadcast. The Georgia NaviGAtor Incident Management Program has reduced annual fuel consumption by 6.83 million gallons, and contributed to decreasing carbon monoxide (CO) emissions by 2457 tons, hydrocarbons by 186 tons, and nitrous oxides (NO) by 262 tons. The evaluation also indicates that average traffic incident delay times were reduced by an average of 46 min in duration. In comparison, the Florida Road Ranger program saved 1.7 million gallons of fuel per month, and Maryland's CHART (Coordinated Highways Action Response Team) program saved 6.4 million gallons of fuel in 2005 and reduced NO emissions by 233 tons. Evaluation of the Maricopa County, Arizona, REACT (Regional Emergency Action Coordinating Team) program indicates that it reduced fuel consumption by 6% and vehicle emissions by 11%.

Traveler information. Traveler information systems, which provide traffic and road conditions and emergency advisories to travelers through dynamic message signs, highway radio, phone messaging, and Internet websites, were reported to reduce CO emissions by 1.5% in Seattle, Washington, and by 2.5% in Oakland County, Michigan. In Cincinnati, Ohio/Kentucky, vehicle emissions were reduced by 3.7–4.6%.

Traffic signals. The *ITE 2007 National Signal Report Card, Technical Report*, showed that poor traffic signal timing accounts for 5–10% of all traffic delays. There are about 300,000 traffic signals in the United States, which, overall, were graded to be operating at level "D" conditions—that is, inefficiently. The vast majority of traffic signal systems across the United States have the potential for greatly improved performance. Signal-system-management improvements, as discussed in the 2007 *National Traffic Signal Report Card, Technical Report*, can help improve traffic signal operations and ultimately reduce delays to travelers. Improved traffic signal operations also can help minimize air pollution by making sure that vehicles are not starting and stopping wastefully and using more fuel than necessary. With increased emphasis on management of these systems, it is expected that we could reduce national fuel consumption by 10% and related vehicle emissions by 22%.

Advanced signal systems in Syracuse, New York, were reported to reduce fuel consumption by 7–14% and related vehicle emissions by 9–13%. In Toronto, Canada, they were reported to reduce fuel consumption by 4–7% while reducing vehicle emissions by 3–6%, and in Los Angeles, California, they reduced fuel consumption by 13% and vehicle emissions by 14%.

Open-road tolling. Open-road electronic toll collection studies have shown that in Baltimore, Maryland, vehicle emissions were reduced by 16–63%; on the New Jersey Turnpike, volatile organic compound (VOC) emissions were reduced by 80%; and the Colorado PrePass® weigh station electronic screening system reduced fuel consumption by 48,200 gallons/month.

In Tucson, Arizona, full ITS deployment was calculated to reduce CO emissions by 10%, NO by 16%, hydrocarbons by 12%, and fuel consumption by 11%.

Traffic engineering applications. Another area that can potentially reduce fuel consumption and vehicle emissions and also improve transportation safety is the construction of modern roundabouts as an option for traffic signal controls. Modern roundabouts improve intersection efficiency by eliminating stopping and queuing patterns. Numerous studies have shown that modern roundabouts reduce greenhouse-gas emissions by 17–65% and fuel consumption by 28%, while also reducing vehicle-related fatalities by 90%, injuries by 75%, and all crashes by 35%. Many agencies, such as the Wisconsin Department of Transportation (DOT), are now requiring that an intersection control evaluation be conducted whenever traffic signals are considered for installation at an intersection. Typically, modern roundabouts have proven to be a more effective control solution based on eight evaluation criteria: (1) operation, (2) safety, (3) pedestrians and bicycles, (4) environmental impacts, (5) right-of-way requirements, (6) operating and maintenance costs, (7) construction costs, and (8) practical feasibility.

The use of light-emitting diode (LED) traffic signal heads has been shown in Portland, Oregon, to save 4.9 kilowatt-hours (kWh) per year. In the United States, the national potential savings is 3 billion kWh. Similarly, in Ann Arbor, Michigan, LED street lighting has produced a 50% energy saving. In Salisbury, South Africa, 53 solar-panel–equipped street lights were reported to reduce carbon dioxide (CO_2) energy-generated emissions by 7 tons.

Other traffic engineering initiatives that have been proven to reduce vehicle emissions and fuel consumption include enhancements for pedestrian and bicycling facilities and the design of complete streets with connected systems. According to a 2004 U.S. Environmental Protection Agency study, increased street network connectivity and pedestrian-friendly design can reduce per-capita vehicle miles traveled and vehicle emissions.

Outlook. Transportation engineers can have a positive impact on reducing greenhouse-gas emissions and fuel consumption through ITS technologies and traffic engineering. Transportation engineers are already having a significant impact on reducing greenhouse gases and fuel consumption, as discussed earlier (see **table**). They should take credit for these efforts and promote increased transportation sustainability actions. They should also increase integration of ITS technologies, design systems for all users, and provide modal choices. In addition, they need to work with land-use planners to create street systems that promote connectivity and reduce urban sprawl. The Institute of Transportation Engineers is developing a certification program for transportation operations that is similar to the U.S. Green Building Council's Leadership in Environmental Excellence in Design (LEED) program. Not only will this program recognize those programs and agencies that are actively managing transportation systems to reduce emissions and fuel consumption, it will also serve to promote the profession's role in environmental stewardship. ITE is well placed, with members in 90 different countries and members of all age groups who are contributing their knowledge and expertise toward meeting the challenges that are before us.

For background information *see* AIR POLLUTION; ENERGY SOURCES; HIGHWAY ENGINEERING; TRAFFIC-CONTROL SYSTEMS; TRANSPORTATION ENGINEERING in the McGraw-Hill Encyclopedia of Science & Technology.

Ken Voigt; Beverly McCombs

Bibliography. *2007 National Traffic Signal Report Card, Technical Report*, Institute of Transportation Engineers, 2007; *Intelligent Transportation Systems Benefits: 2001 Update*, U.S. Department of Transportation, Research and Innovative Technology Administration, 2001; *Intelligent Transportation Systems Benefits, Costs, and Lessons Learned: 2005 Update*, U.S. Department of Transportation, Research and Innovative Technology Administration, 2005; *Regional Emergency Action Coordination Team (REACT) Evaluation*, Maricopa County Department of Transportation, Arizona, 2002.

Summary of transportation engineering operations fuel and vehicle emission reduction benefits

	Annual reduction	
Strategy	Fuel gal	CO tons
* **Traffic signal timing**		
ITE retiming study	10%	22%
Syracuse, NY	7–14%	9–13%
Toronto, Ca	4–7%	3–6%
Los Angles, CA	13%	14%
* **Modern roundabouts**	28%	17–65%
* **Incident management programs**		
Florida road ranger	1.7 million gal.	–
Maryland chart	6.4 million gal.	–
Georgia navigator	6.8 million gal.	2547 tons
AZTECH REACT	6%	11%
* **Traveler information systems**		
Seattle, WA	–	1.5%
Oakland County, MI	–	2.5%
Kentucky/Cincinnati, OH	–	3.7–4.6%
* **Open road tolling**		
Baltimore, MD	–	16–36%
* **Full ITE deployment**		
Tucson, AZ	11%	10%

Tropical glacier monitoring

Mountain glaciers are good indicators of global climate change. Glaciers are masses of ice found on land that are thick enough to flow under their own weight or that show evidence of past movement. Documenting the retreat of tropical glaciers is difficult because they are often located in remote areas with steep terrain and inclement weather. However, today's computerized mapping technologies, including remote sensing and geographic information systems (GIS), can complement field studies in tracking changes of these remote glaciers. In fact, the Intergovernmental Panel on Climate Change (IPCC) has recommended the use of satellite remote sensing to document glacier change.

When most people think of glaciers, they think of the Arctic and Antarctic ice sheets as well as mountain glaciers in the mid-latitudes such as the Alps and the Rockies. However, thousands of small glaciers are found in the tropics in the South American Andes, on Mt. Kenya, Mt. Kilimanjaro, and in the Rwenzori Mountains of East Africa and on Papua (formerly Irian Jaya), Indonesia (**Fig. 1**).

From a glaciological perspective, tropical glaciers exist in three distinct climate regimes shown in Fig. 1: the humid inner tropics, which have abundant precipitation throughout the year; the outer tropics, which have distinct wet and dry seasons; and the semiarid to arid subtropics. In each of these regimes, different climatic factors control glacier growth and retreat. Although they fall geographically outside the glaciologically delimited tropics, a few small glaciers in Mexico and in the northern Chilean Andes face similar challenges with regard to mapping using remote sensing. Glaciers exist throughout the tropics, but most of the world's tropical glaciers—and most of its glacier ice—are found in the South American Andes of Peru and Bolivia, with much smaller numbers and areas of glaciers found elsewhere as illustrated by the varying size of the proportional circles in Fig. 1.

Mass balance of glaciers. Glaciers are mass-balance systems. At any time, they either gain (accumulate) or lose (ablate) mass over their surface. Any net gain or loss of mass causes the glacier to change its size and shape. In most regions of the world, glaciers gain mass through snowfall, avalanching, wind deposition, and other processes during winter and lose mass through melting and sublimation (the direct change of ice to water vapor) in the summer. This situation is more complicated for tropical glaciers, where melting can occur at any time during the year and the wet accumulation season can occur at the warmest time of year.

If the amounts of accumulation and ablation that occur on a glacier during a year exactly balance each other, the glacier is in equilibrium. If there is net accumulation, the glacier will thicken and glacier flow will transfer the additional mass to the glacier's snout, causing it to advance; if, on the other hand, the glacier consistently loses mass each year, it will retreat.

In the field, the mass balance of a glacier is typically determined by measuring the mass gained or lost at a series of locations on the glacier. In the accumulation zones, which are typically found at the highest elevations on a glacier, the depth of the snowpack at the end of the melt season (typically the summer) is measured by probing or by digging snow pits. The density of the snow is also measured to convert the depths into an equivalent amount of water added to the glacier. In the ablation areas on a glacier, stakes are inserted vertically so that the amount of ice lost can be determined by measuring the height of the ice melted away at each stake. These mass-balance measurements are important because they enable scientists to understand the climatic factors that are causing tropical glaciers to retreat. Unfortunately, because of the work involved, the mass balance of only a few glaciers in the tropics is measured each year. Scientists contribute mass-balance information for glaciers worldwide to the World Glacier Monitoring Service (WGMS), which compiles the information and makes it available to the public and to the National Snow and Ice Data Center (NSIDC), which produces and distributes information about all aspects of the Earth's cryosphere to scientists and the public.

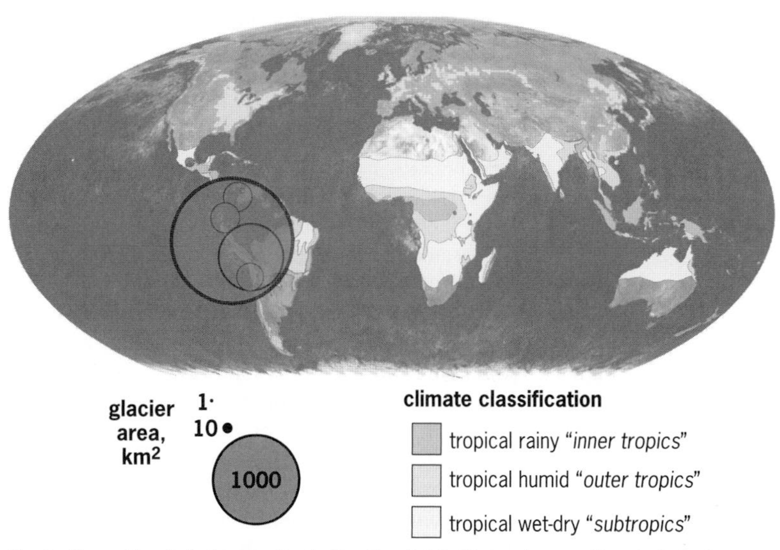

Fig. 1. Proportional circle map illustrating the distribution and area of tropical glaciers (*Glacier data from the World Glacier Inventory and kindly provided by Graham Cogley, Trent University*)

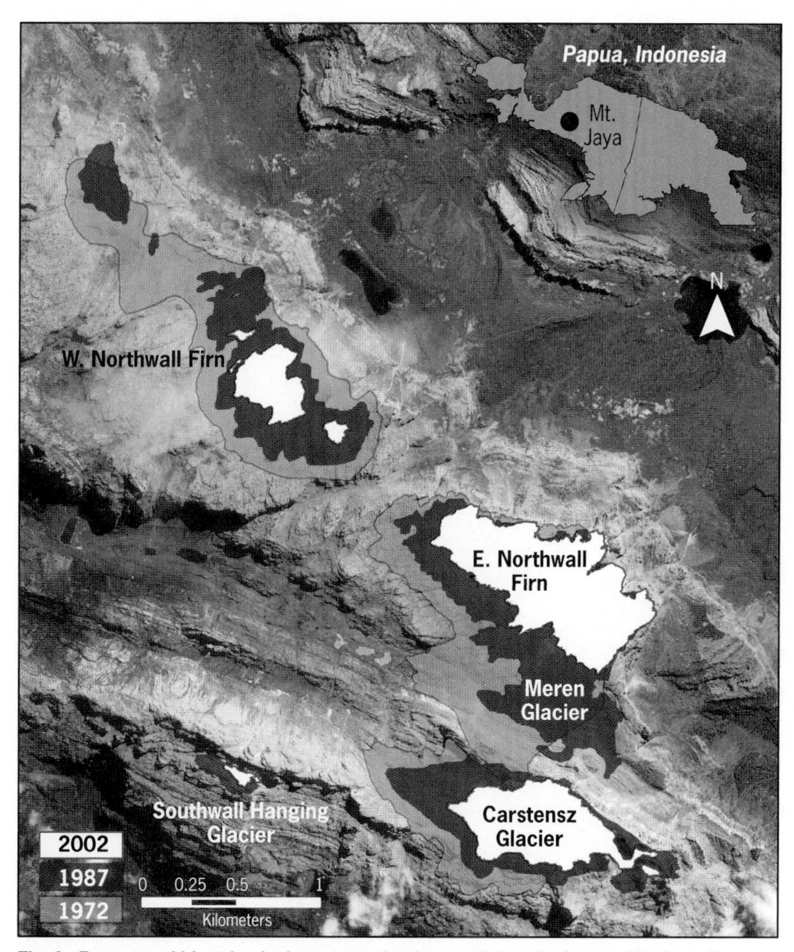

Fig. 2. Recent and historic glacier extents for the remaining glaciers on Mt. Jaya, Papua, Indonesia.

Remote sensing and GIS. Remote sensing and GIS technologies offer a complementary approach for examining the recent retreat of glaciers in the tropics and worldwide as well. GIS can be used to compute the areas of glaciers and to map changes between time periods. Compared to the large glaciers and ice sheets of the high latitudes, the small size of tropical glaciers and the steep terrain in which they are located can be impediments to mapping their retreat accurately. Cloud cover is also a problem, as it can be difficult to capture an image of some tropical glaciers from an aircraft or satellite, because cloud cover often obscures the view.

The easiest glacier property to measure, from either aerial photographs or satellite images, is the glacier area, which is also termed the glacier extent. Although glacier extent changes in response to mass-balance changes and hence climate, changes in glacier extent are also influenced by internal factors, including topography and glacier flow. Depending on the size of the glacier, it can take many years for mass-balance changes to be reflected in changes in a glacier's extent. Fortunately, the small size of tropical glaciers means that their area changes quite quickly in response to mass-balance changes. An exciting new remote sensing technology, lidar (light detection and ranging) can be used to measure changes in the elevation of a glacier's surface and thus measure changes in a glacier's volume directly. lidar has been used on glaciers in the Cordillera Blanca of Peru.

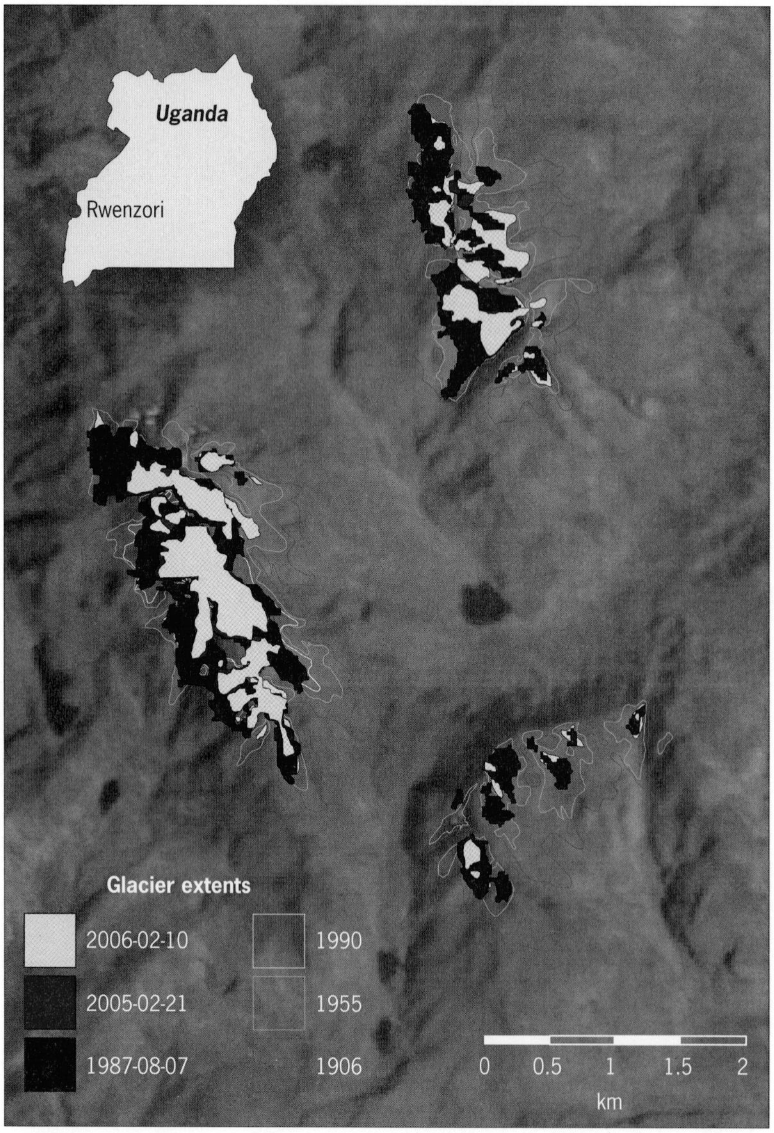

Monitoring changes in glacier extent from satellite images and aerial photographs is a complex process. First, the images must be georegistered, which is a term for making sure that every pixel in the image is in its correct geographic location. This enables glacier extents from different years to be compared. However, it can be difficult to align remote sensing images with glacier extents mapped in the field or taken from historical maps. This can be seen in the misalignment of the Mt. Jaya glacier maps in **Fig. 2**.

Mapping glacier extents. Following georegistration, the extent of the glaciers must be determined. This can be accomplished using two very different approaches. The simplest and often the most accurate approach is manual digitization, in which an analyst visually traces the extent of the glacier from a remotely sensed image on a computer monitor using a GIS. This is a time-intensive approach and requires a knowledgeable analyst to be successful. The extents of the Mt. Jaya glacier shown in Fig. 2 in 2002 were mapped quite accurately using this approach. Depending on the data source, an automated classification can be better than manual digitization, and it is often faster and less subjective in determining glacier area. The types of imagery vary by period and type; for example, it can be difficult to obtain aerial photography for some areas.

The alternative is to use digital image processing techniques to perform a computer classification to determine which pixels in an image represent a glacier and which do not. Different types of computer classification can be performed, but all take advantage of the fact that snow and glacier ice have quite different spectral signatures than other surface features. In the visible wavelengths (0.4–0.7 micrometers) to which the human eye is sensitive, snow and ice are typically much brighter than other materials. In the mid-infrared wavelengths (1.5–2.5 μm), which are outside the range of human vision, snow and ice absorb most electromagnetic energy and appear quite dark in images collected at these wavelengths. This combination is unique among land covers and can make computer classifications of glaciers quite accurate. Examples of using computers to map tropical glacier extents are illustrated in **Fig. 3** for glaciers in the Rwenzori in Uganda.

Mapping glacier extents by either method is hampered when snow occurs not only on the glacier but also on the surrounding terrain. For tropical glaciers, this problem can be minimized by using images collected during the dry seasons of the year. Glacier

Fig. 3. Glacier extents for the Rwenzori Range, Uganda, from satellite and historical mapping.

surfaces may also be covered with a thick layer of rock debris, which makes it difficult, if not impossible, to distinguish accurately between the glacier and surrounding materials in remotely sensed images. Both of these factors must be considered carefully when monitoring glacier extents using remote sensing.

Satellite remote sensing. The historical archive of satellite images and aerial photographs that can be used to map tropical glaciers typically dates back to the 1960s and 1970s. The modern Earth satellite remote sensing era began on July 23, 1972, with the launch of NASA's *Landsat 1* satellite. Although the approximate 80-m resolution of its multispectral scanner (MSS) scanner is not ideal for mapping tropical glaciers, it and more recently declassified satellite images from the U.S. *Corona* satellite program can provide useful information on historical tropical glacier areas from the 1970s and sometimes the 1960s. The earliest aerial photography that is useful for mapping tropical glaciers also dates back to this period.

The launch of NASA's *Landsat 4* and *5* satellites with its 28.5-m resolution Thematic Mapper sensor and the 20-m-resolution SPOT satellites developed at the French Centre National d'Estudes Spatiales (CNES) both provide a better source for information on tropical glacier extents beginning in the mid-1980s. Newer satellites from these series continue to provide useful information today. The ASTER (Advanced Spaceborne Thermal Emission and Reflection Radiometer) instrument on NASA's *Terra* satellite is another important image source for monitoring tropical glacier retreat. Beginning with the launch of Space Imaging's *IKONOS* satellite in 1999, there are now a number of satellites with spatial resolutions of 1 m or greater that match the quality of aerial photography for mapping tropical glacier extent. Aerial photography continues to provide an important source of tropical glacier extents, as is evidenced by recent documentation of glacier shrinkage on Mt. Kilimanjaro.

The Global Land Ice Monitoring from Space Program (GLIMS) program was established to ensure that satellite images suitable for studying glacier change are acquired and to facilitate mapping and analysis of glacier change by researchers around the globe. GLIMS is compiling glacier mapping from many regions of the world into a single database at the U.S. National Snow and Ice Data Center.

Outlook. Since the late 1800s, the world's tropical glaciers have been in a state of disequilibrium. Both remote sensing and field studies have documented their continued retreat. **Figure 4** illustrates the ice loss from locales scattered across the tropics. One of the best-studied tropical glaciers, Chacaltaya Glacier in the Cordillera Real of Bolivia ,which was featured in *National Geographic*, disappeared in 2009. If current retreat rates continue, within a few decades tropical glaciers will probably disappear from Papua, Indonesia, and from the African peaks as well.

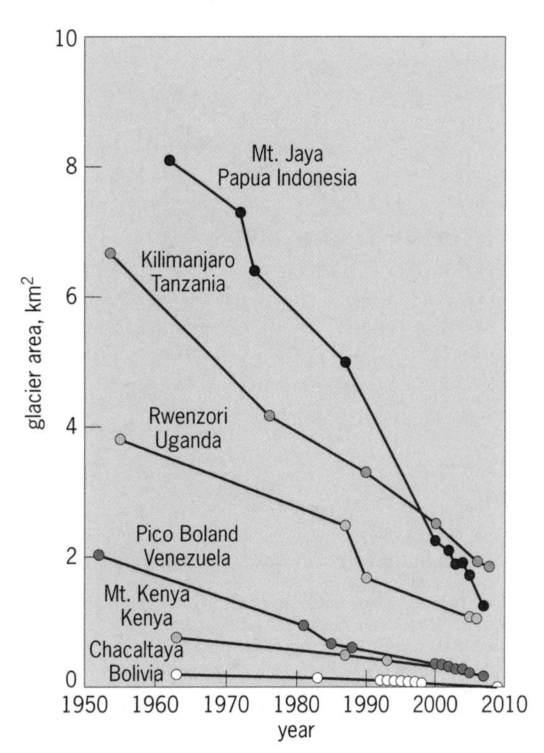

Fig. 4. Changes in the total area of glaciers on selected mountain ranges and for the Chacaltaya Glacier in Bolivia.

For background information *see* LIDAR; GEOGRAPHIC INFORMATION SYSTEMS; GLACIAL GEOLOGY AND LANDFORMS; GLACIAL HISTORY; GLACIOLOGY; GLOBAL CLIMATE CHANGE; HILL AND MOUNTAIN TERRAIN; IMAGE PROCESSING; REMOTE SENSING; SOUTH AMERICA in the McGraw-Hill Encyclopedia of Science & Technology. Andrew Klein

Bibliography. T. Appenzeller, The big thaw, *Natl. Geogr.*, 211(6):56–71, 2007; R. J. Braithwaite, Glacier mass balance: The first 50 years of international monitoring, *Prog. Phys. Geogr.*, 26(1):76–95, 2002; G. Kaser and H. Osmaston, *Tropical Glaciers*, International Hydrology Series, Cambridge University Press, Cambridge, U.K., 2002; L. G. Thompson et al., Glacier loss on Kilimanjaro continues unabated, *Proc. Natl. Acad. Sci. USA*, 106(47):19770–19775, 2009.

Turtle origins

Turtles [order Testudinata (Chelonia)] are characterized by a highly specialized body plan, which renders the analysis of their evolutionary relationships with other reptiles difficult. The most salient feature of the turtle body plan is the shell, composed of a carapace that covers the back of the trunk, and a ventral plastron that covers the belly. The development of the turtle shell is correlated with intricate changes that affect the vertebral column, the associated ribs, and the limb girdles. Discussions of the evolutionary origin of turtles therefore need to address two separate questions: Which ancestral group of reptiles gave rise to turtles; and how did their unique body plan evolve?

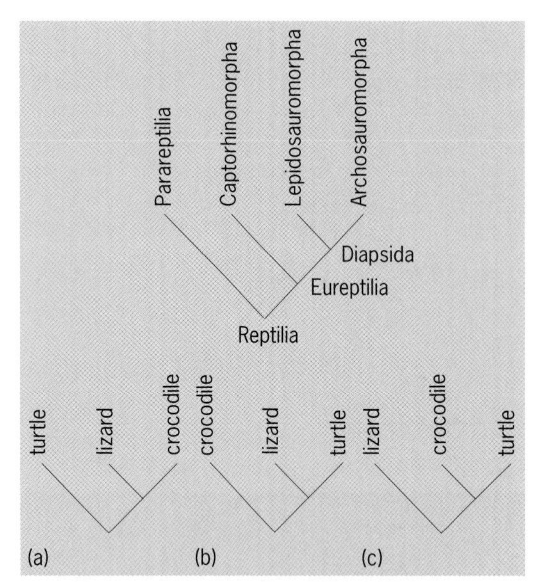

Fig. 1. Top: the major lineages of reptile evolution. In the past, turtles have been related to parareptiles, captorhinomorphs, lepidosauromorphs, and archosauromorphs. Bottom: (*a*) the conventional view of turtle relationships among living reptiles; (*b*) turtle relationships based on anatomical data; (*c*) turtle relationships based on molecular data.

Classification and morphology. Reptile classification (**Fig. 1**) has traditionally been based on the configuration of the temporal (cheek) region of the skull. Reptiles in which the temporal region of the skull is fully roofed over (closed) by bone are called anapsids. They comprise an array of extinct forms, mostly from the late Paleozoic era (320–250 million years before the present), that represent two different lineages of reptile evolution: the Parareptilia and the stem of the Eureptilia (the Captorhinomorpha). In more advanced eureptilians, the cheek region of the skull features two temporal fenestrae (openings) behind the eye socket, for which reason they are called Diapsida. The diapsids in turn split into two major evolutionary lineages. The first is the Archosauromorpha, which includes, among others, the crocodiles, the flying reptiles (pterosaurs), the dinosaurs, and their descendants, the birds. The Lepidosauromorpha is the other diapsid reptile branch, in which evolved the Tuatara, the scaly reptiles such as lizards and snakes, and their fossil relatives.

Until recently, the oldest known turtle was *Proganochelys* from the Late Triassic (210 million years before the present) of Germany, which shows no fenestration of the temporal region of the skull. Turtles were consequently believed to be "living fossils" (living species very closely resembling fossil relatives in most anatomical details), having originated in the Paleozoic era either from the stem of the eureptilian lineage (Captorhinomorpha) or from one (Procolophonia) or another (Pareiasauria) group of parareptiles (Fig. 1*a*). It was only in the late 1990s that a phylogenetic signal (wherein closely related species tend to be more similar to one another than expected by chance) based on anatomical data started to emerge, which linked turtles with the lepidosaurian lineage of diapsid reptiles (Fig. 1*b*). Eager to confirm or disconfirm this phylogenetic signal, systematists turned to molecular (DNA sequence) data to investigate turtle relationships. The results confirmed a diapsid origin of turtles. However, in contrast to the anatomical data, the molecular data link turtles with the archosauromorph branch of diapsid evolution (Fig. 1*c*). Although some lingering doubts persist among paleontologists that turtles are of diapsid descent, zoologists have overwhelmingly accepted that conclusion. Still, the issue of whether turtles are allied with the archosauromorph (crocodile) or lepidosauromoph (lizard) branch of diapsids remains unresolved.

Many species among parareptiles (for example, pareiasaurs), archosauromorphs (for example, crocodiles), and lepidosauromorphs (for example, a variety of lizards) are characterized by the formation of bony plates in deep layers of the skin (in the dermis), which are called osteoderms. It is therefore tempting to assume that the turtle ancestor also developed an osteoderm covering. The individual osteoderms eventually fused to form the larger bony plates that compose the turtle carapace. The ventral body armor of turtles, the plastron, would correspondingly have evolved through the fusion of "gastral ribs." These are riblike structures that develop in the ventral body wall of many reptiles, forming a basketlike structure that supports the belly. Such a scenario of the evolution of the turtle shell was invoked to derive turtles from parareptiles such as pareiasaurs. Indeed, some pareiasaurs had a dense osteoderm covering of their body, and osteoderms are also present on the neck, tail, and limbs of *Proganochelys*.

However, the evolution of the turtle carapace, which also ossifies in the dermis, from osteoderms present in their ancestor fails to explain another unique feature of the turtle body plan. In all tetrapods, the shoulder blade (scapula) lies outside the rib cage. Turtles are unique among tetrapods in that the scapula lies inside the rib cage, which is a consequence of the fact that the trunk ribs themselves are contributing to the formation of the carapace in the form of costal plates. Researchers defending an origin of turtles from pareiasaurs have explained this peculiarity by a backward shift of the pectoral girdle in turtles, such that the scapula moved from a position in front of the rib cage to one inside the latter. There is, however, no evidence for such a backward displacement of the pectoral girdle, neither in fossil turtles nor in the embryonic development of living turtles. This suggests that more complicated evolutionary transformations took place in the origin of the turtle body plan.

Developmental biology. For more than 150 years, discussions of morphological evolution and innovation have been dominated by two paradigms, which can be named the *transformationist* approach as opposed to the *emergentist* approach. The transformationist approach seeks an understanding of morphological evolution through the transformation of ancestral adult structures, such as the evolution of the turtle carapace through the coalescence of osteoderms covering the body of turtle ancestors. The emergentist approach seeks an understanding of morphological evolution as a consequence of changes in embryonic development, which bring about novel anatomical features. It is in the light of this latter paradigm that the evolution of the turtle carapace is best understood.

During the early stages of embryonic development of all vertebrates, cells derived from the middle germ layer (mesoderm) cluster to form a segmented series of somites along either side of the notochord, the axial organ that supports the embryonic body. Cells derived from the somites will, among other things, contribute to the formation of the vertebrae and of the ribs articulating on those. In all tetrapods except turtles, the cells that will form the cartilaginous precursors of the ribs migrate from the somites into the lateral body wall; this is how the ribs come to lie deep to the shoulder blade (scapula).

During early stages of the development of turtles, the deep layer of the skin (dermis) on the embryo's back starts to thicken, forming a disklike structure called the carapacial disk (**Fig. 2**). The carapace will eventually ossify within this carapacial disk. The margin of the carapacial disk is called the carapacial ridge. Very early on during its development, the upper part of the lateral body wall of the turtle embryo folds inward, thus forming the first rudiment of the carapacial ridge. As the carapacial disk continues to develop, the carapacial ridge is completed anteriorly and posteriorly to form the circular perimeter of the carapacial disk. Because of the early infolding of the upper part of the lateral body wall in turtles, the ribs can no longer grow downward into the body wall as they do in all other tetrapods. Instead, the ribs grow outward into the fold of the lateral body wall that forms the carapacial ridge. Therefore, it is not the shoulder blade that changes its position relative to the ribs. Instead, it is the ribs that grow outward into a more superficial position and hence come to lie superficial to the scapula.

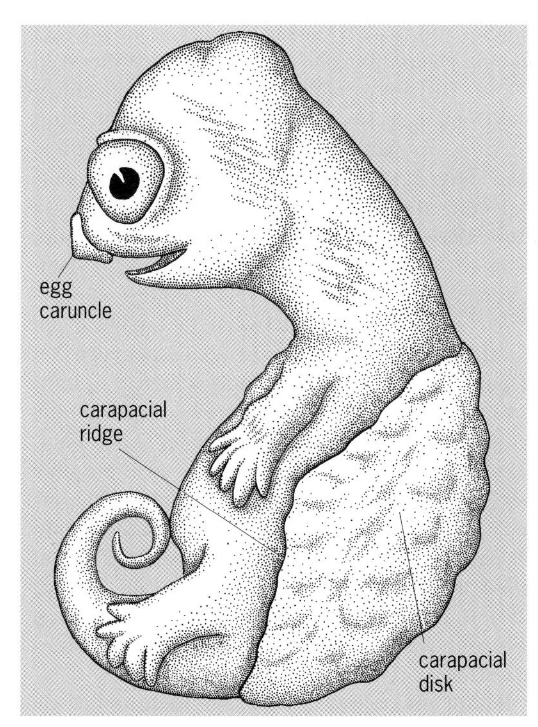

Fig. 2. A turtle embryo showing the carapacial disk and carapacial ridge at a stage before ossification of the carapace begins. (*Redrawn after C. L. Yntema, A series of stages in the embryonic development of Chelydra serpentina, J. Morphol., 125:219–252, 1968*)

Odontochelys semitestacea. These insights, gleaned from developmental biology, were recently complemented by the discovery of the oldest and most primitive turtle, *Odontochelys semitestacea*, from the early Late Triassic (approximately 220 million years before the present) of China. As indicated by its genus name, this turtle had toothed upper and lower jaws, in contrast to all other known turtles, including *Proganochelys*, in which the upper and lower jaws are covered by a beaklike structure, the rhamphotheca. The species name points to the fact that only the plastron was fully developed in this fossil turtle. The carapace remained incomplete. Neural plates ossified on top of the trunk vertebrae and the ribs were broadened, resembling those in a living turtle embryo during early stages of carapace ossification. The ribs are directed laterally, suggesting the presence of the lateral part of the carapacial ridge. However, the anteriormost trunk rib lies behind the scapula, instead of above it as in all other turtles, suggesting that the carapacial disk and hence the carapacial ridge were incompletely developed anteriorly and posteriorly. In fully evolved turtles, the anterior part of the carapacial ridge directs the growth of the anteriormost trunk rib in an anterolateral direction, that is, across the tip of the shoulder blade.

Odontochelys semitestacea thus documents an intermediate stage in the evolution of the turtle shell, which nicely complements insights gained from developmental biology. Completion of the carapace in more advanced turtles, including *Proganochelys*, required the ribs to broaden further to form the costal plates, which establish sutural contact with one another and with the neural plates located above the vertebrae. Moreover, the series of neural and costal plates is complemented by a series of peripheral plates ossifying directly in the dermis. These are the nuchal plate in the neck region, the pygal plate in the tail region, and marginal plates laterally.

Even though the fossil comes from marine sediments, it was not necessarily also living in the sea. *Odontochelys* remained much smaller than modern sea turtles, and its limbs were not transformed to form flippers. The limb proportions, however, indicate aquatic habits, rendering it likely that the remains of these turtles that eventually fossilized were washed into the sea from a nearby river delta system. Aquatic habits would also explain why the plastron evolved before the carapace. It might have provided not only protection from predators attacking from below, but also important ballast for a lung-breathing tetrapod, helping it move through water.

For background information *see* ANAPSIDA; ANIMAL EVOLUTION; CHELONIA; DEVELOPMENTAL BIOLOGY; DIAPSIDA; FOSSIL; ORGANIC EVOLUTION; PALEONTOLOGY; PHYLOGENY; REPTILIA; SYSTEMATICS in the McGraw-Hill Encyclopedia of Science & Technology. Olivier Rieppel

Bibliography. R. L. Carroll, *Vertebrate Paleontology and Evolution*, W. H. Freeman, New York, 1988; H. Nagashima et al., On the carapacial ridge in turtle embryos: Its developmental origin, function and the chelonian body plan, *Development*, 134:2219–2226, 2007; O. Rieppel, Turtles as hopeful monsters, *BioEssays*, 23:987–991, 2001; O. Rieppel and R. R. Reisz, The origin and early evolution of turtles, *Annu. Rev. Ecol. Syst.*, 30:1–22, 1999; J. Wyneken, M. H. Godfrey, and V. Bels (eds.), *Biology of the Turtles*, CRC Press, Boca Raton, FL, 2008.

2010 Haiti earthquake

On January 12, 2010, a magnitude 7.0 earthquake struck the Port-au-Prince region of Haiti (see **illus.** *a*), killing more than 200,000 people, leaving more than 1.5 million homeless, and destroying most of the governmental, technical, and educational infrastructure throughout this region of 3 million people. The earthquake was also strongly felt in the Dominican Republic but caused no damage there because of the significant distance between major urban centers and the epicenter. The event caused an estimated $8 billion in damages, or about 120% of the country's gross domestic product (GDP). No other earthquake of such moderate magnitude has ever caused so many causalities and such extensive damage. This was because the earthquake occurred in a heavily populated region of a very poor country with substandard building practices, and one that had not in any way prepared for such an eventuality. Only limited information on the seismic hazard was available before the event. Efforts were underway to enhance communication and knowledge about earthquakes, but awareness among the public as well as decision and policy makers remained low. As a result, mitigation and preparedness efforts were minimal, as the earthquake threat was not accounted for in construction, land-use planning, or emergency procedures, a situation unfortunately common among earthquake-prone developing countries.

Though earthquake hazards may not have been on the minds of Haitians, the January 2010 earthquake came as little surprise to many geologists and geophysicists. Haiti—and the island of Hispaniola as a whole—lies on the boundary between the Caribbean and North America plates, two major tectonic plates that slide past each other at a speed of about 2 cm/year (illus. *d*). This relative motion causes a buildup of pressure on several faults lines across the island. When the pressure exceeds the mechanical resistance of a fault section, the rock yields and results in a sudden break of the fault and the radiation of seismic waves away from this rupture (that is, an earthquake). Historical archives tell us that this process has occurred a number of times since Columbus first landed on Hispaniola (illus. *a*). Southern Haiti was struck on September 15, 1751, November 21, 1751, and June 3, 1770, by major events, with magnitude estimated between 7.0 and 7.5, which severely affected Port-au-Prince. They were followed on May 7, 1842, by a magnitude 8.0 earthquake that struck the northern part of Haiti and the Dominican Republic, with tremendous damage to the cities of Port de Paix and Cap Haitien. The most recent large events in Hispaniola comprise a series of magnitude 7.5 to 8.1 earthquakes between 1946 and 1953 offshore the northeastern coast of the Dominican Republic.

Fault systems. These large earthquakes highlight the three major fault systems that accommodate the relative motion between the Caribbean and North America plates. Studies based on Global Positioning System (GPS) measurements, a very precise technique that allows geophysicists to track the Earth's deformation with millimeter precision, show that the Caribbean plate moves east-northeastward at a rate of about 2 cm/year relative to North America. This overall plate motion consists mostly of horizontal shear parallel to the plate boundary, with a small amount of contraction perpendicular to it. The plate motion is distributed among a few major faults, with plate boundary-parallel motion accommodated by the Septentrional-Oriente and Enriquillo-Plantain Garden fault zones in Haiti and the Dominican Republic, and plate-boundary normal (contractional) motion taken up offshore to the north of the island (North Hispaniola fault zone) and to the south of the Dominican Republic (Muertos trough). Dense GPS measurements recently completed across the island show a north-south gradient in displacements, reflecting the buildup of elastic strain within the Earth's crust as a result of the overall plate motion (illus. *a*). These measurements, when integrated into kinematic models, indicate that the approximately 2 cm/yr relative motion between the Caribbean and North American plates is partitioned between 2 to 6 mm/yr on the North Hispaniola fault, 6 to 7 mm/yr on the Muertos trough, 12 mm/yr on the Septentrional fault zone, and 5–6 mm/yr on the Southern Peninsula fault zone (uncertainties on these numbers are of the order of ± 2 mm/yr). Additional faults, yet to be identified, may be active throughout Hispaniola. Denser geodetic measurements and more geological investigations (on land and offshore) are needed to obtain the complete inventory of active structures necessary for seismic hazard assessment.

Hazard level. Historical archives indicate that the last major earthquakes to strike the Southern Peninsula fault zone in Haiti occurred in 1751 and 1770, about 250 years ago. Smaller-magnitude earthquakes, which are less well located, occurred in

1701, 1784, 1860, 1864, and 1953, but these were too small to release much of the pressure building up in the fault zone. Assuming that the Enriquillo-Plantain Garden fault zone (EPGFZ) has been accumulating elastic deformation at a constant rate of 6 mm/yr (based on the GPS observations cited earlier), by 2010 it had built up a total amount of 1.5 m (0.06 m/yr × 250 yr) to be released in earthquakes to come. Earthquake scaling laws indicate that the release of 1.5 m of elastic deformation by sudden fault slip in one single earthquake corresponds to a magnitude of 7.1, very close to that of the January 12, 2010, event. This successful forecast of an earthquake magnitude based on geodetic measurements and historical data, while significant and useful, is far from being an earthquake prediction, which would require knowing the date and location of the event as well. It nevertheless demonstrates that scientific studies are critical to place useful bounds on the hazard level posed by seismic faults.

Analysis. Within hours after the January 12, 2010, earthquake, preliminary results became available, in particular, from the U.S. Geological Survey (USGS), which located the epicenter about 25 km (15 mi) WSW of Port-au-Prince and determined a magnitude of 7.0. On the basis of this location, most geologists and geophysicists assumed that the earthquake had ruptured a portion of the EPGFZ, a nearly vertical strike-slip (that is, horizontal motion) fault that runs 900 km (560 mi) from the Enriquillo Valley of the Dominican Republic through the center of the southern Peninsula of Haiti and continues westward across the Jamaica Passage to the Plantain Garden valley of eastern Jamaica. This fault forms the southern boundary of the Gonave microplate, a small elongate plate wedged in between the larger North American and Caribbean plates in the northeastern Caribbean (illus. *d*). Recordings from remote seismological stations were used to rapidly estimate the type of faulting, but with conflicting results regarding the dip of the fault plane; the USGS preferred dip to the south, while the Global Centroid-Moment-Tensor (CMT) Project's solution preferred dip to the north. Regardless of the solution, it quickly became apparent that the earthquake involved a large component of horizontal slip, but that some fault-perpendicular contraction was also present.

In the weeks following the earthquake, several research teams were able to travel to Haiti to study the geological context of the earthquake and assess the short-term threats associated with it. This fieldwork was made possible thanks to the efficient and dedicated participation of the Haitian Bureau of Mines and Energy and the Faculty of Science of the Haiti State University, both of which had been severely affected by the earthquake. The mapping of surface breaks and measurement of ground displacements caused by the earthquake must be done rapidly, as they are quickly overprinted by human activity, erosion, or postearthquake deformation. A first-order priority was to document the surface rupture of the earthquake fault. Field observations from several independent groups, as well as extensive use of high-resolution satellite imagery, found no surface expression of earthquake slip along the prominent trace of the EPGFZ. Therefore, despite the fact that the earthquake focus was relatively shallow, the associated rupture remained at depth and did not reach the surface. These early geological observations also identified uplift along the coast between Leogane and Petit Goave (up to 0.6 m) by mapping dead coral reefs that had emerged above sea level as a result of the earthquake.

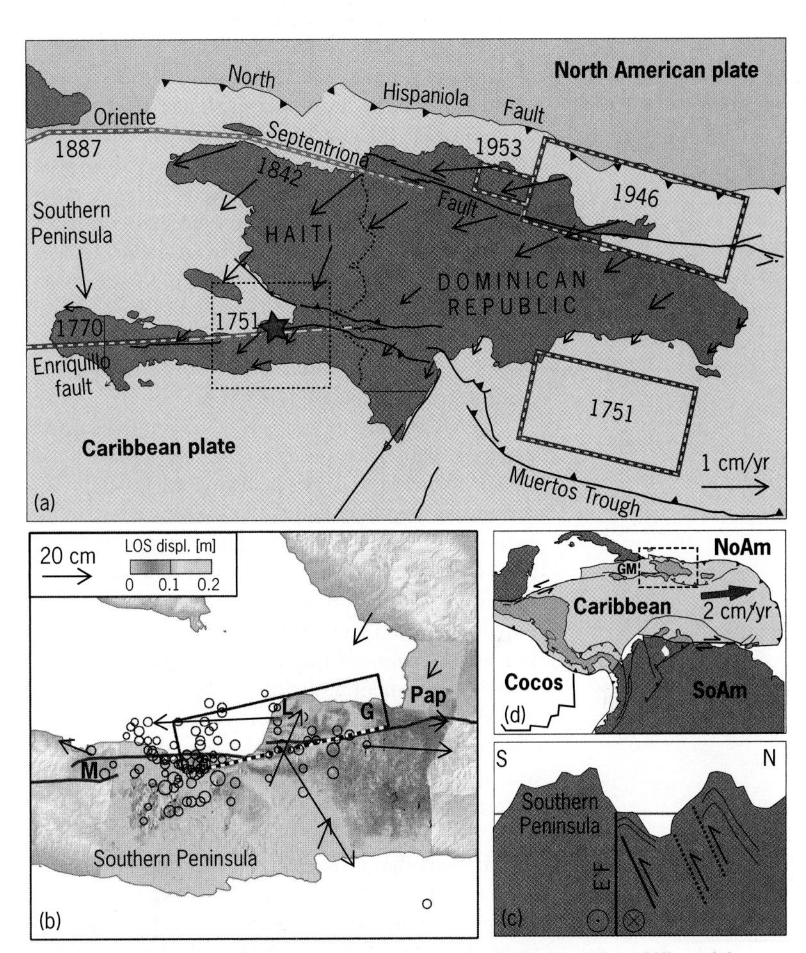

2010 Haiti earthquake. (*a*) Summary of the present-day tectonic setting of Hispaniola. Black lines show major active faults. Gray and white dashed lines delineate recent earthquake ruptures as derived from historical archives (vertical strike-slip events are shown as lines, dip-slip events shown as projected surface areas). Arrows show GPS velocities with respect to the Caribbean plate. Note the combination of shear and contraction across the plate boundary. The star shows the epicenter of the January 12, magnitude 7.0, Haiti earthquake. The dashed rectangle shows area of part *b*. (*b*) Zoom on the epicentral area showing the deformation caused by the earthquake. Background colors (not shown here) represent fringes from a radar interferogram calculated by F. Amelung (University of Miami) using Advanced Land Observing Satellite (ALOS)/Phased Array type L-band Synthetic Aperture Radar (PALSAR) data from the Japan Aerospace Exploration Agency (JAXA) (descending track, constructed from images acquired 9 March 2009 and 25 January 2010). Each fringe represents 0.2 m of displacement in the ground-to-satellite line-of-sight direction. Arrows show coseismic displacements from GPS measurements. Black rectangle shows surface projection of earthquake rupture for a simple one-fault model, black-white dashed line shows its intersection with surface. Open circles show aftershocks as reported by the USGS. G = Greissier, L = Léogane, M = Miragoane, PaP = Port-au-Prince. (*c*) Schematic north–south cross section perpendicular to the earthquake rupture. The thick vertical black line shows the Enriquillo fault (EF), the leftmost line dipping north shows the inferred earthquake rupture (single fault model). (*d*) Overall regional plate tectonic setting of the Caribbean and Central America plates. The Caribbean plate is moving eastward with respect to both the North and South American plates at a speed of about 2 cm/yr. GM = Gonave microplate. The dashed rectangle shows the area of part *a*.

Geodetic studies followed promptly, with the aim of measuring "coseismic" ground displacements—simply put, the difference between the position of the ground before and after the earthquake. This information helps constrain the geometry of the earthquake fault at depth as well as the overall distribution of slip during the earthquake rupture. GPS measurements at sites whose position had been determined before the earthquake showed up to 0.8 m of horizontal displacement in the epicentral area (illus. *b*). Measurable coseismic displacements of a few millimeters were reported up to about 150 km (93 mi) from the epicenter. The spatial distribution of horizontal displacements showed a combination of left-lateral strike-horizontal motion and north-south contraction, in a pattern similar to the long-term (pre-earthquake) deformation pattern across Hispaniola. Coseismic displacements were also well recorded by interferometric synthetic aperture radar (InSAR), a remote sensing technique that compares satellite-to-ground range data from radar images taken before and after the earthquake. InSAR observations showed up to 0.8 m of satellite-to-ground range change over an approximately 50-km-wide (31-mi-wide) region to the north of the EPGFZ and centered on Leogane (illus. *b*).

Taken together, the geological, seismological, and geodetic data indicate that the earthquake was caused by an approximately 30-km-long (19-mi-long) rupture located between Petit Goave and Greissier, extending from about 5 to 20 km (3 to 12 mi) in depth, and dipping 60° to the north. The same data require that the motion on the earthquake fault combined horizontal shear (about 2.6 m) and contractional motion (about 1.8 m), with westward and upward displacement of the crustal block to the north of the fault (illus. *b*). More complex fault models are currently under investigation as more data become available, but all require that most of the slip occurred on a north-dipping fault slightly oblique to the vertical (or steeply south-dipping) EPGFZ. It therefore appears that most of the earthquake slip during the January 12, 2010, event did not occur on the Enriquillo fault, contrary to the original assumption right after the event (illus. *c*). Clearly, more analysis is required to finalize these conclusions and to understand their implications for earthquake hazards in Haiti.

The January 12, 2010, earthquake was followed by a strong sequence of aftershocks (illus. *b*), with magnitudes reaching 5.9 (January 20, 2010) so far. In the absence of a national seismic network in Haiti, it was key to install temporary seismic stations to study the spatial and temporal distribution of the aftershock sequence. A French group (Géoazur, Ifremer, and IPG Paris) was able to install ocean-bottom seismometers (OBS) 4 weeks after the earthquake on both sides of the southern Peninsula of Haiti. Shortly after, they installed five seismic stations on land between Port-au-Prince and Miragoane, which remained in operation until May 2010. Seismologists from the Canadian Geological Survey installed three seismic stations in Port-au-Prince, Leogane, and Jacmel in mid-February, meant to remain in operation for the long term. In March 2010, the USGS deployed eight seismometers outside Port-au-Prince (removed in June 2010) to increase aftershock detection and improve location estimates, and six strong-motion instruments within Port-au-Prince to investigate the variability of shaking associated with variations in local geology and topography (five will stay for the long term). Although work remains to be done to fully characterize the aftershock sequence, preliminary results indicate that they concentrated at the western end of the rupture (between Leogane and Miragoane). Most of them involved faulting on NW-SE trending faults with pure contractional motion, another surprise given the mostly strike-slip nature of the main shock.

Outlook. The January 12, 2010, Haiti earthquake was a tragic reminder of geological reality: the inexorable seismic threat to which the country and the region are exposed. The threat is as old as the active fault lines that cut through Hispaniola, and will remain present as long as the Caribbean and North American plates continue to move past each other at speeds of about 2 cm/yr, that is, for millions of years. This part of the risk equation, driven by the slow but relentless drift of tectonic plates, is nonnegotiable. The remaining part of the risk equation, the population exposure and vulnerability, is the responsibility of governments, international agencies, and, increasingly, actions taken by local communities and individuals. Experience and research have given us a broad range of solutions to reduce these factors. These involve proper land-use planning, appropriate use of construction practices, and raising public and institutional awareness through education and research. However, the effective application of such mitigation strategies requires a strong and sustainable science and engineering foundation in Haiti, so that Haitians themselves can take a lead role in quantifying the threats and designing solutions adapted to their economy and society. The international community has the responsibility of helping Haiti implement these solutions in every aspect of its reconstruction.

For background information *see* EARTHQUAKE; EARTHQUAKE ENGINEERING; FAULT AND FAULT STRUCTURES; GEODESY; SEISMIC RISK; SEISMOGRAPHIC INSTRUMENTATION; SEISMOLOGY; SYNTHETIC APERTURE RADAR (SAR) in the McGraw-Hill Encyclopedia of Science & Technology. Eric Calais

Bibliography. E. Calais et al., Strain partitioning and fault slip rates in the northeastern Caribbean from GPS measurements, *Geophys. Res. Lett.*, 29(18):1856, 2002; C. DeMets et al., GPS geodetic constraints on Caribbean-North America plate motion, *Geophys. Res. Lett.*, 27(3):437–441, 2000; D. M. Manaker et al., Interseismic plate coupling and strain partitioning in the northeastern Caribbean, *Geophys. J. Int.*, 174:889–903, 2008; P. Mann et al., Actively evolving microplate formation by oblique collision and sideways motion along strike-slip faults: An example from the northeastern Caribbean plate margin, *Tectonophysics*, 246(1–3):1–69, 1995; P. Mann. et al., Oblique collision in the northeastern Caribbean from GPS measurements and geological observations, *Tectonics*, 21:1057, 2002.

Type 1 diabetes

Type 1 diabetes (T1D) is an autoimmune disease that poses significant challenges to afflicted individuals, to the development of effective therapeutic interventions, and to public health initiatives at large. Although the precise interactions between inherited susceptibilities and environmental factors that together trigger the disease remain to be elucidated in detail, T1D development is mediated by complex autoimmune processes that eventually destroy insulin-secreting beta cells in the pancreas and lead to elevated blood sugar levels along with serious disturbances of protein, fat, and carbohydrate metabolism. Currently, no cure or effective prevention is available. Despite insulin treatment, severe complications such as kidney failure, heart attack, and stroke are frequent.

Natural history of T1D. The term "diabetes mellitus," literally meaning "honeyed flow-through," refers to a constellation of clinical symptoms, including increased thirst, excessive urination, and elevated sugar content in the urine, that has been known since antiquity, but it was not until 1890 that a specific organ, the pancreas, was implicated in the disease process. The pancreas, a prodigious source of digestive enzymes, also contains clusters of distinct tissues, termed islets, that constitute about 2% of its mass and that harbor specialized hormone-producing cells, including insulin-secreting beta cells. Impaired production or action of insulin, discovered in 1921, is a central feature of diabetes, and, as with many other human diseases, subsequent work established that "diabetes mellitus" is in fact a collection of different disease processes rather than an individual disease. Since the mid-1970s, the concept that a subset of diabetes patients suffers from an autoimmune syndrome has been firmly established. At present, the term T1D describes the form of diabetes resulting from a loss of beta-cell mass (clinical symptoms become manifest only when more than 90% of beta cells are destroyed), with subtype 1A referring to immune-mediated forms and subtype 1B covering the less frequent nonimmune disease forms. For simplicity, type 1A will be referred to here as T1D.

Although the onset of clinical symptoms can occur quite suddenly, T1D is a chronic disease, the cause of which is not fully understood (see **illustration**). What is clear, though, is that a complex interplay between genes and environmental factors is involved in the initiation and perpetuation of destructive autoimmune processes that result in progressive beta-cell loss and ultimately clinical disease. Large-scale genetic studies have documented

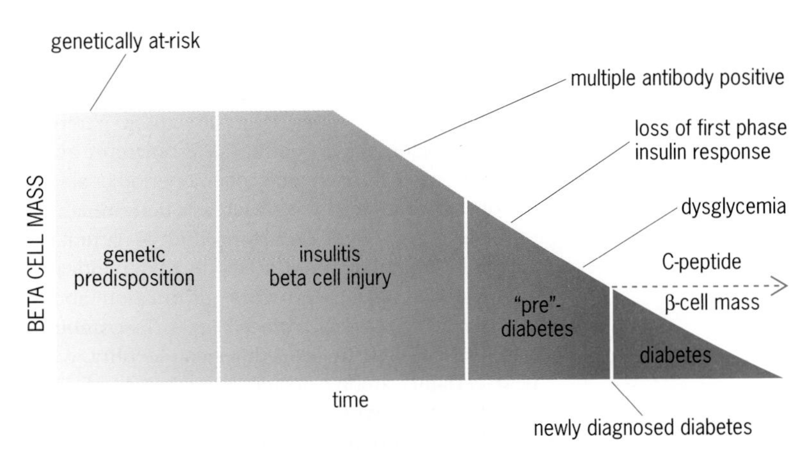

Stages in the development of type 1 diabetes. (*Type 1 Diabetes TrialNet*)

that certain genes are preferentially associated with T1D. Chief among these is a subset of defined major histocompatibility complex II (MHC-II) genes that confer significantly enhanced risk of T1D development. MHC-II genes, which demonstrate great diversity throughout the population, are involved in "flagging" bodily cells for recognition by a subset of immune cells (CD4 T cells), and the preferential association of defined MHC-II genes with T1D thus provides genetic evidence for a role of CD4 T cells in the destructive autoimmune process. The purpose of genetic association studies, which are continually refined and extended, is twofold: identification of individuals who are at risk and identification of disease-associated genes that may provide clues concerning the disease mechanism and hence may point toward new therapeutic targets. These studies also demonstrate that the genetic makeup alone is not sufficient to cause disease: among monozygotic (identical) twins, T1D development may be limited to one twin or occur in both at points in time that are decades apart. Therefore, environmental factors must also contribute to disease manifestation, and potential candidates range from dietary compounds to infectious diseases. On balance, however, it is unlikely that a single specific environmental trigger will ever be identified. Rather, multiple and possibly unrelated insults, including tissue inflammation and viral infections, successively induce changes in the pancreatic microenvironment that, in combination with a defined genetic predisposition, will induce and maintain an autoimmune process in which immune cells such as T cells and macrophages participate in the destruction of beta cells.

Prediction, prevention, and reversion of T1D. Some of the most remarkable progress made in the past decade pertains to the improved accuracy with which the risk for T1D development can now be assessed. A combination of genetic testing (MHC and other genes), serological analyses (detection of blood-borne autoantibodies directed against beta-cell products, including insulin, GAD65, IA-2, and ZnT80), and intravenous glucose tolerance tests

(IVGTT, the measurement of insulin secreted into the blood in response to intravenous administration of glucose) can identify individuals who are at risk for T1D (see illustration). Other metabolic parameters (hemoglobin A1c) permit an estimate of average glucose levels maintained over a period of about 3 months prior to analysis, whereas a determination of the levels of C-peptide, a by-product of natural insulin synthesis that is not a part of therapeutically administered insulin, provides information about the extent of residual beta-cell mass. In combination, these diagnostic procedures can identify, with near certainty among first-degree relatives of T1D patients, those who will eventually develop T1D, thereby establishing a "window of opportunity" for immune intervention therapies aimed at modulating the destructive autoimmune process, preserving the remaining beta cells, and preventing progression to clinical disease (see illustration).

Unfortunately, effective and reliable immune intervention strategies remain an elusive goal at present. The design of specific therapies is largely based on preclinical trials in small animal models. Here, the surprisingly large number of successful experimental protocols that can prevent T1D progression (and the much smaller number of treatments that reverse recent-onset T1D) are testament to the possibilities and limitations inherent in the use of experimental animals. For example, in the nonobese diabetic (NOD) mouse model (a well-established animal model for T1D), development of spontaneous T1D is dependent on autoantibodies, whereas such antibodies are associated with human T1D (and thus can serve as a diagnostic marker), but do not appear to contribute directly to the disease process. It would therefore seem important to evaluate the efficacy of potential treatments in multiple different preclinical models, all of which feature certain commonalities with and divergences from human T1D.

What unites many of the preclinical and clinical trials, though, is a focus on the targeted modulation of destructive autoimmune responses. Early attempts to prevent T1D development by means of generalized immunosuppression demonstrated some promise and confirmed the critical role of autoimmune responses in the disease process. However, any benefits were outweighed by adverse effects of the drugs used, and patients progressed to T1D upon cessation of treatment. The aim of many contemporary therapies is thus to specifically disable subsets of immune cells, in particular T cells, that are directly involved in beta-cell destruction. Interestingly, this goal may be achieved by induction of competing T-cell responses that have the capacity to suppress destructive T-cell responses (so-called regulatory or suppressor T cells). Overall, animal studies have allowed for the generation of a diverse array of novel therapeutic paradigms, but so far most attempts to translate successful treatment strategies from preclinical models to human T1D prevention have met with only modest success.

In addition, the transplantation of islets is a principal possibility for restoring lost beta-cell mass and reestablishing normal blood glucose control. However, although this approach remains challenging for reasons similar to those for other organ transplants (for example, scarcity of suitable donors and quality of donor tissue), long-term survival of functioning islets is further complicated by the fact that the autoimmune process that destroyed the recipient's islets in the first place will also compromise the integrity of the transplanted islets. Thus, islet transplantation will have to be combined with additional treatments similar to those that seek to disrupt the destructive autoimmune responses in prediabetic individuals.

T1D management. In the absence of therapies that effectively prevent, reverse, or cure T1D, management of T1D is centered around diet, exercise, and, most critically, insulin administration to regulate blood glucose levels, prevent metabolic disorders, minimize long-term complications, and avert death. The major challenge for insulin substitution therapy is adjusting the insulin dosage to changing demands (for example, food intake, exercise, and sleep), and it is in the area of novel insulin formulations, administration, and glucose monitoring that T1D patients at large have benefited from recent improvements. External insulin pumps deliver small amounts of rapid-acting insulin every few minutes according to a programmed basal rate (continuous subcutaneous insulin infusion therapy) and are complemented by user-initiated bolus dosages given before meals or to correct high blood glucose levels. The advantages of insulin pumps include convenience, flexibility, and relative freedom to adjust insulin dosages, resulting in a reduction of hypoglycemia. However, these are balanced by expense, skin infections, logistical considerations, psychological factors, weight gain, and potential metabolic complications in the event of pump malfunction (ketonuria or ketoacidosis). Insulin pumps, however, do not obviate the need for frequent manual blood glucose checks, and tighter control can be achieved by devices that measure subcutaneous glucose values and transmit information to a remote monitor in real time (continuous glucose monitoring, CGM). The use and reliability of CGM is likely to increase in the near future, but an ideal system, namely, the combination of an insulin pump and CGM in a closed loop that measures blood glucose and automatically adjusts insulin administration ("artificial pancreas"), remains many years away from more general implementation.

Conclusions. T1D is an exceedingly complex disease, and the development of effective therapies that prevent progression to clinical disease or even reverse recent-onset diabetes is likely to take many years of extended research endeavors and clinical trials. In the immediate future, however, progress in patient care, including the combination of insulin pumps and CGM, will improve the quality of life of affected individuals and reduce T1D-related complications as a result of better fine-tuning of blood glucose control. Here, it should be emphasized that T1D management is not limited to patients and their health-care providers, but intimately involves families and entire communities engaged in education, supervision, and practical assistance.

For background information *see* AUTOIMMUNITY; CLINICAL IMMUNOLOGY; DIABETES; EPIDEMIOLOGY; GENETICS; GLUCOSE; IMMUNOLOGY; INSULIN; PANCREAS; PANCREAS DISORDERS in the McGraw-Hill Encyclopedia of Science & Technology. Dirk Homann

Bibliography. P. Concannon, S. S. Rich, and G. T. Nepom, Genetics of type 1A diabetes, *N. Engl. J. Med.*, 360:1646–1654, 2009; L. K. Shoda et al., A comprehensive review of interventions in the NOD mouse and implications for translation, *Immunity*, 23:115–126, 2005; M. G. von Herrath, R. S. Fujinami, and J. L. Whitton, Microorganisms and autoimmunity: Making the barren field fertile?, *Nat. Rev. Microbiol.*, 1:151–157, 2003; M. von Herrath and D. Homann, Organ-specific autoimmunity, pp. 1331–1374, in W. Paul (ed.), *Fundamental Immunology*, Lippincott Williams & Wilkins, Philadelphia, 2008.

Ultrafast x-ray sources

Sources of x-rays are an essential tool for scientists examining the structure and interactions of matter. X-ray sources already played this role before the scattering of x-rays from DNA led to the first understanding of the double-helix structure. With wavelengths of the order of atomic distances, scattered x-rays produce diffraction patterns of crystal lattices. In microscopy applications, the resolution is proportional to the wavelength of light, so x-rays can be used to see much finer structures than can be seen with visible light, even down to single atoms. In addition, the energy of x-rays is resonant with the core atomic levels of atoms, and so with appropriate photon energies, the placement of specific atoms in a large molecule can be determined. Worldwide, more than 10,000 scientists use synchrotron sources, storage rings of high-energy electrons or positrons (positively charged electrons), each year.

As an example of such use, virtually every picture of a protein or drug molecule that is published in the scientific press is a reconstruction based on x-ray scattering of synchrotron light from the crystallized form of that molecule. However, those pictures are static, and proteins work through configuration (shape) changes in response to energy transfer. To understand how biological systems work requires following the energy flow to these molecules and tracking how shape changes drive their interaction with other molecules. We would like to be able to freeze the action of these molecules at various steps along the way with an x-ray strobe light. How fast would it have to be? Actually getting a picture of a molecule in a fixed configuration requires x-ray pulses as short as 30 femtoseconds (fs; 1/30 of a millionth of a millionth of a second). Capturing the energy flow through changes in electronic levels requires a faster strobe, less than 1 fs. And to acquire such information in smaller samples with higher accuracy demands ever brighter x-rays. Modern synchrotrons (known as third-generation light sources) cannot deliver such short, bright pulses of x-rays. An entirely new approach is required: linear-accelerator- (linac-)based light sources, termed fourth- or next-generation light sources (NGLSs). Although such light sources will not displace synchrotrons, they do offer exciting new capabilities that can be understood from the physics of the light production in each device.

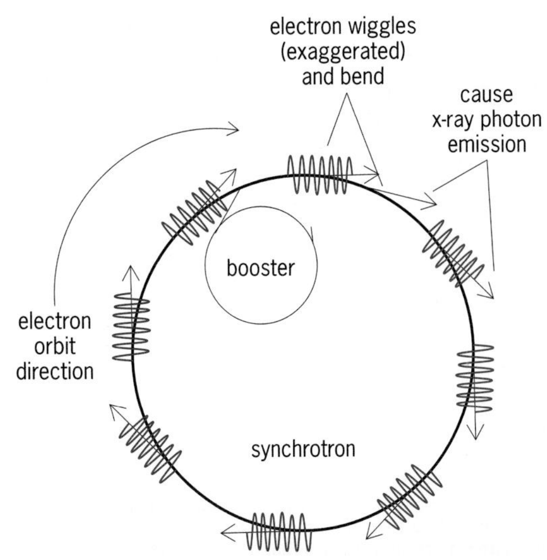

Fig. 1. In a synchrotron, electrons or positrons are boosted to their operating energy and injected into a magnetic ring that confines the electrons to a circular orbit. Numerous bending magnets and undulators around the orbit produce light for research.

X-ray rings. In a modern synchrotron x-ray source, bunches of electrons are accelerated to high energy (up to 7×10^9 eV) in a booster and stored in a magnetic electron/positron storage ring (**Fig. 1**). As the electrons bend in an arc from the magnetic field, they radiate x-rays. The shortest x-ray wavelength produced is determined by the radius of the bend and the cube of the inverse electron energy. The electrons radiate because in bending, they are accelerating transversely, and accelerated charges radiate electromagnetic waves; that is why antennas send out radio waves. Ampere-level currents can be stored in such rings, and the power that is radiated as x-rays must be resupplied in each orbit by passing the electron bunches through a radio-frequency acceleration cavity. Modern machines can store such multibillion-watt-average-power electron beams for many hours. There may be many scientists using the x-rays simultaneously around the circumference; 25 simultaneous experiments would not be unusual (Fig. 1).

The dynamics of the electrons determine the characteristics of the x-rays in terms of their brightness. The interplay between stochastic heating (random energy loss on photon emission) of the electrons emitting the x-rays, the radio-frequency acceleration

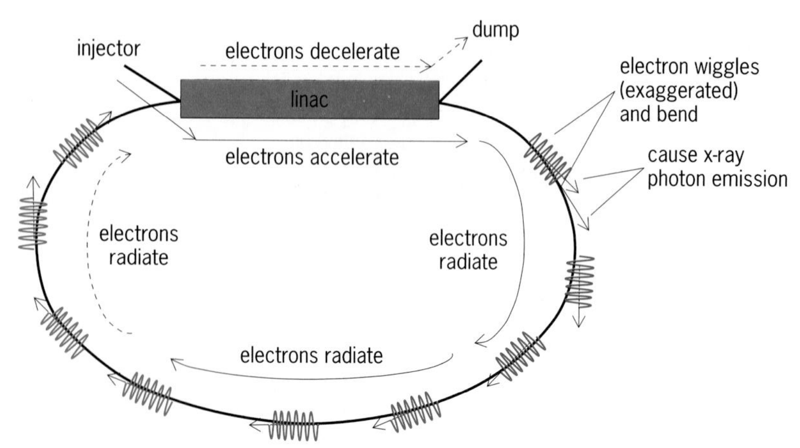

Fig. 2. An energy-recovering linac (ERL) device accelerates electrons from an injector to high energy in a linac, and then sends the beam into a series of bending magnets or undulators to produce light for research. The spent beam passes through the linac out of phase to recover its energy before being deposited in a beam dump. Short pulses can be maintained because of the short residence time of the electrons in the arcs.

fields, the energy spread of the electrons, and the electron bending and focusing by magnetic fields establishes an equilibrium for the electron orbits that sets the shortest pulse that can be produced, typically of the order of 50 picoseconds (millionths of a millionth of a second). This limitation results from a natural coupling between the electrons' energy and their path length around the ring. Large photon power emission yields large electron energy spreads, which result in electron path-length variations around the circular orbit, preventing exact temporal overlap and tight bunching. Although magnetic transport lattice designs can try to minimize this effect, practical limits remain as a result of the rapid heating of the electrons as they orbit many times per second. The only way short pulses can be extracted from such devices is through pulse-slicing techniques, which reduce average brightness.

The radiation from such machines can be produced in a simple magnetic bend forming the arc or enhanced through the use of undulators. These devices consist of a linear array of alternating north/south magnetic poles that repeatedly wiggle the electrons. The radiation fields of each wiggle emission interfere coherently to produce a narrower spectrum (coherent interference of N wiggles narrows the spectrum by the square root of N) and a tighter radiation pattern than in a simple bend. The tightening angular emission occurs because the natural x-ray angular spread of $1/\gamma$ (where γ is the relativistic factor, equal to the ratio of the electron's total energy to its rest energy) is reduced by N in the same way that a series of dipole antennas produces a narrow directed radiation pattern through coherent wave interference.

Linear accelerator sources. Linear accelerators (linacs) can overcome the limitations of synchrotrons because they send the electron beams through the system quickly so that the electrons never come into equilibrium with the competing effects. Very short pulses of electrons can be produced in an injector, accelerated to high energy, and sent through bending magnets to produce the desired x-rays.

The problem faced by machine designers is one of practicality. If a storage ring can hold 1 A at 7000 MV, then a linear accelerator would require an outrageous 10,000 MW of electric power (1 A times 7×10^9 V of acceleration produced at 70% efficiency) to be competitive in average x-ray output—in other words, more power than the output of 10 nuclear power plants. In recent years, two non-mutually exclusive solutions to this dilemma have emerged and are beginning to show practical results.

Energy-recovering linacs (ERLs). The first solution is called an energy-recovering linac (or an energy-recovery linac; ERL). In this system, the electrons are not recycled, but their power is. Modern continuous radio-frequency accelerators utilize superconducting radio-frequency cavities powered by microwaves to accelerate electrons. The cavities are made out of niobium and are bathed in superfluid helium at a temperature of 2 K (−456°F). Because the cavities are superconducting at this temperature, ohmic losses from the microwave fields are very small, and an almost perfect transfer of energy from the microwave field to the electrons occurs. The electron bunches get an increase in energy as they pass through each accelerator cavity in proper phase with the microwave electric field. If, after producing x-rays, the electrons are sent back into the same cavities, but 180° out of phase, they see a decelerating electric field. Their power is transferred back to the microwaves, and is then available to accelerate a subsequent electron bunch (**Fig. 2**). Since the energy transfer process is so efficient, 1 A of electrons can be envisioned accelerating to high energy and back down again with only makeup power for the radiation losses. Because they acquire a large energy spread during the radiating process, the electrons are thrown away at the end, and new "clean" bunches are provided for the next cycle. The concept was first demonstrated at high current by the Thomas Jefferson National Accelerator Facility, based on an idea dating back to 1976. Cornell University, which has operated a storage-ring light source for many years, has proposed the construction of a 5-GeV, 100-mA system based on the ERL concept.

As with the storage-ring concept, many users can be accommodated around the electron beam path. Since the electron beam is not required to go through the system many times, manipulations of the electron bunch length or focus can optimize the output of each radiation generator. Very long undulators for higher brightness can be accommodated because there is no need to bend in a tight circle. This enhanced brightness and capability for the desired ultrashort pulses can more than make up for the somewhat lower average current as compared to that of a storage ring.

Free-electron lasers (FELs). A second concept addresses not so much increasing the electron beam power, but rather enhancing the radiative action through collective stimulated emission of radiation. In synchrotrons, the electrons radiate in random optical phase with each other because they are in long, uniform bunches as measured on x-ray scale lengths. The radiation produced comes from statistical fluctuations in the electron bunch, or "noise," and the intensity is proportional to the number of electrons. If the electrons could be induced to bunch at the x-ray wavelength, then their radiation fields would add in phase, and the emission would be coherent. Moreover, the power would be proportional to the square of the number of electrons, and the bandwidth would be narrowed by the square root of the number of bunches. Since the total number of electrons is typically of the order of 10^9, the square of that value would produce a very intense pulse.

That is what happens in a free-electron laser (FEL): amplification of the optical wave is achieved. The interaction of the optical field from an external seed pulse or simply spontaneous noise coupled with the undulator field causes the electrons to bunch longitudinally at the x-ray wavelength. What was a smooth bunch of electrons at the undulator input becomes density-modulated at the x-ray wavelength into thousands of bunches by the undulator output. Bunching at the x-ray wavelength occurs because the wavelength of the undulator gets one Lorentz contraction and one Doppler shift, allowing a several-centimeter undulator wavelength to produce 0.1-nm bunching (and therefore light) from 10-GeV electron beams. Very high optical gains of 10^8 can be achieved while extracting of the order of 0.1% of the electron beam power in an amplifier (and even more in a long-wavelength oscillator). With a peak current of 1000 A and an energy of 10 GeV, peak powers of 10 GW can be achieved.

Operating in this manner, a device called the Linac Coherent Light Source (LCLS), using an 8-GeV electron beam, produced more than 10^{10} W of saturated output of 0.15-nm-wavelength light in 80 m (260 ft) of undulator in April 2009, becoming the first high-power x-ray laser at this wavelength. The brightness of this radiation exceeds that of previous sources in this range by a factor of 10^{11}. Work is already underway to perform experiments utilizing 30-fs pulse lengths to obtain a flash image of a protein taken in a single shot before the protein violently disassembles from the absorbed energy. To turn such a system into a practical user facility will require rapidly switching the electron beam from a high-repetition-rate linac into multiple undulators, so that more users can be accommodated (**Fig. 3**). George Neil

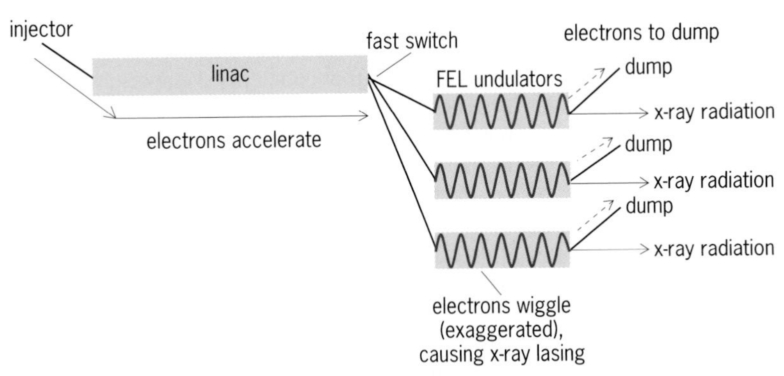

Fig. 3. A free-electron laser (FEL) uses a high-peak-current electron beam from a linac to amplify light to short-pulse, high-intensity radiation in an undulator. The spent beam is then discarded. Higher radiation efficiency compensates for the lower average beam power.

Scientific experiments. The science describing the interactions between electromagnetic radiation and matter has begun to face important challenges in addressing some of today's most urgent and critical needs in such areas as renewable energy sources, climate change, information technology, biological complexity, and biomedicine. Our understanding of some fundamental processes may permit us to convert sunlight into electricity at much higher efficiencies than are obtainable today, synthesize new fuels with no impact upon the global environment, and build magnetic-based memory devices with unprecedented storage density and recording speed. This understanding may also allow us a detailed look at the structure, morphology, and functionality of biological systems, helping to unlock the gates to a sustainable future for humankind.

Success in this endeavor requires realizing advanced radiation sources that have a very high brilliance, a high degree of coherence, and variable polarization, and whose investigators have complete control of the photon flux, together with very innovative detection devices and spectroscopic tools. The impressive progress achieved in the past few decades by new synchrotron sources has resulted in an effective and extraordinary advancement of our knowledge of protein structure, x-ray coherent imaging techniques, and the structure of complex and exotic materials. However, the insights that have been obtained so far are largely relevant to systems in equilibrium. Therefore, the actual understanding of the structure of matter is based mostly upon time-independent theories and experiments in the energy (frequency) domain; by contrast, our knowledge of systems that are out of equilibrium or under extreme conditions is very limited. Nonetheless, access to the physics and structure of states evolving from one equilibrium state to another is of paramount importance to expanding our understanding to mechanisms that play fundamental roles in critical technologies requiring photochemical and biochemical reactions, collective excitations, phase transitions, imaging of ultrafast processes of nanosystems, and handling of matter under extreme conditions.

In recent years, a significant effort has been focused on producing femtosecond-duration x-ray pulses by slicing a synchrotron electron bunch with an ultrashort laser beam, opening the way for low-photon-flux experiments, that is, experiments with up to 10^4–10^5 photons per second, on the dynamics of systems that are out of equilibrium. Unfortunately, the technology of the synchrotron-based radiation sources cannot meet the general need for femtosecond time-resolution experiments, mostly because of the time structure of the electron bunches (with durations of several tens of picoseconds) in storage rings. In addition, the nanometer spatial resolution needed for nanoscience experiments is out of the range of current synchrotron sources because of their limited output coherence. Conversely, the limitations on the intensity, coherence, and time structure of the x-ray sources that are available today—in particular, the limitations on generating coherent light in the x-ray region using conventional laser sources—have made it necessary to consider new sources, such as the free-electron laser (FEL), or insertion devices powered by electron beams from recirculating linacs or energy-recovery linacs (ERLs). These new radiation sources, also known as fourth-generation light sources, can produce x-ray pulses with an unprecedented peak brilliance [measured in photons/(s mm^2 mrad^2 0.1% bandwidth)], orders of magnitude higher than those of third-generation synchrotron radiation sources; pulse durations from a few femtoseconds to tens of femtoseconds with high transverse and longitudinal coherence, approaching values associated with conventional lasers, but in the soft- and hard-x-ray spectral regions; variable (linear and circular) polarization; and significant photon energy tunability. [Although the photon energy ranges for the extreme-ultraviolet (EUV), soft-x-ray, and hard-x-ray regions in the electromagnetic spectrum cannot be defined precisely, the following definitions are used here: EUV region ≈20–300 eV, soft-x-ray region ≈300 eV–3 keV, and hard-x-ray region is above 3 keV.]

Range of experiments. FELs, operating both currently or in the near future, and ERL-based light sources will make it possible to extend experimental investigations to dynamic phenomena through ultrafast stroboscopic (pump-probe) experiments, single-pulse coherent diffraction imaging with spatial resolution in the nanometer domain, or studies of matter under extreme thermodynamic conditions (such as warm dense matter, a state of matter intermediate between solid and plasma, which is expected to exist, for example, in the cores of large planets). This will lead, for example, to the control of several degrees of freedom, such as spin and charge, opening the possibility of gaining new insights into the physics of superconductivity, magnetism, and complex materials, and enabling future developments in information science, energy production, energy storage, and energy saving. Moreover, by taking advantage of the ultrabright and ultrashort pulses, single-photon-pulse experiments can collect images at time scales faster than that at which radiation damage takes place, thereby opening the possibility of studying the morphology and structure of biosystems that are unstable under exposure to x-ray radiation. For these reasons, this new generation of light sources will have a broad and deep impact on experiments involving the temporal evolution in the femtosecond time domain of the formation and breaking of chemical bonds; the dynamics of magnetic moments, spin, and phase-transition processes; and the imaging of objects down to nanoscale sizes, with an ultimate resolution of the order of the nanometer.

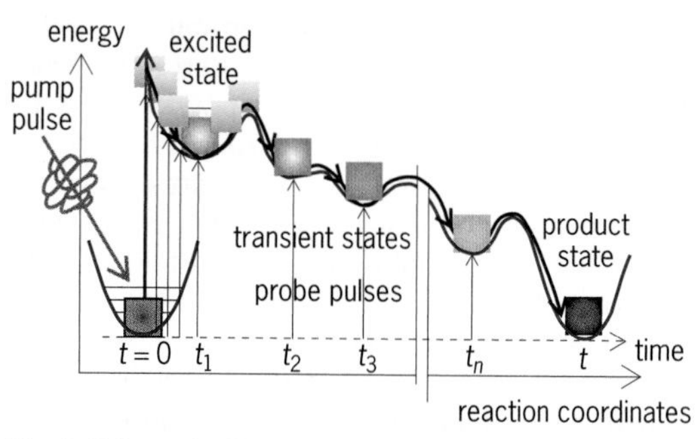

Fig. 4. Schematic diagram of an optical pump-probe experiment. A first light pulse (pump) excites the system from the ground state to a higher energy state. A second pulse probes the system at different and controlled delay times (t_1, t_2, . . .) in order to observe the progress of the system over time along a reaction pathway (reaction coordinates). The purpose of these experiments is to observe the time evolution (transient states) of the matter (atoms, molecules, condensed matter, and biological systems) after an ultrashort excitation pulse. (*Image courtesy of Lin Chen, Argonne National Laboratory*)

Pump-probe experiments. Among the various spectroscopies in the time domain, pump-probe experiments with lasers are well established. When Fourier-transform-limited pulses are used in these experiments, the energy resolution ΔE or the time resolution $\Delta\tau$ will be limited only by the energy-time uncertainty relation ($\Delta E \Delta\tau \geq h/2\pi$, where h is Planck's constant). The principle of these experiments is illustrated in **Fig. 4**. An ultrashort light pulse excites (pumps) an atom, a molecule, or a condensed system, while a second pulse, properly delayed in time, probes the excited matter, making it possible to follow its evolution back to the ground state. By using femtosecond (or subfemtosecond) pulses in the EUV, soft-x-ray, or hard-x-ray wavelength regimes, it will be possible to observe the atomic motion in real time, while electronic excitations, metastable states, phase transitions, and decay processes can be followed over time (**Fig. 5**).

Studies of atom clusters. By taking advantage of the very high brightness of advanced radiation sources, time-resolved spectroscopy and coherent imaging

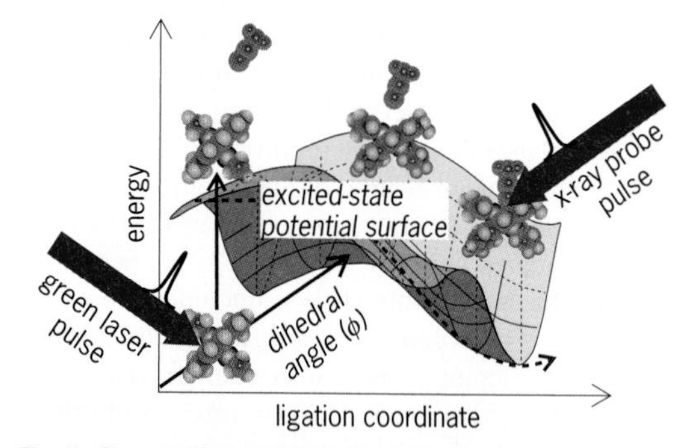

Fig. 5. Structural dynamics of a photoactive copper-complex excited state created by green laser pulses, followed by x-ray probe pulses to capture both dihedral angle movements on a 100-fs time scale and ligation coordinate changes on picosecond to nanosecond time scales. (*From G. B. Shaw et al., Ultrafast structural rearrangements in the MLCT excited state for copper(I) bis-phenanthrolines in solution, J. Am. Chem. Soc., 129:2147–2160, 2007*)

on single molecules or on systems consisting of a few atoms (clusters) in mass-selected cluster beams will be possible. This capability will make it possible to perform experiments on individual nanometer-scale objects, and thereby bridge the knowledge gap with regard to the structural, electronic, and magnetic properties of systems intermediate between single atoms and solids.

Control of experimental parameters. Although the high peak brilliance is of paramount importance for single-photon-pulse techniques or studies on warm dense matter, other experiments in electron-emission spectroscopy and microscopy will need lower peak brilliance and a high repetition rate to avoid radiation damage or space-charge effects. Therefore, the ability of the investigator to control the repetition rate, along with the peak and spectral brightness, represents a key property characterizing fourth-generation radiation sources.

Ultrashort pulses. Fourth-generation light sources have the possibility of generating x-ray pulses of a few femtoseconds or even a few hundred attoseconds (1 as = 10^{-18} s). This capability would make feasible experiments on electron dynamics as well as on subfemtosecond processes in atoms, molecules, and condensed matter that are not possible with existing laser-based high-order-harmonic-generation sources because of their limited peak and spectral brilliance.

End-station equipment and design. Besides the radiation parameters achievable with FELs or ERLs, femtosecond-time-resolution and nanometer-spatial-resolution experiments will need sophisticated end stations, tailored and highly innovative detectors, and advanced x-ray optical components and devices, such as x-ray beam splitters, auto- and cross-correlators, and the Kirkpatrick-Baez (KB) submicrometer-focusing x-ray telescope. For the system shown in **Fig. 6**, the experimental stations are designed for (1) experiments on systems with a very low concentration (10^2–10^6 atoms/cm^3) of atoms, molecules, and clusters of atoms, where the probability of interaction between the light and the matter requires a very high photon density in the radiation pulse (10^{12} or higher photons per pulse); (2) experiments for ultrafast (from a few femtoseconds up to tens of femtoseconds) coherent imaging; and (3) time-resolved studies of inelastic scattering and experiments under extreme conditions.

Furthermore, some experiments will require the storage and analysis of a huge amount of data. In particular, this will be the case for coherent diffraction imaging experiments (**Fig. 7**), where several sets of diffraction patterns over extended photon energy ranges are collected and analyzed in real time, in order to drive and control experimental parameters.

In many cases, the end stations for FEL experiments will be based on the design and technology of the experimental stations that operate on current synchrotron storage rings. Such technologies include coherent diffraction x-ray imaging and time-resolved x-ray imaging, used mostly for biomedical and biostructural studies; time-resolved x-ray diffraction, for revealing the dynamics of the atomic structure of hard and soft matter; time-resolved x-ray absorption and emission spectroscopy, for studying the dynamics of the occupied and unoccupied orbitals and bands in systems that are out of equilibrium; time-resolved inelastic x-ray scattering, for measuring the collective excitations of electrons, phonons, and spin in the energy domain; and time-resolved photoelectron spectroscopy, for observing the dynamics of core levels and valence-band states in momentum space. Although incomplete, this ensemble of experiments will access the unexplored world of the nonequilibrium and transient states of materials, as well as generate and control new and exotic states of the matter present in the universe under extreme conditions, such as matter in planetary and stellar cores.

Prospects. The scientific and technological impact of these new radiation sources will arise from the unprecedented capability for exploring and manipulating matter at the nanoscale, combined with the

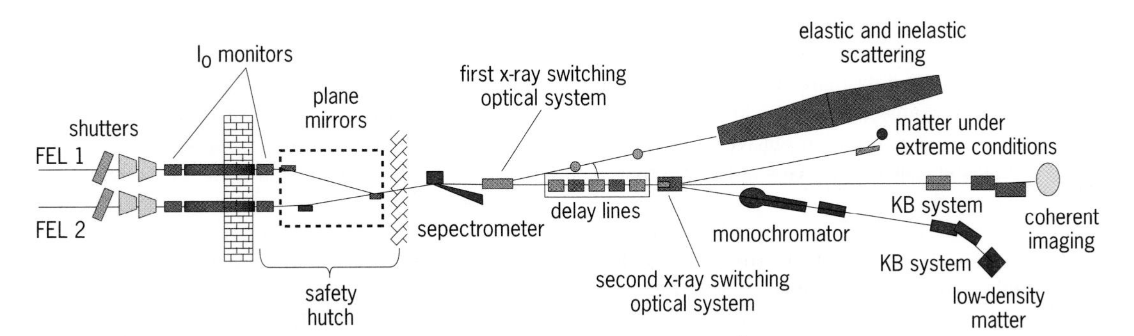

Fig. 6. Optical design of the FEL beamline system for the FERMI@Elettra FEL facility. The FEL pulses generated by FEL1 or FEL2 are brought by the photon beam-transport system (beamline) onto the samples inside the experimental stations. The I_0 monitors measure the intensity of the incoming FEL pulse (I_0). The beamline optical components are designed for focusing, splitting, and monochromatizing the photon beam. For FEL pulses, the optical components must be designed to preserve the ultrashort pulse time structure, while avoiding radiation damage that could be induced by the extremely high photon flux (up to 10^{15} photons per pulse) and the accompanying huge amounts of energy carried by the radiation pulses impinging on the mirrors, diffraction gratings, and beam splitters on the beamline. (*From E. Allaria et al., The FERMI@Elettra free-electron-laser source for coherent x-ray physics: Photon properties, beam transport system and applications, New J. Phys., 12:075002, 2010; Beamline conceptual design by D. Cocco, Sincrotrone Trieste*)

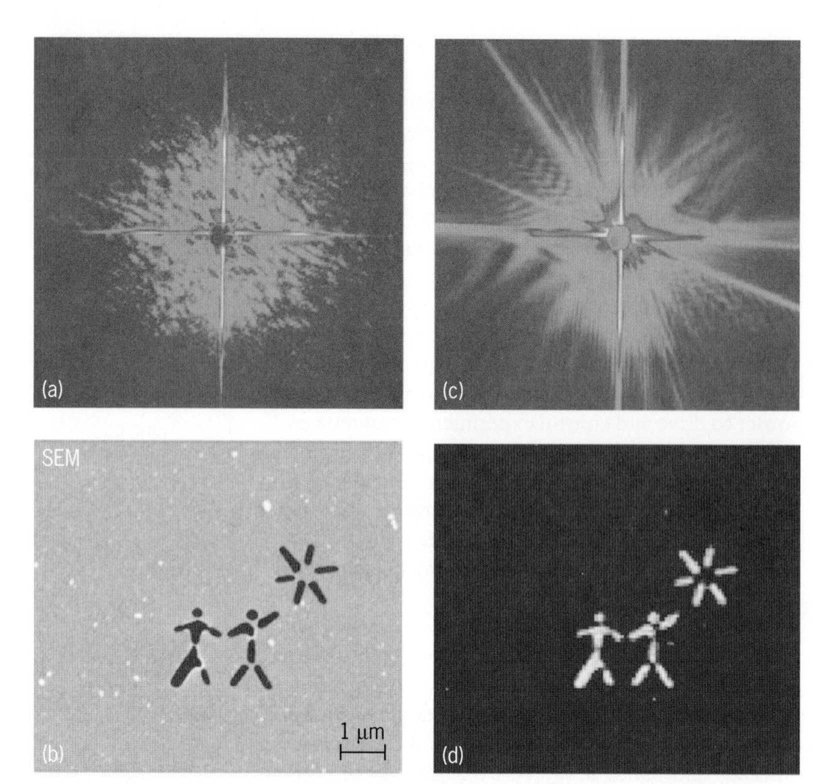

Fig. 7. Coherent diffraction images of an artificial structure imprinted on a Si_3N_4 window. (*a*) Electron diffraction of an image of the structure, as obtained by a scanning electron microscope (SEM). (*b*) The image of the structure whose electron diffraction is shown in *a*, as seen by the same SEM. (*c*) Optical diffraction image of the structure obtained with a single femtosecond EUV-wavelength pulse from the FLASH FEL facility at DESY (Deutsches Elektronen-Synchrotron in Hamburg, Germany). (*d*) The real image, as reconstructed by suitable algorithms from the optical diffraction pattern shown in *c*. As can be seen, the resolutions obtained by the SEM in *b* and by the FEL single pulse in *d* are comparable. This technique, together with proven synchronization methods, could be extended to spatial resolutions below 1 nm and temporal resolutions of a few femtoseconds, allowing ultrafast imaging with a resolution comparable to that obtained by conventional electron microscopes. (*From A. Barty et al., Ultrafast single-shot diffraction imaging of nanoscale dynamics, Nat. Photon., 2:415–419, 2008*)

possibility of observing dynamic processes in the femtosecond and possibly attosecond time domains. In particular, ultrashort soft-x-ray pulses, with photon energies below approximately 3 keV, will allow us to expand the horizons of our knowledge about the dynamics of the motion of electrons in atoms and molecules, or while forming a bond, transiting from one state to another, or forming collective excited states. The extension to ultrashort hard x-rays (photon energies greater than 3 keV) will permit observation of the motion of atoms in hard and soft matter (biomaterials) under impulsive excitations.

However, the first and perhaps most effective results will come from the use of these novel radiation sources as supermicroscopes, capable of combining the spatial resolution of electron microscopes with femtosecond dynamic-image resolution of in vivo samples.

For background information *see* ATOM CLUSTER; DIFFRACTION; DIRECTIVITY; LASER; OPTICAL PULSES; PARTICLE ACCELERATOR; SCANNING ELECTRON MICROSCOPE; SYNCHROTRON RADIATION; ULTRAFAST MOLECULAR PROCESSES; X-RAY DIFFRACTION; X-RAYS in the McGraw-Hill Encyclopedia of Science & Technology.

Fulvio Parmigiani

Bibliography. E. Allaria et al., The FERMI@Elettra free-electron-laser source for coherent x-ray physics: Photon properties, beam transport system and applications, *New J. Phys.*, 12:075002 (17 pp.), 2010; D. H. Bilderback, Review of third and next generation synchrotron light sources, *J. Phys. B Atom. Mol. Opt. Phys.*, 38:S773–S797, 2005; C. A. Brau, *Free-Electron Lasers*, Academic Press, London, 1990; P. Emma, First lasing of the LCLS x-ray FEL at 1.5 Å, paper TH3PBI01, *Proc. Particle Accelerator Conference*, Vancouver, Canada, May 4–8, 2009; E. L. Saldin, E. A. Schneidmiller, and M. V. Yurkov, *The Physics of Free Electron Lasers*, Springer-Verlag, Berlin, 2000.

Ultralong-range Rydberg molecules

The physical world around us is a result of atoms that have been bound together by various interaction mechanisms. The most prominent of these are covalent bonds and ionic bonds, which are responsible for the structure of almost all matter around us. Humankind has examined these bonding mechanisms for many centuries and has made remarkable progress in combining matter at will for specific applications. A crucial prerequisite for designing arbitrary molecules is a detailed understanding of the interaction mechanisms among the binding partners and how to control them. Nowadays this can be done with exactly two atoms with almost complete control over their internal and external degrees of freedom. Their physical nature is therefore determined by quantum mechanics, and with this, a new research field in quantum chemistry is established. In quantum chemistry, the reaction dynamics are no longer described by thermodynamic properties of ensembles; they are described by well-controlled quantum states with deterministic reaction channels.

In the case of Rydberg molecules, two neutral atoms are fused together by absorption of a photon by one of the atoms, and the resulting diatomic bond is stable only as long as the molecule remains in an excited state. In a Rydberg atom, at least one electron is excited to a state close to the ionization continuum. The weakly bound electron lives far away from the remaining core, behaving almost like a free electron. Therefore, the size of a Rydberg atom, which also determines the size of the Rydberg molecule, can be extremely large. The observed molecules have internuclear distances of a tenth of a micrometer, the size of a small virus, and are 1000 times larger than the N_2 molecule.

Binding mechanism. The assembly of a ground-state atom and a Rydberg atom into a diatomic molecule is an elegant example of quantum chemistry, with rich physics to be explored. The binding mechanism relies on the ultralow-energy scattering between the weakly bound electron of the Rydberg atom and a nearby ground-state atom, which produces an attractive force. Therefore, the

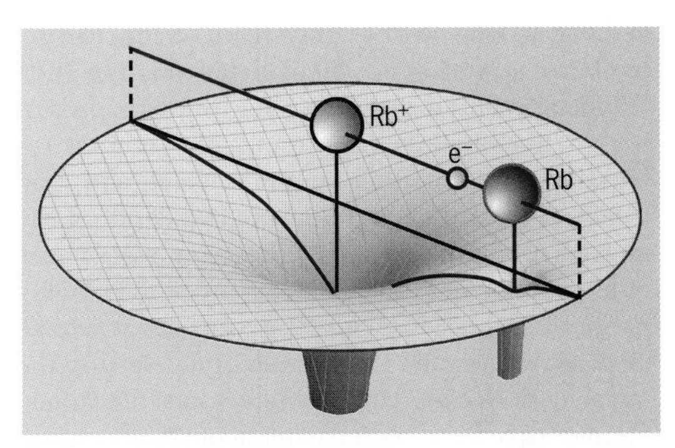

Fig. 1. Diagram of a Rydberg molecule. The electron of a highly excited Rydberg atom has caught a nearby atom in its ground state. The ground-state atom is polarized by the orbiting electron, which results in an attractive force. The surface represents the potential energy of the electron in the molecule. For a symmetric *S* state of the Rydberg atom, the angular momentum of the electron vanishes, and the classical analog of the electron motion in the 1/*r* Coulomb potential is linearly through the core.

ground-state atom is "caught" by the electron, and the electron is still attached to the core of the Rydberg atom (**Fig. 1**).

To a good approximation, the remaining core establishes a $1/r$ Coulomb potential at the position of the distant electron (where r is the distance to the nucleus of the Rydberg atom), just as in the hydrogen atom. The core itself produces only small corrections to the electronic wave functions, and these can be included precisely by applying quantum defect theory, in which the binding energies are adjusted compared to those of the hydrogen atom, based on experimental data. On the other hand, the electron also brings along an electric field, which is seen by a ground-state atom that happens to be nearby. This neutral atom is now polarized by the electric field of the electron, which leads to an α/r^4 interaction potential, where α is the polarizability of the ground-state atom and r is now the distance of the electron from the ground-state atom. Because of the low kinetic energies during the scattering of the electron close to the classical turning point in the $1/r$ Coulomb potential and the ground-state atom almost at rest, the scattering process must be calculated fully quantum mechanically.

Assuming a collision of a neutral atom at rest and an electron with momentum $\hbar k$, where $\hbar$ is Planck's constant divided by 2π, the momentum-dependent scattering length is given in atomic units in the low-energy limit by Eq. (1), which is a good first approximation.

$$a(k) = a_0 + (\pi/3)\alpha\, k + \ldots \qquad (1)$$

The scattering length parameterizes the scattering event by the range of a hard-sphere potential, which is in accordance with the classical elastic collision process of two spheres that each have radius a. Surprisingly, this length scale can also become negative in the quantum world. In the limit of zero velocities, the scattering length is given by a_0, which is negative in the case of rubidium.

Although there exists an attraction between the slow electron and the polarized atom at rest, given by the $1/r^4$ potential, the electron is still much too fast for the much heavier atom to follow. The effective interaction potential seen by the ground-state atom results from the time average of many scattering events of the electron. In the framework of a Dirac pseudopotential, the interaction potential of the electron with the ground-state atom is approximated by a pointlike delta function (which is infinite at the origin and zero everywhere else). This leads to an effective potential given by Eq. (2), where now

$$V_{\text{eff}} = 2\pi a(k)\, |\psi_e|^2 \qquad (2)$$

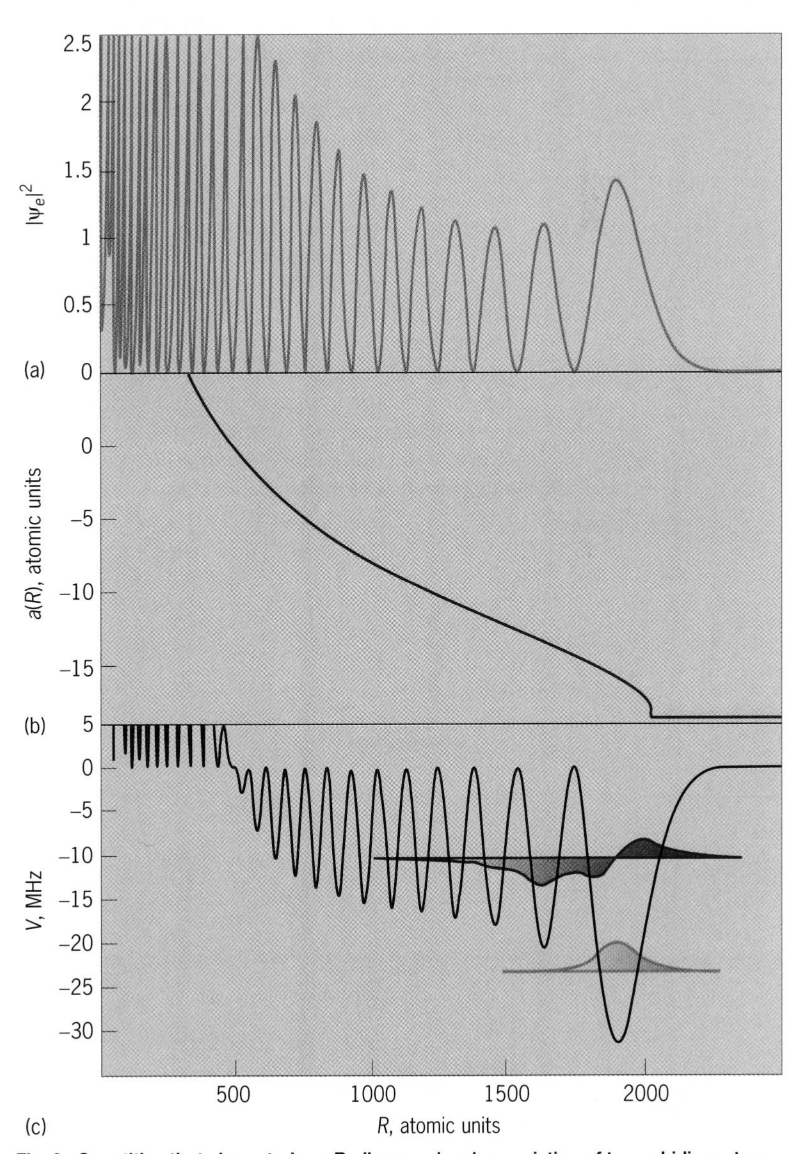

Fig. 2. Quantities that characterize a Rydberg molecule consisting of two rubidium atoms, with the Rydberg atom in the *S* state with principal quantum number $n = 35$. (*a*) Squared density, $|\psi_e|^2$, of the excited electron of the Rydberg atom, plotted as a function of the distance, *r*, of the electron from the nucleus of the Rydberg atom in atomic units (Bohr radii). (*b*) Velocity-dependent electron-atom scattering length, *a*(*R*). (*c*) Effective molecular potential, *V*, which is proportional to the product of the squared electron density and the scattering length. In the low-energy limit of purely *S*-wave scattering, the depth of the potential supports two bound states with an interatomic spacing of almost 0.1 μm.

the scattering length $a(k)$ and the wave function of the electron ψ_e determine entirely the nature of the bond.

What remains to be determined is an expression for the scattering length as a function of position. In a semiclassical picture, one can assume that an electron moves in a $1/r$ Coulomb potential according to Newton's laws, and compute classically the motion of the electron in the Coulomb potential of the remaining core. The extension (size) of the Rydberg atom, and thereby of the effective potential, are then given by the position of the classical turning point at $\hbar k = 0$, which is determined by the quantum-mechanical binding energy of the Rydberg state. As an example, the corresponding potential is shown for rubidium in **Fig. 2** for the principal quantum number $n = 35$. The scattering length is negative for small momenta, and therefore the ground-state atom sees an attractive potential close to the classical turning point. Closer to the core, the velocity of the electron increases, the scattering length changes its sign, and the effective potential becomes repulsive. Nevertheless, the depth of the potential at large distances is sufficient to support two vibrational bound states.

This very simplified picture is valid only for bound states that are localized close to the classical turning point in order to be in agreement with the low-energy scattering limit. A more complete description must also include higher kinetic energies in the scattering process as well as the back action of the ground-state atom on the wave function of the Rydberg electron. This is quite a demanding task, since many other Rydberg levels have to be included in the scattering process. The improved effective potential for $n = 35$ of rubidium is shown in **Fig. 3**.

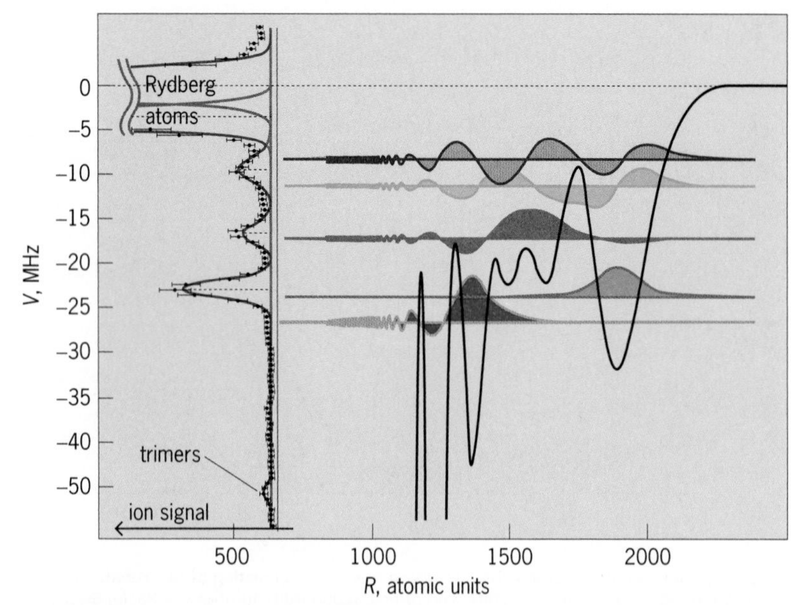

Fig. 3. Results of a full calculation of the effective molecular potential, V, for $n = 35$. There are more bound states than in the simplified model shown in Fig. 2. A steep drop of the potential at roughly 1200 Bohr radii leads to a localization of the two highest-lying states by quantum reflection. The positions of the spectroscopic lines that are seen in the ion signal, which is plotted at the left end of the figure, nicely agree with the calculations, and the existence of trimer states is well supported.

Experimental preconditions. The physical properties of the ultralong-range molecules require extreme starting conditions for the reactants and good shielding from the environment. The depth of the molecular potential is of the order of only a few tens of megahertz, and the kinetic energy of the binding partners or collisions with other particles at submillikelvin temperatures will immediately destroy the molecules. To observe the bound states, the atomic samples have to be cooled to ultracold temperatures and placed in an ultrahigh vacuum to avoid collisions with molecules of the background gas. On the other hand, the distances between the cooled atoms must be smaller than the size of the Rydberg atoms in order for a nearby atom to be caught by the electron, which results in rather high densities. With the development of laser cooling, atom trapping, and evaporative cooling, such cold clouds of atoms can be generated and are used in many experiments dealing with quantum-degenerate gases. After preparation of the sample in a purely magnetic trapping potential under ultrahigh-vacuum conditions, the excitation into Rydberg states is carried out by narrowband, continuous-wave lasers. To obtain the number of excited states, a strong electric field is applied, and the weakly bound electron is pulled out. The resulting ion is then guided into a multichannel plate for efficient detection.

Spectroscopy. Chemists have developed many methods to confirm the production of a certain molecule, such as mass spectrometry, absorption or fluorescence spectroscopy, and analytical reactions, none of which are applicable here. The actual generation and detection of ultralong-range Rydberg molecules is achieved by direct photoassociation of two atoms into the bound state, subsequent field ionization, and the detection of ions on a multichannel plate. Initially, the cold atomic sample contains on the order of 10^8 spin-polarized rubidium atoms in the ground state at a temperature of a few microkelvins and densities of the order of 10^{19} atoms/m^3. The density corresponds to a mean interatomic distance of well below 1 μm, and the probability of finding a pair of atoms within the classical Kepler radius of the Rydberg atom is quite high. The excitation of only one atom to the Rydberg state is ensured by the large van der Waals interaction between two Rydberg atoms. At small distances, only one atom can be excited into a Rydberg state, whereas the other one is shifted out of resonance and therefore blocked.

To excite the molecular state directly, the laser is red-detuned by the binding energy with respect to the position of the atomic excitation. For the principal quantum number around $n = 35$, typically up to five molecular states are found per quantum number within a frequency range of a few tens of megahertz. The positions of the lines agree nicely with the theoretical predictions of the full Green's function approach (Fig. 3).

The molecular states have lifetimes of many microseconds, which is a factor of 3 shorter than the lifetimes of the atomic Rydberg states, but still long enough to allow high-resolution spectroscopy, that is, at resolutions well below 1 MHz. The shorter lifetime is actually not a result of distortion of the electron wave function of the Rydberg atom by the presence of the ground-state atom; instead, it indicates the presence of additional decay channels. The details of such decay processes are unknown, but the observation of Rb_2^+ ions has already given some hint of the underlying reaction dynamics.

Trimers and more. At even lower energies of the excitation laser, signals at twice the energy of the most prominent molecular lines suddenly appear. They can be assigned only to a bound state consisting of one Rydberg atom and two ground-state atoms. This is the first demonstration of a direct photoassociation of a trimer state, a molecule consisting of three atoms, and at moderately higher densities, polymer states with even more atoms should be within reach.

Another peculiarity in some of the dimer states is an additional localization of the ground-state atom by quantum reflection. For example, the two highest-lying states in Fig. 3 live energetically above the potential landscape in a metastable state. The steep drop of the effective potential toward the core would not hinder a classical particle from entering the much deeper potential well, but in the quantum world, the wave function of the ground-state atom is not able to adjust fast enough to the sudden increase of the kinetic energy and therefore gets reflected. The measured lifetimes indicate that the ground-state atom is reflected at the edge at least ten times.

There is much more to be discovered, since so far the rotational structure of the molecules, their behavior in electric fields, the dynamics of the bound states, and so forth have not been examined. One can envisage these molecules as a microscopic laboratory for low-energy scattering of electrons and rubidium atoms. Actually, the binding mechanism shown here is not restricted to rubidium, and in principle any polarizable particle (including other atomic species or even molecules) could be trapped in the wave function of a high-lying Rydberg state of an atom and a molecule.

For background information *see* CHANNEL ELECTRON MULTIPLIER; CHEMICAL BONDING; INTERMOLECULAR FORCES; LASER COOLING; LASER SPECTROSCOPY; MOLECULAR STRUCTURE AND SPECTRA; NONRELATIVISTIC QUANTUM THEORY; PARTICLE TRAP; QUANTUM CHEMISTRY; QUANTUM MECHANICS; RYDBERG ATOM; SCATTERING EXPERIMENTS (ATOMS AND MOLECULES) in the McGraw-Hill Encyclopedia of Science & Technology. Robert Löw

Bibliography. T. F. Gallagher, *Rydberg Atoms*, Cambridge University Press, Cambridge, U.K., 2005; C. H. Greene, Quantum chemistry: The little molecule that could, *Nature*, 458:975-976, 2009, doi:10.1038/458975a; S. D. Hogan and F. Merkt, A new perspective on the binding power of an electron, *ChemPhysChem*, 10:2931-2934, 2009, doi: 10.1002/cphc.200900499; P. van der Straten and H. J. Metcalf, *Laser Cooling and Trapping*, Springer, Berlin, 2001.

Using melt inclusions for understanding crustal melting processes

The Earth's continental crust may start to melt when its temperature exceeds about 650°C. This process, called anatexis, occurs in anomalously hot areas of the crust, such as young orogenic belts or continental rifts. By producing a melt of broadly granitic composition and a silica-poor, mafic solid residue, crustal anatexis is of paramount importance in shaping the continental lithosphere. On the one hand, it determines the geochemical layering (differentiation) of the crust; on the other, it allows—by the lubricating and weakening effects of the melt—much easier and faster deformation of the crust, with important tectonic and geodynamic implications. Just as for any other geological process occurring in an inaccessible area of the Earth, several aspects of crustal melting are still poorly understood by scientists, whose studies are mostly focused on either the experimental reproduction of anatexis in the laboratory or the characterization of the most common natural products of crustal melting: migmatites. Migmatite is a rock that has partially melted. It often consists of two parts: a clear leucosome, representing the crystallized melt, and a dark melanosome, representing the residual, solid material. One major unknown is the composition of the natural melts produced during anatexis, as both leucosomes in migmatites and allochthonous crustal granites appear to have been modified and/or contaminated to variable degrees after the melts have formed.

In recent years, the need to fill this gap in knowledge has prompted a new approach to the characterization of the composition of natural crustal melts: the study of melt inclusions.

The perspective: entrapment during incongruent melting. Building on the extensive work and literature on fluid inclusions and melt inclusions in mafic rocks, we realized that tiny droplets of the melt phase produced during crustal anatexis can be trapped by, and preserved within, those minerals that grow simultaneously with the melt—that is, the peritectic phases produced during incongruent melting reactions. For example, a garnet crystal that forms during the incongruent melting of biotite has the potential for trapping primary inclusions of the melt that it is in contact with. While straightforward, this perspective has found little application to migmatites and granulites in the past. This was probably due to technical obstacles; in fact, the study of micrometer-sized melt inclusions requires changing the scale of observation of rocks and using magnifications that are not routine.

Melt inclusions in enclaves and xenoliths. The inclusions of anatectic melt trapped during incongruent

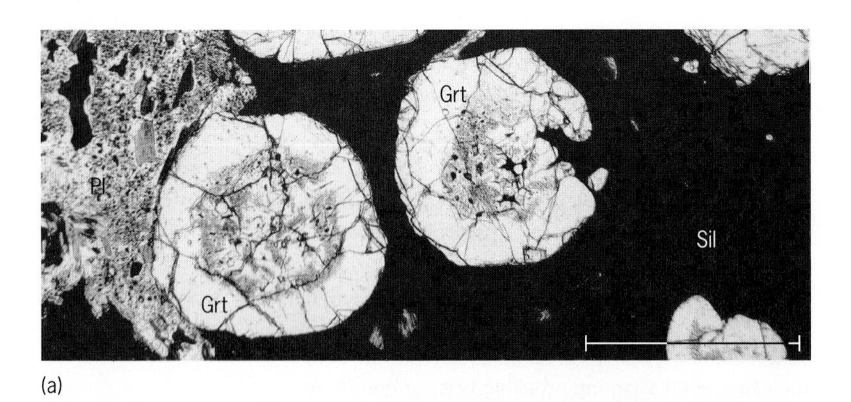

(a)

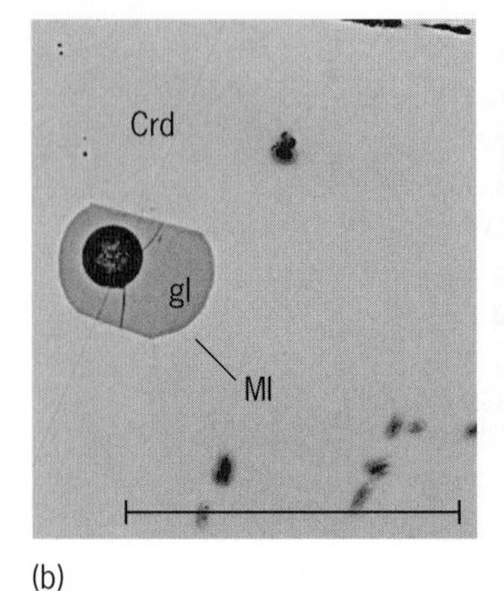

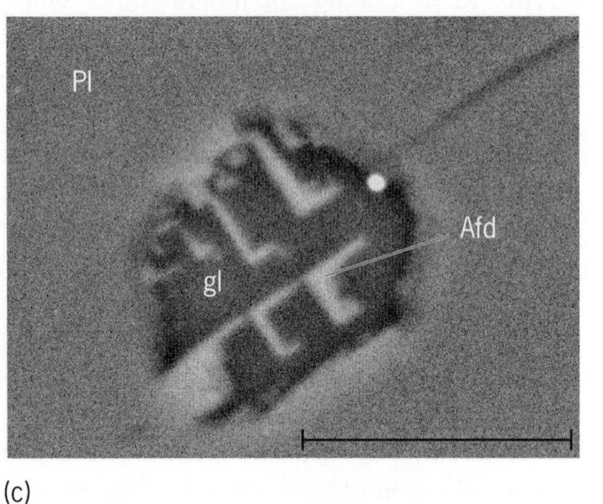

(b) (c)

Fig. 1. Glass inclusions in the enclaves from El Hoyazo. (*a*) Photomicrograph showing a thin-section view of a typical garnet-biotite-sillimanite enclave, with euhedral garnet (Grt) porphyroblasts surrounded by plagioclase (Pl) and a sillimanite-glass intergrowth (Sil). Glass inclusions are concentrated in the cores of garnet, giving crystals a "dusty" appearance. (*b*) Backscattered scanning electron microscope (SEM) image of a melt inclusion (MI) in cordierite (Crd), consisting of undevitrified glass (gl) with a shrinkage bubble. (*c*) Backscattered SEM image of a partly crystallized melt inclusion in plagioclase, showing dendritic crystallization of alkali feldspar (Afd). Scale bars correspond respectively to 5 mm, 100 μm, and 15 μm. (Images *b* and *c* from *A. Acosta-Vigil et al., Microstructures and composition of melt inclusions in a crustal anatectic environment, represented by metapelitic enclaves within El Hoyazo dacites, SE Spain, Chem. Geol., 237:450–465, 2007*).

melting can solidify to, and be preserved as, glass if the partially melted rock is cooled very rapidly from the high temperatures at which it was residing. This may occur in fragments of crustal rocks (called xenoliths or enclaves) entrained in lavas that rapidly ascend to and extrude on the Earth's surface. An example of this geological context is the small volcano of El Hoyazo (southeast Spain), where dacitic lava is rich in crustal enclaves consisting of garnet, biotite, sillimanite, plagioclase, graphite, and glass. The mineralogical and chemical composition of these rocks is comparable to that of melanosomes in migmatites, but here abundant glass is present, particularly as melt inclusions in all the minerals. The inclusions range in size from 5–50 micrometers (rarely up to 100 μm), have a primary texture, contain fresh and undevitrified glass, and show very little evidence of melt crystallization upon cooling (**Fig. 1**). Along with textural evidence, the peraluminous leucogranitic and close-to-eutectic compositions of glasses (see below) support the conclusion that they represent natural anatectic melts. In addition, the evidence of extensive deformation coeval with melting implies that anatexis of these rocks took place in a regional metamorphic setting, and not through melting induced by contact with the hot lavas.

The study of enclaves has one major drawback; that is, most of the currently known examples of anatectic enclaves do not represent common anatectic contexts such as regionally metamorphosed migmatites and granulites, but instead represent contact metamorphic scenarios. Therefore, further data are needed in order to understand how representative enclaves are of the major petrologic processes occurring in the continental crust, and how relevant is the information they bear to the petrogenesis of the widespread migmatites and granulites. Nonetheless, this type of occurrence has boosted an important development of melt inclusion studies by raising the questions: Why shouldn't similar inclusions also occur in migmatites and granulites? And if they do occur, what will they look like?

Nanogranites and glass inclusions in migmatites. Based on the experience acquired with glass inclusions in enclaves, a systematic search was started for the existence of inclusions of anatectic melt in garnet and other peritectic phases in crustal anatectic migmatites. Several migmatite and granulite localities worldwide were targeted, and the first occurrence of melt inclusions that was found was in the slowly cooled granulites of the Kerala Khondalite Belt (KKB) in southern India. There, the inclusions are hosted within garnet, where they are distributed toward the core of the crystals and have a tendency to group and form clusters. Those in the range of 15–25 μm are fully crystallized to a cryptocrystalline aggregate of quartz, alkali feldspar, biotite, and plagioclase (**Fig. 2*a***). Owing to their very small size, these inclusions could easily be overlooked or mistaken for dust or the result of bad sample preparation. A tip for their identification is through optical analysis under crossed polarizers, where their polycrystalline, birefringent nature is revealed. The inclusions generally have a negative crystal shape, and the minerals in the polycrystalline aggregates form crystals with similar dimensions, only in part with well-formed shapes, and ranging in size from hundreds of nm to a few μm. In places, inclusions contain granophyric to micrographic intergrowths of quartz and feldspars. Given the microstructural and chemical features (see below), the cryptocrystalline aggregate found within these inclusions was named "nanogranite." Another exceptional and intriguing discovery is that, within the same cluster of nanogranite inclusions, some inclusions <15 μm are still completely glassy despite the very slow cooling rate of the host rocks (Fig. 2*b*). The solidification of melt to glass and its preservation in regionally metamorphosed migmatitic rocks, confirmed by subsequent findings in other localities, can be explained by the kinetics of crystallization, in particular by the inhibition of nucleation in the inclusions with the smallest volumes, by analogy with observations of the behavior of aqueous solutions in sediments or of glass in films and pores of contact metamorphic rocks.

Since the findings in the KKB rocks, nanogranite and glassy inclusions also have been identified in regional metamorphic migmatites from other geological settings with various pressure-temperature (P–T) conditions, such as the Ivrea Zone and the Ulten Zone (Italy), Ronda (Spain), the Barun Gneiss (Nepal), and south-central Massachusetts (United States). Inclusions are hosted in garnet at all these localities. In addition, migmatites from Ronda also show melt inclusions in ilmenite, another peritectic mineral that grows during the melting of titanium-rich biotite (Fig. 2*c*).

Composition of melt inclusions. The existence of primary melt inclusions, such as those described here, allows the direct analysis of crustal anatectic melts. Major element concentrations can be analyzed by electron microprobe (EMP), whereas trace element contents are obtained by laser ablation-inductively coupled mass spectrometry (LA-ICP-MS). These techniques can be applied directly to melt inclusions that solidified to glass in enclaves or migmatites, but not to those crystallized to a cryptocrystalline nanogranite. The latter ones need to be re-homogenized to the original melt by heating the inclusions in an experimental device. Experimental remelting of nanogranite has been successfully done in samples from the KKB and Ronda, by means of both a high-temperature heating stage and a piston-cylinder apparatus (Fig. 2*d*). Once it has been remelted and quenched to a glass through fast cooling, the inclusion can be analyzed by the above analytical methods.

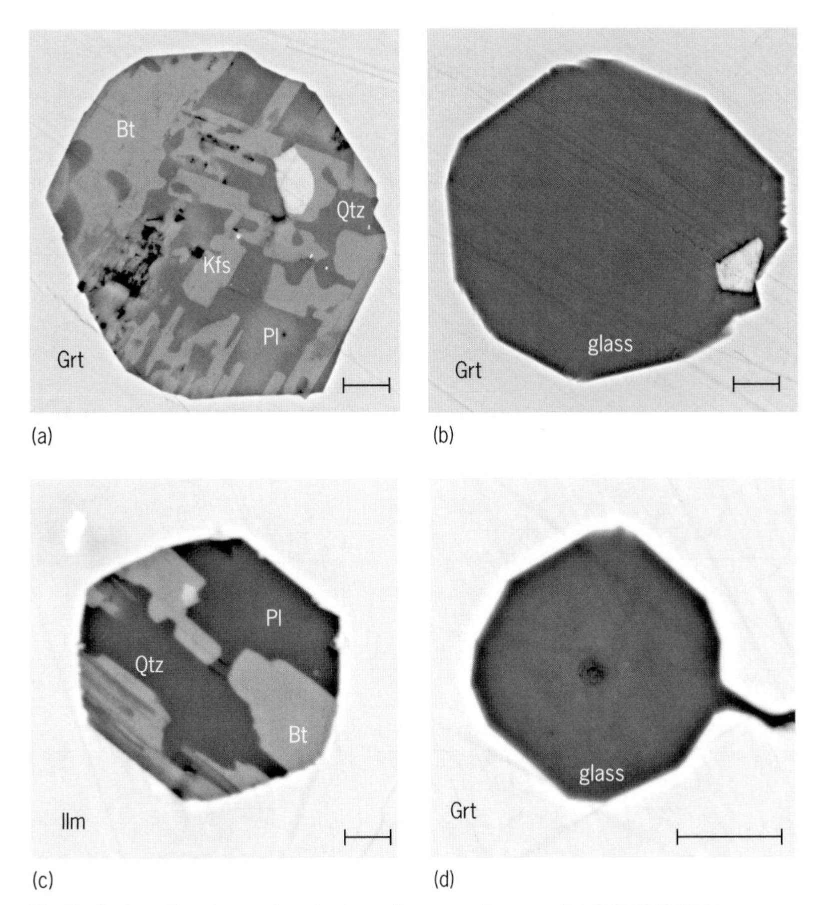

Fig. 2. Backscattered scanning electron microscope images of melt inclusions in migmatites. (*a*) Nanogranite inclusion in garnet (Grt) from a granulite of the KKB, consisting of quartz (Qtz), K-feldspar (Kfs), biotite (Bt), and plagioclase. (*b*) Glassy inclusion in the same rocks. (*c*) Nanogranite inclusion in ilmenite (Ilm) from a migmatite of Ronda. (*d*) Remelted nanogranite inclusion in garnet from a migmatite of Ronda. Scale bar in all photos corresponds to 2 μm.

The **table** and **Fig. 3** summarize the major element chemical compositions and some trace element concentrations of the melt inclusions described here. Inclusions from all contexts (enclaves and migmatites) always have a peraluminous, granitic composition consistent with experimental results on the partial melting of aluminum-rich metasedimentary parent rocks. Interestingly, the analyses show systematic compositional patterns and variations. Melt inclusions in several host phases from El Hoyazo enclaves (plagioclase, garnet, cordierite, and ilmenite) show different compositions. Comparison of textural and chemical data suggests the progressive entrapment of chemically evolving melts during prograde

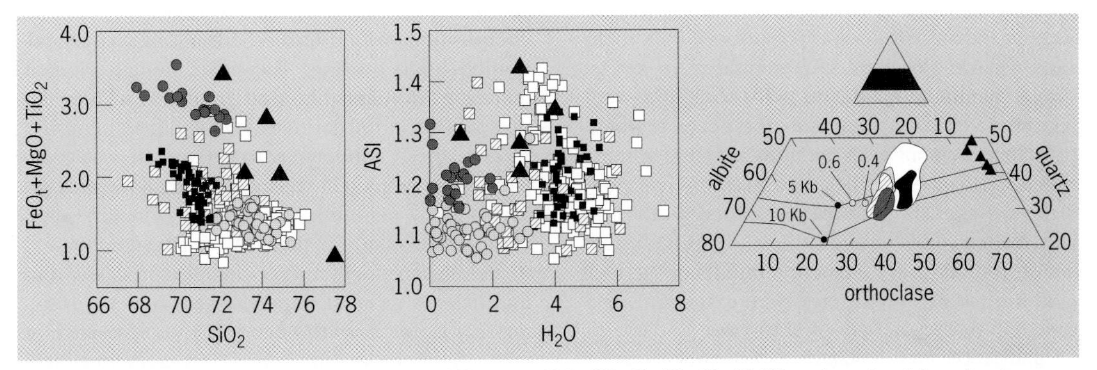

Fig. 3. Major element concentrations, molar ratio ASI = mol. $Al_2O_3/[(CaO)+(Na_2O)+(K_2O)]$, and quartz-albite-orthoclase normative compositions (in wt%) of glasses from El Hoyazo enclaves and KKB migmatites. Glasses from El Hoyazo: white squares or field = MI in plagioclase; stippled squares or field = MI in garnet; light gray circles or field = MI in Crd; dark gray circles or field = MI in Ilm; black squares or field = matrix glasses. Glasses from the KKB migmatites: black triangles. Black dots and lines in the quartz-albite-orthoclase diagram refer to eutectic points and cotectic lines, respectively, of the H_2O-saturated haplogranite system at 5 and 10 kbar. Gray dots in the quartz-albite-orthoclase diagram refer to the eutectic points of the H_2O-undersaturated haplogranite system at a_{H2O} = 0.6 and a_{H2O} = 0.4.

Mean electron microprobe and laser ablation ICP-MS analyses of glasses from El Hoyazo enclaves and the KKB migmatites (37 and 7)

Ocurrence Microstructure	Enclave MI in Pl	Enclave MI in Grt	Enclave MI in Crd	Enclave MI in Ilm	Enclave Matrix Gl	Migmatite Nanogr MI	Migmatite Glassy MI
Electron microprobe concentrations, wt%							
No. analyses	145	43	51	18	48	37	7
SiO_2	72.90	71.51	73.56	70.68	70.60	73.08	75.41
TiO_2	0.08	0.10	0.07	0.31	0.17	0.08	0.11
Al_2O_3	12.87	14.45	14.07	15.69	13.99	13.31	12.09
FeO*	1.14	1.66	1.28	2.58	1.55	3.03	1.58
MnO	0.01	0.07	0.04	0.07	0.01	0.03	0.03
MgO	0.15	0.04	0.04	0.13	0.14	0.76	0.25
CaO	0.22	0.60	0.94	0.96	0.36	0.60	0.08
Na_2O	2.86	3.62	3.41	3.55	2.96	1.14	0.62
K_2O	5.10	4.92	4.87	4.92	5.54	6.76	7.73
P_2O_5	0.19	0.34	0.20	0.31	0.29	0.09	0.04
F	0.06	0.07	0.05	0.07	0.06	n.d.	n.d.
Cl	0.01	0.01	0.45	0.00	0.01	0.25	n.d.
O=F	−0.02	−0.03	−0.02	−0.03	−0.02	–	–
O=Cl	0.00	0.00	−0.10	0.00	0.00	−0.05	–
H_2O by diff	4.42	2.61	1.10	0.82	4.32	0.82	2.06
ASI	1.22	1.17	1.12	1.22	1.22	1.30	1.27
#K	0.54	0.47	0.48	0.48	0.55	0.79	0.89
#Mg	0.19	0.04	0.04	0.08	0.13	0.30	0.15
Laser ablation ICP-MS analyses concentrations, ppm							
No. analyses	18 to 22	8 to 12	4	3	20 to 28		
Rb	212	237	250	206	211	n.d.	n.d.
Sr	28	112	135	242	82	n.d.	n.d.
Ba	80	235	543	512	355	n.d.	n.d.
V	0.72	0.16	2.0	9.7	2.3	n.d.	n.d.
Zr	24	28	7.7	34	37	n.d.	n.d.
Th	1.53	1.1	0.83	3.3	3.5	n.d.	n.d.
U	4.0	4.2	3.9	20	2.9	n.d.	n.d.
Sum REE	18	21	35	69	44	n.d.	n.d.
Eu/Eu*	0.69	2.5	1.2	1.2	1.4	n.d.	n.d.
$(La/Lu)_N$	11	32	4.6	2.2	13	n.d.	n.d.

* Total Fe as FeO.
#K = mol. $K_2O/(K_2O + Na_2O)$
#Mg = mol. MgO/(MgO + FeO*)
Sum REE refers to total rare earth element concentration, and Eu/Eu* to the Europium anomaly; $(La/Lu)_N$ refers to the ratio Lanthanum/Lutetium, normalized to the chondritic values

heating of these rocks. For instance, the composition of melt inclusions in plagioclase and garnet indicates relatively low (700–750°C) temperatures of formation and supplies information on the prepeak anatectic history of the rock. Conversely, the ultrapotassic and anhydrous compositions of melt inclusions in garnets from the KKB granulites are very far from a "minimum melt" and point to melting temperatures well above minimum or eutectic temperatures, in agreement with the ultra high-temperature (>900°C) origin of the rock. In our experience, most of the petrogenetic information gained on the partially melted rocks through the study of inclusions comes from the trace element contents of the melt inclusions, which so far have been extensively analyzed only in the enclaves of El Hoyazo.

Significance and problems. Glass inclusions in these metasedimentary enclaves and migmatites are evidence of the melt phase that was produced during crustal melting and trapped by the peritectic minerals growing simultaneously during incongruent melting reactions. The inclusions of nanogranite represent their crystallized counterpart. The inclusions were trapped during a prograde event of magma formation, and hence they differ from other examples of granitic inclusions found in minerals from plutonic and volcanic rocks of granitoid composition, which formed during magma crystallization upon cooling. The novel finding of melt inclusions in migmatites and granulites widens the perspective in crustal petrology and geochemistry. In fact, these nanometer- to micrometer-scale objects, so far studied in volcanic and shallow plutonic rocks, allow for the direct analysis of natural melt compositions and for their use in constraining and modeling petrologic and geochemical processes during anatexis. As an example, along with demonstrating that garnet growth occurred in the presence of melt, the inclusions from the KKB granulites highlight the pitfalls of considering almost binary normative quartz (Q)–K-feldspar (Or) compositions as corresponding to fractionated anatectic melts, or assuming that anatectic melt has a minimum melt composition.

There are some problems associated with the study of melt inclusions. A major problem is analytical and relates to the small size of these objects. On the one hand, this implies that the size of the beam may be larger than the crystal or glass portion that is being analyzed and therefore may include some of the surrounding material; on the other, the use of focused beams enhances the loss of alkalis [in particular sodium (Na)] during analysis. Another problem is methodological and resides in the necessity of determining, case by case, the extent to which inclusions preserve their primary features, including the degree of chemical interaction with the host and the degree of crystallization upon cooling. Even processes taking place during entrapment have to be carefully evaluated in order to assess the extent to which boundary-layer phenomena affect the composition of the trapped melt so that it deviates in composition from the bulk melt in the system at the time of entrapment. A solution to this problem requires the acquisition of, and comparison among, large and high-quality analytical datasets on many occurrences worldwide. Up to now, such a dataset exists only for the El Hoyazo enclaves, and work is in progress for a dataset for migmatitic rocks from the Kerala Khondalite Belt (India) and from Ronda (Spain).

The trapping of melt inclusions during anatexis depends on parameters such as the amount of melt, the stress field acting on the rock, the growth rate of the peritectic host mineral, and the presence of very fine-grained crystals in the rock matrix. In addition, the preservation of inclusions depends on the extent of chemical interaction with, and mechanical behavior of, the host mineral during the subsequent history, as microfracturing would allow the access of fluids and alteration of the primary melt composition or nanogranite assemblage. Therefore, melt inclusions in the most chemically inert and mechanically strong mineral hosts (for example, spinel, ilmenite, and zircon) from the least deformed rock domains should be targeted. Not least, our studies also show that finding anatectic melt still "frozen" as glass is possible even in slowly cooled regional migmatites and granulites, although we expect it to be restricted to inclusions $<10–15\ \mu m$.

Outlook. Our discovery implies that from now on, the composition of anatectic melts potentially can be analyzed rather than assumed, and hence we foresee that this investigation will stimulate research on anatectic enclaves in volcanic rocks and nanogranite inclusions in migmatites and granulites, with important impacts on a wide spectrum of disciplines, including metamorphic and igneous petrology, geochemistry and isotope geology, mineralogy, and volcanology. We also believe that many occurrences of melt inclusions have been overlooked because they simply were not searched for, and that they will be uncovered by careful reinvestigation of migmatite and granulite samples worldwide. In addition, the small size of glassy inclusions (often $<10\ \mu m$) and crystals within nanogranite (often $<1\ \mu m$) offers new challenges and applications for cutting-edge microanalytical techniques, such as field emission gun-based scanning electron microscopy (SEM) and EMP, LA-ICP-MS, nanoSIMS (secondary ion mass spectrometer), synchrotron-based micro-XRF (x-ray fluorescence) and micro-XRD (x-ray diffraction) spectroscopy. Rapid technological development is likely to eliminate all analytical obstacles in a few years.

For background information *see* BIREFRINGENCE; DACITE; GRANITE; IGNEOUS ROCKS; LAVA; MIGMATITE; XENOLITH in the McGraw-Hill Encyclopedia of Science & Technology.

Bernardo Cesare; Antonio Acosta-Vigil

Bibliography. A. Acosta-Vigil et al., Mechanisms of crustal anatexis: A geochemical study of partially melted metapelitic enclaves and host dacite, SE Spain, *J. Petrol.*, 51:785–821, 2010; M. Brown, Melting of the continental crust during orogenesis: The thermal, rheological, and compositional consequences of melt transport from lower to upper continental crust, *Can. J. Earth Sci.*, 47:655–694, 2010; B. Cesare, Crustal melting: Working with enclaves, pp. 37–55 in E. W. Sawyer and M. Brown (eds.), *Working with Migmatites*, Mineralogical Association of Canada, Quebec City, Quebec, Short Course 38, 2008; B. Cesare et al., Nanogranite and glassy inclusions: The anatectic melt in migmatites and granulites, *Geology*, 37:627–630, 2009; E. Roedder, Fluid inclusions, *Rev. Mineral.*, vol. 12 (644 pp.), 1984; E. W. Sawyer, *Atlas of Migmatites: The Canadian Mineralogist Special Publication 9*, Mineralogical Association of Canada, Quebec City, Quebec; NRC Research Press, Ottawa, 2008.

VASIMR propulsion system

Electric propulsion systems have the possibility of reducing the propellant mass needed for long-range planetary missions and drag compensation in orbiting satellites, as compared to chemical rocket systems. Electric propulsion reduces the mass needed for a given transfer of momentum to the space vehicle and thus its acceleration by exhausting the mass at a higher velocity. Electric propulsion systems are not used to launch payloads from the Earth's surface to low-Earth orbit because the force and energy required are too large. Instead, they are used in a space environment for drag compensation of space stations and for long space trips.

The thrust (measured in newtons) to the rocket is the product of exhaust velocity measured relative to the spaceship and the rate of propellant flow. A high exhaust velocity can produce more thrust with a lower mass flow, but at a cost of more energy. Momentum scales as mass times velocity, while kinetic energy scales as mass times velocity squared. In a chemically fueled rocket, the exhaust velocity is limited by the energy stored in chemical bonds to a velocity characteristic of a temperature of about 2000 K (3140°F). In contrast, an electric propulsion system can have an exhaust velocity characteristic

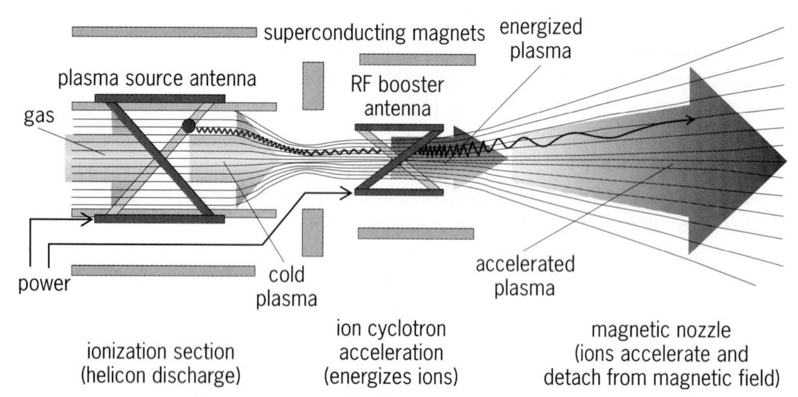

Schematic of a VASIMR plasma rocket. The first section ionizes a neutral gas that flows into the second section. In the second section, additional energy is added to the ions by a second radio-frequency (RF) antenna. The energetic ions are accelerated in the expanding magnetic field and ultimately detach from the confining magnetic field. The solid lines are the magnetic field lines. The wavy line that follows a magnetic field line is the trajectory of a plasma ion. (Ad Astra Rocket Company).

of temperatures greater than 100,000 K (180,000°F), depending only on the energy used to accelerate the ions. Another possible advantage is that an electric propulsion system can provide a controlled and variable thrust for a long period of time, whereas chemical rockets are generally explosive and operate only for a short period of time. The energy for electric propulsion can come from either a solar array or a nuclear reactor on board the rocket.

VASIMR (Variable Specific Impulse Magnetoplasma Rocket) is a plasma rocket that is designed to operate in either a low-thrust, high-specific-impulse mode for maximum fuel efficiency, or a high-thrust, low-specific-impulse mode when a higher thrust is needed. (The specific impulse is the thrust force times its duration divided by the weight of propellant used, measured in seconds.) The VASIMR concept was created by Franklin Chang Diaz at NASA's Johnson Space Center, where Chang Diaz was an astronaut. Development has since been privatized.

Principles of operation. The VASIMR engine is made up of three sections (see **illustration**).

Ionization section. The first section, a helicon discharge, transforms the propellant gas (made up of neutral atoms) into a plasma composed of electrons and ions, using radio-frequency (RF) power and an antenna in a confining magnetic field. In a helicon discharge, the currents in the antenna drive electron currents (waves) in the plasma that transport the electromagnetic radio-frequency energy into the plasma, where it is absorbed. The absorption of the wave radio-frequency energy is due to electron-ion collisions in the plasma. The magnetic field is important because it minimizes the interactions between the plasma and the material surfaces of the engine. In contrast to other electric propulsion techniques, there is no electrode in direct contact with the hot plasma. The magnetic fields are provided by either current-carrying magnetic field coils or high-temperature superconducting magnets.

The process of ionizing an atom (knocking an electron from a neutral atom) is usually done by a collision between an energetic electron and a neutral atom. The electrons are energized by absorption of the energy in the radio-frequency waves. The energy cost to create an electron-ion pair is the ionization energy of the atom plus any radiated energy, usually in the form of line radiation from the neutral atom or ion generated by collisions with the free electrons in the plasma. If any ions travel across the magnetic field and collide with the wall, they will recombine with an electron on the surface, causing a further energy loss, and increase the cost of an electron-ion pair in the exhaust of the rocket. The cost of an electron-ion pair depends on the atomic characteristics of the fuel and on the transport mechanisms operating in the helicon and is typically about 5–10 times the ionization energy of the fuel atom. In recent tests, the costs to extract an electron-ion pair in VASIMR (VX-100) with argon fuel were 78 ± 11 eV, as compared to an ionization energy of 15.8 eV for argon.

Typical electron temperatures in the ionization (plasma-producing) section are of the order of 50,000 K (90,000°F), depending on the propellant gas and the mode of operation. In this section, the ions are accelerated by the electric field in the plasma up to a velocity near the speed of sound before they enter the radio-frequency acceleration section.

Ion cyclotron acceleration. The second section, again in a strong magnetic field, accelerates the plasma ions further using a single-pass ion cyclotron absorption process to achieve the desired ion energy. A second radio-frequency antenna generates electromagnetic waves that resonate with the perpendicular motion of charged ions around the magnetic field. The absorption of radio-frequency energy increases the kinetic energy of the charged ions around the confining magnetic field lines (cyclotron motion). This kinetic energy is perpendicular to the magnetic field but is converted to kinetic energy parallel to the magnetic field when the magnetic field drops. Some of the wave energy also goes into plasma electrons.

By varying the neutral gas flow into the helicon and the relative power into the helicon and the ion acceleration section, the VASIMR engine can operate in either a high-thrust, low-specific impulse mode or a low-thrust, high-specific-impulse mode. For efficient operation, it will probably be necessary to accelerate the ions in the second section to an energy somewhat greater than the cost to create an electron-ion pair in the helicon section. The acceleration of ions in single-pass heating has been demonstrated in an experiment by E. A. Bering and colleagues.

Magnetic nozzle. In the third section, a magnetic nozzle, the confining magnetic field drops and plasma ion energy is transferred from rotational cyclotron motion perpendicular to the magnetic field to motion parallel to the magnetic field because of the conservation of the magnetic moment of the orbiting ion. That is, the ions are accelerated along the direction of the magnetic field by the reduction of the magnetic field. This process is also at work in the Earth's magnetosphere. The energetic plasma ions will then detach from the magnetic field and provide thrust to the rocket when the kinetic pressure of the ions is greater than the magnetic pressure, [$B^2/(2\mu_0)$, where B is the magnetic field and μ_0 is the magnetic constant, also known as the permeability of vacuum]. The detachment of the plasma from the magnetic field has been demonstrated in an experiment at Marshall Space Flight Center by C. A. Deline and colleagues.

Capabilities. An important capability of the VASIMR rocket is the possibility of varying its exhaust parameters depending on the mission requirements. To minimize trip time, the rates of fuel consumption and power usage can be controlled separately. The use of radio-frequency electromagnetic waves to transfer energy to the plasma also minimizes plasma-surface interactions, which should yield a robust and long-lived plasma rocket. Almost any gas can be used as a fuel for VASIMR, since the energy for the rocket comes from radio-frequency fields instead of the chemical bonds of the fuel. Another advantage of the VASIMR rocket is the possibility of scaling to larger sizes or powers. That is, there do not appear to be fatal flaws in going to larger systems.

Current status. A demonstration rocket in the Ad Astra Laboratory, the VX-200, with superconducting magnets has been operated at a total power of 200 kW for a time of 15 s using argon as the propellant gas. It produces a thrust of the order of 10 N (2 lb). Preparations for the VF-200, the flight version of VASIMR, are underway, with expectations for a launch in late 2013 to the *International Space Station*. In addition to testing VASIMR technology in space, the VF-200 will have the capability of maintaining the *International Space Station* in a stable orbit by compensating for drag. Further possible applications include lunar cargo delivery and satellite operations.

For background information *see* ELECTROTHERMAL PROPULSION; ION PROPULSION; PARTICLE ACCELERATOR; PLASMA PROPULSION; ROCKET PROPULSION; SPACECRAFT PROPULSION; SPECIFIC IMPULSE in the McGraw-Hill Encyclopedia of Science & Technology.

Roger Bengtson

Bibliography. E. A. Bering III et al., Observations of single-pass ion cyclotron heating in a trans-sonic flowing plasma, *Phys. Plasma*, 17:043509 (19 pp.), 2010; L. D. Cassady et al., VASIMR technological advances and first stage performance results, *45th AIAA/ASME/SAE/ASEE Joint Propulsion Conference & Exhibit*, Denver, Colorado, August 2-5, 2009; F. R. Chang Diaz, The VASIMR rocket, *Sci. Amer.*, 283(5):90–97, November 2000 E. Y. Choueiri, A critical history of electric propulsion: The first 50 years (1906–1956), *J. Propul. Power*, 20(2):193–203, 2004, doi:10.2514/1.9245; C. A. Deline et al., Plume detachment from a magnetic nozzle, *Phys. Plasma*, 16:033502 (9 pp.), 2009.

Water on the Moon

The Moon is constantly bombarded by debris, which includes the cores of comets, water-bearing asteroids, and interplanetary dust, all of which contain varying amounts of water. The fate of this water added to the Moon from external sources has always been a mystery. Now new data suggest that this mystery is on its way to a solution.

Water is not stable in the vacuum of the typical lunar surface. The rock and soil samples returned by the Apollo missions completely lack any hydrous mineral phases or water-bearing weathering products. If water is present during normal geological conditions, it will oxidize most minerals. The lack of oxidized minerals (for example, those containing ferric iron, Fe^{3+}) in lunar samples suggests that they formed under anhydrous conditions and that water is largely absent from the Moon's surface and interior.

Previous attention in the search for lunar water has focused on the poles of the Moon. Because the Moon's spin axis is nearly perpendicular to the ecliptic plane (1.5° obliquity), the floors of deep craters near the poles are in permanent shadow. These dark areas receive heat only from the interior of the Moon and are extremely cold, calculated to be about 50 K (50° above absolute zero). Recent measurements indicating even colder temperatures (25–35 K) have been directly measured by the current *Lunar Reconnaissance Orbiter* (*LRO*) spacecraft of the National Aeronautics and Space Administration (NASA). These areas are so cold that molecular motion is slowed to rates that exceed the age of the solar system. Thus, water molecules cannot "escape" from these cold areas and are "trapped." Over the extreme age of the Moon (more than 4.5 billion years), significant amounts of water molecules could have accumulated in these "cold traps."

Early evidence for lunar polar ice. The first hint that there could be ice in these polar cold traps came from a radio experiment on the Defense Department/NASA Clementine mission in 1994. High values of the circular-polarization ratio of radio waves reflected from the lunar surface can result from either high degrees of surface roughness or high internal (volume) scattering from ice deposits. The circular-polarization ratios of radio echoes from the south polar region were elevated and consistent with the presence of ice in the crater Shackleton. Four years later, NASA's *Lunar Prospector* spacecraft carried an instrument designed to measure the amount and energy of neutrons given off the Moon's surface when cosmic rays collide with atoms in the lunar crust. Hydrogen absorbs neutrons, so when the *Lunar Prospector* investigators saw a decrease in the flux of medium-energy neutrons near the lunar poles, they concluded that excess amounts of hydrogen were present there. This observation is consistent with the presence of polar water ice.

A problem with the *Lunar Prospector* data is that its maps of hydrogen concentration are low in spatial

Fig. 1. In this M^3 image of the Moon, the water absorption feature at 2800 nm is shown in gray, increasing in brightness with increasing water content. Both poles display water absorption, which becomes stronger as the pole is approached. This water is being created or deposited all over the Moon, but is preserved only in the colder areas that permit its longer survival.

resolution; it is impossible to identify any structure in the data smaller than about 40 km (25 mi). Thus the *Lunar Prospector* neutron data show a large, smeared-out area of enhanced hydrogen content and we cannot tell if this excess is confined solely to the dark floors of permanently shadowed craters (consistent with the presence of water ice) or just an overall enrichment of hydrogen near the poles (consistent with implanted solar-wind protons). Models of the distribution of water ice near the lunar poles indicate that if the hydrogen is present as water ice, concentrations exceed 1% by weight in some dark areas. This might sound like a small amount, but when totaled over a large area, it could constitute hundreds of millions of tons of water ice on the Moon. Moreover, because the neutron instrument only senses the upper 30–40 cm (12–16 in.) of the surface, total concentrations of water could be up to ten times greater than these results.

New studies and missions. Several studies and missions in recent years have provided additional data about the presence of water on the Moon. Study of volcanic glass from the *Apollo 15* landing site demonstrated that tiny amounts of water (about 50 parts per million) are present in the interiors of these glasses, suggesting that the lunar mantle contains about ten times this amount—a startling result, considering the extreme dryness of other lunar samples. The Indian *Chandrayaan-1* mission carried two instruments provided by NASA that are revolutionizing our view of lunar water, the Moon Mineralogy Mapper (M^3) and the Mini-SAR (SAR = synthetic aperture radar). The M^3 instrument collected reflectance spectra for most of the Moon and found evidence for the presence of both water (H_2O) and hydroxyl (OH) molecules either as a monolayer on lunar dust grains or bound into the mineral structures in surface materials poleward of 65° latitude at both poles (**Fig. 1**). Moreover, this surface water is temporally variable, being present in greater quantity in both local early morning and late evening and increases in abundance with increasing latitude. These results were verified by NASA's *Cassini* and *EPOXI* (the former *Deep Impact*) spacecraft during separate flybys of the Moon. New lunar surface observations indicate significant quantities of water moving toward areas with lower mean surface temperatures and increasing in abundance with latitude. Taken together, the results indicate that water is being deposited (for example, by comet impact) or created (for example, by reduction of metal oxides in the surface by solar-wind protons) and transported to the poles. By this process, significant quantities of water ice could accumulate at the poles over geological time.

Both poles were well covered by the Mini-SAR radar data. Much of the north polar region displays backscatter properties typical for the ordinary Moon, with circular-polarization ratio values in the range of 0.1–0.3, increasing to greater than 1 for young primary impact craters (**Fig. 2**). These high circular-polarization ratio values likely reflect a high degree of surface roughness associated with these fresh features. Another group of craters in the region show elevated circular-polarization ratios (between 0.6 and 1.7) in their interiors, but no enhanced circular-polarization ratios in deposits exterior to their rims (typical circular-polarization ratio values of about 0.2 to 0.4). Almost all of these features are in permanent Sun shadow and correlate with proposed locations of polar ice modeled on the basis of *Lunar Prospec-*

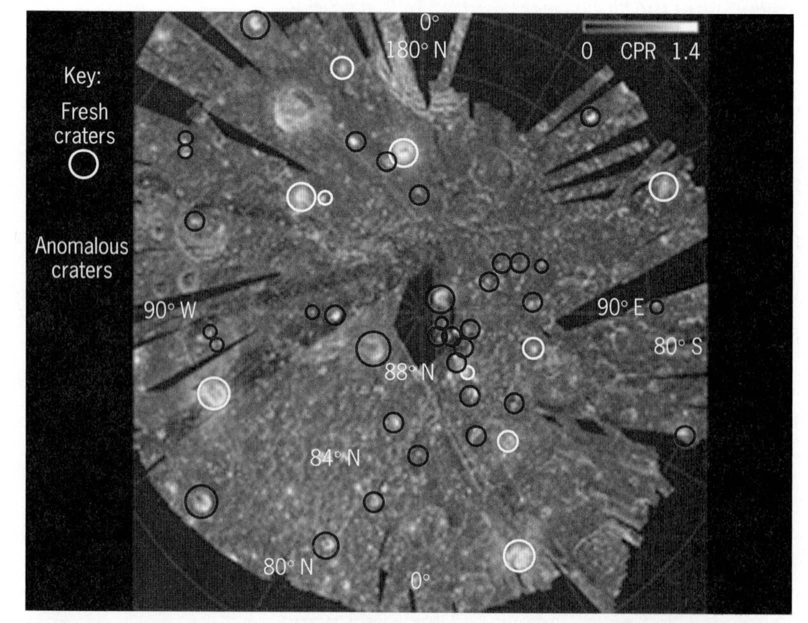

Fig. 2. Mini-SAR (synthetic aperture radar) mosaic of the north pole of the Moon, showing the circular-polarization ratio (CPR) of surface materials. Fresh craters that display high CPR due to surface roughness are circled in white, while the anomalous craters that have high CPR only inside their shadowed rims are circled in color. Over 40 of these craters have been identified and are interpreted to contain nearly pure water ice deposits. There are over 600 million metric tons of water ice in these craters.

tor neutron data. These relations are consistent with deposits of nearly pure water ice inside these shadowed craters, with approximately 600 million metric tons of nearly pure water ice present in over 40 small craters within 10° of the pole. The south polar region shows similar relations, except that it has more extensive terrain with low circular-polarization ratio and fewer anomalous craters than the north pole. Small areas of high circular-polarization ratio are found in some southern craters, notably Shoemaker, Haworth, and Faustini; these areas might be deposits of water ice. The crater Shackleton has also been partly mapped and new results show relations similar to and consistent with the previous interpretations of the presence of water ice.

The companion satellite to the *Lunar Reconnaissance Orbiter*, the *Lunar Crater Observation and Search Satellite* (*LCROSS*), slammed the upper stage of its launch vehicle into the Moon's south pole on October 9, 2009, and observed the ejected material for evidence of lunar water. Results indicate that both water vapor and ice were part of the ejected materials from the *LCROSS* impact crater; initial analyses indicate that water is present in these deposits at the 5–10% level by weight. The *LCROSS* impact site exhibits no anomalous radar behavior, suggesting that such water content cannot be detected by radar. However, the *LCROSS* results do indicate that significant amounts of lunar polar water may be present even in the absence of specific radar evidence for it. Spectra from this impact event show evidence for other volatile substances, including ammonia and simple carbon compounds. The presence of such material may indicate a cometary source for these volatile materials.

Conclusions. Results from the recent and ongoing lunar missions, as well as continued study of lunar samples, indicate that water is present in large quantity with multiple occurrences on the Moon. Water was present in the deep lunar interior 3.3 billion years ago, at concentration levels of a few hundred parts per million. This water would have been released during eruption of lunar magma and could have made its way to sequestration in the polar cold traps. Water is also being either made or deposited all over the Moon, nearly continuously. Most of this water is subsequently lost to space (for example, by sputtering, ionization, or thermal escape) but some of it is being retained. Any water arriving at a cold trap near the pole will be captured permanently. This picture of a constantly migrating stream of water molecules to the poles is supported by direct observation of surface spectral features. Water, once in the polar areas, is stable as ice in the permanent darkness where sublimation is prevented when the ice is buried by a thin layer of soil. Significant quantities can accumulate there; the *LCROSS* results suggest that several percent to tens of percent proportion by weight of water ice may exist in the polar soils. Finally, some of this migrating water apparently collects at a rate high enough that significant soil cannot mix with it during normal impact bombardment, as shown by the presence of relatively "pure" water ice deposits in selected lunar craters imaged by radar.

Water is present on the Moon in significant amounts; at least several billion metric tons of water are at the poles, a volume comparable to that of the Great Salt Lake of Utah. This water is more than enough to support a permanent, sustainable human presence on the Moon and for export as rocket fuel (hydrogen and oxygen are the two most powerful chemical propellants known), energy storage, and life support in cislunar space. These new discoveries fundamentally alter our knowledge of the Moon's processes and history, and highlight its scientific value and utilization potential.

For background information *see* COMET; MOON; NEUTRON; RADAR ASTRONOMY; SOLAR WIND; SPACE FLIGHT; SPACE PROBE; SYNTHETIC APERTURE RADAR (SAR) in the McGraw-Hill Encyclopedia of Science & Technology.

Paul D. Spudis

Bibliography. P. G. Lucey, A lunar waterworld, *Science*, 326:531–532, 2009, doi:10.1126/science.1181471; C. M. Pieters et al., Character and spatial distribution of OH/H_2O on the surface of the Moon seen by M^3 on *Chandrayaan-1*, *Science*, 326:568–572, 2009, doi:10.1126/science.1178658; P. D. Spudis, *The Once and Future Moon*, Smithsonian Institution University Press, 1996; P. D. Spudis et al., Initial results for the north pole of the Moon from Mini-SAR, *Chandrayaan-1* mission, *Geophys. Res. Lett.*, 37:L06204, 2010, doi:10.1029/2009GL042259; R. R. Vondrak and D. H. Crider, Ice at the lunar poles, *Amer. Sci.*, 91:322–329, 2003.

Wave slamming

Wave slamming occurs when a ship operating in a seaway is subjected to forces due to the presence of ocean waves. For most of a ship's life, these forces (typically referred to as ordinary wave-induced loads) will be of low to moderate magnitude. However, when a ship travels at high speed in moderately high seas or when operating in heavy seas, these hydrodynamic forces, in conjunction with the ship's own rigid-body motions, can cause the ship's bow to come out of the water (**Fig. 1***a*).

As the ship reenters the sea surface, the ship's structure may be subject to a high reentry velocity, which results in a rapid increase in the wetted area (Fig. 1*b–d*). The result is an impulsive pressure acting on a localized area of the ship's structure for an extremely short period of time (milliseconds). An example of this type of pressure measurement is shown in **Fig. 2**.

Effects of slamming. The effect of pressure loadings is important for both the local and global structure of ships. Slamming of the fore or aft body of a ship can result in significant stresses and deformations on local structural details, such as the bow flare sections of a ship or the cross deck (wetdeck)

(a)

(b)

(c)

(d)

Fig. 1. Example of wave slamming of a scaled model of a ship. (*a*) Ship bow is out of water. (*b*) Ship reenters sea surface with high vertical velocity. (*c*) Rapid increase in wetted area moments after initial impact. (*d*) Rapid increase in wetted area at end of impact event.

regions of a catamaran. The results of local deformations can lead to cracks (**Fig. 3**).

In addition to local structural damage, slamming events can affect the global ship structure as well. For such an event to occur, the slamming force must be of sufficient magnitude that it causes the entire ship to vibrate. This vibratory response of the ship's hull girder, also known as whipping, can result in loads that are of equal magnitude to the ordinary wave-induced loads already present and acting on the ship's hull girder.

Slam-induced whipping events are the result of either keel slamming or momentum slamming. In the first case, large impact pressures are generated when the ship is lifted out of the water and the impact area is characterized by a low deadrise angle (that is, the angle of the boat's bottom relative to the horizontal). In the case of momentum slamming, it is not necessary for the keel to emerge from the water surface. Rather, the ship's motions are of sufficient magnitude that the bow region is quickly submerged. This rapid immersion leads to increased fluid pressures operating over a large area which, in turn, results in forces capable of causing a whipping response. In either case, this type of loading event could affect the ship's structural integrity.

Aside from structural integrity issues, the presence of slamming affects both ship passengers and crew alike. At a minimum, the presence of ship vibration due to a slamming event will affect passenger comfort. More importantly, the associated vertical accelerations reduce performance skills as attention must be devoted to keeping one's balance.

Designing for slamming loads. In order to mitigate the effects of slam loads, several options are available to the naval architect. One option is to use weather routing (that is, weather forecasting) to avoid adverse wave conditions. However, this approach has several shortcomings. First, as good as such forecasts may be, they are not infallible. Hence, the ship may still end up operating in conditions conducive to slamming. Another problem is that a ship may need to operate in adverse conditions due to mission requirements or financial considerations.

As a result, it is best to make allowance for slamming loads in the actual design of the ship. During the preliminary stages of design, naval architects can use nonlinear ship motion and load programs to simulate the dynamic behavior of a ship operating in select sea conditions. However, in order to perform such simulations the design must be developed sufficiently such that the hull form, basic ship attributes, and structural details are known.

As the design matures, another option available to the naval architect is to perform model tests to determine how a ship will behave in severe operating conditions. Before doing such tests, care must be taken to ensure that the model hull form is geometrically similar to the actual ship. In addition, it is also necessary to properly scale the ship's mass properties, centers of gravity, pitch and roll, radius of gyration, and structural properties.

In order to facilitate this, a dimensional analysis based on Froude scaling laws is performed to ensure that gravitational forces are properly scaled. With this approach, naval architects are assured that dynamic similarity exists between the model and the full-scale ship.

In order to measure the global effects of slamming, the model is outfitted with a longitudinal beam (or backspline), typically made out of aluminum, which represents the actual hull girder of the ship. The model is then segmented into sections at longitudinal locations where structural loads are to be measured.

Finally, with respect to local slamming loads, a model can be outfitted with either pressure gauges or pressure panels (that is, structurally scaled sensors, typically made out of polyvinyl chloride). While pressure gages will measure the hydrodynamic pressure, the use of pressure panels is preferred, as they can be designed to respond exactly at the natural frequency of the local structure of the ship. Once the design is completed, the pressure panel will behave in the same way dynamically as the actual local full-scale section. As a result, the measured pressure effectively represents the structural response to the slam event.

A typical scaled-model arrangement consists of the hull form, backspline, pressure panels, propeller shafts, and rudders (**Fig. 4**). The model is then tested in selected wave conditions (Fig. 1). The resulting structural loads (both the ordinary wave-induced loads and the slam-induced whipping loads) are recorded for additional analysis (**Fig. 5**).

Extreme-value analysis. Since the ocean's waves are random, the results obtained from either numerical simulation or model tests represent only a snapshot in time of the possible loading events that can occur within a given operational condition. Although the basic statistical properties would not change if the experiment were repeated, the observed maximum response during each test would be different. This is because the waves being generated during a test are random. Consequently, if the same test were repeated 100 times, 100 different maximum responses would be recorded. The distribution of these observed maximum responses defines what is called the extreme distribution.

Although it is possible to perform repeat tests, the cost and time involved are prohibitive. As a result, designers utilize various statistical methods to determine design slamming loads. One popular extreme-value prediction method is Weibull analysis. With this approach, measured events (or amplitudes) are accounted for and applied to the Weibull distribution function. The general three-parameter Weibull (cumulative) distribution function is expressed in the equation below, where x is the data ($x \geq x_0$),

$$P(x) = 1 - \exp\left[-\left(\frac{x - x_0}{\theta - x_0}\right)^{\beta}\right]$$

$P(x)$ is the (cumulative) distribution of x (that is, the probability that a data point is less than x), x_0 is the threshold value below which there is no measurable data, β is the Weibull shape parameter or slope, θ is the characteristic value, corresponding to the x value with a cumulative probability of 0.632, and $\theta - x_0$ is the scale factor.

The advantage of using this type of statistical method is that other statistical distributions such as the normal, exponential, and Rayleigh distributions are subsets of the Weibull. Hence, designers do not need to know, prior to the analysis, what type of distribution will best fit the test data.

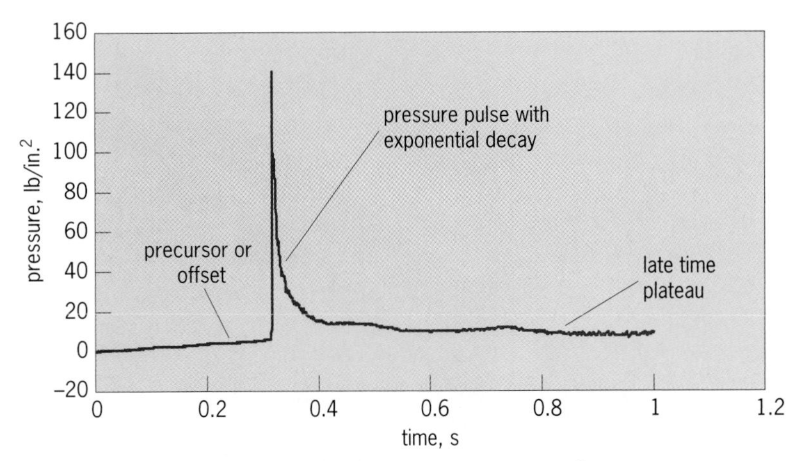

Fig. 2. Example of a slam-induced pressure measurement. 1 lb/in.² = 6.9 kPa.

(a)

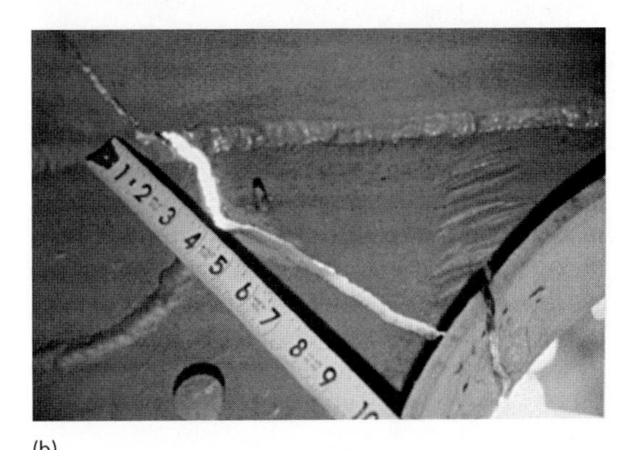

(b)

Fig. 3. Example of (*a*) local structural damage (or "plate dishing") and (*b*) localized cracking due to a slam event.

Results obtained from a Weibull analysis allow designers to determine an acceptable level of risk based on such factors as consequence of failure, increased structural weight, cost to repair, mission criticality, and so on. For example, failure of the shell plating of a ship, due to a slamming event, usually means that the plate will locally deform (or "dish in"). While this deformation is not aesthetically pleasing, dishing does little harm to the structural integrity of the ship. As a result, one may use Weibull analysis results to design such a structure to a probability of nonex-

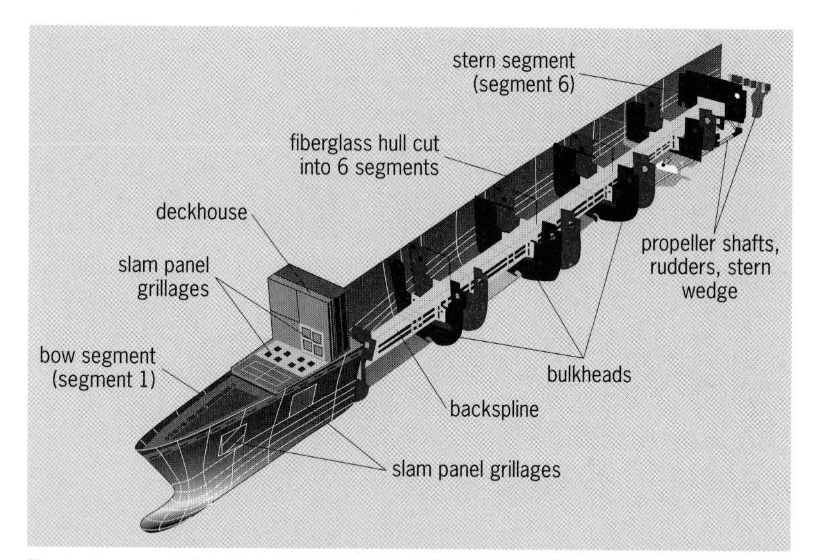

Fig. 4. Representation of a structural loads model (a scaled ship model that is tested in selected wave conditions to determine structural loads).

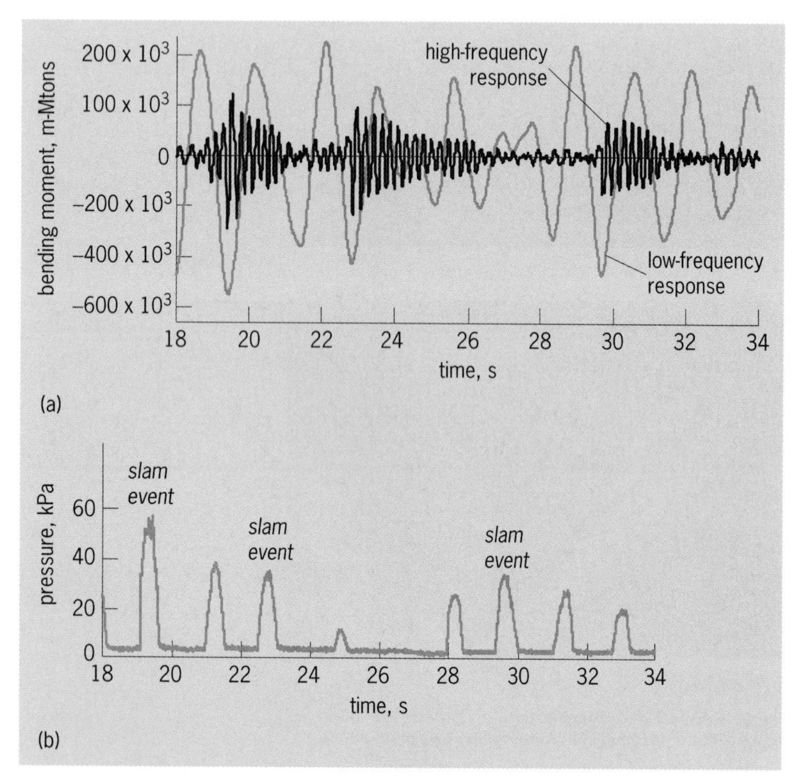

Fig. 5. Examples of responses of a scaled model, recorded during testing and scaled to a full-size ship. (*a*) Measured global response, showing high-frequency whipping events and low-frequency response to ordinary wave-induced loads. Bending moment is expressed in units of meter-metric tons. (*b*) Measured local slamming response.

ceedence consistent with that of the most probable maximum load. However, it can be shown through extreme value statistics, that there is a 63% chance that such a load can be exceeded. Hence, when taking into consideration slamming events related to the ship's hull girder or other primary ship structure, designing to a probability of nonexceedence equal to 99.9% may be suitable.

For background information *see* DIMENSIONAL ANALYSIS; DISTRIBUTION (PROBABILITY); DYNAMIC SIMILARITY; NAVAL ARCHITECTURE; OCEAN WAVES; PROBABILITY; SHIP DESIGN; SHIP POWERING, MANEUVERING, AND SEAKEEPING; STOCHASTIC PROCESS in the McGraw-Hill Encyclopedia of Science & Technology.

Allen Engle

Bibliography. R. E. D. Bishop and W. G. Price, *Hydroelasticity of Ships*, Cambridge University Press, cloth ed. 1979, paper. 2005; J. C. Daidola, V. Mishkevich, and A. Bromwell, *Hydrodynamic Impact Loading On Displacement Ship Hulls, An Assessment of the State of the Art*, Ship Structure Committee Report SSC-385, 1995; O. A. Hermundstad and T. Moan, Efficient methods for direct calculation of slamming loads on ships, *Society of Naval Architects and Marine Engineers Transactions*, 117:156–181, 2009; M. K. Ochi, *Applied Probability and Stochastic Processes in Engineering and Physical Sciences*, Wiley-Interscience, 1990.

Wide-scale infrastructure for electric cars

Most cars are powered by gasoline or diesel fuel. There is a massive infrastructure of service stations to provide these fuels in most regions of the world, especially in areas where there are many cars. In the United States alone there are approximately 120,000 service stations, often clustered in larger cities and distributed along interstate and other highways.

The range of gasoline-powered cars is typically 250 mi (400 km) or more on a single tank of fuel. The mean distance between service stations is generally much less than the mean range. Thus, the range of gasoline-powered cars allows them to be refueled at a service station without concern about running out of fuel.

Battery-powered electric cars offer many advantages over gasoline-powered cars, including the possibility of charging cars with electricity from a green (renewable sources of energy) grid. However, there are a number of issues for charging (refueling) batteries that are particular to electric cars, most of which involve the limited range and the lack of charging infrastructure. These issues do not pertain to hybrid gas-electric cars, which carry gasoline onboard as a range extender. Enormous infrastructure development is required for battery-powered electric cars to be feasible on a large scale.

Range. The energy density of a battery, both by volume and by weight, is not sufficient to power a car over a long distance. The energy density of a typical lithium-ion battery is of the order of 0.7 MJ/L and 0.5 MJ/kg. The energy density of gasoline is 34.6 MJ/L and 47.5 MJ/kg. The much lower energy storage density of a lithium-ion battery compared to that of gasoline is the reason that a battery-powered electric car has a range considerably lower than that of a gasoline-powered car. Battery-powered electric cars that will be on the market in 2011–2012, with rare exception, will have an advertised range of 100 mi (160 km) or less.

Approximately 70% of vehicle miles traveled in a day by drivers in the United States are on trips of 100 mi (160 km) or less. This means that an electric car with a range of 100 mi (160 km) could handle most of the daily trips of many such drivers. Drivers in Europe and Asia will typically have an even higher percentage of their vehicle miles traveled per day in trips of less than 100 mi (160 km). Thus, for many users, an electric car could be charged at home at night, assuming charging capability exists at the home location.

Charging locations. The situation for charging cars depends on location: home, street, work, parking garage, hotel, roadside, and so on. Charging in all cases should be done at night rather than during the day, as grid load and generating capacity are more stressed during daytime, for example, when air conditioners are generally operating.

Charging at home. Charging in the driveway or garage will work for many drivers who live in a house with available charging capability. However, not all houses have a garage or driveway, and not all of these will be equipped with an adequate and safe outlet for charging a car.

Charging for apartment dwellers. This will have to be available in either a parking lot or a parking structure, equipped with safe outlets, adequate electricity supply, and a means of paying for the electricity used.

Charging in public places. Infrastructure must be available for charging cars on public streets. The outlet will have to be weatherproof, tamper- and fraud-resistant, and have a means of paying, likely by credit card or perhaps mobile phone.

Charging at work. Charging at places of employment will require the necessary infrastructure at the office, store, or other place of work. Charging could take place in a parking structure or in a parking lot if parking is provided at the place of employment. The issues of how to pay for the infrastructure and the electricity are to be determined.

Charging at locations in transit is described below. Although innovative solutions like solar charging in parking structures may be imagined, the sun does not always shine and other electricity sources must be available.

Charging time. The time to refill the tank of a gasoline-powered car is measured in minutes, with the flow rate limited primarily by pump and hose capacity. The time needed to recharge the battery of an electric car, however, is measured in hours, with the number of hours depending on battery design and the electrical capacity of the charging circuit. Batteries tend to overheat if the charging rate is too high. Thus, recharging an electric car will not be done in a few minutes.

Only specially designed circuits can operate at high voltage (higher than 220 V) or high current (40 A or more) to allow a charging time of just a few hours. There is rarely a circuit with sufficient capacity available in the garage, driveway, or near the street in most homes. Although a car can be charged on an ordinary circuit (110 V and 20 A in the United States, 220 V in Europe and elsewhere), many hours will be required.

Need for guaranteed recharging capacity. A driver must have a guaranteed location to recharge his or her car to avoid "range anxiety." If the driver can complete a day's driving needs on a single charge, and the car can be recharged at home overnight, then there is no need for recharging capability elsewhere. However, all other cases will require infrastructure: electric charging stations at work and in public places and electric charging service stations along the roads and highways. Since charging time is generally longer than a customer will wait, one can imagine an infrastructure for battery swapping, in which a nearly depleted battery is quickly swapped for a fully charged battery. However, there are also serious issues with battery swapping, as discussed below.

Charging en route. The biggest challenge will be daytime recharging when time is limited, for example, when one takes a trip for a distance that exceeds the range of the car. This could be daily, in the case of a long commute, or episodic, in the case of travel to a distant place. The short range of electric cars and the long charging time for batteries limit the available options. Although it is conceivable that quick-charging batteries will be developed, such batteries are presently not on the horizon.

In the near term it will be necessary to have a different plan for powering electric cars to travel long distances without an overnight stop for recharging. One alternative to quick-charging batteries (not yet developed) is changing batteries at "swapping stations." Batteries would have to be interchangeable among cars of different makes and models, and charged batteries of the right style would have to be readily available. Batteries would also need to be rapidly swapped, in about the same amount of time it now takes to refuel with gasoline.

Battery swapping. Battery swapping stations would have to be even more numerous than present-day gasoline service stations, as the range of electric cars is less than that of gasoline-powered cars. Several hundred thousand swapping stations would be required in the United States, and developing the network would require massive capital. The battery (or battery pack) on an electric car will likely have an integrated computer, which will almost certainly be brand and model specific, much as automobile engines are brand and model specific. There is little likelihood that automobile manufacturers will all agree on one or a few battery and computer styles to allow battery packs to be interchangeable among brands and models. One has only to shop for a headlight or taillight for a car or another trivial replacement part to see that automobile manufacturers do not standardize on a few styles.

Battery swapping might work for fleet vehicles that return to their garage periodically, as models can be standardized. It might work in limited locations, as in small countries like Denmark and Israel. At least one company is planning to implement battery-swapping stations, although no cars presently made have swappable batteries. It is unlikely that battery swapping will be implemented nationwide in a large country like the United States, at least in the foreseeable future, due to the lack of cars designed to swap batteries, the enormous capital costs of the stations, the large capital cost of the batteries themselves, and the limited lifetime of present-day batteries. It is more likely that new generations of batteries will have higher storage capacity and shorter charging times, so that eventually charging stations will supplement or replace gasoline service stations.

Battery issues. Present-day batteries have a limited life span, that is, their charge capacity diminishes with time and use. Swapping batteries will mean that sometimes one might get a better battery than the one being swapped out, and sometimes one might get one that is worse, with lower performance and more limited range. Customers may resist swapping for this reason.

Automotive battery packs are expensive, currently costing between $10,000 and $20,000. A swapping station would need to have a considerable stockpile of charged batteries available. The cost of the electricity to charge the batteries is very low. However, the capital cost of owning batteries, which are expensive and which degrade with time, is very high. Cars would be sold without batteries. The rental cost of a battery will be high due to the high capital costs of the batteries and charging stations. It is not certain that drivers will be prepared to pay the high rental cost of a charged battery.

Safety issues. There are a number of safety and liability issues which arise when swapping batteries is considered, assuming that such batteries are standardized to work in many different vehicles. One issue is that a swappable battery will likely mount underneath the car. Latch failure could lead to a battery being dropped onto the roadway or being propelled forward through the car. Latches will obviously be designed to not unlatch accidentally, but experience with strategic systems such as nuclear weapons suggests that it would be extremely difficult to prevent such accidents entirely. Batteries and the connections to the car will have to be weather resistant and highly reliable, and swapping will have to work reliably with connectors that wear and degrade with time. Swapping will have to work extremely well to avoid damage to batteries or cars.

The outlet used for charging a car also poses potential safety issues, as a high-voltage high-current outlet in most cases will be exposed to the public. The outlets must be tamper- and childproof, and designed to withstand vandalism and other threats.

Electric grid and generation issues. As noted above, cars must be charged at night, when electricity demand is low and generating and grid capacity are available. The grid in the United States could handle a large fleet of electric cars, on the assumption that there is a means of enforcing night-time charging of cars by controls or by differential pricing.

Hybrid-electric solution. A present-day solution to the range problem for electric cars is to have a gasoline engine and a supply of gasoline together with an electric motor and a battery. A car with a gasoline engine and an electric motor is the familiar hybrid-electric car if all the energy is provided by gasoline. One can drive long distances because all the energy comes from the gasoline on-board, but the car is highly efficient due to the electric motor. A variant of this is the plug-in hybrid-electric car—a car in which the battery can be charged by the electric grid, but there is gasoline on-board. The gasoline engine is often called a "range extender" by the automobile industry. The car can be powered by electric power from the grid, which is stored in the battery until the battery is exhausted. The energy to power the car then comes from gasoline.

Two-car solution. Given the limited range of battery-powered electric cars at present, it is possible that some drivers or families will choose to have at least two cars: one a conventional gasoline-powered car or hybrid, the second a battery-powered electric car for use as a "city car." This option will be appealing for some subset of the populace. However, implementation would likely increase the total number of cars on the road and require resources to build and maintain the additional cars and places to park the additional cars when not in use. Some families may require two or more cars with range greater than that offered by a "city car." Thus this solution will be useful only in limited situations.

For background information *see* AUTOMOBILE; AUTOMOTIVE ENGINE; BATTERY; GASOLINE; ELECTRIC VEHICLE; MOTOR in the McGraw-Hill Encyclopedia of Science & Technology.

Alfred S. Schlachter

Bibliography. D. M. Lemoine, D. M. Kammen, and A. E. Farrell, An innovative and policy agenda for commercially competitive plug-in hybrid electric vehicles, *Environ. Res. Lett.*, 3:014003 (10 pp.), 2008; D. Sperling and D. Gordon, *Two Billion Cars: Driving Toward Sustainability*, Oxford University Press, 2009.

Wobbling motion in nuclei

A body with a triaxial shape has three perpendicular axes of symmetry whose lengths are all different. In the 1970s, Aage Bohr and Ben Mottelson proposed that some rotating atomic nuclei could undergo a stable triaxial deformation, and showed that a unique signature of such a deformation would be an exotic excitation mode, called a quantized wobbling-phonon excitation, whose classical analog is the wobbling motion of an asymmetric top. Although this excitation mode was predicted some 35 years ago, it was finally observed only in recent years in several lutetium and tantalum isotopes. A new dimension thus has been added to the rotational motion of atomic nuclei, the best-established mode of collective motion in nuclei.

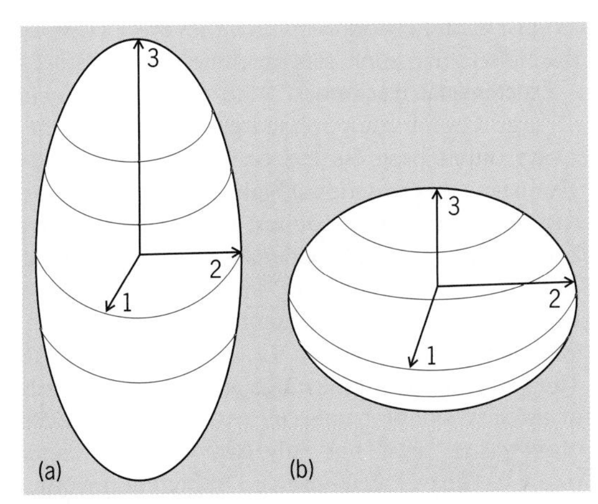

Fig. 1. Axially symmetric nuclear shapes with (a) prolate and (b) oblate deformations. The three principal axes are perpendicular to each other. The symmetry axis-3 is the longest and shortest axis for prolate and oblate nuclei, respectively.

Deformation of atomic nuclei. The nucleus of any atom is composed of certain combinations of neutrons and protons, collectively called nucleons. The nucleons are bound together by a nuclear force that exists only inside the nucleus. Some nuclei are more rigid and have stable shapes, others are softer and their shapes are more malleable. The deformation is determined by the intrinsic shell structures of nuclei. The most important deformation of nuclei is the axially-symmetric deformation corresponding to either prolate or oblate shapes. The prolate shape is similar to that of an American football, while the oblate shape is similar to that of a pancake (**Fig. 1**). In both cases the principal axis 3 is the symmetry axis of rotation; the lengths of the other two principal axes, 1 and 2, are equal. The difference between the radii along the longest and shortest axes can be up to 30% of the averaged radius.

Depending on the nuclear shell structure associated with the given numbers of neutrons and protons, other nuclear shapes are obtained for excited nuclear states. When a nucleus rotates rapidly, the intrinsic shell structure can stabilize to some more exotic nuclear shapes, such as a much larger deformation (super-deformation) with a long-to-short axis ratio of 2:1, or an axially asymmetric (triaxial) deformation, where the lengths of the three axes are all different. Here we are concerned with a stable triaxial shape, rather than a possible quantum-fluctuation about an axially-symmetric shape, and one that is stable in a sufficiently wide region of rotational angular momentum. For years, clear experimental evidence for the presence of the predicted stable triaxial shape has been sought, because nuclei with different deformations have different excitation properties.

Collective rotation. Unlike classical rotors, nuclei are quantum-mechanical many-body systems. Axially symmetric deformed nuclei, like linear molecules, cannot rotate collectively about the symmetry axis-3. Any rotation about an axis perpendicular to the symmetry axis is equivalent to any other such rotation. The nuclei have lowest energy when they are in their ground state. When gaining additional energy and angular momentum, the nuclei will be excited to higher-spin states, rotating faster. The rotational energy is expressed in terms of the moment of inertia, $\Im_1$, about the perpendicular axis-1 as Eq. (1), where $\hbar$ is the unit of angular momentum and

$$E(I) = \frac{\hbar^2}{2\Im_1} I(I+1) \qquad (1)$$

I the spin quantum number for the rotation about axis-1. The level energy is inversely proportional to the moment of inertia. Nuclei in higher-spin states decay to lower-spin states in a collective rotational band, and eventually back to their ground states, by emitting a series of gamma rays.

Wobbling mode. Nuclei with a triaxial shape can, however, rotate about any one of the principal axes, showing much richer collective rotation spectra. Assuming that axis-1 is the shortest axis and axis-3 the longest, we have moments of inertia $\Im_1 > \Im_2 > \Im_3$. The rotation about axis-1 is most favorable because it takes the least energy to excite the nucleus to a given spin I. For a certain intrinsic nuclear structure (the numbers of protons and neutrons occupying each particular quantum state), a family of rotational bands can be built by transferring some angular momentum to the other two axes. All bands in the family have identical intrinsic structure and other properties, such as deformation, moment of inertia, and level lifetime. The level energies of this family of rotational bands are formulated in terms of so-called phonon excitations as in Eq. (2), where n_w is the

$$E(I) = \frac{\hbar^2}{2\Im_1} I(I+1) + \hbar\omega_w \left(n_w + \frac{1}{2}\right) \qquad (2)$$

wobbling phonon number, ω_w is the wobbling frequency depending on all three moments of inertia, and $I(I+1) = I_1^2 + I_2^2 + I_3^2$. [The excitations are identified with phonons because the second term on the right-hand side of Eq. (2) is also just the expression for the energy of an elastic wave in a solid with n_w quanta of energy or phonons.] It should be noted that the total angular momentum I consists of components along the three principal axes; thus its direction does not coincide with a particular principal axis. The electromagnetic decay properties of the transitions between the bands in a wobbling family depend on the triaxial shape as well as the intrinsic structure. The characteristic pattern of wobbling bands is depicted in **Fig. 2**.

A classical analog of the wobbling mode is the motion of an asymmetric top which executes a precessional motion while spinning, describing a cone with a variable apex angle. The wobbling motion is also well known in chemical molecules. For example, it is used to explain the nature and dynamics of water molecules in a nanoscopically confined environment, such as dielectric relaxation and solvation

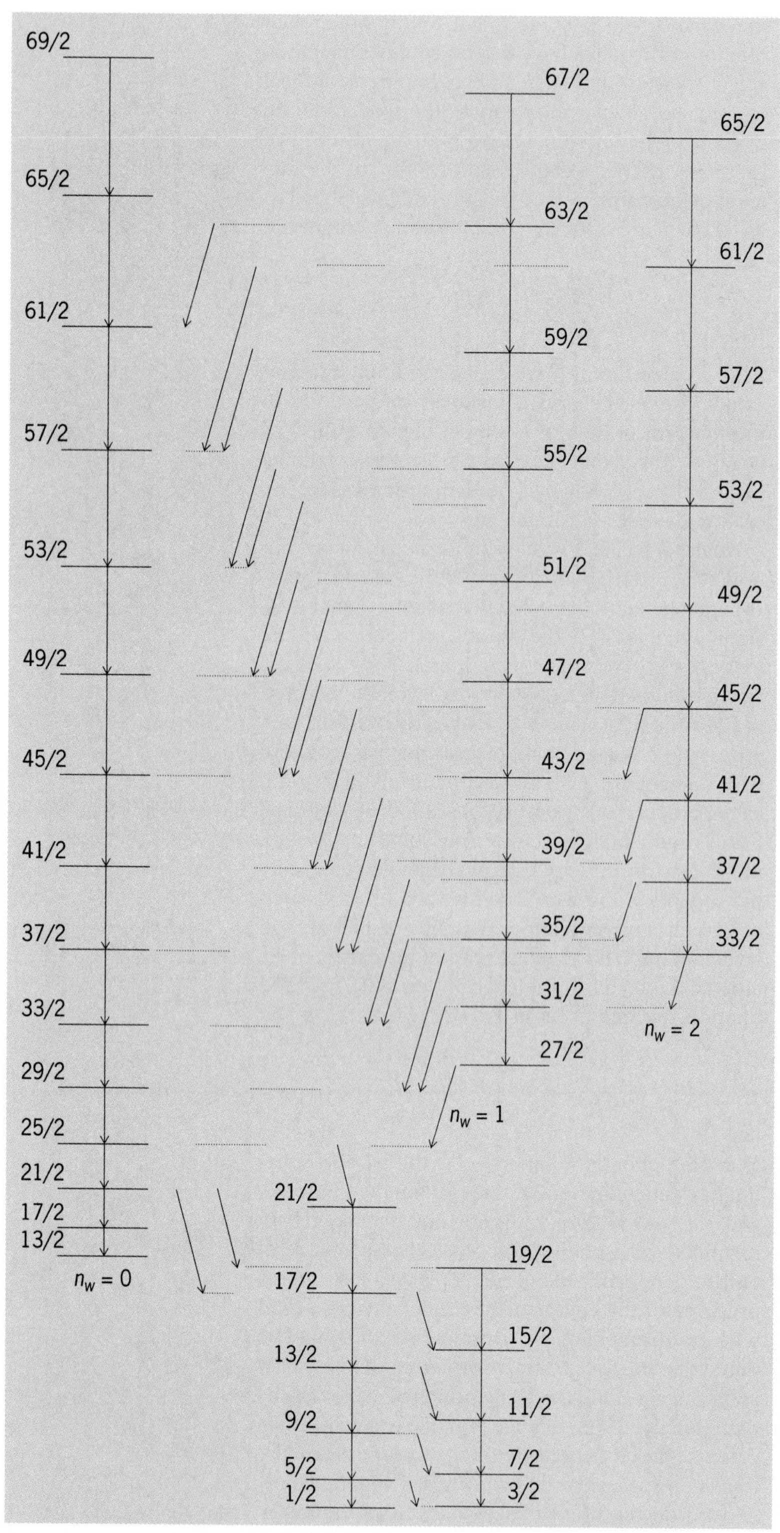

Fig. 2. Energy-level diagram for wobbling bands, showing that the bands with higher phonon numbers n_w decay to bands with lower phonon numbers. The 0-phonon band decays to the lower-lying ground-state band. The levels are labeled by their spins.

dynamics in complex and biological systems. However, the wobbling motion in the nuclear system is quantized. The quantized wobbling phonon picture was proposed as a unique fingerprint of triaxial deformation in 1975 by Bohr and Mottelson, in an extension of the work for which they were awarded the Nobel Prize in physics the same year.

Experimental discoveries. With the development of large Compton-suppressed germanium detector arrays, rapid progress has been made in nuclear gamma-ray spectroscopy at high-spin states. After extensive experimental searches, wobbling bands have been identified in the odd-mass isotopes lutetium-161, -163, -165, and -167 (^{161}Lu-^{167}Lu, proton number $Z = 71$ and neutron numbers $N = 90$, 92, 94, 96) and in tantalum-167 (^{167}Ta, $Z = 73$ and $N = 94$). The bands reach spins as high as (97/2) $\hbar$. A family of three wobbling bands ($n_w = 0, 1, 2$) have been observed in ^{163}Lu, but only two wobbling bands ($n_w = 0, 1$) have been observed in each of the other isotopes. The labels $n_w = 1{,}2$ indicate their interpretation as one- and two-phonon excitations built on the $n_w = 0$ band. The spin difference between the corresponding levels in the $n_w = 1$ and $n_w = 0$ bands (also in the $n_w = 2$ and $n_w = 1$ bands) is $1\hbar$, and the bands with higher phonon numbers have higher excitation energy than bands with lower phonon numbers. The measured properties of bands in each wobbling family, such as their moments of inertia, level lifetimes, and quadrupole moments, are strikingly similar, indicating that they are built upon identical intrinsic structures. The inter-band transitions from the $n_w = 1$ band to the $n_w = 0$ band (and from the $n_w = 2$ band to the $n_w = 1$ band in ^{163}Lu) have also been identified. The electromagnetic properties of these linking transitions have been investigated. Electric quadrupole (E2) transitions are, surprisingly, the dominant components even though these transitions carry only $1\hbar$ angular momentum ($\Delta I = 1$). At high spins the strength of these E2 components are approximately proportional to $1/I$ and not to $1/I^2$, as is usually observed for $\Delta I = 1$ transitions in high-spin structures. The calculated transition strengths are determined almost exclusively by the amount of triaxiality, while the wobbling frequencies and the phonon excitation energies depend on the moments of inertia. The observed configurations correspond to triaxial nuclear shapes with axis ratios of approximately 0.8:1:1.2. Overall, the measured electromagnetic transitions in these wobbling families are in good agreement with the wobbling phonon picture and theoretical calculations. The wobbling mode, one of the most exotic properties of rotating nuclei, has now been proved to be a more general phenomenon.

Crucial role of a single aligned proton. Protons and neutrons generally exist within nuclei in pairs. The two nucleons in a pair have spins in opposite directions and cancel with each other; thus the pair has zero spin. All five lutetium and tantalum isotopes in which the wobbling mode has been observed have a single unpaired proton. In the wobbling bands this proton is found to be in the $i_{13/2}$ orbital, an excited quantum state, and it aligns its intrinsic angular momentum (about $6\hbar$) predominantly along the axis with the largest moment of inertia (the axis-1). In the textbook example of wobbling motion and in previous theoretical investigations, the angular momentum of the intrinsic motion of single particles does not appear. The rotation-aligned $i_{13/2}$ proton orbital plays a significant role in odd-Z nuclei. It favors a strongly deformed triaxial shape. Furthermore, a smaller rotational energy is needed for building a given total angular momentum because of this additional aligned angular momentum from the single proton, and thus, the wobbling excitations may appear more easily at lower excitation energy. When going to higher excitation energies, nucleons from broken pairs will be excited into different orbitals, forming different "quasiparticle" configurations upon which the rotational bands can be developed. Experiments for neighboring nuclei with even numbers of protons and neutrons in this mass region indicate that high-spin excitations based on quasiparticle configurations are generally more favorable than the wobbling excitations. Consequently, the wobbling mode has not been observed in any even-even nucleus.

Measurements with large detector arrays. The study of the characteristic cascades of gamma transitions through rotational bands, and in particular the weaker transitions between bands, requires advanced detector arrays with high resolution and sensitivity. The aim is to register simultaneously as many as possible of the coincident gamma rays in order to establish the decay pattern. The observation of the wobbling mode became possible only after the construction of the powerful germanium spectrometer arrays: GAMMASPHERE in the United States and EUROBALL in Europe. A few cases of wobbling excitations have now been identified. Discrepancies between measured and calculated quantities exist. Triaxiality and wobbling may also be found in other mass regions. The purity of the wobbling mode may be studied by its interaction with coexisting quasiparticle excitations. All these questions, as well as the further study of higher phonon excitations, will require an experimental sensitivity level beyond what is available today. The new "gamma-ray energy tracking arrays," the GRETINA in the United States and the AGATA in Europe, will reach a sensitivity level that is orders of magnitude higher than that of the current arrays. Other, even more fascinating collective modes of the atomic nucleus may be expected.

For background information *see* ANGULAR MOMENTUM; GAMMA-RAY DETECTORS; GAMMA RAYS; MOMENT OF INERTIA; NUCLEAR STRUCTURE; NUTATION (ASTRONOMY AND MECHANICS); PHONON; PRECESSION; RIGID-BODY DYNAMICS in the McGraw-Hill Encyclopedia of Science & Technology. Wenchao Ma

Bibliography. A. Bohr and B. Mottelson, *Nuclear Structure*, vol. 2, W. A. Benjamin, Inc., Reading, MA, 1975; G. Hagemann, Triaxiality and wobbling, *Acta Physica Polonica B*, 36:1043–1054, 2005; D. J. Hartley et al., Wobbling mode in ^{167}Ta, *Phys. Rev. C*, 80:041304(R) (5 pp.), 2009; R. Janssens and F. Stephens, New physics opportunities at Gammasphere, *Nucl. Phys. News*, 6(4):9–17, 1996.

Wood supply and demand

At times in history, there have been concerns that demand for wood (timber) would be greater than the ability to supply it, but that concern has recently dissipated. The wood supply and demand situation has changed because of market transitions, economic downturns, and continued forest growth. This article provides a concise overview of this change as it relates to the United States, looking at nationwide trends in the pulp and paper sector and the solid wood and composite wood product sectors (lumber and wood panels) and also nationwide trends in timber growth, timber removals, and real prices for timber.

Pulp and paper trends. Until the 1990s, wood pulp production in the United States trended more or less continuously upward. **Figure 1** shows annual U.S. wood pulp production from the beginning of the twentieth century through 2009. The upward trend through most of the twentieth century may have suggested until recently that pulpwood demand might lead to unsustainable pressure on timber supply. For many decades, the United States was by far the world's leading producer of wood pulp, and it still is, but the trend in U.S. wood pulp production transitioned after peaking in the mid-1990s and has generally declined over the past 15 years.

Wood pulp is used mainly for paper and paperboard production, so the transition in wood pulp output reflects a transition in paper and paperboard markets. Economic recessions negatively impacted paper and paperboard demand in 2001-2002 and more recently in 2008–2009, which resulted in U.S. paper and paperboard consumption declining by 25% from 1999 to 2009. Furthermore, since the 1990s, U.S. paper and paperboard consumption deviated from the trend in overall U.S. economic growth, represented by real gross domestic product (GDP), as shown in **Fig. 2**.

Historically, U.S. paper and paperboard demand was closely correlated with real U.S. GDP growth, and thus the transition in paper and paperboard

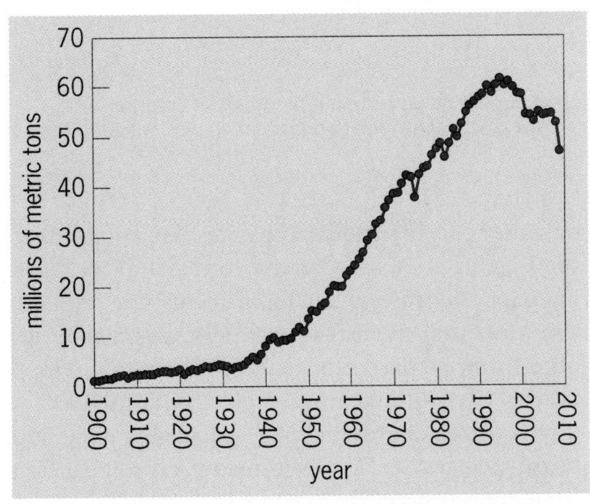

Fig. 1. U.S. annual wood pulp production, 1900–2009. *(Sources: Forest Service, Commerce Department, American Forest & Paper Association, International Woodfiber Report)*

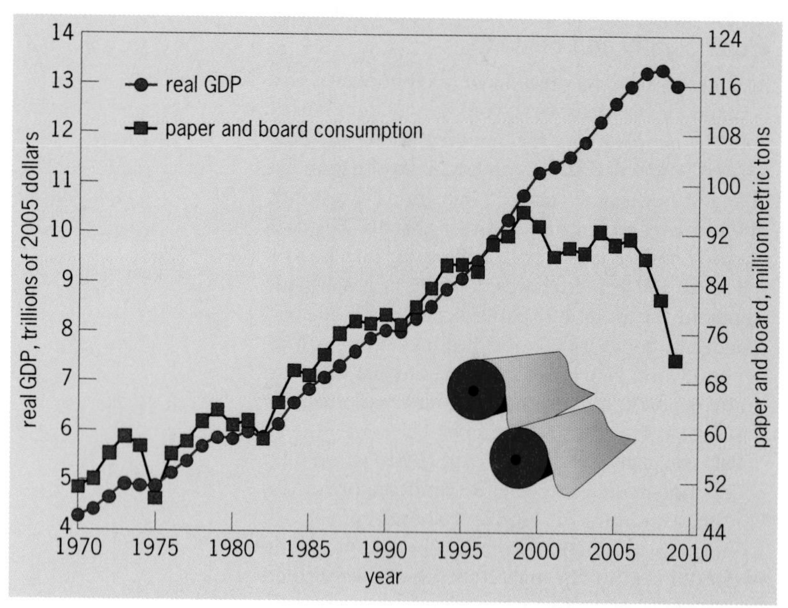

Fig. 2. U.S. annual paper and paperboard consumption, and real gross domestic product (GDP), 1970–2009. [*Sources: U.S. Bureau of Economic Analysis (GDP); American Forest & Paper Association (apparent consumption, including wet machine board and construction paper)*]

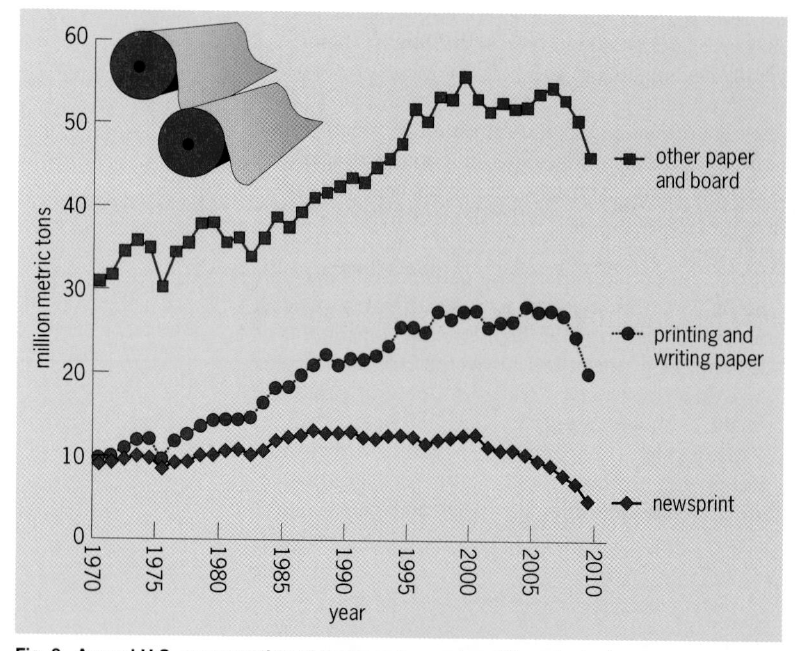

Fig. 3. Annual U.S. consumption of paper and paperboard by principal category, 1970–2009. [*Source: American Forest & Paper Association (excluding wet machine board and construction paper)*]

demand reflects structural changes. Important structural changes included reduced demand for printing paper because of declining advertising expenditures for print media (newspapers, magazines, and other print media), especially as advertising expenditures have shifted toward electronic media, and displacement of growth in U.S. packaging paper and board demand as domestic manufacturing of consumer goods was displaced by growth in goods imported from other countries. Not surprisingly, China recently became the leading global producer of paper and paperboard products (edging out the United States in total tonnage, although not necessarily in product value).

Figure 3 illustrates trends in U.S. paper and paperboard consumption for several principal product categories—newsprint, printing and writing paper, and other paper and paperboard (including all packaging paper and board products, tissue and sanitary products, and other industrial paper and board products). The trends indicate that the decline in demand has been more significant for printing paper products from 1999 to 2009 (with newsprint consumption down by 61%, and printing and writing paper down by 26%), whereas other paper and board consumption has declined by 17%.

As stated previously, the United States is the world's leading producer of wood pulp. Nevertheless, the unabated growth in U.S. pulp and paper output throughout most of the twentieth century has transitioned to declining output levels and declining pulpwood demands since the late 1990s. In addition to paper and paperboard, pulpwood is also used in the manufacture of oriented strandboard (OSB) and miscellaneous other lesser products. These uses are small in comparison to the demands on pulpwood for wood pulp. Certainly, any view that pulpwood demands will exert unsustainable pressure on U.S. timber supply should give way to a less ominous perspective, given the declining trends in U.S. pulp and paper production and the much lesser demands for other pulpwood-based products.

Solid wood and composite wood product trends. Solid wood products (chiefly, lumber and plywood) and composite wood products (chiefly, other wood panels, such as particleboard and OSB) are important materials for the construction, manufacturing, and shipping segments of the U.S. economy. Nearly all new single-family houses and low-rise multifamily residential structures are framed with lumber and sheathed with wood panels. Large amounts of such products are used also in the construction of low-rise nonresidential buildings and in the upkeep and improvement of existing structures. Solid wood and composite wood products are used extensively in various manufactured products, and nearly all manufactured products are shipped on wooden pallets.

At the turn of the new millennium, total U.S. consumption of solid wood and composite wood products was growing steadily, but consumption declined in recent years. Lumber consumption increased from 120 million cubic meters (m^3) in 2000 to nearly 136 million m^3 in 2005 (**Table 1**). Consumption of all other wood products increased similarly. In 2005, a record high of 204 million m^3 of solid wood and composite wood products were consumed in the United States. However, during 2005, the rumblings of an impending economic downturn were beginning to be felt. In 2006, consumption of wood products began a steady and dramatic decline. By 2009, total solid wood and composite wood products consumption was just 110 million m^3, the lowest total level of consumption since the 1980s.

TABLE 1. Solid wood and composite wood products consumption in the United States, 2000–2009

Year	Lumber* (million m^3)	Structural panels† (million m^3)	Nonstructural panels‡ (million m^3)	Engineered wood§ (million m^3)	Total (million m^3)
2000	120.4	32.3	22.5	4.4	179.6
2001	117.8	31.6	21.1	4.8	175.4
2002	121.5	33.1	21.6	5.1	181.3
2003	121.4	33.8	21.7	5.6	182.5
2004	133.6	35.4	24.1	6.3	199.3
2005	135.9	37.5	23.8	6.5	203.6
2006	128.1	34.3	23.0	6.4	191.8
2007	113.8	32.8	20.6	5.1	172.3
2008¶	96.0	23.6	16.7	3.6	139.9
2009¶	75.1	19.1	13.6	2.4	110.2

*Includes softwood and hardwood lumber.
†Includes softwood plywood and oriented strandboard.
‡Includes hardwood plywood, particleboard, medium-density fiberboard, hardboard, and insulation board.
§Includes prefabricated wood I-joists, glued laminated timbers, laminated veneer lumber, and other structural composite lumber.
¶Preliminary data.
Sources: *C. Adair, Market Outlook: Structural Panels and Engineered Wood Products 2009–2014, American Plywood Association, Tacoma, WA, 2009; J. L. Howard, U.S. Timber Production, Trade, Consumption, and Price Statistics 1965–2005, USDA Forest Service/Forest Products Laboratory, Madison, WI, 2010.*

Much of the recent decline in wood products consumption can be linked directly to the residential housing market. Between 2000 and 2005, single-family housing starts increased from 1.2 to 1.7, before falling to less than 0.5 million in 2009. The result was a drop in total wood products consumption in residential construction from 116 million m^3 in 2005 to 45 million m^3 in 2009 (**Table 2**). This 60% decline in total wood products use for residential housing construction accounted for nearly 80% of the total drop in solid wood and composite wood products consumption from 2005 to 2009.

Despite the recent economic downturn, wood products remain vital to the strength and well-being of the U.S. economy and will continue to be important as the economy recovers and grows in the future. Construction is by far the largest consumer of wood products. Historically, it has accounted for about two-thirds of total solid wood products consumption, about half of which is new residential construction. Information on consumption in nonresidential end-use markets provides important insights into the strength of the market and into how factors affecting the market can impact consumption.

For example, nonresidential construction was fairly unscathed by the initial impact of the recent recession. Nonresidential construction typically takes years to complete rather than months for residential construction. Many projects were in progress during the beginning of the recession. Consumption for nonresidential construction peaked in 2008 and now appears to be in a slow decline, but not as precipitous as the decline in residential construction (Table 2).

The United States has been a net importer of solid wood and composite wood products for decades, but imports have also dropped precipitously as

TABLE 2. Solid wood and composite wood products consumption in the United States by end-use market, 2000–2009

Year	Construction: Residential* (million m^3)	Construction: Nonresidential (million m^3)	Construction: Total (million m^3)	Manufacturing (million m^3)	Packaging and shipping (million m^3)	Other† (million m^3)	Total (million m^3)
2000	92.0	14.3	106.4	27.9	16.6	24.2	175.1
2001	93.1	13.9	107.0	24.3	15.2	24.0	170.6
2002	97.9	12.8	110.7	22.7	15.7	27.1	176.3
2003	101.2	12.2	113.4	21.1	15.4	27.0	176.9
2004	108.8	12.5	121.3	20.8	17.1	33.9	193.0
2005	115.8	12.9	128.7	19.8	17.9	30.7	197.2
2006	106.8	14.5	121.3	18.8	18.9	26.4	185.4
2007	89.4	15.5	105.0	17.9	20.2	24.1	167.2
2008‡	61.1	15.8	76.9	16.9	20.2	22.3	136.3
2009‡	44.8	14.4	59.2	16.5	20.2	11.9	107.8

*Includes new construction, mobile homes, and improvements and repairs.
†Includes activities not captured in specific end-use markets, and adjustments to compensate for inherent differences between total consumption and the sum of its parts.
‡Preliminary data.
Sources: *C. Adair, Market Outlook: Structural Panels and Engineered Wood Products 2009–2014, American Plywood Association, Tacoma, WA, 2009; D. B. McKeever, Estimated Annual Timber Products Consumption in Major End Uses in the United States, 1950–2006, No. FPL-GTR-181, USDA Forest Service/Forest Products Laboratory, Madison, WI, 2009; U.S. Census Bureau: Manufacturing, Mining, and Construction Statistics, 2010.*

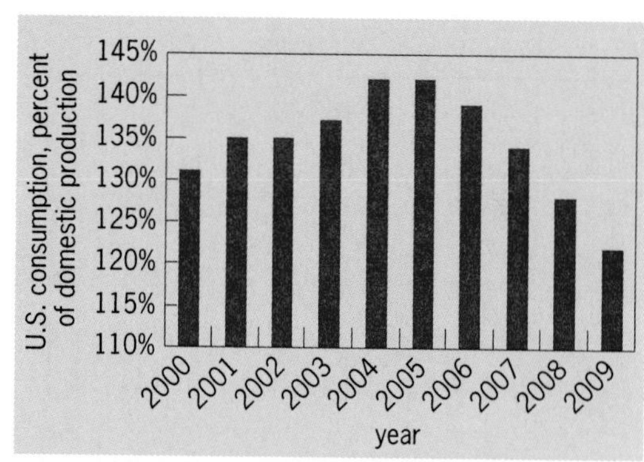

Fig. 4. U.S. solid wood and composite wood consumption as a percentage of domestic production.

consumption declined in recent years. In periods of robust economic growth, domestic demands far exceed domestic production, allowing imports to increase. However, during a slow economy, domestic production reclaims a larger share of consumption. Whereas total consumption of solid wood and composite wood products exceeded domestic production by 31% in 2000 and by 42% in 2005, this percentage dropped to just 22% in 2009 (**Fig. 4**). Thus, domestic producers have reclaimed a larger share of the U.S. wood products market.

Trends in timber growth and removals. Just as timber market trends are never static but rather constantly changing, so too are trends in timber growth and timber supply. As U.S. timber demand and harvest levels have moderated in recent years, U.S. timber growth and timber inventories have increased. Timber inventories are the volumes of standing timber on commercial forest land. **Figure 5** shows trends since the 1980s in net annual growth of timber growing stock (average net change in the volume of forest growing stock inventory) and timber growing stock removals on U.S. timberland (total private and public timberland), as reported by the U.S. Forest Service. [Note that U.S. timberland is defined as all forest land producing or capable of producing industrial wood (at least 20 cubic feet per acre per year) and not withdrawn from production by statute or regulation, including both private and public timberland. Timber growing stock is a classification of timber inventory that includes live commercial species of trees meeting specified standards of quality and vigor (it excludes dead trees or cull trees). Removals include the growing stock removed by timber product harvest (including logging residues) and other removals such as land clearing and stand thinning (harvest and logging residues accounted for 92% of removals in 2006).]

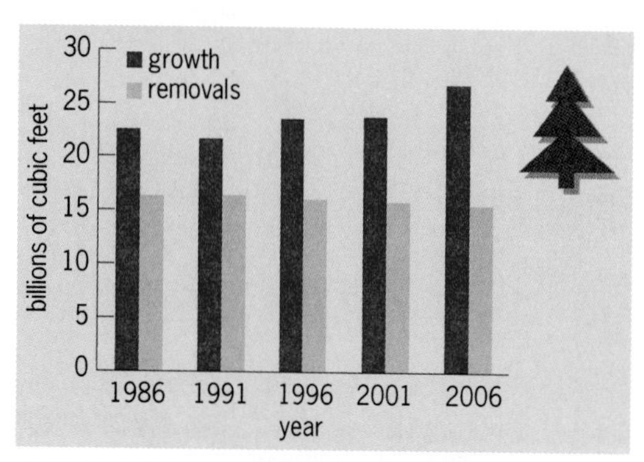

Fig. 5. U.S. timber growing stock net annual growth and removals, billions of cubic feet, 1986–2006.

If net annual growth equaled annual removals, there would be an exact balance between tree removal volumes, tree mortality, and tree growth. However, since net annual growth is increasing and is much larger than annual removals, the data indicate that timber inventory is increasing, as timber growth is far ahead of timber removals. In fact, U.S. timber growing stock inventory has increased from 755,935 million cubic feet in 1987 to 932,096 million cubic feet in 2007, an increase of 23% in 20 years, at an average rate of 9 billion cubic feet per year, and the rate of accumulation has been increasing in recent years. It should be noted that this increase is largely the result of increased volumes per acre, not increased acreage of forest land. Total forest land has been fairly stable over the past 25 years or more. The U.S. timber growing stock inventory is now 51% higher than in 1953, when it was 615,884 million cubic feet.

Trends in real prices for timber. Real (inflation-adjusted) price indexes are indicators of relative economic scarcity or economic abundance for any commodity. [Real price indexes are computed by adjusting nominal price indexes for inflation, in this case by dividing producer price indexes for timber commodities by the all-commodity producer price index (using nationwide price index data as reported by the U.S. Bureau of Labor Statistics).] **Figure 6** shows U.S. nationwide real price indexes for several timber commodities, including softwood logs and bolts, hardwood logs and bolts, and pulpwood. Although timber commodity prices have experienced some volatility in recent decades, the overall trends in real prices have been generally declining since the mid-1990s, which tends to indicate abundance of timber supply relative to timber demand. Timber supply indicates the volumes of timber that owners are willing to sell at differing price levels, whereas timber demand indicates the volumes of timber that buyers are willing to purchase at differing price levels.

Softwood logs experienced a sharp price spike in the late 1980s and early 1990s, largely because of administrative reductions in timber harvests from national forests and other public lands (particularly in the U.S. West). However, after peaking in the early 1990s, the real price index for softwood logs and bolts declined, and the index is currently at a level nearly the same as in the early 1980s. Meanwhile, real price indexes did not increase as much for hardwood logs and bolts, or for pulpwood, and those price indexes have also substantially declined with declining demands since the 1990s. In fact, they are now at levels well below price levels of the 1980s in real (inflation-adjusted) terms (Fig. 6). These changes indicate that timber supply growth has far outpaced domestic demand growth since the early 1990s, which resulted from a declining paper and paperboard sector, a declining manufacturing sector (particularly furniture manufacturing, which has moved largely offshore), and increased importation of softwood lumber. These factors all contributed to increased accumulation of timber inventories (a contributing factor in supply growth).

Conclusions. The United States is the world's leading producer of primary forest products, including wood pulp, lumber, and wood panel products. In 2008, the United States accounted for 17% of worldwide lumber, plywood, and particleboard production, down from 25% a decade earlier, and 21% of worldwide paper and paperboard production, down from 29% a decade earlier. Those products serve a wide array of end users in commercial activities ranging from packaging, shipping, warehousing, publishing, personal care, and hygiene to housing construction, manufacturing, and transportation. As such, sustaining adequate supply of timber for those primary products is economically and commercially an important element in sustaining the economic vitality, competitiveness, and future growth of the U.S. economy. However, growth in timber demand has diminished in recent years, and concern that wood supply is out of balance with wood demand should give way to a new perspective. Real prices for timber commodities have been declining for many years (at least since the 1990s), a clear reflection of the fact that timber supply and timber growth are outpacing timber demand. From a resource perspective, trends indicate that U.S. timber demands have become much more sustainable in relation to timber supply, with annual timber growth expanding its lead over annual timber harvest volume, and standing timber inventories reaching levels higher than at any time in at least the past 60 years or more.

For background information *see* FOREST AND FORESTRY; FOREST MANAGEMENT; FOREST MEASUREMENT; FOREST TIMBER RESOURCES; LUMBER; PAPER; REFORESTATION; WOOD PROCESSING; WOOD PRODUCTS in the McGraw-Hill Encyclopedia of Science & Technology.

Peter J. Ince; David B. McKeever

Bibliography. C. Adair, *Market Outlook: Structural Panels and Engineered Wood Products 2009–2014*, American Plywood Association, Tacoma, WA, 2009; American Forest and Paper Association (AF&PA), *Statistics of Paper, Paperboard and Wood Pulp*, AF&PA, Washington, D.C., 2008; J. L. Howard, *U.S. Timber Production, Trade, Consumption, and Price Statistics 1965–2005*, USDA Forest Service/Forest Products Laboratory, Madison, WI, 2010; D. B. McKeever, *Estimated Annual Timber Products Consumption in Major End Uses in the United States, 1950–2006*, No. FPL-GTR-181, USDA Forest Service/Forest Products Laboratory, Madison, WI, 2009; D. S. Powell et al., *Forest Resources of the United States, 1992*, General Technical Report RM-234, USDA Forest Service/Rocky Mountain Forest and Range Experiment Station, Fort Collins, CO, 1993; W. B. Smith et al., *Forest Resources of the United States, 1997*, General Technical Report NC-219, USDA Forest Service/North Central Research Station, St. Paul, MN, 2001; W. B. Smith et al., *Forest Resources of the United States, 2002*, General Technical Report NC-241, USDA Forest Service/North Central Research Station, St. Paul, MN, 2003; W. B. Smith et al., *Forest Resources of the United States, 2007*, General Technical Report WO-78, USDA Forest Service/Washington Office, Washington, D.C., 2009; K. L. Waddell et al., *Forest Statistics of the United States, 1987*, Resource Bulletin PNW-RB-168, USDA Forest Service/Pacific Northwest Research Station, Portland, OR, 1989.

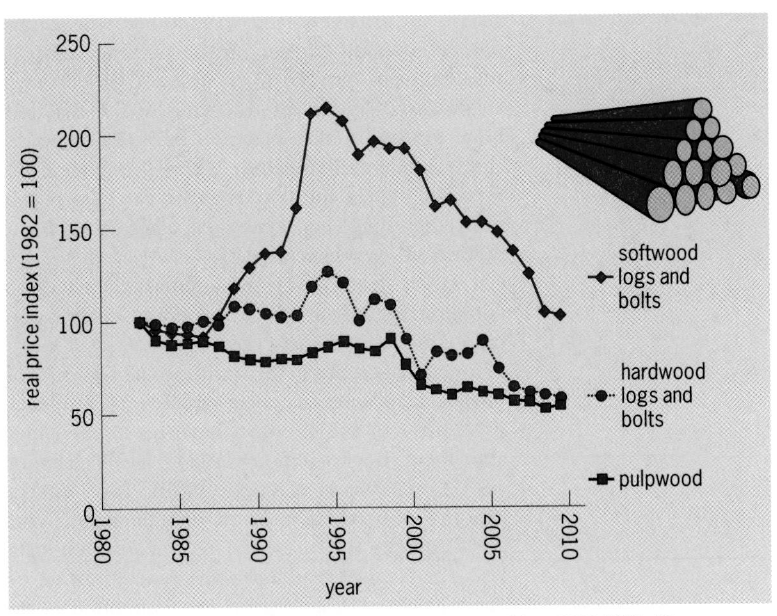

Fig. 6. U.S. real price indexes for delivered timber commodities, 1982–2009. [*Source: Bureau of Labor Statistics, Producer Price Indexes (deflated using all-commodity producer price index)*]

Nobel Prizes for 2010

The Nobel Prizes for 2010 included the following awards for scientific disciplines.

Chemistry. The chemistry prize was awarded to Richard F. Heck of the University of Delaware, Ei-ichi Negishi of Purdue University, and Akira Suzuki of Hokkaido University, Sapporo, Japan, for palladium-catalyzed cross-coupling reactions in organic synthesis.

Cross-coupling is a carbon-carbon bond-forming reaction. In organic synthesis, cross-coupling reactions are used to produce organic compounds such as natural products that have useful properties, biologically active molecules used in medicine and agriculture, and organic molecules used in electronic devices.

The mechanisms of palladium cross-coupling reactions are quite complex. Simply, however, cross-coupling involves the reaction of an organohalide (the electrophilic coupling partner) with either an alkene or organometallic compound (the nucleophilic coupling partner) in the presence of a zerovalent palladium [$Pd_{(0)}$] catalyst and proceeds via an organopalladium intermediate. In the reaction, bond formation takes place on Pd by bringing the carbon atoms close together, where they couple and form a bond. In the last step, the catalyst—a substance that speeds up a chemical reaction but is not consumed—is regenerated.

The key features of palladium-catalyzed cross-coupling reactions are the reagents and reaction conditions are mild, the reactions tolerate a wide range of functional groups in the starting materials, and the products are stereoselective, with few by products.

All three of the scientists have reactions named after them: Heck reaction (1972), Negishi reaction (1977), and Suzuki reaction (1979). The Heck reaction is the reaction of an organic halide with an alkene in the presence of a palladium catalyst. The Negishi reaction is the reaction of an organic halide with an organozinc compound in the presence of a palladium catalyst. The Suzuki reaction is the reaction of an organic halide with an organoboron reagent in the presence of a palladium catalyst. Today, all three reactions are used widely in organic synthesis at laboratory through multiton scales.

In organic synthesis, green chemical reactions reduce or eliminate waste and hazardous substances. Among the principles of green chemistry are using mild reagents, avoiding protecting groups, using less solvent, and using catalysts. Palladium-catalyzed cross-coupling reactions attain these principles, and recent research has shown that Heck reactions, for example, can be run in ionic liquids or water instead of organic solvents, potentially extending their application and usefulness.

For background information *see* ALKENE; CATALYSIS; GREEN CHEMISTRY; ORGANIC CHEMISTRY; ORGANIC SYNTHESIS; ORGANOMETALLIC COMPOUND; PALLADIUM in the McGraw-Hill Encyclopedia of Science & Technology.

Physiology or medicine. The prize in physiology or medicine was awarded to Dr. Robert G. Edwards, Emeritus Professor of Human Reproduction at the University of Cambridge, U.K., for the development of human in vitro fertilization (IVF).

Human infertility has been estimated to affect about 10% of couples throughout the world. In women, it is often due to the impairment of fertilization by damage to the Fallopian tubes, the oviducts through which eggs travel from the ovaries to the uterus. In males, infertility is often due to poor quantity or quality of sperm. The research conducted by Edwards was not only instrumental in overcoming infertility but also contributed much to our basic understanding of human developmental biology.

In humans and other mammals, eggs (ova) develop from primordial cells known as oocytes within individual follicles in the ovaries. Prior to the release of mature eggs (ovulation), the oocytes remain in a resting stage of the first meiotic division (prophase). Then, in humans during an approximately monthly cycle, hormones and other factors induce the resumption of growth and the continuation of meiosis to the metaphase stage. Following rupture of a follicle, a mature egg is released and enters a Fallopian tube. If it is met by sperm ascending from the uterus, a single sperm may penetrate the egg and initiate a series of events including further progress of meiosis leading to two haploid (half) sets of chromosomes, one of which fuses with the haploid set from the sperm, and the other discarded. The process of fertilization results in an embryo with the full (diploid) complement of chromosomes, which then proceeds to divide, descend the Fallopian tube, and eventually implant in the wall of the uterus. Edwards's work required research on and a full understanding of all of the steps in this sequence of events, and the development of methods and techniques to implement in vitro fertilization.

Research on in vitro fertilization had roots in studies done in the nineteenth century on marine invertebrates and nonmammalian vertebrates, many of which achieve union of egg and sperm outside the body in the aquatic environment. In the mid-1930s, Gregory Pincus (who later on was a developer of the contraceptive pill) reported experimental maturation of rabbit oocytes in vitro, and in 1959, his colleague Min Chueh Chang demonstrated successful in vitro fertilization of matured rabbit oocytes that could develop into viable offspring after implantation of the embryos in adult females. His technique, however, was based on a contemporary belief that sperm needed to be activated in vivo prior to in vitro fertilization by exposure to the uterus of a pregnant female, a process known as capacitation. Chang later developed experimental conditions that showed this was not necessary, which set the stage for much further work.

Successful in vitro fertilization of human gametes and subsequent implantation of resulting embryos required overcoming several technical hurdles: control of oocyte maturation; ability to retrieve oocytes at a stage appropriate for IVF; activation of sperm in vitro; providing conditions conducive to fertilization and embryo development; and a technique for successful transfer of the embryo to the uterus for prenatal development.

Edwards, working at the University of Cambridge, started studying the maturation of human eggs using ova obtained from pieces of ovary that had been removed during unrelated surgical procedures. His initial attempts at initiating fertilization were unsuccessful until he utilized a culture medium developed by Barry Bavister, a student at Cambridge, for experiments being done on hamsters fertilization. It was about this time in the late 1960s that Edwards started a crucial collaboration with Patrick Steptoe, an obstetrician and gynecologist at Oldham General Hospital in London, who had developed techniques for laparoscopy, the direct visualization of internal organs by means of an optical instrument inserted into the body through a small incision, thus eliminating the need for major surgery. Working with volunteer patients, they pharmacologically induced maturation of eggs in the ovaries and then collected them using a modified laparoscope. Using fertilization techniques they had been perfecting, the eggs were exposed to sperm provided by the prospective father. With time, they were able to have embryos develop normally to the blastocyst stage, at which point embryos normally implant themselves in the uterine wall. Attempts at implantation failed, however, until they changed their procedure to use eggs removed just at the time of normal ovulation rather than induced maturation. In 1977, a success implantation was achieved, and on July 25, 1978, Louise Brown, the first person conceived through IVF, was born.

Edwards successfully continued his collaboration with Steptoe until the latter's death in 1988. Edwards's work was met with criticism at the outset from those who felt it was morally objectionable or politically incorrect. He was very careful to meet these objections and develop ethical guidelines for his procedures. Researchers in many other laboratories and countries expanded on the original techniques, and today more than 4,000,000 babies have been born worldwide through in vitro fertilization.

For background information *see* ANIMAL REPRODUCTION; FERTILIZATION (ANIMAL); INFERTILITY; PREGNANCY; REPRODUCTIVE SYSTEM DISORDERS; REPRODUCTIVE TECHNOLOGY in the McGraw-Hill Encyclopedia of Science & Technology.

Physics. The physics prize was awarded to Andre K. Geim and Konstantin S. Novoselov for groundbreaking experiments regarding the two-dimensional material graphene. Geim and Novoselov are both Russian-born physicists working at the University of Manchester in the United Kingdom; Geim is a citizen of the Netherlands, while Novoselov is a British and Russian citizen.

Graphene is a flat, two-dimensional sheet of carbon atoms arranged in a honeycomb structure. Graphite, such as found in a pencil, consists of millions of such sheets stacked on top of one another. The covalent chemical bonds between neighboring carbon atoms in a sheet are very strong, while the inter-sheet bonds in graphite are extremely weak. In fact, the writing ability of a pencil results from the consequent ease with which thin layers of graphite are cleaved from a larger crystal by rubbing it against a surface such as paper. Two other carbon materials, carbon nanotubes and fullerenes, can be thought of as graphene sheets that have been rolled up into tubes and balls, respectively.

It was long thought that it would be impossible to isolate stable graphene sheets, but in 2004 Geim, Novoselov, and their colleagues accomplished precisely that. They used adhesive tape to cleave a graphite crystal repeatedly, 10 to 20 times, to produce successively thinner graphite flakes on the tape. Then they transferred the flakes to an oxidized silicon wafer by pressing the tape against the wafer, and were able to identify single graphene layers with an optical microscope, using the transparent silicon dioxide layer as a kind of interference filter.

Besides the exfoliation method, other ways of fabricating very thin carbon films soon followed. In particular, the fabrication of epitaxial graphene on crystalline silicon carbide has been studied by a group led by Walter de Heer.

Graphene is the first truly two-dimensional crystalline material. Its electronic structure, which was described theoretically in 1947, is quite different from that of three-dimensional materials, and gives rise to properties that were studied by Novoselov, Geim, and colleagues in their initial work, with the help of observations of the Hall effect in the material. They varied the voltage across a graphene sheet to transform it from an electron conductor to a hole conductor and back, and to alter its carrier concentration, thereby radically changing its conductivity. Thus, they were able to demonstrate a metallic field-effect transistor. The ease of altering the carrier density is the basis of graphene's predicted usefulness in many possible device configurations. *See* QUANTUM ELECTRONIC PROPERTIES OF MATTER.

Graphene's electronic structure also gives rise to an unusual form of the quantum Hall effect, which may lead to a more accurate standard of electrical resistance. Moreover, it allows electrons to travel relatively large distances in graphene without scattering (ballistic transport), giving rise to large electron mobilities and to a conductivity greater than that of copper. This and other electronic properties, as well as the prospect of fabricating graphene circuits that are smaller than those that can be manufactured from silicon, have led to the prediction that graphene transistors will be faster and more efficient than current silicon transistors; the high mobility may be particularly useful in high-frequency applications. Thus, as noted in the initial paper by Novoselov, Geim, and colleagues, graphene technology could play a critical role at a time when the semiconductor industry is nearing the limits of performance for technologies dominated by silicon (though graphene computers are still a distant dream). Finally, electrons in graphene behave as if they were massless; like massless photons they travel at a constant speed, which is, however, only about 1/300 that of light. It is, therefore, possible that the study of certain phenomena that would otherwise require large particle accelerators can be conducted more easily on a smaller scale using graphene.

Graphene has a number of other unusual properties that will probably be the basis of future applications. The strong bonding of the carbon atoms gives rise to a nearly perfect crystalline structure, which makes graphene suitable for extremely sensitive detectors of pollutants and other materials. Graphene conducts heat more than 10 times better than copper, and is more than 100 times stronger than the strongest steel. It is almost transparent, and this property, together with its large conductivity, renders it suitable for use in touch screens, light panels, and perhaps even solar cells. Graphene might be mixed into plastics to make them electrically conducting, more heat resistant, and more mechanically robust. The new composite materials could be used in the manufacture of automobiles, aircraft, and spacecraft.

For background information *see* CARBON NANOTUBES; CHEMICAL BONDING; COMPOSITE MATERIAL; EPITAXIAL STRUCTURES; FULLERENE; GRAPHITE; HALL EFFECT; INTEGRATED CIRCUITS; INTERFERENCE FILTERS; TRANSISTOR in the McGraw-Hill Encyclopedia of Science & Technology.

Contributors

The affiliation of each Yearbook contributor is given, followed by the title of his or her article. An article title with the notation "coauthored" indicates that two or more authors jointly prepared an article or section.

A

Acosta-Vigil, Dr. Antonio. *Consejo Superior de Investigaciones Científicas (CSIC), Instituto Andaluz de Ciencias de la Tierra, Granada, Spain.* USING MELT INCLUSIONS FOR UNDERSTANDING CRUSTAL MELTING PROCESSES—coauthored.

Adams, Dr. David J. *Wellcome Trust Genome Campus, Experimental Cancer Genetics Laboratory, Wellcome Trust Sanger Institute, Hinxton, United Kingdom.* DNA SEQUENCING—coauthored.

Ailor, Dr. William H. *Director, Center for Orbital and Reentry Debris Studies, The Aerospace Corporation, El Segundo, California.* ASTEROID EXTINCTION.

AlAbdulkarim, Layla O. *Faculty of Technology, Policy and Management, Delft University of Technology, The Netherlands.* SMART METERING—coauthored.

Angielczyk, Dr. Kenneth D. *Department of Geology, Field Museum of Natural History, Chicago, Illinois.* DICYNODONTIA.

Annese, Dr. Jacopo. *The Brain Observatory, University of California, San Diego.* NEUROPSYCHOLOGY OF MEMORY REVISITED.

Apsel, Dr. Alyssa B. *School of Electrical and Computer Engineering, Cornell University, Ithaca, New York.* LOW-POWER RADIO LINKS—COAUTHORED.

B

Baird, Prof. Donald G. *Department of Chemical Engineering, Virginia Polytechnic Institute & State University, Blacksburg.* FLOW BEHAVIOR OF VISCOELASTIC FLUIDS.

Barger, Prof. Amy J. *Department of Astronomy, University of Wisconsin-Madison.* SUPERMASSIVE BLACK HOLES.

Batten, Dr. Sonia. *Sir Alister Hardy Foundation for Ocean Science, Plymouth, United Kingdom.* CONTINUOUS PLANKTON RECORDERS IN THE NORTH PACIFIC OCEAN.

Beard, Ronald L. *Space Applications Branch, U.S. Naval Research Laboratory, Washington, District of Columbia.* COORDINATED UNIVERSAL TIME (UTC) SCALE.

Beckstead, Dr. Robert B. *Department of Poultry Science, University of Georgia, Athens.* GENE DISCOVERY IN DROSOPHILA MELANOGASTER.

Bellantoni, Jeffrey. *Chairperson of Graduate Art Design, Pratt Institute, New York.* KINETIC TYPOGRAPHY—coauthored.

Belov, Dr. Katherine. *Faculty of Veterinary Science, University of Sydney, New South Wales, Australia.* CONTAGIOUS CANCERS.

Bengston, Dr. Roger. *Physics Department, University of Texas, Austin.* *VASIMR* PROPULSION SYSTEM.

Bickel, Dr. Sharon E. *Department of Biological Sciences, Dartmouth College, Hanover, New Hampshire.* SISTER CHROMATID COHESION—coauthored.

Bielory, Dr. Leonrad. *Department of Environmental Sciences, Rutgers University, Piscataway, New Jersey.* MODELING EFFECTS OF CLIMATE CHANGE ON ALLERGIC ILLNESS—coauthored.

Binder, Prof. Dr. Wolfgang H. *Institute of Chemistry/Chair of Macromolecular Chemistry, Martin-Luther University Halle-Wittenberg, Germany.* CLICK CHEMISTRY IN POLYMER SCIENCE—coauthored.

Bjorhovde, Dr. Reidar. *The Bjorhovde Group, Tucson, Arizona.* SUSTAINABLE DEVELOPMENT IN STEEL CONSTRUCTION.

Bluestein, Dr. Howard B. *School of Meteorology, University of Oklahoma, Norman.* RECENT ADVANCEMENT IN TORNADO OBSERVATION.

Boubli, Dr. Jean Philippe. *Wildlife Conservation Society, WCS Brasil, Rio de Janeiro, Brazil.* NEW VERTEBRATE SPECIES—coauthored.

Bovenkerk, Jens. *Institut für Kraftfahrzeuge, RWTH Aachen University, Body Department, Aachen, Germany.* PEDESTRIAN PROTECTION.

Boyden, Prof. Edward S. *Synthetic Neurobiology Group, MIT Media Lab, Cambridge, Massachusetts.* OPTOGENETICS.

Brayard, Dr. Arnaud. *Laboratoire Biogéosciences, Université de Bourgogne, Dijon, France.* PELAGIC ECOSYSTEM RECOVERY AFTER END-PERMIAN MASS EXTINCTION.

Brennesholtz, Matthew. *Insight Media, Norwalk, Connecticut.* PICOPROJECTORS.

Brouns, Dr. Stan J. J. *Laboratory of Microbiology, Department of Agrotechnology and Food Sciences, Wageningen University, Gelderland, The Netherlands.* CRISPR-BASED IMMUNITY IN PROKARYOTES—coauthored.

Bruynseraede, Prof. Yvan. *Department of Physics and Astronomy, Katholieke Universiteit, Leuven, Belgium.* ANDERSON LOCALIZATION—coauthored.

C

Calais, Prof. Eric. *Department of Earth & Atmospheric Sciences, Purdue University, West Lafayette, Indiana.* 2010 HAITI EARTHQUAKE.

Callegaro, Dr. Luca. *Electromagnetism Division, Istituto Nazionale di Ricerca Metrologica (INRIM), Torino, Italy.* ESTABLISHING THERMODYNAMIC TEMPERATURE WITH NOISE THERMOMETRY.

Campbell, Dr. Kevin L. *Department of Biological Sciences, University of Manitoba, Winnipeg, Canada.* STAR-NOSED MOLE—coauthored.

Cannon, Prof. James W. *Department of Mathematics, Brigham Young University, Provo, Utah.* CANNON'S CONJECTURE.

Cesare, Dr. Bernardo. *University of Padova, Department of Geosciences, Padova, Italy.* USING MELT INCLUSIONS FOR UNDERSTANDING CRUSTAL MELTING PROCESSES—coauthored.

Coates, Dr. John D. *Department of Plant and Microbial Biology, University of California, Berkeley.* MICROBIAL FUEL CELLS—coauthored.

Crne, Dr. Matija. *Procter & Gamble Service GmbH, Darmstadt, Germany.* SCARAB BEETLE IRIDESCENCE—coauthored.

Cunliffe, Dr. Vincent T. *MRC Centre for Developmental and Biomedical Genetics, Department of Biomedical Science, University of Sheffield, South Yorkshire, United Kingdom.* EPIGENETIC MECHANISMS IN DEVELOPMENT.

Curtis, Dr. Kevin. *InPhase Technologies, Longmont, Colorado.* HOLOGRAPHIC DATA STORAGE.

D

Devenport, Prof. William J. *Aerospace and Ocean Engineering, Virginia Polytechnic Institute and State University, Blacksburg.* AEROACOUSTICS.

Dokania, Rajeev K. *School of Electrical and Computer Engineering, Cornell University, Ithaca, New York.* LOW-POWER RADIO LINKS—coauthored.

Duprez, Dr. Delphine. *Biologie Moléculaire et Cellulaire du Développement, Université Pierre et Marie Curie, Paris, France.* MUSCLE DEVELOPMENT AND REGENERATION.

E

Ebel, Dr. Rainer. *Marine Biodiscovery Centre, University of Aberdeen, United Kingdom.* MARINE MYCOLOGY.

Engle, Allen. *Naval Surface Warfare Center, Carderock Division, West Bethesda, Maryland.* WAVE SLAMMING.

Epstein, Dr. Richard I. *Los Alamos Natiional Laboratory, New Mexico.* LASER COOLING OF SOLIDS—coauthored.

Etemad, Dr. Kamran. *Mobile Wireless Group, Intel Corporation, Potomac, Maryland.* MOBILE WIMAX.

F

Ferroir, Dr. Tristan. *Laboratoire des Sciences de la Terre, Ecole Normale Superieure de Lyon, Lyon, France.* HARDER-THAN-DIAMOND CARBON CRYSTALS.

Floodeen, Robert. *Software Engineering Institute/CERT, Carnegie Mellon University, Pittsburgh, Pennsylvania.* CYBER DEFENSE—coauthored.

Frauendorf, Prof. Stefan. *Department of Physics, University of Notre Dame, Indiana.* CHIRALITY IN ROTATING NUCLEI.

G

Gaudet, Dr. Justin J. *Department of Biological Sciences, Dartmouth College, Hanover, New Hampshire.* SISTER CHROMATID COHESION—coauthored.

Gegear, Dr. Robert J. *Department of Neurobiology, University of Massachusetts Medical School, Worcester.* MONARCH BUTTERFLY MIGRATION—coauthored.

Gelenbe, Prof. Erol. *Department of Electrical and Electronic Engineering, Imperial College London.* SELF-AWARE NETWORKS.

Georgopoulos, Dr. Panos G. *Environmental and Occupational Health Sciences Institute, Piscataway, New Jersey.* MODELING EFFECTS OF CLIMATE CHANGE ON ALLERGIC ILLNESS—coauthored.

Geschwind, Dr. Daniel H. *Departments of Neurology and Human Genetics, UCLA School of Medicine, Los Angeles, California.* FOXP2 GENE AND HUMAN LANGUAGE—coauthored.

Ghose, Dr. Shohini. *Department of Physics and Computer Science, Wilfrid Laurier University, Waterloo, Ontario, Canada.* QUANTUM SIGNATURES OF CHAOS—coauthored.

Giurgiutiu, Prof. Victor. *Department of Mechanical Engineering, University of South Carolina, Columbia.* ACTIVE MATERIALS AND SMART STRUCTURES.

Goodkind, Prof. John M. *Department of Physics, University of California-San Diego, La Jolla.* THE SEARCH FOR A SUPERSOLID.

Gooßen, Dr. Käthe. *Fachbereich Chemie/Organische Chemie, Technische Universität Kaiserslautern, Germany.* Decarboxylative coupling—coauthored.

Gooßen, Prof. Dr. Lukas J. *Fachbereich Chemie/Organische Chemie, Technische Universität Kaiserslautern, Germany.* Decarboxylative coupling—coauthored.

Graves, Dr. Jennifer A. Marshall. *ARC Centre for Kangaroo Genomics and Research School of Biology, Australian National University, Acton, Australian Capital Territory.* Genomics and relationships of Australian mammals—coauthored.

Grupe, Daniel W. *Department of Psychology, Waisman Laboratory for Brain Imaging and Behavior, University of Wisconsin-Madison.* Anxiety disorders and the amygdala—coauthored.

H

Hageskal, Dr. Gunhild. *Section of Mycology, National Veterinary Institute, Oslo, Norway.* Fungi in drinking water—coauthored.

Halzen, Prof. Francis. *IceCube Research Center, University of Wisconsin, Madison.* IceCube neutrino observatory.

Hamilton, Prof. Joseph H. *Department of Physics and Astronomy, Vanderbilt University, Nashville, Tennessee.* Asymmetric, pear-shaped nuclei—coauthored.

Harrell, Ashley. *Department of Sociology, University of South Carolina, Columbia.* Human altruism—coauthored.

Harvati, Dr. Katerina. *Department of Paleoanthropology, Eberhard Karls University of Tübingen, Baden-Württenberg, Germany.* Homo heidelbergensis.

Herbst, Florian. *Institute of Chemistry/Macromolecular Chemistry, Martin-Luther University Halle-Wittenberg, Germany.* Click chemistry in polymer science—coauthored.

Hewitt, Dr. Patrick W. *Aerojet, Gainesville, Virginia.* Ramjet engines.

Homann, Dr. Dirk. *Barbara Davis Center for Childhood Diabetes, University of Colorado-Denver, Aurora, Colorado.* Type 1 diabetes.

Hou, Dr. Tuo-Hung. *Department of Electronics Engineering and Institute of Electronics, National Chiao Tung University, HsinChu, Taiwan, Province of China.* Metal-oxide resistive-switching RAM technology.

Hsieh, Prof. Shang-Hsien. *Deputy Dean, Office of International Affairs & Professor, Department of Civil Engineering, National Taiwan University, Taipei, Taiwan.* Internet-based simulation for earthquake engineering.

Huber, Dr. Alan. *Institute for the Environment, University of North Carolina at Chapel Hill.* Air toxics computational fluid dynamics.

I

Ince, Dr. Peter J. *USDA Forest Service, Forest Products Laboratory, Madison, Wisconsin.* Wood supply and demand—coauthored.

Irmis, Dr. Randall B. *Utah Museum of Natural History, and Department of Geology and Geophysics, University of Utah, Salt Lake City.* Reappraisal of early dinosaur radiation—coauthored.

Isukapalli, Dr. Sastry S. *Environmental and Occupational Health Sciences Institute, Piscataway, New Jersey.* Modeling effects of climate change on allergic illness—coauthored.

J

Jentschel, Dr. Michael. *Nuclear and Particle Physics Group, Institute Laue-Langevin, Grenoble, France.* Einstein's mass-energy equivalence principle.

Jessen, Prof. Poul S. *College of Optical Sciences, University of Arizona, Tucson.* Quantum signatures of chaos—coauthored.

K

Kanak, Dr. Katharine M. *School of Meteorology, University of Oklahoma, Norman, Oklahoma.* Multimoment microphysical parameterizations in cloud models—coauthored.

Kang, Dr. Julie. *Department of Plant Biology, University of California, Davis.* Regulation of leaf shape—coauthored.

Keane, Dr. Thomas M. *Wellcome Trust Sanger Institute, Hinxton, United Kingdom.* DNA sequencing—coauthored.

Klein, Dr. Andrew G. *Department of Geography, Texas A&M University, College Station.* Tropical glacier monitoring.

Konopka, Dr. Genevieve. *Department of Neurology, UCLA School of Medicine, Los Angeles, California.* FOXP2 gene and human language—coauthored.

Kun, Dr. Luis. *Senior Research Professor of Homeland Security, National Defense University, IRM College, Washington, District of Columbia.* Health and environmental interoperability—coauthored.

Kwan, Dr. Mei-Po. *Department of Geography, Ohio State University, Columbus.* Three-dimensional geographic information systems.

L

Labandeira, Dr. Conrad C. *Department of Paleobiology, National Museum of Natural History, Smithsonian Institution, Washington, District of Columbia.* Pollination mutualisms by insects before the evolution of flowers.

Levy, Dr. Salomon. *Levy & Associates, Campbell, California.* Innovative nuclear fuel cycles.

Li, Prof. Yuanqi. *Department of Building Engineering, Tongji University, Shanghai, China.* High-speed rail—coauthored.

Liu, Dr. Edwin. *Section of Pediatric Gastroenterology, Hepatology, and Nutrition, The Children's Hospital, University of Colorado Health Sciences Center, Aurora, Colorado.* CELIAC DISEASE.

Lively, Mark B. *Utility Economic Engineers, Gaithersburg, Maryland.* ELECTRIC GENERATION ANCILLARY SERVICES.

Löw, Dr. Robert. *Institute of Physics, University of Stuttgart, Germany.* ULTRALONG-RANGE RYDBERG MOLECULES.

Lukszo, Dr. Zofia. *Faculty of Technology, Policy and Management, Delft University of Technology, The Netherlands.* SMART METERING—coauthored.

Lumbsch, Dr. H. Thorsten. *Department of Botany, Field Museum of Natural History, Chicago, Illinois.* PHYLOGEOGRAPHY AND BIOGEOGRAPHY OF FUNGI.

Luo, Dr. Yixiao. *Department of Physics and Astronomy, Vanderbilt University, Nashville, Tennessee.* ASYMMETRIC, PEAR-SHAPED NUCLEI—coauthored.

M

Ma, Prof. Wenchao. *Department of Physics and Astronomy, Mississippi State University.* WOBBLING MOTION IN NUCLEI.

Malasri, Kriangsiri. *Department of Computer Science, University of Memphis, Tennessee.* SECURING WIRELESS IMPLANTABLE HEALTH-CARE DEVICES—coauthored.

McCombs, Beverly. *Cofounder and a director emeritus of Traffic Design Group Ltd, New Zealand.* TRANSPORTATION'S ROLE IN SUSTAINABILITY—coauthored.

McDowell, David Q. *Standards Consultant, Penfield, New York.* REFERENCE PRINTING CONDITIONS.

McKeever, David B. *USDA Forest Service, Forest Products Laboratory, Madison, Wisconsin.* WOOD SUPPLY AND DEMAND—coauthored.

Melnyk, Ryan A. *Department of Plant and Microbial Biology, University of California, Berkeley.* MICROBIAL FUEL CELLS—coauthored.

Merlin, Dr. Christine. *Department of Neurobiology, University of Massachusetts Medical School, Worcester.* MONARCH BUTTERFLY MIGRATION—coauthored.

Mostoslavsky, Dr. Gustavo. *Department of Medicine, Section of Gastroenterology, and Center for Regenerative Medicine (CReM), Boston University School of Medicine, Massachusetts.* INDUCED PLURIPOTENT STEM CELLS.

N

Nagano, Dr. Takashi. *Laboratory of Chromatin and Gene Expression, Babraham Institute, Cambridge, United Kingdom.* LONG NONCODING RNAS (LONG NCRNAS).

Nägerl, Prof. Urs Valentin. *Bordeaux Neuroscience Institute, Université Victor Segalen Bordeaux 2, France.* FLUORESCENCE MICROSCOPY: HIGH-RESOLUTION METHODS.

Nawrot, Dr. Mark. *Center for Visual Neuroscience, Department of Psychology, North Dakota State University, Fargo.* PARALLAX AND THE BRAIN.

Neil, Dr. George. *Thomas Jefferson National Accelerator Facility (Jefferson Lab), Newport News, Virginia.* ULTRAFAST X-RAY SOURCES—in part.

Nesbitt, Dr. Sterling J. *Jackson School of Geosciences, University of Texas, Austin.* REAPPRAISAL OF EARLY DINOSAUR RADIATION—coauthored.

Nitschke, Dr. Jack B. *Department of Psychology, Waisman Laboratory for Brain Imaging and Behavior, University of Wisconsin-Madison.* ANXIETY DISORDERS AND THE AMYGDALA—coauthored.

O

Ostrowski, Dr. Daniela. *Department of Biological Sciences, University of Missouri, Columbia.* GENETICS OF MEMORY—coauthored.

Otnes, Dr. Roald. *Maritime Systems Division, Norwegian Defence Research Establishment (FFI), Horten, Norway.* INTERFERENCE FROM POWER-LINE TELECOMMUNICATIONS.

P

Park, Dr. Jung Ok. *School of Materials Science and Engineering, Center for Advanced Research on Optical Microscopy, Georgia Institute of Technology, Atlanta.* SCARAB BEETLE IRIDESCENCE—coauthored.

Parker, William G. *Division of Resource Management, Petrified Forest National Park, Petrified Forest, Arizona.* REAPPRAISAL OF EARLY DINOSAUR RADIATION—coauthored.

Parmigiani, Prof. Fulvio. *Department of Physics, Università degli Studi di Trieste, Italy.* ULTRAFAST X-RAY SOURCES—in part.

Pask, Dr. Andrew J. *Department of Molecular and Cellular Biology, University of Connecticut, Storrs.* GENOME OF THE PLATYPUS.

Patel, Dr. Hardip R. *ARC Centre for Kangaroo Genomics and Research School of Biology, Australian National University, Acton, Australian Capital Territory, and Molecular Genetics Division, Victor Chang Cardiac Research Institute, Darlinghurst, New South Wales, Australia.* GENOMICS AND RELATIONSHIPS OF AUSTRALIAN MAMMALS—coauthored.

Pierce, Dr. Marcia M. *Department of Biological Sciences, Eastern Kentucky University, Richmond.* CANINE INFLUENZA; H1N1 INFLUENZA.

Pines, Prof. Daryll J. *Farvardin Professor and Dean, A. James Clark School of Engineering, University of Maryland, College Park.* PULSAR NAVIGATION—coauthored.

Plackett, Dr. David V. *Risø National Laboratory for Sustainable Energy, Technical University of Denmark, Roskilde.* BIOREFINERY (WOOD).

Platt, Dr. Donald. *Micro Aerospace Solutions, Inc., Melbourne, Florida.* SPACE FLIGHT, 2009.

Pleijel, Dr. Fredrik. *Department of Marine Ecology, Tjärnö Marine Biological Laboratory, University of Gothenburg, Strömstad, Sweden.* DEEP-SEA ANNELID WORMS.

Politis, Dr. Gustavo G. *CONICET, Facultad de Ciencias Sociales, Universidad Nacional del Centro de la Provincia de Buenos Aires, Argentina.* ETHNOARCHEOLOGY.

Price, Nadine T. *Department of Biological Sciences, University of Manitoba, Winnipeg, Canada.* STAR-NOSED MOLE—coauthored.

Priola, Dr. Suzette A. *Laboratory of Persistent Viral Diseases, Rocky Mountain Laboratories, National Institute of Allergy and Infectious Diseases, Hamilton, Montana.* ANIMAL PRIONS.

R

Raguso, Dr. Robert A. *Department of Neurobiology and Behavior, Cornell University, Ithaca, New York.* FLORAL VOLATILES.

Rahman, Prof. M. Aziz. *Faculty of Engineering and Applied Science, Memorial University of Newfoundland, St. John's, Canada.* PERMANENT-MAGNET SYNCHRONOUS MACHINES.

Reppert, Dr. Steven M. *Department of Neurobiology, University of Massachusetts Medical School, Worcester.* MONARCH BUTTERFLY MIGRATION—coauthored.

Rieppel, Dr. Olivier. *Rowe Family Curator of Evolutionary Biology, Department of Geology, The Field Museum of Natural History, Chicago, Illinois.* TURTLE ORIGINS.

Ritchie, Dr. Michael F. *Department of Biochemistry, Temple University School of Medicine, Philadelphia, Pennsylvania.* STIM FAMILY OF PROTEINS—coauthored.

Robertson, Dr. Jyothi V. *Koret Shelter Medicine Program, School of Veterinary Medicine, University of California, Davis.* INFECTIOUS DISEASE CONTROL IN ANIMAL SHELTERS.

Rodriguez, Dr. Nuria. *Fachbereich Chemie/Organische Chemie, Technische Universität Kaiserslautern, Germany.* DECARBOXYLATIVE COUPLING—coauthored.

Röhe, Dr. Fabio. *Wildlife Conservation Society, WCS Brasil-Amazon, and Departamento de Biologia, Instituto de Ciências Biológicas, Universidade Federal do Amazonas, Manaus, Brazil.* NEW VERTEBRATE SPECIES—coauthored.

Rotenberg, Dr. Eli. *Advanced Light Source, Lawrence Berkeley National Laboratory, Berkeley, California.* QUANTUM ELECTRONIC PROPERTIES OF MATTER.

Rowell, Dr. Roger M. *Professor Emeritus, Department of Biological Systems Engineering, University of Wisconsin, Madison.* CONTAMINANTS IN RECYCLED WOOD.

S

Sakai, Dr. Sharleen T. *Department of Psychology and Neuroscience Program, Michigan State University, East Lansing.* SOCIAL INTELLIGENCE AND THE BRAIN IN THE SPOTTED HYENA.

Schetz, Prof. Joseph A. *Department of Aerospace and Ocean Engineering, Virginia Polytechnic Institute & State University, Blacksburg.* SKIN FRICTION MEASUREMENT.

Schlachter, Dr. Alfred S. *Advanced Light Source, Lawrence Berkeley National Laboratory, Berkeley, California.* WIDE-SCALE INFRASTRUCTURE FOR ELECTRIC CARS.

Schlueter, Dr. John A. *Materials Science Division, Argonne National Laboratory, Argonne, Illinois.* MOLECULE-BASED SUPERCONDUCTORS.

Schoch, Dr. Conrad L. *National Center for Biotechnology Information, National Library of Medicine, National Institutes of Health, Bethesda, Maryland.* DNA BARCODING IN FUNGI—coauthored.

Scholes, Prof. Gregory D. *Department of Chemistry, University of Toronto, Ontario, Canada.* QUANTUM COHERENCE IN PHOTOSYNTHESIS.

Seaton, Dr. Daniel B. *Solar Influences Data Analysis Center, Royal Observatory of Belgium, Brussels.* STEREO mission.

Seifert, Dr. Keith A. *Biodiversity (Mycology and Botany), Eastern Cereal and Oilseed Research Centre, Agriculture and Agri-Food Canada, Ottawa.* DNA BARCODING IN FUNGI—coauthored.

Shapiro, Dr. Kristen L. *Department of Astronomy, University of California, Berkeley.* GALAXY EVOLUTION IN THE EARLY UNIVERSE.

Sharma, Dr. Vivek. *Department of Mechanical Engineering, Massachusetts Institute of Technology, Cambridge.* SCARAB BEETLE IRIDESCENCE—coauthored.

Sheik-Bahae, Prof. Mansoor. *Department of Physics and Astronomy, University of New Mexico, Albuquerque.* LASER COOLING OF SOLIDS—coauthored.

Sheikh, Dr. Suneel I. *President and CEO, ASTER Labs, Inc., Shoreview, Minnesota.* PULSAR NAVIGATION—coauthored.

Shen, Prof. Zuyan. *Department of Building Engineering, Tongji University, Shanghai, China.* HIGH-SPEED RAIL—coauthored.

Siegel, Dr. Jay A. *Professor and Chair, Department of Chemistry and Chemical Biology, Director, Forensic and Investigative Sciences Program, Indiana University-Purdue University Indianapolis.* NATIONAL ACADEMY OF SCIENCES REPORT ON FORENSIC SCIENCE.

Silva, Claudia Regina. *Instituto de Pesquisas Científicas e Tecnológicas do Estado do Amapá (IEPA), Macapá, Brazil.* NEW VERTEBRATE SPECIES—coauthored.

Simpson, Dr. Brent T. *Department of Sociology, University of South Carolina, Columbia.* HUMAN ALTRUISM—coauthored.

Singer, Ronald L. *Tarrant County Medical Examiner's Office Criminalistics Laboratories, Fort Worth, Texas.* FORENSIC FIREARMS IDENTIFICATION.

Sinha, Dr. Neelima R. *Department of Plant Biology, University of California, Davis.* REGULATION OF LEAF SHAPE—coauthored.

Skaar, Dr. Ida. *Section of Mycology, National Veterinary Institute, Oslo, Norway.* FUNGI IN DRINKING WATER—coauthored.

Sloan, Dr. Steven R. *Joint Program in Transfusion Medicine and Department of Laboratory Medicine, Children's Hospital and Harvard Medical School, Boston, Massachusetts.* BLOOD GROUP GENOTYPING.

Soares, Dr. Daphne. *Department of Biology, University of Maryland, College Park.* SENSORY STRUCTURES IN CROCODILIANS.

Soboloff, Dr. Jonathan. *Department of Biochemistry, Temple University School of Medicine, Philadelphia, Pennsylvania.* STIM FAMILY OF PROTEINS—coauthored.

Souza, Dr. Sergio Marques. *Coleções Zoológicas, Instituto Nacional de Pesquisas da Amazônia, Manaus, Amazonas, Brazil.* NEW VERTEBRATE SPECIES—coauthored.

Spencer, Prof. Michael G. *School of Electrical and Computer Engineering, Cornell University, Ithaca, New York.* BETAVOLTAICS.

Spudis, Dr. Paul D. *Lunar and Planetary Institute, Houston, Texas.* WATER ON THE MOON.

Srinivasarao, Dr. Mohan. *School of Materials Science and Engineering, School of Chemistry and Biochemistry, Center for Advanced Research on Optical Microscopy, Georgia Institute of Technology, Atlanta.* SCARAB BEETLE IRIDESCENCE—coauthored.

Stanton, Prof. Anthony P. *Carnegie Mellon University, Tepper School of Business, Pittsburgh, Pennsylvania.* AUTOMATED INK OPTIMIZATION SOFTWARE FOR THE PRINTING INDUSTRY.

Stemp, Dr. James. *Department of Sociology and Anthropology, Keene State College, Keene, New Hampshire.* LITHIC USE-WEAR.

Stephan, Dr. Douglas Wade. *Department of Chemistry, University of Toronto, Ontario, Canada.* FRUSTRATED LEWIS ACID AND BASE PAIR REACTIONS.

Stone, Marcia. *Science and Medical Writer, New York, New York.* BACTERIAL SYMBIONTS OF FARMING ANTS.

Straka, Dr. Jerry M. *School of Meteorology, University of Oklahoma, Norman, Oklahoma.* MULTI-MOMENT MICROPHYSICAL PARAMETERIZATIONS IN CLOUD MODELS—coauthored.

Sullivan, Prof. Richard. *Section of Research Oncology, King's Health Partners Integrated Cancer Centre, King's College London, and Guy's and St Thomas' NHS Foundation Trust, London, United Kingdom; and European Institute of Oncology, Milan, Italy.* MEDICINAL MUSHROOMS AND BREAST CANCER.

T

Teicher, Dr. Martin H. *Developmental Biopsychiatry Research Program and Laboratory of Developmental Psychopharmacology, McLean Hospital, Harvard Medical School, Belmont, Massachusetts.* EFFECT OF ABUSE ON THE BRAIN.

Tentzeris, Prof. Manos M. *School of Electrical and Computer Engineering and Georgia Electronic Design Center, Georgia Institute of Technology, Atlanta.* INKJET-PRINTED PAPER-BASED ANTENNAS AND RFIDs—coauthored.

Tjaden, Dr. Brett C. *Department of Computer Science, James Madison University, Harrisonburg, Virginia.* CYBER DEFENSE—coauthored.

Tse, Terry. *Federal Railroad Administration Research & Development, Washington, District of Columbia.* POSITIVE TRAIN CONTROL.

V

Vallero, Dr. Daniel. *Adjunct Professor of Engineering Ethics, Pratt School of Engineering, Duke University, Durham, North Carolina.* HEALTH AND ENVIRONMENTAL INTEROPERABILITY—coauthored.

van der Oost, Prof. John. *Laboratory of Microbiology, Department of Agrotechnology and Food Sciences, Wageningen University, Gelderland, The Netherlands.* CRISPR-BASED IMMUNITY IN PROKARYOTES—coauthored.

Van Haesendonck, Prof. Chris. *Department of Physics and Astronomy, Katholieke Universiteit, Leuven, Belgium.* ANDERSON LOCALIZATION—coauthored.

Vekilov, Prof. Peter G. *Department of Chemical and Biomolecular Engineering, University of Houston, Texas.* PROTEIN CRYSTALLIZATION.

Voigt, Kenneth H. *Department of Civil Engineering, University of Wisconsin-Milwaukee.* TRANSPORTATION'S ROLE IN SUSTAINABILITY—coauthored.

W

Waikel, Patricia A. *Oceanographic Center, Nova Southeastern University, Dania, Florida.* METATRANSCRIPTOMICS—coauthored.

Waikel, Dr. Rebekah L. *Department of Biological Sciences, Eastern Kentucky University, Lexington.* METATRANSCRIPTOMICS—coauthored.

Wang, Dr. Lan. *Department of Computer Science, University of Memphis, Tennessee.* SECURING WIRELESS IMPLANTABLE HEALTH-CARE DEVICES—coauthored.

Wang, Xiao. *School of Electrical and Computer Engineering, Cornell University, Ithaca, New York.* LOW-POWER RADIO LINKS—coauthored.

Wegst, Dr. Ulrike G. K. *Department of Materials Science and Engineering, Drexel University, Philadelphia, Pennsylvania.* RELATIONS AMONG STRUCTURE, PROPERTIES, AND FUNCTION IN BIOLOGICAL MATERIALS.

Wen, Dr. Hua-An. *Key Laboratory of Systematic Mycology and Lichenology, Institute of Microbiology, Chinese Academy of Sciences, Beijing, China.* MEDICINAL FUNGAL RESEARCH IN CHINA—coauthored.

Westra, Edze R. *Laboratory of Microbiology, Department of Agrotechnology and Food Sciences, Wageningen University, Gelderland, The Netherlands.* CRISPR-BASED IMMUNITY IN PROKARYOTES—coauthored.

Woolman, Prof. Matthew. *Graphic Design, Virginia Commonwealth University, Richmond, Virginia.* KINETIC TYPOGRAPHY—coauthored.

X

Xie, Dr. Huikai. *Department of Electrical and Computer Engineering, University of Florida, Gainesville.* IN VIVO 3D OPTICAL MICORENDOSCOPY.

Y

Yang, Dr. Li. *Texas Instruments, Dallas, Texas.* INKJET-PRINTED PAPER-BASED ANTENNAS AND RFIDs—coauthored.

Z

Zars, Dr. Troy. *Department of Biological Sciences, University of Missouri, Columbia.* GENETICS OF MEMORY—coauthored.

Zhu, Dr. Ping. *Institute of Materia Medica, Chinese Academy of Medical Sciences, Beijing, China.* MEDICINAL FUNGAL RESEARCH IN CHINA—coauthored.

Zhu, Prof. Sheng-Jiang. *Department of Physics, Tsinghua University, Beijing, China.* ASYMMETRIC, PEAR-SHAPED NUCLEI—coauthored.

Zielke, Dr. Olaf. *School of Earth and Space Exploration, Arizona State University, Tempe.* CONSTRAINING SLIP AND TIMING OF PAST EARTHQUAKE RUPTURES.

Index

Asterisks indicate page references to article titles.

A

D

E

F

H

M

O

S

T

W

X

Y

Z